WIND TURBINES

Edited by **Ibrahim Al-Bahadly**

INTECHOPEN.COM

Wind Turbines
Edited by Ibrahim Al-Bahadly

CBS, Edition 2016

Published by InTech

Janeza Trdine 9, 51000 Rijeka, Croatia

Publishing Process Manager Ana Nikolic

Technical Editor Teodora Smiljanic

Cover Designer InTech Design Team

Image Copyright 2010. Used under license from Shutterstock.com

Additional hard copies can be obtained from orders@intechopen.com

Wind Turbines, Edited by Ibrahim Al-Bahadly
p. cm.
ISBN 978-953-307-221-0

INTECH

open science | open minds

Contents

Preface

The need for energy consumes our society. As technology has advanced in certain areas the ability to produce power has had to keep pace with the ever increasing demands. There always seems to be energy-crisis whether contrived or real, and society allows the pollution of our environment in the name of power production. Power production with traditional means has polluted our planet. Hydro power dams release carbon that was locked up in the trees and plants that were drowned during the filling of the dam. Any sort of fossil fuel powered plant releases carbon into the environment during the combustion process. Renewable, environmentally friendly, clean, safe, even wholesome, are the types of adjectives we should be using to describe power production. Wind energy is the closest we may have at present that may be considered to fit into these criteria.

There is a tremendous amount of free energy in the wind which is available for energy conversion. The use of wind machines to harness the energy in the wind is not a new concept. The early machines were used for pumping water for irrigation purposes and later developed as windmills for grinding grain. The power in the wind at any moment is the result of a mass of air moving at speed in a particular direction. To capture this power or should we say part of it, it is necessary to place in the path of the wind a machine, a wind turbine, to transfer the power from the wind to the machine. It has only really been in the last century that the intensive research and development have gone into the use of wind energy for electricity generation. A number of different types of wind machines, or wind turbines, exist today. They can generally be categorized into two main categories. Turbines, whose rotor shaft rotates around a horizontal axis and those whose rotor rotates around a vertical axis.

The area of wind energy is a rapidly evolving field and an intensive research and development has taken place in the last few years. Therefore, this book aims at providing an up-to-date comprehensive overview of the current status in the field to the research community. The research works presented in this book could be divided into three main groups. One group deals with the different types and design of the wind mills aiming for efficient, reliable and cost effective solutions. The second group deals with works tackling the use of different types of generators for wind energy. The third group is focusing on improvement in the area of control. Each chapter of the book offers detailed information on the related area of its research with the main objectives of the works carried out as well as providing a comprehensive list of references which should provide a rich platform for research of the field.

The editor has been privileged by the invitation of INTECH to act as editor of the book "Wind Turbines" which encompasses high quality research works form internationally renowned researchers in the field. The editor is glad to have had the opportunity of acknowledging all contributing authors and expresses his gratitude for the help and support of INTECH staff particularly the Publishing Process Manager Ms Ana Nikolic.

Dr. Ibrahim Al-Bahadly,
Massey University,
Palmerston North,
New Zealand

Part 1

Windmills

1

Special Issues on Design Optimization of Wind Turbine Structures

Karam Maalawi
National Research Centre, Mechanical Engineering Department, Cairo Egypt

1. Introduction

A wind turbine is a device that exploits the wind's kinetic energy by converting it into useful mechanical energy. It basically consists of rotating aerodynamical surfaces (blades) mounted on a hub/shaft assembly, which transmits the produced mechanical power to the selected energy utilizer (e.g. milling or grinding machine, pump, or generator). A control system is usually provided for adjusting blade angles and rotor position to face the wind properly. All units are supported by a stiff tower structure, which elevates the rotor above the earth's boundary layer. There are two common types: horizontal-axis and vertical-axis wind turbines. In the former, which dominate today's markets, the blades spin about an axis perpendicular to the tower at its top (see Fig.1), while in the latter they spin about the tower axis itself. In fact, wind turbines have been used for thousands of years to propel boats and ships and to provide rotary power to reduce the physical burdens of man. From the earliest

Fig. 1. Offshore horizontal-axis wind turbine

times of recorded history, there is evidence that the ancient Egyptians and Persians used wind turbines to pump water to irrigate their arid fields and to grind grains (Manwell et al., 2009). The technology was transferred to Europe and the idea was introduced to the rest of the world. Early wind turbines were primitive compared to today's machines, and suffered from poor reliability and high costs. Like most new technology, early wind turbines had to go through a process of "learning by doing", where shortcomings were discovered, components were redesigned, and new machines were installed in a continuing cycle. Today, Wind turbines are more powerful than early versions and employ sophisticated materials, electronics and aerodynamics (Spera, 2009). Costs have declined, making wind more competitive clean energy source with other power generation options. Designers apply optimization techniques for improving performance and operational efficiency of wind turbines, especially in early stages of product development. It is the main aim of this chapter to present some fundamental issues concerning design optimization of the main wind turbine structures. Practical realistic optimization models using different strategies for enhancing blade aerodynamics, structural dynamics, buckling stability and aeroelastic performance are presented and discussed. Design variables represent blade and tower geometry as well as cross-sectional parameters. The mathematical formulation is based on dimensionless quantities; therefore the analysis can be valid for different wind turbine rotor and/or tower sizes. Such normalization has led to a naturally scaled optimization models, which is favorable for most optimization techniques. The various approaches that are commonly utilized in design optimization are also presented with a brief discussion of some computer packages classified by their specific applications. Case studies include blade optimization in flapping and pitching motion, yawing dynamic optimization of combined rotor/tower structure, and power output maximization as a measure of improving aerodynamic efficiency. Optimization of the supporting tower structure against buckling as well as the use of the concept of material grading for enhancing the aeroelastic stability of composite blades have been also addressed. Several design charts that are useful for direct determination of the optimal values of the design variables are introduced. This helps achieving, in a practical manner, the intended design objectives under the imposed design constraints. The proposed mathematical models have succeeded in reaching the required optimum solutions, within reasonable computational time, showing significant improvements in the overall wind turbine performance as compared with reference or known baseline designs.

2. General aspects of wind turbine design optimization

Design optimization seeks the best values of a set of n design variables represented by the vector, $\underline{X}_{nx1}$, to achieve, within certain m constraints, $\underline{G}_{mx1}(\underline{X})$, its goal of optimality defined by a set of k objective functions, $\underline{F}_{kx1}(\underline{X})$, for specified environmental conditions. Mathematically, design optimization may be cast in the following standard form (Rao, 2009): Find the design variables $\underline{X}_{nx1}$ that minimize

$$F(\underline{X}) = \sum_{i=1}^{k} W_{fi} F_i(\underline{X}) \tag{1a}$$

subject to

$$G_j(\underline{X}) \leq 0 \,,\, j=1,2,........I \tag{1b}$$

$$G_j(\underline{X}) = 0 \text{ , } j=I+1,I+2,\ldots.m \tag{1c}$$

$$0 \leq W_{fi} \leq 1$$
$$\sum_{i=1}^{k} W_{fi} = 1 \tag{1d}$$

If it is required to maximize $F_i(\underline{X})$, one simply minimizes $-F_i(\underline{X})$ instead. The weighting factors W_{fi} measure the relative importance of the individual objectives with respect to the overall design goal. Fig. 2 shows the general scheme of an optimization approach to design.

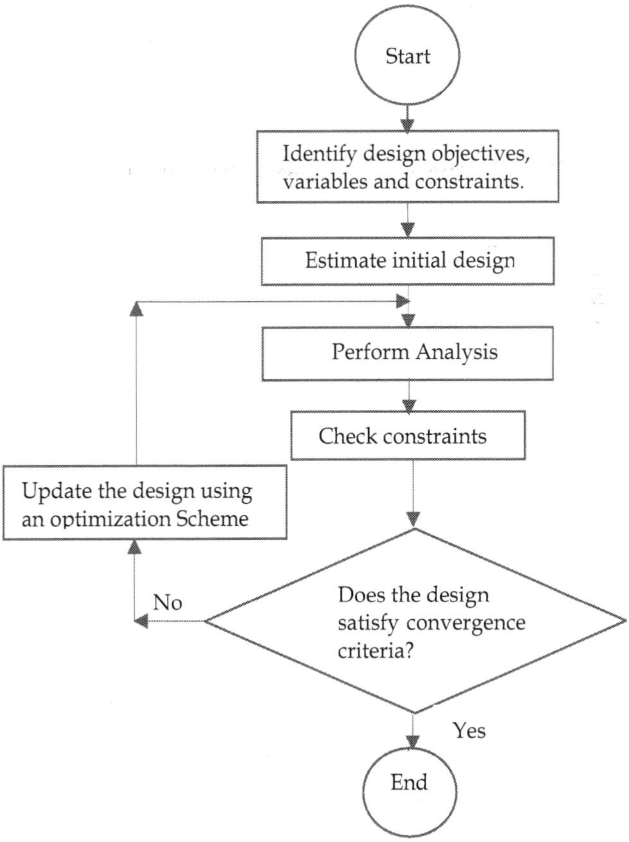

Fig. 2. Design optimization process

Several computer program packages are available now for solving a variety of design optimization models. Advanced procedures are carried out by using large-scale, general purpose, finite element-based multidisciplinary computer programs, such as *ASTROS* (Cobb et al., 1996), *MSC/NASTRAN* and *ANSYS* (Overgaard and Lund, 2005). The *MATLAB* optimization toolbox (Vekataraman, 2009) is also a poweful tool that includes many routines for different types of optimization encompassing both unconstrained and constrained minimization algorithms. Design optimization of wind energy conversion systems involve many objectives, constraints and variables. This is because the structure of the wind turbine contains thousands of components ranging from small bolts to large, heavyweight blades and spars. Therefore, creation of a detailed optimization model incorporating, simultaneously, all the relevant design features is virtually imposible. Researchers and engineers rely on simplified models which provide a fairly accurate approximation of the real structure behaviour. In the subsequent sections, the underlying concepts of applying optimization theory to the design of a conventional wind turbine will be applied. The relevant design objectives, constraints and variables are identified and discussed.

2.1 Design objectives of a wind turbine

A successful wind turbine design should ensure efficient, safe and economic operation of the machine. It should provide easy access for maintenance, and easy transportation and erection of the various components and subcomponents. Good designs should incorporate aesthetic features of the overall machine shape. In fact, there are no simple criteria for measuring the above set of objectives. However, it should be recognized that the success of structural design ought to be judged by the extent to which the least possible cost of energy production can be achieved (Wei, 2010). For a specified site wind characteristics, the analysis of the unit energy cost (Euro/Kilowatt.Hour) involves many design considerations such as the rotor size, rated power, fatigue life, stability, noise and vibration levels.

2.2 Design variables

The definition of design variables and parameters is of great importance in formulating an optimization model. Design variables of a wind turbine include layout parameters as well as cross-sectional and spanwise variables. The main variables of the blades represent the type of airfoil section, chord and twist distributions, thickness of covering skin panels, and the spacing, size and shape of the transverse and longitudinal stiffeners. If the skin and/or stiffeners are made of layered composites, the orientation of the fibers and their proportion can become additional variables. Tower variables include type (truss- tubular), height, cross sectional dimensions, and material of construction.

2.3 Design constraints

There are many limitations that restrict wind turbine design, manufacturing and operation. The most significant among these are: (a) type of application (e.g. electricity generation), (b) site condition (location - wind speed characteristics - wind shear – transportation - local electricity system-.....), (c) project budget and financial limitations, (d) technological and manufacturing limitations, (e) manpower skills and design experience, (f) availability of certain material types, (g) safety and performance requirements. An optimal design for a wind turbine must achieve the system objectives and take into consideration all aspects of the design environments and constraints.

3. Basic aerodynamic optimization

The aerodynamic design of a wind turbine rotor includes the choice of the number of blades, determination of blade length, type of airfoil section, blade chord and twist distributions and the design tip-speed ratio (TSR=rotational speed x rotor radius/design wind speed at hub height). Concerning blade number (N_B), a rotor with one blade can be cheaper and easier to erect but it is not popular and too noisy. The two-bladed rotor is also simpler to assemble and erect but produces less power than that developed by the three-bladed one. The latter produces smoother power output with balanced gyroscopic loads, and is more aesthetic. The determination of the blade length (or rotor size) depends mainly on the needed energy for certain application and average wind speed of a specific site. The choice of the type of airfoil section may be regarded as a key point in designing an efficient wind rotor (Burger & Hartfield 2006). Other factors that can have significant effects on the overall rotor design encompass the distribution of wind velocity in the earth boundary layer as well as in the tower shadow region (see Fig. 3).

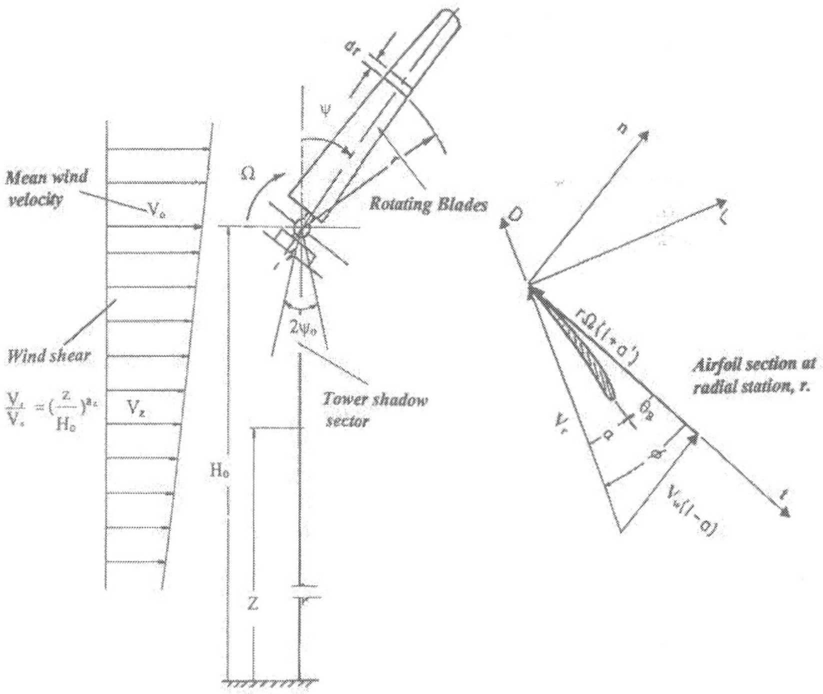

Fig. 3. Wind rotor geometry and velocity components

The various symobles in Fig. 3 are defined as follows: a=axial induction factor, a′=angular induction factor, a_z=wind shear exponent, D=aerodynamic drag, H_0=hub height, L=aerodynamic lift, r=local blade radius, V_r=resultant wind velocity, Z=height above ground or sea level, α=angle of attack, θ_B=blade twist, φ=inflow angle, ψ=azimuth angle, Ω=rotor rpm. More definitions can be found in (Maalawi & Badr, 2003). The following formula is used to calculate the output rated power P_r, or generator capacity of a wind turbine:

$$P_r = (C_p \eta)\frac{1}{2}\rho_{air}(\pi R^2)V_r^3 \text{ (Watts)} \quad (2)$$

Where:
C_p=power coefficient, depending on blade geometry , airfoil section and tip-speed ratio.
η = Transmission and generator efficiency.
R = Rotor radius (meter).
V_r =rated wind speed (m/s).
ρ_{air}= air density (kg/m^3).
An optimized wind rotor is that operates at its maximum power coefficient at the design wind speed, at which the design tip-speed ratio is set. This defines the rotor rpm and thus required gear ratio to maximize the energy production. The calculations of the annual energy productivity are accomplished by an iterative computer calculation based on the Weibul wind representation and the specified power performance curve (Wei, 2010). Operating the wind turbine at the design TSR corresponding to the maximum power point at all times may generate 20–30% more electricity per year. This requires, however, a control scheme to operate with variable speed. Several authors (Kusiak, et al, 2009), (Burger & Hartfield 2006), (Maalawi & Badr, 2003) have studied optimum blade shapes for maximizing C_p. Important conclusions drawn from such studies have shown that the higher the lift/drag ratio, the better the aerodynamic performance of the turbine. Analytical studies by (Maalawi & Badawy, 2001) and (Maalawi & Badr, 2003) indicated that the theoretical optimum distributions of the blade chord and twist can be adequately determined from an exact analysis based on Glauert's optimum conditions. The developed approach eliminated much of the numerical efforts as required by iterative procedures, and a unique relation in the angle of attack was derived, ensuring convergence of the attained optimal solutions. Based on such analytical approach, the theoretical optimum chord distribution of the rotating blade can be determined from the following expression:

$$C(r) = \frac{8\pi rF\sin\varphi}{N_B C_L\{[(\lambda_r + \tan\varphi)/(1-\lambda_r\tan\varphi)] - (C_D/C_L)\}} \quad (3)$$

where N_B is the number of blades, C_D/C_L the minimum drag-to-lift ratio of the airfoil section, F tip loss factor, λ_r (=$\Omega r/V_o$) local speed ratio at radial distance *r* along the blade, Ω rotational speed of the blade and V_o wind velocity at hub height (refer to Fig. 3). Having determined the best blade taper and twist, the aerodynamic power coefficient can be calculated by integrating all of the contributions from the individual blade elements, taking into account the effect of the wind shear and tower shadow. Fig. 4 shows variation of the optimum power coefficient with the design tip-speed ratio for a blade made of NACA 4-digit airfoil families. Both cases with (lower curves) and without (upper curves)wind shear and tower shadows were investigated. It is seen that Cp increases rapidly with TSR up to its optimum value after which it decreases gradually with a slower rate. The optimum range of the TSR is observed to lie between 6 and 11, depending on the type of airfoil. The effect of wind shear and tower shadow resulted in a reduction of the power coefficient by about 16%. The value of the design TSR at which $C_{p,max}$ occurs is also reduced by about 9%. It is also observed that blades with NACA 1412 and 4412 produce higher power output as compared with other airfoil types.

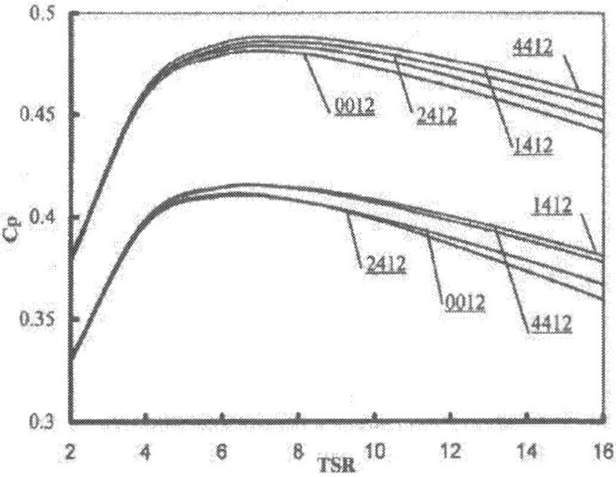

Fig. 4. Variation of the optimum power coefficent with tip-speed ratio for a three-bladed rotor made of NACA-4 digit airfoils

4. Frequency optimization models

Large horizontal-axis wind turbines (HAWT) utilized for electricity generation are characterized by their slender rotating blades mounted on flexible tall towers. Such a configuration gives rise to significant vibration problems, and assesses the importance of analyzing blade and tower dynamics in the design of successful wind generators. A major issue for reducing vibration levels is to avoid the occurrence of resonance, which plays a central role in the design of an efficient wind turbine structure. Vibration reduction fosters other important design goals, such as long fatigue life, high stability and low noise levels. It is one of the main emphasises of this chapter to optimize the system frequencies and investigate their variation with the stiffness and mass distributions of the rotating blades or the supporting tower structures. These frequencies, besides being maximized, must be kept out of the range of the excitation frequencies in order to avoid large induced stresses that can exceed the reserved fatigue strength of the materials and, consequently, cause failure in a short time. Expressed mathematically, two different design criteria are implemented here for optimizing frequencies:

Frequency-placement criterion: Minimize $$\sum_i W_{fi}(\omega_i - \omega_i^*)^2 \quad (4)$$

Maximum-frequency criterion: Minimize $$-\sum_i W_{fi}\,\omega_i \quad (5)$$

In both criteria, an equality constraint should be imposed on the total structural mass in order not to violate other economic and performance requirements. Equation (4) represents a weighted sum of the squares of the differences between each important frequency ω_i and its desired (target) frequency ω_i^*. Appropriate values of the target frequencies are usually chosen to be within close ranges (called frequency windows) of those corresponding to a reference or baseline design, which are adjusted to be far away from the critical exciting frequencies. The main idea is to tailor the mass and stiffness distributions in such a way to make the objective function a minimum under the imposed mass constraint. The second alternative for reducing vibration is the direct maximization of the system natural frequencies as expressed by equation (5). Maximization of the natural frequencies can ensure a simultaneous balanced improvement in both of stiffness and mass of the vibrating structure. It is a much better design criterion than minimization of the mass alone or maximization of the stiffness alone. The latter can result in optimum solutions that are strongly dependent on the limits imposed on either the upper values of the allowable deflections or the acceptable values of the total structural mass, which are rather arbitrarily chosen. The proper determination of the weighting factors W_{fi} should be based on the fact that each frequency ought to be maximized from its initial value corresponding to a baseline design having uniform mass and stiffness properties (Negm & Maalawi, 2000).

4.1 Yawing dynamic optimization of combined rotor/tower structure

For wind turbines of horizontal-axis type, the rotation of the nacelle/rotor combination at the top with respect to the tower axis, called yawing motion, is an important degree of freedom in the system dynamics. Such a rigid body motion is produced by the yawing mechanism to direct the rotor towards the wind in order to maximize energy capture. Usually this is accomplished actively with an electrical or hydraulic yaw servo. A wind vane, placed on top of the nacelle, senses the wind direction. The servo is activated when the mean relative wind direction exceeds some predefined limits. Therefore, the wind turbine spends much of its time yawed in order to face the rapidly changing wind direction, so it would seem reasonable to expect that designers should have a sufficient understanding of the turbine response in that condition to take it properly into account. There are frequent yaw system failures world-wide on wind turbines, where some statistical studies indicated that such failures accounts for about 5-10% of breakdowns in any given year the wind plant is in operation. This fact emphasizes the need to improve the design of yaw mechanisms in order to increase the availability of turbines and reduce their maintenance overheads. One of the most cost-effective solutions in designing efficient yaw mechanisms and reducing the produced vibrations is to separate the natural frequencies of the tower/nacelle/rotor structure from the critical exciting yawing frequencies. An optimization model was developed by (Maalawi, 2007) showing the necessary exact dynamical analysis of a practical wind turbine model, shown in Fig. 5, for proper placement of the system frequencies at their target values. The rotor/nacelle combination was considered as a rigid body with mass polar moment of inertia I_N spinning about the vertical axis x at an angular displacement $\psi(t)$ relative to the top of the tower,where t is the time variable. The yawing mechanism was assumed to have a linear torsional spring with a stiffness K_y. The tower is in the state of free torsional vibration about its centroidal axis with an absolute angular displacement denoted by $\varphi(x,t)$. The associated eigenvalue problem is cast in the following:

$$\frac{d}{dx}[GJ(x)\frac{d\Phi}{dx}]+\rho I_p(x)\omega^2\Phi(x)=0 \quad (6)$$

The boundary conditions are:

$$\text{at } x=0 \qquad \Phi(x)=0 \quad (7a)$$

$$\text{at } x=H \quad \frac{GJ}{\psi_o}\frac{d\Phi}{dx} - K_y = 0; \quad \psi_o = \left[\frac{GJ}{\omega^2 I_N}\frac{d\Phi}{dx} - \Phi\right] \tag{7b}$$

where GJ is the torsional rigidity of the tower structure and Φ space-dependent angular displacement.

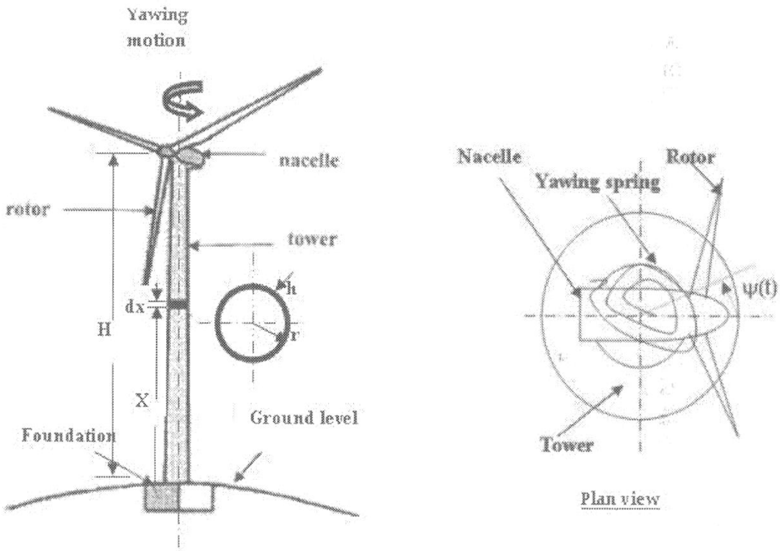

Fig. 5. Horizontal-axis wind turbine in free yawing motion

Considering a tapered, thin-walled tower, the various cross sectional parameters are defined by the following expressions:

$$\text{mean radius} \qquad r = r_o(1-\beta\hat{x}) \tag{8a}$$

$$\text{wall thickness} \qquad h = h_o(1-\beta\hat{x}) \tag{8b}$$

$$\text{torsional constant } J = I_p = 2\pi r^3 h \tag{8c}$$

The variables r and h are assumed to have the same linear distribution, and $\hat{x}$ and β are dimensionless quantities defined as:

$$\hat{x} = \frac{X}{H}, \; \beta = (1-\Delta), \; \Delta = r_H / r_o \tag{9}$$

where Δ denotes the taper ratio of the wind turbine tower, r_H raduis at height H and r_o at tower base. Quantities with the hat symbol (^) are dimensionless quantities obtained by dividing by the corresponding parameters of a baseline tower design having uniform properties and same structural mass, height and material properties of the optimized tower design (refer to Table 1). Introducing the transformation

$$\hat{x} = \frac{1}{\beta} - \frac{1}{\hat{\omega}} y \text{ ; } (\beta \neq 0) \tag{10}$$

Equation(6) takes the form:

$$\frac{d^2\Phi}{dy^2} + \frac{4}{y}\frac{d\Phi}{dy} + \Phi = 0 \tag{11}$$

which can be further transformed to the standard form of Bessel's equation by setting $\Phi = (y)^{3/2}\theta$, to get

$$y^2\frac{d^2\theta}{dy^2} + y\frac{d\theta}{dy} + (y^2 - \frac{9}{4})\theta = 0 \tag{12a}$$

which has the solution

$$\theta(y) = C_1 J_{3/2} + C_2 J_{-3/2} \tag{12b}$$

where C_1 and C_2 are constants of integration and $J_{3/2}$ and $J_{-3/2}$ are Bessel's functions of order $k=\pm 3/2$, given by:

$$J_{3/2}(y) = \sqrt{\frac{2}{\pi y^3}}(\sin y - y\cos y) \tag{13a}$$

$$J_{-3/2}(y) = \sqrt{\frac{2}{\pi y^3}}(\cos y + y \sin y) \tag{13b}$$

Quantity	Notation	Dimensionless expression
Circular frequency	ω	$\hat{\omega} = \omega H\sqrt{\rho/G}$
Spatial coordinate	x	$\hat{x} = x/H$
Tower mean radius	r	$\hat{r} = r/r_b$
Tower wall thickness	h	$\hat{h} = h/h_b$
Tower torsional constant	J	$\hat{J} = J/J_b \; (= \hat{r}^3\hat{h})$
Nacelle/rotor polar moment of inertia	I_N	$\hat{I}_N = I_N/\rho H J_b$
Yawing stiffness coefficient	K_y	$\hat{K}_y = \frac{K_y}{(GJ_b/H)}$
Structural mass	M	$\hat{M} = M/M_b$
Baseline design parameters: M_b=structural mass=$2\pi H r_b h_b$, J_b=torsional constant of tower cross section (=$2\pi r_b^3 h_b$), where r_b=mean radius, h_b=wall thickness; $\hat{\omega}_b$ is the circular frequency=$\pi/2$.		

Table 1. Definition of dimensionless quantities

Finally, the exact analytical solution of the associated eigenvalue problem can be shown to have the form:

$$\Phi(y) = A[\frac{y\cos y - \sin y}{y^3}] + B[\frac{y\sin y + \cos y}{y^3}] \tag{14}$$

where *A* and *B* are constants depend on the imposed boundary conditions:

at $\underline{y = \zeta\ (= \hat{\omega}/\beta)}$ $\Phi(y)=0$

$$\text{at } \underline{y = \xi\ (= \zeta\Delta)} \quad \frac{\hat{\omega}(d\Phi/dy)}{\Psi_o} + (\hat{K}_y/\hat{J}) = 0\,; \quad \Psi_o = [\frac{\hat{J}}{\hat{\omega}\hat{I}_N}\frac{d\Phi}{dy} + \Phi] \tag{15}$$

Applying the boundary conditions, and considering only nontrivial solution of Eq. (14), that is A≠0 and B≠0, it is straightforward to obtain the frequency equation in the following compacted form:

$$\frac{\hat{\omega}^2}{\alpha\hat{\omega} + (\hat{J}/\hat{I}_N)} - (\hat{K}_y/\hat{J}) = 0\,; \quad \alpha = \frac{\xi[\hat{\omega} - (1+\zeta\xi)\tan\hat{\omega}]}{(\xi^2 - 3)(\zeta - \tan\hat{\omega}) + 3\xi(1+\zeta\tan\hat{\omega})} \tag{16}$$

It is to be noticed that in the above equations $\hat{J}$ is the dimensionless torsional constant at the top of tower and is equal to $\hat{r}_o^3\hat{h}_o\Delta^4$ (refer to Eq.8c). The frequency equations for the special cases of the limiting conditions are summarized in Table 2.

Condition	Reduced frequency equation
Locked yawing mechanism ($\hat{K}_y \to \infty$)	$\alpha\hat{\omega} + \hat{J}/\hat{I}_N = 0$
Uniform tower with no taper ($\Delta = 1;\ \beta = 0$)	Apply Eq.(16) with $\alpha = -\tan\hat{\omega}$
Stand-alone tapered tower with no attached masses at the top ($\hat{I}_N = 0$) .	$[(3-\xi^2) + 3\zeta\xi]\tan\hat{\omega} - [3\hat{\omega} - \zeta\xi^2] = 0$

Table 2. Frequency equation for limiting conditions

Once the exact dimensionless natural frequencies have been determined the associated mode shapes can be obtained analytically from:

$$\Phi(y)=A[(\frac{y\cos y - \sin y}{y^3}) - (\frac{\zeta - \tan\zeta}{1+\zeta\tan\zeta})(\frac{y\sin y + \cos y}{y^3})] \tag{17}$$

Several case studies were investigated and discussed in (Maalawi, 2007). A specific design case for locked yawing mechanism is shown in Fig. 6. It is seen that the objective function is well behaved in the selected design space (Δ, h_o). Actually, the developed chart represent the fundamental frequency function augumented with the imposed mass constraint so that the problem may be treated as if it were an unconstrained optimization problem. Each point inside the chart corresponds to different mass and stiffness distribution along the tower height, but the total mass is preserved at a constant given value. Now, it is possible to choose the desired frequency, which is far away from the excitation frequencies, and obtain the corresponding optimum variables directly from the chart. It is observed that there is a distinct optimum zone which encompasses the global optimal solution. Results have also indicated that as the rotor/nacelle inertia increases the attained global optimum configuration changes from that one having higher thicknes at tower base with low tapering ration to a configuration with lower thickness and higher tapering ratio. The maximum attainable frequencies for several design cases are summarized in Table 3. The associated optimum values of the chosen design variables, which satisfy the imposed mass equality constraint, are also presented. Fig. 7 shows variation of the optimal fundamental frequency with the yawing stiffness for a practical range of the mass polar moment of inertia of the nacelle/rotor combination in the unlocked condition. As a remarkable observation, the frequency increases for increasing yawing stiffness (K_y) and decreasing the nacelle/rotor inertia (I_N) with a higher rate. The fundamental frequency can be shifted sufficiently from the range which resonates with the excitation yawing frequencies. Other factors that should be considered, as a natural extension of the given formulation, include minimization of the system response to a sudden yaw motion, which can cause severe bending and shearing stresses within the blades and tower structures.

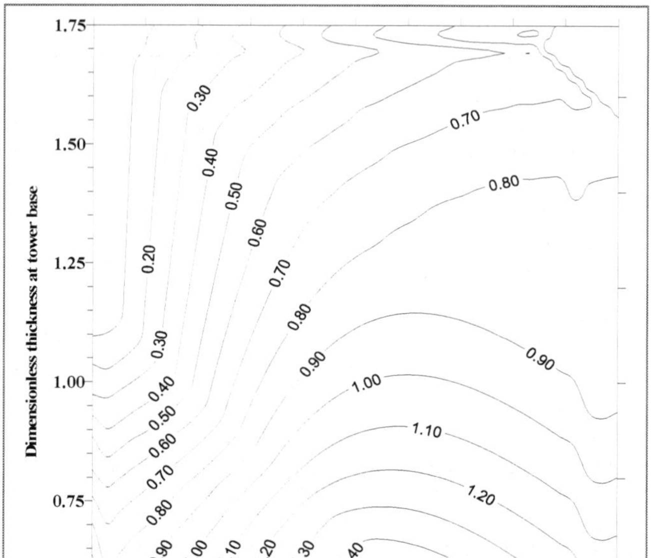

Fig. 6. Frequency chart for a locked yawing mechanism ($\hat{M} = 1$, $\hat{I}_N = 1.0$)

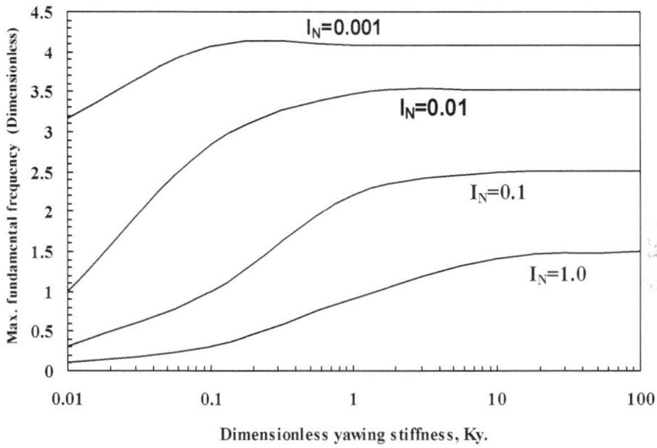

Fig. 7. Variation of the maximum fundamental frequency with yawing stiffness for different inertias of unlocked nacelle/rotor combination, ($\hat{M}$ = 1.0)

$\hat{I}_N$ / $\hat{K}_y$	0.001	0.01	0.1	1.0
0.01	3.1625 (0.688,0.6, 2.315)	0.9988 (0.863,0.463,2.489)	0.316 (1.0, 0.4, 2.5)	0.099 (1.0,0.538,1.861)
0.1	4.0663 (0.088,1.10,2.9)	2.833 (0.30,0.875,2.467)	0.9879 (1.0, 0.4, 2.5)	0.3137 (1.0, 0.4, 2.5)
1.0	4.0851 (0.075,1.113,2.495)	3.476 (0.188,0.988,2.485)	2.2105 (0.438,0.738,2.497)	0.9036 (1.0, 0.4, 2.5)
10.0	4.0909 (0.075,1.113,2.495)	3.5191 (0.175,1.0,2.488)	2.4859 (0.375,0.8,2.474)	1.4122 (0.763,0.513,2.497)
(∞) Locked	4.0915 (0.075,1.113,2.495)	3.5241 (0.175,1.0,2.49)	2.515 (0.375,0.8,2.474)	1.50 (0.7,0.55,2.491)

Table 3. Maximum yawing frequency, $\hat{\omega}_{1,max}$ and optimum design variables ($\Delta, \hat{h}_o, \hat{r}_o$)

4.2 Frequency optimization of blades with variable pitch

In strong wind conditions it is necessary to waste part of the excess energy of the wind in order to avoid damaging the wind turbine structure. All wind turbines are therefore designed with some sort of power control. There are different ways of doing this safely on modern wind turbines: pitch, active stall and passive stall controlled wind turbines (Manwell et al., 2009). On a pitch controlled wind turbine the turbine's electronic controller checks the power output of the turbine several times per second. When the power output becomes too high, it sends an order to the blade pitch mechanism which immediately pitches (turns) the rotor blades slightly out of the wind. Conversely, the blades are turned back into the wind whenever the wind drops again. The rotor blades thus have to be able to turn around their longitudinal axis (to pitch) as shown in Fig. 8a. The pitch mechanism is usually operated using hydraulics or electric stepper motors. Fig. 8b shows the optimal operational conditions of a pitch-controlled 2 MW wind turbine. During normal operation the blades will pitch a fraction of a degree at a time, and the rotor will be turning at the same time. The computer will generally pitch the blades a few degrees every time the wind changes in order to keep the rotor blades at the optimum angle o maximize output power for all wind speeds. (Maalawi & Badr, 2010) formulated an optimization model for avoiding resonance due to blade pitching motion. They considered reduction of vibration either by a direct maximization of the natural frequencies or by separating the natural frequencies of the blade structure from the harmonics of the exciting torque applied from the pitching mechanism at the rotor hub. The derived exact frequency equations for both cases of active and inactive pitching motion are given in the following:

- Baseline design with rectangular planform (Δ=1):

$$\hat{\omega}\tan\hat{\omega} = \hat{K}_S/(\hat{h}_o\hat{C}_o^3) \tag{18a}$$

- Active pitch:

$$\tan\hat{\omega} = \frac{3\hat{\omega}(3+\gamma\delta)}{(\gamma\delta)^2 - 3\gamma^2(1+\Delta^2) + 9(1+\gamma\delta)} \tag{18b}$$

- Inactive pitch:

$$\tan\hat{\omega} = \frac{(1-3\theta)(3\hat{\omega}-\gamma\delta^2) - 3\gamma^2\delta\theta}{3\gamma\delta(1-3\theta) + (3-\delta^2)(1-3\theta+\theta\gamma^2)} \tag{18c}$$

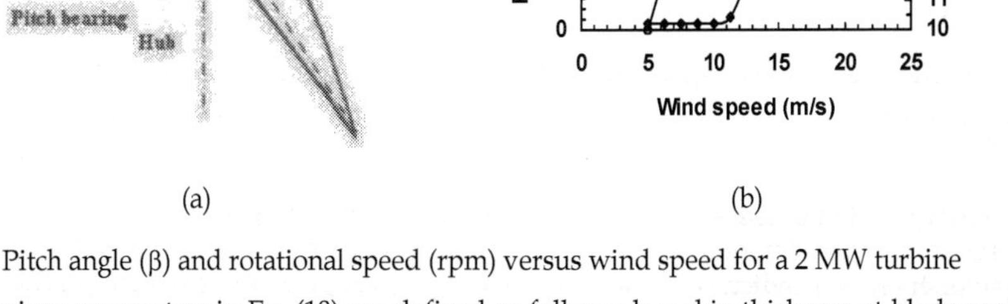

Fig. 8. Pitch angle (β) and rotational speed (rpm) versus wind speed for a 2 MW turbine

The various parameters in Eq. (18) are defined as follows: h_o= skin thickness at blade root, C_o= airfoil chord at root, K_s=stiffness coefficient at root, Δ=blade tapering ratio, α=(1-Δ), $\gamma = \hat{\omega}/\alpha$, $\delta=\gamma\Delta$ and $\theta = \alpha\hat{h}_o\hat{C}_o^3/\hat{K}_S$. More details can be found in (Maalawi & Badr, 2010), where several case studies were considered. As a demonstration, Fig. 9 depicts the developed frequency chart for the design case of locked pitching mechanism with a flexibile root having a stiffness coefficient $\hat{K}_S = 10$. The chart can be seen to have a banana-shaped profile bounded by two curved borderlines; the one from above represents a triangular blade geometry ($\Delta=0$) while the lower one represents a rectangular blade geometry ($\Delta=1$). It is not allowed to penetrate these two lines in order not to violate the imposed mass equality constraints. Each point inside the feasible domain in the middle corresponds to different mass and stiffness distributions along the blade span, but the total structural mass is preserved at that value corresponding to a uniform baseline design. The lower and upper empty regions represent, respectively, infeasible blade designs with structural mass less or greater than that of the baseline design. The global optimal design is too close to the design point $\{\hat{C}_o, \hat{h}_o, \Delta\} = \{1.202, 2.011, 0.207\}$ with $\hat{\omega}_{1,max} = 2.6472$. If it happened that such global optima violates frequency windows, another value of the frequency can be taken near the optimum point, and an inverse approach is utilized by solving the frequency equation for any one of the unknown design variables instead.

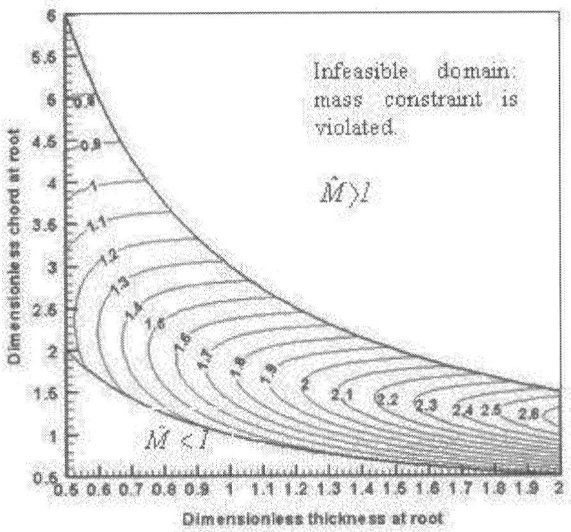

Fig. 9. Augmented pitching frequency-mass contours ($\hat{\omega}_1$) for a blade with flexible root: $\hat{K}_S = 10$ ($\hat{M} = 1$)

4.3 Frequencies of teetered and hingeless blades in flapping motion

Perhaps, the most important degree of freedom in the case of HAWT is the elastic bending deformation perpendicular to the plane of rotation; called flapping motion. Usually, for a typical blade configuration, the chordwise bending (lead-lag) and torsional stiffnesses are much higher than that in the spanwise direction. Therefore, the main focus of this section will be given to flapping dynamics as an important consideration in wind turbine structural design. There are two commonly used types of blade-to-hub connection; namely, teetered and hingless. The teetered type allows rigid body motion of the blades normal to the plane of rotor disk. It is employed only in one- and two-bladed wind turbines. The teetering rotor has substantial advantages over the hingless rotor with respect to blade shank stresses, fatigue life, and tower loading. The teetering motion of the blades parallel to the power transmission shaft reduces the internal bending stresses in the blades and the vibratory loads transmitted to the supporting tower structure. The only option for the three-bladed rotor is the hingeless type, which is cheaper in construction than the teetered one. However, for wind turbines with power output greater than 6 MW, the decision to whether to select two- or three-bladed rotor is still questionable when considering the overall system design objectives (e.g. cost and fatigue). (Maalawi & Negm, 2002) considered the optimal blade design by maximizing a weighted sum of the flapping frequencies for both teetered and hingeless types. Constraints included avoidance of aeroelastic instabilities, limitations imposed on the total structural mass and capability of starting rotation at the specified cut-in speed. Results showed that the approach used can produce improved designs as compared with known baseline designs. Figs. 10 and 11 show the developed isomerits of the second mode rotating frequency of hingeless and teetered blades, respectively.

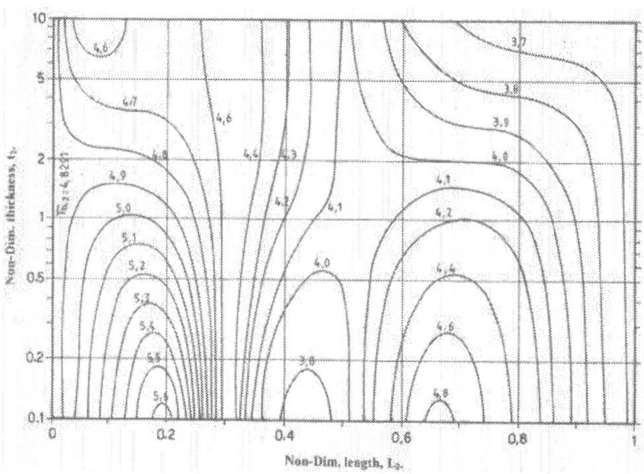

Fig. 10. Second-mode flapping frequency of a hingeless blade

As seen in Fig. 10 the design space contains three distinct frequency branches. The middle one has a local minimum while the one to the left has the global maximum solution. For the teetered type depicted in Fig. 11, there are only two distinguished frequency branches. The one to the left encompasses the global maximum while the one to the right contains a local minimum. The central zone contains a well behaved frequency contours which separate the two branches from each other. It is also observed that, for both types, the global maximum always occurs at the design point which corresponds to the minimum acceptable value of the outboard shear wall thickness.

5. Optimization against buckling

With the development of huge wind turbines, the consideration of inbuckling stability of the supporting tower structure or the rotating blades can be another crucial factor in designing

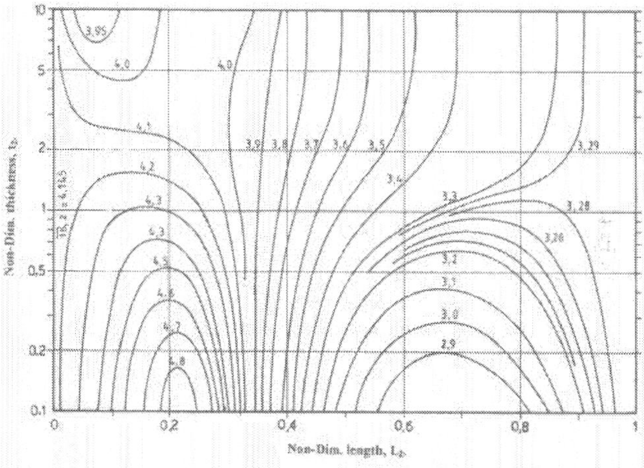

Fig. 11. Second-mode frequency of a teetered blade

efficient wind generators. The tower constructions are usually manufactured as tubular steel sections, which are more economical than solid sections in resisting compressive loads. (Maalawi, 2002) developed an optimization model for maximizing the critical buckling load of elastic columns under equality mass constraint. The given formulation considered columns made of uniform segments with the design variables defined to be the radius of gyration (r_k), wall thickness (t_k) and length (L_k) of each segment. The exact structural analysis ensured the attainment of the absolute maximum critical buckling load for any number of segments, type of cross section and type of boundary conditions. Fig. 12 shows the developed contours of the dimensionless critical buckling load (P_{cr}) for columns having the wall thickness held at its design value in order to avoid the possibility of local buckling of the cylindrical shell structure. The buckling load was found to be very sensitive to variation in the segment length. Investigators who use finite elements have not recognized that the length of each element can be taken as a main variable in addition to the cross-sectional properties.

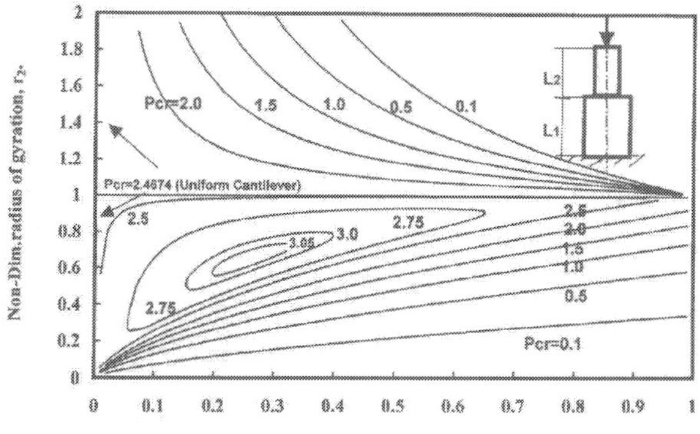

Fig. 12. Optimal buckling design of tubular cantilevers constructed from two segments

Table 4 gives the attained optimal solutions for cantilevers constructed from different number of segments (N_s). About 36% optimization gain was obtained for a cantilever only made of five segments not equally spaced, which represents a truly strongest column design. It is to be noticed that all variables are given in dimensionless form with respect to a known baseline design having uniform properties with the same total length, material properties and cross-sectional type and shape. For example, the critical buckling load P_{cr} shown in Fig. 9 is normalized by the quantity (EI/L^2) of the baseline design, where E=Young's modulus, I=second moment of area and L=total column's length.

Ns	(r_k, L_k), k=1,2,.....,N_s	$P_{cr,max}$	Gain (%)
1	(1,1)	2.467	0
2	(1.117,0.75), (0.65,0.25)	3.072	24.5
3	(1.185,0.507), (0.935,0.333), (0.555, 0.161)	3.248	31.62
4	(1.195,0.495), (0.988, 0.283), (0.687, 0.146), (0.395, 0.077)	3.348	35.70
5	(1.214,0.392), (1.098,0.207), (0.916,0.217), (0.629, 0.109),(0.399, 0.075)	3.368	36.5

Table 4. Optimum tower patterns under mass and local buckling constraints

Another work by (Maalawi, 2009) considered optimization of columns made of composite materials with the design variables taken to be the volume fraction distribution (V) of the constituent materials along the column's length. Results for the optimum columns made of unidirectional composites having the largest possible resistance against buckling are depicted in Fig. 13. It is seen that the increase in the number of segments would, naturally, result in higher values of the critical buckling load. However, care ought to be taken for the corresponding increase in cost of manufacturing procedures.

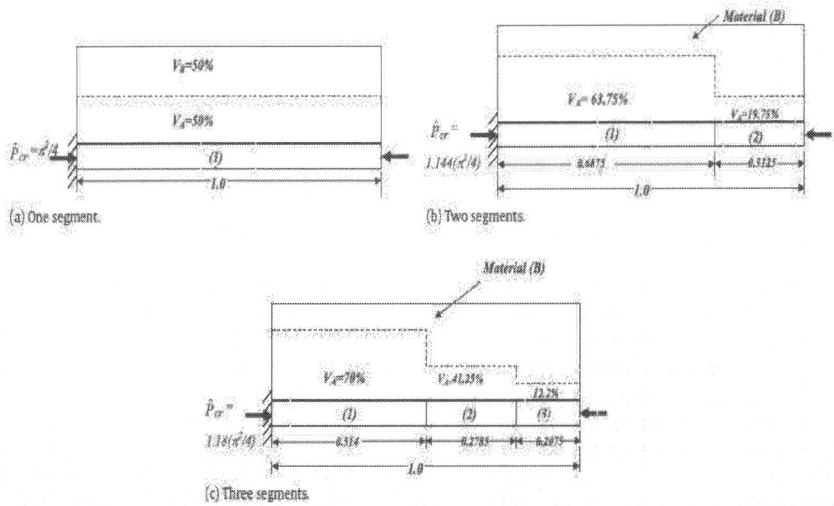

Fig. 13. Strongest columns with axial material grading: Material (A)=E-glass fibers, material (B)=epoxy matrix

6. Material grading for improved aeroelastic stability of composite blades

Wind turbine blades in parked position can experience aeroelstic instability condition for winds coming from all possible directions with speeds ranging from the cut-in up to the survival wind speed. A solution that can be promising to enhance aeroelastic stability of composite blades is the use of the concept of functionally graded materials (FGMs), in which the mechanical and physical properties vary spatially within the structure. FGMs may be defined as advanced composite materials that fabricated to have graded variation of the relative volume fractions of the constituent materials. (Librescu & Maalawi, 2007) introduced the underlying concepts of using material grading in optimizing subsonic wings against torsional instability. They applied an exact mathematical approach allowing the material properties to change in the wing spanwise direction, where both continuous and piecewise structural models were successfully implemented. The enhancement of the torsional stability was measured by maximization of the critical air speed at which divergence occurs with the total structural mass kept at a constant value in order not to violate other performance requirements. Fig. 14 shows a rectangular composite blade model constructed from uniform piecewise panels, where the design variables are defined to be the fiber volume fraction (V_f) and length (L) of each panel. It was shown by many investigators in the field that the use of piecewise models in structural optimization gives excellent results and can be promising for many engineering applications (Negm & Maalawi, 2000).

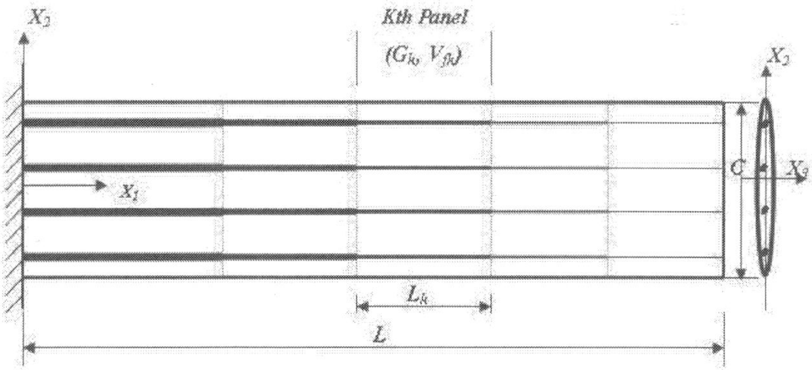

Fig. 14. Composite blade model with material grading in spanwise direction

Fig. 15 shows the isomerits for a blade composed from two panels made of carbon-AS4/epoxy-3501-6 composite. The selected design variables are (V_{f1}, L_1) and (V_{f2}, L_2). However, one of the panel lengths can be eliminated, because of the equality constraint imposed on the blade span. Another variable can also be discarded by applying the mass equality constraint, which further reduces the number of variables to only any two of the whole set of variables. Actually the depicted level curves represent the dimensionless critical air speed augmented with the imposed equality mass constraint. It is seen that the function is well behaved, except in the empty regions of the first and third quadrants, where the equality mass constraint is violated. The final constrained optima was found to be (V_{f1},L_1)= (0.75, 0.5) and (V_{f2},L_2)= (0.25, 0.5), which corresponds to the maximum value of the critical speed of 1.81, representing an optimization gain of about 15% above the reference value $\pi/2$.

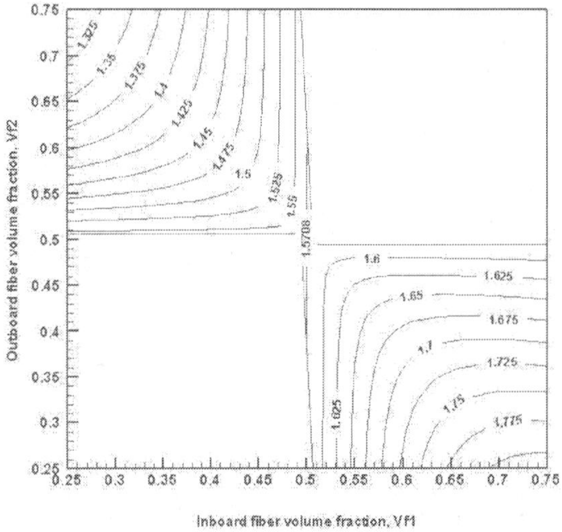

Fig. 15. Isomerits of the normalized divergence speed ($\hat{V}_{div}$) for a two-panel blade model

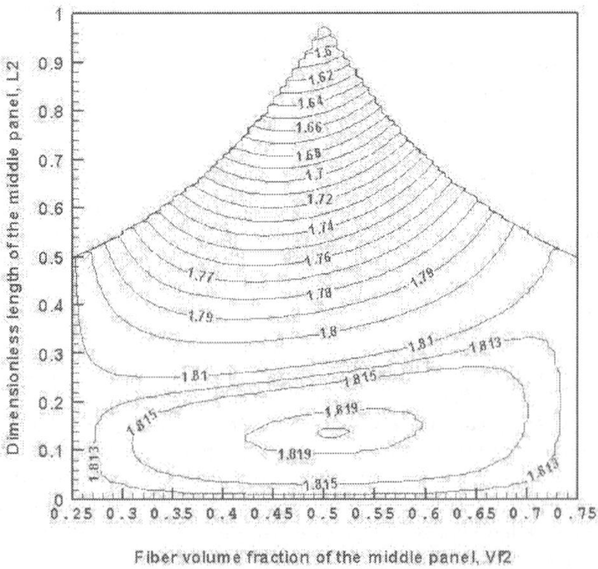

Fig. 16. Isomerits of $\hat{V}_{div}$ in (V_{f2}-L_2) design space for a three-panel model

The functional behavior of the critical air speed $\hat{V}_{div}$ of a three-panel model is shown in Fig. (16), indicating conspicuous design trends for configurations with improved aeroelastic performance. As seen, the developed isomerits has a pyramidal shape with its vertex at the design point (V_{f2}, L_2)=(0.5, 1.0) having $\hat{V}_{div}$ =π/2. The feasible domain is bounded from above by the two lines representing cases of two-panel blade, with V_{f1}=0.75 for the line to the left and V_{f3}=0.25 for the right line. The contours near these two lines are asymptotical to them in order not to violate the mass eqaulity constraint. The final global optimal solution, lying in the bottom of the pyramid, was calculated using the MATLAB optimization tioolbox routines as follows: $(V_{fk}, L_k)_{k=1,2,3}$ = (0.75, 0.43125), (0.5, 0.1375), (0.25, 0.43125) with $\hat{V}_{div}$ =1.82, which represents an optimization gain of about 16%. Actually, the given exact mathematical approach ensured the attainment of global optimality of the proposed optimization model. A more general case would include material grading in both spanwise and airfoil thickness directions.

7. Conclusion

Wind power is now growing rapidly in the world. Although it currently supplies little of the world's electricity needs, the amount of proposed new wind plants is significant, and could soon make wind the largest source of new power supply generating zero-emissions electricity at an affordable cost. It is the main concern of this chapter to consider system optimization analysis for improving performance and operational efficiency of wind turbines, especially in early stages of product development. Simplified realistic optimization models focusing on both aerodynamic and structural efficiencies of the main structural components, namely; the rotating blades and the supporting tower structures are presented and discussed. Design variables encompass blade and tower geometry as well as cross-sectional parameters. Different strategies have been addressed, including power output maximization, blade optimization in flapping and pitching motion and yawing dynamic optimization of combined rotor/tower structure. Optimization of the supporting tower structure against buckling as well as the use of the concept of material grading for enhancing the aeroelastic stability of composite blades have been also addressed. It has been shown that normalization of all terms results in a naturally scaled objective functions, constraints and design variables, which is recommended when applying different optimization techniques. Several design charts that are useful for direct determination of the optimal values of the design variables are given. The various approaches that are commonly utilized in design optimization are also presented with a brief discussion of some computer packages classified by their specific applications. Another promising area of future research is the development of topology optimization methods to simultaneously design smart wind turbine structures. Actually, the most economic design that will perform its intended function with adequate safety and durability requires much more than the procedures that have been described in this chapter. Further optimization studies must depend on a more accurate analysis of constructional cost. This combined with probability studies of load applications and materials variations, should contribute to further efficiency achievement. Finally, much improved and economical designs for wind turbines may be obtained by considering multi-disciplinary design optimization, which allows designers to incorporate all relevant design objectives simultaneously.

8. References

Burger, C. & Hartfield, R. (2006). Wind turbine airfoil performance optimization using the vortex lattice method and genetic algorithm. 4th AIAA Energy Conversion Engineering Conference, June 2006, AIAA 2006-4051, 26-29.

Cobb, R., Canfield, R. and Liebst, B. (1996). Finite element model tuning using automated structural optimization system software. *AIAA Journal*, Vol. 34, No.2, 392-399, ISSN: 0001-1452.

Kusiak A., Zheng H. and Song Zhe (2009). Power optimization of wind turbines with data mining and evolutionary computation. *Renewable energy*, Vol.35, 695-702.

Librescu, L. & Maalawi, K. (2007). Material grading for improved aeroelastic stability in composite wings. *Journal of Mechanics of Materials and Structures*, Vol.2, No.7, 1381-1394.

Maalawi, K. & Badawy, M. (2001). A direct method for evaluating performance of horizontal axis wind turbines. *Renewable and Sustainable Energy Reviews*, Vol.5, No.2, 175-190.

Maalawi, K. & Negm, H. (2002). Optimal frequency design of wind turbine blades. *Journal of Wind Engineering and Industrial Aerodynamics*, Vol.90, No.8, 961-986.

Maalawi, K. (2002). Buckling optimization of flexible columns. *International Journal of Solids and Structures*, Vol.39, 5865-5876.

Maalawi, K. & Badr, M. (2003). A practical approach for selecting optimum wind rotors. *International Journal of Renewable Energy*, Vol.28, No.5, 803-822.

Maalawi, K. (2007). A model for yawing dynamic optimization of a wind turbine structure. *International Journal of Mechanical Sciences*, Vol.49, No.10, 1130-1138.

Maalawi, K. (2009). Optimization of elastic columns using axial grading concept. *Engineering Structures*, Vol.31, No.12, 2922-2929.

Maalawi, K. & Badr, M. (2010). Frequency optimization of a wind turbine blade in pitching motion. *Proceedings of the Institution of Mechanical Engineers, Part A: Journal of Power and Energy*, Vol.224 (A4), No.2, 545-554, Professional Engineering Publisher, ISSN: 0957-6509JPE907.

Manwell, J.; McGowan, J. & Rogers, A. (2009). *Wind Energy Explained: Theory, Design & Application*, 2nd edition, Wiley, ISBN-13: 978-0470015001, United Kingdom.

Negm, H. & Maalawi K. (2000). Structural design optimization of wind turbine towers. *Computers and Structures*, Vol.74, No.6, 649-666.

Overgaard, L.C.T. & Lund E. (2005). Structural design sensitivity analysis and optimization of Vestas V52 wind turbine blade, *Proceedings of the 6th World Congress on Structural and Multidisciplinary Optimization*, CD-ROM, 10 pages, ISBN: 85-285-0070-5, Rio de Janeiro, 30 May-03 June 2005, Brazil.

Rao, S. (2009). *Engineering Optimization: Theory and Practice*, 4th edition, John Wiley & Sons, ISBN: 978-0470183526, New York.

Spera, D. (2009). *Wind Turbine Technology: Fundamental Concept in Wind Turbine Engineering*, 2nd edition, ASME Press, ISBN-13: 978-0791802601, New York.

Venkataraman, P. (2009). *Applied Optimization with MATLAB Programming*, 2nd edition, John Wiley & Sons, ISBN: 978-0470084885, New York.

Wei Tong (2010). *Wind Power Generation and Wind Turbine Design*, WIT Press, ISBN: 978-1845642051, United Kingdom.

2

Productivity and Development Issues of Global Wind Turbine Industry

Ali Mostafaeipour
Industrial Engineering Department, Yazd University, Yazd, Iran

1. Introduction

Renewable energies (REs) represent a cornerstone to steer our energy system in the direction of sustainability and supply security and a broad set of different RE technologies and resources exist today [1].Wind is one of the world's fastest growing renewable energy sources. The rapid growth in wind power is a result of improvements accomplished in technology [2].The recent focus on finding solutions for mitigating global warming has resulted in renewable energy technologies gaining importance. Among the renewable energy technologies, wind power is one of the fastest growing technologies globally at an average annual growth rate of more than 26% since 1990 [3, 4].

The worldwide demand for renewable energy is increasing rapidly because of the climate problem, and also because oil resources are limited. Wind energy appears as a clean and good solution to cope with a great part of this energy demand [5]. Developing a utility-scale wind project is a complicated and time-consuming process involving developers, landowners, utilities, the public and various local authorities. Although each wind energy project is unique and have different characteristics, basic features and related steps are common [2].In practice, the steps are iterative and overlap one another depending on the specific project circumstances. The key steps of development and planning for a wind farm are site selection, detailed wind assessment, feasibility, construction and operation [2, 6].

Estimates of the global technical potential of solar energy range from 15 to 4300 EJ, and for on-shore wind 3-600 EJ [7, 8, 9]. Over three quarters of global wind turbine sales come from only four turbine manufacturing companies: Vestas, GE Wind, Enercon, and Gamesa [10].

The Intergovernmental Panel on Climate Change (IPCC) in 2001 [11], for example, estimated the annual global theoretical terrestrial potential of wind as 1728 EJ from all land with mean annual wind speeds >5.1 m/s at 10m above the ground. This amount was then reduced to give a technical potential of only 72 EJ based on the experience of the Netherlands and the USA [7].As a resource, wind energy offers a number of advantages over solar energy. Peak energy intensity for wind turbines is currently around 400 W/m^2, more than twice that for solar PV [7].

Energy is one of the essential inputs for economic development and industrialization. Fossil fuels are the main resources and play a crucial role to supply world energy demand. However fossil fuel reserves are limited and usage of fossil fuel sources have negative environmental impacts. Therefore, management of energy sources, rational utilization of energy, and renewable energy source usage are vital .Since the first oil crisis, renewable energy sources have gained a great importance due to their inexhaustibility, sustainability, ecological awareness and supply of energy security. So, renewable energy sources are expected to play an important role especially in electrical energy generation [12].

Among the renewable energy sources wind energy is currently viewed as one of the most significant, fastest growing, commonly used and commercially attractive source to generate electrical energy because of the mature and cost effective energy conversation system technology. So, electricity generation cost from wind energy system has become competitive with fossil fuel systems. Installed total wind power capacity has reached over 93 GW and installed wind power capacity generates more than 1% of the global electricity consumption .In recent years Weibull distribution has been one of the most commonly used, accepted, recommended distribution to determine wind energy potential and it is also used as a reference distribution for commercial wind energy softwares such as Wind Atlas Analysis and Application Program (WAsP) [12,13,14,15].Germany is a leader in Europe on shifting from conventional to renewable sources of energy. As its land-based sites of wind energy are built to capacity [16,17], Germany looks to the sea for further production possibilities .In the United States and Germany, offshore areas are generally considered public space, which makes offshore renewable energy development public in nature [18].

Many countries and sub-national governments are looking not only to expand their domestic use of renewable energy, but also to develop accompanying local renewable energy technology manufacturing industries to serve that demand [18]. Cost is a central issue in meeting greenhouse gas emission reduction goals [19, 20]. Some clean energy technologies are relatively costly today but costs may decrease over time as technological improvements occur, equipment is standardized, and economies of scale take hold [20].

Electricity generated from wind power currently represents only 0.5 percent of global electricity production, and about a 7 billion (US) dollar annual industry [23]. The market is expected to double over the next 4 years [10], and it is this perceived potential for future growth and the rapid growth rates to date that are causing many governments to look toward developing domestic wind technology manufacturing industries. Countries and sub-national governments around the world—in both developed and developing countries—are therefore establishing policies to promote the construction of new wind power installations, and some have developed targeted policies to specifically encourage local manufacturing of large wind turbine technology [25]. The reasons why it is difficult to quantify the exact number of jobs coming from wind-related activities have to do with the lack of detail in the official statistics, which does not allow researchers to exploit the data, and with the variety of company profiles that make up the sector, which can be hard to identify and examine. The wind energy sector has grown exponentially since the end of the 1990s, especially within the European Union (EU), and this has affected the employment levels of the regions involved [26].

2. Global installed wind turbines and energy potential

A reliable supply of energy is essential to maintain and to improve human being's living conditions. Compared to the conventional coal-fired approach, renewable energy (RE) helps to mitigate the impacts of greenhouse gas emissions to a large extent .According to the Global Wind Energy Council (GWEC), the global cumulative installed capacity has reached 94 GW in 2007, which increased 31% than the previous year [24,27]. Today, the world's energy supply is largely based on fossil fuels and nuclear power. These sources of energy

will not last forever and have proven to be contributors to our environmental problems. The environmental impacts of energy use are not new but they are increasingly well known; they range from deforestation to local and global pollution. In less than three centuries since the industrial revolution, mankind has already burned roughly half of the fossil fuels that accumulated under the earth's surface over hundreds of millions of years. Nuclear power is also based on a limited resource (uranium) and the use of nuclear power creates such incalculable risks that nuclear power plants
Cannot be insured [28].

Based on a survey amongst the WWEA (World Wind Energy Association) member associations, a double digit growth for the wind energy market is expected despite the general economic crisis. Based on available figures from 11 of the top 15 countries representing over 80% of the world market, WWEA recorded 5374 MW new installed capacity in the first quarter of 2009, equaling an increase of 23% compared with last year in the same countries. WWEA keeps its previous prevision of a total installed capacity of 152000 MW worldwide by the end of 2009 (Fig. 1), which will mean a new record of over 30,000 MW newly installed capacity within one year. This represents a market growth of 25% compared with last year [29].

Based upon the World Wind Energy Report in 2008, following global highlights is important [30]:

- Worldwide capacity reaches 121188 MW, out of which 27261 MW were added in 2008.
- Wind energy continued its growth in 2008 at an increased rate of 29 %.
- All wind turbines installed by the end of 2008 worldwide are generating 260 TWh per annum, equaling more than 1.5 % of the global electricity consumption.
- The wind sector became a global job generator and has created 440'000 jobs worldwide.
- The wind sector represented in 2008 a turnover of 40 billion Euros.
- For the first time in more than a decade, the USA took over the number one position from Germany in terms of total installations.
- China continues its role as the most dynamic wind market in the year 2008, more than doubling the installations for the third time in a row, with today more than 12 GW of wind turbines installed.
- North America and Asia catch up in terms of new installations with Europe which shows stagnation.
- Based on accelerated development and further improved policies, a global capacity of more than 1,500,000 MW is possible by the year 2020.

The status of wind energy is as a stable, profitable and low-risk investment. Although some wind energy projects are postponed due to financing challenges, the overall market development can still compensate such delays showing great signs of vitality. A substantial share of the slow down in some regions are a consequence of new regulations and bureaucratic delays that undermine the development of new wind parks rather than of financing difficulties [29].

Areas of potential wind turbine improvements include [21, 22]:

- Advanced tower designs, including taller towers, new materials, and telescoping towers that are easier to install.
- Larger rotors made from lighter materials and having improved aerodynamics.
- More efficient gear boxes, drive trains, generators, and electronics.

Renewable sources of energy are essential parts of an overall strategy of sustainable development. They help reduce dependence of energy imports, thereby ensuring a sustainable supply. Furthermore renewable energy sources can help improve the competitiveness of industries over the long run and have a positive impact on regional development and employment. Renewable energy technologies are suitable for off-grid services, serving those in remote areas of the world without requiring expensive and complicated grid infrastructure [28].

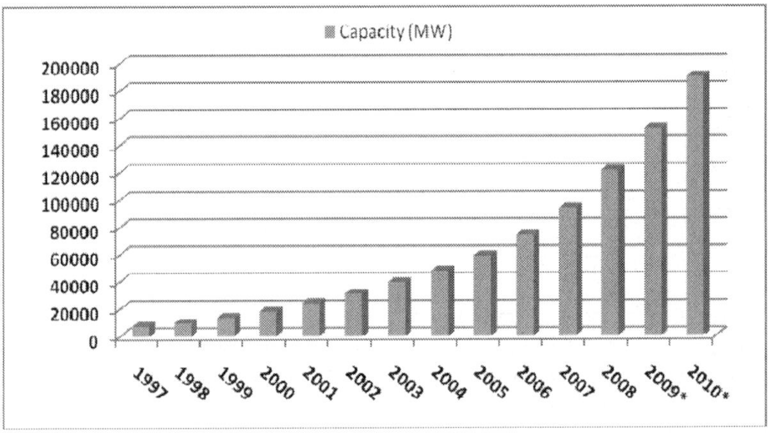

*= Prediction

Fig. 1. Total world installed capacity [30].

2.1 Energy potential

The theoretical potential of wind energy as illustrated in Table 1 amounts to 6,000 EJ (or more than 12 times current energy needs), what seems to be enormously high when compared to its current use. A technical potential is estimated to be 10% of the theoretical one. The ultimate potential of wind-generated electricity worldwide could indeed be very large: Other assessments state figures of up to 50 times current global final electricity consumption [1, 33, 34, and 35].

Height limitations of wind converters, coast distance of offshore sites, insufficient wind velocities and land use, and the feasible grid integration all limit the realizable potential of this promising technology option.

2.1.1 Technical potential

If technical boundary conditions (i.e. efficiencies of conversion technologies, overall technical limitations as, e.g. the available land area to install wind turbines) are considered the technical potential can be derived.

2.1.2 Theoretical potential

For deriving the theoretical potential general physical parameters have to be taken into account (e.g. based on the determination of the energy flow resulting from a certain energy

resource within the investigated region). It represents the upper limit of what can be produced from a certain energy resource from a theoretical point-of-view—of course, based on current scientific knowledge [1].
Potential for geothermal energy is higher than other resources of energy, but its use was only 2 EJ in 2004 which is very low.
Potential for biomass energy is 2900 EJ, but its use was 50 EJ in 2004 which was highest if we compare with other resources of energy in the world.

Resource	Use in 2004 (EJ)	Technical Potential	Theoretical Potential
Biomass	50	250	2900
Geothermal	2	5000	140000000
Hydropower	10	50	150
Ocean	-	-	7400
Solar	0.2	1600	3900000
Wind	0.2	600	6000
Total	62.4	7500	143916450

Table 1. Technical and theoretical potentials and usage for various renewable energy sources (in terms of primary energy) at global scale [33].

The top five countries in terms of total installed wind capacity at the end of 2004 were Germany, Spain, USA, Denmark, and India; wind turbine manufacturers from these top five countries sold 94 percent of all wind turbines installed globally in 2004. Germany clearly stands out as having maintained the most sizable and stable market [25].
The US and Indian markets have been much less stable than those of Germany, Spain, and Denmark. Annual installations in the US were highest in 2001 and 2003 with over 1,600MW installed each year. However, annual installations dipped well below 200MW per year between 1995 and 1998, and again in 2000. The year 2004 was also slow in the US, as the on-again, off-again nature of the federal production tax credit (PTC) has created significant uncertainty in the market in recent years. India's market has also been unstable, with initial growth in the mid 1990s, a slowdown in the late 1990s, and some resurgence in recent years. From 2001 to 2004, however, India has been able to maintain annual installations of over 200MW per year. Though the instability of the Indian and US markets has not stopped local investments in wind manufacturing (in part because the long-term market potential is so large in both countries), it has often complicated the process of developing successful local wind manufacturing industries [25].
The Brazilian government has also pursued policies governing wind farm development that include stringent local content requirements, primarily through the recent Proinfa legislation (the Incentive Program for Alternative Electric Generation Sources) that offers fixed-price electricity purchase contracts to selected wind projects. Starting in January 2005, the Proinfa legislation requires 60 percent of the total cost of wind plant goods and services to be sourced in Brazil; only companies that can prove their ability to meet these targets can take part in the project selection process. In addition, from 2007 onwards, this percentage increases to 90 percent [25, 31].
China has also been using local content requirements in a variety of policy forms. China's 1997 "Ride the Wind Program" established two Sino-foreign joint venture enterprises to domestically manufacture wind turbines; the turbines manufactured by these enterprises

under technology transfer arrangements started with a 20 percent local content requirement and a goal of an increase to 80 percent as learning on the Chinese side progressed [25,32]. China's recent large government wind tenders, referred to as wind concessions, have a local content requirement that has been increased to 70 percent from an initial 50 percent requirement when the concession program began in 2003. Local content is also required to obtain approval of most other wind projects in the country, with the requirement recently increased from 40 to 70 percent [25].

3. Wind market growth rates

An important indicator for the vitality of the wind market is the growth rate in relation to the installed capacity of the previous year. The growth rate went up steadily since the year 2004, reaching 29.0 % in 2008, after 26.6 % in 2007, 25.6 % in the year 2006 and 23.8 % in 2005. However, this increase in the average growth rate is mainly due to the fact that the two biggest markets showed growth rates far above the average: USA 50 % and China 107 % (Fig. 2). Bulgaria showed the highest growth rate with 177 %, however, starting from a low level. Also Australia, Poland, Turkey and Ireland showed a dynamic growth far above the average [30]. Figure 3 shows that world wind market growth rate in 1999 was the highest and then was decreased to the year 2004 which was the lowest. Since 2004 it has had a slight increase. It is also interesting to know that growth rate for Turkey in 2007 was a lot more than 2008.

Between 1991 and 1995 both the average list price of wind turbines and turnkey investment costs of wind farms in Germany have declined steadily by about 8-9% per year. However, average prices remained rather stable since 1995. In fact, the price of the cheapest turbine available even increased during 1995-1999. There are a number of possible explanations for these trends. In Germany, more and more wind parks are situated in inland areas with

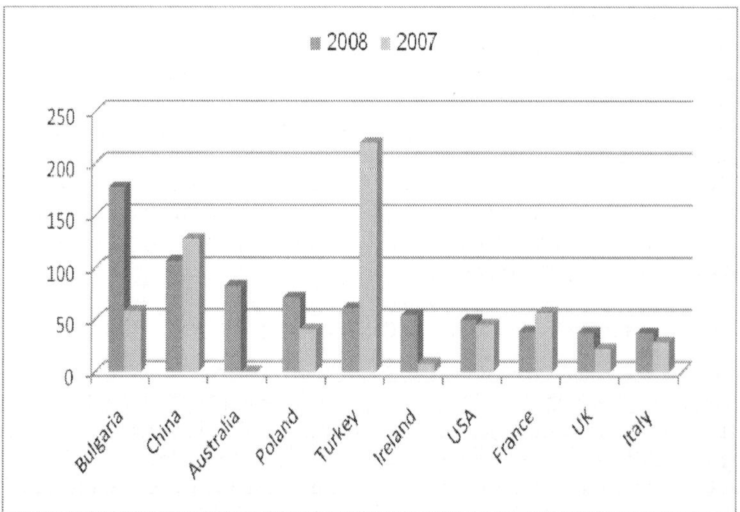

Fig. 2. Top ten countries with highest growth rate [30].

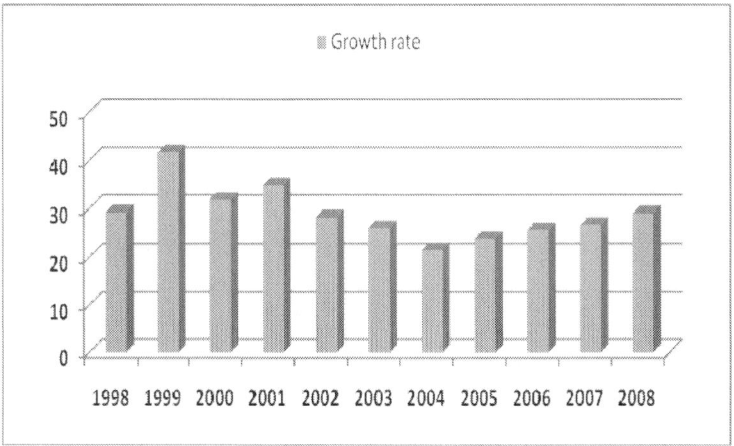

Fig. 3. World market growth rate in % [30].

lower wind speeds, due to a lack of appropriate sites near the coast. While in 1993, 70% of all new wind parks (in terms of capacity) were installed in coastal regions, this share has dropped to a mere 10-15% in 1999 [36,37]. The wind energy sector is one of the fastest-growing energy sectors in the world. From 1991 until the end of 2002, global installed capacity has increased from about 2GW [38] to over 31GW [39], with an average annual growth rate of about 26%. During this period, both prices of wind turbines and cost of wind-generated electricity have been reduced. In spite of these developments, electricity derived from wind is not yet able to fully compete with electricity produced from fossil fuel. However, this may change in the near future [36, 40].

In terms of countries, the 'big five' (Germany, Spain, Denmark, the USA and India) have been at the top for the last decade (from 1995 to 2005). In these countries over 80% of the worldwide wind-based power generation capacity was installed in 2005 [36, 39]. The expansion of renewable energies requires additional investments into production facilities as well as into the transportation and distribution grid .Since the majority of renewable energy technologies is not profitable at current energy prices, its furtherance is not only associated with production and employment effects but with increasing cost as well. It is apparent that the cost disadvantage of renewable compared to conventional energies is crucially dependent on future prices of energies used in power plants as well as on the amount of CO_2 emission permits [63, 64].

4. Installed wind turbines worldwide (2007- 2008)

The share of new installed capacity for USA and China with 31.62% and 23.83% respectively accounts for more than half of the other countries in the world for 2008 (Fig 4).

The USA and China took the lead, USA taking over the global number one position from Germany and China getting ahead of India for the first time, taking the lead in Asia. The USA and China accounted for 50.8 % of the wind turbine sales in 2008 and the eight leading markets represented almost 80 % of the market for new wind turbines. One year ago, still only five markets represented 80 % of the global sales. The pioneer country Denmark fell back to rank 9 in terms of total capacity, whilst until four years ago it held the number 4 position during several years (Fig. 5). However, with a wind power share of around 20 % of the electricity supply, Denmark is still a leading wind energy country worldwide [30].For the year 2008, USA was in top position following by Germany, Spain, China, India, Italy, France, UK, Denmark and Portugal. But Germany was in top position in 2007.

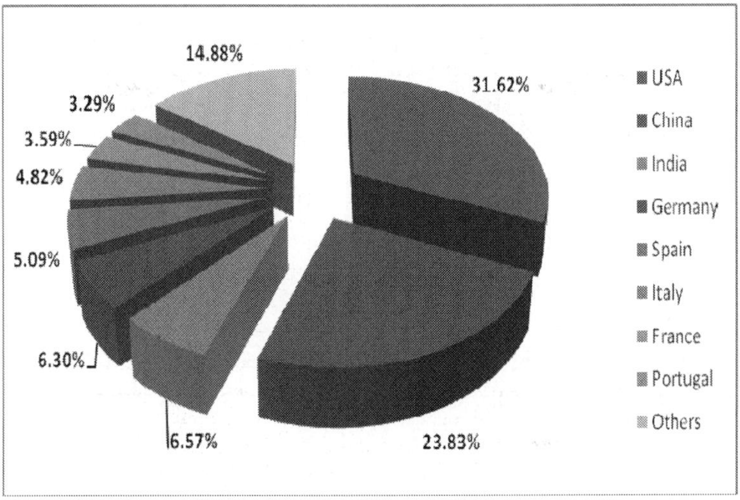

Fig. 4. Share of countries new installed capacity for 2008 [30]

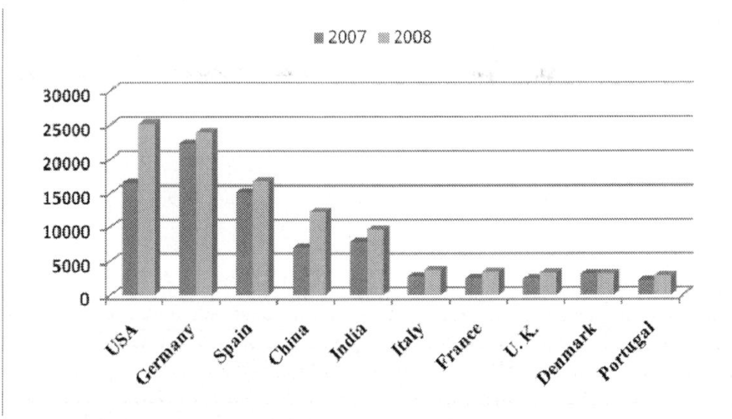

Fig. 5. Top ten wind turbine installed countries (MW) [30].

5. World installation of wind turbines for 2006

The global wind energy market experienced yet another record year in 2006, demonstrating a growth of 32% over 2005 figures. According to the statistics (Table 2) issued by the Global Wind Energy Council (GWEC), 2006 saw the installation of 15197 megawatts (MW) of new capacity, taking total installed wind energy capacity to 74,223 MW. In terms of economic value, the wind energy sector is now established as an important player in the energy market, the GWEC says. The total value of new generating equipment installed in 2006 was worth €18 billion (US$23 billion). The countries with the highest total installed capacity are Germany (20,621 MW), Spain (11,615 MW), the USA (11,603 MW), India (6,270 MW) and Denmark (3,136 MW). Thirteen countries around the world have now passed 1000 MW level for installed capacity. In terms of new capacity added in 2006, the USA led the way with 2,454 MW, followed by Germany (2,233 MW), India (1,840 MW), Spain (1,587 MW), China (1,347 MW) and France (810 MW). These figures show that new players such as China and France are gaining ground [41]. By the end of 2006, cumulative installed wind capacity of China had reached 2.6GW; the average annual growth rate over the past ten years has been 46%. Between 2004 and 2006, China's ranking in the world wind energy league moved up from the top 10 to the top 6, and the country is planning to host some of the biggest wind farms in the world. At the present growth rate, the 2010 target will be reached two years earlier. Wind power has not just contributed to supplying electricity but has lowered supply costs, reduced carbon emissions and helped to limit air pollution [42].

New capacity	MW	Market share (%)
USA	2454	16.1
Germany	2233	14.7
India	1840	12.1
Spain	1587	10.4
China	1347	8.9
France	810	5.3
Canada	776	5.1
Portugal	694	4.6
UK	634	4.2
Italy	417	2.7
Top 10 total	12 792	84.2
Rest of world	2405	15.8
World total	15 197	

Table 2. Installed capacity of top countries for January-December 2006 [41].

5.1 Europe

Europe was the leading player in the market, with 48,545 MW of installed capacity at the end of 2006 - 65% of the global total. In 2006, European wind capacity grew by 19%, producing approximately 100 TWh of electricity, equal to 3.3% of total European Union (EU) electricity consumption in an average wind year. "While Germany and Spain still represented 50% of the EU market, there was healthy trend towards less reliance on these two countries," says Christian Kjaer, the European Wind Energy Association's (EWEA) CEO. "In the EU, 3,755 MW was installed outside of Germany, Spain and Denmark in 2006.

In 2002, this figure still stood at only 680 MW [41].The figures show that most of the European countries were serious about investing into wind market.

Following the agreement reached in March 2008 by the Heads of State [43], the European Union has committed itself to achieving, by 2020, that 20% of the energy it consumes comes from renewable energies and that its CO_2 emissions are cut by 20% in comparison with 1990 levels (30% if other developed countries join the effort) [44].Wind is the most dynamic renewable energy in Europe and in the world; it already covers 3% of electricity demand in the EU—up to 23% in Denmark and around 8% in Spain and Germany [45] and is the second largest attractor of energy investments after natural gas [44,46]. Germany with total amount of 20,622 MW, Spain with 11,615 MW and Denmark with 3,136 MW installed wind power capacity were in top positions in Europe. It shows that there was a tremendous need for renewable energies like wind in order to combat high price of fossil fuel. Europe with 48,545 MW of installed wind power capacity in 2006 was in top position which is admirable.

5.2 Asia

Asia experienced the strongest increase in installed capacity outside of Europe, with an addition of 3,679 MW. This took the continent's total to over 10,600 MW. In 2006, wind capacity in Asia grew by 53% and accounted for 24% of new installations. The strongest market remains India, which installed over 1,840 MW of new capacity in 2006, increasing its total to 6,270 MW. China more than doubled its total installed capacity in 2006, taking it up to 2,604 MW by installing 1,347 MW of capacity, making it the sixth largest market worldwide. The Chinese market was boosted by the country's new Renewable Energy Law, which entered into force on 1 January 2006 [41].

In 2006, the burning of coal produced two-thirds of the primary energy consumed in China. Even with improvements in end-use energy efficiency, energy demand continues to grow and so does the air pollution. In China, pollution is causing serious health problems; crop damage and acid rain, all of which are taking a social and economic toll [42]. Air pollution has been a very serious problem in China, therefore government has implemented new regulations toward using renewable energies in order to decrease co_2 .They plan to have 5,000 MW of wind energy by the year 2010. India with total amount of 6,270 MW, China with 2604 MW and Japan with 1394 MW installed wind power capacity were in top positions in Asia. It shows that there was a great effort and attention in these countries toward using wind energy. The reason might be high cost of fossil fuel which was imported from Persian Gulf countries.

5.3 North America

North America accounted for 22% of the world's new installed wind capacity in 2006. For the second year running, the US wind energy industry installed nearly 2,500 MW, making it the country with the most new wind power. "Wind's exponential growth reflects the nation's increasing demand for clean, safe and domestic energy, and continues to attract both private and public sources of capital," comments Randy Swisher, president of the American Wind Energy Association (AWEA). "New generating capacity worth US $4 billion was installed in 2006, billing wind as one of the largest sources of new power generation in the country - second only to natural gas - for the second year in a row." Canada also had a record year, with the installed capacity more than doubling from 683 MW in 2005 to 1459 MW at the end of 2006. "Wind energy is an emerging Canadian success story and 2006 will be remembered as the year that our country first began to seriously capture its economic and environmental benefits," according to Robert Hornung, president of the Canadian Wind Energy Association (CanWEA). "Canada's on the cusp of a wind energy boom as provincial governments are now targeting to have a minimum of 10,000 MW of installed wind energy capacity in place by 2015" [41]. USA with total amount of 11,603 MW and Canada with 1,459 MW installed wind power capacity were only countries in North America.

5.4 Latin America and Caribbean

Brazil with total amount of 237 MW, Mexico with 88 MW and Costa Rica with 74 MW installed wind power capacity were in top positions in Latin America & Caribbean. It shows that there was not tendency for wind turbine installation in this part of the world. Reason could be high resources of fossil fuel in countries like Mexico and also great attention toward manufacturing of methanol in Brazil.

5.5 Rest of the world

According to table 3, growth in the relatively young African and Middle Eastern market picked up considerably in 2006, with 172 MW of new installed capacity, bringing the total up to 441 MW. This represents a 63% growth. The main countries experiencing increases are Egypt, Morocco and Iran. Compared to previous years, the Australian market only experienced slow growth in 2006 [41].Egypt with total amount of 230 MW, Morocco with 124 MW and Iran with 48 MW installed wind power capacity were in top positions in Africa and Middle East. It shows that there was not too much attention in other countries in theses regions toward using wind energy. Australia with total amount of 817 MW, New Zealand with 171 MW and Pacific Island with 12 MW installed wind power capacity were in top positions in Pacific Region. Australia has been active in field of wind energy.

Country	Total end 2005	New 2006	Total end 2006
Africa & middle east			
Egypt	145	85	230
Morocco	64	60	124
Iran	23	27	48
Tunisia	20	0	20
Other	11	0	11
Total	271	172	441
Asia			
India	4430	1840	6270
China	1260	1347	2604
Japan	1061	333	1394
Taiwan	104	84	188
South Korea	98	75	173
Philippines	25	0	25
Other	13	0	13
Total	6990	3679	10667
Europe			
Germany	18415	2233	20622
Spain	10028	1587	11615

Country	Total end 2005	New 2006	Total end 2006
Denmark	3128	12	3136
Italy	1718	417	2123
UK	1332	634	1963
Portugal	1022	694	1716
France	757	810	1567
Netherland	1219	356	1560
Austria	819	146	965
Greece	573	173	746
Ireland	496	250	745
Sweden	510	62	572
Norway	267	47	314
Belgium	167	26	193
Poland	83	69	153
Rest of Europe	364	192	556
Total Europe	40898	7708	48545
Out of which UE- 27	40512	7611	48062
Latin America & Caribbean			
Brazil	29	208	237
Mexico	3	85	88
Costa Rica	71	3	74
Caribbean (w/o Jamaica)	35	-	35
Argentina	27	-	27
Colombia	20	-	20
Jamaica	20	-	20
Other	7	-	7
Total	212	296	508
North America			
USA	9149	2454	11603
Canada	683	776	1459
Total	9832	3230	13062
Pacific region			
Australia	708	109	817
New Zealand	169	3	171
Pacific island	12	-	12
Total	889	112	1000
Word total	59091	15197	74223

Table 3. Global installed wind power capacity (MW)- regional distribution[41].

6. World installation of wind turbines for 2008

In terms of continental distribution, a continuous diversification process can be watched as well: In general, the focus of the wind sector moves away from Europe to Asia and North America. Europe (Fig 6) decreased its share in total installed capacity from 65.5 % in 2006 to 61 % in the year 2007 further down to 54.6 % in 2008. Only four years ago Europe dominated the world market with 70.7 % of the new capacity. In 2008 the continent lost this position and, for the first time, Europe (32.8 %), North America (32.6 %) and Asia (31.5 %) account for almost similar shares in new capacity. However, Europe is still the strongest continent while North America and Asia are increasing rapidly their shares. The countries in Latin America and Africa counted for respectively only 0.6 % and 0.5 % of the total capacity and fell back in terms of new installations down to respectively only 0.4 % and 0.3 % of the additional capacity installed worldwide in the year 2008[30]. Wind energy generating capacity in the US increased from about 2,500 MW in 1999 to about 21,000MW in mid 2008 and about 28,000MW in early 2009. At the same time, the costs of installed utility- scale wind projects (in constant $/kW) declined until the early 2000s and then generally increased [21, 47, 48]. Mass production is likely to play a significant role for future cost reductions. In the last 5 years, wind farms of several hundred MW capacities have been realized in Spain and the USA [36]. Since the majority of renewable energy technologies are not profitable at current energy prices, its furtherance is not only associated with production and employment effects but with increasing cost as well. It is apparent that the cost disadvantage of renewable compared to conventional energies is crucially dependent on future prices of energies used in power plants as well as on the amount of CO_2 emission permits [19, 20].Australian share in this regards is more than share of both Latin America and Africa.

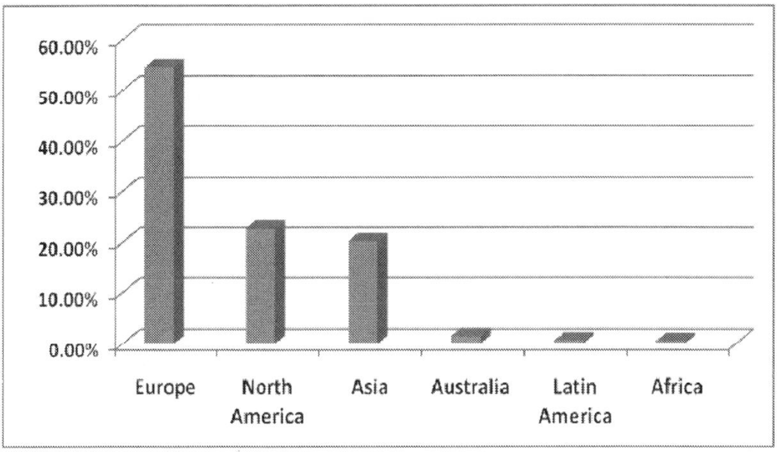

Fig. 6. Continental share of total installed capacity 2008[30].

6.1 Europe

Europe lost its dominating role as new market but kept its leading position in terms of total installation with 66'160 MW. Germany and Spain maintained as leading markets, both showing stable growth. The most dynamic European markets were Ireland (adding 440 MW, 55 % growth) and Poland (196 MW added, 71 % growth), the first Eastern European country with a substantial wind deployment. All in all, the European wind sector showed almost stagnation with a very small increase in added capacity from 8,607 MW to 8,928 MW. The biggest market Germany is expected, after the amendment of the renewable energy law EEG, to show bigger market growth in 2009. An encouraging change happened in the UK where the government announced the introduction of a feed-in tariff for community based renewable energy projects. However, the cap of 5 MW represents a major hurdle so that the UK wind market will still grow at moderate rates. However, without additional incentives for wind power in more EU member states, such as improved feed-in legislation, the European Union may not be able to achieve its 2020 targets for renewable energy [30].It goes without saying that most of the European countries were in top positions in 2008. Germany and Spain were in second and third position with total capacity installed of 23,902.8 and 16740.3 MW respectively. But Germany with 22,247.4 MW and Spain with 15,147.4 MW of total capacity installed for 2007 were in first and third positions. Italy, France, United Kingdom, Denmark and Portugal were in position of six to ten respectively for 2008.It shows great effort of European countries toward using wind energy for electricity production. Recently, because of the global economic crisis, some wind turbine manufacturing companies in Europe dismissed the workers and decreased production lines in order to combat the crisis.

6.2 Asia

Asia with the two leading wind countries China and India and 24,439 MW of installed capacity is in a position of becoming the worldwide locomotive for the wind industry. China has again doubled its installations and Chinese domestic wind turbine manufacturers have started for the first time to export their products. It can be expected that in the foreseeable future Chinese and Indian wind turbine manufacturers will be among the international top suppliers.

The Indian market has shown robust and stable growth in the year 2008. It has already a well-established wind industry which already plays a significant and increasing role on the world markets. Further countries like South Korea (already with 45 % growth rate in 2008) start investing on a larger scale in wind energy and it can be observed that more and more companies are developing wind turbines and installing first prototypes. In parallel with the market growth in the country, it can be expected that also new manufacturers will be able to establish themselves. The World Wind Energy Conference held on Jeju Island in June 2009 is expected to push the development in the region. Pakistan installed its first wind farm in the year 2008 and the Government of the country aims at further wind farms in the near future [30]. China has chosen wind power as an important alternative source in order to rebalance the energy mix, combat global warming and ensure energy security. Supportive measures have been introduced. In order to encourage technical innovation, market expansion and commercialization, development targets have been established for 2010 and 2020, concession projects offered and policies Introduced to encourage domestic production [42].

China with 12,210.0 MW and India with 9,587.0 MW of total capacity installed in 2008 were in positions of fourth and fifth in the world. Japan, South Korea and Iran with total installed capacities of 1,880.0 MW, 278 MW and 823 MW respectively were in positions of 13, 27 and 35 in the world for 2008(Table 3). The positions of Japan and South Korea for year 2007 were same as 2008, but Iran had position of 34 in 2007. Philippine, Israel, Pakistan, Jordan, Indonesia, Mongolia, Kazakhstan, Syria and South Korea were among the Asian countries with wind turbine activities in 2008.

6.3 North America

North America showed very strong growth in the year 2008, more than doubling its capacity since 2006 to 27,539 MW. Breaking two world records, the USA became the new number one worldwide in terms of added as well as in terms of total capacity. More and more US states are establishing favorable legal frameworks for wind energy and try to attract investors in manufacturing facilities. It can be expected that the new President Obama administration will improve substantially the political frameworks for wind power in the country, especially for those types of investors that have practically been excluded from the production tax credit scheme, like farmers, smaller companies or community based projects. The credit crunch, however, may lead to delays in project development in the short term. The Canadian government has rather been hesitating. However, among the Canadian provinces Quebec and Ontario are showing increasing commitment towards an accelerated deployment of wind energy. In Quebec, contracts for new projects were signed for a total of 2000 MW, the first to be operational by 2011[30].USA with total installed capacity of 25,170 MW and Canada with 2369 MW in 2008 were in positions of 1st and 11th in the world. But for the year 2007, USA was in position of 2nd and Canada in position of 11th.

6.4 Latin America

Many Latin American markets still showed stagnation in the year 2008 and the overall installed capacity (667 MW) in the region accounts for only 0.5 % of the global capacity. Only Brazil and Uruguay installed major wind farms in the year 2008. This slow wind deployment is especially dangerous for the economic and social prospects of the region as in many countries people are already suffering from power shortages and sometimes do not have access to modern energy services at all. However, in some countries like Argentina, Brazil, Chile, Costa Rica or Mexico many projects are under construction thus putting lights in the forecast for 2009[30]. Brazil with 338.5 MW and Mexico with 85.0 MW of total capacity installed in 2008 were in positions of 24th and 34th in the world. Costa Rica, Argentina, Uruguay and Chile with total installed capacities of 74.0 MW, 29.8 MW, 20.5 MW and 20.1 MW respectively were in positions of 37,41, 46 and 47 in the world for 2008(Table 4).

6.5 Africa

In spite of the huge potentials all over the continent, with world's best sites in the North and South of the continent, wind energy plays still a marginal role on the continent with 563 MW of total capacity. Several major wind farms can be found in some of the North African countries like Morocco, Egypt or Tunisia. In the year 2009 and 2010, substantial increases can be expected from projects which are already in the development stage. However, so far, the emergence of domestic wind industry in African countries is only in a very early stage. However, it is interesting to see that companies from the region are showing an increasing interest and have started investing in the wind sector. In Sub-Saharan Africa, the installation of the first wind farm in South Africa operated by an Independent Power Producer can be seen as a major breakthrough. The South African government prepares the introduction of a feed-in tariff which would create a real market, enable independent operators to invest and thus play a key role in tackling the country's power crisis. In the mid-term, small, decentralized and stand-alone wind energy systems, in combination with other renewable energies, will be key technologies in rural electrification of huge parts of so far unserved areas of Africa. [30]. Egypt with 390.0 MW and Morocco with 125.2 MW of total capacity installed in 2008 were in positions of 21st and 32nd in the world (Table 4). South Africa, Tunisia, Nigeria, Eritrea and Namibia with total installed capacities of 21.8 MW, 20.0MW , 2.2 MW, 0.8 MW and 0.5 MW respectively were in positions of 43, 48 , 64, 69 and 72 in the world for 2008.

6.6 Australia and Oceania

The region showed encouraging growth rates, reaching 1,819 MW by the end of 2008, most of it thanks to Australia. Commitments made by the Australian government to increase their efforts in climate change mitigation and expansion of renewable energies create the expectation that the Australian wind energy market will show further robust growth also in the coming years. New Zealand, after a change in government, may, however, face major delay in its switch to renewable energy [30]. Australia with 1494 MW and New Zealand with 325.3 MW of total capacity installed in 2008 were in positions of 14thand 26th in the world. Australia was in position of 16th and New Zealand was in position of 20th for the year 2007.

Position 2008	Country	Total Capacity installed end 2008	Added Capacity 2008	Growth Rate 2008	Position 2007	Total Capacity installed end 2007	Total Capacity installed end 2006	Total Capacity installed end 2005
		[MW]	[MW]	[%]		[MW]	[MW]	[MW]
1	USA	25170.0	8351.2	49.7	2	16818.8	11603.0	9149.0
2	Germany	23902.8	1655.4	7.4	1	22247.4	20622.0	18427.5
3	Spain	16740.3	1595.2	10.5	3	15147.4	11630.0	10027.9
4	China	12210.0	6298.0	106.5	5	5912.0	2599.0	1266.0
5	India	9587.0	1737.0	22.1	4	7850.0	6270.0	4430.0
6	Italy	3736.0	1009.9	37.0	7	2726.1	2123.0	1718.0
7	France	3404.0	949.0	38.7	8	2455.0	1567.0	757.2
8	United kingdom	3287.9	898.9	37.6	9	2389.0	2123.4	1353.0
9	Denmark	3160.0	35.0	1.1	6	3125.0	1567.0	3128.0
10	Portugal	2862.0	732.0	34.4	10	2130.0	1962.0	1022.0
11	Canada	2369.0	523.0	28.3	11	1846.0	3136.0	638.0
12	The Netherlands	2225.0	478.0	27.4	12	1747.0	1716.0	1224.0
13	Japan	1880.0	352.0	23.0	13	1528.0	1460.0	1040.0
14	Australia	1494.0	676.7	82.8	16	817.3	1559.0	579.0
15	Ireland	1244.7	439.7	54.6	17	805.0	1309.0	495.0
16	Sweden	1066.9	235.9	28.4	18	831.0	817.3	509.0
17	Austria	994.9	13.4	1.4	14	981.5	746.0	819.0
18	Greece	989.7	116.5	13.3	15	873.3	964.5	573.3
19	Poland	472.0	196.0	71.0	24	276.0	757.6	73.0
20	Norway	428.0	95.1	28.5	19	333.0	153.0	268.0
21	Egypt	390.0	80.0	25.8	21	310.0	230.0	145.0
22	Belgium	383.6	78.3	33.7	22	286.9	194.3	167.4
23	Chinese Taipei	358.2	96.7	28.0	23	297.9	187.7	103.7
24	Brazil	338.5	91.5	37.0	25	247.1	236.9	28.6
25	Turkey	333.4	126.6	61.2	26	206.8	64.6	20.1
26	New Zealand	325.3	3.5	1.1	20	321.8	171.0	168.2
27	Korea (south)	278.0	85.9	44.7	27	192.1	176.3	119.1
28	Bulgaria	157.5	100.6	176.7	33	56.9	36.0	14.0
29	Czech Republic	150.0	34.0	29.3	28	116.0	56.5	29.5
30	Finland	140.0	30.0	30.3	29	110.0	86.0	82.0
31	Hungary	127.0	62.0	95.4	35	65.0	60.9	17.5
32	Morocco	125.2	0.0	0.0	36	125.2	64.0	64.0
33	Ukraine	90.0	1.0	1.1	30	89.0	85.6	77.3
34	Mexico	85.0	0.0	0.0	31	85.0	84.0	2.2
35	Iran	823.0	15.5	23.3	34	66.5	47.4	31.6
36	Estonia	78.3	19.7	33.6	37	58.6	33.0	33.0

Position 2008	Country	Total Capacity installed end 2008	Added Capacity 2008	Growth Rate 2008	Position 2007	Total Capacity installed end 2007	Total Capacity installed end 2006	Total Capacity installed end 2005
37	Costa Rica	74.0	0.0	0.0	32	74.0	74.0	71.0
38	Lithuania	54.4	2.1	4.0	38	52.3	55.0	7.0
39	Luxembourg	35.3	0.0	0.0	39	35.3	35.3	35.3
40	Latvia	30.0	2.6	9.5	41	27.4	27.4	27.4
41	Argentina	29.8	0.0	0.0	40	27.8	27.8	27.8
42	Philippines	25.2	0.0	0.0	42	25.2	25.2	25.2
43	South Africa	21.8	5.2	31.4	49	16.6	16.6	16.6
44	Jamaica	20.7	0.0	0.0	43	20.7	20.7	20.7
45	Guadeloupe	20.5	0.0	0.0	44	20.5	20.5	20.5
46	Uruguay	20.5	19.9	3308.3	68	0.6	0.2	
47	Chile	20.1	0.0	0.0	46	20.1	2.0	2.0
48	Tunisia	20.0	0.0	0.0	45	20.0	20.0	20.0
49	Colombia	19.5	0.0	0.0	47	19.5	19.5	19.5
50	Croatia	18.2	1.0	5.8	48	17.2	17.2	6.0
51	Russia	16.5	0.0	0.0	50	16.5	15.5	14.0
52	Switzerland	13.8	2.2	19.2	53	11.6	11.6	11.6
53	Guyana	13.5	0.0	0.0	51	13.5	13.5	13.5
54	Curacao	12.0	0.0	0.0	52	12.0	12.0	12.0
55	Romania	7.8	0.0	0.0	54	7.8	2.8	0.9
56	Israel	6.0	0.0	0.0	55	6.0	7.0	7.0
57	Pakistan	6.0	0.0	New	New	0.0	0.0	0.0
58	Slovakia	5.1	6.0	2.8	56	0.5	5.0	5.0
59	Faroe Islands	4.1	0.1	0.0	57	4.1	4.1	4.1
60	Ecuador	4.0	0.9	3.7	58	3.1	0.0	0.0
61	Cuba	3.2	5.1	242.9	61	2.1	0.5	0.5
62	Cape Verde	2.8	0.0	0.0	59	2.8	2.8	2.8
63	Mongolia	2.4	2.4	new	New	0.0	0.0	0.0
64	Nigeria	2.2	0.0	0.0	60	2.2	2.2	2.2
65	Jordan	2.0	0.0	0.0	62	2.0	1.5	1.5
66	Indonesia	1.2	0.2	20.0	65	1.0	0.8	0.8
67	Martinique	1.1	0.0	0.0	63	1.1	1.1	1.1
68	Belarus	1.1	0.0	0.0	64	1.1	1.1	1.1
69	Eritrea	0.8	0.0	0.0	66	0.8	0.8	0.8
70	Peru	0.7	0.0	0.0	67	0.7	0.7	0.7
71	Kazakhstan	0.5	0.0	0.0	69	0.5	0.5	0.5
72	Namibia	0.5	0.0	6.4	70	0.3	0.3	0.3
73	Netherland Antilles	0.3	0.0	0.0	71	0.0	0.0	0.0
74	Syria	0.3	0.0	0.0	72	0.03	0.03	0.03
75	North Korea	0.2	0.2	2010.0	73	0.01	0.01	0.01
76	Bolivia	0.01	0.0	0.0	74	0.0	0.0	0.0
	Total	121187.9	27261.1	29.0		93926.8	74150.8	59024.1

Table 4. Total capacity installed and position of countries [30].

7. Employment issues regarding wind energy

Wind energy is often said to have positive effects on employment, but few studies have systematically dealt with this matter [26]. The development of renewable energy industries and saving energy technologies became a way to achieve environmental objectives and a means of increasing energy self-sufficiency and employment (e.g. [49 and 50 to 55]. The use of renewable energies offers the opportunity to diminish energy dependence, reduce the emission of CO2 and create new employment. The involvement of local agents is highly important for the future development in this field, especially in regions whose industrial mix was based on traditional energy sources [49]. Wind industry in Europe is a predominantly male business with 78% employment, where men make up majority of the labor in fields of construction, production and engineering.

One fundamental advantage of wind energy is that it replaces expenditure on mostly imported fossil or nuclear energy resources by human capacities and labor. Wind energy utilization creates many more jobs than centralized, non-renewable energy sources. The wind sector (Fig. 7) worldwide has become a major job generator: Within only three years, the wind sector worldwide almost doubled the number of jobs from 235,000 in 2005 to 440'000 in the year 2008. These 440,000 employees in the wind sector worldwide, most of them highly skilled jobs, are contributing to the generation of 260 TWh of electricity [30].

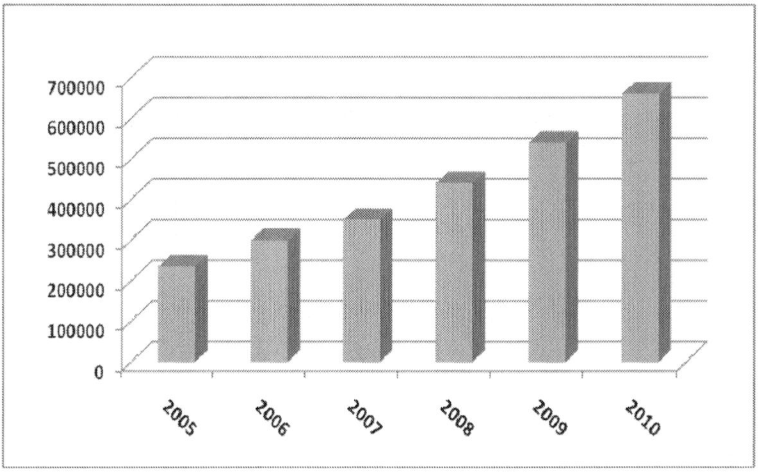

Fig. 7. Wind energy jobs worldwide [30].

The wind energy sector has grown exponentially since the end of the 1990s, especially within the European Union (EU), and this has affected the employment levels of the regions involved[26]. The expansion of renewable energies requires additional investments into production facilities as well as into the transportation and distribution grid [19]. Unemployment rates around 10% shifted the focus of the analysis of the economic effects of the German Renewable Energy Sources Act (EEG) on labor market effects, and several studies have analyzed these effects [56 to 59] These earlier studies either focused on the effects of electricity only, or modeled the end of the German feed-in tariff system and focus on the development until 2010 [60]. Wind energy represents an attractive source of employment in Europe. Since a number of activities (construction, O&M, legal and environmental studies) are best dealt with at local level, there will always be a positive co-relation between the location of the wind farm and the number of jobs it creates. The decision of where to locate large manufacturing centers, however, seems to rely on other, often microeconomic factors, and this is where regional and municipal authorities have a role to play. Another relevant point is that wind energy employment is following the opposite trend to the general energy sector, particularly coal extraction and electricity generation, and measures that encourage the transfer of workers from general energy to wind energy will be highly beneficial from both social and economic point of view [26].Manufacturers and component manufacturers (Fig. 8) with 37% and 22% respectively make up the highest share of direct jobs in wind energy .Service companies are the third largest category, followed by project developers. Operation and Maintenance (O&M) with 11% is in next category.

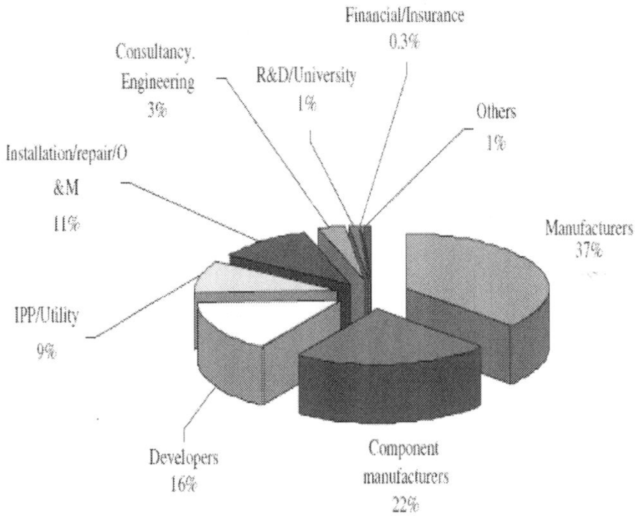

Fig. 8. Direct employment by type of company in EU [26].

The development of any new industry, including wind power, can create new domestic job opportunities, and wind development is often credited with creating more jobs per dollar invested and per kilowatt-hour generated than fossil fuel power generation [61]. Direct jobs are typically created in three areas: manufacturing of wind power equipment, constructing and installing the wind projects, and operating and maintaining the projects over their lifetime [25].

In addition, there are limited global locales possessing a skilled labor force in wind power, with Denmark still representing a unique hub of skilled laborers and an experienced network of key components suppliers to support turbine manufacturers. Suzlon recently decided to base its international headquarters in Denmark to take advantage of this knowledge base, even though it has stated that it is unlikely to sell its turbines to the Danish market [25, 62].Wind energy companies in the EU employed around 104,350 people in 2008. This represents a growth of 226% with respect to 2003 [26]. Germany with total No. of 38,000 persons employed directly in wind industry is leader in Europe (Table 5). Spain and Denmark are also countries with high employment rates in wind energy business too.

Country	No. of Direct Jobs
Austria	750
Belgium	2000
Bulgaria	100
Czech	100
Denmark	17000(23500)
Finland	800
France	6000
Germany	38000
Greece	1800
Hungary	100
Ireland	1500
Italy	3000
Netherlands	2000
Poland	800
Portugal	3000
Spain	20500
Sweden	2000
UK	4500
Rest of world	400

Table 5. Summary of employment profiles (direct jobs) in different EU member states [26, 50 to 55]

8. Implementation of wind turbines in buildings

A new design of a Darrieus turbine in buildings is known as Crossflex [Fig. 9 & 10] which has an innovative system for the blades. This turbine can be located on corners and ridges of the buildings which creates an interesting aesthetic view.

Most iterations of the Darrieus form have placed the turbine on a mast. Its disadvantage is requirement for a rigid foundation, because it causes bending stress on the shaft. Also, it causes high localized loads on the building structure when mounted on buildings [65].

To maximize the number of potential locations that may be exploited, and to enable variable positioning to exploit augmented airflows, the design of the cowl also allows considerable flexibility in the positioning of the turbine. Fig. 11 shows a variety of positions on a 90° corner. This could be horizontal mounting on roof pitches from flat to 45°; horizontal mounting on parapet edges; or vertical mounting on building corners in plan. This enables placement where concentration of wind occurs, for example, rising flow up vertical surfaces, or toward the prevailing wind direction on building corners or ridges. A significant development of the Crossflex concept is the new design and placement of the turbine within a cowling and the general arrangement is shown in Fig. 12. Omitting the shaft is an advantage of this system [65]. There are numerous advantages of Crossflex over conventional Darrieus turbines in terms of performance and usability. This system is at its early stage, but needs more future work. It is a promising technique for future buildings.

Fig. 9. Architectural integration corner [65].

Fig. 10. Architectural integration parapet and ridge [65].

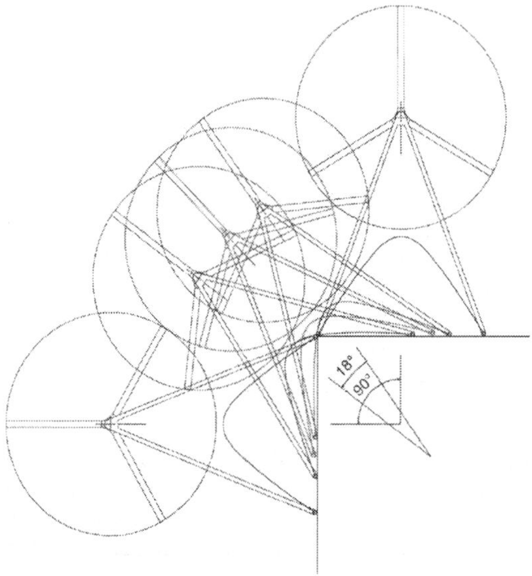

Fig. 11. Variable placement options [65].

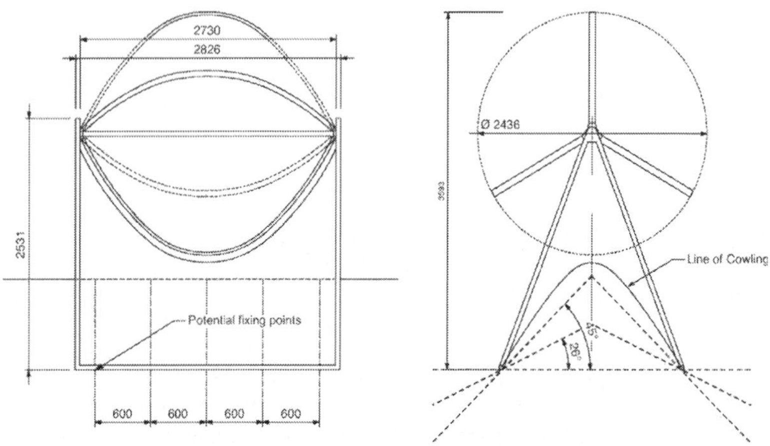

Fig. 12. Front and side elevations [65].

9. Conclusion

Renewable energy sources have been facing a growing attention in global energy markets due to many benefits associated with their importance. During past few years, a great attention was paid toward using wind energy in many countries around the world.USA; Germany, India and China were among the countries which were more successful in order to install wind turbines in recent years. It should be noted that other countries like Bulgaria and Turkey had the highest growth rate for 2008 and 2007 respectively. In general, the focus of the wind sector moves away from Europe to Asia and North America. Europe decreased its share in total installed capacity from 65.5 % in 2006 to 61 % in the year 2007 further down to 54.6 % in 2008. Only four years ago Europe dominated the world market with 70.7 % of the new capacity. In 2008 the continent lost this position and, for the first time, Europe (32.8 %), North America (32.6 %) and Asia (31.5 %) account for almost similar shares in new capacity. Europe lost its dominating role as new market but kept its leading position in terms of total installation with 66,160 MW. Asia with the two leading wind countries China and India and 24,439 MW of installed capacity is in a position of becoming the worldwide locomotive for the wind industry. In spite of the huge potentials all over the Africa, with world's best sites in the North and South of the continent, wind energy plays still a marginal role on the continent with 563 MW of total capacity. Australia showed encouraging growth rates, reaching 1,819 MW by the end of 2008. Many Latin American markets still showed stagnation in the year 2008 and the overall installed capacity (667 MW) in the region accounts for only 0.5 % of the global capacity. North America showed very strong growth in the year 2008, more than doubling its capacity since 2006 to 27,539 MW. The wind sector worldwide has become a major job generator. Within only three years, the wind sector worldwide almost doubled the number of jobs from 235,000 in 2005 to 440,000 in the year 2008.Wind energy represents an attractive source of employment in many countries in the world. There are some activities like operation and maintenance (O&M), research and development (O&M), manufacturing and construction which are able to create jobs in wind industries.

10. References

[1] Held A, Faber T, Panzer C, Toro F, Haas R, Resch G. Potentials and prospects for renewable energies at global scale. Energy Policy 2008; 36: 4048-56.

[2] Ozer S, Tosun M. Feasibility study of wind farms: A case study for Izmir, Turkey Baris Ozerdem. Journal of Wind Engineering and Industrial Aerodynamics 2006; 94: 725-43.

[3] Usha Rao K, Kishore VVN. Wind power technology diffusion analysis in selected states of India. Renewable Energy 2009; 34: 983-88.

[4] Pullen A. Global wind 2005 report. Global Wind Energy Council. <www.gwec. net>; 2005; 1-50.

[5] Breton SP, Moe G. Status, plans and technologies for offshore wind turbines in Europe and North America. Renewable Energy 2009; 34: 646-54.

[6] National Wind Energy Coordinating Committee, Permitting Wind Energy Facilities—A Handbook, NWCC Siting Subcommittee. <www.nationalwind.org>; March 1998.

[7] Honnery D, Moriarty P. Estimating global hydrogen production from wind. International journal o f hydrogen energy 2009; 34: 727-36.

[8] De Vries BJ, Van Vuuren D, Hoogwijk MM. Renewable energy sources: their potential for the first half of the 21st century at a global level: an integrated approach. Energy Policy 2007; 35: 2590-610.

[9] Moriarty P, Honnery D. Can renewable energy avert global climate change?. International clean air conference, Hobart, Tasmania. Clean Air Society of Australia and New Zealand; May 2005.

[10] BTM Consult ApS. International Wind Energy Development, World Market Update 2004; March 2005.

[11] Intergovernmental Panel on Climate Change (IPCC). Climate change 2001: mitigation. Cambridge, UK: Cambridge University Press; 2001.

[12] Seyit A, Akdaga, Dinler A. A new method to estimate Weibull parameters for wind energy applications. Energy Conversion and Management 2009; 50: 1761–66.

[13] World Wind Energy Association. Wind turbines generate more than 1% of the global electricity. <http:// www.wwindea.org>: Press Release: February 2008.

[14] Carta JA, Ramírez P, Velázquez S. A review of wind speed probability distributions used in wind energy analysis. Renewable and Sustainable Energy Reviews 2009; 13(5): 933–55.

[15] Petersen LE. Wind power meteorology. Roskilde (Denmark): Risoe National Laboratory Press; 1997.

[16] Ohlhorst D, Bruns E, Schon S, Koppel J. Wind energie boomin Deutsch land. Peter Lang publishing; 2008, p. 5–60

[17] Nitsch J. "Leitstudie 2008: Weiterentwicklung der "Ausbaustrategie Erneuer- bare Energien" vor dem Hintergrund der aktuellen Klimaschutzziele Deutsch- lands und Europas,". Berlin; 2008.

[18] Portman ME, Duff JA, Koppel J, Reisert J, Higgins ME. Offshore wind energy development in the exclusive economic zone: Legal and policy supports and impediments in Germany and the US. Energy Policy 2009; 37: 3596–607.

[19] Hillebrand_B, Buttermann HG, Behringer JM, Bleuel M. The expansion of renewable energies and employment effects in Germany. Energy Policy 2006; 34: 3484–94.

[20] European Parliament. Directive 2004/101/EC of the European Parliament and the Council of 27 October 2004 amending directive 2003/87/EC establishing a scheme for greenhouse gas emission allowance trading within the Community, in respect of the Kyoto Protocol's project mechanisms. L 338/18. Brussels; 2004.

[21] Berry D. Innovation and the price of wind energy in the US. Energy Policy 2009, doi: 10.1016/ j.enpol.2009.05.071.

[22] Thresher R, Robinson M, Veers P. Wind Energy Technology: Current Status and R&D Future. National Renewable Energy Laboratory. Golden, CO (NREL/CP- 500-43374); 2008.

[23] International Energy Agency (IEA). Wind energy annual report 2003. < www.ieawind.org/iea >; 2004.

[24] Zhao ZU, Hua J, Zuo J. Performance of wind power industry development in China: A Diamond Model study. Renewable Energy 2009; 34: 2883–91.

[25] Lewisa JI, Wiserb RH. Fostering a renewable energy technology industry: An international comparison of wind industry policy support mechanisms. Energy Policy 2007; 35: 1844–57.

[26] Blanco MI, Rodriguez G. Direct employment in the wind energy sector: An EU study. Energy Policy 2009; 37: 2847–57.

[27] Global wind energy outlook 2008. Green Peace International, DLR and Ecofys; October 2008.

[28] Investing in Renewable Technologies: Wind, Solar, Geothermal, Hydro, Biomass. Renewable Energy Reports <www.energybusinessreports.com>; September 9, 2009.

[29] The 8th World Wind Energy Conference 2009 (WWEC2009) on Jeju island/South Korea the World Wind Energy Association (WWEA); 2009.
[30] World Wind Energy Association (WWEA); 2008.
[31] Cavaliero C, DaSilva E. Electricity generation: regulatory mechanisms to incentive renewable alternative energy sources in Brazil. Energy Policy 2005; 33: 1745-52.
[32] Lew D. Alternatives to coal and candles: wind power in China. Energy Policy 2000; 28: 271-86.
[33] Rogner H H. Energy resources. In: World Energy Assessment—2004 update. United Nations Development Programme, United Nations Department of Economic and Social Affairs, World Energy Council, chapter 5; 2004.
[34] Grubb M, Meyer N. Resources, systems and regional strategies. Renewable Energy. Sources for Fuels and Electricity. Island Press, Washington, DC. Wind energy 1993: 157-212.
[35] Haefele W. Energy in a finite world: a global systems analysis. Report by the Energy Systems Group of the International Institute for Applied Systems Analysis. Ballinger Publishing Company, Vol. 2, Cambridge, MA: 1981.
[36] Junginger M, Faaij A, Turkenburg WC. Global experience curves for wind farms. Energy Policy 2005; 33: 133-150.
[37] Durstewitz M, Hoppe-Kilpper M. Bericht zur Markt- und Kosten entwicklung bei Windenergie anlagen. Kassel, Institut fur Solare Energie versorgungs technik e.V. (ISET); 2002.
[38] EWEA (European Wind Energy Association). Wind Energy: The Facts. London; 1997.
[39] Milborrow D, Tishler C, Harrison L, O'Bryant M. The Windicator. Wind power Monthly, April 2003.
[40] Turkenburg WC, Beurskens J, Faaij A, Fraenkel P, Fridleifsson I, Lysen E, Mills D, Moreira JR, Nilsson LJ, Schaap A, Sinke WC. World Energy Assessment. Chapter 7: Renewable energy technologies. World Energy Assessment. Goldemberg J. Washington, DC, UND; 2000, P. 220-72.
[41] Global Wind Energy Council (GWEC). <www.gwec.net >; April 2007.
[42] China Renewable Energy Market Outlook. <http://www.energybusinessreports.com >; May 2008.
[43] Council of the European Union, 7224/1/07 Rev: Brussels European Council, 8-9 March 2007. Presidency Conclusions. <www.consilium.europa.eu >; 2007.
[44] Blanco MI, Rodriguez G. Can the future EU ETS support wind energy investments? .Energy Policy 2008; 36: 1509-20.
[45] European Wind Energy Association (EWEA). Wind Map in Europe, 2006. < www.ewea.org>; 2007.
[46] Platts. Platts Power Vision; March 2007.
[47] Lemming J. Cost reduction potentials in wind. Presentation at the IEA Technology Learning and Deployment Workshop, Paris; 2007.
[48] Wiser R, Bolinger M. Annual report on US wind power installation, cost, and performance trends: 2007. Energy Efficiency and Renewable Energy, US Department of Energy, Washington, DC; 2008.
[49] Moreno B, Lopez AJ. The effect of renewable energy on employment. The case of Asturias (Spain). Renewable and Sustainable Energy Reviews 2008; 12: 732-51.

[50] Connor PM. UK renewable energy policy: a review. Renewable and Sustainable Energy Reviews 2003; 7:65–82.
[51] Dincer I. Renewable energy and sustainable development: a crucial review. Renewable and Sustainable Energy Reviews 2000; 4:157–75.
[52] Hillebrand B, Buttermann HG, Behringer JM, Bleuel M. The expansion of renewable energies and employment effects in Germany. Energy Policy 2006; 34:3484–94.
[53] Laitner S, Bernow B, Cicco JD. Employment and other macroeconomic benefits of an innovation-led climate strategy for the United States. Energy Policy 1998; 26: 425–32.
[54] Lenzen M, Dey CJ. Economic, energy and greenhouse emissions impacts of some consumer choice, technology and government outlay options. Energy Econom 2002; 24: 377–403.
[55] Thothathri R. The wind brought jobs and prosperity. New Energy 1999; 4: 28–30.
[56] Pfaffenberger W. Wertscho¨ pfung und Bescha¨ ftigung durch gru¨ ne Energieproduktion? Energiewirtschaftliche tagesfragen ; 2006, p. 22–6.
[57] Pfaffenberger W, Nguyen K, Gabriel J. Ermittlung der Arbeitspla¨ tze und Bescha¨ ftigungswirkungen im Bereich erneuerbarer Energien. Studie des bremer energie instituts im Auftrag der Hans- Bo¨ ckler-Stiftung; 2003.
[58] Hillebrand B, Buttermann HG, Behringer JM, Bleuel M. The expansion of renewable energies and employment effects in Germany. Energy Policy 2006; 34 (18): 3484–94.
[59] Fahl U, Kuster R, Ellersdorfer I. Jobmotor O¨ kostrom? Bescha¨ ftigungseffekte der Fo¨ rderung von erneuerbaren Energien in Deutschland. Energiewirtschaftliche Tagesfragen 2005; 55 (7):476–81.
[60] Lehr U, Nitsch J, Kratzat M, Lutz C, Edler D. Renewable energy and employment in Germany. Energy Policy 2008; 36: 108–17.
[61] Singh V, Fehrs J. The Work that Goes into Renewable Energy. Renewable Energy Policy Project Research Report 14, November 2001.
[62] Wind power Monthly (WPM). Denmark picked for global headquarters. Wind power Monthly News Magazine A/S, Denmark. October 2004:25.
[63] Stern N. The economics of climate change. American Economic Review 2008; 98 (2): 1–37.
[64] Mc Kinsey Global Institute. The carbon productivity challenge; 2008.
[65] Sharpea T, Proven G. Crossflex: Concept and early development of a true building integrated wind Turbine. Energy and Buildings 2010; 42: 2365–2375.
[66] US Wind power < http://EnergyBusinessReports.com >; 2010.

3

Adaptive Bend-Torsional Coupling Wind Turbine Blade Design Imitating the Topology Structure of Natural Plant Leaves

Wangyu Liu and Jiaxing Gong
South China University of Technology
China

1. Introduction

In the wind turbine system, the size of the blade is determined by the level of single output power. With the rise of offshore wind turbines (Breton & Moe, 2009), the output power of commercial blade has reached 5MW, and the length of the blade is over 100m. The design and manufacture limit of large-scale wind turbine is facing severe challenges, and as a result, wind turbine blade has become a research focus of scholars from all over the world. In a poor working environment, the problems of large blades in the following two aspects, which occur in the process of operation, will become more and more prominent.

1. In the operation, the blade should bear a good rigidity in order to minimize the destruction that random wind load and gust may cause to the blade (Bishop et al, 1999). Knut (1999) points out that due to the increasing length of wind turbine blade, the blade becomes more vulnerable to the unpredictable destruction caused by random gusts and ultimate wind load. And since the fatigue test of the blade has its limitations, therefore, he proposed a kind of random probability model based on bad working conditions to predict the fatigue life and reliability of the series of blades (Ronold & Larsen, 2000). Christoph (2006) points out that the length and weight of large-scale blade have an increasing impact on the bending load withstood internally. Meanwhile, it becomes more and more difficult for the large blades, which are subjected to wind, rain, moisture, and other adverse environmental effects, to meet the design requirements of a 20-year basic fatigue life. In order to predict the fatigue life of blade more accurately, apart from the unidirectional fiber, he has also made some research on the S-N curve of the off-axis fiber which bears the shear load. In addition, the research on the blade fatigue and damage mechanism has also been attached importance to by domestic and foreign researchers. For example, Daniel (2008), Raif (2008) and other scholars have delved into the fatigue and destruction data and the interlayer destruction mechanism of glass fiber and carbon fiber. They point out that both the technology of blade fiber manufacture and the adaptability of the blade have an important impact on the fatigue and destruction of the blade.

   Thus with the wind turbine blade becoming lager and larger, it becomes more and more difficult to maintain the rigidity of the blade. Even within the rated wind speed, the instability of the wind speed produces a serious varied load to the blade, increasing the possibility of cyclic fatigue damage of the blade. When subjected to gusts or ultimate wind load, the blade, failing to adapt to the change immediately, becomes more vulnerable to invalid breakage, and thus its fatigue life is reduced. Therefore, it is very important to increase the blade flexibility as well as to improve its unloading effects.

2. The longer the blade is, the higher the tower holder is, and the more unstable the direction of the wind speed will be. As a result, the attack angle of the blade will vary in accordance with the change of the wind speed more easily, which in turn results in the increase of the instability of the power output. At the same time, with the increase of the blade cost, to increase wind turbine efficiency in order to lower the cost of wind power by designing adaptive blade has become the research focus (Lobitz & Veers,1999).

   As mentioned above, thus to reduce the impact of the instability of wind speed such as gusts on the blade, to improve the reliability and antifatigue merits of the large-scale blade as well as the stability of power output, and to broaden the scope of running wind speed have become the research hotspots that the current wind power field is concerned about. The existing technology is to improve the stability (Bao et al.,2007; Lin,2005) of the power output of the wind turbine blade by stalling, varying pitch angle and other methods, and some researchers also try to realize this in the ultra-large blade by reducing the area of the blade trailing edge strip (Lackner & Kuik, 2010). However, for those blades which are similar to slender cantilever beam, when the wind load is not fixed and the inertia force becomes larger and larger, the blade structure and the instability of the power output can hardly be regulated and controlled by the electromechanical system. And the feedback effect and the governing response speed can not meet the requirements of real-time control. As a result, the cost will increase accordingly. Therefore, to improve the adaptivity of the blade and reduce the reliance on the control system to achieve the stability of output power can meanwhile enhance the unloading function of the blade and improve its fatigue life, which is thus of research value.

2. Review of the development of adaptive blade

In fact, the tailorability and designability of the composite aeroelasticity has long been widely used (Büter & Breitbach, 2000) in the military and aviation fields, etc. Till 1990s, relevant researchers had begun to try to develop an intelligent blade which bears a good adaptability to the wind load from outside by designing a laminated blade material. Karaolis and other researchers have realized the blade twisting and the coupling of the relevant acting force through the mirror symmetry laminated design of the FRP composite of small blades. That is to say, with the change of the acting force, the twist angle of each section of the blade will change accordingly thus to unload some force and control the power output (Jeronimidis & Musgrove, 1989) of the blade. Joose and others have designed a kind of structure in which axial tensile deformation and twist deformation are coupled to adjust the rotation angle of the blade tip, and have analyzed the stability of power output and its protective capability for the blade in the cases of over-rated wind speed. DON (Lobitz & Veers, 1999) discussed about the tension, shear and twist coupling algorithm of the linear beam units and verified it through a set of combined experiments, figuring out the coupling factors of the blade tip in different twist angles. They have done a preliminary research on the contribution (Joosse et al., 1996) of coupling factors to the stability of the power. Andrew for the first time applied twist coupling effect design to the 50kW blade systematically, and proposed a design idea of intelligent blade to enhance the blade's power control and lower the blade's fatigue loss (Andrew et al.,1999). So far, the blade with twist coupling effect is still at the stage of preliminary design, with some verifying experiments on some small blades.

Till early 21th century, as great importance had been attached to renewable energy by the countries all over the world, technology of wind power generation and the blade design theory has been developed rapidly. Researchers in this field have reached a common view to design an intelligent blade that bears a good adaptability to the bad environment. The Sandia National Laboratories of the United States had also begun to do some independent research as well as to fund developing the adaptive blade with torsion coupling property.

Don applied torsion coupling design to medium-sized blades of 300kW, and proved in detail in his report that adaptive blades are improved blades in terms of wind capturing efficiency, which are therefore able to increase the annual wind power catch. Although in the case of stalling, the chances of fatigue and destruction may be increased for adaptive blades, in the process of operation, its antifatigue property (Lobitz et al., 2001) is actually increased. Due to the complexity of the structure and the shape of wind turbine blade, it is difficult to obtain an accurate solution to the mechanical problem of wind turbine structure by applying the normal numerical analysis theory. However, with the maturity of the finite element analysis technology, it is also widely applied in the mechanical calculation of wind turbine blade structure. For instance, Ladean attains the bending stiffness and the shearing rigidity of the blade through the finite element technology, and also gains the static and dynamic mechanical property and the buckling frequency (McKittrick et al., 2001) under the circumstance of rated wind speed and ultimate wind load, which is of great reference value for further study.

The increase of the size of the blade calls for some material better than glass fibers which can now hardly meet the requirements of structure reliability. While carbon fibers, with its light weight and good comprehensive mechanical property, have become the first choice to replace glass fiber. In recent years, some scholars are dedicated to the study (Griffin & Ashwill, 2003) of the hybrid fibers mixed by glass fibers and carbon fibers, and have applied (Mohamed & Wetzel, 2006) them to the large scale commercial blade constantly. However, in terms of the hybrid fibers, there is still a problem of manufacture and cost constraints. And due to the complexity of the mechanism of fracture of the composite material itself, it will certainly be more difficult to predict and grasp the failure mode of hybrid fibers. In order to better grasp the fatigue and destruction mechanism of the hybrid fibers, John has made about 10^{10} experimental research on the fatigue property and strength reliability of different glass fibers and hybrid fibers samples in different stress ratio, which provides a good reference for the fracture mechanism of the fibers. The research results show that sewed epoxy hybrid fibers is better than the knitted hybrid fibers in terms of compression ratio intensity and antifatigue property (Mandell et al., 2003). Don (2001), under fatigue load, delves into the interlayer destruction mechanism of the hybrid fibers of glass fibers with variable cross-section and carbon fibers, and the result shows that interlayer stress and strain have great impact on the destruction of the fiber layer. Compared with glass fibers, under the maximum stress and strain in the bottom layer, carbon fibers are more susceptible to interlayer separation failure. Selwin used the finite element method to analyze the cause for the fatigue failure of the broken blade, and discovered that the estimated results agreed with the improved fatigue failure criteria (Rajadurai et al., 2008), which applies to all kinds of composite materials. This provides a good reference for the optimization design of the wind turbine material. James, based on the NPS blade model, has introduced carbon fibers to the torsion coupling design of the blade, and established the model of the torsion coupling design of hybrid fibers. He has also checked the static and dynamic property under the ultimate load, and compared it with standard glass fibers blade, proving the advantages and feasibility (Locke & Valencia, 2004) of applying hybrid carbon fibers to the torsion coupling design.

In recent years, researchers have begun to apply torsion coupling design to the large scale blade. Dayton has begun to apply torsion coupling design to a large blade of 1.5Mw, and has studied the airfoil design of adaptive blade. At last he chose an off-axis skin airfoil, and evaluated it from the aspect of coupling effect. The results show that although the blade airfoil he designed has the potential (Griffin, 2004) to be manufactured as large scale adaptive blade, in his documents, due to the restrictions on the using of hybrid fibers considering cost and manufacture factors, he has only considered to use the hybrid fibers of the combination of 20° and 70°, which in turn makes it difficult to design an adaptive blade with optimal coupling effect. Based on this, Kyle put all off-axis carbon fibers in the beam cap of a long blade of 1.5MW, and used parametric methods to analyze the influence of different angles and volume ratios of the offset fibers on the torsion coupling effect. He proposed to optimally analyze the offset fiber angel (Wetzel, 2005) in which the optimal coupling effect of the blade occurs considering the stress, strain, and in-plane stress and strain as well as other comprehensive evaluation factors of the cross fibers of the blade. But Kyle also put too much emphasis on the cost of the blade manufacture, and ignored the possibility of balancing the cost through the improved power and fatigue property.

With the maturity of the design technology of torsion coupling blade, Alireza and other researchers have developed a comprehensive procedure combining aerodynamics and the coupling of the blade structure to accurately predict the angle of attack and bending moment of the adaptive blade during the process of design as well as the output power (Maheri et al., 2006; Maheri et al., 2007), which has reduced the heavy finite element calculation task in the process of optimization design thus saving the time and efforts in calculation. After that, they soon started to develop a set of design tools (Maheri & Isikveren, 2009a, 2009b) of adaptive blade. This procedure has applied the aerodynamic - structure iteration coupling algorithm, and is able to design an intelligent blade in which the maximum annual wind power capturing is realized through the alteration of the coupling design factors in different wind speeds according to the requirements of designers. This design procedure is still in the research stage, and has great reference value.

Some researchers, for example, Rachel (2007) and Nicholls-Lee (2009) have even applied the torsion coupling design idea of wind turbine to the design of tidal turbine blade as to improve the annual power capturing capability of the tidal turbine, approaching the Betz limit. The research shows that the adaptive blades of tidal power generation can also improve the annual power capturing capability, as well as reduce the load-carrying capability to enhance the reliability of the blade.

3. Bionic research of wind turbine blade

Martin Ryle, astronomer of Cambridge University, Nobel Prize winner, had predicted that if the blades can work like the palm tree in high wind speed, then the energy conversion efficiency can be increased by 50%, and the material can be saved by 2/3 (Platts & Liu,

2008). Ryle thus had tried to study the flexible wind turbine. As early as in the early 80s of last century, the concept of flexible wind turbine has aroused great attention, and small wind turbines represented by Carter flexible wind turbine have been developed and put into use. However, since the traditional theory applied to the wind turbine design is no longer suitable for the flexible wind turbine in which the plane of rotation has changed, at that time the research and development of small flexible wind turbine mainly adopted the trial and error method. As the wind turbine is developed towards the tendency of large size, the trial and error method is no longer suitable for the development of large equipments. And flexible wind turbine, due to its inadequacy in the reliable design theory, was a mere flash in the pan, failing to become the mainstream of the development of wind turbine. Nevertheless, flexible wind rotor, due to its excellent cost performance and the wide adaptability to wind load changes, is still the research field that some scholars would like to devote their whole life to. In the 90s of the 20th century, Peter Jamieson, British engineer, proposed the concept of coning blade, and conducted some fruitful research, gaining the relevant aerodynamic performance curves of coning blade. The results show that compared with conventional design, using the wind blade with the same length, the minimum start-up wind speed can be decreased from 4m/s to 3m/s, and the ability to withstand hurricanes can reach 70m/s. A research team led by Jim Platts of Cambridge University, based on the research of Peter Jamieson, established a BEM coning blade theory based on trimming and revision. In the theory, articulated juncture method is adopted to control the open and close of the wind blade, so as to adjust the solidity of the wind blade to control the output power. In this way, the complete performance curve of the coning blade is gained, the power capture coefficiency of wind power is increased, and the cost of wind power generation is lowered (Curran & Platts, 2006; Curran, 2006).

Sarikaya (1990) and Gordon (1980) have delved into biological composite materials and its bionic application, and both of them believe that the organisms have optimized their structure, shape and function after a long-term evolution of hundreds of millions of years in the natural environment. Steele, Somerville and other scholars have verified that there is a necessary connection (Steele, 2006; Chris et al., 2004) between the production of the biological shape and tissue and the environmental stress. The research of Jones and others shows that the optimization results of biological tissues present Michell's truss topological structure, and match trajectory of principal stress (Jones & Platts, 1998). Medial axis exists in a lot of organisms (Blum & Nagel, 1978) as an optimized pattern, such as the common vein tissues or grass tissues, as well as the skeleton structure of many animals. There must be some links between the pattern and structure and the growth of the organisms.

In recent years, the author's group is mainly devoted to the research on the enhancement mechanism (Liu & Hou, 2007; Liu et al., 2009) of the topological structure of plant leaves. The research shows that the mesophyll of the leaves and the marrow in the skeleton form a multiphase and natural composite structure, and the medial axis network undoubtedly bears the function of enhancing the tissues. Therefore, in order to explore the mechanical enhancement mechanism of the leaf skeleton on the part of the topological structure of plant leaves, some experimental studies and numerical simulations on many groups of leaf samples were carried out. The analysis shows that the plant leaf can always maintain relatively smaller internal strain energy. According to the topological graph of plant leaves in single load field and multi-load field, stress is one of the inducing factors for vein growth, and apart from the realization of its physiological function, vein distribution, in order to adapt to the changes of the complicated environmental stress, its network has evolved into a stable and adaptive medial axis structure. Subsequently, our study again focuses on the wind resistant plants in the coastal area, for example, the palm tree series. The study finds that palm tree trunks and branches are the natural fibrous composite which forms a good system combining rigidity and flexibility. And the leaves also bear a good flexible property. From our point of view, this is not only related to the flexible structure of itself, but also has some links with its material structure, which bears good rigidity, strength, and toughness.

3.1 The bend-torsion coupling design of wind turbine blade

3.1.1 Blade material and coupling effect

3.1.1.1 Composite blade

In recent decades, high performance composite material develops rapidly. The typical composite material are glass fibers reinforced polymers and carbon fibers, which are widely applied in engineering. The advantages of composites are (Jiang et al., 1990):

1. High ratio strength and stiffness, which means stronger and stiffer with the same mass.
2. Good fatigue performance.
3. Design ability, which means the engineer can customize the performances of the composite products.
4. High safety coefficient.

In wind turbine machine, the turbine blade is the key structural component. Therefore, reasonable structure, superior and light material, advanced manufacturing technique and etc must be required for the blade to withstand the adverse stress environment, such as the bend and torsion moment, the pulling force induced by the wind load, the gravity and inertial force. Composites material becomes the first choice owning to its advanced performances mentioned above. Currently, hybrid composites, mixed with glass/carbon fiber reinforced polymers, are utilized in large-scale wind turbine blade. With the increase of blade size, more carbon fiber reinforced composites are used in order to reduce the weight and raise its reliability.

3.1.1.2 Fundamentals of the layering theory of composite laminate

The turbine blade is manufactured by composite layering technique. As the good designable characteristic of fiber reinforced composites, some complex coupling effect, such as tensile-shear, bend-torsion, can be achieved through composite layering method. The following section will give a brief introduction to the coupling design of composite laminates. The context of this part can be found in many textbooks (Jiang et al., 1990).

Laminate is made up of unidirectional thin plates layered according to the designed sequence and angles. In order to evaluate the laminate stiffness, the following hypothesis is made:

1. Made-in-art bond strength and no gap between layers, discounting the thickness of the bond layer.
2. The thickness is very small, and the line perpendicular to the middle plane keeps its shape and position (perpendicular to the middle plane) after deformation.

The stress in the *kth* layer can be represented as,

$$\begin{bmatrix} \sigma_x \\ \sigma_y \\ \tau_{xy} \end{bmatrix}_k = \begin{bmatrix} \overline{Q_{11}} & \overline{Q_{12}} & \overline{Q_{16}} \\ & \overline{Q_{22}} & \overline{Q_{26}} \\ sym. & & \overline{Q_{66}} \end{bmatrix}_k \left\{ \begin{bmatrix} \xi_x^0 \\ \xi_y^0 \\ \gamma_{xy}^0 \end{bmatrix} + z \begin{bmatrix} K_x \\ K_y \\ K_{xy} \end{bmatrix} \right\} \tag{1}$$

Where $[Q]_k$ is the stiffness matrix of layer *k*, *z* is the coordinate variable in the direction of thickness. ξ_x^0, ξ_y^0, γ_{xy}^0 are the normal and shear strain in the middle plane of the laminate, K_x, K_y, K_{xy} are the bend and torsion curvatures, respectively. *Sym* means the elements in the matrix is symmetrical to the main diagonal elements. If the displacement of the middle plane are μ_0, v_0, ω_0, then

$$\begin{bmatrix} \xi_x^0 \\ \xi_y^0 \\ \gamma_{xy}^0 \end{bmatrix} = \begin{bmatrix} \frac{\partial_{\mu 0}}{\partial x} \\ \frac{\partial_{v0}}{\partial y} \\ \frac{\partial_{\mu 0}}{\partial x} + \frac{\partial_{v0}}{\partial y} \end{bmatrix}; \begin{bmatrix} K_x \\ K_y \\ K_{xy} \end{bmatrix} = \begin{bmatrix} -\frac{\partial^2 \omega_0}{\partial x^2} \\ -\frac{\partial^2 \omega_0}{\partial y^2} \\ -2\frac{\partial^2 \omega_0}{\partial x \partial y} \end{bmatrix} \tag{2}$$

Suppose the resultant inner force per unit of the laminate are N_x (pulling force), N_y (compressive force), and N_{xy} (shear force), and the resultant inner moment (bend or torsion moment) are M_x, M_y and M_{xy}, shown as Fig.1 and Fig.2.

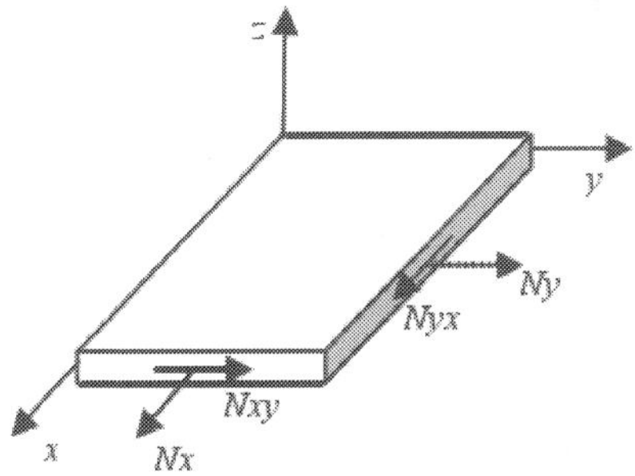

Fig. 1. The resultant inner force

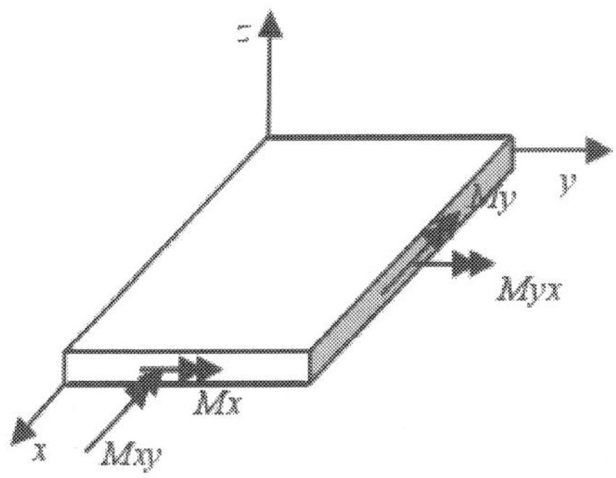

Fig. 2. The resultant inner moment

The constituent equations of the laminate are,

$$\begin{bmatrix} N_x \\ N_y \\ N_{xy} \end{bmatrix}_k = \begin{bmatrix} A_{11} & A_{12} & A_{16} \\ & A_{22} & A_{26} \\ sym. & & A_{66} \end{bmatrix} \begin{bmatrix} \xi_x^0 \\ \xi_y^0 \\ \gamma_{xy}^0 \end{bmatrix} + \begin{bmatrix} B_{11} & B_{12} & B_{16} \\ & B_{22} & B_{26} \\ sym. & & B_{66} \end{bmatrix} \begin{bmatrix} K_x \\ K_y \\ K_{xy} \end{bmatrix} \tag{3}$$

$$\begin{bmatrix} M_x \\ M_y \\ M_{xy} \end{bmatrix}_k = \begin{bmatrix} B_{11} & B_{12} & B_{16} \\ & B_{22} & B_{26} \\ sym. & & B_{66} \end{bmatrix} \begin{bmatrix} \xi_x^0 \\ \xi_y^0 \\ \gamma_{xy}^0 \end{bmatrix} + \begin{bmatrix} D_{11} & D_{12} & D_{16} \\ & D_{22} & D_{26} \\ sym. & & D_{66} \end{bmatrix} \begin{bmatrix} K_x \\ K_y \\ K_{xy} \end{bmatrix} \tag{4}$$

$$A_{ij} = \int_{-\frac{h}{2}}^{\frac{h}{2}} Q_{ij} dz \,,\; B_{ij} = \int_{-\frac{h}{2}}^{\frac{h}{2}} Q_{ij} dz \,,\; D_{ij} = \int_{-\frac{h}{2}}^{\frac{h}{2}} Q_{ij} dz \tag{5}$$

In equation (5), A_{ij} are the stiffness coefficients between the stress and middle plane strain, called as the pulling stiffness; D_{ij} are the stiffness coefficients related with the inner moment, the curvature, and torsion curvature, generally called as flexural stiffness; B_{ij} is a coefficient not only coupled with the pulling deformation but also with the flexure, generally called as the coupling stiffness. When the layer of the laminate is decided, all the stiffness mentioned can be calculated in the first place, then the stress and strain analysis can be carried out.
It can be observed from formula (3) and (4) that the in-plane inner force in the laminate will result in flexural deformation (bend and torsion), and the flexural inner force (bend and torsion moment) will result in the in-plane deformation, which are called the coupling effect of pulling and flexure. The complicated physical relationship of laminate can be explained from the coupling stiffness matrix B. Matrix A denotes the coupling effect of pulling and shearing, and matrix D gives the coupling effect of bend and torsion. In adaptive wind turbine blade design, the main task is to design proper matrix D to realize the adaptive coupling of bend and torsion.
The performances of composites not only relate with the matrix and reinforced material, but also with the fibers volume ratio and arrangement. The stiffness of the laminate can be designed by changing the fibers arrangement. Their relation is shown in table 1, where the definition of the layer specification can be found in literatures about composites mechanics.
It can be found from Table 1 that, in order to achieve the bend-torsion coupling effect given by matrix D and avoid the influence of coupling matrix B, the off-axis fiber arrangement in symmetrical means must be followed.

3.1.2 Example validation of coupling design

In this section, a simple example is taken to verify the bend-torsion coupling effect of the theory mentioned above, and to provide evidence for the following computing method. Supposing that there is a piece of simple laminated plate with 10m length, 1m width and 30 layers, 10mm thickness of each layer. The material property is shown as the followings: axial modulus $E_x = 36.47\text{GPa}$, radial modulus E_y=12.6GP$_a$, poisson ratio U_x=0.22, shear modulus G_{xy}=3.94GPa. Fix the root and put a concentrated force, F=10KN, at the midpoint of the end.

stiffness		Pulling stiffness	Coupling stiffness	Flexural stiffness
Symmetrical arrangement	Normal isotropic inhomogeneous	$A_{ij} \neq 0$	0	$D_{ij} \neq 0$
	Special isotropic inhomogeneous	$A_{16} \neq 0$ $A_{26} \neq 0$	0	$D_{16} \neq 0$ $D_{25} \neq 0$
	Off-axis arrangement	$A_{ij} \neq 0$	0	$D_{ij} \neq 0$
Anti-symmetrical arrangement	Isotropic arrangement	$A_{11} = A_{12}$ $A_{16} = A_{26} = 0$	$B_{22} = -B_{11}$ the other $B_{ij} = 0$	$D_{11} = D_{12}$ $D_{16} = D_{26} = 0$
	Off-axis arrangement	$A_{16} = 0$ $A_{26} = 0$	$B_{16} \neq 0$ and $B_{26} \neq 0$ the other $B_{ij} = 0$	$D_{16} = 0$ $D_{26} = 0$
Non-symmetrical arrangement	Normal isotropic inhomogeneous	$A_{ij} \neq 0$	$B_{ij} \neq 0$	$D_{ij} \neq 0$
	Special isotropic inhomogeneous	$A_{16} = 0$ $A_{26} = 0$	$B_{16} = 0$ $B_{26} = 0$	$D_{16} = 0$ $D_{26} = 0$
Bi-unidirectional		$A_{16} = 0$ $A_{26} = 0$	$B_{16} = 0$ $B_{26} = 0$	$D_{16} = 0$ $D_{26} = 0$

Table 1. The stiffness of different fiber arrangement

Theoretical method is firstly used to deduce the stress and strain of the laminated plate, and then finite element method is carried out to verify the correctness of the computing method. Supposing that the laminate plate is unidirectional and the fiber placement follows the lengthwise direction of the plate. As the laminated plate is in regular shape, the length, width and height are l=10m, b=1m, h=0.3m respectively. The laminated plate could be equivalent to a beam, and the root of which is fixed. According to the beam theory, the bend moment at the beam root is,

$$M_{\max} = F \times l = 100000 N.m$$

The bend inertial moment of the beam is,

$$W = \frac{b * h^2}{6} = \frac{1 \times 0.3 \times 0.3}{6} = 0.015 \ m^3$$

Then the maximum tensile stress at the root is,

$$\sigma_{\max} = \frac{M_{\max}}{W} = \frac{100000}{0.015} = 66.7 \ \text{MPa}$$

The strain is,

$$\xi_x = \frac{\sigma_{\max}}{Ex} = 0.00183$$

Use finite element method to solve the problem. Fig.3 and Fig.4 show the results. The root average tensile stress is about 67MPa, and the maximum strain is 0.00187, which is close to the result with beam theory method.

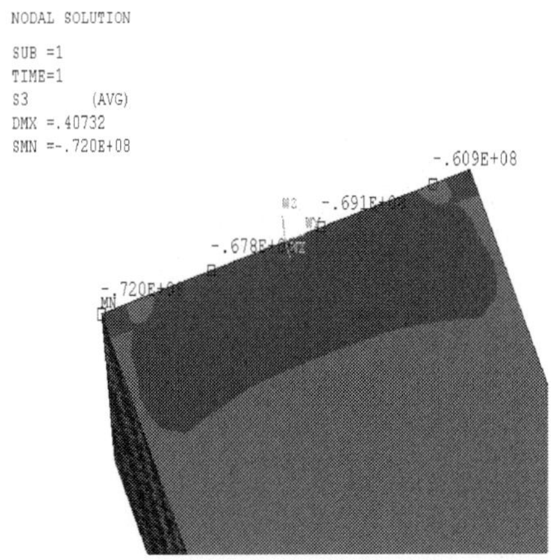

Fig. 3. Tensile stress at root

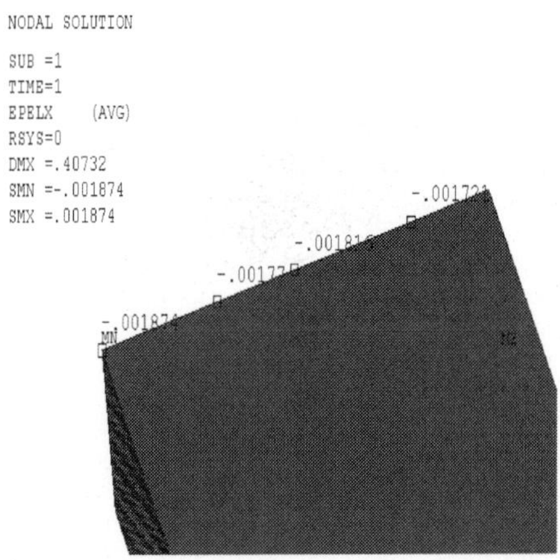

Fig. 4. Compressive strain at root

The following content takes the laminated plate for an example to discuss the coupling effect of the laminated plate. Firstly, under the unidirectional layer condition mentioned above, the result of the displacement of the laminated plate end is shown in Fig.5. It can be seen that the displacement of the two end-points is equal, and no torsion displacement happens. However, when the layer pattern is changed into angular layer, for example, the fibers are in 20° off the lengthwise direction, the torsion deflection, say 2.73° drawn from the result, will be achieved shown as Fig.6. In the context, a group of off-axis fibers placed in different angles among 0°~40°, could make the similar bend-torsion coupling effect.

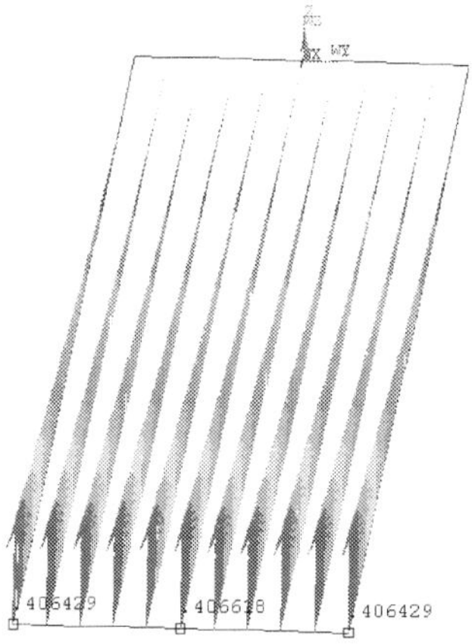

Fig. 5. End displacement of unidirectional laminated plate

Fig. 6. End displacement of off-axis laminated plate

It can be seen from the example mentioned above, when the layer angle of the fiber in the laminated plate does not deviate from the coordinate axis, no coupling effect would be achieved. And when it does, the bend-torsion coupling effect would be generated. It just verifies that the bend-torsion coupling theory of laminated plate is correct, and it is reasonable and accurate to use finite element method to calculate the coupling effect.

3.1.3 The layer and coupling effect in blade

The coupling design method of composite laminate is widely used in several fields, for example, the aviation. While in the wind turbine machine design, the designers usually try to eliminate the coupling effect caused by the blade layer. Traditionally, unidirectional fibers (0°) and ±45° triaxial fibers are used in blade layering. With the increase of the blade size, coupling design of the blade shows potential advantages in some aspects. According to previous content, laminated plate with off-axis fibers placement would achieve favorable bend-torsion coupling effect. Whereas, in wind turbine blade design, it would work well if the fibers are placed in mirror way (not 0°) in the coupling area of the two chord faces up and down, shown as Fig.10. Where, θ means the off-axis angle.

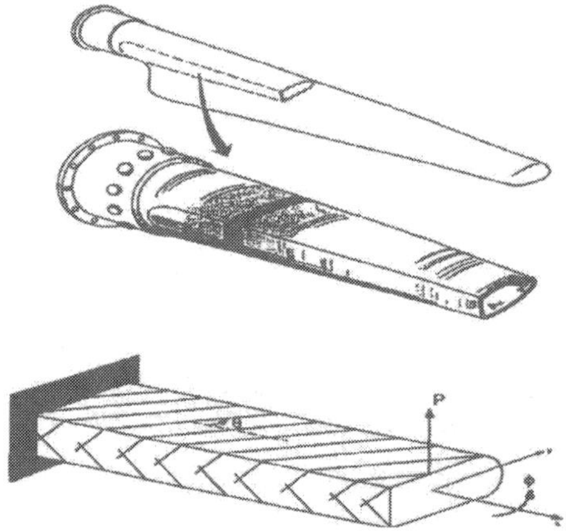

Fig. 7. Bend-torsion coupling design method in the blade spar (Ong, 2000)

3.1.3.1 The control of coupling ratio

In finite element, the 3-D beam element provides an efficient and short-cut computing method for the coupling effect design. In order to analyze the coupling effect of fiber layer more accurately, Cheng conducts two part of modification to beam element theory (Cheng, 2000), (1) Considering about the in-plane lateral shear stress and strain; (2) Considering about the out-of-plane bending effect relative to torsion, and conducting intensive study on the bend-torsion coupling effect through theory analysis.

The bend-torsion coupling effect of wind turbine blade is realized through the laminate layer design method to achieve the stiffness D mentioned above, and at the same time, certain torsion effect could be achieved when the blade bends. Whereas, how to evaluate the coupling effect with a parameter indicator? Kyle. K Wetzel. (Wetzel, 2005) have made much investigations. Carried on his work and based on cantilever beam theory, the authors made some calculations for the selected blade model.. Since the blade spar is usually made up with symmetrical section, it is equal to the cantilever beam of symmetric distribution. Its bend-torsion coupling control equation is,

$$\begin{bmatrix} EI & -K \\ -K & GJ \end{bmatrix} \begin{bmatrix} \dfrac{\partial \theta}{\partial z} \\ \dfrac{\partial \varphi}{\partial z} \end{bmatrix} = \begin{bmatrix} M_b \\ M_t \end{bmatrix} \tag{6}$$

In the equation above: EI is the bending stiffness in the blade flapping direction; GJ is the torsion stiffness in the blade spanwise direction; K is the coupling term; θ is the bend angle

of the cross section; φ is the torsion angle; M_b is bending moment of the cross section; M_t is the torsional moment of the cross section.
According to equation(6)：

$$EI*GI\frac{\partial\theta}{\partial z}-K*GJ\frac{\partial\varphi}{\partial z}=M_b*GJ \tag{7}$$

$$-K^2\frac{\partial\theta}{\partial z}+K*GJ\frac{\partial\varphi}{\partial z}=M_t*K \tag{8}$$

Using the two equations above, it can be deduced that,

$$\frac{\partial\varphi}{\partial z}=\frac{M_b*K+M_t*EI}{EI*GJ-K^2};\ \frac{\partial\varphi}{\partial z}=\frac{M_b*GJ+M_t*K}{EI*GJ-K^2} \tag{9}$$

Supposing that the blade length is l, and there is a concentrated force working on the spar end. At the same time, the bending and torsional moment at any section x in spanwise direction are $M_b=F(l-x)$, $M_t=0$. Substitute them into equation (9), an integral calculation is conducted on the equation above in the beam spanwise direction, and the bending deformation and torsion angle at section x can be obtained as the followings:

$$\theta=\frac{P(2lx-x^2)}{2(EI-K^2/GJ)} \tag{10}$$

$$\varphi=\frac{P*K(2lx-x^2)}{2(EI*GJ-K^2)} \tag{11}$$

Then the maximum torsion angle at the blade tip is:

$$\varphi_l=\frac{FKl^2}{2(\mathrm{EI}*\mathrm{GJ}\text{-}\mathrm{K}^2)} \tag{12}$$

Now, supposing that:

$$\alpha^2=\frac{K^2}{EI{\cdot}GJ} \tag{13}$$

Substitute equation (13) into equation (12), then

$$\varphi_l=\frac{F\alpha l^2}{2\sqrt{\mathrm{EI}*\mathrm{GJ}}(1\text{-}\alpha^2)}=\frac{Fl^2}{2}\frac{1}{\sqrt{\mathrm{EI}*\mathrm{GJ}}}\frac{\alpha}{1\text{-}\alpha^2} \tag{14}$$

It can be seen from equation (13) and (14) that, when $\alpha/1\text{-}\alpha^2$ becomes the maximum, α tends to be 1 and $EI*GJ$ tends to be the minimum, and the biggest torsion angle at the blade tip would become the maximum. This demonstrates that, to get the biggest torsional deformation, $EI*GJ$ should be minimized, and α should be maximized. Here, α is called as the blade coupling control ratio. The following section will give more details about the factors that affects α, and how the coupling ratio relates with the off-axis angle of the fibers.

3.1.3.2 The factors that affect the control ratio

1. Material factor

It is necessary to discuss the factors that affect coupling control ratio before designing the coupling airfoil profile of the blade. In this section, a 2-D composite plate is chosen as an example to make the explanation. The constitutive equations of material stress and strain is:

$$\begin{pmatrix} \sigma_x \\ \sigma_y \\ \tau_{xy} \end{pmatrix} = \begin{pmatrix} Q_{11} & Q_{12} & Q_{16} \\ Q_{21} & Q_{22} & Q_{26} \\ Q_{61} & Q_{62} & Q_{66} \end{pmatrix} \begin{pmatrix} \varepsilon_x \\ \varepsilon_y \\ \gamma_{xy} \end{pmatrix} \tag{15}$$

In the equation above: σ_i , τ_{ij} are tensile stress and shearing stress respectively; ε_i , γ_{ij} are tensile strain and shearing strain; When the material principal axis and the coordinate axis tend to be the same, Q_{ij} is the flexible ($S_{16} = S_{26} = S_{61} = S_{62} = 0$) component; When the material principal axis is deviated from the coordinate axis with an angle θ, S_{ij} is off-axis flexibility component.
Now resolve equation (15) into 3 terms,

$$\begin{cases} \sigma_x = Q_{11}\xi_x + Q_{12}\xi_y + Q_{16}\gamma_{xy} \\ \sigma_y = Q_{21}\xi_x + Q_{22}\xi_y + Q_{26}\gamma_{xy} \\ \sigma_x = Q_{61}\xi_x + Q_{62}\xi_y + Q_{66}\gamma_{xy} \end{cases} \tag{16}$$

Supposing that σ_y =0 in plane stress state, then,

$$\xi_y = -\frac{Q_{21}\varepsilon_x + Q_{26}\gamma_{xy}}{Q_{22}} \tag{17}$$

Substitute equation (17) into equation (16), and the following equations can be obtained,

$$\sigma_y = \left(Q_{11} - \frac{Q_{12} \bullet Q_{21}}{Q_{22}} \right) \xi_x + \left(Q_{16} - \frac{Q_{12} \bullet Q_{26}}{Q_{22}} \right) \gamma_{xy} \tag{18}$$

$$\tau_{xy} = \left(Q_{61} - \frac{Q_{62} \bullet Q_{21}}{Q_{22}} \right) \xi_x + \left(Q_{66} - \frac{Q_{62} \bullet Q_{26}}{Q_{22}} \right) \gamma_{xy} \tag{19}$$

Then turn the two equations above into matrix form:

$$\begin{pmatrix} \sigma_y \\ \tau_{xy} \end{pmatrix} = \begin{pmatrix} \overline{Q_{11}} & \overline{Q_{12}} \\ \overline{Q_{21}} & \overline{Q_{22}} \end{pmatrix} \begin{pmatrix} \xi_x \\ \gamma_{xy} \end{pmatrix} \tag{20}$$

Where,

$$\overline{Q_{11}} = Q_{11} - \frac{Q_{12}Q_{21}}{Q_{22}}$$

$$\overline{Q_{12}} = Q_{16} - \frac{Q_{12}Q_{26}}{Q_{22}}$$

$$\overline{Q_{21}} = Q_{61} - \frac{Q_{62}Q_{21}}{Q_{22}}$$

$$\overline{Q_{22}} = Q_{66} - \frac{Q_{62}Q_{26}}{Q_{22}}$$

And according to the classical composite laminate theory, the relationship among force, moment, stress and strain are,

$$\begin{pmatrix} N_x \\ N_{xy} \\ M_x \\ M_{xy} \end{pmatrix} = \begin{pmatrix} \overline{A_{11}} & \overline{A_{16}} & \overline{B_{11}} & \overline{B_{16}} \\ \overline{A_{61}} & \overline{A_{66}} & \overline{B_{61}} & \overline{B_{66}} \\ \overline{B_{11}} & \overline{B_{61}} & \overline{D_{11}} & \overline{D_{16}} \\ \overline{B_{16}} & \overline{B_{66}} & \overline{D_{61}} & \overline{D_{66}} \end{pmatrix} \begin{pmatrix} \xi_x^0 \\ \gamma_{xy}^0 \\ \kappa_x \\ \kappa_{xy} \end{pmatrix} \tag{21}$$

Where,

$$\overline{A_{ij}} = \int \overline{Q_{ij}}dz \; ; \; \overline{B_{ij}} = \int \overline{Q_{ij}}xdx \; ; \; \overline{D_{ij}} = \int \overline{Q_{ij}}xdx$$

ξ_x^0, γ_{xy}^0 are the tensile stress and shearing strain of the middle plane, κ_x and κ_{xy} are the bending and torsional ratio of the middle plane, x is the vertical distance between the laminate and the middle plane. As to unidirectional laminated plate with symmetrical layer, equation (21) could be changed into:

$$\begin{pmatrix} N_x \\ N_{xy} \\ M_x \\ M_{xy} \end{pmatrix} = \begin{pmatrix} \overline{A_{11}} & \overline{A_{16}} & 0 & 0 \\ \overline{A_{61}} & \overline{A_{66}} & 0 & 0 \\ 0 & 0 & \overline{D_{11}} & \overline{D_{16}} \\ 0 & 0 & \overline{D_{61}} & \overline{D_{66}} \end{pmatrix} \begin{pmatrix} \xi_x^0 \\ \gamma_{xy}^0 \\ \kappa_x \\ \kappa_{xy} \end{pmatrix} \tag{22}$$

According to the equation above, only tensile-shear and bend-torsion coupling effect exist, which is as exactly as what is expected.

If dissymmetry layer method is used in laminate layer, the coupling effect would be more complicated. Equation (22) can be turned into the following form,

$$\begin{pmatrix} N_x \\ N_{xy} \\ M_x \\ M_{xy} \end{pmatrix} = \begin{pmatrix} \overline{A_{11}} & 0 & 0 & \overline{B_{16}} \\ 0 & \overline{A_{66}} & \overline{B_{61}} & 0 \\ 0 & \overline{B_{61}} & \overline{D_{11}} & 0 \\ \overline{B_{16}} & 0 & 0 & \overline{D_{66}} \end{pmatrix} \begin{pmatrix} \xi_x^0 \\ \gamma_{xy}^0 \\ \kappa_x \\ \kappa_{xy} \end{pmatrix} \tag{23}$$

According to equation (23), the laminated plate contains all of the coupling terms, and it is too complicated to control.

Because most of the blade spar of wind turbine machine is made up of unidirectional off-axis fibers, the coupling control ratio could be expressed as:

$$\alpha = \frac{-\overline{Q_{16}}}{\sqrt{\overline{D_{11}} \bullet \overline{D_{16}}}} \tag{24}$$

Substitute each terms of equation (20) into the equation above, then

$$\alpha = \frac{Q_{16} \bullet Q_{22} - Q_{12} \bullet Q_{26}}{\sqrt{(Q_{11} \bullet Q_{22} - Q_{16}^2) - (Q_{66} \bullet Q_{22} - Q_{26}^2)}} \tag{25}$$

$$\alpha^2 = \frac{(Q_{16} \bullet Q_{22} - Q_{12} \bullet Q_{26})^2}{(Q_{11} \bullet Q_{22} - Q_{16}^2) - (Q_{66} \bullet Q_{22} - Q_{26}^2)} \tag{26}$$

Define,

$$\nu_{16} = \frac{Q_{12} \bullet Q_{26} - Q_{16} \bullet Q_{22}}{Q_{11} \bullet Q_{22} - Q_{12}^2} \tag{27}$$

$$\nu_{61} = \frac{Q_{12} \bullet Q_{26} - Q_{16} \bullet Q_{22}}{Q_{66} \bullet Q_{22} - Q_{26}^2} \tag{28}$$

Then the coupling parameter equation could be simplified as:

$$\alpha = \sqrt{\frac{\nu_{16}}{\nu_{61}}} \tag{29}$$

So far, a simple equation through theory deduction of coupling control ratio is clear. According to equation (29), the coupling control ratio is only relative to Q_{ij} of the material, the layer flexible component of laminated plate. It means that the coupling ratio is only relative to the material itself and the layer angle.

2. Other factors

In theoretical calculation, a simplified model, say that a regular unidirectional one layer laminate plate with common modified beam theory, is chosen as the example to make the validation. Whereas, it is necessary to study the effect of other important factors, such as the geometry shape, material property, layer ratio and so on, on the coupling control ratio. Ong (1999) took composite materials spar of D-SPAR shape as an example, and analyzed the effect of several factors on the coupling ratio,

1. Geometry factor. Through the calculation of several samples of D-SPAR and aerofoil NACA0012, it was found that the coupling control ratio between the samples made little difference.
2. Layer factor. The layer factor mainly includes the effect of layer thickness and layer arrangement. The research showed that the layer thickness and layer material gave little influence on the coupling control ratio.

3. Fiber volume ratio. The coupling control ratio magnifies as the fiber volume ratio does.
4. Inner web stiffness. If the web is fixed inside the sample, its existence would increase the bending and torsional stiffness of the blade, which results in the reduction of coupling effect.
5. Layer method. There are several layer method of composite materials mentioned above, and wind turbine machines usually employ unidirectional and dissymmetrical laminate as the choice of fiber arrangement in composite. Moreover, mirror arrangement of fibers is necessary to provide bend-torsion coupling effect. As Fig.8(a) shows, the stiffness matrix from the layer is bend-torsion coupling stiffness and tensile-shear coupling stiffness. But if dissymmetry mirror layer arrangement is employed, as Fig.8(b) shows, the resultant stiffness matrix contains not only the bend-torsion coupling stiffness and tensile-shear stiffness, but also the tensile-shear-bend-torsion coupling stiffness, making the coupling behavior of the D-SPAR too complicated to control, as well as the coupling ratio of the laminated plate.

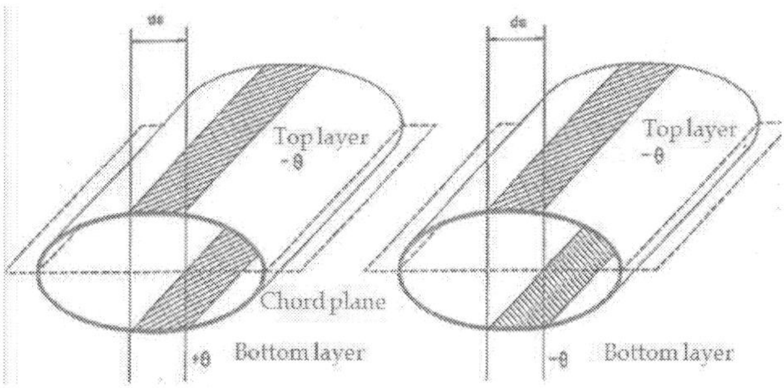

(a) Mirror symmetry layer (b) Dissymmetry layer

Fig. 8. Layer method

3.2 The adaptive blade design based on the compliant structure of plant leaf

Medial axis pattern is a general phenomenon almost for all biology, such as the plant leaf vein, the skeleton of creature observed from the macro-level. From the viewpoint of material science, the medial axis skeleton, combined with the other phase of the organism, formed a kind of composite structure with multiple phases. Obviously, the network of medial axis pattern takes the function of reinforcement (Gibson & Ashby, 1997). Whereas, how does this kind of topology structure adapt to the environmental stress? Does it have any meaning in guiding the structural design? In this section, we firstly achieved the mechanical performances of plant leaf vein through experiment, then the pattern evolution process from the viewpoint of topology optimization was carried out. Finally, a case study was given to validate the effect of bionic design methodology in wind turbine blade design.

3.2.1 The mechanical property and bionic study of compliant structure of plant leaf (Liu & Zhang, 2010a)

3.2.1.1 Sample preparation and measurement method

In order to explore the compliant structure and adaptability of the vein pattern of plant leaf, five fresh and mature leaf samples, which represent the typical leaf network in nature, are collected. 6 samples, which are ficus altissima, pineapple, kemirinoten, madagascar palm, ficus viren, and royalplam are picked in the campus shown in Fig.9. Royalplam and madagascar palm are monocotyledons plants with parallel pulse, the others are dicotyledonous plants with network veins. Royalplam is one of the most wind resistant plant in offshore, whereas, ficus viren has poor ability in carrying high wind load. In addition to the different abilities in carrying loads, the 6 plants are also quite different in thickness, shape outline, leaf texture and vein distribution. Therefore, the samples chosen are typically representative.

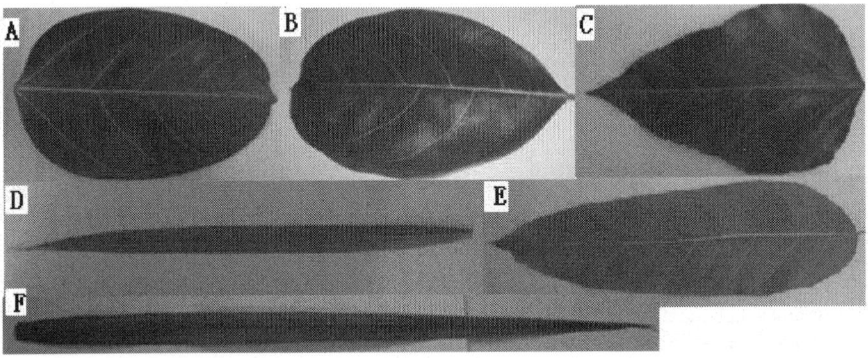

Fig. 9. The fresh plant leaf

For each kind of plants, 15 fresh samples with similar shape and size are collected. Firstly, a digimatic calliper is used to measure the leaf thickness. As the actual thickness of plants leaf is different from the root to the tip, the thickness measured was processed into an average value along the leaf spanwise direction. Secondly, in order to prevent the water loss of leaf from influencing the density result, the leaves with the polythene bags on are placed in the refrigerator, and keep the refrigerator temperature around 4℃. Before experiment, the leaves are wiped dry and each leaf mass is measured by using electronic balance with the accuracy of 0.001g. Finally, the mechanical performances are measured on a universal material tester in Fig.10, the results are shown in Table 2.

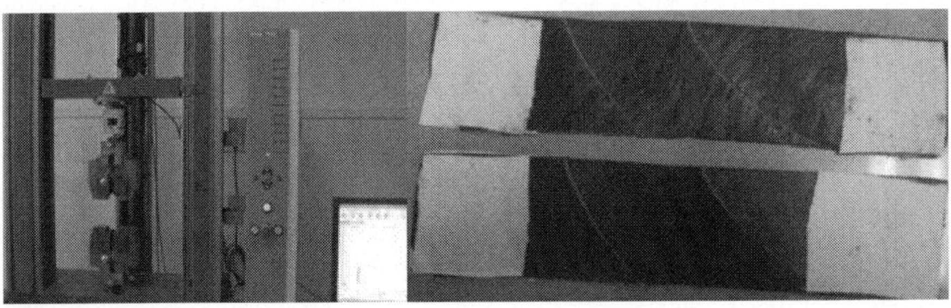

Fig. 10. The tension test of plant leaf

Property	Symbol	Ficus altissima	Pineapple	Kemirinoten	Ficus viren	Royalplam
Thickness	$T\ /\ mm$	0.43 ±0.023	0.37 ±0.029	0.29 ±0.016	0.17±0.019	0.25 ±0.029
Density	$\rho\ /\ kg\ /\ m^3$	55.87 ±7.63	63.05 ±19.44	40.23 ±5.11	703.9 ±37.4	0.25 ±0.029
Mesophyll modulu	$E_m\ /\ MP_a$	55.87 ±7.63	63.05 ±19.44	40.23 ±5.11	45.31 ±10.14	938 ±120.24
Leaf vein modulu	$E_v\ /\ MP_a$	229.7 ±55.13	199.9 ±35.85	157.6 ±28.70	99.8 ±33.82	4833 ±913.93
Possion ratio[13]	ν	0.33	0.33	0.33	0.33	0.33

Table 2. The material property of the plant leaves

3.2.1.2 Numerical simulation and topology optimization of the samples

LH65 WENZEL trilinear coordinates measuring instrument is used to measure the plant leaf data. In this study, only the coordinate data of the main vein and lateral vein in the setup system are measured, meanwhile the minor veins are ignored, considering that it plays an insignificant role in the mechanical performances. In order to build the mechanical model of leaf, the practical leaf structure is simplified. In practice, the cross section and thickness of the veins becomes smaller from the root (or the medial axis) to the tip (or the leaf rim). In this paper, the cross-section is hypothesized as round shape, and the outer diameter changes from the maximum to the minimum with linear function, whilst, the thickness keeps the same. The geometry model of leaves is built with the tool of ANSYS APDL Then, the model is imported to the commercial software Hypermesh to make the topology optimization. The target function is the minimum weighted compliance, and the constraint equation is volume fraction, which is set to be 0.3. In addition, the leaves are assumed to withstand loads from different directions, such as wind loads, rain and gravity. The topological optimum results are showed in Fig.11. The result gave an explanation that plant leaf vein is a kind of compliance effective structure with vein network pattern. A phenomenon was found by viewing all kinds of plant leaves in nature that different leaves growing in different conditions present various configurations, that is, the included angles between the major vein and lateral vein varied from 0° to 90° etc, while most of them varied in the range between 45° and 60°. What differs is the size of the cross section of the vein. Two main reasons cause the result, (1) in order to make the result less influenced by the error of numerical algorithm, the applied load is much bigger than the practical load that ordinary plants could withstand; (2) the reserved material based on the constraint condition in topology optimization is much more than the practical condition. Whereas, It is still deserved to notice that the material density is correlated with the load vector, that is, when the load takes place, the material appears where the stress needs, and the topology pattern of material elements corresponds well to the loading orientation, which explains that the mid axis pattern of plant vein is adaptive to the environmental stress. The main vein consistently remains unchanged, the lateral vein changes slightly along different load direction, which shows that the vein network of leaf has robust adaptability. The analysis shows that the plant leaf helps to keep the internal strain energy at a small value, and the environmental stress is one of the inducing factors for vein growth. It deserves to notice that the vein network of the plant leaf not only lends itself the physiological functions but also adapts to the complex environmental stress by evolving itself into a steady network pattern.

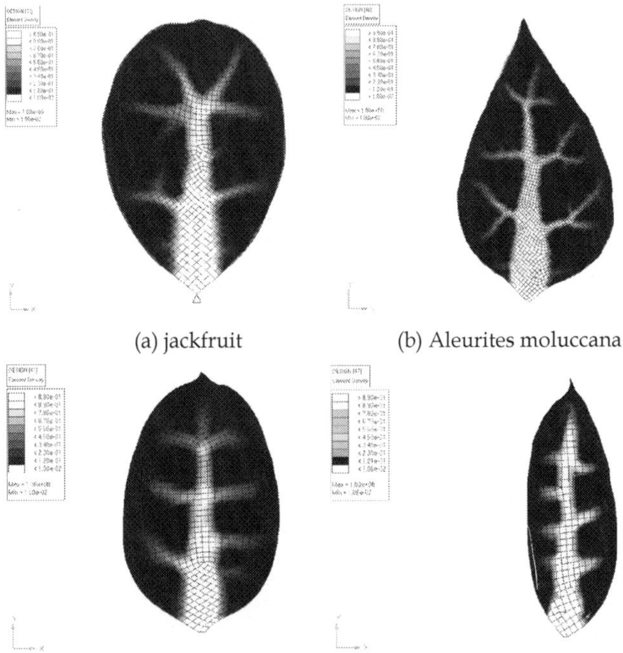

(a) jackfruit (b) Aleurites moluccana

(c) Ficus altissima (d) ficus viren

Fig. 11. The topology network of plant leaf

3.2.2 Adaptive blade design based on vein structure of plant leaf

The similarity between the plant leaf vein and the wind turbine blade can be explained as followings:

1. Structure and environment: both of the plant leaf and the blade are cantilever structure working in natural environment, and mainly suffer from wind load.
2. The Inner topology structure: the plant leaf has the principal vein and the lateral vein, and the lateral vein locates symmetrically at bi-lateral sides of the principal vein; Contrastively, large wind turbine blade is usually designed or configured as spanwise spar, a set of shear webs and composite skin structure, which is similar to the topology pattern of plant leaf. The design intention of large wind turbine blade is also obvious, that is, the spanwise spar is mainly used to carry the centrifugal force and the self weight, and the shear webs are used to carry the shear wind force.

It is not hard to see that the adaptive growth of plant leaf is driven by stress environment, and it can be used as a general guide to design wind turbine blade because of the similar working environment and structure requirement.

3.2.2.1 Blade optimization and bionic design for wind turbine

The baseline blade is originally developed by the institute of renewable energy research of Shan Tou University for 1.5 MW wind turbine, which is illustrated in Fig.12(a) (Xin, 2005). The length of the blade is 34m, and the airfoils are derived from Wortmann FX77/79Mod airfoil series, where the first airfoil profile position begins at 8.15m from root part, whose detailed profile parameters and design operating case have been thoroughly documented by Han (2008). In order to prevent big deformation of blade tip from influencing the accuracy of calculation, the tip part is magnified slightly.

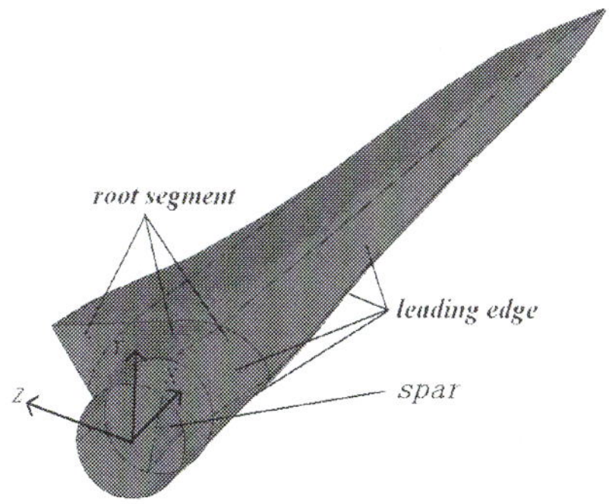

(a) The 1.5MW blade model and its key parts

(b) Blade topology optimization result

Fig. 12. The model and topology optimization result of 1.5MW blade

We select the blade segment from 12 to 20m along the blade spanwise direction, and suppose this part is made of homogeneous material. The finite element model of the blade is established in HyperMesh, where, PSOLID elements and Solid Isotropic Material with Penalization (SIMP) method are used for structural topology optimization. The target function is the minimum weighted compliance, and the constraint equation is the volume fraction, which is set to be 0.3. Topological optimization results are shown in Fig.12(b) with the load of critical wind of 50 m/s and gravity. As shown in Fig.12(b), the blade topological structure suggests a rough impression of the blade material distribution, which are, the spar and web configuration. If the blade topology pattern compares with the plant leaf, much more clear impression could be achieved, shown as Fig.13. The blade spar cap and webs correspond to the main vein, and the blade skin corresponds to the lateral vein, which would change with the wind load direction.

It is of indubitability that the most adaptive structure in the world comes from natural design. The authors were highly inspired by the similar cantilever structure between plant leaf and wind turbine blades, as well as the similar stress environment. Therefore, it could be expected that wind turbine blade imitating the plant leaf structure could achieve the excellent adaptive performances. The authors main work mainly focus on the fiber orientation design imitating the plant leaf skeleton. As it is known that different plant leaves have different morphological structure and side vein angles, in order to explore which leaf skeleton pattern are more suitable for the performance requirement of wind turbine blades,

Wind directions in 30°, 45°, 60° and 90° , respectively

Fig. 13. Comparison of the topology structure between plants and turbine blade[]

different plying angles changed in the range of [0,90] are chosen to make the calculation of different performances. Referring literature, the particular angle 20° is specially considered. The main vein angles, calculated from the medial axis of the blade, are [10°/0°] (Liu et al., 2009). Traditionally, the stiffen spar is chosen as the coupled design region. Whereas, in our work, the design region is expanded to the skin part of the blade considering the leaf vein structure. First of all, the uncoupled e-glass blade with small modification is chosen as a baseline model for coupling structure design, which is illustrated in Fig.12. The parameters of the blades and material properties are listed in Table 3 (Hermann T & Locke).

Parameters	E-glass fiber/epoxy		T600 carbon fiber/epoxy		Foam
	Experiment	safety	Experiment	safety	
Flap module E11[GPa]	36.47	-	127.3	-	0.61
Radial module E12[GPa]	12.62	-	8.78	-	-
Shear module G12[GPa]	3.94	-	5.07	-	-
Possion ratio	0.22	-	0.24	-	0.2
Density [Kg/m3]	1880	-	1520	-	120
Tensile strain [%]	2.00%	0.82%	1.21%	0.49%	-
Compressive strain [%]	2.09%	0.85%	0.97%	0.40%	-
Shear stress F12[MPa]	62.5	25.5	73.6	30.04	-

Table 3. Material properties

Parameters	Baseline	Prototype	Error
weight(Kg)	5808	5800	0.10%
Tip deflection(m)	1.12	-	-
1st natural freq(Hz)	1.05	1.08	3.40%
2^{nd}natural freq(Hz)	1.87	1.82	2.70%
3rd natural freq(Hz)	3.8	3.39	12%

Table 4. Comparison between baseline blade and prototype

The baseline blade is made up of outer shells and two internal shear webs with the same width as the spar cap, which is designed as variable section, from 20% to 85% of spanwise direction, with thickness tapered via ply drops to reduce the blade weight, and the exterior skins and internal shear webs are both sandwich construction with triaxial fiberglass laminate separated by foam core, whose material property is shown in table 3. For the baseline model, the unidirectional glass fibers are used for uncoupling effect, and the shells are overlaid inside and out with bidirectional ±45° glass fabric normally.

The baseline blade, named as Model A, uses all glass/epoxy fibers shown in table 3. In order to examine if the baseline blade matches well with the prototype model chosen from document (Lobitz, 2000), the comparison is made by numerical simulation for the blade tip deflection under critical wind load and the previous two order natural frequencies. The results shown in table 4 indicate that the baseline model matches well with the prototype model, and only small deviations exist within the acceptable scope. It means that the approaches of modeling and simulating are accurate and practical.

In addition, two more configurations are built for bend-torsion coupling design. One is assumed to be the same as the baseline blade except that the unidirectional fibers in spar cap are replaced with T600 carbon material shown in table 3, named as model B; another model is adapted from model A except that the skin of the blade is replaced with T600 carbon fibers in order to make the effect of coupling.

3.2.2.2 Evaluation of performances considering different design models

Based on the blade FEM calculation method (McKittrick et al., 2001), the equations from (30) to (32) were used to calculate the coupling controlled factor and stiffness of each blade section (Griffin, 2002), whilst, If all the parameters of each section are used to evaluate the overall aerodynamic performances of the wind turbine blade, it is too complex and inconvenient to fulfill the evaluation. Literature (Wetzel, 2005) reported that the aerodynamic performances in the region closer to the blade tip are more important. Therefore, a full blade involved the equivalent coupling factor based on the weighted average result of each spanwise section could be established with the following formula:

$$\alpha^* = \frac{2}{z_{tip}^2}\sum_{i=1}^{n}(\alpha_1 l_1 + \alpha_2 l_2 + \cdots + \alpha_i l_i) \tag{30}$$

where, z_{tip} is the overall length of the blade, α_i is the coupling factor of each section, l_i is the station of each section. Referring to this method, the full blade involved the equivalent flapping and torsion stiffness can be defined as：

$$EI^* = \frac{2}{z_{tip}^2}\sum_{i=1}^{n}(EI_1 l_1 + EI_2 l_2 + \cdots + EI_i l_i) \tag{31}$$

$$GJ^* = \frac{2}{z_{tip}^2}\sum_{i=1}^{n}(GJ_1 l_1 + GJ_2 l_2 + \cdots + GJ_i l_i) \tag{32}$$

The off-axis fibers orientation in the coupling region is the design variable. The random variable, namely the off-axis fiber angle in the coupling region, is supposed to follow the uniform distribution in the range of 0°~ 45°, because the related study shows that, when the off-axis fibers angles change in the range of 0°~±45°, the best coupling effect can be obtained. 22 samples are randomly chosen, and the static and dynamic performances of the blade are evaluated for each change of the off-axis orientation. The static performance is achieved through 2 load steps in ANSYS. In the first step, the parameters of the blade working in flapping and torsion moment including the blade stiffness and the coupling factor are respectively calculated; In the second step, the stress and strain of the blade working at extreme wind speed 50m/s are calculated, including two groups of stress that play the main role in deciding the blade failure: one group involves the interlaminate shear stress and in-plane Von Mises stress; another group involves the maximum tensile strain and compressive strain. In the mean time, the dynamic performance is calculated to figure out the 1st out-of-plane and in-plane frequency. All the calculation job is realized with ANSYS random calculation module by the user-subroutine language APDL, and the results are shown in Fig.14, Fig.15 till Fig.19.

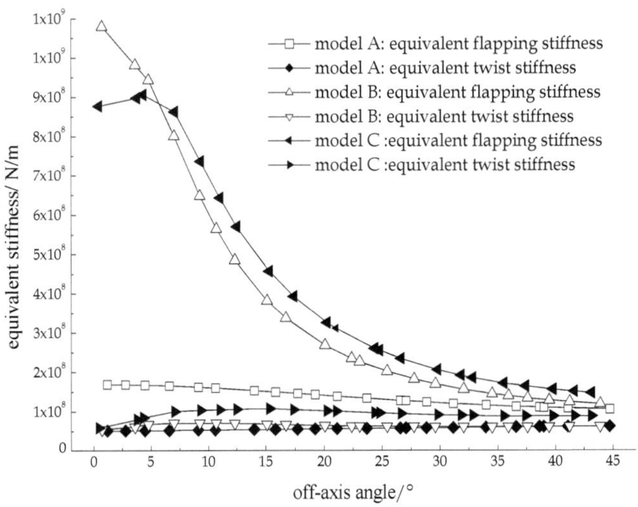

Fig. 14. The equivalent flapping and torsional stiffness

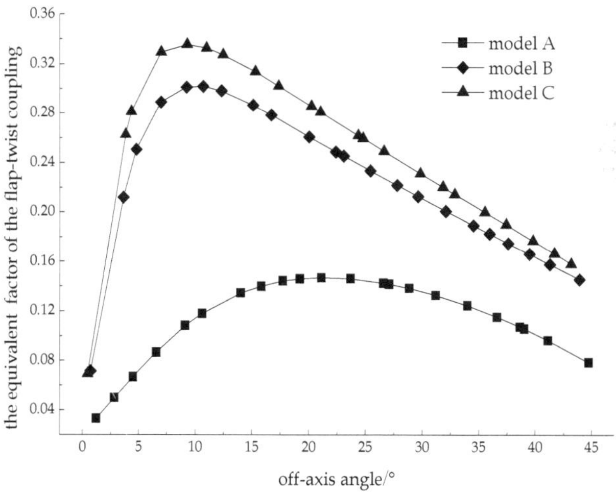

Fig. 15. The equivalent coupling factor -torsion

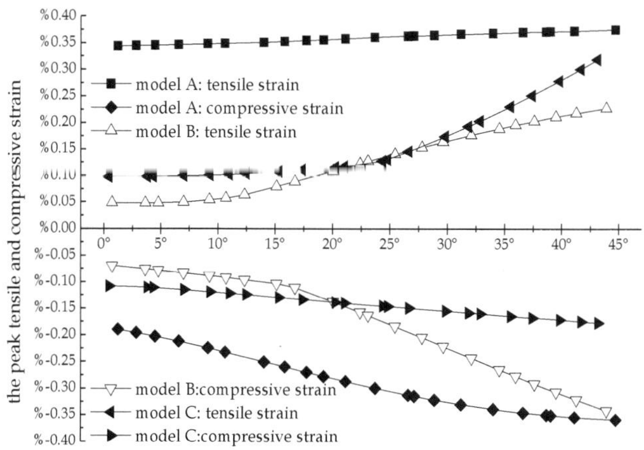

Fig. 16. The peak fibers tensile and compress strain

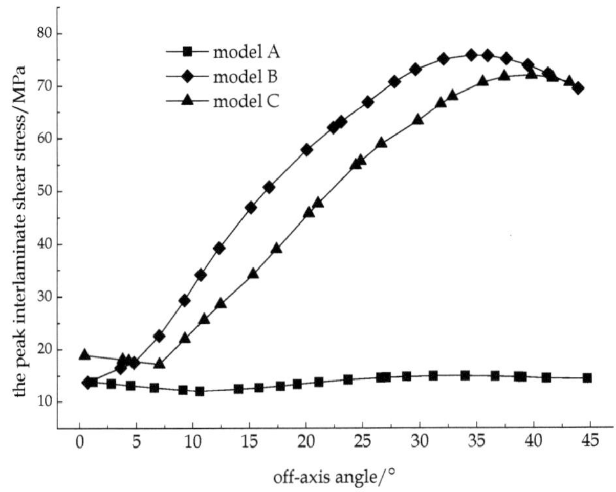

Fig. 17. The interlaminar shear stress

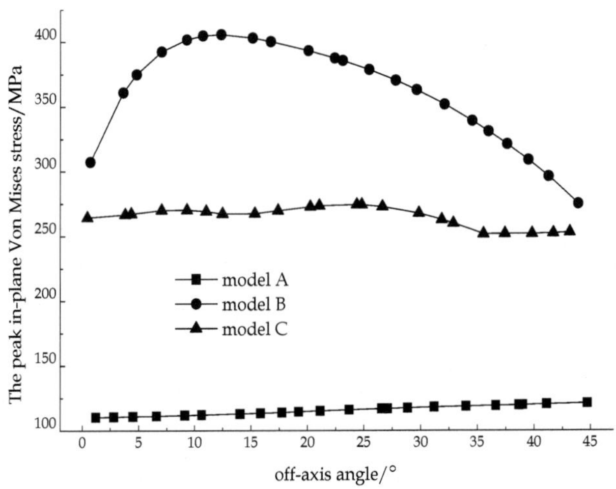

Fig. 18. The in-plane Von Mises stress

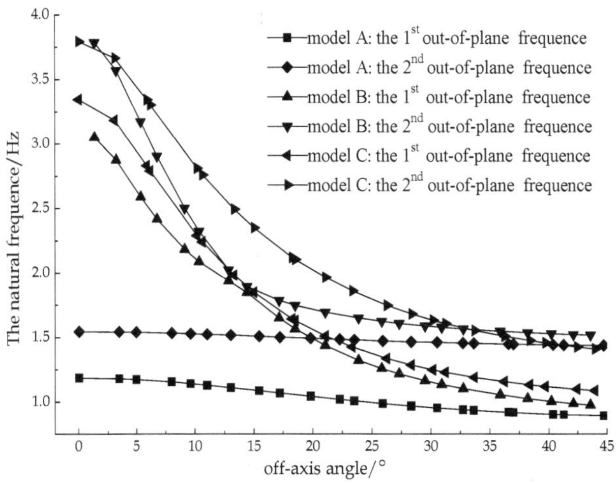

Fig. 19. The natural frequence

It can be observed from Fig.14 that, the overall trend of the equivalent flapping stiffness of each model is decreased with the increase of off-axis angle. Because the carbon fibers axial module is several times of the glass fibers, the flapping stiffness of model B and model C will be proportionally increased other than model A near 0°. But obviously, the carbon fibers axial module is dramatically decreased with the increase of the off-axis angle, which leads to the rapid decrease of the flapping stiffness. Comparatively, the torsion stiffness of the 3 models have little difference. The result achieved in paper (Liu &Zhang, 2010b) accords well with the theory introduced in the context, and it also agrees well with the result in the literature mentioned before. Meanwhile, it validates that the definition of the equivalent parameters in previous section is reasonable.

It can be known from Fig.15 that model B and model C achieve better coupling effect than model A, and off-axis fibers in blade skin achieve better coupling effect than off-axis spar cap. The off-axis angle which generates the maximum coupling effect of the three models is also different. The off-axis angle of model B and model C is about 11°, whereas, model A is about 20°. In addition, it is clear that, near the off-axis angle which achieves the maximum coupling effect, the flapping stiffness is still large. From this point, the spar thickness or the skin thickness can be also reduced so as to reduce the blade weight.

It can be observed from Fig.16 that the maximum tensile and compressive strains are increased with the increase of off-axis angle. The reason is that the decrease of the flapping stiffness leads to the increase of the deflection, which results in the increase of the corresponding tensile and compressive strain. Therefore, the fibers volume fraction and off-axis angle should ensure that the strains are within the safe range. It can be deduced that, when the glass fibers are replaced with carbon fibers, if the designed stiffness is expected to be equivalent with the reference stiffness, it can be reduced by diminishing the layer thickness. As the maximum compressive strain of carbon fibers is smaller than that of glass fibers, it should be careful not to make the tensile and compressive strain of carbon fibers exceed the safe value. The blade structure in this chapter is not exactly equal to the practical blade, and all the tensile and compressive strain do not exceed the safe range.

In Fig.17, it can be observed that the interlaminate shear stress is increased considerably with the increase of off-axis fiber angle in model B and model C. It climbs to about 74MPa, which is close to the dangerous situation for the carbon fibers used in this chapter. Large interlaminate shear stress will increase the possibility of transverse breakage of the interlaminate fibers if it exceeds the reference value. Definitely, it also becomes an important factor to constrain the blade design.

In Fig.18, it can be observed that the maximum in-plane Von Mises stress for model A has a small increasing trend with the increase of off-axis angle. Before 15°, there is an obvious decrease for model B, then decrease rapidly. Whereas, for model C, we can see the peak in-plane stress drops with the increase of angle in a way of slight fluctuation. According to the cumulate principle of Palmgren Miner's fatigue damage, low in-plane stress would be helpful to increase the blade fatigue life. In this sense, model B and model C may not be good to achieve longer fatigue life. Actually, this phenomena is caused by the properties of material itself. If the models are made up of the same material, and the coupling design is achieved only through regulating the off-axis fibers arrangement, it is found that the model imitating the compliant structure of plant leaf has better fatigue performances(Liu & Zhang, 2010a)

It can be known from Fig.19 that, the first order frequency of the out-of-plane and in-plane are all decreased with the increase of off-axis angle in three models. This is because the blade natural frequency is proportional to its stiffness, especially for model B and model C. The decline trend of the natural frequency is much more dramatically owing to the rapid drop of the stiffness, but carbon fibers can highly raise the natural frequency, which can be clearly seen that the design stiffness for model B and model C are always higher than that of the baseline blade. Therefore, it will not influence the dynamic performance of the blade.

3.2.3 Adaptive blade design based on stress trajectory

It is well known that in fiber reinforced composites, the fibers take the main function of carrying the load, and the matrix takes the function of bonding material and spreading the stress. Thus, it is commonly acceptable and understandable to match the fiber orientation with the principal stress orientation. Following this thought, at each point in the structure, three orthogonal sets of fibers, each subjected to an essentially uni-directional load, would carry all of the three principal stresses by utilizing the immense longitudinal strength and stiffness of the fibers. As fibers are orientated with the three principal stresses, the further advantage for a composite component is that it leads to minimal secondary stresses in the resin (Liu & Platts, 2008).

Wind turbine blades are critical components carrying the bending and torsional moment caused by wind and other force source. The general failure mode of wind generator is fatigue failure happened on some fracture-critical components in wind turbine system. Using ANSYS APDL and the special composite element Shell99, a 1.5MW wind turbine blade model, whose data was originated from (Li et al., 2005), was created and shown as Fig.12. The principal stress field in different wind load cases was processed, and the streamlines of the principal stress are shown in Fig.20(a)-(c), plotted with the shadow lines.

It can be observed from Fig.20(a)-(c) that when the incoming wind flow is perpendicular to the blade windward surface, the streamlines of the principal stress would go along with the

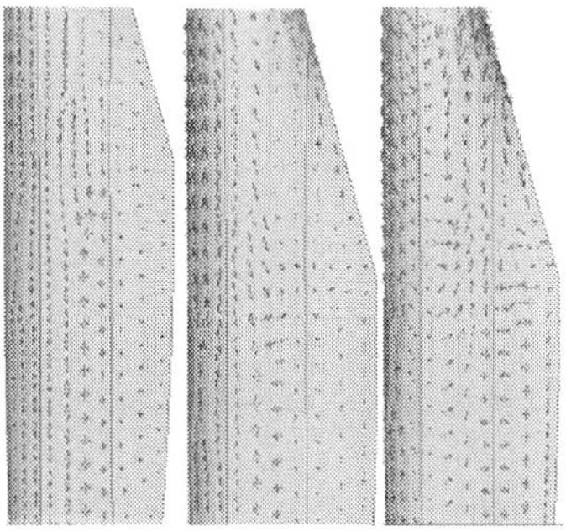

(a)0° (b)45° (c)-45°

Fig. 20. The blade principal stress field distribution

blade span direction; When the incoming flow is tilted to the blade windward surface and the rotational axis or rotational plane, the streamlines of the principal stress would change their orientation according to the force source change. Usually, the blade is made of GFRC (or CFRC) in UD(Unidirectional) type(the volume ratio could be 7:1) and lateral type(the volume ratio could be 1:1), where the UD plies carry most of the bending moment and centrifugal force, and the lateral plies bear most of the torsional moment and shearing force. In practice, the orientation of the incoming flow changes continuously, which leads to the principal stress field in the blade change homologous. This kind of complex stress environment requires that the blade material could be able to adapt to the change of the principal stress field in order to make use of its maximum loading effect. According to the similar working environmental analysis and similar structural requirement, the adaptive topology structure of plant leaf could be used as a guide to design a kind of adaptive blade structure. Traditionally, the blade plying orientations are usually 0° and 45°, whereas, according to medial axis pattern of plant leaf, the newly designed blade plying orientation are 0°, 22.5°, 45° and 67.5°, where, the angles are defined as the included angles between the UD plies and the lateral plies. Two kinds of 1.5MW blade model were designed in different plying orientations, and their dynamic and static performances are analyzed. Complying with the IEC61400- 1999 standards, the random wind field mode was taken as the random load sample, and the wind velocity complies with Weibul distribution, whose shape parameter is 2 and the scaling parameter is 15. The incoming flow orientation is regarded as uniform distribution. Now supposing two kinds of blades are working in the same wind field mode mentioned above, and the Monte Carto sampling method in ANSYS was used to calculate the blade performances. According to the change trend of the standard deviation for output values specified, 200 samples are reasonable. The maximum Von Mises stress and the standard deviation for traditional type and medial axis type are listed respectively as followings: 2.5594MP_a, 2.5750MP_a; 2.4631MP_a, 2.4007MP_a; It denotes that the maximum stress of the modified medial axis type is less discrete and more stable than the traditional one, which is helpful to improve the blade fatigue lifetime.

3.2.4 Material design of the blade

The length and weight of the blade would be considerably increased with the increase of power capacity of individual wind turbine generator. The confliction between power increase and blade length and weight increase leads to the problem that GFRP can not fully meet high performance requirement of the large-scale blade(>50meters). Instead, lightweight CFRP with high performance becomes the trend of blade material application. Whereas, high cost of CFRP makes it impossible to fully use it in the whole blade. Therefore the application of hybrid composites of GFRP and CFRP in large and medium-sized blades becomes the development trend.

According to design methods of large-scale blades, usually the airfoil leading edge and spar should be taken as the key design regions because of carrying the most moment, and at the same time the root segment of the blade where stress concentration easily appears also becomes focus of attention, especially for large-scale seashore wind turbines which are liable to be damaged from the roots caused by torsional vibration in typhoon. Therefore, the general guideline is to add a small amount of carbon fibers in the leading edge, spars and root segment of the blade to improve its comprehensive performance.

Taking the 1.5MW blade as an example, the model and each key part is defined as Fig.12, the material used is listed in Table 5.

Material	E1(MP_a)	E2(MP_a)	G12(MP_a)	ν	ρ (Kg/m^3)
GFRP(7:1)	42600	16500	5500	0.22	1950
GFRP(1:1)	17500	17500	3600	0.14	1950
CFRP(7:1)	142500	24200	10170	0.24	1580
CFRP(1:1)	26500	26500	8600	0.15	1580
Urethanefoam	60.8	59.8	19.18	0.20	119.7

Table 5. Material mechanical properties

The blade performance is discussed considering the following cases.

Case 1. the analysis results of prototype model in document (Liu & Platts, 2008), which only have the frequency values.

Case 2. the model used above in this article. Which exists somewhat inaccuracy compare with prototype model.

Case 3. based on Case2, 2mm CFRP (7:1) layers are added in leading edge as UD laminates.

Case 4. based on Case3, 2mm CFRP (1:1) layers are added in leading edge as lateral laminates.

Case 5. based on Case4, 2mm CFRP (1:1) layers are added in spar sandwich structure.

Case 6. based on Case5, 2mm CFRP (7:1) layers are added in root segment.

In Case 3, 5 and 6, each CFRP laminate locates at the same layer.

40m/s wind speed in the wind field site is taken as the extreme load, and a series of computations considering above cases were carried out with ANSYS. Fig.21 shows the results of dynamic performance, where the first-order waving and second-order swing natural frequency were obtained, the sixth torsional frequency, which has significant effect on the blade torsional vibration. Fig.22 shows the static performance, where the maximum displacement and Von Mises stress is obtained.

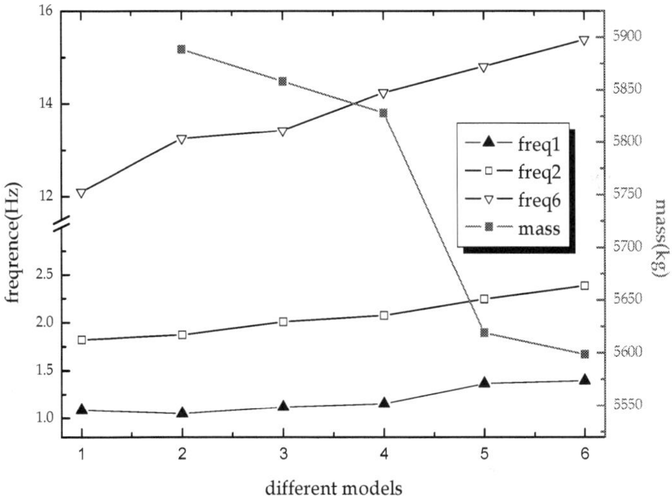

Fig. 21. The dynamic performance in different cases

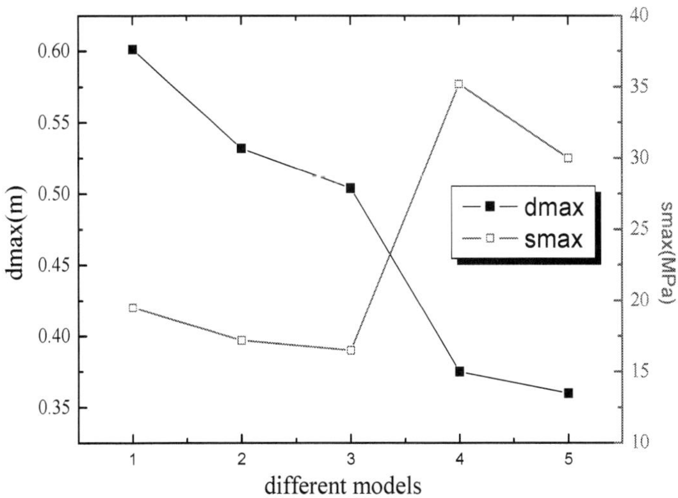

Fig. 22. The static performance in different cases

In Fig.21, number 1-6 on the X-axis represent Cases 1-6, and number 1-5 in Fig.22 represent cases 2-6. It can be observed from the figures that when the CFRP in the leading edge is added, and then in spar, and root segment in turn, each order frequency can be improved obviously and the blade weight was reduced. The blade tip displacement becomes smaller, but Von Mises stress becomes greater with the increasing of CFRP amount. These can be explained as followings, where:

$$\omega = \sqrt{\frac{K}{I}} \tag{33}$$

$$\sigma = E\varepsilon \tag{34}$$

In formula (33), *K*, *I* represent stiffness coefficient and inertia of the blade. It is obvious that the frequency ω is in direct proportion to *K*. Formula (34) shows that stress σ will increase with *E*. Whereas the strain ε is in inverse proportion to *EI*, which was named as bending stiffness in the spar theory. Here, *K* and *E* become bigger as the CFRP is added, and the frequency ω and stress σ increase at the same time.

Above all, when hybrid composites are utilized in the blade, the dynamic performance can be improved and the deformation and the weight of the blade can be reduced. Whereas, the stress will increase unavoidable. Necessary measures must be taken to balance the contradictory, which needs further study.

4. Conclusion

In this chapter, the design and evaluation of adaptive blade based on tend-torsion coupling effect is explored, incorporated the bionic design method from the flexible topological structure of plant leaf. Three models for coupling design are built referring to a 1.5MW baseline blade. The investigations on parametric design for off-axis fiber angle of the coupled blade are conducted respectively. The results show that, the glass/carbon hybrid fibers are the best choice for coupling design, which can provide high coupling coefficient between 15°~25° of off-axis fiber angles, rather than single glass fibers. For the three models, the maximum tensile and compressive strain and stress increase with the increase of off-axis fibers angle, whereas, the in-plane stress in hybrid fiber blades decreases. It is recommended to place the fibers before 25° to ensure the blade structure safety. Following that, a kind of bionic design method is integrated into the coupled blade design, and the result shows that it can improve the blade bend-torsion coupling effect further.

5. Acknowledgement

The authors are thankful to the support of national natural science foundation of China (the Grant No. 50675067 and 50975090).

6. Reference

Simon Philippe Breton & Geir Moe. (2009). Status, plans and technologies for offshore wind turbines in Europe and North America. *Renewable Energy*. 34, pp. 646-654.

N W M Bishop & L W Lack, L Li. (1999). Analytical fatigue life assessment of vibration induced fatigue damage. *IPENZ Transactions*. 26, pp. 1-17.

Knut O Ronold, Jakob Wedel-Heinen & Carl J Christensen. (1999). Reliability-based fatigue design of wind-turbine rotor blades. *Engineering Structure*. 21, pp. 1101-1114.

Knut O Ronold & Gunner C Larsen. (2000). Reliability-based design of wind-turbine rotor blades against failure in ultimate loading. *Engineering Structures*. 22, pp. 565–574.

Christoph W Kensche. (2006). Fatigue of composites for wind turbines. *International Journal of Fatigue*. 28, pp. 1363–1374.

Daniel D Samborsky, et al. (2008). Delamination at Thick Ply Drops in Carbon and Glass Fiber Laminates Under Fatigue Loading. *Journal of Solar Energy Engineering*. 130, pp. 1-8.

Raif Sakin & Irfan Ay. (2008). Statistical analysis of bending fatigue life data using Weibull distribution in glass-fiber reinforced polyester composites. *Materials and Design*. 29, pp. 1170-1181.

Don W. Lobitz & Paul S. Veers. (1998). Eroelastic Behavior of Twist-Coupled Hawt Blades. *American Institute of Aeronautics and Astronautics*. AIAA-98-0029, pp. 1-9.

Bao N S, et al. (2007). Dynamic characteristics of large-scale stall wind turbine system. *Acta Energiae Solaris Sinica*. 28(12), pp. 1329-1334.

Lin Y G. (2005). Study on the Technology of Pitch-control for Large Scale Wind Turbine. Doctor Dissertation. Zhejiang, Zhe Jiang University.

Matthew A Lackner & Gijs A M. van Kuik. (2010). The Performance of Wind Turbine Smart Rotor Control Approaches During Extreme Loads. *Journal of Solar Energy Engineering*. 132, pp. 1-8.

Andreas Büter & Elmar Breitbach. (2000). Adaptive blade twist calculations and experimental results. *Aerosp. Sci. Technol* . 4, pp. 309–319.

Karaolis N M, G Jeronimidis & P J Musgrove. (1989). Composite Wind Turbine Blades: Coupling Effects and Rotor Aerodynamic Performance, *EWEC'89, European Wind Energy Conf*, Glasgow, pp. 10-13.

Joosse, P. A. & R. M. van den Berg. (1996). Development of a TenTorTube for Blade Tip Mechanisms, Part 1: Feasibility and Material Tests, Proc, *European Union Wind Energy Conf. and Exhib*, GF6teborg, pp. 20-24.

Andrew, T.L & Richard, G. J. (1999). The compliant blades for wind turbines. *IPENZ Transactions*. 26, pp. 7 -12.

Don W. Lobitz, et al. (2001). The Use of Twist-Coupled Blades to Enhance the Performance of Horizontal Axis Wind Turbines. SAND 2001-1303.

Ladean R. McKittrick, Douglas S. Cairns, & John Mandell. (2001). Analysis of a Composite Blade Design for the AOC 15/50 Wind Turbine Using a Finite Element Model. SAND 2001-1441.

Dayton A Griffin & Thomas D Ashwill. (2003). Alternative composite materials for megawatt-scale wind turbine blades: design considerations and recommended testing. *Wind Energy*. 125, pp. 515-521.

Mansour H Mohamed & Kylc K Wetzel. (2006). 3D Woven Carbon/Glass Hybrid Spar Cap for Wind Turbine Rotor Blade. *Journal of Solar Energy Engineering*. 128, pp. 562-573.

John F. Mandell, Daniel D. Samborsky & Lei Wang. (2003). New fatigue data for wind turbine blade materials. *Wind Energy*. 125, pp. 506-514.

J Selwin Rajadurai, et al. (2008). Finite element analysiswith an improved failure criterion for composite wind turbine blades. *Forsch Ingenieurwes*. 72, pp. 193–207.

James Locke & Ulyses Valencia. (2004). Design Studies for Twist-Coupled Wind Turbine Blades[R]. SAND 2004-0522.

Dayton Griffin. (2002). Evaluation of Design Concepts for Adaptive Wind Turbine Blades[R]. SAND2002-24 24.

Kyle. K Wetzel. (2005). Utility Scale Twist-Flap Coupled Blade Design. *Journal of Solar Energy Engineering*. 127, pp. 529-537.

Alireza Maheri, Siamak Noroozi & Chris A Toomer. (2006). WTAB, a computer program for predicting the perfor-mance of horizontal axis wind turbines with adaptive blade. *Renewable Energy*. 31, pp. 1673–1685.

Alireza Maheri, Siamak Noroozi & John Vinney. (2007). Application of combined analytical FEA coupled aero structure simulation in design of wink turbine adaptive blades. *Renewable Energy*. 32, pp. 2 011-2018.

Alireza Maheri & Askin T Isikveren. (2009). Design of Wind Turbine Passive Smart Blades, *European Wind Energy Conference*, France.

Alireza Maheri & Askin T. Isikveren. (2009). Performance prediction of wind turbines utilizing passive smart blades: approaches and evaluation. *Wind Energy*. (in press)

Rachel F Nicholls-Lee, et al. (2007). Enhancing Performance of a Horizontal Axis Tidal Turbine using Adaptive Blades, *OCEANS 2007 Europe*, UK, pp. 1-6.

R F Nicholls-Lee, S W Boyd & S R Turnock. (2009). Development of high performance composite bend-twist coupled blades for a horizontal axis tidal turbine, *17th International Conference on Composite Materials* , UK.

Jim Platts & W Y Liu. (2008). Developing Chinese Wind Energy Technology, *Young Scholars Symposium for Design and Manufacture of New Energy of National Natural Science Foundation*, Guangzhou.

Crawford Curran & Platts Jim. (2006). Updating and optimization of a coning rotor concept. Collection of technical papers, *44th AIAA Aerospace Science Meeting*, Reno, Nevada, 10, pp. 7265-7280.

Crawford Curran. (2006). Re-examination the precepts of the blade element momentum theory for coning rotors. *Wind Energy*. 9(5), pp. 457-478.

Sarikaya M, Gunnison K.E & Yasrebi M. (1990). See shells as a natu-ral model to study laminated composites, *Lancster Penn sylvania: Tech Pub Co Inc*, pp. 176-183.

Gordon J E, Jeronimidis G. M & O W Richardson. (1980). Composites with high work of fracture. *Mathematical and Physical Sciences*. 294,1411(A), pp. 545- 550.

Steele C R. (2000). Shell stability related to pattern formation in plants. *Journal of Applied Mechanics*. 67(2), pp. 237-247.

Somerville Chris, Bauer Stefan & Brininstool Ginger. (2004). Review: Toward a system approach to understanding plant cell Walls. *Science*. 306(24), pp. 2206-2211.

S E Jones & M J Platts. (1998). Practical Matching of Principal Stress Field Geometries in Composite Components, *97 International Conference on Automated Composites*, Oxford, Elsevier, 29, pp. 821-828.

Blum H & Nagel R. (1978). Shape description using weighted symmetric axis features. *Pattern Recognition*, 10(3), pp. 67-180.

W Y Liu, et al. (2007). Relationship between medial axis pattern of plant leaf and mechanics self-adaptability(I): experimental investigation and numerical simulation. *Journal of South China University of Technology: Natural Science Edition*, 35(3), pp. 42-46.

W Y Liu, J X Gong, W F Hou. (2009). Relationship between medial axis pattern of plant leaf and mechanics self-adaptability(II): vein structures with different vector angles and topological pattern of plant leaf. *Journal of South China University of Technology: Natural Science Edition*, 37(8), pp. 12-16.

Y Q Jiang, F S Lu & Gu Z J. (1990). *Mechanics of Composite Materials*, Xi'an Jiaotong University Press, Xi'an .

Cheng Huat Ong. (2000). Composite wind turbine blades. Doctor Dissertation. *Stanford University*, USA.

Cheng Huat Ong & Stephen W Tsai. Design, manufacture and testingofa bend-twist spar. SAND99-13.

Lornal J. Gibson & Michael F. Ashby. (1997). Cellular solids: structure and properties, *Cambridge university press*, England.

W Y Liu & Y Zhang. (2010a). Network study of plant leaf topological pattern and mechanical property and its application, *The 3rd International Conference of Bionic Engineering*, Zhuhai

W P Xin. (2005). Analysis of dynamic characteristics and response for rotating blades of wind turbine, Master Dissertation , *Shantou University*, China.

Xinyue, Han. (2008). Muti-objective optimization design and structure dynamic analysis of HAWT blade, Master Dissertation, *Shantou University*, China.

W Y Liu, J X Gong, & W F Hou. (2009). Relationship between medial axis pattern of plant leaf and mechanics self-adaptability (II): vein structures with different vector angles and topological pattern of plant leaf. *Journal of South China University of Technology (Natural Science Edition)*. 37, pp. 12-16.

Hermann T & Locke J. (2005). Failure Analysis of Anisotropic Laminate Composites *Utilizing Commercial FEA Software*. in: AIAA-2005-0975 (2005), pp. 370–381.

W Y Liu & Y Zhang. (2010b). Bend-Twist Coupling Design and Evaluation of Spar Cap of Wind Turbine Compliance Blade, *International Conference on Manufacturing Engineering and Automation 2010*, Guangzhou. (in press)

W Y Liu & M . J. Platts, (2008). Concept representation, practical topology decision and analysis in composites lug design, *ICFDM*, Tianjin, 8, pp. 636-643.

D Y Li, Z Q Ye, & Chen Y. (2005). Mutli-body dynamics numerical analysis of rotating blade of horizontal axis wind turbin. *Acta Energiae Solaris Sinica*. pp. 475-781.

W Y Liu, J X Gong & X F Liu. (2009). A kind of innovative design methodology of wind turbine blade based on natural structure.in: Information and Computing Science, *ICIC '09. Second International Conference*, Manchester, England ,pp. 350-354.

4

A Ducted Horizontal Wind Turbine for Efficient Generation

I.H. Al-Bahadly and A.F.T. Petersen
Massey University
New Zealand

1. Introduction

This chapter investigates ducted turbines for the use of wind power generation. The interest for this grew from the ever increasing demand for energy. After investigating the nature of the three bladed wind turbines, it became apparent that the machines were not very efficient, expensive and have a limited fatigue life.

The ducted twin turbine wind power generator is proposed in this chapter and a comparison in performance has been carried out between the ducted turbine and the conventional turbine. The ducted turbine has the ability to accelerate the air flow through a converging intake thereby increasing the power that can be extracted from the air flow. As the wind passes through a converging duct the velocity increases while the pressure decreases. The power extracted has a cubic relationship to wind velocity where as the relation to pressure is linear.

The need for energy consumes our society. As technology has advanced in certain areas the ability to produce power has had to keep pace with the ever increasing demands. There always seems to be an energy-crisis weather contrived or real, and society allows the pollution of our environment in the name of power production.

Power production with traditional means has polluted our planet. Hydro power dams release carbon that was locked up in the trees and plants that were drowned during the filling of the dam. Any sort of fossil fuel powered plant releases carbon into the environment during the combustion process. Nuclear plants are generally unpopular and will not be accepted in New Zealand for a very long time.

Renewable, environmentally friendly, clean, safe, even wholesome, are the types of adjectives we should be using to describe power production. Wind energy is the closest we may have at present that may be considered to fit into those criteria. Certain aspects such as noise and blade flash are a concern. The ducted twin turbine is proposed in this report as an environmentally friendly, safe alternative method of power production from renewable sources.

The ducted twin turbine wind power generator is proposed in this chapter and a comparison in performance has been carried out between the ducted turbine and the conventional turbine. The ducted turbine has the ability to accelerate the air flow through a converging intake thereby increasing the power that can be extracted from the air flow. As the wind passes through a converging duct the velocity increases while the pressure decreases. The power extracted has a cubic relationship to wind velocity where as the relation to pressure is linear. This is exploited in the ducted turbine and gives an advantage of a factor of 17 (improvement) over the conventional turbine in theoretical calculations not including coefficients of power transformation.

This improvement in power that can be extracted from the wind is vital to the implementation of this power generation system. The system is proposed for installation and location in urban areas thereby reducing transmission losses. Blade flash is reduced and noise production is low when the formulas of (Grosveld, 1985) are applied to the situation. These factors will be vital if the generators are to be mounted close to dwellings.

To reduce complexity of design the mechanism controlling angle of attack is not built into the turbine blades as with conventional wind turbines. The ducted turbine uses Variable Inlet Guide Vanes (VIGVs) mounted in the air stream prior to the first stage turbine; this controls angle of attack maintaining optimum performance, while the mechanisms do not have to be mounted in confines of a hub. An annular arrangement is proposed that houses the pitch change mechanism in the nacelle or inner ducting reducing inertia on the rotating mechanism.

The sensors for the control of the VIGVs will be a set of twin pitot static tubes measuring wind speed. As the turbine is constant speeding the only variable that will affect angle of attack is wind velocity. This is accounted for by the VIGVs as the adjustments are made the air flow is offered to the first stage turbine at the optimum angle of attack; this is the induced swirl motion from the VIGVs.

Overall this work provides an alternative to the power production systems that are available on the market. There have been similar systems produced and some early patients filed, yet the ducted system is not common, perhaps due to cost. But as more people object to the construction and expansion of wind farms full of conventional turbines, then other systems will have to be investigated. Perhaps the quieter, smaller, more efficient twin ducted wind turbine power generator.

Clean, renewable, affordable, safe, efficient, and non-invasive are the terms we need to describe our power generation systems. The environment cannot keep absorbing the industrial poisons that are the by products of modern humanities lifestyle. Coal fired power plants place carbon into the atmosphere, hydro dams cover and drowns forests that release carbon into the atmosphere during decomposition; LPG/LNG fired gas turbine power generators discharge carbon into the atmosphere, and nuclear power generation is politically unacceptable in New Zealand. The reasonable options that are left are wind and wave (tidal) power generation.

Wind turbine generation is a system that will provide clean power at an affordable rate from renewable resources. Safety and efficiency are the two areas of wind power generation that are debatable at present.

Wind energy has been touted as the energy generating saviour. It is branded renewable; that is it will be available for future generations and the 'reserve' does not diminish with use. Wind turbines also create carbon credits as the CO_2 displaced by each kilowatt of energy generated that would have otherwise been generated by a fossil fuel powered station. This has lead to the increase in profitability of existing turbines in their operation.

The use of wind power dates back as long as history itself; with the use of sail harnessing the wind to power boats and ships alike. Holland would not be as it is without the use of wind to power the water pumps that held back the sea and the stones that milled the flour.

Dutch settlers took the wind mill to the United States where, in the mid 1800's, it evolved into the multi-bladed wind turbine that was synonymous with the older generation cowboy movies. According to (Mathew, 2006), between 1850 and 1930 over six million of these wind turbines were sold, these were primarily used to lift water from ground wells.

(Mathew, 2006) recognises Denmark as the first place where a purpose built wind turbine power generator was built, this occurred about 1890. At the same time a twelve kilowatt generator was built in Cleveland, Ohio, this sported a set of step-up gears to the generator. By 1910, hundreds of generators were supplying electricity to villages in Denmark, and by 1925 small generators, 0.2 - 3 kW, were available in the United States to charge batteries. These wind turbines were of a two or three blade configuration with an aerodynamic consideration to the shape of the cross section.

At the beginning of 2007 the total installed and operational capacity of wind turbines in New Zealand was 170.8 MW (The New Zealand Wind Energy Association, 2007). The installed capacity at the beginning of 2004 was 36.3 mega watts (MW): In other words this is equivalent to an increase in installed wind energy capacity of 470% - most of this occurred during 2004.

151 MW of new capacity is currently under construction and when this is commissioned during the year the total installed capacity in New Zealand will nearly double to 321.8 MW.

Today there are 7 wind farms in North and South Island (or 9 if you count the staged Tararua and Hau Nui developments as two separate sites) and with the growth of this industry the generation capacity will only increase. Although, there is a need for renewable power generation in the future, public objection to the erection of large scale wind farms has increased.

The lobby group against the expansion of the wind farm expansion cited:

- Low frequency noise pollution
- Blade flash
- Spoiling the natural aesthetics

As the factors that would depress real estate prices, raise health issues and create a general annoyance. A ducted turbine may be a possible solution for some of these problems (Dannecker and Grant, 2002; Hu and Cheng, 2008).

2. The proposed system

The optimisation of the wind turbine and ducting assembly is the focus of this work and concentrates on the interface between the wind and the ducting assembly. The specifications for the generator were set as a goal for us to attain or beat, these are:

Rotational speed	300 rpm
Power	3 kW
Efficiency	95 %
Diameter of the generator housing	250 mm (fixed)
Generator drive	gearless

Setting these specifications provided a foundation from which to build the design. Other specifications were not constrained, such as torque, thereby allowing any extra power captured from the wind to be realised as current. Typically the wind strength is not constant and allowing an unconstrained parameter allows the design to harness various levels of power, when available.

A ducted turbine with the generator mounted on a common shaft is the arrangement investigated in this work. Investigation into this system indicates a possibility of reducing the affects created by the larger wind generators.
Another prospect of this turbine will be the ability to site this item close to built-up areas. As noise and blade flash is reduced these items will be mounted directly onto buildings and in urban areas where the power is used, and any surplus sold to the power distribution company. This will; also have the affect of reducing transmission losses as power production will occur on site.
The ducted turbine has not been a popular choice with wind turbine designers. The cheaper option has been the three bladed, low rotational speed turbines commonly seen in the large commercial wind farms. With micro generation, a smaller more efficient system of a ducted turbine is proposed.
Efficiencies can be gained with a ducted system as this gives the ability of accelerating the airflow through a convergent path much like a venturi. The shape of the structure becomes critical as this determines the velocity of the airflow and partial pressures that will be required to maximise the efficiencies. Consideration as to the surface finish will be critical also as this affects the size or depth of the vena contractor and the characteristics of the airflow.
The ducted twin screw design chosen, has the ability to greatly increase the power that can be extracted from the airflow. Simply put the theoretical power that can be extracted is:
Power (W) = ½ * (ρ air(kg/m^3)) * (turbine swept area(m^2)) * (wind velocity(m/sec))^3

$$\text{Power} = \tfrac{1}{2} * \rho * TSA * V^3 \tag{1}$$

This shows that power is directly proportional to half the cube of the wind velocity.

2.1 Acceleration of airflow velocity

Giovanni Battista Venturi (1746 - 1822), accredited with the effect named after him, is an example of Bernoulli's principle of incompressible gasses where an acceleration of airflow must occur through a constriction to satisfy the equation of continuity.

$$Pa + \tfrac{1}{2}\rho V^2 + \rho g h = \text{constant} \tag{2}$$

From this we can derive,

$$Pa_1 - Pa_2 = \tfrac{1}{2}\rho(V^2{}_2 - V^2{}_1) \tag{3}$$

And,

$$A_1 * V_1 = A_2 * V_2 \tag{4}$$

Where:

A	=	area,
V	=	velocity,
ρ	=	density of fluid,
TSA	=	turbine swept area,
g	=	acceleration constant,
h	=	height,
Pa	=	fluid pressure.

The acceleration of the airflow was achieved with the use of a converging duct. This is illustrated in the diagram of Fig.1.

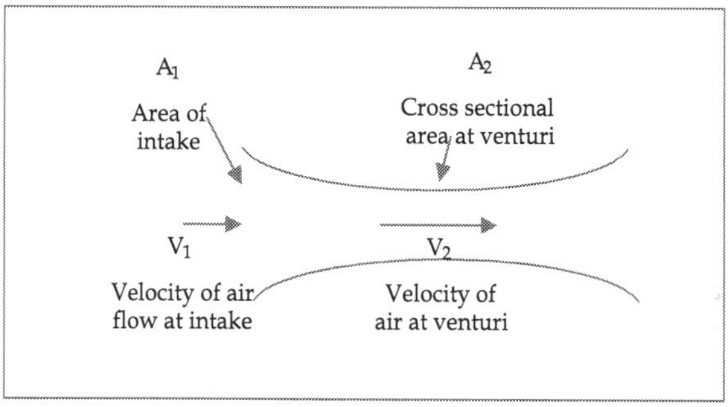

Fig. 1. Illustration of the venture equation maintaining continuity

Therefore,

$$V_2 = (A_1 * V_1)/A_2 \quad (5)$$

Then by placing the wind turbine inside a convergent duct we should see increased efficiencies as the airflow is accelerated through the venturi. Aside from this advantage, gains will also be made with the reduction of span-wise flow and the elimination of blade tip vortices. According to (Spera, 1994) a six percent reduction in drag can be gained by the elimination of blade tip vortices.

Noise will be reduced as more effective control of airflow over the turbine blades will be possible and acoustic relationships will be examined. By controlling the angle of attack of the airflow entering the ducting the design of the blades can be optimised for a narrow range of aerodynamic conditions, this can provide the opportunity to reduce the compromises that are normal with aerofoil design. This will enable positioning of these wind generators closer to populated areas.

An iterative approach to the development of this turbine was pursued. The software package used was EDF.lab. EDF.lab is a program designed to simulate fluid flows through a structure. It contains a solidworks modeler and may prove affective with the design process.

3. System design

A relationship between shape and airflow was envisaged and the areas in which the advantages to efficiency were to be sought. These were:

- Acceleration of airflow velocity
- Reduction/elimination of blade tip vortices
- Reduction of span-wise flow
- Gearless transfer of power to the generator core.

Gearless transfer of power is being achieved through the efforts of incorporating a switched reluctance generator (Chen, 2008; Zhao et al, 2009). This should enable efficient power generation without the need to incorporate step-up gear systems.

According to (Mathew, 2006), the power available in the wind can be calculated from its kinetic energy. This shall be examined at this stage although losses occur transitioning from wind kinetic energy to turbine rotational energy this can be described by a coefficient. Actual power will be discussed later in this report.

$$E = ½ * m * V^2 \quad (6)$$

If we consider the mass in terms of density and volume
Dimensional analysis gives:

$$E = \tfrac{1}{2} * \rho * V * V^2$$

$$E = \tfrac{1}{2} * \rho * V^3 \qquad (7)$$

The air flow that interacts with the turbine is offered up to the front face of the rotor, this is referred to as the Turbine Swept Area (TSA). Therefore (Mathew, 2006) describes the parcel of air offered as having a cross sectional area of that of the TSA and thickness equivalent to the wind velocity. Hence energy per unit time is:
Power (W) = ½ * (ρ air (kg/m^3)) * (turbine swept area (m^2)) * (wind velocity (m/sec))^3 as shown in equation (1).
If we look at the relationships between the factors of the power equation we notice a linear relation between density and power, where as, there exists a cubic relation between velocity and power. This will be exploited with the ducted turbine design.
The density will be affected by the acceleration of the air flow due to the venturi effect. To investigate the change in density to the air flow we must assume dry as being an ideal gas then apply the ideal gas law equation:

$$Pa * Vg = n * R * T \qquad (8)$$

Where, Pa = pressure, Vg = volume of gas, n = number of kilo moles, R = gas constant, T = Temperature Kelvin, m = mass.
If we consider the density of air as:

$$\rho = \frac{m}{Vg} \qquad (9)$$

Then from equations 8 and 9 we can calculate the density of air in the turbine.

$$\rho = \frac{n \times Pa}{R \times T} \qquad (10)$$

From the section view shown in Fig. 2, the minimum pressure gradient is 101221 Pa. From this we can calculate the density drop in air flow through the duct using equation (10) where:

n	=	29	kg/kilo mol
Pa	=	101221	pascal's
R	=	8.314	J/K/mol
T	=	293	K

$$\rho = \frac{29 \times 101221}{8.314 \times 293}$$

ρ = 1.205 kg/m^3

Density of air at standard temperature and pressure, ρ = 1.293 kg/m^3
As proof of the advantage of wind velocity over density we shall demonstrate this through a mathematical example. According to NIWA the average wind velocity in New Zealand is 6 m/sec. This value will be used as a benchmark for the calculations.

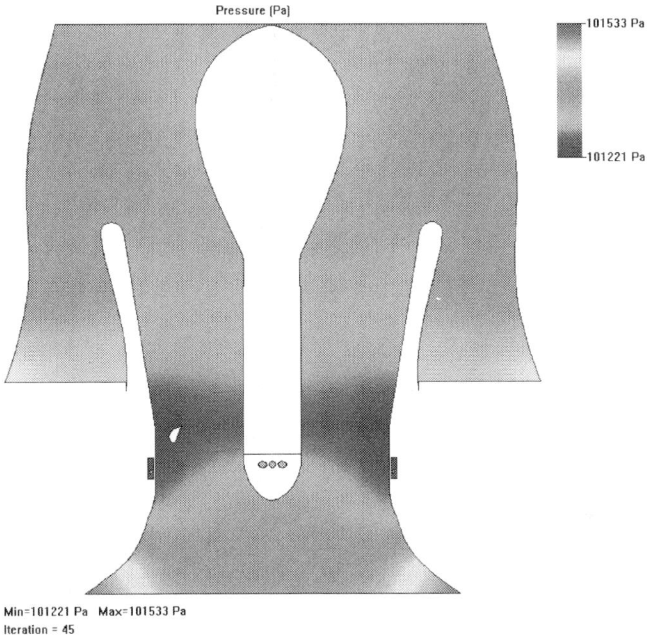

Fig. 2. Section view of turbine ducting with pressure gradients

3.1 Calculation for condition 1

No augmented acceleration of air flow through ducts, this condition uses the standard pole mounted turbine on a horizontal axis as shown in Fig. 3. Turbine diameter of 2m with an unusable hub diameter of 0.25m.

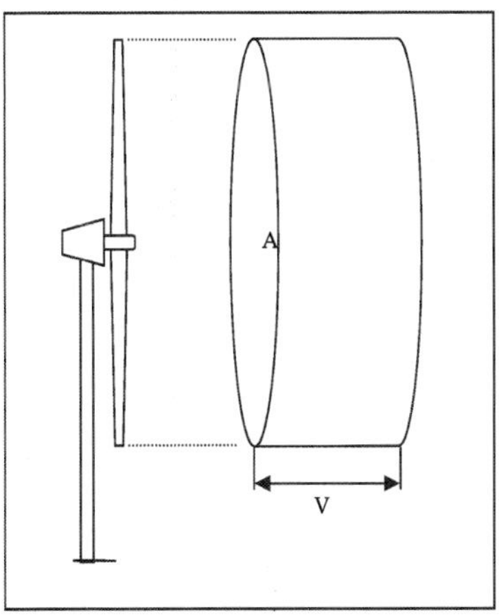

Fig. 3. Illustration of first example, standard turbine in open air

Using equation (1). Power = ½ * ρ * TSA * V^3
Power = ½ * 1.293 * 3.09252 * 6^3
Power = 431.9 W

3.2 Calculation for Condition 2

A ducted turbine with acceleration of airflow due to the venturi affect aligned with Bernoulli's equation of continuity, as shown in Fig. 4. A drop in density is observed and will be included in the calculation.

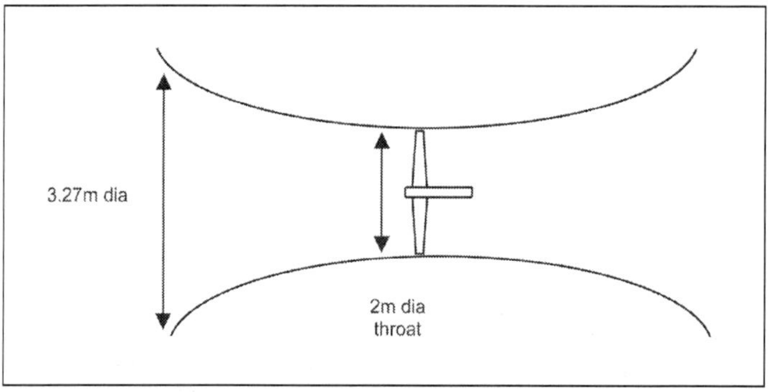

Fig. 4. Illustration of the ducted turbine, with dimensions provided

Wind velocity at the turbine. Use equation (5), $V_2 = (A_1 * V_1)/A_2$

$$V_2 = \left(\pi \times 1.633^2 \times 6\right) \div \pi$$

V_2 = 16 m/sec

From equation (10) air density at the turbine is 1.205 kg/m^3.

Using equation (1). Power = ½ * ρ * TSA * V^3
Power = ½ * 1.205 * 3.09252 * 16^3
Power = 7631.8 W.

While condition 1 gave an output of 431.9 W, condition 2 gave an output of 7631.8 W. This provides evidence of the increase in efficiency available from a ducted wind turbine. The calculation showed an increase in efficiency of a factor of 17, this was from the same ambient wind velocity of 6 m/sec and included the reduction in density as the air accelerates through the venturi. This is a significant increase in efficiency and can be exploited in a number of ways, either make more energy with the same size diameter turbine or make the turbine smaller. Before calculations are completed on turbine sizing, consideration must be allowed for losses during the transfer of energy from the wind to the shaft.

3.3 Betz limit

In 1919 Albert Betz concluded that no more than 16/27 (59.3%) of the wind's kinetic energy could be captured by a turbine, (Gorban et al, 2001). For all the energy to be removed from the air flow and stop the wind the turbine would be a solid plate, this would not allow any transfer of kinetic energy. If the turbine some how contained one small thin blade, the kinetic energy would be carried away with the wind going through the rest of the swept area. The Betz limit has nothing to do with the inefficiencies of the generator and bearings, just the conversion of energy from the wind to the shaft.

As the Betz limit is the maximum efficiency that may be harnessed by the turbine, the actual efficiency is estimated between 0.35 and 0.45, it must be noted this is turbine efficiency and does not include losses in bearings or generation. Actually it is believed that only 10 - 30% of the energy is converted to useful electricity.

As this project does not include the design of the turbine blade we must choose a Coefficient of Performance (C_p).

A conservative figure of 0.35 is to be taken but (Spera, 1994) postulated 6% savings in the elimination of blade tip vortices. This is significant and should be included in this calculation. (Mathew, 2006) gives figures of between 0.38 and 0.46 for a ducted turbine depending on the tip speed ratio (λ), this supports the claim by (Spera, 1994) of a 0.06 increase in efficiency.

3.4 Blade tip vortices

Blade tip (wing tip) vortices are created when air rushes from the high pressure area 'beneath' the lifting surface to the low pressure area 'above' the lifting surface. This occurs without the air travelling cord-wise from leading edge to trailing edge. Fig. 5 shows illustration of a simple aerofoil and position of pressure gradients. Fig. 6 shows the direction of airflow.

As the air rushes from the high to the low pressure area two losses occur, reduction in the efficiency of the lifting body from span-wise flow and extra interference drag from the generation of blade tip vortices.

Energy is transferred from the motion of the blade and is used to accelerate the vortex airflow that spirals from the tip thereby creating drag.

The ducted turbine does not generate any vortices and the tip clearances are small enough to eliminate tip vortices and large enough to clear the vena contractor area. Considering the affect of the ducted turbine with respect to span-wise flow and blade tip vortices, the 6 % (Spera, 1994) attributes to the deterioration of efficiency will be added to the 0.35 Cp to give a value of 0.41Cp.

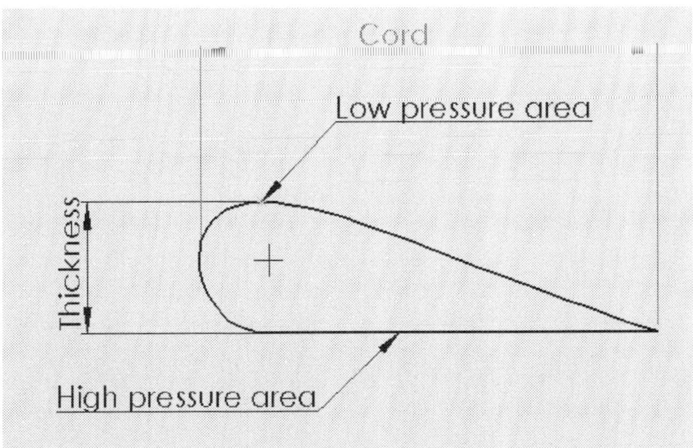

Fig. 5. Illustration of a simple aerofoil and position of pressure gradients

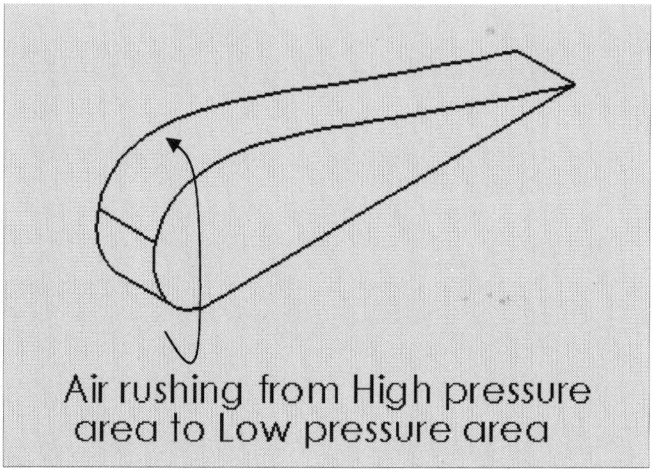

Fig. 6. Direction of airflow

3.5 Actual power

As discussed so far power contained in a parcel of wind is vastly different to that extracted by a wind turbine. The ducted turbine has an estimated C_p of 0.41.
According to (Kenttield, 1996), the expression between coefficient of power an actual power is:

$$Cp = \frac{2 \times P_T}{\rho \times TSA \times V^3} = \frac{2 \times T_T \times \Omega}{\rho \times TSA \times V^3} \tag{11}$$

Where, Ω = angular velocity, T_T = actual torque
Therefore, rearranging,

$P_T = ½ * \rho * TSA * V^3 * C_p$

Where, P_T = Total actual power.

$P_T = ½ * 1.205 * 3.09252 * 16^3 * 0.41$

$P_T = 3129.06$ W

Assume actual power at 3 kW.

3.6 Actual torque

To calculate torque first we must investigate the importance of tip speed ratio. The relative speed between the wind and turbine is important. If the wind is at high velocity and the turbine is at relatively low rotational speed, then the wind may be able to pass through the rotor without interacting with it. This is vital as any wind that does not interact is kinetic energy that is wasted. As the design parameters for this project set the rotational speed of the turbine at 300 rpm the turbine was designed with 8 blades, although this may not optimise the relation between wind velocity and 'obstructing' the duct it ensures the removal of energy from the stream.
Inlet guide vanes improve the efficiency of power extraction by ensuring the optimum angle of attack of the air flow onto the turbines. Using the guide vanes ensures our calculations are

conservative and if this project goes from theory to design, there is confidence that design performance will be exceeded.
The tip speed ratio (λ) is the ratio between the peripheral speed of the turbine and the wind velocity. According to (Mathew, 2006) is calculated:

$$\lambda = \frac{R \times \Omega}{V} = \frac{2 \times \pi \times N \times R}{V} \tag{12}$$

Where, R = radius of the rotor, Ω = angular velocity (rad/sec), N = rotational speed of the rotor,
Also the coefficient of torque is expressed,

$$C_T = \frac{2 \times T_T}{\rho \times TSA \times V^2 \times R} \tag{13}$$

If we divide equation (11) by equation (13), then we get,

$$\lambda = \frac{C_P}{C_T} = \frac{R \times \Omega}{V} \tag{14}$$

Calculating Ω,

$$\Omega = \frac{2 \times \pi \times 300}{60} = 31.42 \text{ rads/sec}$$

Tip speed ratio

$$\lambda = \frac{1 \times 31.42}{16} = 1.964$$

Therefore the torque coefficient is,

$$C_T = \frac{0.41}{1.964} = 0.2087576$$

To calculate torque we rearrange equation (13)

T_T = ½ * ρ * TSA * V^2 * R * C_T

Therefore actual torque is:

T_T = ½ * 1.205 * 3.09252 * 16^2 * 1 * 0.2087576

T_T = 99.575 kg.m.sec^{-1}

or

T_T = 976.83 N.m.sec^{-1}

4. Prototype

The results from the calculations for the ducted turbine are encouraging and lead us to the design. This is to be a horizontal axis wind turbine, but this will be ducted. This was a basic Venturi with a 3.3 m diameter opening to the convergent duct that has its smallest diameter at the turbine, this measures 2m. See Fig. 7.

The shape embodies the Venturi theory and through Bernoulli's theory of incompressible gasses and application through the equation of continuity. Applying equation (4) enables acceleration of the air flow through the turbine that is mounted 1200mm from the leadingedge of the duct. This shape has been optimised to accelerate the air through the ducting.

This configuration contains two turbines on the same shaft, the second has less blades as there is not so much energy left in the airflow.

The tapering of the duct in the last 1100mm to the rear of the turbine is to accommodate the stream tube.

As the air enters the turbine it obviously slows as some of the kinetic energy is removed. This implies that the air flow is slower leaving the turbine than when it entered. But the same amount of air that enters the turbine, in a given amount of time, must leave the turbine in that same time period, therefore the cross sectional area to the rear of the rotor must be greater to accommodate the slower moving scavenge air. Fig. 8. shows section view of the inner and outer shrouds.

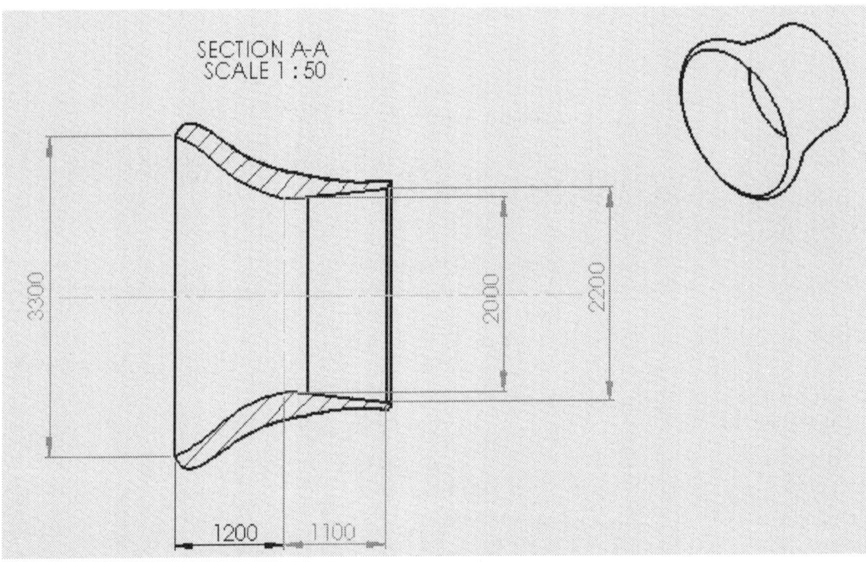

Fig. 7. Section view of the inner turbine ducting, all dimensions in mm

A second shroud is fitted to the rear of the inner duct and draws air in through an annular orifice, this is to aid the scavenging of the air from behind the turbine. As low pressure air exists in the turbine ducting after transferring energy to the rotor, another venturi is added to lower the pressure at the rear to help scavenge the exhaust air.

As the centre of drag of the second fairing is to the rear of the pivot point this also aids weather cocking as shown in Fig. 9.

Applying equation (4) to the area where the air flow leaves the inner duct and mixes with the air from the outer duct, there is an observed narrowing of the passage. The conclusion when analysing the relationship between A_1, V_1 and A_2, V_2 is that as A_2 gets smaller relative to A_1 then V_2 increases. This principle (the venturi principle) is used where the two airflows mix.

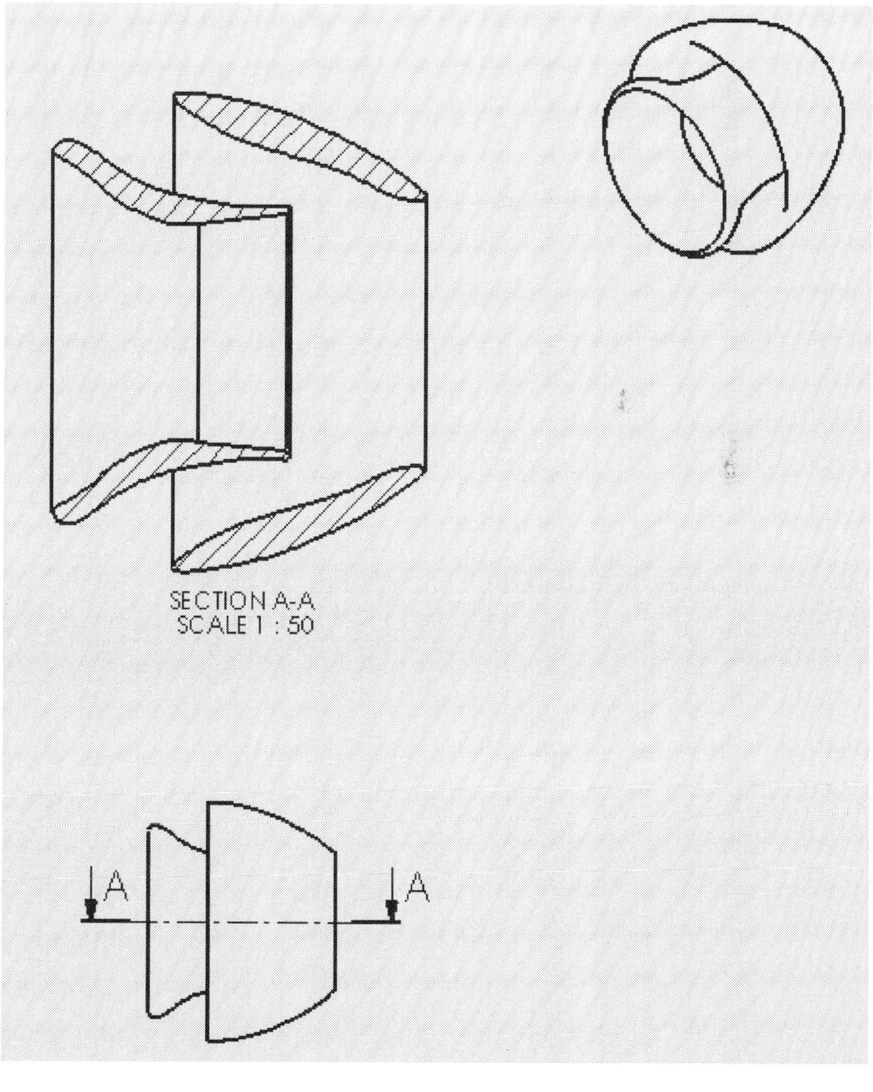

Fig. 8. Section view of the inner and outer shrouds

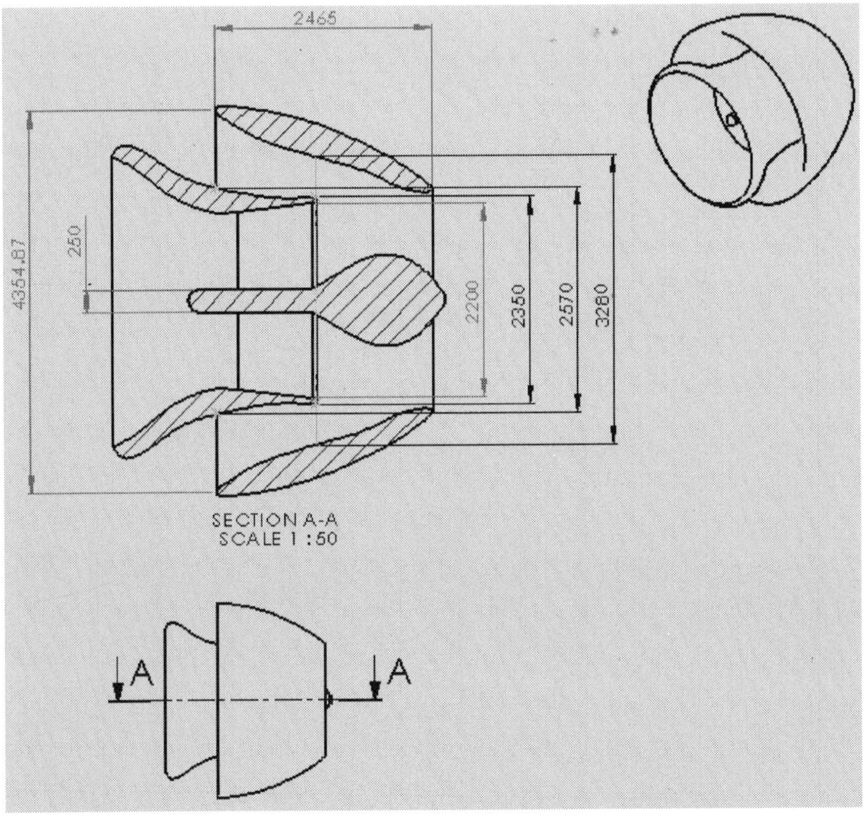

Fig. 9. Dimensioned inner and outer shrouds complete with generator fairing

4.1 Surface finish

The surface finish of all areas exposed to the airflow is critical. A smooth finish is required to minimise the size of the vena contractor and to reduce turbulent flow. The air travelling through the ducting is turbulent; this is defined by a dimensionless value called the Reynolds Number (Re) and is calculated:

$$Re = \frac{\rho \times V \times D}{\mu}$$

Where, D = pipe diameter, μ = fluid viscosity.

The Reynolds number for fluid flow through the ducted turbine is from 1.21×10^6 to 2.14 x 106. This is outside the lamina flow range, therefore surface finish is critical as it affects the Friction Factor. The easy way to determine the friction factor is with the use of the Moody chart as shown in Fig. 10.

The e/D ratio must first be calculated where, e = the average height of the bumps on the surface, D = the diameter of the pipe.

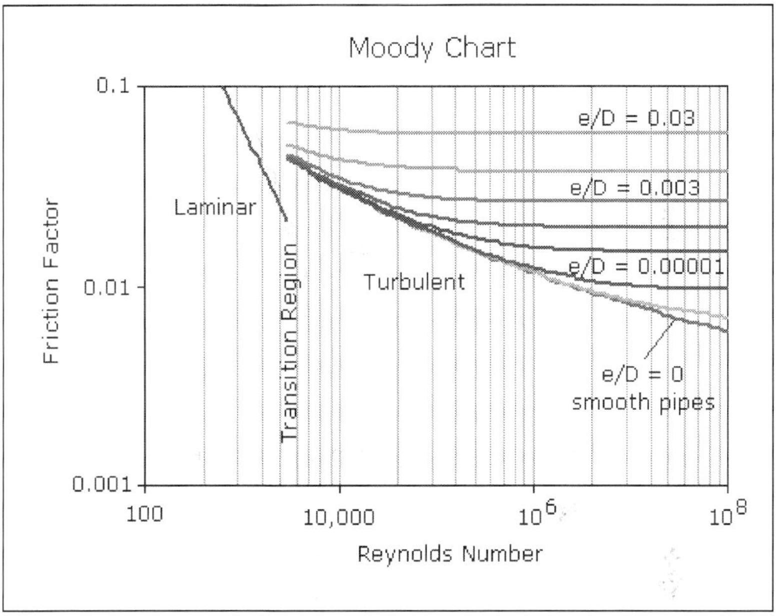

Fig. 10. Moody chart, courtesy of (eFunda, 2001)

Assume:

Plastic moulding with a surface finish of	3 µm
Smaller diameter of duct	2 m
This gives an e/D factor of:	1.5×10^{-6}

Applying this to the Moody chart we get a friction factor of: 0.0098

The friction factor is the multiplier that calculates the pressure drop per meter length of pipe. The pressure drop is directly proportional to the wind velocity through the ducting, hence a direct indication of kinetic energy lost. As the e/D ratio is the basis of the friction factor calculation, the only way a reduction in losses can be achieved is, either make the surface finish smoother and/or the hole larger.

4.2 Angle of attack

Inlet guide vanes and stators have been incorporated into the design to ensure the airflow isoffered to the turbine stages at an optimum angle of attack. For any aerofoil cross section to be efficient it has to be offered to the air flow at the correct angle of attack [see Fig. 11].

Inlet guide vanes and stators act to direct a fluid in a particular direction. In a situation where there are areas of low pressure turbulence that are present due to a change in profile or cross sectional area; then a device that will precisely alter the flow to compensate for these changes will increase efficiency.

Variable inlet guide vanes (VIGVs) can be adjusted to apply a favourable spin to the mass airflow thereby offering the stream at the most optimum angle of attack for the particular wind speed. The VIGVs can also be used in high winds where the vanes can be turned as this alleviates the need for a variable pitch turbine blade and reduces the complexity of the design. However, a variable pitch system is still required for the VIGVs, but this becomes less complex than the variable pitch blade as the actuating mechanism does not need to mounted on a shaft and is better suited to positioning within the nacelle.

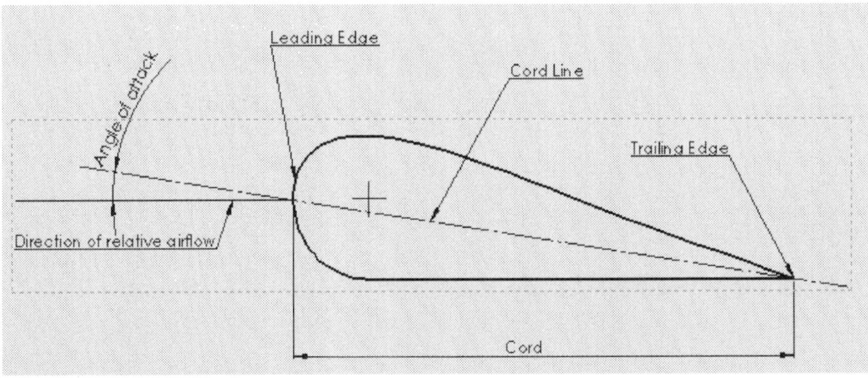

Fig. 11. Angle of attack

Stators are situated between the two turbines within the stream tube. These have a dual purpose: as a means to reduce turbulent flow by offering the airflow up to the turbine at the optimum angle of attack and, as a set of cooling fins to dissipate heat from the generator.
V_{Tur} is the component of the wind velocity induced by the rotation of the turbine, V_{Wind} is the component induced by the wind. $V_{Resultant}$ is the resultant of the previous two vectors this has its coefficients of drag and lift. Φ is the angle of attack.
Aero foils have a critical angle of attack where the maximum efficiency is attained. Depending on the aerofoil cross section ± ½ a degree can be greater than the tolerance allowable to maintain maximum efficiency. Therefore it is import to maintain the optimum angle of attack.
Consider the parallelogram of forces shown on Fig. 12. For the same rotational speed of the turbine, the linear velocity changes for different points of the blade from the hub to the tip. Close to the hub of the blade V_{Tur} will be short representing the slower speed, this will move the $V_{Resultant}$ thereby increasing the angle of attack Φ. This will mean the aerofoil is not at its optimum at certain points. The rectification of this is to design a twist into the blade; this ensures a constant angle of attack at a fixed rotational speed.
With VIGVs, the optimum angle of attack can be maintained through a wide range of rotational speeds and wind speeds. Contemplating the parallelogram of forces in figure 13, if the vector V_{Wind} was to greatly increase, the angle of attack Φ would also change. By altering the angle at which the wind was offered to the turbine, $V_{Wind,}$ would solve the angle of attack Φ to the optimum angle for greater efficiency.
Turbine blades can also be part reaction and part aerodynamic. Close to the hub blade cross sections are either reaction in cross section or, if the gain is not optimal, structural to take the high loads from centrifugal forces and fatigue from harmonics. If a section is taken close to the blade tip a more aerodynamic lifting surface will be observed. The aerodynamic shape is to optimise the aerodynamic forces to remove the kinetic energy from the airflow (Schreck et al, 2007, Schreck and Robinson, 2007).

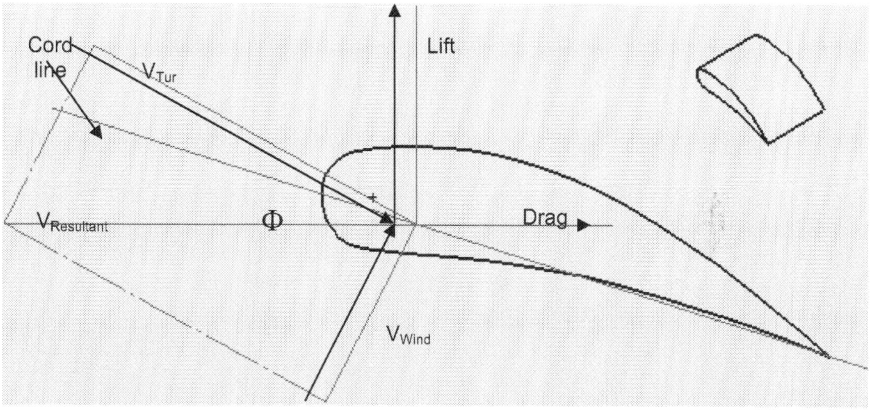

Fig. 12. Vector of forces on an aerofoil

4.3 Vector flow

Figure 13 depicts the vector flow through the ducted turbine. The length of the vector arrows represents relative wind speed and direction. This illustration was produced using EDF.Lab™ in conjunction with Solidworks™. The purpose of the lids depicted in yellow, orange, and green, are to define the boundaries around the design and are only used within the mathematical modelling of the turbine assembly. The vectors depict smooth flow through the system with no turbulence evident.

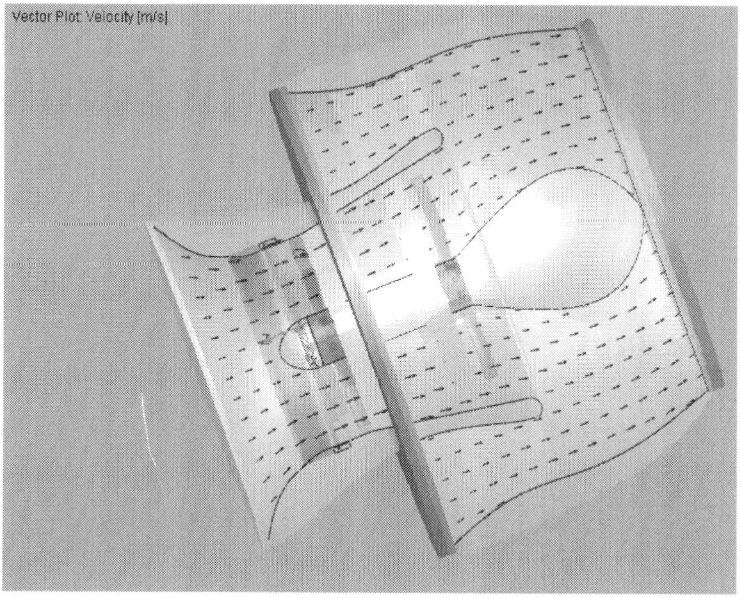

Fig. 13. A sectioned representation of vector flow through the Turbine ducting

The turbine assembly comprises a convergent duct assembly that accelerates the air flow to the VIGVs. The VIGVs are adjustable and will guide the airflow on to the first stage turbine at the optimum angle of attack. As shown on page 34 this the angle between the resultant vector and the cord line of the aerofoil and this varies with wind velocity and rotational speed of the turbine. By maintaining the optimum angle of attack greater efficiencies can be maintained.

Control of the VIGVs will be proportional to wind speed only. As the turbine is constant speeding, V_{tur} is constant, therefore V_{wind} is the only variable. Wind velocity can be sensed using twin pitot static tubes that can be mounted diametrically opposite each other on the outside edge of the turbine intake as shown in Fig. 14. The wind speed can be read as the highest of the results received from either pitot tube.

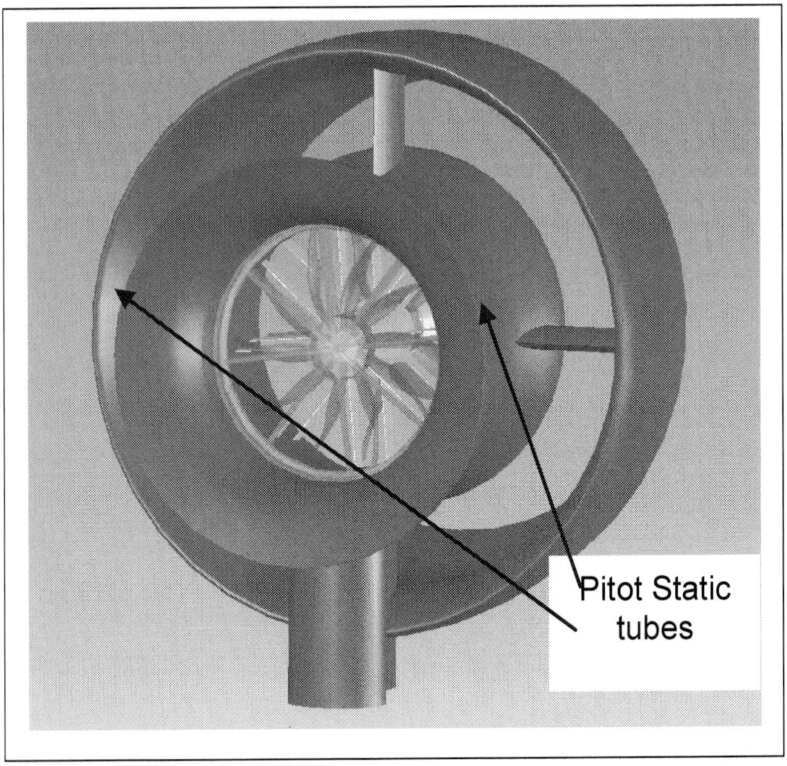

Fig. 14. Position of pitot static tubes

5. Conclusion

The specifications were set as a goal for us to achieve with our design. From the specification a rough design was envisaged and the first of the basic principles were

mathematically modelled, this provided evidence that the design was feasible. Further expansion to the design was carried out and more of the fundamental mathematics was discovered that would describe the Air flow through the duct.
Power calculations were provided for the ducted turbine and the conventional turbine. These provided a direct comparison that was referred to as condition 1 and condition 2. This showed a theoretical power rating difference of a factor of 17, it must be shown here that the theoretical power shows the amount of power that is available in the wind flow.
The Betz limit describes that maximum amount of power that is able to be extracted from the wind flow this is 16/27 (59.3%). A more conservative approach was taken and a coefficient of 0.41 was chosen. This provided evidence of a 3 kW out put.
The Reynolds number shows turbulent flow through the ducting as the number was in the order of 10^6 (anything above 2000 can be considered turbulent flow). Therefore VIGVs and stators were added to reduce the possibility of vortex generation. The VIGVs also provide control of the wind flow over the turbine blades; this allows the optimum power to be extracted while reducing the boundary layer over the blade and minimising the possibility of turbulent boundary layer flow interacting with the blade trailing edge.

6. References

Chen H. Implementation of a three-phase switched reluctance generator system for wind power applications. In: Proceeding of the 14th Symposium on electromagnetic launch technology, 2008.

Dannecker R, Grant A. Investigations of a building-integrated ducted wind turbine module. Wind Energ. 2002; 5:53-71.

eFunda. Smart Measure, Pipe Friction. Retrieved from, 2001; www.efunda.com/ formulae/smc_fluids/calc_pipe_friction.cfm, on 19 Jul 07.

Gorban AN, Gorlov AM, and Silantyev VM. 2001. Limits of the Turbine Efficiency for Free Fluid Flow. Journal of Energy Resources Technology, 2001; 123: 311 - 317.

Grosveld FW. Prediction of Broad Band Noise from Horizontal Axis Wind Turbines. Journal of Propulsion and Power, 1985; Vol 1, No 4.

Hu SY, Cheng JH. Innovatory designs for ducted wind turbines. J. Renewable Energy 33 (2008):1491-1498.

Kentfield JAC. The Fundamentals of Wind-Driven Water Pumps. Gordon and Breach Science Publishers. Amsterdam, Netherlands, 1996.

Mathew S. Wind Energy, Fundamentals, Resource, Analysis and Economics. Springer, Berlin, 2006.

Schreck SJ, Robinson MC. Horizontal axis wind turbine blade aerodynamics in experiments and modelling. IEEE Trans. on Energy Conversion 2007;22(1):61-70.

Schreck S, Sorenson N, and Robinson M. Aerodynamic structures and processes in rotationally augmented flow fields. Wind energy, vol. 1, no. 2, 2007.

Spera, DA. Wind Turbine Technology, Fundamental Concepts of Wind Turbine Engineering. ASME Press, New York, 1994.

The New Zealand Wind Energy Association. Current Wind Farms and Projects Under Construction. Retrieved from, 2007;www.windenergy.org.nz/FAQ/map.htm, on the 21 Mar 07.

Zhao H, Lingzhi Y, Hanmei P, and Kunyan Z. Reseach and control of SRG for variable-speed wind application. Proceeding of the 6th IEEE international conference IPEMC, 2009;2238-2243.

5

Small Wind Turbine Technology

Oliver Probst[1], Jaime Martínez[1,2], Jorge Elizondo[1,2] and Oswaldo Monroy[1]
[1]*Instituto Tecnológico y de Estudios Superiores de Monterrey, Monterrey*
[2]*Diseño Eólico y Solar, Monterrey*
Mexico

1. Introduction

Small wind turbines are an attractive alternative for off-grid electrification and water pumping, both as stand-alone applications and in combination with other energy technologies such as photovoltaic, small hydro or Diesel engines. Under these conditions, the cost of energy alone is often not the only criterion to consider, and aspects like system performance, suitability for a given wind regime, reliability under normal and extreme wind conditions, and overall system life are often equally important. Where no grid connection is available or the grid is unreliable, it is the energy-providing service that matters, not its precise cost. In grid-connected situations, the actual vs. the rated performance may be of more interest in order to achieve the cost saving benefits proposed in the design of the project. In either case, an uninterrupted service with a performance close to the one specified by the provider is a key requirement for a successful small wind project.

While all wind turbines, both MW-class utility turbines and small wind generators, are subject to the fluctuating nature of the wind, there are several reasons why it is more difficult to guarantee the performance of a small wind turbine. First, the smaller inertia of rotor/generator leads to significant transient effects in response to changing wind speeds. Moreover, often small wind turbines rely on passive mechanisms for aligning the rotor with the wind direction, such as lifting forces acting on a tail vane in the case of an upwind rotor or axial (drag and lift) forces acting on the rotor in the case of a downwind turbine. Therefore, under conditions of varying wind directions, an incomplete alignment of wind turbine and wind direction may occur, and the alignment error can be expected to be a function of the turbulent time scales present at the site. In the case of furling systems, a technology used by many manufacturers to passively protect the turbine from overspeeding and generator overheating, the situation is still far more complex since the mechanism can be triggered both by sustained high winds and gusts, also exhibiting a hysteresis depending on the specific design parameters chosen.

Another important aspect relates to maintenance. While large wind turbines are routinely serviced as part of an ongoing service agreement during the lifetime of a wind farm and are constantly monitored by means of a SCADA (supervisory control and data acquisition) system, small wind generators are expected to function largely unsupervised and with only occasional maintenance. Therefore, the only indication for a typical owner of a small wind turbine is the charge level of the battery, which is only very indirectly related to the (integrated) wind turbine system performance in the period prior to the observation.

Systematic measurement campaigns are therefore essential to assess the system performance under quasi-steady state conditions and compare this performance to the predictions of the theoretical design model, study the turbine under transient conditions, such as during furling and start-up, and detect energy losses due to storage or inappropriate matching of turbine capacity and battery size, among others. All these issues will be addressed in the present chapter.

Another issue yet has to do with the very concept of performance certification and verification. While large wind turbines are often certified according to international standards, generally the IEC61400-12, only recently a performance standard for small wind turbines has been issued (AWEA 2009). While certification may be a high financial burden, especially for small companies with a modest output of units, a unit-by-unit on-site performance verification is generally prohibitive because of the cost and complexity of the associated measurement equipment. More importantly still, due to the generally low output voltage level Joule losses on the transmission line from the hub height to the battery bank, end-use application or grid-intertie inverter significantly impact on the true amount of energy captured. Small wind turbine providers therefore often specify the expected power or energy output at the nacelle, i.e. before the transmission line. While this is a sound practice, the end user has to be aware of the fact that the true energy output may be substantially less than announced. This and related issues will be addressed in the chapter.

2. Conceptual design of small wind turbine systems

While no strict definition of a small wind turbine exists in literature, wind turbines with a rated capacity of 10kW or less are generally considered *small*; this definition is sometimes extended up to about 50kW due to the recent appearance of higher rated machines suitable for servicing more energy demanding applications, including agricultural tasks such as water pumping for irrigation or livestock watering. As it will explained below, rated capacity is not a very well defined parameter for a small wind turbine, so the rotor diameter or, equivalently, the swept rotor area are often preferred for classifying small wind turbines, where rotor diameters of about 10m can be taken as the dividing line. Another means of distinguishing *small* from *large* wind turbines is by requiring a small wind turbine to have a tail vane (see discussion below). While in principle many of the aspects discussed in this chapter can be applied to vertical- and horizontal-axis wind turbines equally we will limit ourselves to the latter only.

A small wind turbine generally consists of the following minimal components: (1) A rotor with a variable number of blades (section 3), (2) an electric generator (section 4), and (3) passive or active electronic components (section 6) for feeding electricity into a battery bank, the public grid or, occasionally, into a direct application such as a water-pump. Upwind wind turbines are generally equipped with a tail vane to assure the rotor is facing the wind, while downwind turbines rely on the self-orienting effect of the axial forces acting on the rotor, albeit at the expense of a periodic tower shading effect acting on the turbine blades. Many upwind turbines rely on a furling mechanism for overspeed and output power control at high wind speeds, although other mechanisms such as load-induced stall are used occasionally. Passively pitching blades, generally triggered by the action of the centrifugal forces acting on the rotor, have been used in the past but are currently less common.

Most small wind turbines are variable-frequency devices, allowing for an optimal operation at all wind speeds below the threshold for the onset of the overspeed and power control mechanism. While in the case of battery-charging applications the use of a passive rectifier together with the selection of an appropriate voltage level may be sufficient to maintain the operating point close to the system optimum, especially when the aerodynamic efficiency curve (section 6) is broad (Elizondo et al., 2009, Probst et al., 2006), some commercial systems rely on the use of an active load control in order to maintain the system at the optimal operating point for each wind speed (Martínez et al., 2006). In the case of a coupling to the electric grid, a full *wild* AC/fixed frequency AC conversion is generally feasible through the use of a *back-to-back* AC/DC/AC converter, as opposed to large wind turbines where a direct full conversion is still rather the exception than the rule and most commercial large wind turbines rely on doubly fed induction generators (DFIG) where only a fraction of the total power is passed through a converter.

3. Rotor aerodynamics and loads

Just like large wind turbines most modern small wind turbines use a three-bladed rotor with aerodynamic sections (airfoils), although designs with two or four blades are occasionally encountered. Two-bladed rotors develop their optimal aerodynamic performance at higher tip speed ratios (TSR) and therefore have to rotate faster than three-bladed rotors. This allows for the use of high-speed electric generators which are smaller and less expensive to manufacture, although at the expense of potentially greater noise problems. Multi-bladed rotors, on the other hand, have a higher starting torque which favors starting at low wind speeds.

Due to the smaller dimensions, the Reynolds numbers (Re) at sections of small wind turbine blades are considerably smaller than for large wind turbine blades. While typical Reynolds numbers for large wind turbines are in excess of 10^6, where the aerodynamic lift and drag coefficients vary little with Re, at small wind turbine blades the aerodynamic performance of a given airfoil may be substantially poorer at the inboard sections, where the Reynolds number may be of the order of 100,000 or less. Low Reynolds number operation is highly dependent on the behavior of the laminar boundary layer (Selig and McGranahan, 2004), so ambient turbulence and surface roughness have a pronounced effect on the wind turbine behavior. Surface roughness is generally affected by the manufacturing technique, but may also change considerably over time due to soiling. In Fig. 1 two examples illustrating the effect of soiling are shown for two airfoils (the E387 and the S822) designed for the use with small wind turbines (data from Selig and McGranahan, 2004). In their experiments, Selig and McGranahan simulated the effect of airfoil soiling by attaching a zigzag shaped boundary layer trip to the airfoil surface.

In the graphs, the ratio between the lift and the drag coefficient has been plotted as a function of the angle of attack for both clean and soiled surfaces. It is conspicuous from the figure that lift/drag quotient is substantially reduced in the angle-of-attack range in which the rotating airfoils will be operating most of the time. The maximum C_L/C_D value is reduced from 80 to about 55 in the case of the E387 airfoil and from 65 to 40, a 31% and 38% reduction, respectively. Also, the difference in aerodynamic performance between attached flow and stall conditions is decreased since the performance in stall is little affected by soiling, so the effectiveness of turbine power control schemes based on active stall regulation, either by load control or by pitching the blades towards higher angles of attack, is greatly reduced if soiling is not controlled.

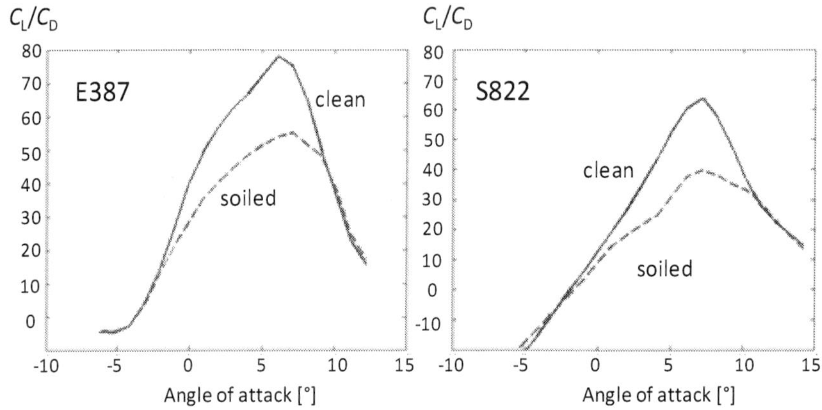

Fig. 1. Effect of airfoil soiling on the ratio of lift and drag coefficient for two airfoils designed for small wind turbines (Reynolds number=100,000). Data from Selig and McGranahan (2004)

A characteristic element of low Reynolds number flow is the appearance of a laminar separation bubble caused by the separation of the laminar flow from the airfoil with a subsequent turbulent reattachment (Selig and McGranahan, 2004). This phenomenon leads to a considerable increase in the drag coefficient at low angle of attack. This quite dramatic drag increase is illustrated in Fig. 2 where the measured C_L-C_D diagram (drag polars) for the Eppler airfoil E387 (data from Selig and McGranahan, 2004) has been drawn for Reynolds numbers in the 100,000 to 500,000 range. While a moderate increase in drag occurs for any given lift coefficient upon decreasing the Reynolds number from 500,000 to 200,000, the drag at 100,000 is substantially higher.

Low Reynolds number flow also has higher associated uncertainties, as shown by Selig and McGranahan (2004, chapter 3) in their comparisons of their aerodynamic force measurements with those obtained at the NASA Langley in the Low-Turbulence Pressure Tunnel (McGhee et al., 1988). While an excellent agreement between the two sets of measured drag polars is obtained for Reynolds numbers of 200,000 and higher, substantial differences arise at 100,000. Although the same shape of the drag polars was observed in both cases, showing the appearance of the laminar separation bubble, the drag coefficients for a given lift coefficient were found to be higher in the measurements by Selig and McGranahan (2004). Interestingly, a similar discrepancy, limited to the low Reynolds number case of 100,000, was found in a theoretical analysis of small-scale wind turbine airfoils (Somers and Maughmer, 2003), including the Eppler airfoil E387 mentioned above. In their study, the authors use two different airfoil codes, the XFoil and the Eppler Airfoil Design and Analysis Code (Profil00), finding similar results for drag polars, except for the low Reynolds number case of 100,000, where the experimentally observed drag is better reproduced by the Profil00 code. From the above it should have become clear that the uncertainty in the prediction of the lift and drag coefficients at low Reynolds is larger than at higher Reynolds, making predictions of rotor performance and energy yield less accurate.

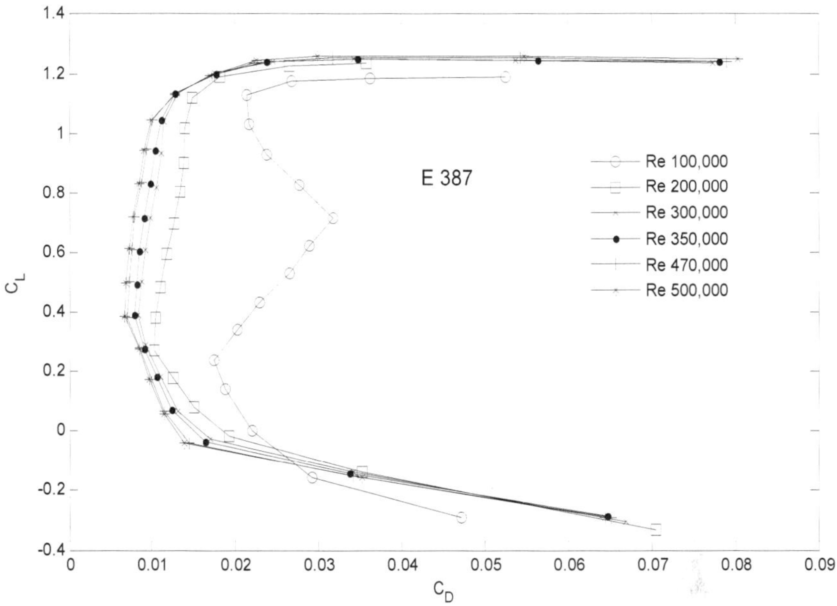

Fig. 2. Aerodynamic lift vs. drag coefficient for the Eppler airfoil E387 designed for the use with small-scale wind turbines. After Selig and McGranahan (2004)

A direct consequence of the lower aerodynamic performance at low Reynolds numbers is a generally somewhat lower aerodynamic power coefficient (C_p< 0.46-0.48 for a well-designed rotor at peak efficiency, as opposed to >0.50 for large wind turbines) and a dependence of C_p on both the tip speed ratio (TSR) and the wind speed, as opposed to large rotors, where to a good approximation the power coefficient is a function of TSR only. This effect is illustrated in Fig. 3, where the experimental results of the aerodynamic power coefficient C_p vs. the tip speed ratio (TSR) λ of a turbine rated at 1.4 kW (swept diameter 3m), obtained from a field characterization, have been plotted together with the predictions of a mathematical model of the turbine. The experimental data was obtained by operating the turbine under different controlled load conditions, including direct connection to a battery bank with a voltage of 48V, 24V, or 12V; to provide higher load conditions, the 12V battery bank was shunted with a resistance whose value was varied from 2.1Ω to 1.1Ω (Elizondo et al., 2009). It can be seen that for low values of the tip speed ratio all power coefficient values fall onto a universal curve, while for higher TSR values a greater spread between the recorded values exsist, as predicted by the mathematical model based on a combination of a Blade Element Momentum (BEM) and an electromechanical model of the generator/rectifier. The appearance of different C_p - λ curves at high values of TSR can be traced back to the lower aerodynamic performance of the blade sections at low wind speeds (and therefore low Reynolds numbers), as illustrated by the difference of the lower curve in Fig. 3 (corresponding to a wind speed of 6 m/s) and the higher curve (valid for 12 m/s).

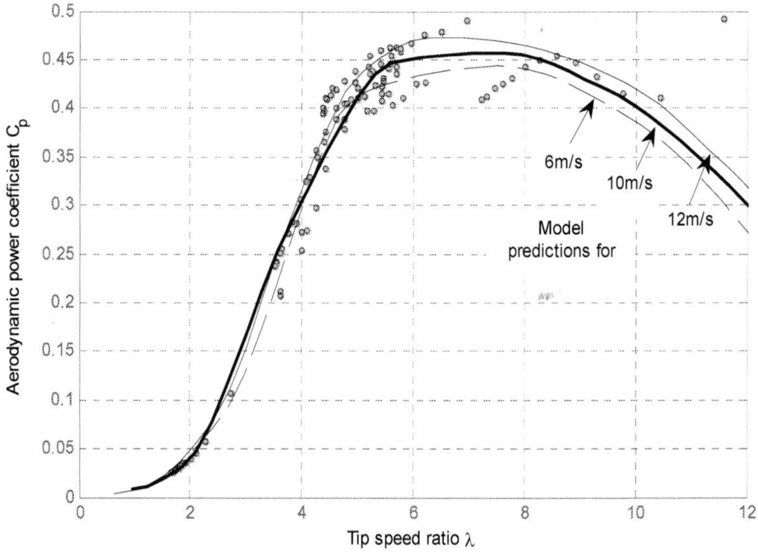

Fig. 3. Measured C_p-λ curve for a wind turbine rated at 1.4 kW and comparison with the predictions of a mathematical model of the turbine.

Another important point refers to the influence of the average blade aspect ratio; while long slender blades can be often well described with Blade Element Momentum (BEM) models using aerodynamic lift and drag coefficients determined in the wind tunnel under two-dimensional flow conditions, three-dimensional effects become important for blades with low aspect ratios, especially under conditions of flow separation or *stall*. As shown by Martínez et al. (2005), a good prediction (as opposed to a parametric fit) of the output power curves was obtained by modeling several research wind turbines with rated capacities in the range of 10-20kW by combining a 2D with a 3D-stall model. For that purpose, the lift coefficient as a function of the angle of attack of a given blade section was modeled according to

$$C_{L,\text{eff}} = \begin{cases} C_{L,\text{wind tunnel}}(\alpha) & \alpha \le \alpha_S \\ \frac{1}{2}\left(C_{L,\text{VC}}(\alpha;\kappa=50) + C_{L,\text{VC}}(\alpha;\kappa=\kappa_0)\right) & \alpha > \alpha_S \end{cases} \tag{1}$$

where $C_{L,\text{wind tunnel}}$ refers to the lift coefficient of the blade section measured under 2D flow conditions, and $C_{L,\text{VC}}$ is the lift coefficient as determined by the Viterna-Corregan stall model (Martínez et al., 2005):

$$C_L = \tfrac{1}{2} C_{D,\max} \sin 2\alpha + K_L \cos^2\alpha / \sin\alpha \tag{2}$$

$$C_D = C_{D,\max} \sin^2\alpha + K_D \cos\alpha\ , \tag{3}$$

where

$$K_L = \left(C_{L,S} - C_{D,\max} \sin\alpha_S \cos\alpha_S\right) \sin\alpha_S / \cos^2\alpha_S \tag{4}$$

$$K_D = \left(C_{D,S} - C_{D,\max} \sin^2\alpha_S\right) / \cos\alpha_S \tag{5}$$

Here, $C_{L,S}$ and $C_{D,S}$ denote the lift and drag coefficient at the stall angle α_S, respectively, and $C_{D,\max}$ is the maximum drag coefficient (at α=90°). It depends on aspect ratio (defined as the ratio of the blade span L and the mean chord of the blade) as follows:

$$C_{D,\max} = \begin{cases} 1.11 + 0.018\kappa & \kappa \le 50 \\ 2.01 & \kappa > 50 \end{cases} \tag{6}$$

Another important difference between large and small wind turbines has to do with hub height. Small wind turbines are usually placed at heights around 20 to 30 meters, as compared to 60 to 80 meters for utility-scale units. Therefore, small wind turbines usually operate at wind with higher turbulence intensity on its blades, as shown in the following approximate expression (Burton et al., 2001) for the turbulence intensity TI, defined as the ratio of the standard deviation σ_u of the fluctuations of longitudinal component of the wind velocity and the wind speed U:

$$\mathrm{TI} \equiv \frac{\sigma_u}{U} = \frac{2.4u^*}{U(z)} \approx \frac{1}{\ln(z/z_0)} \tag{7}$$

where z_0 is roughness length of the site and u^* is the friction velocity. For a typical utility-scale wind turbine with a hub height of z=80m and a roughness length of 0.1m, the turbulence intensity is about 15%, whereas for a small wind turbine hub height, say, z=20m, the corresponding figure is 19%

Increased turbulent intensity has a predominantly detrimental effect on turbine performance, mostly due to increased transient behavior, causing frequent acceleration and deceleration events, increased yawing movement and vibrations on most components. Wind shear, on the other hand, can generally be neglected due to the small dimensions of the rotor.

Although structural aspects cannot be neglected in the design process of small rotor blades, their impact on the design is less pronounced compared to the large wind turbine case. Structural properties are generally analyzed after the aerodynamic design stage has been completed, as opposed to large wind turbines where the structural design precedes the aerodynamic design. In the following we will discuss the main aspects to be considered in the structural design of small-scale rotors.

Three types of main operational loads can be distinguished: (1) Inertial, (2) aerodynamic, and (3) gravitational loads. Loads on small wind turbines blades are the same as on blades of utility size wind turbines, but their relative importance is different. If we assume the tip speed ratio (λ) to be constant, the three principal forces on the blades scale can be discussed as follows: The centrifugal force on the blade root (Fig. 4) can be calculated from

$$F_c = \omega^2 \rho A \int_0^R r \, \mathrm{d}r = \tfrac{1}{2}\omega^2 \rho A R^2 \tag{8}$$

where A is the cross-sectional area of the blade. We then have

$$A \propto c \times \gamma \propto R^2, \tag{9}$$

with c and γ being a typical chord and thickness dimension, respectively, each of which scale approximately with the blade radius R. Introducing the tip speed ratio λ by setting $\omega = \lambda U/R$ we see that the centrifugal force at the blade root scales with the square of the blade length:

$$F_c \propto R^2. \tag{10}$$

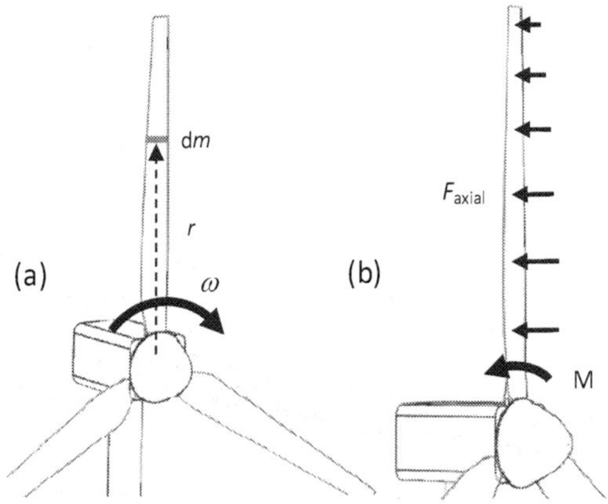

Fig. 4. Forces on a wind turbine rotor. (a) Centrifugal force on the blade root. (b) Axial force and root moment.

The aerodynamic forces create an axial load which translates into a root bending moment:

$$M = \int_0^R \mathrm{d}F_{ax}(r)\, r \;\propto \int_0^R c\, W^2\, r\, \mathrm{d}r \tag{11}$$

where r is the distance from the rotational axis. The precise axial force variation along the blade has to be calculated through numeric methods such as Blade Element Momentum Theory. However, supposing that aerodynamic performance is not affected by the blade length (i.e. supposing the same C_L and C_D for any blade length), we know that the axial force is proportional to chord c and effective velocity W squared, as expressed in equation (11). We observe that the effective wind speed squared can be calculated from

$$W^2 = U^2\left((1-a)^2 + \left(\lambda\frac{r}{R}\right)^2 (1+a')^2\right) \propto \left(\lambda\frac{r}{R}\right)^2 \quad \text{for } \lambda\frac{r}{R} >> 1 \tag{12}$$

Where a, a' are the axial and tangential induction factors, respectively, and are assumed to be independent of the scaling process, i.e. are assumed equal for small and large wind turbines. It should be noted that the approximate proportionality in equation (13) is valid at most radial positions, except for the blade root. Therefore, if the tip speed ratio λ is taken to be unchanged upon scaling the blade length, then

$$M \propto R^3 \tag{13}$$

Gravitation, finally, gives rise to an oscillating force on the blade that acts alternatively as compressive, tensile or shear force, depending on the azimuthal blade position. Its magnitude depends directly on blade total mass:

$$F_g = mg = \rho g V \propto \rho g (c \times t \times R) \tag{14}$$

If we assume again that the chord c and wall thickness t scale linearly with the blade radius, then

$$F_g \propto R^3 \tag{15}$$

i.e., the gravitational force scales as the cube of the blade radius. It has been proposed in literature (Burton et al., 2001) that blade mass can be scaled as $R^{2.38}$ with a proper engineering design to optimize blade material. In either case, the axial force bending moment and gravitational force become dominant as the blade gets larger, while with small wind turbines centrifugal forces usually dominate. A direct effect of the dominant role of centrifugal forces at small blades is that blades have greater stiffness (due to centrifugal stiffening) and are only lightly bent due by the axial force.

The discussion above directly translates into guidelines for the materials selection and the manufacturing process. While mechanical properties are highly dependent on the materials used in the manufacturing process, a typical blade material is glass fiber reinforced plastic (GFRP), although wood (either as the blade material or for interior reinforcement) and carbon fiber are also used by some manufacturers. With GFRP, manufacturing methods vary widely from hand lay-up to pultrusion (e.g. Bergey), depending on whether the precise blade geometry or a high production volume are the major concern.

The observed mechanical properties for blades are usually lower than the expected properties for the material, usually due to the following causes:

a. Air bubbles can form inside the material, concentrating stress and reducing overall resistance. This is a typical situation in hand lay-up manufacturing processes.
b. Material degradation due to weathering in operating blades. Usually UV radiation and water brake polymer chains, while wind acts as an abrasive on the surface. (Kutz, 2005).
c. In small wind turbines both centrifugal and axial forces can lead to failure in the following way: Centrifugal force failures occur as a direct consequence of surpassing the tensile strength of the reinforcement and usually occur near the root where centrifugal force is maximum, and close to the bolts fixing the blade due to stress concentration. Failures due to axial force bending moments usually occur due to buckling in the inboard section of the blade.

4. Generators

The generator is the center piece of a small wind turbine. The advent of powerful permanent magnets based on Neodymium has opened the door to compact permanent magnet synchronous generator designs (Khan et al., 2005) with potentially high efficiencies. Radial flow generators are still the predominant choice, but axial flow designs (Probst et al., 2006) are becoming increasingly popular because of their modular design and relatively low manufacturing requirements. Currently, axial flow designs are typically limited to smaller-scale turbines with rated capacities of 10 kW or less due to the strong increase in structural material requirements for larger machines. Induction generators are occasionally used because of the abundance and low cost of induction machines which can be configured as generators, but are suitable only for grid-coupled applications.

Designing an efficient generator requires an understanding of the different loss mechanisms prevailing in such generators. Often, Joule losses occurring at the armature winding of the stator coils (often referred to as *copper losses*) are by far the greatest source of losses, so care has to be taken to avoid overheating, either by using high-voltage designs, allow for a large wire cross section to reduce armature resistance, provide efficient passive cooling mechanisms, or a combination of the former. Clearly, higher magnetic field strengths lead to higher induction voltages which in turns allow for lower currents, hence the need for powerful magnets. Iron cores instead of air cores can be used to increase the magnetic flow and therefore the induction voltage, albeit at the expense of a cogging torque (detrimental at startup) and higher stator inductivity (Probst et al., 2006). Wiring several stator coils in series is often a simple and efficient measure to increase the system voltage and diminish copper losses. Peak efficiencies of about 90% can be achieved with such a scheme even in a modest manufacturing environment (Probst et al., 2006). Under more stringent manufacturing conditions, where a small and stable air gap between the stator and the rotor can be assured, efficiencies of the order of 95% can be achieved routinely (Khan et al., 2005).

4.1 Common generator topologies

As described above, the electric generators of modern small wind turbines are generally designed to use permanent magnets and a direct coupling between rotor and generator. The following common topologies can be encountered:

1. Axial flow air-cored generators
2. Axial flow generators with toroidal iron cores
3. Axial flow generators with iron cores and slots
4. Radial flow generators with iron cores and slots
5. Transverse flow generators with slotted iron core

In the topologies above the type of flow refers to the direction of the magnetic flow lines crossing the magnetic gap between the poles with respect to the rotating shaft of the generator. Once the flow lines reach the iron core (in practical realizations actually laminated steel), the flow lines may change their direction according to the geometry of the core. Two of the most common topologies are shown in Fig. 5 and Fig. 6, respectively. Fig. 5 shows a typical radial magnetic flow topology, whereas Fig. 6 exhibits the conceptual design and magnetic flow field of an axial flow generator. Similar magnetic flux densities can be achieved in the magnetic gap for both topologies, but the axial flow geometry has the advantage of a modular design, since the two rotor disks and the stator disk (not shown) can be simply stacked on the rotor axis, making this design conceptually attractive for small-scale wind turbines, where often less sophisticated manufacturing tools are available than for large wind turbines.

Each topology has specific advantages and disadvantages (Dubois et al., 2000; Yicheng Chen et al., 2004; Bang et al., 2000), which makes it difficult to define a clearly preferred choice; in most cases the topology chosen will depend on the design preference. An overview of the most important up- and downsides is given in Table 1.

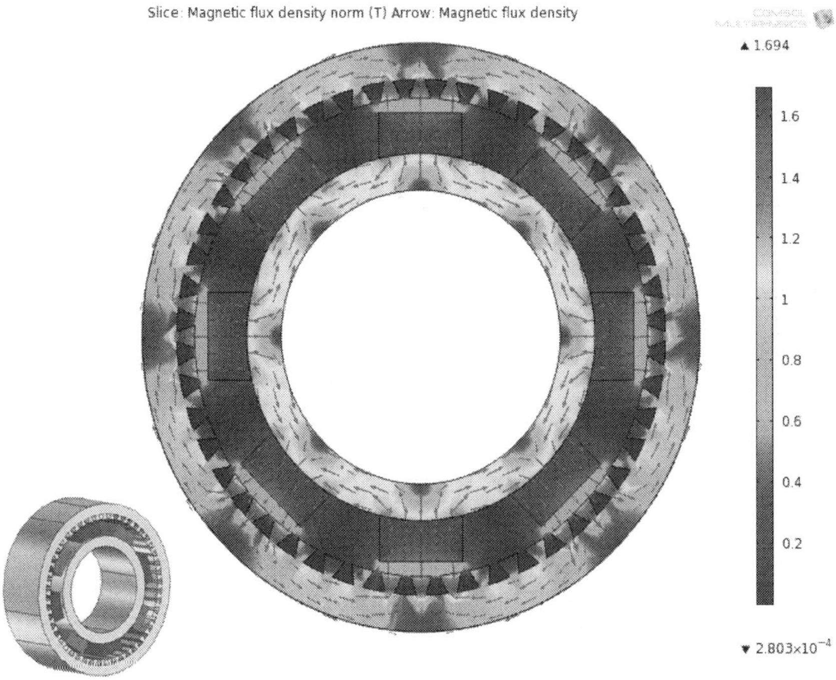

Fig. 5. Typical radial flow permanent magnet generator with iron core and slots. Small figure: Perspective view of general arrangement. Main figure: Color map: Magnetic flux density in T. Arrows: Magnetic flux density vector field.

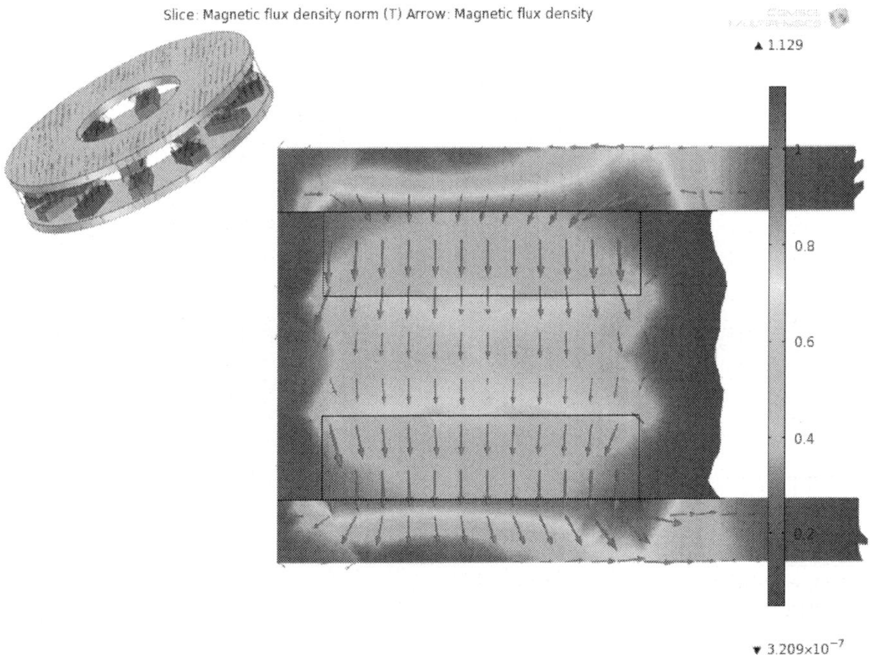

Fig. 6. Typical axial flow permanent magnet generator with iron core. Small figure: Perspective view of general arrangement. Main figure: Color map: Magnetic flux density in T. Arrows: Magnetic flux density vector field.

Topology	Advantages	Disadvantages
Axial flow with air core	Simple design and manufacture No cogging torque Quiet operation Low risk of demagnetization of permanent magnets No core losses Stackable and therefore scalable generators Multi-phase operation can be implemented easily	Structural challenges for maintaining a constant air gap for larger diameters Possible thermal instability of the polymer resin encapsulation Large amount of neodymium required Large external diameter Eddy losses in copper windings
Axial flow with toroidal iron core	Simple design and manufacture No cogging torque Quiet operation Low risk of demagnetization of permanent magnets No core losses Stackable and therefore scalable generators Multi-phase operation can be implemented easily Short end-coil connections	Structural challenges for maintaining a constant air gap for larger diameters Large amount of neodymium required Large external diameter Eddy losses in copper windings and in iron core
Axial flow with iron core and slots	Very high torque density Stackable and therefore scalable generators	Structural challenges for maintaining a constant air gap for larger diameters Complex manufacture Presence of cogging torque Relatively noisy Eddy core losses
Radial flow with iron core and slots	Structurally more robust tan axial flow generators; therefore less structural material required More widely used and well-known topology Smaller exterior diameter Diameter can be defined without considering the axial length	Presence of cogging torque Relatively noisy Eddy core losses Large amount of magnetic material due to laminated core.
Transverse flow with slots in iron core	Generally needs the least amount of neodymium Low copper losses Simple coil winding High torque density, if properly designed	Complex design and manufacture Uncommon topology so far Potentially high magnetic dispersion Potentially low power factor Presence of cogging torque Needs a stator for each electric phase Eddy core losses

Table 1. Comparative table of different generator topologies commonly used in small-scale wind turbines

4.2 Mechanical loads

Independently of the generator topology chosen, a structural analysis is indispensable before settling on a specific generator design. In small wind turbines it is common to directly couple the rotor and the generator; therefore mechanical loads on the rotor are directly transferred to the generator. Several extreme conditions should be considered when evaluating a generator design:

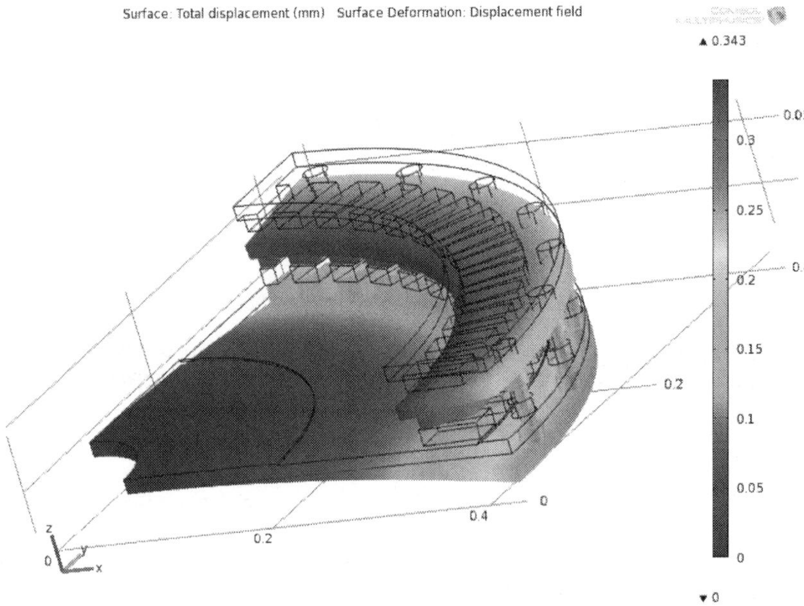

Fig. 7. Axial displacement field for the deformation of the rotor disks of an axial flow permanent magnet generator due to magnetic forces between the magnets

Electromagnetic forces. Since Neodymium magnets are particularly strong, their forces on the structural design have to be considered carefully. These forces are especially important for axial flow designs where either the axial forces between the magnets and the iron core or the forces between magnets (in the case of air cores) have to be considered. A common consequence is a deformation of the rotating disks on which the magnets are mounted, thereby reducing the clearance between stator and rotational disks. This reduction affects the magnetic flow distribution at best, but can ultimately lead to a collision between the disks and therefore the destruction of the generator, if not properly accounted for. Fig. 7 shows an example of the axial deformation field simulated for an axial flow permanent magnet generator with a free gap between the magnets and toroidal stator (not shown) of 3mm. Independent assessments determined that the tolerance of gap width should not be larger than 5% of this value, i.e. 0.15mm. For the design shown in Fig. 7, however, where the separation of the rotor plates is controlled by a ring of bolts distributed at the outer perimeter of the disks, axial displacements over 0.3mm are observed at the inner perimeter, leading to a considerable reduction in free gap width.

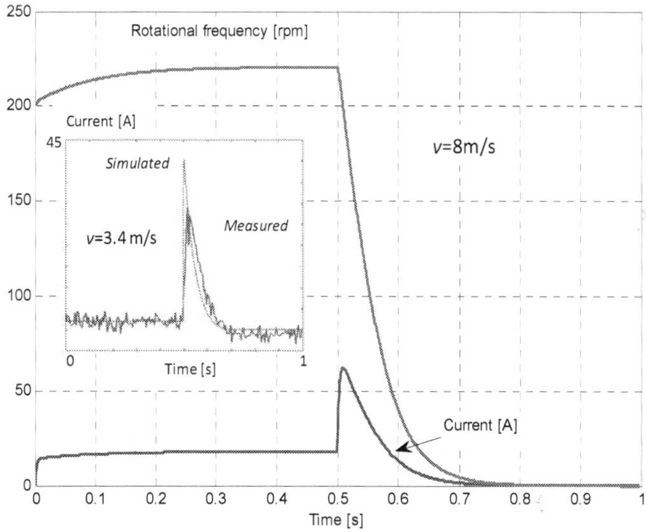

Fig. 8. Simulation of a short-circuit event. *Main figure*: Rotational frequency and stator current as a function of time. The short-circuit occurs at 0.5s. The wind speed was taken as 8 m/s. *Inset:* Simulated vs. measured stator current for a wind speed of 3.4 m/s.

Abrupt braking. It is important to consider different failure modes, such as the occurrence of a continuous short circuit at the generator terminals. This kind of failure leads to an abrupt braking of the rotor and may cause severe damage if not contemplated at the mechanical design stage. In the first place, excessive mechanical stress may occur at the structural elements due to the high braking torque. Moreover, thermal stress may occur due to the high electric current flowing under the short-circuit conditions, which is only limited by the internal resistance and inductance of the stator coils. Overheating occurring under these conditions may damage the wiring of the stator and electronic components. En Fig. 8 a simulation of the effect of such a short-circuit event is shown for the variables rotor frequency and stator current (main figure, wind speed = 8 m/s). In the inset of the figure, an example of the validation of the dynamic model is given for a wind speed of 3.4 m/s. In either case it can be observed that upon short-circuiting the rotor a steep rise in stator current is obtained, followed by a slower decay once the rotor-generator slows down as the result of the strong opposing torque. It is conspicuous from Fig. 8 that the rotation is brought to a complete halt in less than 0.2 seconds, after an initial condition of about 225rpm, with most of the braking occurring during the first 0.1 seconds. Such a spike both in mechanical torque and in current creates strong mechanical and thermal stresses, respectively, and can inflict severe damage to the rotor-generator (including the complete destruction), if not accounted for properly.

Blade fracture. Small-scale wind turbines rotate at a relatively high frequency compared to large turbines. As pointed out in the rotor section, centrifugal forces are generally the largest design concern, even under normal conditions, but in the case of a blade failure a severe imbalance of the rotor may occur. This imbalance gives rise to an eccentric force in the rotating shaft because of the remaining blades. Since many small wind turbines may reach high rotational speeds, centrifugal force can be several tons even for a turbine rated at 1kW, which can ultimately damage the generator shaft or the structure. Since the generator accounts for a significant part of the overall cost of the turbine, a damaged generator will generally lead to a total loss of the turbine.

Blade forces during extreme and turbulent wind events. Small turbines generally align themselves with the wind direction by means of passive yawing, so large and rapid changes in turbine orientation are common. Moreover, some wind turbines use furling mechanisms, which rotate the complete wind turbine abruptly and deviate it from the prevailing wind direction. These conditions induce gyroscopic forces in the blade root and the clamping supports. Those forces sum up with the aerodynamic bending moment; the gyroscopic forces are not axially symmetric because each blade experiences a particular force according to its angular position. This imbalance tries to bend the generator shaft and in extreme cases can lead to air-gap closure in the generator, the ultimate consequence of which is a magnet collision with the stator.

5. Control mechanisms

Control and protection mechanisms are peripheral elements that are necessary to ensure the reliability and long-term performance of a wind turbine. These mechanisms vary significantly with wind turbine size. While large turbines rely on active blade pitch and mechanical brakes, small wind turbines frequently use passive mechanism and controlled short circuits. The most common control mechanisms in small wind turbines are discussed below.

5.1 Furling systems

Furling is a passive mechanism used to limit the rotational frequency and the output power of small-scale wind turbine in strong winds. While other mechanisms, such as passive blade pitching or all-electronic control based on load-induced stall can occasionally be encountered, furling is the most frequently used mechanism. The basic idea is the turn the rotor out of the wind once a critical wind speed value has been reached. This principle is illustrated in Fig. 9 where photographs of an operating commercial wind turbine (Aeroluz Pro, rated at 1.4kW) are shown for normal operation (a) and under furled conditions (b).

Fig. 9. Furling mechanism operating in a commercial wind turbine rated at 1.4kW. (a) Normal (unfurled) operation. (b) Furled turbine.

The basic operating principle is sketched in Fig. 10. The mechanism is based on the interplay of three torques caused by the aerodynamic forces on the rotor and the tail vane, respectively, as well as a force of restitution, often provided by gravity in conjunction with an appropriate inclination of the tail axis. Due to an eccentric mounting of the turbine the axial force creates a moment around the vertical turbine axis, tending to turn the turbine out of the wind (counter-clockwise rotation in Fig. 10). At low or normal wind speeds this rotation is avoided by two opposing moments working in conjunction. Firstly, the aerodynamic torque on the tail vane tends to realign the vane with the wind direction, thereby causing a clock-wise rotation of the vane with respect to the generator structure. Now, if the tail rotation axis (generally referred to as the furl axis) is chosen not be non-perpendicular to the horizontal plane, then this rotation results in an increase in gravitational potential energy, which translates into an opposing torque for the turbine rotation. If the wind speed is strong enough, however, the opposing torques will be overcome and the turbine furls (Fig. 10 (b)). If the wind speed is reduced, then the moment of restitution prevails and operation of turbine in alignment with the wind direction is reestablished.

The transition into the furling regime and back to normal operation for a selected case is shown in Fig. 11, where both the yaw (a) and the furl angle (b) have been plotted as a function of the steady-state wind speed for (i) entering and (ii) exiting the furling regime.

These results were obtained by feeding a constant wind speed into a dynamic model of the furling mechanism of a small wind turbine (Audierne et al. 2010) and observing the

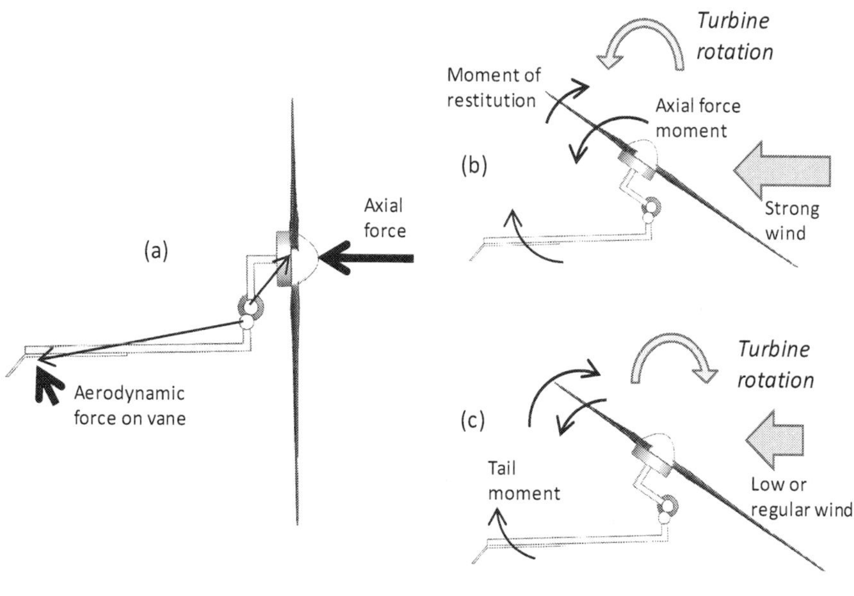

Fig. 10. Overview of the operating principles of a furling system. (a) Aerodynamic forces. (b) Furling movement in strong winds. (c) Restitution of normal (aligned) operation upon reduction of the wind speed.

asymptotic value of the yaw and furl angle, respectively. It can be seen from Fig. 11 that in this particular case the onset of furling, characterized by a steep transition of the angles, occurs at about 12.25 m/s. In order to return to normal operation the wind speed has to be lowered below that value, in this case to about 12.15 m/s, i.e. some hysteresis occurs.

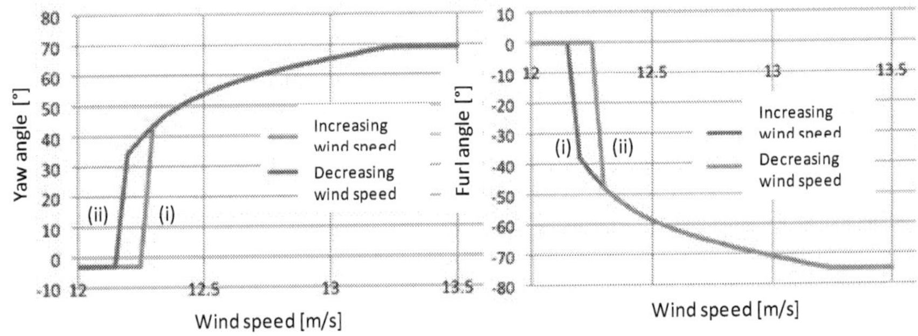

Fig. 11. Simulation results for steady state transition for (i) Entering, (ii) exiting the furling regime. (a) Yaw angle, (b) furl angle.

This hysteresis is quite common in furling mechanisms and can be traced back to the different variation of the torque components with the yaw angle. This allows the designer to fine-tune the system according to his or her requirements; see Audierne et al. (2010) for details. It should be noticed that in the case exhibited in Fig. 11 a relatively smooth transition to the asymptotic values of the angles occurs after the initial steep transition, allowing for a relatively smooth variation of the power curve beyond the onset of furling, as opposed to situations where the rotors jumps to its stop position (about 70° in this case) abruptly.

Due to this hysteresis it can be anticipated that that additional complexity will be present in the case of a dynamically varying wind speed. To explore this dynamics, stochastic wind speed time series (Amezcua et al., 2011) with a given turbulence intensity, a defined Kaimal turbulence spectrum and specified gust values was fed into the simulator developed for furling system (Etienne et al., 2010). 15 realizations of each stochastic process were simulated and the results of furling calculations for these 15 runs were averaged. An initial hypothesis was that short gusts beyond the steady-state wind speed for the onset of furling might trigger a transition into the furling regime where the system would be trapped due to hysteresis. In Fig. 12 a phase diagram identifying the system phases (non-furled, furled, transitioning) has been plotted for two cases of the standard deviation of the wind speed. The variable plotted on the horizontal axis is the mean wind speed of the time series, whereas the vertical axis shows the difference between the gust and the mean wind speed, i.e.

$$\Delta v = v_{\max} - \langle v \rangle = \langle v \rangle G \qquad (16)$$

where G is the gust factor. In the case of a low standard deviation of 1m/s (Fig. 12 (a)) a clear dividing line can be seen between the two main regimes, i.e. the normal operation regime with yaw angles of up to about 15°, and the furling regime, where the yaw angle is in excess of 60°. The transition between the two regimes occurs in a thin range around the dividing line the width of which is essentially constant for all ($\langle v \rangle$, Δv) combinations. It is intuitively clear that for smaller average wind speeds $\langle v \rangle$ the required gust Δv (measured relative to

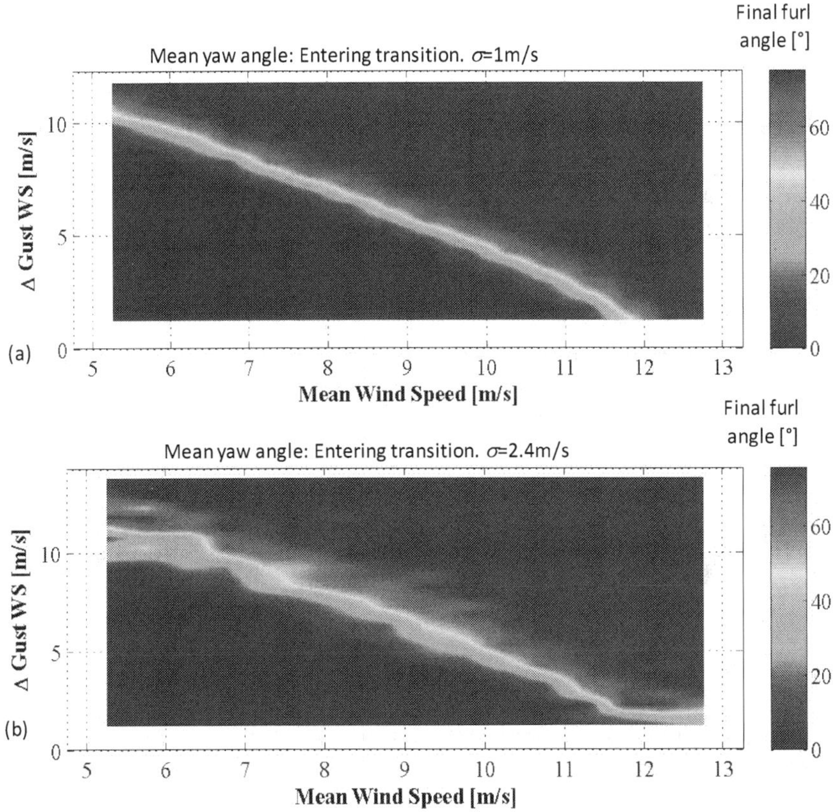

Fig. 12. Phase diagram for dynamic transitions into the furling regime: Simulated asymptotic value of the yaw angle as a function of the mean wind speed and the difference between the gust and the mean wind speed. a) Wind speed standard deviation = 1 m/s, b) standard deviation = 2.4 m/s.

the mean wind speed) for triggering a transition into furling has to be higher than at high wind speeds. As expected, for zeros gust (G=0) transition to furling occurs approximately at the critical steady-state wind speed (about 12 m/s in this example).

The fact that the transition boundary is approximately a straight line with negative slope can be stated in terms of the following simple equation

$$\alpha\ \Delta v + \langle v \rangle = \frac{\tau}{T}\ \Delta v + \langle v \rangle = v_{ss}, \tag{17}$$

Where T is the simulation time frame and τ the effective gust duration and v_{ss} is the steady-state wind speed value for the transition to furling. Using the graph in Fig. 12 (a) an effective relative gust duration τ/T can be estimated. While $\tau/T < 1$ in all cases simulated, the effective gust duration determined this way is considerably larger than the real gust duration (typically 15s), indicating that a short gust that takes the total wind speed over the threshold value may be sufficient to trigger the transition to furling, even if the average wind speed in the interval is low. As mentioned earlier, the hysteresis between the transitions to and from the furling regime is a plausible qualitative explanation of this phenomenon.

Not unexpectedly, a higher turbulence intensity gives rise to a more stochastic behavior near the transition boundary, as shown in Fig. 12 (b), where the phase diagram in the $<v>$- Δv plane has been shown for a value of the standard deviation of 2.4m/s. While still the linear relationship between the critical $<v>$ and Δv values can be seen, the boundary is now blurred out, indicating a more chaotic movement of the rotor angle near the threshold. Interestingly, higher fluctuations can be seen for low mean wind speeds, where the turbulence intensity is higher than for high $<v>$ values, given the fact that the standard deviation and not the turbulence intensity was held constant in the simulations.

While more complex phase diagrams can be produced by choosing extreme value of the geometric parameters of the system, for plausible design parameters it can be seen that the system behavior remains relatively predictable (in a statistical sense), so that a stationary analysis provides a useful guidance for the design of furling systems.

5.2 Load-induced stall control

Stall control is a common practice in fixed rotational speed wind turbines and was used in utility-size turbines until relatively recently, when multi-megawatt turbines became the standard for commercial wind farms. Some utility-scale turbines, such as the NEG-Micon 1.5MW (later upgraded to 1.65MW under the label Vestas), also used active stall control, where the blade is pitch in the opposite direction as compared with regular pitch control.

In small wind turbines stall control had been used by different means, either by rotating the blade through a mechanism activated by centrifugal forces (Westling et al., 2007) or by changing the rotational speed. This last approach has the advantage of reducing the moving parts and hence increasing the reliability. Variations on rotational speed can be caused by changing the load on the generators terminals. If a smaller load is connected, higher current will be demanded and therefore higher mechanical torque. This will lead to a reduction in the rotational speed and higher angle of attack. Stalled blades will reduce output power of the wind turbine as shown in Fig. 13 (Elizondo 2007). These results were obtained with an experimental wind turbine built based on a Bergey XL.1 commercial wind turbine generator, but equipped with a specially designed rotor and tail vane (Elizondo, 2007). A simple load control was implemented based on a switchable resistor bank, where the value of the load was changed as a function of the measured wind speed.

For wind speeds from 0 to 6.0m/s the load was a resistance of 3Ω, which was reduced to 1 Ω at 6.1m/s, 0.3 Ω at 7.1m/s, and 0.25 Ω at 8.1m/s. Without the intervention of the control system (i.e. for a constant load of 3 Ω) the system would be expected to operate in near-optimal conditions for wind speeds of up to about 7m/s; consequently, power should increase as the cube of the wind speed. Due to the change in load from 3 Ω to 1 Ω at 6.1m/s power increases a little slower, but the change is not obvious from the graph. Actually, instead of operating at a tip speed ratio (TSR) slightly above the TSR for optimum system power coefficient $C_{p,system}$ at a wind speed of 7m/s, due to the change of load the system now operates under mild stall conditions with a slightly lower $C_{p,system}$. The difference becomes more conspicuous at higher wind speeds of about 7 to 8m/s where a clear change of curvature in the P(v)-curve is apparent. When the resistance is lowered to 0.3 Ω at 8.1m/s, a noticeable change occurs, with a clear reduction in slope of the P(v) curve, leading to a plateau of the power production. At a wind speed of 9.1m/s a further reduction of load to 0.25Ω is induced by control system, but this small increase in load is incapable of coping with the increase of available power density; consequently, power production increases again. It should be noted that this behavior is a specific limitation of the Bergey XL.1 generator which was designed to operate with a different control system and whose intrinsic (armature) resistance is too large to allow for a further reduction of the load resistance, as required for a full-scale active-stall control.

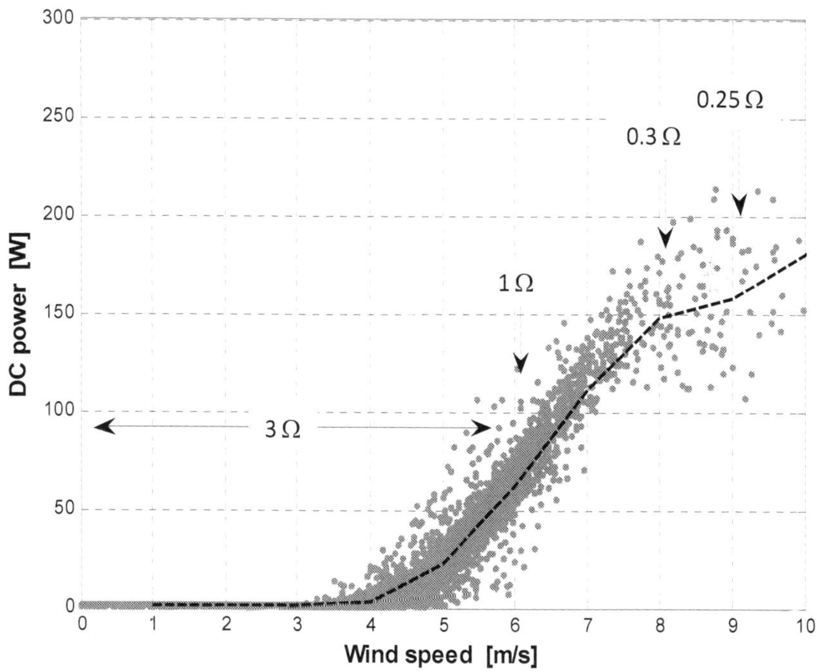

Fig. 13. Proof-of-concept demonstration of active stall regulation

5.3 Passive blade pitch

Blade pitching is very common among mega-watt size wind turbines where a motor is used to rotate the blade along its axis depending on measured wind speed and desired performance. This principle has also been used in small wind turbines but usually with passive mechanisms that convert an existing force into the blade rotation.

Mechanisms vary significantly depending on the manufacturer and the purpose of the pitch. Some turbines use the pitch mechanism as a means of power control, however, since the system is passive, a precise control is hard to achieve and therefore most turbines just implement it as a protection system against high wind speed or high blade rotational speed.

In any case, the operational principle is to reduce the accelerating wind force on the blades, by changing the angle of attack in the airfoil sections. Examples of blade rotations both to increase angle of attack and reach stall conditions (pitch to stall), or to decrease angle of attack (pitch to feather) can be encountered in commercial wind turbines. The activation of the mechanism commonly uses the centrifugal force, where the radial movement of the blades induces the rotation. However, in some cases the aerodynamic torque caused by pressure difference along the airfoil chord has also been used. Restoration of the unpitched blade orientation is usually achieved using springs attached to the blades that overcome the activation force when non-operational conditions finish. In principle, any other force such as gravity may be used to restore the blade orientation.

One critical aspect to consider when implementing a passive blade pitch system is to ensure that all blades rotate at the same time in order to prevent aerodynamic or inertial imbalance that may damage components of the wind turbine. Furthermore, a careful balance between the activation and restoration force of the mechanism has to be considered in order to avoid system oscillations that will cause damage in the long-tem.

6. System behavior

6.1 Power flows and efficiency

In general terms, power losses in small-scale wind systems are significant, given the fact that small wind turbines generally operate at low voltages; this is particularly notorious in battery-charging systems where typical system voltages are in the range of 12V to 48V. The different loss mechanisms cover a quite wide range for each loss component, given the range of electric power and rotational frequencies encountered. At low frequencies, rotational friction at the shaft bearings is the dominant loss mechanism, where at nominal output power normally Joules losses in coils and cables play the most important role. Generally, the point of maximum efficiency does not occur at maximum output power due to the interplay between the extraction of aerodynamic power (peaking at a given rotational frequency for a given wind speed) and the losses of power in the electromechanical system (which generally increase with rotational power).

An overview of the power conversion stages occurring in a small wind system, together with the main loss mechanisms is given in Fig. 14. The power delivered to the rotor-generator shaft by the extraction of power from the wind is first transformed into electromagnetic power, with losses corresponding mainly to friction at the bearings and ventilation losses. In order to produce power at the generator terminals, energy has to be converted at the generator, involving Joule losses at the armature resistances of the stator coils, eddy losses in coils, as well as losses at the ferromagnetic core, if present. Joule losses at the coils are generally the dominant loss mechanism at small wind turbine generators, especially at higher power values. If an electronic converter is used, such as for maximum power point tracking and related functions, then additional losses occur at this stage, although these losses are generally more than compensated by the corresponding gain in aerodynamic efficiency of the rotor. Before getting to the load, a transmission line is required. Even if the load is placed at the foot of the tower (an unlikely situation), the minimum length of the line is still of the order of 20m to 30m, generating substantial losses if the power is transmitted in 12V or 24V. While self transformers can in principle be used to raise the level of the transmission voltage, the compact size of the nacelles of small wind turbines often does not allow for such a measure. Moreover, the system complexity increases by requiring a second transformer near the load in order to return to the system voltage. In the case of battery-charging systems, the rectifier has to be placed after the second transformer, so the whole transmission line has to be three-phase, which increases the system cost beyond the additional expenses for the transformers. After delivering the power to the load, there are still losses occurring at the cabling of the load, as well as internal losses, so selecting an efficient end-use device is important to obtain an efficient overall energy service.

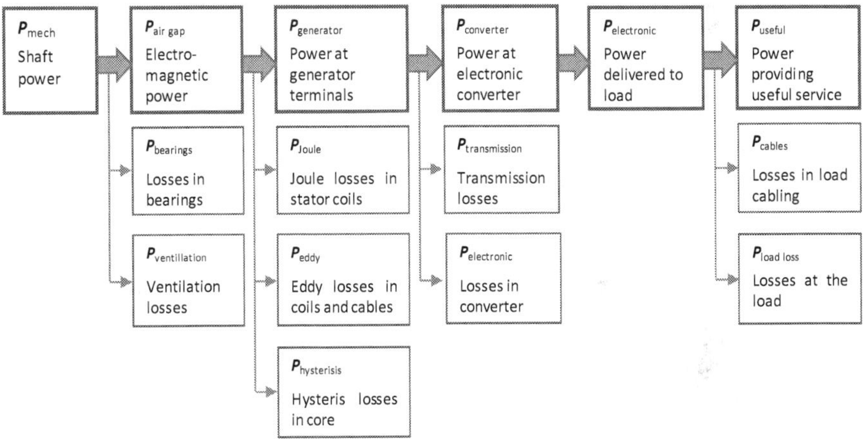

Fig. 14. Power conversion chain at a small wind system. The upper chain identifies the main conversion stages, while the lower boxes describe the losses occurring at each stage.

Typical component efficiencies are the following: (1) Generator efficiencies may vary widely, depending on the design, the manufacturing accuracies, and the operating point. Commercial generators generally have efficiencies of 90% and beyond, while the efficiency of home-built generators may be as low as 65%. Similar efficiencies (65%-95%) apply for the transmission line, where higher transmission voltage (such as 48V) helps keeping the losses at an acceptable level. The downside of a 48V system voltage is the fact that few appliances are available for this voltage level, as opposed to 12V where an abundance of products exists due to the use of this system voltage in the automotive sector. The efficiency for delivering power to the load is in the range of 60% to 85% if a battery bank and a subsequent DC/AC converter is used.

The total system efficiency, based on these values, is in the range of 25% to 75% if batteries are used for energy storage, and 45% to 85% for grid-connected systems. From the discussion above it is clear that a high-quality electric system and installation are essential. This is particularly true in small systems with a rated output power of 1kW or less, and even more so in home-built systems where the overall available power may be deceptively low and hardly enough to light a few light bulbs.

6.2 Steady-state system behavior

In order to determine the steady-state operating point of the wind system it is necessary to specify two variables on both ends of the generator: (1) On the upstream side either the shaft frequency or the mechanical torque can be specified. (2) Possible variables on the downstream side include the output voltage or the electric current. A useful tool in the design of the system operation is a diagram where the power supply and demand curves are shown as a function of the rotational frequency n (Fig. 15).

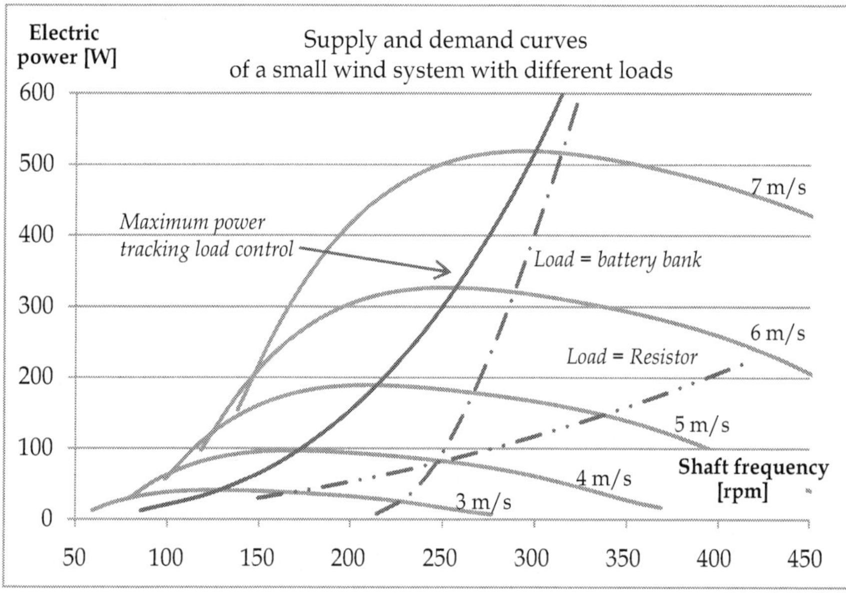

Fig. 15. Supply and demand curves for different wind speeds and load types.

In Fig. 15 the supply curves (converted into electric power curves by incorporating all inefficiencies and their respective dependencies on n in the conversion from shaft to load power) have been drawn for a wind turbine with a swept diameter of 3.0 m for different wind speeds in the range of 3 m/s to 7 m/s, together with the load curves for (a) a simple fixed electric resistor (such as a light bulb), (b) a battery bank, and (c) a load equipped with an electronic load control, in this case designed to track the maximum electric power point. It can be seen that an electric resistor generally represents a load which is inappropriately matched to the source, since only at certain wind speeds an operation close to optimal conditions is possible, even for the relatively broad supply curves shown in this example. Clearly, a direct connection of the wind turbine to a resistor is not an efficient means of utilizing the power extracted from the wind and will not be used in other than the simplest systems. One possible example is electric space heating in cold climates with strong winds where an abundance of the wind resource may minimize concerns about overall efficiency. Under these conditions, also the fluctuating nature of the wind resource which directly translates into a fluctuating output power is of less importance, since the heated space itself acts as a storage medium.

If a battery bank is used as a load, the source-load match improves substantially. If the supply curves and the system voltage are appropriately matched, a certain self-regulation occurs in this case, since the power draw from the battery matches the supply from the rotor-generator quite well for all wind speed values other than the very low ones. The poor match at low wind speeds is due to the simple fact the a minimum rotational speed is required for a synchronous generator in order to produce at least the battery voltage at its terminals, so no charging (and the corresponding power flow) will occur at shaft frequencies below the critical value. Battery-charging systems are quite common for remote electrification, and a direct connection of the wind turbine to the battery bank may be a viable alternative for cost-sensitive applications.

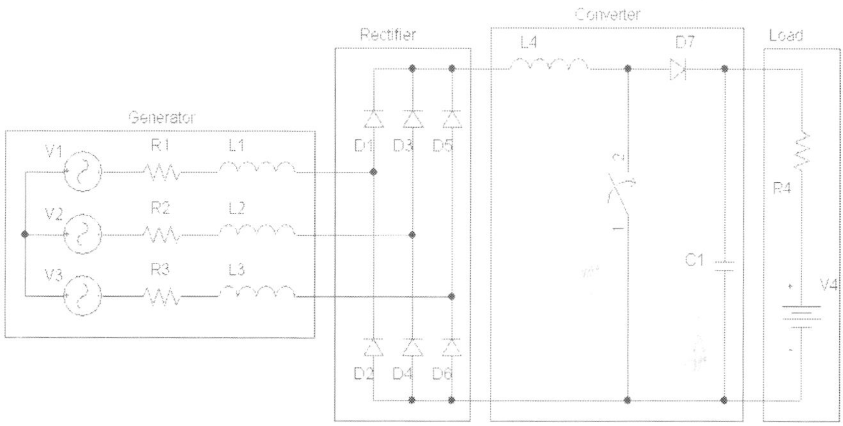

Fig. 16. Simple boost controller operating with a three-phase generator and rectifier, as well as a battery bank.

In the case of grid-connected applications where a reduction of the electric bill is the main concern, a load control to ensure an optimal extraction of the available wind power may be appropriate. As shown in Fig. 15 a maximum power point tracking load may significantly increase the output power at low wind speeds and improve the starting behavior of the turbine. As mentioned in the preceding paragraph on battery charging, the problem to be overcome at low rotational frequencies (required to match the supply curves at low wind speeds) is the low voltage produced by the generator under these conditions. An electronic device which dynamically changes the apparent load impedance can solve this problem. A simple scheme designed to boost the voltage is shown in Fig. 16 where a three-phase generator, operating with a passive six-pulse rectifier, is connected to a battery bank through a switching device. The switch can be a Mosfet or an IGBT device and is fed by a pulse width modulated (PWM) signal generated by a control unit, often a micro-controller. Opening and closing the switch generates a modulated current and therefore an induction voltage at the inductor which is smoothed out by the capacitor. The average output voltage can be adjusted by varying the duty cycle of the switching process. The inductor shown in Fig. 16 is often not necessary, since the self inductance of the generator can be used for that purpose.

This simple device can be used in different ways to control the operation of the small-scale wind system: (1) As mentioned before, boosting the voltage at low wind speeds allows charging of the battery bank at wind speeds well below the critical value for constant load voltage. (2) At high wind speeds the current can be increased in order to operate the turbine under partially stalled conditions and slow down the rotor. This is a convenient way of controlling the operation of the turbine at high wind speeds. If appropriately designed, the rotor will dissipate most of the excess power available at high wind speeds by entering the stall regime, limiting the electric power dissipation in the generator due to the increased current.

6.3 Mathematical modeling

Despite their apparent simplicity, small-scale wind turbines are complex systems. This makes it necessary to establish a mathematical model which couples all the physical processes occurring in the system and can provide accurate guidance during the design, analysis, and continuous improvement. A good starting point is a conceptual model of the electric generator which can be used to identify all the physical processes and develop a more detailed model of the magnetic flow topology, the effective electric parameters and the heat transfer. Often, the electric load characteristics can be studied in a lumped circuit model, but the values of the effective parameters such as armature resistance and stator inductance have to be determined from the more detailed model, if the costly construction of a series of prototypes for measurement and optimization purposes is to be avoided. Detailed mathematical models have been developed for certain generator topologies (see, e.g., Bumby et al., 2005; Gieras et al., 2008), but numerical multi-physics models can also be conveniently constructed with tools like Comsol (see examples in section 5).

Once the generator characteristics are sufficiently well known, effective load curves can be constructed for typical situations, such as a battery or a resistive load. Also, the total generator efficiency (including the rectifier if a DC application is sought) as a function of the rotational frequency should be known in order to narrow the range of the steady-state operating points, the exact location of which will of course depend on the detailed rotor characteristics. The rotor is generally modeled using Blade Element Momentum theory with suitable expressions for the aerodynamic lift and drag coefficients, as discussed earlier. As it was mentioned before, the available wind speed and the instantaneous electrical load connected are the principal factors determining the operating point of the generator. For most purposes, a steady-state operating model of the turbine will be enough, although the assessment of failures will generally require a transient model of the system, as discussed in section 5. A control model which appropriately maps the aerodynamic mechanisms such as furling or passive pitching and the electronic control elements such as load control (if present) will conclude the analysis of the operating characteristics of the wind turbine.

Thermal characteristics finally, such as the heat transfer mechanisms, are important to know in order to define the generator's loading characteristics. A proper thermal design should also the consider the ventilation properties of the rotor-generator, since a wind turbine has some intrinsic thermal self-regulation characteristics, given the fact that high wind situations, where efficient cooling by forced convection is available, generally correlate with high power output.

The results of the modeling process can be conveniently summarized by a performance map, defined as a set of tables showing the values of the different variables of interest as a function of both the wind speed and the rotational frequency of the generator. Variables of interest include the electric output power, the temperature of the stator coils, the terminal voltage and the electric current. The information of the different tables can be combined by marking table cells where critical conditions for any of the variables occur. The performance

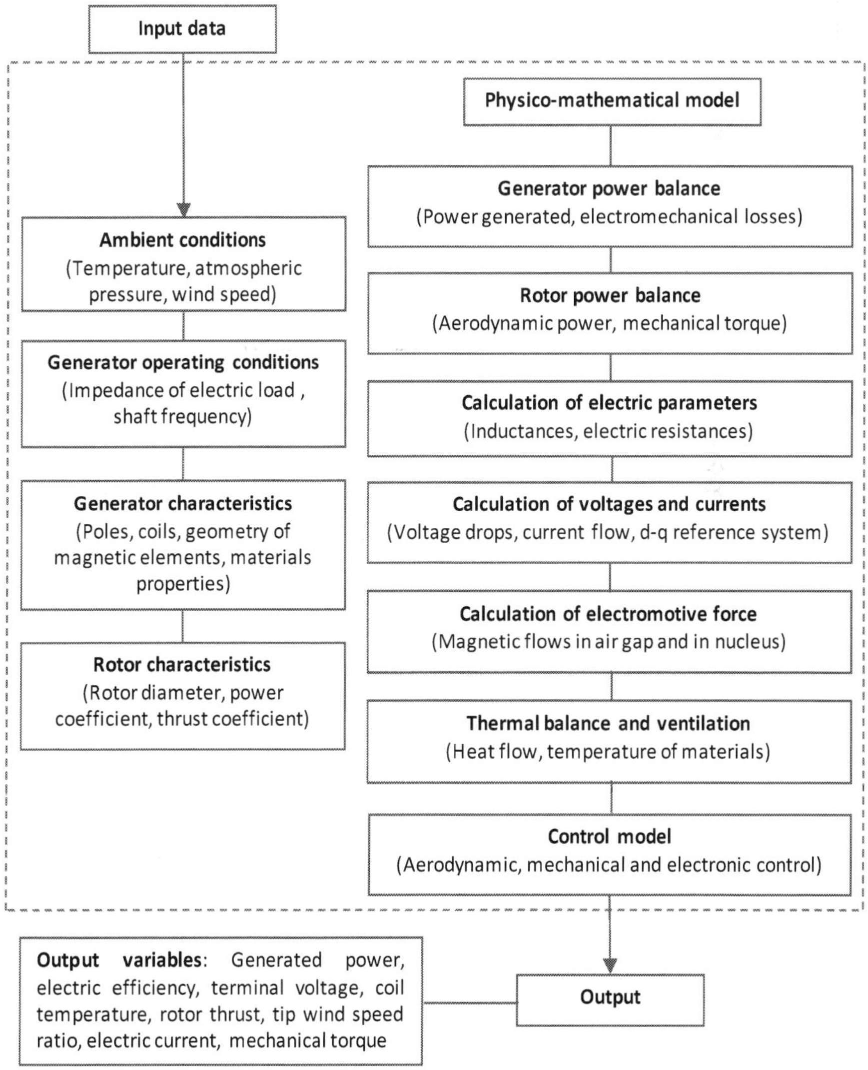

Fig. 17. Schematic representation of the system modeling process

map can then be used to establish operating limits, desirable operation points or zones, and methods of regulation, control, and protection. In Table 2 an example of such a performance map (shown for the variable output power) for a small wind turbine rated at 1.4 kW.

	U_{wind} (m/s)					
	4	6	8	10	12	14
n (rpm)	P_{gen} (W)					
50	4	7	10	14	19	24
100	55	54	54	63	71	82
150	**84**	186	200	177	186	204
200	78	281	417	457	437	429
250	52	312	606	735	768	777
300	16	**313**	714	1,009	1,021	**940**
350	0	291	775	1,239	**1,251**	723
400	0	249	**786**	**1,405**	1,615	86

Table 2. Performance map for a wind turbine rated at 1.4 kW. Dark-shaded cells: Optimal output power. Light-shaded cells: Regulated high-wind speed path. Hatched cells: Overheating coils.

At low wind speeds reaching the maximum output power point is desirable in order to make an efficient use of the relatively small available power density. For wind speeds of up to 8 m/s the required rotational frequency remains within a window where centrifugal forces on the blades are tolerable on structural grounds and noise due to air-cutting blades is not a serious problem. At still higher wind speeds, following the maximum output power point would lead to very high shaft frequencies and produce excessive heating of the stator coils. Therefore, limiting the rotor frequency by an appropriate control scheme is necessary. In the present example, limiting the rotor frequency to 400rpm, 350rpm, and 300rpm at 10m/s, 12m/s, and 14m/s, respectively, leads to a path where all system variables of interest remains within their allowed boundaries. Such a frequency control may be achieved using electronic load control, e.g. by actively inducing stall at (part of) the blades, by passively veering the rotor out of the wind (furling, see section 4), by passively pitching the rotor blades, or a combination of the former. The vertically hatched cells in Table 2 correspond to situations where overheating of the coils occur (due to high dissipation due to Joule losses and others) and have to be avoided.

7. Summary and conclusions

Small-scale wind turbines are an attractive option for a host of remote applications, such as rural electrification, water pumping or telecommunication, and also provide an option for saving energy and mitigating greenhouse gases in grid-tied situations. While conceptually simple, small wind turbines are quite complex systems and require a professional design, construction and operation. While motivated individuals with a sufficient engineering background may be capable of designing and building a functional wind turbine, its performance may be deceptively low, unless great care is taken to design, construct and operate the system based on best engineering practice. As pointed out above, the rotor design and performance evaluation have specific issues related to low Reynolds number flow and 3D aerodynamic effects. The generator, on the other hand, in general has to be designed for a given application, which precludes the use of off-the-shelf appliances in most cases, unless suboptimal operation is tolerable. Manufacturing of the generator is critical due to the small gap widths required for efficient magnetic flow, and potentially hazardous events have to be considered early in the design stage. The control mechanisms, finally, are key to the efficient operation of the turbine under normal operation and will keep the turbine safe in extreme winds. And last, but certainly not least, design for safety should guide the whole process, and certification of the performance and safety characteristics (e.g. according to AWEA (2009)) should be obtained for any turbine intended for commercial sales.

8. References

Amezcua, J.; Muñoz, R.; Probst, O. (2011). Reconstruction of gusty wind speed time series from autonomous data logger records, *Wind & Structures* (accepted)

Audierne, E.; Elizondo, J.; Bergami, L.; Ibarra, H.; Probst, O. (2010). Analysis of the furling behavior of small wind turbines, *Applied Energy* 87(7): 2278-2292

AWEA Small Wind Turbine Performance and Safety Standard AWEA Standard AWEA 9.1 - 2009. Available at http://www.smallwindcertification.org/.

Bang, D.; Polinder, H.; Shrestha, G.; Ferreira, J.A. (2008). Review of Generator Systems for Direct-Drive Wind Turbines. Proceedings of the European Wind Energy Conference (EWEC) 2008.

Baroudi, Jamal A.; Dinavahi, Venkata; Knight, Andrew M. A review of power converter topologies for wind generators. University of Alberta, Edmonton, AB, Canada, 2007

Bumby, J.R.; Martin, R. Axial-flux permanent-magnet air-cored generator for small-scale wind turbines. IEE Proc. Electr. Power Appl., Vol. 152, No. 5, September 2005

Burton, T.; Sharpe, D.; Jenkins, N., Bossanyi, E. (2001). *Wind Energy Handbook*, John Wiley & Sons

Dubois, M.R.; Polinder, H. Ferreira. J.A. (2000) Comparison of generator topologies for direct-drive wind turbines. Delft University of Technology 2000.

Elizondo, J. (2007). Diseño, manufactura y caracterización experimental de aspas y controlador de carga resistiva para una turbina de viento Bergey BWC XL.1. Master Thesis, Instituto Tecnológico y de Estudios Superiores de Monterrey, Monterrey, Mexico (in Spanish)

Elizondo, J.; Martínez, J.; Probst, O. (2009). Experimental study of a small wind turbine for low- and medium-wind regimes. *International Journal of Energy Research* 33:309–326, John Wiley & Sons, Ltd

Gieras, Jacek F. Performance Characteristics of a Permanent Magnet Transverse Flux Generator. IEEE, 2005

Gieras, Jacek F.; Wang, Rong-Jie; Kamper, Maarten J. Axial Flux Permanent Magnet Brushless Machines. 2nd Edition, Springer 2008

Khan, M.A.; Pillay, P.; Visser, K.D. (2005). On Adapting a Small PM Wind Generator for a Multiblade, High Solidity Wind Turbine. IEEE Transactions on Energy Conversion Vol. 20, No. 3, 685-692

Kutz, M. (2005). Handbook of Environmental Degradation of Materials. William Andrew Publishing.

McGhee, R.J., Walker, B.S., and Millard, B.F. (1988). Experimental Results for the Eppler 387 Airfoil at Low Reynolds Numbers in the Langley Low-Turbulence Pressure Tunnel. NASA TM 4062

Martínez, J.; Morales, A.; Probst, O., Llamas, A., Rodríguez, C. (2006). Analysis and simulation of a wind-electric battery charging system. *International Journal of Energy Research* 30:633-646 John Wiley & Sons, Ltd

Martínez, J.; Bernabini, L.; Probst, O; Rodríguez, C. (2005). An improved BEM model for the power curve prediction of stall-regulated wind turbines. *Wind Energy,* vol. 8, pp. 385-402, John Wiley & Sons, Ltd

Probst O., González A., Roidl M., Llamas A. (2006). A small wind–electric system based on an improved version of the ITDG axial flux permanent magnet generator. *Wind Engineering* 2006; 30(5):385–400.

Wastling, M.A; Balson, J.Ch.; Irving, D.; Cann, R.J. (2007). United States Patent US 7,172,392 B2

Yicheng Chen; Pragasen Pillay; Khan, A.. (2004) PM Wind Generator Comparison of Different Topologies. Industry Applications Conference, 2004. 39th IAS Annual Meeting. Conference Record of the 2004 IEEE

Yicheng Chen; Pillay, P.; Khan, A.; PM wind generator topologies. IEEE Transactions on Industry Applications 2005; 41:1619-1626

Selig, M.S., McGranahan, B.D. (2004). Wind Tunnel Aerodynamic Tests of Six Airfoils for Use on Small Wind Turbines. NREL/SR-500-34515. National Renewable Energy Laboratory, Golden, Colorado, USA

Somers, D.M., Maughmer, M.D. (2003). Theoretical Aerodynamic Analyses of Six Airfoils for Use on Small Wind Turbines. NREL/SR-500-33295. National Renewable Energy Laboratory, Golden, Colorado, USA

6

Innovative Concepts in Wind-Power Generation: The VGOT Darrieus

Fernando Ponta, Alejandro Otero*, and Lucas Lago
Michigan Technological University
USA

1. Introduction

Considerable progress in wind-power technology during the last decades has made it an important supplier of grid-connected electricity in the world energy picture. Nowadays, wind power is the most rapidly growing and most widely utilized renewable energy technology, with a total of 159.2 GW installed worldwide at the end of 2009 producing 340 TWh per year, which is about 2% of worldwide electricity usage (WWEA, 2010).

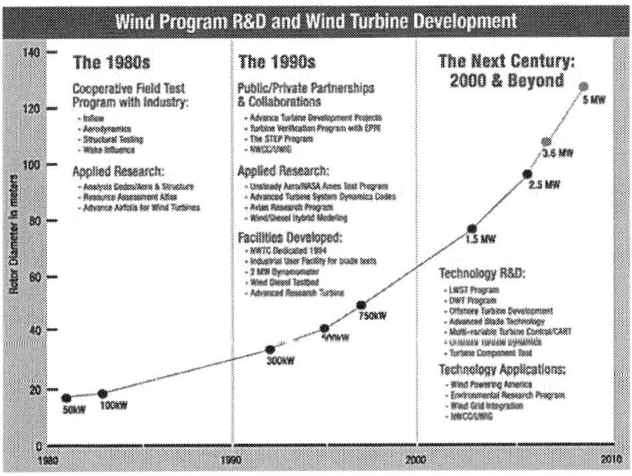

Fig. 1. The wind-turbine upscaling phenomenon (from NREL, 2005).

Over the last 30 years, there has been a spontaneous tendency in the wind-turbine industry to increase the size of the state-of-the-art machine. Figure 1 shows a chart published by NREL-DOE where this upscaling phenomenon is clearly depicted (NREL, 2005). This tendency is driven by economies-of-scale factors that substantially reduce the cost of wind energy. Starting from the typical 50 kW machine of the early 1980s, output power of the state-of-the-art wind turbines is now in the range of 3.6 to 6 MW, with rotor diameters up to

*Also at: FIUBA/CONICET, Buenos Aires, Argentina

127 meters. Commercial models within this range are available from several manufacturers like GE, RE-Power, Enercon, Vestas, and Siemens. In recent years industry insiders have been talking about a next-generation of giant offshore turbines of 7.5 to 12 MW with rotor diameters up to 200 meters (de Vries, 2005). If this generation of superturbines is successfully developed, wind-energy costs would be reduced substantially.

There is still a wide margin for improvement within the current horizontal-axis wind-turbine technology (HAWT), which looks as the most effective way to go in the short- to medium-term future. The development of adaptive-blades with aeroelastic tailoring the flexo-torsional modes of the blade structure to reduce aerodynamic loads, the introduction of a modular-concept blade that could be split into easy-to-handle segments, and the use of advanced materials and manufacturing techniques in blades and other components are only some examples of how to improve the economics of design, manufacturing, transport-logistics, and operation of HAWTs (NREL, 2008). But the HAWT technology would ultimately reach its limits in the long-term future. As sizes increase, a number of related problems would appear in rotor design, transport and assembly. Ultimately, the low rotational speed associated with huge radii would complicate the coupling with the electrical generator to the point that the size and weight of the drive train required would become impractical.

On the other hand, there are geographical regions around the world (e.g. Alaska, British Columbia, Labrador, Patagonia) characterized by vast wind resources. Mean speeds in some areas almost double those recorded at the typical locations for which commercially-available wind turbines were designed. Those regions, classified as outstanding and superb, offer an enormous potential in terms of energy resource where, in the future, it might even be possible to produce hydrogen as a substitute fuel in competitive terms, helping to close the gap between global fuel needs and the maximum amount of biofuel that may be produced sustainably. Hence, it is worthwhile to explore innovative concepts in extra-large wind-power plants to be able to exploit the renewable energetic potential that those regions offer.

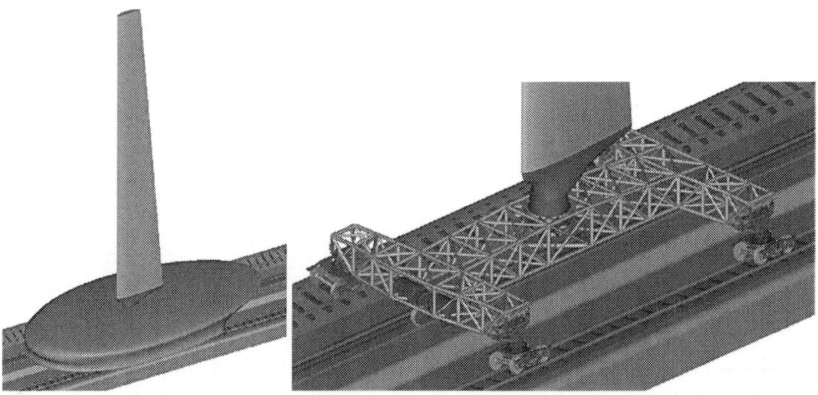

(a) (b)

Fig. 2. Perspective view of a blade-wagon (a). Structural configuration of a blade-wagon showing bogies, suspension system, and the electrical generators coupled to their driving wheels. (b).

In this chapter we are going to explore an innovative concept that, by going beyond the classical evolutive-design process based on the HAWT concept, may help address those challenges in the long-term future of wind-turbine technology. The Variable-Geometry Oval-Trajectory (VGOT) is an innovative type of wind turbine based on the Darrieus-type rotor which has been introduced by Ponta & Luna Pont (1998). In a traditional Darrieus, the blades rotate around a central vertical axis. In the VGOT, instead, each blade slides over rails mounted on a wagon formed by a reticulated structure supported by standard train bogies (see Figure 2). Each wagon contains its own electrical generation system coupled to the power-wheels and the electricity is collected by a classical third rail system (Ponta et al., 2004; 2001; 2002). With the VGOT design, if we keep constant the velocity of the wagons (i.e. the tangential speed of the blades), we can increase the area swept by the blades (and hence the rated power of the plant) without the low-rotational-speed problems associated with a classical Darrieus rotor of large diameter.

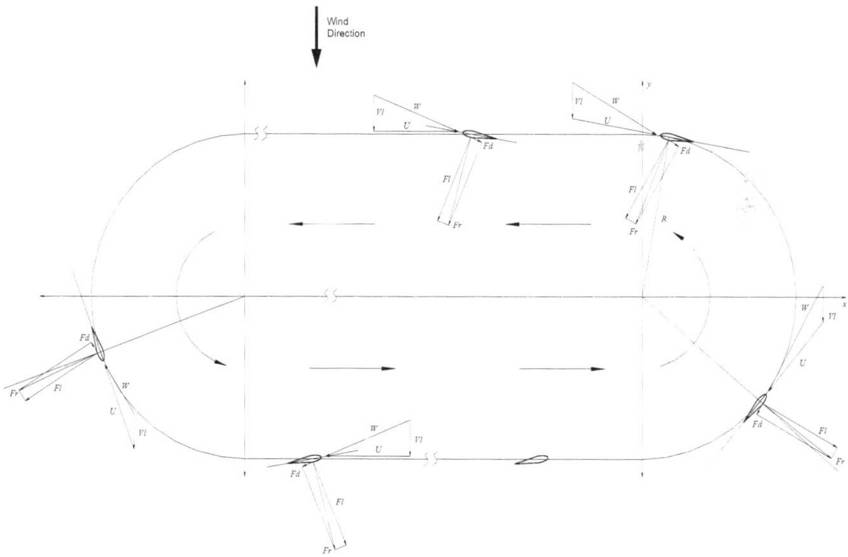

Fig. 3. Schematic view of the path followed by the blade-wagons of the VGOT Darrieus showing the velocity composition of the incoming flow and the aerodynamic forces acting on the blades.

The blade-wagon elements of a VGOT Darrieus, not being solidly affixed to a central axis, could move following a non-circular trajectory (see Figure 3). For certain locations where the compass rose shows a preferential bearing it is possible to optimize the energy-conversion efficiency of the entire plant by increasing the portion of transit perpendicular to the bearing of the incoming wind. Along the perpendicular tracks, the blade generates the higher output-power, while along the portions where the trajectory is in-line with the incoming wind the blade-wagon not only does not produce energy but actually consumes it, taking power from the rest of the plant to keep moving. Thus, extending those portions of the path perpendicular to the wind by the addition of straight tracks the overall energy-conversion efficiency of the plant increases (Ponta et al., 2007). This configuration also allows for increasing the area swept by the blades (which depends on the blade height and the trajectory width across the direction of the incoming wind), with a smaller increment in the path length compared to a circular trajectory. It reduces investment in rails and blades to get the same output-power. On the other hand, along these straight tracks the direction of inflow on the blades would remain practically constant, contributing to aerodynamic and structural stability of the system. This situation is different from a traditional Darrieus where the blades are subjected to an inflow variable in both magnitude and angle of incidence all along the path.

1.1 Technical features of the VGOT Darrieus

In our design, we looked for the use of normal railway material as much as possible. Rails, sleepers, supporting wheels, and other mechanical devices are intended to be chosen among the standards. As well as electrical generators, third-rail electricity-collecting lines, and other electrical devices are intended to be standard too. Besides, other stock pieces of equipment could be adapted to our purposes and some other devices should be designed specially.

When confronting the study of the structural behaviour of the VGOT Darrieus we found several particularities that distinguish it from other three-dimensional reticulated structures. As mentioned above, each blade is mounted on a chassis supported by standard train bogies and covered by a streamlined case designed to decrease aerodynamic drag (see Figure 2(a)). The chassis is formed by a tubular reticulated structure that should be designed to absorb the efforts in the vertical and traverse directions of the railroad due to the aerodynamic loads, the weight of the components and the centrifugal acceleration along the curved tracks. The effects of the link between the blade and the suspension should be considered, together with the added mass effects of the components and the ballast placed to improve the stability of the wagon. To this end, we developed a computer model to simulate the integral behaviour of the reticulated structure, the suspension, and the blade itself. It involves several subroutines including a finite element code to solve the three-dimensional beam reticulated assemble. Due to its complexity, this code is treated comprehensively in a separate paper, Otero & Ponta (2004).

Each blade-wagon contains its own electrical generation system. The generators are connected to the power wheels by a gear transmission. The wheels are placed at the center of the front part of the chassis (see Figure 2(b)) and draw the power by the friction of the wheels on a ground rail. For this device we propose a twin design based upon a pair of oscillating-lever gearboxes coaxial with the electrical generators. This layout not only drives the torque from the wheels to the electrical machines, but also adjusts the rotational speed to perform a suitable mechanical coupling and keeps a uniform distribution of the weight. The electricity generated is collected by a classical third rail system.

Another interesting feature of the VGOT concept is its relatively long cycle time. It allows the use of a blade-positioning control system that operates continuously during the cycle without the fatigue and mechanical-inertia problems associated with variable-geometry attempts in classical Darrieus rotors. With the addition of an orienting device, the blades could set themselves at an optimal angle with a further increment in the energy-conversion efficiency. Several components of this control system could be taken directly from the standard technology used for nacelle-orientation in classical horizontal-axis wind-turbines. The possibility of a blade-positioning control could help remarkably with the issue of controlling the speed of the wagons and the distance between them, as well as acting as an effective emergency-halt system.

There are three alternative strategies for the combined generation and blade-positioning system. First is to use *asynchronous generators linking the blade-wagons electrically*. Here, the speed of each generator is fixed by the frequency of the electric line, and due to the steep torque/rotational-speed characteristic curve of the asynchronous generator, it remains close to the synchronous speed. The control system must keep the distance among the generating elements, and also takes charge of the optimum positioning of the blades to maximize the energy-conversion efficiency. Second is to use *asynchronous generators but linking the blade-wagons mechanically*. In this strategy, the speed of the generator of the whole set of wagons is fixed by the frequency of electric line and it remains close to the synchronous speed. The control system takes charge of only the optimum positioning of the blades to maximize efficiency, working as speed-controller only in case of an emergency. Finally, a third alternative would be to use *synchronous generators*. In this case the speed of the wagons is variable and is adjusted with the position of the blade to optimise the conversion efficiency in each case according to the wind conditions. Here the control system must keep the distance among the blade-wagons. Electricity must be converted to DC before transmission through the common third-rail in order to avoid problems of loss of synchronicity between the individual generators while they are adjusting their respective speeds during normal operation.

The suspension system of the VGOT deserves particular attention. Given the particularities of this application, it was necessary to conceive a non-conventional design. The blade-wagon is subject to fluctuating aerodynamic loads in the direction perpendicular to the rails. This means that the suspension system should absorb not only the vertical oscillating loads due to the weight and the irregularities of the rails, but also the cross-rail oscillations induced by the fluctuating aerodynamics. If these loads were transmitted directly to the bogies and the rails, they would compromise the stability of the blade-wagon. The suspension must also adapt itself to follow the imperfections and misalignments in the rails layout, minimizing their effects on the structure and the components. To this end we propose a design based upon a two-axis deformable parallelogram which actions in the lateral and vertical directions by means of a set of spring/shock-absorber units (see Figure 4). This device is complemented with a special cushioned-pin spherical joint that connects the parallelogram to the bogie. This joint allows limited motion in order to let the bogie adjust itself to the eventual misalignments on the rails while its rubber cushion absorbs the inevitable vibrations produced by the imperfections on the railroad.

The concept of mounting a blade on a wagon with the aim of generating electricity has been proposed before (see for example Fig. 1.6, Sec. 1.1.3 Manwell et al., 2002, among others), but no systematic study of its practical possibilities has ever been attempted. With the VGOT concept, we intend to carry on a systematic study of the actual feasibility of such a design. We can schematize the dynamical response of the whole VGOT Darrieus as a combination of three blocks, each one with its own dynamical transference. Namely: the aerodynamic system, the structural system of the wagon's chassis, and the suspension system. In the following sections we are going to analyze each one of these systems in detail.

2. Aerodynamic study of the VGOT Darrieus

When undertaking the study of the aerodynamic behavior of a VGOT Darrieus, several particularities arise that make it different from other studies in vertical-axis wind-turbine aerodynamics. The aerodynamic loads depend on both the position of the blade-wagon

(a) (b)

Fig. 4. Perspective view of the suspension system and cushioned-pin spherical joint that connects the parallelogram to the bogie.(a). Details of the shock absorption mechanism (b).

along the path and the height from the ground. Even though it is composed of a number of individual blades that control their setting angle independently, this plant works as a whole and it is not possible to analyze the aerodynamic behavior of each blade-wagon separately due to the interaction of the adjacent blades. Strictly speaking in terms of its aerodynamics, the VGOT is a vertical-axis multi-blade Darrieus turbine with adaptive setting-angle control.
Since the 1970s several aerodynamic prediction models have been formulated for Darrieus machines. Modelling of these turbines follows two different schools: the streamtube-model approach based upon equating the forces on the rotor blade to the change in streamwise momentum through the rotor, and the vortex-model approach based upon vortex representation of the blades and their wakes. The streamtube approach needs much less computation time but the vortex approach is more accurate. Nevertheless, for a multi-blade machine like the VGOT, the use of a vortex-model approach may become prohibitive in terms of computational cost. The VGOT's number of blades (about a hundred or so), is substantially bigger than the conventional rotors. Representing all those blades and their wakes by a vortex-lattice arrangement would make the computer code extremely slow. On the other hand, the presence of a number of blades uniformly distributed over the cross section of the rotor helps to close the gap between the streamtube model and physical reality. Streamtube models compute the time-average of the forces acting on the blades on the portion of trajectory across the streamtube. This is the equivalent of assuming that there exist an infinite number of blades of infinitesimal chord length in such a way that the product of the number of blades times the chord length (i.e. the solidity) of the real rotor is preserved. In a conventional rotor, the aerodynamic forces are concentrated on the few blades. Conversely, in the multi-blade VGOT they are more uniformly distributed in the many blades along the path, hence the VGOT layout is closer to the ideal assumed by the streamtube-model approach. Another issue to take into account is that the VGOT's big size implies large radii on the curved portions of the trajectory and so low values of the angular speed (ω). That minimizes the so-called curved-flow effect (which depends on ω) which induces error in the streamtube-model calculation of conventional rotors.
Streamtube models have evolved with time and we can distinguish several categories. The first of them is the *single streamtube model* proposed by Templin (1974). It uses a single

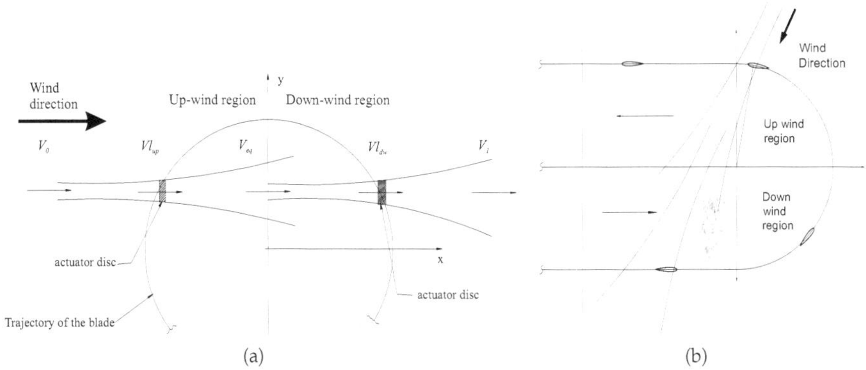

Fig. 5. Double-multiple streamtube model. Schematic view of the tandem arrangement of a double-multiple streamtube (a). Perspective view of the suspension system and cushioned-pin spherical joint that connects the parallelogram to the bogie. Diagram of the dual-scheme streamtube arrangement combining straight and curved tracks. (b).

tube that covers the entire span of the rotor. The rotor's interference is represented by an actuator disk. As the entire rotor is represented by only one tube with one actuator disk, this model predicts a uniform flow for the entire cross section and can not take into account variations of the flow parameters between the upwind and downwind halves of the rotor. To improve the predictive capacities, Strickland (1975) proposed the *multiple streamtube model*, which was also studied by Wilson et al. (1976). It uses an array of adjacent tubes to cover the rotor's span, which takes into account flow variations on the rotor's cross section. A more sophisticated approach called *double-multiple streamtube model* was proposed by Paraschivoiu (1982; 1983a;b). As it uses two actuator disks placed in tandem into each tube of the multiple array (see Figure 5(a)), it can also predict differences between the upwind and downwind halves. Advanced versions of Paraschivoiu's model incorporate dynamic-stall simulation. Streamtube models (specially Paraschivoiu's) can predict effectively the general performance of the rotor (power coefficient vs. tip-speed ratio, power output vs. wind speed, etc.) and their relatively low computational cost makes them a useful practical tool for design. In order to deal with the particular features of the VGOT, we developed a new computational model (see Figure 5(a)) based on the double-multiple streamtube concept that incorporates the capacities of dealing with rotors whose blades follow oval-trajectories at variable setting-angles (Ponta et al., 2007).

As it was mentioned above, the streamtube models are based on the principle of conservation of momentum. The local velocity for each streamtube is computed by equating the average forces on the blades (calculated by the blade element theory using the lift and drag coefficients) to the change in streamwise momentum on the streamtube. From the scheme in Figure 5(a), we have the following relation for the magnitude of the velocity along a double streamtube:

$$V_0 > Vl_{up} > V_{eq} > Vl_{dw} > V_1 \tag{1}$$

where V_0 is the free-stream speed, Vl_{up} and Vl_{dw} the local upstream and downstream velocities, V_{eq} the equilibrium velocity in the coupling of both semitubes, and V_1 is the final velocity in the pseudo-infinite downstream.

The local velocities for each half of the rotor may be written in terms of the velocity at the entrance of each semitube as:

$$Vl_{up} = \xi_{up} V_0, \qquad Vl_{dw} = \xi_{dw} V_{eq} \tag{2}$$

where

$$0 < \xi_{up} < 1, \qquad 0 < \xi_{dw} < 1, \tag{3}$$

are the interference factors for each half of the rotor. From the streamtube theory (see Manwell et al. (2002)), the expression for the local velocity in terms of the inflow and outflow velocities is $Vl = \frac{1}{2}(V_{in} + V_{out})$, then

$$Vl_{up} = \frac{1}{2}\left(V_0 + V_{eq}\right), \tag{4}$$

which gives

$$V_{eq} = \frac{1}{2}\left(2\,\xi_{up} - 1\right) V_0, \tag{5}$$

then

$$\xi_{up} = \frac{Vl_{up}}{V_0}, \qquad \xi_{dw} = \frac{Vl_{dw}}{(2\,\xi_{up} - 1)\,V_0}. \tag{6}$$

The tip speed ratio is defined as $\lambda_0 = V_p / V_0$, where V_p is the tangential velocity of the moving blade. We may define expressions for the local tip speed ratios:

$$\lambda_{up} = \frac{V_p}{Vl_{up}} = \frac{\lambda_0}{\xi_{up}}, \tag{7}$$

$$\lambda_{dw} = \frac{V_p}{Vl_{dw}} = \frac{\lambda_0}{\xi_{dw}(2\,\xi_{up} - 1)}. \tag{8}$$

From the scheme in Figure 6(a), we have that the inflow velocity on the blades may be expressed as,

$$\left(\frac{W}{Vl}\right)^2 = \left(\frac{Vp}{Vl} + \cos\theta\right)^2 + (\sin\theta)^2 = (\lambda + \cos\theta)^2 + (\sin\theta)^2. \tag{9}$$

We can also obtain the angle of the inflow,

$$\tan\gamma = \frac{\sin\theta}{\frac{Vp}{Vl} + \cos\theta} = \frac{\sin\theta}{\lambda + \cos\theta}, \tag{10}$$

which in terms of the local values gives,

$$\left(\frac{W}{Vl_{up}}\right)^2 = (\lambda_{up} + \cos\theta)^2 + (\sin\theta)^2, \tag{11}$$

$$\left(\frac{W}{Vl_{dw}}\right)^2 = (\lambda_{dw} + \cos\theta)^2 + (\sin\theta)^2, \tag{12}$$

and

$$\tan\gamma_{up} = \frac{\sin\theta}{\lambda_{up} + \cos\theta}, \qquad \tan\gamma_{dw} = \frac{\sin\theta}{\lambda_{dw} + \cos\theta}. \tag{13}$$

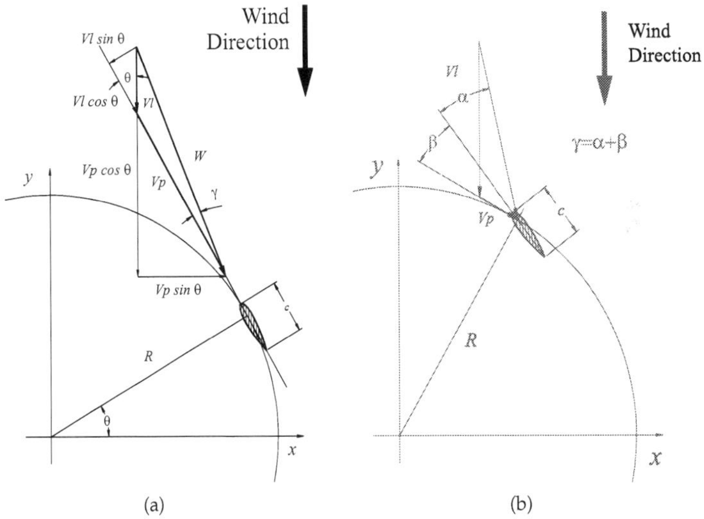

Fig. 6. Comparison of the inflow-velocity diagram for a fixed-geometry Darrieus (a), and a variable-geometry one (b).

For a variable-geometry rotor we have that $\gamma = \alpha + \beta$, where β is the variable setting angle and α is the effective angle of attack on the blade section. Figure 6 illustrates the differences between the velocity diagram for a fixed-geometry Darrieus and a variable-geometry one.

Now, an oval-trajectory Darrieus has two axes of symmetry, thus we have to introduce new concepts which were absent in the standard model for the classical Darrieus rotor. Our model must take into account the angle of the incident wind with respect to the main axis of the oval. We chose to measure the incident-wind angle (φ) from the minor axis of the oval. Earlier versions of our model (Ponta & Seminara, 2000; Seminara & Ponta, 2000) only contemplated four-point compass roses. That is, they just solve the cases where the incidence-wind angle was 0 or 90 degrees (i.e. perpendicular and parallel respectively to the major axis of the oval). It was satisfactory for a first attempt, but it produced too conservative of a prediction of the general performance for winds blowing from a broad range of directions around each side of the mayor axis of the oval. It is of capital importance to determinate the ideal layout of trajectory for a particular compass-rose configuration because the cost-effectiveness of the plant depends on it. Thus, we developed an improved version of the model (Ponta & Seminara, 2001; Seminara & Ponta, 2001) that could deal with those winds blowing from intermediate directions which the previous versions underestimated. To this end, we introduced a dual-scheme streamtube arrangement (see Figure 5(a)) which has the capacity of combining curved and straight tracks for every double-streamtube in order to deal with incident winds blowing from any bearing of the compass.

We also had to introduce a shape coefficient (CF) which defines a proportional relation between the radius of the curved portions of the trajectory and the length of the straight tracks. In this way, we may do changes in the trajectory layout and evaluate the resulting performance under different wind conditions. Another particular aspect that we had to contemplate was the definition of a new system of coordinates to describe the motion of the blades. For an oval

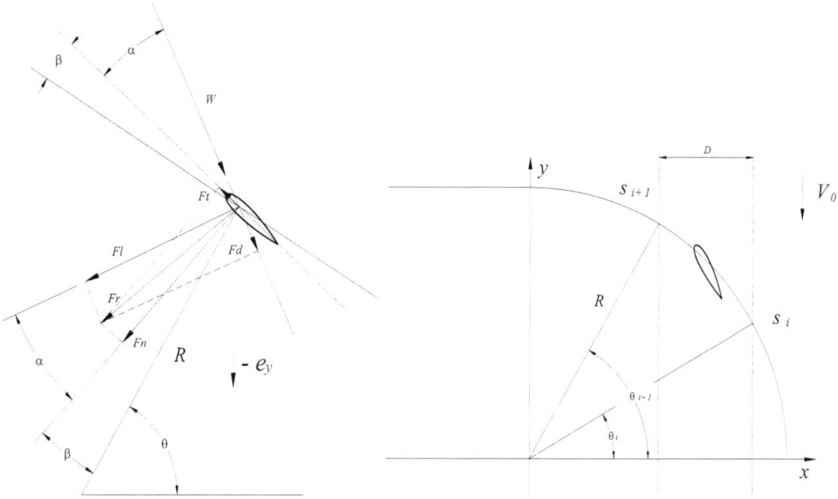

Fig. 7. Diagram of the aerodynamic forces acting on the blades (a). An example of calculation of the width for an upwind tube at $\varphi = 0$ (b).

machine it is no longer possible to simply refer the blade motion to the instant angle along the cycle as in a conventional Darrieus. Hence, we define a new parametric coordinate (s), which describes the motion of the blade and from which its position and the instant angle of its trajectory are computed. Figure 7(b) shows a diagram of the oval trajectory described by the parametric coordinate s.

Figure 7(a) shows a diagram of the aerodynamic forces acting on the blades. From the blade-element theory we calculate that the average value of the aerodynamic force acting on a set of blades passing across a generic streamtube is:

$$F = \frac{N\Delta s}{s_{total}} \frac{1}{\Delta s} \int_{s_i}^{s_{i+1}} \int_{z_j}^{z_{j+1}} \frac{1}{2} \rho W^2 c \left(C_n \sin\left(\theta - \beta + \varphi\right) - \right. \tag{14}$$

$$\left. - C_t \cos\left(\theta - \beta + \varphi\right) + C_{d_o} \cos\left(\theta + \varphi\right) \right) \, dz \, ds \tag{15}$$

where N is the total number of blades, $s_{total} = 2(\pi + CF)R$ is the total length of the trajectory, $\Delta s = s_{i+1} - s_i$ is the length of the portion of trajectory across each tube, z_{j+1} and z_j are the lower and upper heights of the streamtube, C_t and C_n are the Lilenthal's coefficients, and C_{d_o} is the drag coefficient of the wagon (i.e. the aerodynamic drag exerted on the wagon's chassis itself).

On the other hand, from the classical theory of fluid mechanics we have that the axial force acting on a generic streamtube due to the change of momentum is:

$$F = \rho \, A \, Vl \left(V_{in} - V_{out} \right) = 2\rho \, D \, b \, Vl^2 \, \frac{(1 - \xi)}{\xi} \tag{16}$$

where A is the area of the actuator disc placed in the middle of the streamtube, which can be written in terms of its height $b = z_{j+1} - z_j$ and its width D (see an example for an upwind tube at $\varphi = 0$ in Figure 7(b)).
Equating expressions (15) and (16), and after some algebra, we have the general expression of the interference factor for a generic streamtube:

$$\frac{(1-\xi)}{\xi} = \frac{N\, c_m}{8(\pi+CF)\, R\, D} \int_{S_i}^{S_{i+1}} \left(\frac{W}{Vl}\right)^2 (C_n \sin(\theta - \beta + \varphi) - \\ - C_t \cos(\theta - \beta + \varphi) + C_{d_o} \cos(\theta + \varphi))\, ds, \tag{17}$$

where c_m is the average value of the blade-chord length along the height of the streamtube. Combining expressions (8), (11), (12) and (17), we have the following expression for the upstream interference factor:

$$\frac{(1-\xi_{up})}{\xi_{up}} = \frac{N\, c_m}{8(\pi+CF) R\, D} \int_{S_i}^{S_{i+1}} \left(\left(\frac{\lambda_0}{\xi_{up}} + \cos(\theta + \varphi) \right)^2 + (\sin(\theta + \varphi))^2 \right) \\ \left(C_n \sin(\theta - \beta + \varphi) - C_t \cos(\theta - \beta + \varphi) + C_{d_o} \cos(\theta + \varphi) \right) ds, \tag{18}$$

and for the downstream one:

$$\frac{(1-\xi_{dw})}{\xi_{dw}} = \frac{N\, c_m}{8(\pi+CF) R\, D} \int_{S_i}^{S_{i+1}} \left(\left(\frac{\lambda_0}{\xi_{dw}(2\xi_{up}-1)} + \cos(\theta + \varphi) \right)^2 + (\sin(\theta + \varphi))^2 \right) \\ \left(C_n \sin(\theta - \beta + \varphi) - C_t \cos(\theta - \beta + \varphi) + C_{d_o} \cos(\theta + \varphi) \right) ds, \tag{19}$$

We may compute the local interference factors solving 18 and 19 numerically for each streamtube. We use a combination of subroutines: cubic-spline for interpolation of the aerodynamic coefficients, Newton-Cotes for integration and Newton-Rapson to solve the nonlinear functions. From the interference factors we computed the instantaneous value of the upstream and downstream local velocities. Then, the power coefficient may be calculated as,

$$Cp = \frac{N \lambda_0}{2(\pi+CF)(2+CF\cos\varphi)R^2} \sum_{j=1}^{nth} c_m{}^j \sum_{i=1}^{nt} \int_{S_i}^{S_{i+1}} \left(\left(\lambda_0 + \xi_{up}{}^{ij} \cos(\theta + \varphi) \right)^2 + \\ + \left(\xi_{up}{}^{ij} \sin(\theta + \varphi) \right)^2 + \left(\lambda_0 + \xi_{dw}{}^{ij} \left(2\xi_{up}{}^{ij} - 1 \right) \cos(\theta + \varphi) \right)^2 + \\ + \left(\xi_{dw}{}^{ij} \left(2\xi_{up}{}^{ij} - 1 \right) \sin(\theta + \varphi) \right)^2 \right) \left(C_n \sin\beta + C_t \cos\beta - C_{d_o} \right) ds, \tag{20}$$

where nth is the total number of streamtubes in height and nt the total number of streamtubes in width.

2.1 Especial parameters

Due to the especial geometric characteristics of the VGOT Darrieus, it was necessary to propose three new non-dimensional parameters to quantify its performance under different wind conditions: the equivalent power coefficient, the equivalent solidity coefficient and the trajectory efficiency. The two former are the corresponding extensions to the VGOT case of the classical solidity and power coefficients of standard turbines. For the equivalent power coefficient we have

$$Cp_{eq} = \sum_{j=1}^{npcr} f_{\%}{}^{j}\, Cp_j, \tag{21}$$

where $f_{\%}{}^{j}$ is the frequency of occurrence of the wind blowing from the jth bearing in the compass rose, Cp_j is the power coefficient for an incident-wind angle corresponding to the jth bearing, and $npcr$ is the total number of bearings in the wind compass rose. For the equivalent solidity coefficient we have

$$\sigma_{eq} = \frac{N\, c_m}{\left(1 + \frac{CF}{\pi}\right) R}. \tag{22}$$

Note that expression (22) converges to its classical counterpart for conventional Darrieus rotors when $CF \to 0$ (i.e. for a circular-trajectory layout).
The third parameter is a completely new conception exclusive for VGOT machines. The trajectory efficiency is an indicator of the economic efficiency of a particular configuration (i.e. a trajectory layout). It relates the total efficiency of energy conversion with the investment on rails and blades. The former is given by the product of the frequency of occurrence of a certain bearing, times the correspondent power coefficient, times the width of the respective swept area, and the latter is proportional to the total length of the path. The expression for the trajectory efficiency is

$$\mathrm{Ef} = \sum_{j=1}^{npcr} \frac{f_{\%}{}^{j}\, Cp_j\, (2 + CF\cos(\varphi_j))}{2(\pi + CF)}. \tag{23}$$

2.2 Numerical Results

In this section we include some numerical results of the application of our model. We first tested different configurations of oval-trajectory rotors with a fixed trajectory layout of $CF = 8$. Figure 8(a) shows the power-coefficient curves at $\varphi = 0$ for different values of equivalent solidity obtained by changing the number of blades.
Next, we tested several rotor configurations changing CF (i.e. the trajectory layout) and the number of blades in such a way of keeping constant the equivalent solidity. Figure 8(b) shows the corresponding power-coefficient curves. We repeated the test for both extreme cases of incident-wind angle $\varphi = 0$ and $\varphi = 90$ (i.e. when the wind blows, perpendicular and parallel to the mayor axis of the oval trajectory).
To study the aptitude of a particular shape under specific wind conditions, we have computed the equivalent power coefficient and the trajectory efficiency for different compass roses. Three artificially-constructed wind conditions that illustrate the extreme cases at which a VGOT Darrieus with its mayor axis oriented in a North-South direction could be subjected. Compass Rose 1 corresponds to winds with a preferential bearing aligned with the minor axis of the oval, Compass Rose 3 to winds with no preferential bearing, and Compass Rose 4 to winds with a preferential bearing aligned with the mayor axis. This series is completed with Compass Rose 2, which corresponds to the real case of the region of Comodoro Rivadavia in Patagonia, which has a strong west-east directionality.
Figures 9(a) and 9(b) show the values of equivalent power coefficient and the trajectory efficiency for a series of VGOT rotors of different shape. All the rotors have a fixed solidity $\sigma_{eq} = 0.6767$ (which is a typical value for this kind of machine), and work at a tip speed ratio $\lambda = 2.2$ which gives the optimum value for that solidity.

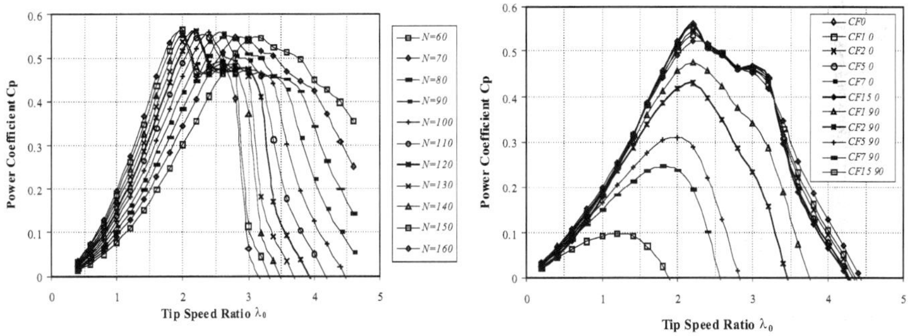

(a) (b)

Fig. 8. Power-coefficient curves at $\varphi = 0$ for VGOT rotors with different number of blades and $CF = 8$ (a). Power-coefficient curves at $\varphi = 0$ and $\varphi = 90$ for VGOT rotors with different trajectory layout but constant solidity (b).

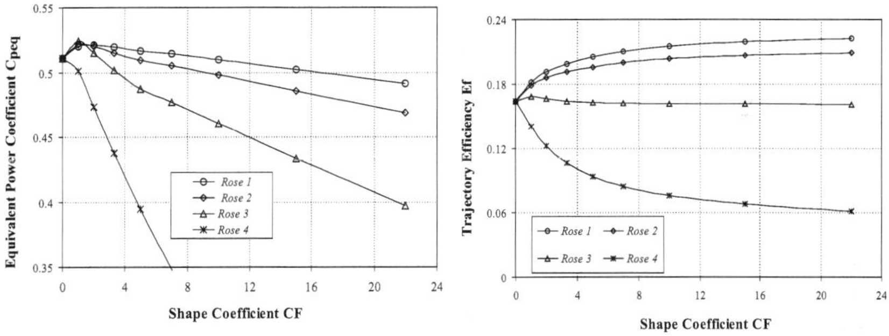

(a) (b)

Fig. 9. Power-coefficient curves for a series of VGOT rotors of different shape under four representative wind conditions (a). Trajectory-efficiency curves for a series of VGOT rotors of different shape under four representative wind conditions (b).

Finally, we computed the aerodynamic loads which were applied to the blade as a distributed load per unit-length. These loads varied in function of both the wagon position along the path and the height from the ground, Figures 10(a) and 10(b) show the aerodynamic load per unit-length in the chord-wise and chord-normal directions (f_{chws}, f_{chnor}) for different heights along the blade in function of the parametric position along the path (i.e. s goes from 0 to 1 to complete the cycle). These data are used as input for a forthcoming study of the structural behavior of the blade-wagon.

3. The structural problem

For the structural study of the blade-wagon, we used a linear analysis approach (i.e. small displacements, small deformations and linear-elastic homogeneous material were assumed). This analysis will be very precise in normal operational conditions at rated power where real

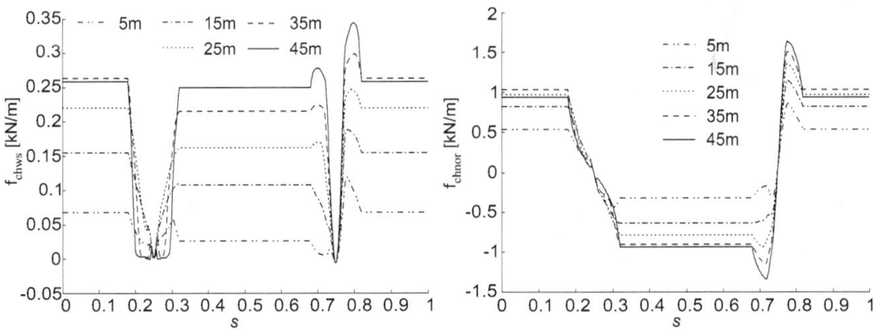

(a) (b)

Fig. 10. Aerodynamic load for different heights along the blade span: (a) in the chord-wise direction and (b) in the chord-normal direction.

work conditions fulfil the proposed hypothesis. This linear analysis provides an essential tool for project purposes and serves as the first step for a future study on the non-linear behaviours that are likely to appear when the plant is working at extreme operational conditions.

We used beams and bars to represent the reticulated structure of the wagon, the blade and the suspension. The blade was modelled by 50 variable-section beam elements; the blade-section chord length varies from 8 meters at the bottom to 4 meters at the tip. Each tubular beam of the three-dimensional reticulated structure of the wagon was modelled by one beam element of constant section. Depending on which portion of the structure the beam belonged to, the exterior and interior diameters differ according to design. The details of the suspension system mechanism are going to be studied in the following section. For the purpose the structural study, the behaviour of the suspension system mechanism can be satisfactorily modelled by an assembly of four two-node bar elements. One assembly was located at each one of the four ends of the wagon in place of the actual suspension mechanism. This helps us determine the overall stiffness required from the suspension system in order to keep the stability of the wagon and the aerodynamic configuration. Another mechanism that should be modelled to study the whole structural group of the generating wagon is the blade attachment. This device should link the blade bottom with the reticulated structure and also include the positioning mechanism. It was modelled by beam elements of extremely high stiffness which is quite realistic considering that stiffness is a characteristic inherent to the functionality of this device.

The structure of the VGOT-Darrieus is mainly subject to loads of aerodynamic origin. As mentioned above, aerodynamic loads were calculated by means of a Double-Multiple Streamtube Model and were applied to the blade as a distributed load per unit-length. These loads varied in function of both the wagon position along the path and the height from the ground, figures 10(a) and 10(b) show the aerodynamic load per unit-length in the chord-wise and chord-normal directions (f_{chws}, f_{chnor}) for different heights along the blade in function of the parametric position along the path s. The distributed loads acting on the blade are obtained by projecting f_{chws} and f_{chnor} onto a global system of coordinates aligned with the rails. We also considered loads due to the weight of the chassis, the blade, and the mechanical devices, and also inertial loads due to the centrifugal acceleration. The geometrical boundary conditions apply onto the suspension support points where displacement is restricted in vertical and transverse directions. Being this i a support bond, bond reactions act only in one sense (i.e. pressing the wheels against the rails) and it was necessary to verify that contact was always preserved. In those cases where this condition was not fulfilled, the ballast was modified to increase the wagon's stability. To take into account the effects of eventual imperfections and misalignment of the rails due to ageing, we introduced randomly-simulated displacements of the points of the structure where the wheels are attached. Displacements in the vertical and transverse directions may be assumed to have a normal statistical distribution with well-known mean and standard deviation. The combination of random displacements that produced the highest stress at each position along the path was selected for evaluation of the effect of rail imperfections.

In order to characterize the structural behaviour of the VGOT Darrieus Rotor, we defined a set of representative parameters: The *Von Misses-yielding stress ratio* (σ_{VM}/σ_y), which indicates the load state, measured in six witness beams including the beams where the maximum and minimum σ_{VM}/σ_y were observed. *Blade-tip transverse displacement* (Δ_{trav}), computed and decomposed in three components: one due to the action of the suspension system, a second due to the deformation of the chassis, and the third due to the lateral bending of the blade. This parameter is useful to ponder the effect of the structure/suspension response on the setting of the blade and check that the aerodynamic configuration is not substantially altered. Finally, the *Blade-tip torsion angle* (ϕ), computed in order to check that the angle of attack of the inflow onto the blade (hence, the aerodynamic load) is not substantially altered.

3.1 Finite element implementation

As mentioned above, we used beams and bars to represent the reticulated structure of wagon chassis, blade, and suspension. We used 3-node isoparametric finite elements with quadratic interpolation assuming Timoshenko beam hypothesis to deal with shear and bending. Torsional and axial effects were included following the classical theory for bars.

The basic expression for the Hellinger-Reissner Functional (see Bathe, 1996, section 4.4.2) leads to a mixed formulation with displacement and strain as independent variables. For the particular case of Timoshenko beams with linear elastic isotropic material, we have the strain tensor $\boldsymbol{\varepsilon} = [\varepsilon_{zz}\ \gamma_{zx}^{AS}\ \gamma_{yz}^{AS}]^T$, where coordinate z is aligned with the axis of the beam. γ_{yz}^{AS} and γ_{zx}^{AS} represent the distortion due to shear effects in yz and zx planes (the superscript AS denotes that the distortions due to shear will be "assumed" with linear variation along the element length and constant on each cross section). The actual strains given by the strain-displacement relations are $\partial_{\varepsilon}\mathbf{u} = [\varepsilon_{zz}\ \gamma_{zx}\ \gamma_{yz}]^T$, with $\gamma_{zx} = \frac{du_1}{dz} - \theta_2$ and $\gamma_{yz} = \frac{du_2}{dz} + \theta_1$, where θ_1 and θ_2 are the angles of rotation of the cross section of the beam in the yz and zx planes respectively, and u_1 and u_2 are the displacements in x and y. $\boldsymbol{\theta} = [\theta_1\ \theta_2\ \theta_3]$ and $\mathbf{u} = [u_1\ u_2\ u_3]$ together form the so-called *generalized displacements* which are the primitive unknowns to be interpolated quadratically. The stress-strain relations involve the Young and shear moduli of the material, E and G respectively. Then, the expression for the Hellinger-Reissner functional reduces to

$$\Pi^{**} = \int_V \left(\frac{1}{2}\varepsilon_{zz}\, E\, \varepsilon_{zz} - \frac{1}{2}\gamma_{zx}^{AS}\, G\, \gamma_{zx}^{AS} - \frac{1}{2}\gamma_{yz}^{AS}\, G\, \gamma_{yz}^{AS} + \gamma_{zx}^{AS}\, G\, \gamma_{zx} + \gamma_{yz}^{AS}\, G\, \gamma_{yz} \right) dV - \int_V \left(\mathbf{u}^T \mathbf{f}^B \right) dV - \int_{S_f} \left(\mathbf{u}^{S_f} \right)^T \mathbf{f}^{S_f}\, dS. \tag{24}$$

Under the linear hypothesis we started from, it is possible to add to (24) the contribution of the axial and torsional loads, arriving to the final expression

$$\Pi^{**} = \frac{E}{2}\int_L \Bigg[\underbrace{I_x\left(\frac{d\theta_1}{dz}\right)^2 + I_y\left(\frac{d\theta_2}{dz}\right)^2}_{(i)} + \underbrace{A\left(\frac{du_3}{dz}\right)^2}_{(ii)}\Bigg]\,\mathrm{dz}$$
$$+\frac{G}{2}\int_L \Bigg[\underbrace{-A\left(k_x\left(\gamma_{zx}^{AS}\right)^2 + k_y\left(\gamma_{yz}^{AS}\right)^2\right)}_{(iii)}$$
$$\underbrace{+\,2\,A\left(k_x\,\gamma_{zx}^{AS}\left(\frac{du_1}{dz}-\theta_2\right) + k_y\,\gamma_{yz}^{AS}\left(\frac{du_2}{dz}+\theta_1\right)\right)}_{(iii)} + \underbrace{I_p\left(\frac{d\theta_3}{dz}\right)^2}_{(iv)}\Bigg]\,\mathrm{dz}$$
$$\underbrace{-\int_L \mathbf{u}^T\mathbf{p}\,\mathrm{dz} - \int_L \boldsymbol{\theta}^T\mathbf{m}\,\mathrm{dz} - \sum_i\left(\mathbf{u}^i\right)^T\mathbf{F}^i - \sum_j\left(\boldsymbol{\theta}^j\right)^T\mathbf{M}^j}_{(v)} \tag{25}$$

where I_x, I_y, I_p and A are respectively the inertia and polar moments and the area of the section. k_x and k_y are the shear correction factors (in this case we assumed $k_x = k_y = 1$); $\mathbf{p}$, $\mathbf{m}$, $\mathbf{F^i}$ and $\mathbf{M^j}$ are respectively the distributed and concentrated loads and moments, and L is the length of the beam. Terms in (25) marked as (i) are associated to bending, term (ii) is associated to axial loads, those marked as (iii) are associated to shear, term (iv) is associated to torsion, and the last terms marked as (v) correspond to the external loads and moments.

We discretized the generalized displacements using 1D isoparametric 3-node-element interpolation (see Bathe (1996); Kwon & Bang (1997)). The interpolated displacements and rotations in the jth direction in terms of displacements u_j^i and rotations θ_j^i and the interpolation functions $h_i(r)$ corresponding to node i are $u_j(r) = h_i(r)\;u_j^i$ and $\theta_j(r) = h_i(r)\;\theta_j^i$, where the repeated index indicates summation on the 3 nodes and r is the intrinsic coordinate along the beam element. For the displacement and rotation derivatives with respect to the local coordinate z, we have $\frac{du_j}{dz}(r) = J^{-1}\frac{dh_i}{dr}u_j^i$ and $\frac{d\theta_j}{dz}(r) = J^{-1}\frac{dh_i}{dr}\theta_j^i$.

These magnitudes are then re-written in matrix form as $u_j(r) = \mathbf{H}_{u_j}\;\hat{\mathbf{u}}$, $\frac{du_j}{dz}(r) = \mathbf{B}_{u_j}\;\hat{\mathbf{u}}$, $\theta_j(r) = \mathbf{H}_{\theta_j}\;\hat{\mathbf{u}}$ and $\frac{d\theta_j}{dz}(r) = \mathbf{B}_{\theta_j}\;\hat{\mathbf{u}}$, where $\hat{\mathbf{u}}$ is the array of nodal values of the generalized displacements, and $\mathbf{H}$ and $\mathbf{B}$ are the arrays of interpolation functions and their derivatives in matrix form respectively.

We used 3-point Gaussian integration for the terms interpolated by quadratic functions Bathe (1996); Burden & Faires (1998).

In order to avoid locking problems we used discontinuous linear interpolation for γ_{zx}^{AS} and γ_{yz}^{AS} with 2-point Gaussian integration and condensation at element level. A detailed description of this technique can be found in Bathe (1996). Distortion interpolation can be expressed in matrix form as $\gamma_{zx}^{AS} = \mathbf{H}_{\gamma_{zx}}\gamma^{AS}$ and $\gamma_{yz}^{AS} = \mathbf{H}_{\gamma_{yz}}\gamma^{AS}$, where γ^{AS} is the array with the values for the distortion at the integration points while $\mathbf{H}_{\gamma_{zx}}$ and $\mathbf{H}_{\gamma_{yz}}$ are the corresponding arrays of interpolation functions.

Substituting the variables in (25) by their discretized counterparts and invoking the stationarity of the functional, we have

$$\begin{bmatrix} \mathbf{K}_{uu} & \mathbf{K}_{u\gamma} \\ \mathbf{K}_{\gamma u} & \mathbf{K}_{\gamma\gamma} \end{bmatrix} \begin{bmatrix} \hat{\mathbf{u}} \\ \gamma^{AS} \end{bmatrix} = \begin{bmatrix} \mathbf{P} \\ \mathbf{0} \end{bmatrix} \tag{26}$$

where

$$\mathbf{K}_{uu} = E \int_{-1}^{1} \left(I_x \, \mathbf{B}_{\theta_1}^T \, \mathbf{B}_{\theta_1} + I_y \, \mathbf{B}_{\theta_2}^T \, \mathbf{B}_{\theta_2} + A \, \mathbf{B}_{u_3}^T \, \mathbf{B}_{u_3} \right) |J| \, dr$$

$$+ G \int_{-1}^{1} I_p \, \mathbf{B}_{\theta_3}^T \, \mathbf{B}_{\theta_3} \, |J| \, dr,$$

$$\mathbf{K}_{\gamma u} = \mathbf{K}_{u\gamma}^T = G \int_{-1}^{1} A \left[\mathbf{H}_{\gamma_{zx}}^T \, (\mathbf{B}_{u_1} - \mathbf{H}_{\theta_2}) + \mathbf{H}_{\gamma_{yz}}^T \, (\mathbf{B}_{u_2} + \mathbf{H}_{\theta_1}) \right] |J| \, dr,$$

$$\mathbf{K}_{\gamma\gamma} = - G \int_{-1}^{1} A \left(\mathbf{H}_{\gamma_{zx}}^T \, \mathbf{H}_{\gamma_{zx}} + \mathbf{H}_{\gamma_{yz}}^T \, \mathbf{H}_{\gamma_{yz}} \right) |J| \, dr,$$

$$\mathbf{P} = \int_{-1}^{1} \left[(\mathbf{H}_{u_1} + \mathbf{H}_{u_2} + \mathbf{H}_{u_3})^T \, \widehat{\mathbf{p}} + \left(\mathbf{H}_{\theta_1} + \mathbf{H}_{\theta_2} + \mathbf{H}_{\theta_3} \right)^T \widehat{\mathbf{m}} \right] |J| \, dr$$

$$+ \sum_i \widehat{\mathbf{F}}^i + \sum_j \widehat{\mathbf{M}}^j,$$

$\widehat{\mathbf{p}}$, $\widehat{\mathbf{m}}$, $\widehat{\mathbf{F}}^i$ and $\widehat{\mathbf{M}}^j$ are the corresponding nodal values for the distributed and concentrated loads and moments.

The degrees of freedom associated with γ^{AS} can be condensed at element level. From the second row of (26), we have $\gamma^{AS} = -\mathbf{K}_{\gamma\gamma}^{-1} \, \mathbf{K}_{\gamma u} \, \hat{\mathbf{u}}$, and substituting for γ^{AS} in the first row of (26), it yields

$$\underbrace{\left(\mathbf{K}_{uu} - \mathbf{K}_{u\gamma} \, \mathbf{K}_{\gamma\gamma}^{-1} \, \mathbf{K}_{\gamma u} \right)}_{\mathbf{K}_{el}} \, \hat{\mathbf{u}} = \mathbf{P}. \tag{27}$$

Now, matrix $\mathbf{K}_{el}$ and array $\mathbf{P}$ are transformed from the local coordinates of the beam element to the global coordinates of the structure and assembled into a global matrix $\tilde{\mathbf{K}}$ and load array $\tilde{\mathbf{P}}$ by the standard procedure used in finite-element theory, arriving to the final system

$$\tilde{\mathbf{K}} \, \tilde{\mathbf{U}} = \tilde{\mathbf{P}}, \tag{28}$$

where $\tilde{\mathbf{U}}$ is the global array of nodal values of the generalized displacements. We then follow the classical procedure to impose the geometrical boundary conditions (see Bathe (1996)) and solve the system of equations to obtain the generalized displacements.

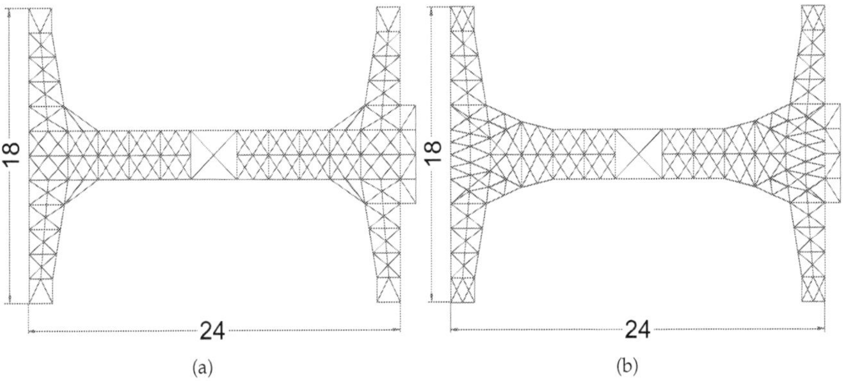

Fig. 11. Top views of (a) Configuration A and (b) Configuration B. Geometrical coordinates given in meters.

3.2 Numerical Results

After a preliminary study on the basic structural outline (Otero & Ponta, 2002; Ponta & Otero, 2002), we systematically applied our computational code simulating three different configurations for the complete structure of the wagon (chassis, blade and suspension). The combined structural response of each configuration for different positions along the path was analyzed and compared, and the design evolved to improve its performance. We assumed for the three configurations that the curved tracks have a 350-meters radius. We started from Configuration A (see figure 11(a)) deriving the other two in order to improve different aspects of the structural behaviour.

One of the aspects to improve was the stress state of the beams at different zones of the structure. Figure 12(a), shows σ_{VM}/σ_y for 6 witness beams (including the beams which show the maximum and minimum). Maximum σ_{VM}/σ_y was about 60 %, several beams exceed 30 % on some point along the path while there were many that did not even reach 20 % at any point. This dispersion indicates an inadequate distribution of material for the different portions of the chassis and a redesign of the structure was recommendable.

Figure 11(b) shows the modified design of Configuration B. By redesigning the thickness of the beams according to the results obtained for the stress distribution in Configuration A, we achieved a substantial reduction in the maximum stress without increasing the total weight. Figure 12(b) shows a comparison of the maximum stress for the three configurations studied; the reduction in maximum stress between configurations A and B is clearly depicted.

A second aspect to consider during the redesign was the reduction of the transverse displacement shown by the blade tip in Configuration A. We started by analyzing the contribution of each major structural component (the blade, the suspension and the chassis) to the total transverse displacement of the blade tip Δ_{trav}. Figure 13(a) shows the total value of Δ_{trav}, together with the contribution of the three major structural components. We reduced the contribution of the suspension by modifying the stiffness of the springs in the front and back wheels. The deflection of the blade was reduced by redesigning the upper blade structure in order to reduce the top-mass affected by the centrifugal force. To reduce the chassis contribution, we reinforced the zones of the structure where the transverse arms are attached to the longitudinal body of the chassis, which could be noticed by comparing figures 11(a)

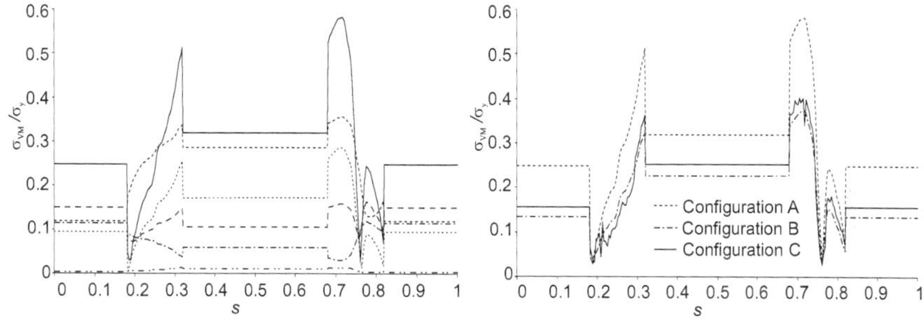

(a) (b)

Fig. 12. Von Misses yielding stress ratio in function of the parametric position along the path. (a) Configuration A. (b) Comparison of the maximum stress for Configurations A, B, and C. Data for Configuration C also include the oscillating stress component due to rail imperfections.

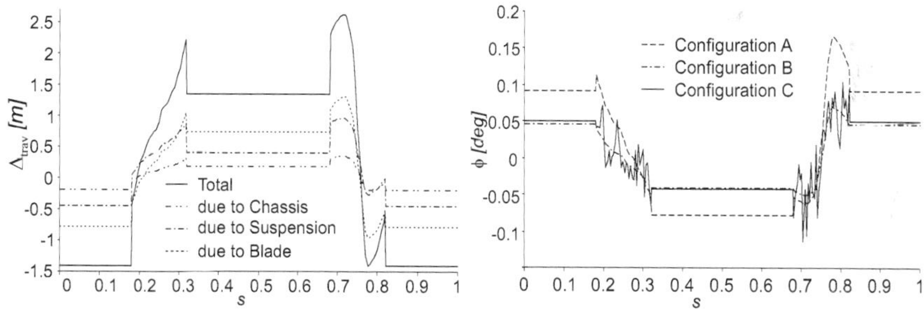

(a) (b)

Fig. 13. Blade-tip transverse displacement in function of the parametric position along the path. Configuration A (a). Comparison of blade-tip torsion angle in function of the parametric position along the path for configurations A, B and C. Data for Configuration C include the oscillatory effect induced by rail imperfections (b).

and 11(b). The latter modification substantially increased the torsional stiffness of the chassis. This reduction of the wagon's torsion translates into a reduced roll, and then decreases the chassis contribution to blade-tip displacement. The combined effect of these modifications to the three major structural components reduced the total transverse blade-tip displacement by 20%.

The global behaviour shown by Configuration B was satisfactory, but we were looking for a more compact design for the chassis in order to reduce the investment in materials and especially the cost of civil works. To this end, we reduced the distance between the railroads in 3 meters by cutting the 1.5-meter stubs that connect the transverse arms with the suspensions on each side of Configuration B. Thus, we arrived to Configuration C, which combines the satisfactory global behaviour of its predecessor with compactness of construction, and constitutes a somehow definitive design. At this point, we introduced in our simulations the effect of rail imperfections. Figure 12(b) shows the maximum stress along the path for Configuration C when rail imperfections are present. It is clear that stress fluctuations induced by the imperfections are relatively small compared with the overall stress, without load peaks that may compromise structural integrity by fatigue.

Another important aspect substantially improved by the new configuration was the reduction of the blade-tip torsion angle, which can be seen in figure 13(b) where a comparison of this parameter among the designs is shown. The absolute value of ϕ for Configuration B is smaller than 0.07 degrees; and when the effect of rail imperfections is included, this value does not exceed 0.12 degrees. This result proved to be important because, before starting this study, we considered the possibility of feeding-back the torsion angles at each point along the blade to recalculate aerodynamic forces. Now, in view of the fact that the fluctuations in blade torsion angle are very small in terms of the optimum angle of attack (which is the angle of attack in normal operation), we may discard the effects of blade torsion in future calculation of the aerodynamic forces.

4. Analysis of the dynamical response of the suspension system

In this section, we shall focus on the problem of the suspension system, considering its interaction with the other two systems according to the following hypothesis: One, the reticulated structure of the wagon chassis acts as a rigid body (i.e. its stiffness is high compared with the suspension's); two, the link between the bearing of the blade and the wagon is rigid; three, the mass of the springs and dampers is negligible compared to the mass of the whole blade-wagon.

In order to compute the inertia tensor and mass of the blade-wagon, we first generate a three-dimensional mesh of isoparametric finite elements, each one representing one beam of the reticulated structure of the chassis. The same meshing code was used to discretize the blade as a series of variable-section beam finite elements. This provides the necessary data to obtain the inertia tensor and the loads for the chassis and the blade by the classical process of numerical integration used in the finite element method. We did not solve a finite element problem, but used the finite-element interpolation functions and integration techniques. By rotating and relocating each single element in the structure, we were able to calculate its inertia tensor and applied load, and referred them to a global coordinate system. We chose the point where the blade is linked to the chassis as the reference point because of its very high stiffness compared to the rest of the structure. To obtain the mass of the blade-wagon, we simply integrated the volume of each element applying the corresponding density according to a materials database.

Given a mass system in which the position of its particles is referred to a local coordinate system (x_1, x_2, x_3), the inertia tensor for the mass system, referred to the origin, is represented by the matrix:

$$I = \begin{bmatrix} I_{11} & I_{12} & I_{13} \\ I_{21} & I_{22} & I_{23} \\ I_{31} & I_{32} & I_{33} \end{bmatrix}. \tag{29}$$

The elements on the diagonal I_{11}, I_{22}, I_{33} are the axial moments of inertia referred to the x_1, x_2, x_3 axes. The elements outside of the diagonal are called the centrifugal moments of inertia with respect to each pair of axes. Being the inertia tensor a symmetric matrix, we have: $I_{12}=I_{21}$, $I_{23}=I_{32}$, $I_{13}=I_{31}$. The expresion for each element is:

$$
\begin{aligned}
I_{11} &= \int \left(x_2^2 + x_3^2\right) dm & I_{12} &= \int \left(x_1\, x_2\right) dm \\
I_{22} &= \int \left(x_1^2 + x_3^2\right) dm & I_{23} &= \int \left(x_2\, x_3\right) dm \\
I_{33} &= \int \left(x_1^2 + x_2^2\right) dm & I_{13} &= \int \left(x_1\, x_3\right) dm
\end{aligned}
\tag{30}
$$

Alternatively, each element can be also expressed by:

$$
I_{ij} = \int \left[\left(\sum_{k=1}^{3} x_k^2 \right) \delta_{ij} - |x_i x_j| \right] dm \tag{31}
$$

where δ_{ij} is a scalar of the form

$$
\delta_{ij} = 1 \quad \text{if } i = j, \qquad \delta_{ij} = 0 \quad \text{if } i \neq j.
$$

The evaluation of these integrals in the inertia tensor and the mass of the blade-wagon was done by using the interpolation functions and numerical integration rule for isoparametric three-node one-dimensional finite elements. A detailed description of these functions and the isoparametric technique may be found in Bathe (1996). Once the inertia tensor of each element has been calculated on its local coordinate system, we relocate it by using Steiner's Theorem with the blade-chassis link as the reference point. Once we have the element in its corresponding location in the structure and referred to a global coordinate system, we obtain the inertia tensor of the complete structure by applying the same process to the rest of the elements.

The new components on the inertia tensor can be defined as:

$$
\begin{aligned}
I_{11st} &= I_{11} + m\left({y'}^2 + {z'}^2\right) & I_{12st} &= I_{12} + m\left(x'y'\right) \\
I_{22st} &= I_{22} + m\left({x'}^2 + {z'}^2\right) & I_{23st} &= I_{23} + m\left(y'z'\right) \\
I_{33st} &= I_{33} + m\left({x'}^2 + {y'}^2\right) & I_{13st} &= I_{13} + m\left(x'z'\right)
\end{aligned}
\tag{32}
$$

$$
x' = |x_{global} - x_{ref}| \qquad y' = |y_{global} - y_{ref}| \qquad z' = |z_{global} - z_{ref}|
$$

From the aerodynamic study, described in section 2, we have data about the optimum value for the blade pitch angle for a discrete number of positions along the trajectory. These data were used to compute the exact value of the inertia tensor in those positions, which depends on the relative angle between the blade and the chassis. Finally, we obtain a continuous expression for each component of the inertia tensor along the trajectory using a spline interpolation between those discreet positions.

We compute a solution for the time-dependent dynamics by solving the system of ordinary differential equations (ODE) for the blade-wagon as a body. Once we have the solid body modeled with its correspondent inertia tensor and the loads, we proceeded to solve the conservation of the linear momentum in axis y and z, and the angular momentum in the three dimensions. It involved the solution of an ODE system of five equations, which gives us

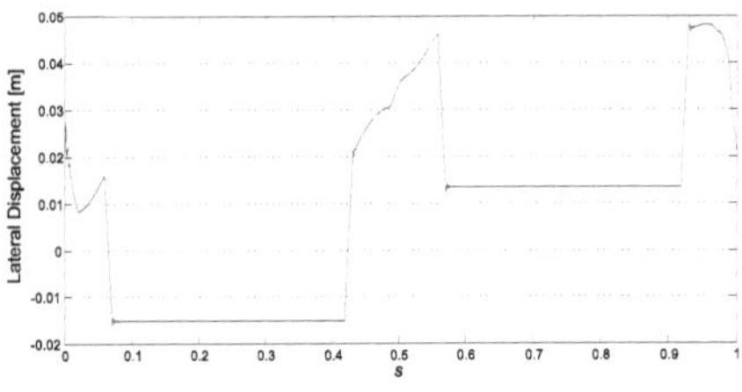

(a)

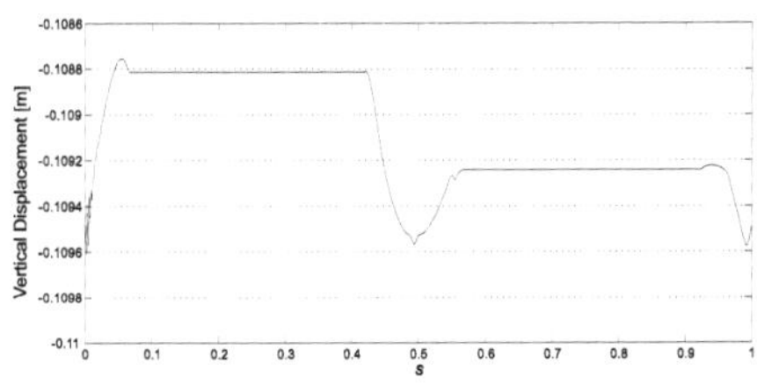

(b)

Fig. 14. Displacements of the blade wagon at the point of the blade-chassis link: lateral direction (a), vertical direction (b).

the displacements and rotations of the blade wagon referred to the blade-chassis link (i.e. the *reference point*). Once we have these displacements and rotations, computing the loads on the springs and shock absorbers in the suspension system is straightforward. Finally, these will give us the crucial information about the normal and tangential loads exerted on the rails as for each position along the trajectory.

The original ODE system of five second-order equations was first transformed into an equivalent system of ten first-order equations using the change-of-variables technique. Then, the transformed system was solved by a multivalue variable-order predictor-corrector solver with adaptive stepsize control. The whole set of three subroutines was written in the MATLAB language.

4.1 Numerical results

In this section we show the numerical results of a series of tests we conducted on a typical VGOT configuration (Ponta & Lago, 2008). Vertical spring stiffness is $4.9370 * 10^6$ N/m, lateral spring stiffness is $3.5737 * 10^6$ N/m, and shock-absorber stiffness in both vertical and horizontal directions is 10^5 N/m. The pre-load of the lateral springs is 250 kN. We included a 110 kN ballast at the rear of the chassis to compensate the pitch moment induced by the aerodynamic *pushing* force applied at the center of pressure of the blade. This permanently acting pitch moment is inherent to the normal operation of the blade-wagon and the use of the ballast is a simple and practical solution to compensate it. The blade height is 50 m; its airfoil section has a chord length of 8 m at the base and 4 m at the top. The thickness of the fiberglass composite shell that forms the blade structure is 0.1 m at the base and 0.01 m at the top. We used a mesh of 1453 finite elements to model the reticulated structure of the chassis, and a mesh of 50 beam elements of variable section to discretize the blade.

Figure 14 shows the displacements of the blade wagon at the point of the blade attachment onto the chassis: In the lateral direction 14(a), and in the vertical direction 14(b). Figure 15 shows the angular motions of the blade-wagon: Roll 15(a) and Pitch 15(b). Figure 16 shows the loads on the springs along the trajectory: In the lateral direction 16(a), and in the vertical direction 16(b). Finally, Figure 17 shows the loads on the shock-absorbers along the trajectory: for the lateral 17(a), and vertical direction 17(b).

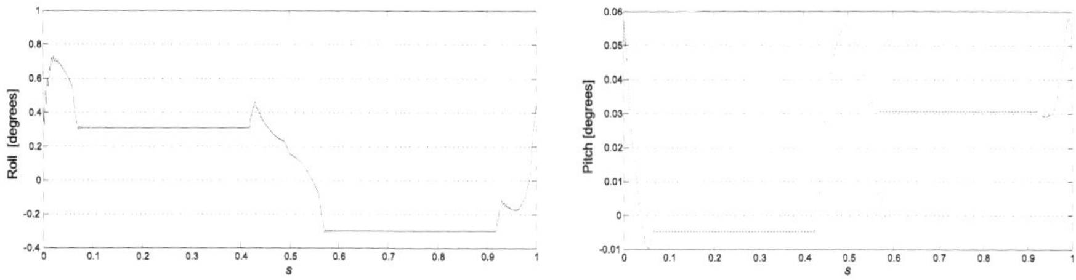

(a) (b)

Fig. 15. Angular motions of the blade-wagon: roll (a), and pitch (b).

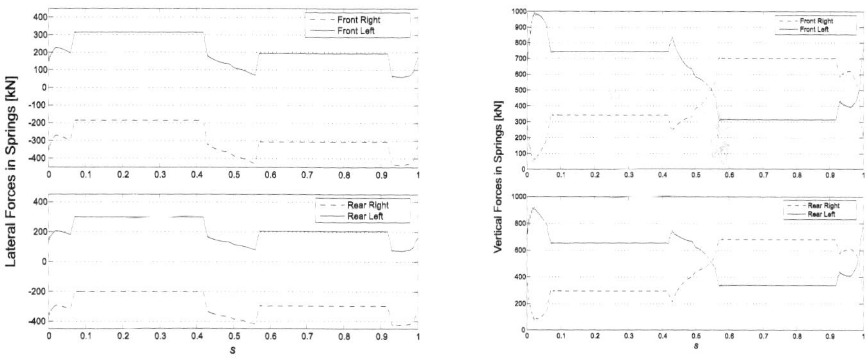

(a) (b)

Fig. 16. Loads on the springs along the trajectory: lateral (a), and vertical direction (b).

5. Concluding remarks

A good evaluation of the actual economic efficiency of the VGOT as an energy-conversion system must take into account the savings in investments due to the substantial reduction in the number of blades and railway length due to the adoption of an oval shape. As it is logical to expect, for the unfavorable cases where the wind shows a preferential bearing aligned with the oval's mayor axis, efficiency is appreciably smaller than for a standard rotor and it keeps decreasing with CF. However, this last situation will never occur in reality because nobody would design a trajectory layout in such a way that its mayor axis is aligned with the wind's preferential bearing. Hence, the worst possible case of all reduces to a compass rose with no preferential bearing which could be dealt with simply by adopting an almost-circular trajectory.

We also have to keep in mind that, even in those cases when we were compelled to use a circular-trajectory layout due to the characteristics of the compass-rose, the advantages mentioned in Section 1 regarding the low-rotational-speed problems associated to a classical

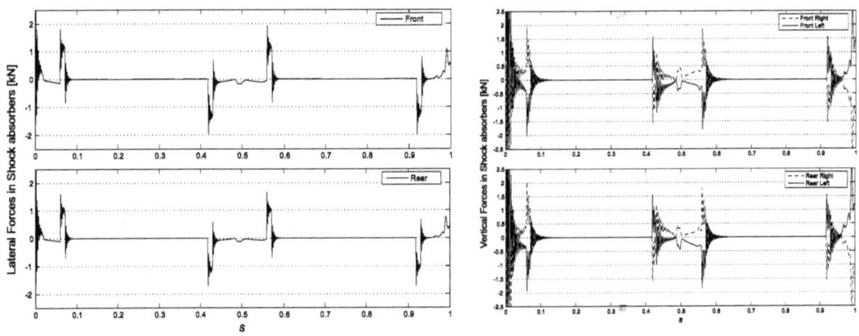

(a) (b)

Fig. 17. Loads on the shock-absorbers along the trajectory: lateral (a), and vertical direction (b).

Darrieus rotor of large diameter are still valid, as well as the possibility of using a blade-positioning control system that operates continuously during the cycle without the fatigue and mechanical-inertia problems associated to variable-geometry attempts in classical Darrieus rotors.

Results shown here indicate that our model possesses the capability to simulate the behavior of this unconventional layout, which expands the spectrum of analysis for Darrieus-turbine design. With this tool, we may determine the optimum values for the layout, the structural configuration, and the stiffness of the springs and shock-absorbers that would be adopted for the suspension system. This process would give us the right balance between the energy-conversion efficiency, the weight and cost of the structure, the stability of the blade-wagon during its normal operation, and the loads exerted on the rails. Excessive stiffness and weight would not only increase the costs of manufacturing but would also induce high loads on the rails and other components of the drive train (like wheels and bogies) compromising their durability.

In the present study, the rails were assumed as brand new (i.e. misalignments and surface degradation due to wearing were not taken into account), which would represent the original design conditions. In an extended version of our model, we shall include an additional subroutine that introduces random displacements on the wheels with the same statistical regularity of the misalignments and/or increased roughness due to natural ageing of the rails. This will allow us to simulate more realistic conditions in further stages of the plant's operational life.

6. Acknowledgements

F.P. is very grateful for the financial support made available by the National Science Foundation through grants CEBET-0933058 and CEBET-0952218.

7. References

Bathe, K. J. (1996). *Finite element procedures*, Prentice Hall, Englewood Cliffs, New Jersey, USA.
Burden, R. L. & Faires, J. D. (1998). *Numerical analysis*, Brooks Cole.

de Vries, E. (2005). Thinking bigger: Are there limits to turbine size?, *Renewable Energy World* 8(3).

Kwon, Y. W. & Bang, H. (1997). *The finite element method using Matlab*, CRC Press.

Manwell, J. F., McGowan, J. G. & Rogers, A. L. (2002). *Wind energy explained: Theory, design and application*, Wiley.

NREL (2005). Wind power today, *Report DOE/GO-102005-2115*, U.S. Department of Energy.

NREL (2008). 20% wind energy by 2030: Increasing wind energy's contribution to U.S. electricity supply, *Report DOE/GO-102008-2567*, U.S. Department of Energy.

Otero, A. D. & Ponta, F. L. (2002). Numerical results for the structural behavior of a blade element of a V.G.O.T. Darrieus, *VIIth World Renewable Energy Congress, Cologne*, Pergamon, p. 223.

Otero, A. D. & Ponta, F. L. (2004). Finite element structural study of the VGOT wind turbine, *Int. J. Global Energy Issues* 21: 221–235.

Paraschivoiu, I. (1982). Aerodynamics loads and performance of the Darrieus rotor, *J. Energy* 6: 406–412.

Paraschivoiu, I. (1983a). Double multiple streamtube model with recent improvements, *J. Energy* 7: 250–255.

Paraschivoiu, I. (1983b). Predicted and experimental aerodynamics forces on the Darrieus rotor, *J. Energy* 7: 610–615.

Ponta, F. L. & Lago, L. I. (2008). Analysing the suspension system of VGOT-Darrieus wind turbines, *Energy for Sustainable Development* 12: 5–16.

Ponta, F. L. & Luna Pont, C. A. (1998). A novel technique for high-power electricity generation in high-speed wind regimes, *Vth World Renewable Energy Congress, Florence*, Pergamon, pp. 1936–1939.

Ponta, F. L. & Otero, A. D. (2002). A 3-node isoparametric finite element model for structural analysis of the V.G.O.T. Darrieus, *VIIth World Renewable Energy Congress, Cologne*, Pergamon, p. 252.

Ponta, F. L., Otero, A. D. & Lago, L. (2004). The VGOT Darrieus wind turbine, *Int. J. Global Energy Issues* 21: 303–313.

Ponta, F. L., Otero, A. D., Luna Pont, C. A. & Seminara, J. J. (2001). Mechanical, structural and electrical concepts for the engineering of a V.G.O.T. Darrieus turbine, *2001 European Wind Energy Conference and Exhibition, Copenhagen*, WIP - Renewable Energies and ETA, pp. 599–601.

Ponta, F. L., Otero, A. D., Seminara, J. J. & Lago, L. I. (2002). Improved design for the structure and gear system of a blade element of a V.G.O.T. Darrieus, *VIIth World Renewable Energy Congress , Cologne*, Pergamon, p. 222.

Ponta, F. L. & Seminara, J. J. (2000). Double-multiple streamtube model for variable-geometry oval-trajectory Darrieus wind turbines, *VIth World Renewable Energy Congress, Brighton, U.K.*, Pergamon, pp. 2308–2311.

Ponta, F. L. & Seminara, J. J. (2001). Double-multiple streamtube model for V.G.O.T. Darrieus turbines with recent improvements, *2001 European Wind Energy Conference and Exhibition, Copenhagen*, WIP - Renewable Energies and ETA, pp. 410–413.

Ponta, F. L., Seminara, J. J. & Otero, A. D. (2007). On the aerodynamics of variable-geometry oval-trajectory Darrieus wind turbines, *Renewable Energy* 32: 35–56.

Seminara, J. J. & Ponta, F. L. (2000). Numerical experimentation about an oval-trajectory Darrieus wind turbine, *VIth World Renewable Energy Congress, Brighton, U.K.*, Pergamon, pp. 1205–1209.

Seminara, J. J. & Ponta, F. L. (2001). Numerical results for a V.G.O.T. Darrieus turbine for different wind compass-rose conditions, *2001 European Wind Energy Conference and Exhibition, Copenhagen*, WIP - Renewable Energies and ETA, pp. 406–409.

Strickland, J. H. (1975). The Darrieus turbine: a performance prediction model using multiple stream tubes, *Report SAND75-0431*, Sandia Laboratory.

Templin, R. J. (1974). Aerodynamic performance theory for the NRC vertical axis wind turbine, *Report LTR-LA-160*, National Research Council of Canada.

Wilson, R. E., Walker, S. N. & Lissaman, P. (1976). Aerodynamics of the Darrieus rotor, *J. Aircraft* 13: 1023–1024.

WWEA (2010). World wind energy report 2009, *Report*, World Wind Energy Association. www.wwindea.org.

7

Wind Turbine Simulators

Hossein Madadi Kojabadi[1] and Liuchen Chang[2]
[1]Sahand University of Technology
[2]University of New Brunswick
[1]Iran
[2]Canada

1. Introduction

Electricity generation using wind energy has been well recognized as environmentally friendly, socially beneficial, and economically competitive for many applications. Wind turbine simulators (WTS) are important equipments for developing wind energy conversion systems and are used to simulate wind turbine behavior in a controlled environment without reliance on natural wind resources, for the purpose of research and design into wind turbine drive trains, especially energy conversion systems. Wind turbine simulators can significantly improve the effectiveness and efficiency of research and design in wind energy conversion systems. The simulator can be used for research applications to drive an electrical generator in a similar way as a wind turbine, by reproducing the torque developed by a wind turbine for a given wind velocity. Also, it can be used as an educational tool to teach the behavior, operation and control of a wind turbine. Usually induction motor, separately excited DC motor and permanent magnet synchronous motor can be used to reproduce the static and dynamic characteristics of real wind turbines.

In the past few years, there have been many studies on wind turbine simulators. Authors of (Nunes et al., 1993), (Battaiotto et al., 1996), (Rodriguez et al., 1998) and (Monfared et al., 2009) utilized separately excited dc motors with controlled armature current. (Nunes et al., 1993) presented a wind turbine simulator using the electromagnetic (EM) torque equation of a dc machine. The armature and the field circuits were controlled so that the dc machine generated the static characteristics of a constant pitch wind turbine. (Monfared et al., 2009) , (Rabelo, et al., 2005) and (Guangchen, et al., 2010) utilized separately excited dc motor to reproduce the static and dynamic characteristics of real wind turbines. The wind turbine simulator described in (Monfared et al., 2009) aims to reproduce the mechanical behavior of wind turbine rotor, drive train and a gear box under dynamic and static conditions. The reference for the torque developed by the wind turbine includes the gradient and tower shadow effects that result in pulsating torques. The larger inertia effect and the steady-state behavior of real wind turbine are also included in the paper. In (Nichita et al., 1998) a more general approach of WTS was presented by the application of a dc motor. In (Jian & xu, 1987) a microcomputer-controlled SCR-DC motor was used to supply the shaft torque. A dc machine, although is ideal from the standpoint of control, is, in general, bulky and expensive compare with an AC machine and it needs frequent maintenance due to its commutators and brushes. Inverter driven PMSM can also work like a real wind turbine (Weihao Hu, et al., 2009), (Xu, Ke, et al., 2007). The wind turbine simulator consists of a microcontroller, an intelligent power module inverter, a PMSM and a control desk on the computer. The control desk obtains the wind speed and calculates the theoretical torque of a real wind turbine by using the wind turbine characteristics and the rotation speed of PMSM. Then the output torque of the PMSM can be regulated by controlling the stator current and frequency. Hence, the inverter driven PMSM can work like a real wind turbine. Inverter driven IM can also reproduce the dynamic and static characteristics of real wind turbine (Madadi, et al., 2004), (Madadi, & Chang, 2005) and (Yun, et al., 2009). A control program is developed that obtains wind profiles and, by using turbine characteristics and rotation speed of induction motor(IM), calculates the theoretical shaft torque of a real wind turbine. Comparing with this torque value, the shaft torque of the IM is regulated accordingly by controlling stator current demand and frequency demand of the inverter.

In this chapter, an IGBT inverter-controlled squirrel cage induction motor was used instead of a dc motor as a WTS. A dc machine, although is ideal from the standpoint of control, is, in general, bulky and expensive compare with an AC machine and it needs frequent maintenance due to its commutators and brushes. This drive is controlled using the measured shaft torque directly, instead of estimating it as conventional drives do. The proposed structure for WTS is achieved in two closed loops of control: speed control and torque control. In this chapter we present the working principles, structures, and test results of wind turbine simulator.

The structure of the wind turbine simulator is depicted in Fig. 1. In this figure, wind simulator, three phase IGBT inverter and IM behaves like a real wind turbine in steady-state. The wind simulator was programmed in C language by utilize of a real wind turbine model, torque PI controller, and IM model. Each of these WTS parts will be discussed in details in the following sections. The structure of the controller for the induction motor is shown in Fig. 2. The rotor speed and the shaft torque are the feedback signals from the torque/speed transducer.

The torque error determines the demand stator current by a proportional integral (PI) regulator. This demand stator current, in turn, determines the required slip frequency based on the function generator in a tabular format as defined by (9).

2. Wind turbine model

According to Betz's law, no turbine can capture more than 59.3 percent of the kinetic energy in wind. The ideal or maximum theoretical efficiency (also called power coefficient, C_p) of a wind turbine is the ratio of maximum power obtained from the wind to the total power available in the wind. The factor 0.593 is known as Betz's coefficient. It is the maximum fraction of the power in a wind stream that can be extracted. So power coefficient, C_p, is the ratio of power output from wind machine to power available in the wind.

$$\text{power output from wind machine} = P = \frac{1}{2}\rho A.u^3 C_p \tag{1}$$

$$\text{power available in wind} = P_{wind} = \frac{1}{2}\rho A.u^3 \tag{2}$$

$$C_p = P / P_{wind} \tag{3}$$

Steady state wind turbine model is given by the power-speed characteristics shown in Fig. 3. The curves in Fig. 3 represent the characteristics of a 3-kW, three-blade horizontal axis wind turbine with a rotor diameter of 4.5 m. These curves can be obtained by wind turbine tests or calculated during the design by manufacturers. At a given wind speed, the operating point of the wind turbine is determined by the intersection between the turbine characteristic and the load characteristic. Usually, the turbine load is an electrical generator, such as an induction generator (IG), synchronous generator (SG), or permanent-magnet synchronous generator (PMSG). From Fig. 3, it is noted that the shaft power (P_m) of the wind turbine is related to its shaft speed (n) and wind speed (u). In practice, the characteristics of a wind turbine can also be represented in a simplified form of power performance coefficient (C_p) and tip speed ratio (λ) as shown in Fig. 4 for the same wind

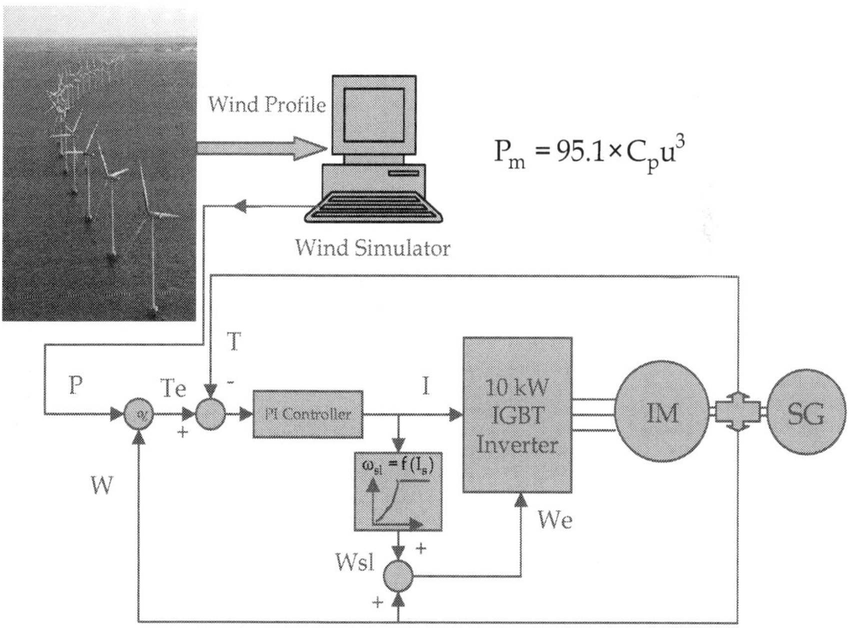

Fig. 1. System stucture of the wind turbine simulator

turbine in Fig. 3. The $C_p - \lambda$ curve is usually used in industry to describe the characteristics of a wind turbine. The tip speed ratio of a turbine is given by:

$$\lambda = \frac{r_m n \pi}{30u} \tag{4}$$

where r_m is the turbine rotor radius in meters, n turbine rotor speed in revolutions per minute (r/min), and u is wind speed in m/s. The turbine output power is given by:

$$P_m = 10.28 \times C_p u^3 \tag{5}$$

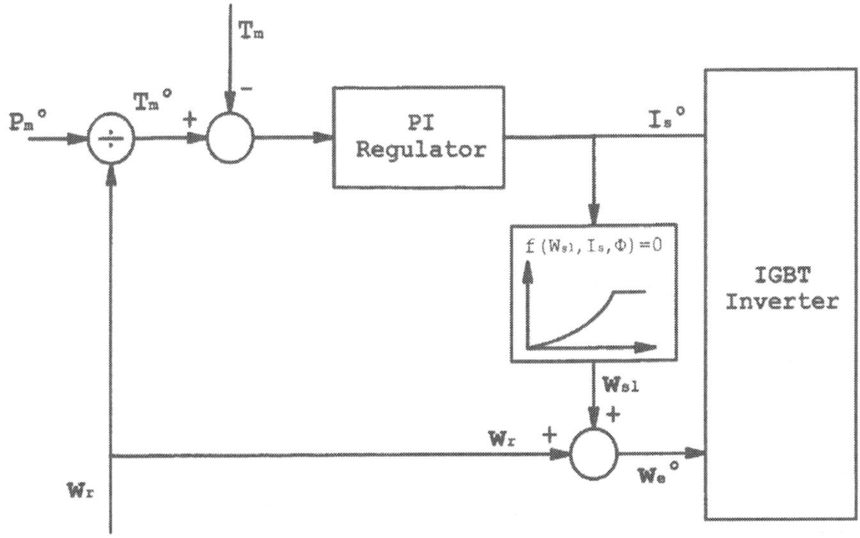

Fig. 2. Controller structure of the induction motor

In a wind turbine simulator, the power-speed characteristics of a wind turbine are physically implemented by an induction motor, dc motor or permanent magnet synchronous motor drive. The shaft power P_m and speed ω_r of the induction motor represent the power and speed of the wind turbine rotor. An inverter fed IM is used to drive a load (i.e., a generator as if it were driven by a real wind turbine). In order to reproduce the turbine characteristics of Figs. 3 and 4 in a laboratory, a microcontroller-and PC-based

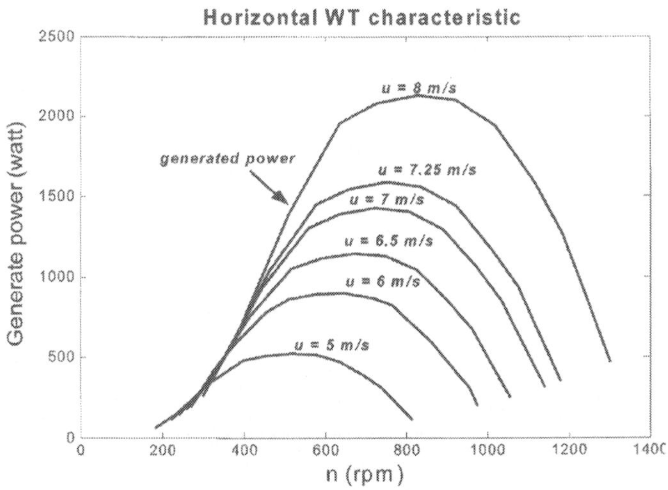

Fig. 3. Power-speed characteristics of a real wind turbine

control system is developed. The wind speed signal needed for the simulator is supplied from wind profiles which can be obtained from measured wind data of a site or can be get in any artificial form by users. Thus, researchers can conduct their studies on wind turbine drive trains in a controllable test environment in replacement of an uncontrollable field one without reliance on natural wind resources.

3. IM model

The model of an induction machine is highly nonlinear and mutually coupled. The complexity of the model is dictated by the purpose for which the model is employed. A simplified equivalent circuit model of an induction machine is shown in Fig. 5. It is used to develop an appropriate controller with an approximate input-output model which relates the stator voltage input V_s, and the outputs, namely the angular speed ω_r and the developed torque T_m . The model is established based on the following assumptions.

- The dynamics of the electrical subsystem are neglected as its time constant is substantially smaller than that of the mechanical subsystem.
- The core loss resistance is ignored.
- The impedance of the magnetizing circuit is much larger than the impedance of the stator, that is

$$(\omega_e L_m)^2 >> (R_s{}^2 + (\omega_e L_s)^2) \Rightarrow V_s \approx V_m \quad (6)$$

where $L_m, R_s, \omega_e, L_s, V_m$ and V_s are the magnetizing inductance, stator resistance, electrical angular speed, stator inductance, magnetizing voltage, and stator supply voltage, respectively. The rotor leakage reactance is much smaller than the equivalent rotor load resistance.

$$\frac{R_r}{S} >> \omega_e L_r \quad (7)$$

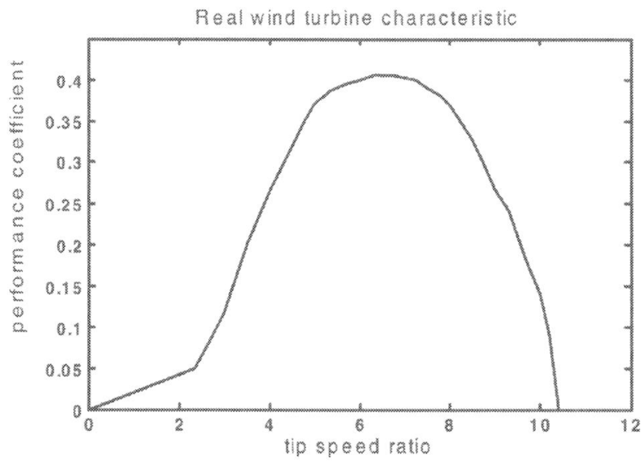

Fig. 4. Cp-Lamda characteristics of a real wind turbine

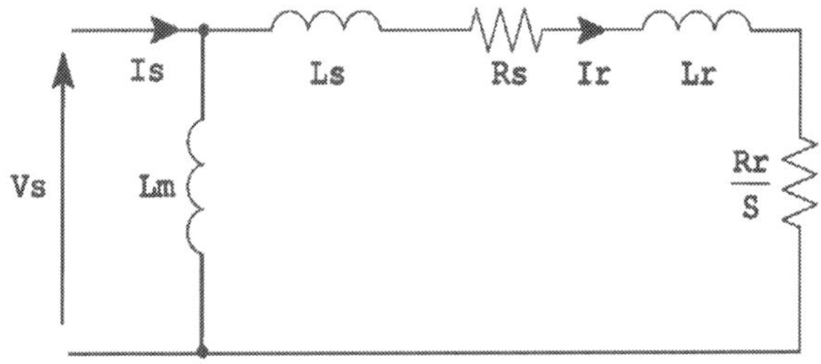

Fig. 5. Induction motor model

where S is the slip, and R_r and L_r are the rotor resistance and inductance, respectively. The stator impedance is much smaller than the reflected rotor impedance at normal operating condition

$$(\frac{R_r}{S})^2 + (\omega_e L_r)^2 >> R_s + (\omega_e L_s)^2 \tag{8}$$

The IM model consists of an algebraic equation which governs the flux linkage , the supply frequency , the output current , and the slip frequency , as given by (9). The equation takes the form of (Bose, 1986), (Rashid, 1993).

$$(I_s)^2 - \varphi^2 \frac{R_r^2 + L_m^2 \omega^2{}_{sl}}{L_m (R_r^2 + L_r^2 \omega^2{}_{sl})} = 0 \tag{9}$$

where $\varphi = V_s / \omega_e$

So based on (9) we get:

$$\omega_{sl} = F(I_s) \tag{10}$$

Function F relates ω_{sl} (induction motor slip speed) to the magnitude of induction motor's stator current in steady-state condition.

4. Inverter control

The three-phase IGBT inverter converts the fixed dc link voltage obtained from a three-phase bridge rectifier into a three-phase variable frequency current source, feeding to the induction machine as the prime mover of a synchronous generator, which acts as the load of the wind turbine. The inverter is controlled by an Intel 80C196KD microcontroller.

The microcontroller accepts current and frequency signals from the output of the wind simulator (refer to Fig. 2) as the demand input and sends out the appropriate triggering pulses to the IGBT driver circuits, based on the errors between the demand current and actual current using the current hysteresis control strategy. The current hysteresis control forces the inverter output currents to track demand current waveforms within a hysteresis upper and lower bands. The output currents are detected by current sensors and compared with the demand current waveforms. When an output current exceeds the upper band, the IGBT gate control signal will be switched to an appropriate state to reduce the actual current. The IGBT gate control state will be properly switched again when the actual output current drops below the lower band. Regulating the magnitude of the sinusoidal demand current waveform will provide an effective control to the inverter output power. The hysteresis bandwidth can be decided by considering the switching frequency limitations and the switching losses of the IGBT modules. Since output currents are always detected and regulated, the current hysteresis control scheme has a fast current response and provides an inherent over current protection. However, the scheme may be more expensive to implement due to the requirements for current sensors and fast A/D conversion devices. The microcontroller system for the IGBT inverter incorporates various hardware and software protection functions such as over current, over voltage, and under voltage and over temperature protections. Three phase inverter is shown in Fig. 6.

5. Experimental results of static wind turbine simulator

The described static wind turbine simulator has been implemented and tested in our laboratory. The wind turbine model and the digital PI controller are realized using a PC interfaced to LAB Windows I/O board. This LAB Windows system is equipped with eight A/D channels and two D/A channels for control and acquisition purposes. The controller generates the current and frequency demands for the IGBT inverter. The inverter control

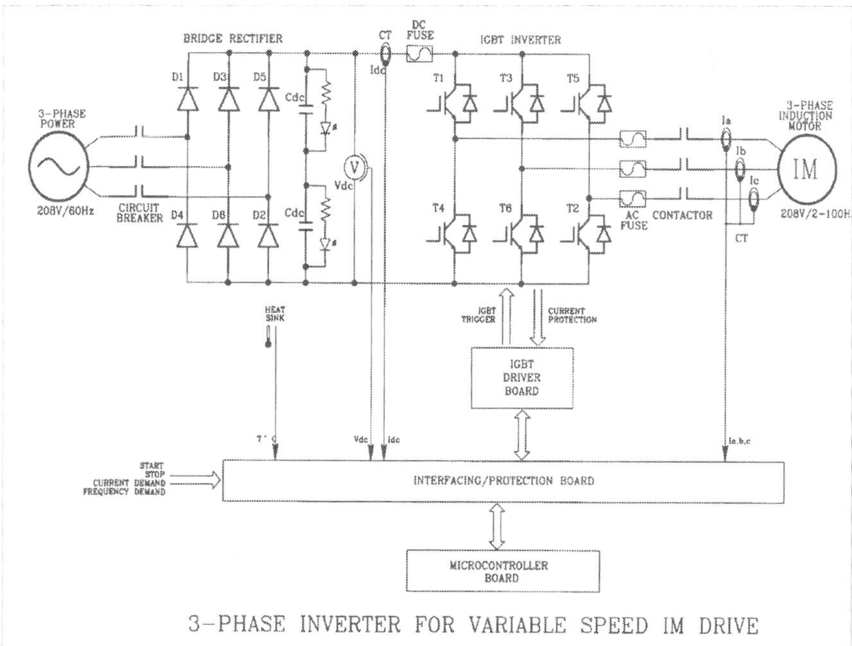

Fig. 6. Configuration of three phase inverterr for WTS

system is implemented by an Intel 80C196KD /KC microcontroller and associated hardware-software. In this research, a horizontal axis wind turbine as described by Figs. 3 and 4 is used. The electrical and mechanical parameters of the IM are listed in Table 1. The power–speed characteristics of WTS at different wind speeds as measured during tests are shown in Fig. 7, and are compared with those of a real wind turbine of Fig. 3. Based on equation (4) and equation (5), the $C_p - \lambda$ characteristic can be derived from the measured $P_m - n$ characteristics of Fig. 7. The measured $C_p - \lambda$ characteristic is compared with that of the wind turbine given in Fig. 4, as presented in Fig. 8. Both Figs. 7 and 8 verify that the developed wind turbine simulator reproduces the steady-state characteristics of a given wind turbine at various wind conditions.

6. Application of steady-state wind turbine simulator

The wind turbine simulator is a valuable test facility to create a controlled test environment for drive trains of various wind turbines without reliance on natural wind resources and actual wind turbines. Several research projects have been conducted using the developed wind turbine simulator at our laboratory. In the development of the "intelligent maximum wind energy extraction algorithm", the developed wind turbine simulator, instead of real wind turbine, is used as the prime mover to drive a synchronous generator. The block diagram of wind power system is shown in Fig. 9 for a stand alone load. A single phase inverter is adopted to extract the generator output and feed to a stand alone load represented by a resistor load bank. The maximum power algorithm is implemented inside

Induction motor parameters and their values	
Rated power = 10 hp	Rr = 0.2032 Ω
B = 0.026 N.m.s/rad.	Ls = Lr = 0.02851 H
P = 4 (number of poles)	Lm = 0.02738 H
Rs = 0.2056 Ω	J = 0.8 $Kg.m^2$

Table 1. Induction motor parameters

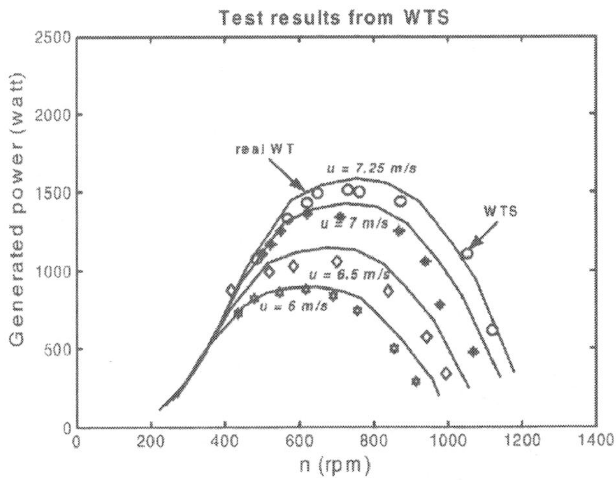

Fig. 7. Tested power–speed characteristics of a WTS.

the single-phase inverter, as part of the control software. The search performance of the developed algorithm based on the WTS as a prime mover, is shown in Fig. 10. In this figure, the first 180 s is the simulator start-up period. In this period, the single - phase inverter is stopped, and no load is added onto the WTS. After the start -up period, the single-phase inverter is started, and begins to search the optimal point for maximum power extraction. This figure confirms that the searching for maximum power is effective, since both C_p and motor power are increased toward the maximum values. In steady-state, the performance factor C_p approaches 0.4, which is the practical limit of most modern wind turbines. These tests were conducted at a constant wind speed of 6 m/s (Wang, 2003). The dynamic response of the maximum power extraction algorithm under the wind speed change has been tested. The test is conducted after the system has two hours of on-line training. Fig. 11 shows the test results when the wind speed was changed from 3 m/s to 6 m/s. Again the first 160 s are for wind simulator start-up. The system operated at a simulated wind speed of 3 m/s until 225 s, when the wind speed was increased to 6 m/s. It can be observed from the figure that the transition time takes about 175 s before the output power and the performance factor reach the new steady state with a high C_p value of 0.39.

7. Dynamic wind turbine simulator

The accurate wind turbine simulator should includes several important components of real wind turbine including, wind shear and tower shadow effects, larger turbine inertia, and turbulent wind speed. The wind turbine simulator described in (Monfared et al., 2009) incorporated all above mentioned components of real WT in designed WTS.

7.1 Tower shadow/wind shear model

As a turbine blade rotates, it is acted upon by wind at various heights. The variation of wind speed with height is termed *wind shear*. Wind speed generally increases with height. Torque pulsations, and therefore power pulsations, are observed due to the periodic variations of

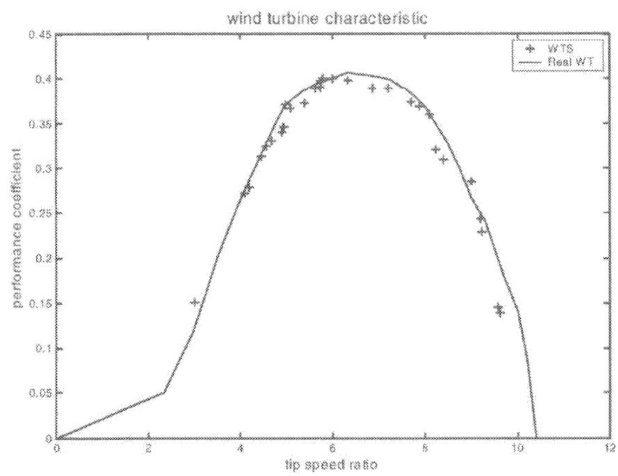

Fig. 8. $C_p - \lambda$ characteristics of a WT and WTS

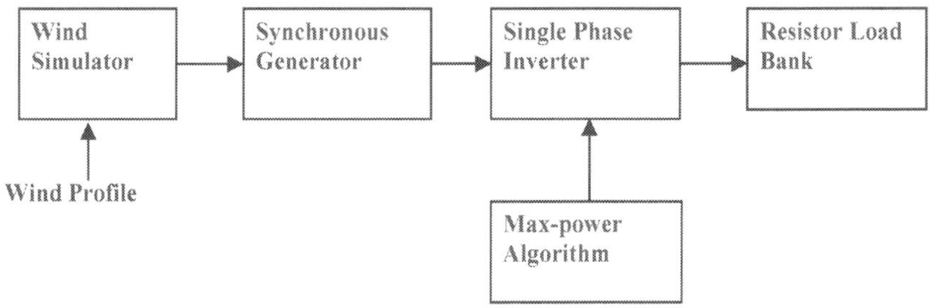

Fig. 9. Block diagram of wind power system

wind speed experienced at different heights. Power and torque oscillate due to the different wind speeds encountered by each blade as it rotates through a complete cycle. For instance, a blade pointing upwards would encounter wind speeds greater than a blade pointing downwards. During each rotation the torque oscillates three times because of each blade passing through minimum and maximum wind. To determine control structures and possible power quality issues, the dynamic torque generated by the blades of a wind turbine must be represented. It is therefore important to model these wind shear and tower shadow induced 1P and 3P torque pulsations for a meaningful WTE as (Lops, 2005).

$$T_{mech} = T_{mill}\left[1 + A_1 \sin(\omega_{mill}t) + A_3 \sin(3\omega_{mill}t)\right] \quad (11)$$

where $A_1 = 0.2$ and $A_3 = 0.4$, T_{mill} and T_{mech} are average and aerodynamic torques of wind turbine, respectively. The average power and torque developed by a wind turbine are a function of the wind speed (u), the rotational speed of the shaft (ω_{mill}), the tip-speed ratio(λ) and the torque and power coefficients (C_q & C_p) as given by (12) and (13) where r_m is the radius of the turbine and ρ is the air density.

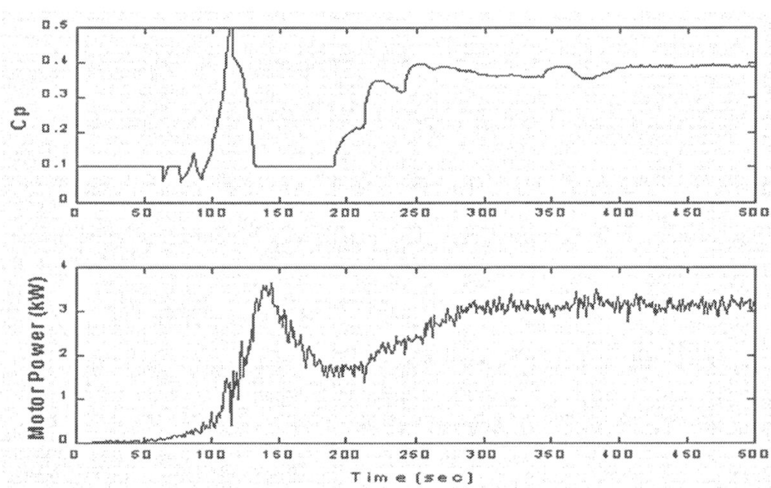

Fig. 10. Searching performance with several hours online.

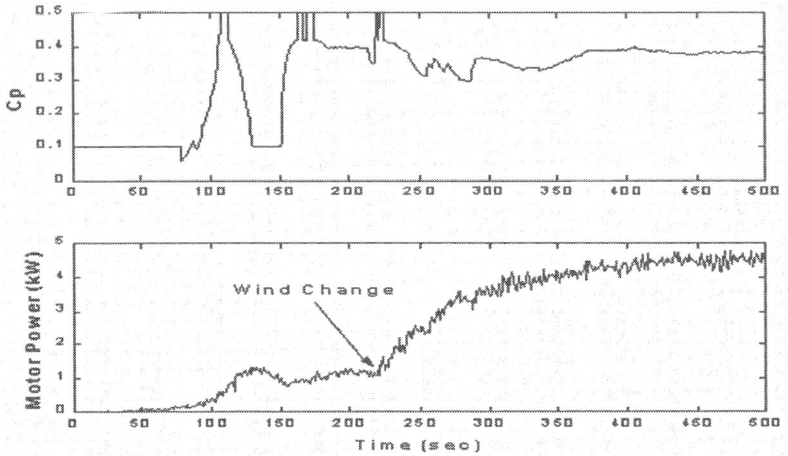

Fig. 11. Dynamic test results of the wind turbine control algorithm on the developed WTS

$$P = 0.5 \times \rho \pi r_m^2 C_p u^3 \tag{12}$$

$$T_{mill} = 0.5 \times \rho \pi r_m^3 C_q u^2 \tag{13}$$

The turbine torque oscillations due to tower shadow and wind shear were calculated based on the turbine specifications. These oscillations were sufficient to cause a resultant fluctuation in the output voltage, such that the effects due to tower shadow could easily be distinguished in the time domain. For this purpose the constant wind speed will be applied such that the generator shaft speed rises to 1500 rpm. If the gear ratio is assumed to be 1:20 so the 1P and 3P frequency of oscillations due to tower shadow and wind shear effects can be calculated as

$$1P = \frac{1500}{60 \times 20} \approx 1.25Hz$$

$$3P = 3 \times 1P \approx 3.75Hz$$

7.2 Inertia model

The inertia model is determined by equating the generator acceleration in the field and laboratory(simulated) systems. The effect is to alter the turbine torque that the dc motor will produce in response to a given wind, such that the effect of the larger turbine rotor inertia is emulated. A mechanical diagram of both the field system and the representative laboratory(simulated) system is shown in Fig. 12, where T_{motor} = dc motor's torque, T_{mech} = aerodynamic turbine rotor torque, T_{gen} = generator torque, J_{motor} = dc motor inertia, J_g = generator inertia, and J_{mill} = turbine rotor inertia. Equations of motion may be written for both systems, real wind turbine (14) and simulator (15) and solved to determine the reference torque (16) required for the dc motor.

$$\frac{T_{mech}}{n} = \left(\frac{J_{mill}}{n^2} + J_g\right)\frac{d\omega_{gen}}{dt} + T_{gen} \tag{14}$$

$$T_{motor} = \left(J_{motor} + J_g\right)\frac{d\omega_{gen}}{dt} + T_{gen} \tag{15}$$

$$T_{motor} = \frac{T_{mech}}{n} + \left(J_{motor} - \frac{J_{mill}}{n^2}\right)\frac{d\omega_{gen}}{dt} \tag{16}$$

where n is gear ratio and the second term of right hand side of (16) represents the compensation torque of wind turbine emulator. Therefore the wind turbine emulator will accurately represent the real wind turbine system if the driving torque T_{motor} is controlled according to (16). Simulation and experimental results from WTS (Monfared et al., 2009) clearly confirm that with larger rotor inertia demonstrates reduced magnitude for 1P and 3P harmonics.

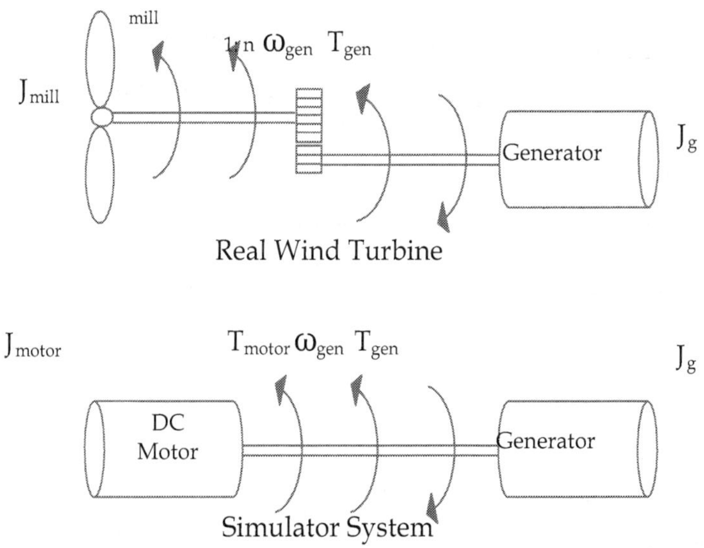

Fig. 12. Mechanical models of real and simulated WT

8. Conclusion

In order to improve the effectiveness and efficiency of research into wind energy conversion systems, a wind turbine simulator has been developed to create a controlled test environment for drive trains of wind turbines. The simulator is realized using an IGBT inverter controlled induction motor drive. The wind turbine model and the digital controller are developed on a C-language platform for easy access, programming, and modifications. Various wind turbines and wind profiles can be incorporated in the control software. The experimental results show that the steady-state characteristics of the WTS are similar to those of the given wind turbine. Various tests conducted on the developed WTS for real wind turbine and the resultant responses of a variety of WTS parameters have confirmed the performance of the wind turbine simulator under the designed digital controllers. The accurate wind turbine simulator should includes several important components of real wind turbine including, wind shear and tower shadow effects, larger turbine inertia, variable wind speed, and steady state characteristics.

9. References

Abelo. B.; Hofmann. W.; Gluck, M. (2004). Emulation of the Static and Dynamic Behavior of a Wind Turbine with a DC Machine, *in Proc. of 35th IEEE Power Electronics Specialists Conference (PESC-05)*, pp. 107 - 2112, Jun. 2004,.

Battaiotto, P. E.; Mantz, R. J.; Puleston, P. f. (1996). A wind turbine emulator based on a dual DSP processor system, *Control Engineering Practice*, Vol. 4, No. 9, 1996, pp. 1261-1266.

Bose, B. K.(1986) *Power Electronics and AC Drives*. Englewood Cliffs, NJ:Prentice-Hall, 1986.

Guangchen, L.; Shengti, W.; Jike, Z. (2010). Design and Realization of DC Motor and Drives Simulator for Small Wind Turbine, Proceedings of *Power and Energy Engineering Conference* (*APPEEC*) pp. 1-4, China, March 2010, Chengdu

Jian. L.; Xu. W. C.(1987). The simulation system of wind turbine performance using a microcomputer controlled SCR-DC motor, *in Proc. Beijing Int. Conf. Electrical Machines*, pp. 865-868. Beijing, China, 1987

Lopes, L. A. C.; Lhuilier, J.; Mukherjee, A.; Khokar, M. F. (2005). A Wind Turbine Emulator that Represents the Dynamics of the Wind Turbine Rotor and Drive Train, *36th IEEE Conf. Power Electronics Specialists*, pp. 2092-2097, 12 Jun 2005

Madadi kojabadi, H.; Chang, L. (2005). A Novel Steady State Wind Turbine Simulator Using an Inverter Controlled Induction Motor, *Wind Engineering*, Vol. 28, No. 4, 2004, pp. 433-443

Madadi Kojabadi, H,. Chang, L., Boutot, T. (2004), Development of a Novel Wind Turbine Simulator for Wind Energy Conversion Systems Using an Inverter- Controlled Induction Motor, *IEEE Trans. On Energy Conversion*, Vol. 19, No. 4., 2004, pp. 547-552.

Monfared, M. Madadi kojabadi, H., Rastegar, H., (2009), Static and dynamic wind turbine simulator using a converter controlled dc motor, *Renewable Energy, Elsevier*, Vol. 34, 2009, pp. 845-848

Nichita, C.; Diop, A. D.; Elhache, G.; Dakyo, B.; Protin, L. (1998). Control structures analysis for a real wind system simulator, *Wind Engineering*, Vol. 22, No. 6, 1998, pp. 275-286,

Nunes, A. A.; Seixa, C. P. F.;. Cortizo, P. C.;. Silva, S. R.(1993). Wind turbine simulator using a dc machine and a power reversible converter, *in Procceedings of the International Conf. On Electrical Machines, ICEMA*, pp. 536-540, Adelaide, 1993

Rashid, M. H. (1993). *Power Electronics: Circuits Devices, and Applications* Englewood Cliffs, NJ: Prentice-Hall, 1993.

Wang, Q. (2003). Maximum wind energy extraction strategies using power electronic converters, *Ph.D. Thesis, University of New Brunswick, Fredericton, Canada, 2003.*

Weihao, Hu, Yue, W., Xianwen, S, Zhaoan, W, (2009), Development of wind turbine simulator for wind energy conversion systems based on permanent magnet synchronous motor, *Electrical Machines and systems,* pp. 2322-2326, Feb. 2009 Wuhan, China

Yun, D.; Han, B.; Choi, N. (2009). Hardware simulator for PMSG wind power system with matrix converter, *International Conference on Telecommunications Energy,* pp. 1-6, Incheon, Oct. 2009, China.

8

Analysis and Mitigation of Icing Effects on Wind Turbines

Adrian Ilinca

Wind Energy Research Laboratory, Université du Québec à Rimouski

Canada

1. Introduction

Precipitation, atmospheric and in-cloud icing affect wind turbine operation in various ways, including measurement and control errors, power losses, mechanical and electrical failures and safety hazard. Anti-icing and de-icing strategies are used to minimize these effects. Many active and passive methods are in development but few are available on the market. Active heating of blades is the most tested, used and reliable way to prevent icing effects. It is used in parallel with passive hydrophobic coating to lower energy consumption. Precise icing evaluation of the site should be done during the assessment phase to evaluate the necessity and benefits of installing anti-icing and/or de-icing system. This evaluation shall continue during operation in order to optimize production and avoid component failure related to icing events. Multiple anemometry in combination with relative humidity measurements is a cheap and reliable icing detection method during assessment while the use of ice sensors and the power curve method is recommended during operation. Most of the wind turbines operating in cold climates are facing icing events, but very few of them are equipped with blade de-icing systems, and few studies were performed and published on the characteristics of these systems.

Technical difficulties due to cold climate conditions have occurred for most of the existing projects in Quebec. Thus, icing simulations were carried out in the refrigerated wind tunnel of the Anti-icing Materials International Laboratory (AMIL) at Université du Québec à Chicoutimi (UQAC). The effect of icing conditions observed at the wind farm in Murdochville, Québec, Canada has been assessed on a 0.2 m NACA63-415 blade airfoil. The shape and mass of the ice deposit on a wind turbine airfoil has been measured, as well as the lift and drag force on the iced airfoil. Scaling was carried out based on the 1.8 MW–Vestas V80 wind turbine technical data, for three different radial positions and two in-fog icing conditions measured at the Murdochville wind farm in the Gaspé Peninsula. For both icing events, the mass of ice accumulated on the blade airfoil increased as we move to the tip of the blade. In wet regime, glaze formed mostly near the leading edge and on the pressure side. It also accumulated by run-off on the trailing edge of the outer half of the blade. In dry-regime, rime mostly accreted on the leading edge and formed horns. For both icing events, when glaze or rime accreted on the blade airfoil, lift decreased and drag increased. A load calculation using the blade element theory shows that drag force on the entire blade becomes too large compared to lift, leading to a negative torque and the stop of the wind turbine. Torque reduction is more significant on the outer third of the blade. Setting up a de-icing system only on the outer part of the blade would enable significant decrease of heating energy costs. In order to optimize the design and power consumption of an electro-thermal de-icing system for wind turbine blades, an experimental setup was built and used to test the system under similar icing conditions. The parameters of the de-icing control system consider only the convective heat loss at the blade surface during ice accretion. The results show the relation among the meteorological conditions, the ridge formed by liquid water runback, the heating power and the airfoil surface temperature. The study provides useful data for the design of electro-thermal de-icing systems for wind turbine blade application.

2. Ice accretion effects on wind turbines

Wind turbines (WT) operating in cold regions or at high altitudes are frequently facing icing conditions during winter operation. At the same time, the best sites for wind farm installation are located at higher altitudes, as wind speed generally increases by 0.1m/s per 100m of altitude for the first 1000m. In regions with cold climate, available wind power is approximately 10% higher than in other regions due to increased air density at lower temperatures (Fortin et al., 2005a). Therefore, wind farms installed in some of the best wind sites around the world are facing possible icing events. Icing affects the wind assessment and the operation of wind farms. The following problems are directly related to icing and cold climate: measurement errors, power losses, overproduction, mechanical failures, electrical failures and safety hazards.

Measurement errors: during the assessment phase, the anemometers, wind vanes and temperature sensors can be affected by ice. In icing conditions, wind speed errors can be as high as 30% (Laakso et al., 2003a). Another study identifies a maximum error of 40% for an ice-free anemometer and 60% for a standard anemometer during icing events (Fortin et al., 2005b);

Power losses: ice accretion changes the shape and roughness of the blade airfoil (consequently affecting their aerodynamic characteristics) and introduces measurement errors from turbine instruments (wrong wind speed or direction, which affects yaw and power controls). Small amounts of ice on the leading edge of airfoils significantly reduce aerodynamic properties of the blade and the resulting power production (Marjaniemi and Peltola, 1998). Power losses may vary from 0.005 to 50% of the annual production, depending on icing intensity and its duration on the site, wind turbine models and the evaluation methodology (Botta et al., 1998; Gillenwater, 2008; Laakso et al., 2005b; Tammelin et al., 2005);

Overproduction: Higher air density related to low temperatures and airfoil modifications can lead to overproduction of the WT. Overproduction of up to 16% has been recorded (Jasinski et al., 1997);

Mechanical failures: ice accretion will increase the load on the blades and on the tower structure, causing high amplitude vibrations and/or resonance as well as mass imbalance between blades. Operation at low temperatures affects oil viscosity and changes the dimensions and mechanical properties of different components of the WT. This results in possible overheating and higher fatigue charges on components; one of the most affected being gearboxes whose lifetime is considerably reduced (Botta et al., 1998; Laakso and Peltola, 2005; Seifert, 2003; Tammelin et al., 2005);

Electrical failure: snow infiltration in nacelle and extreme temperature lead to condensation in the electronics (Laakso et al., 2003a);

Safety hazard: large icing accumulation on blades can be thrown at a distance of up to 1.5 x the combined height of the turbine and the rotor diameter (Tammelin et al., 2000). Using a Monte-Carlo simulation, Battisti et al. (Battisti et al., 2005b) have shown that the odds to be hit by a piece of ice (between 0.18 and 0.36kg), on a site with moderate icing conditions (5 days per year), is 1 in 10. This is valid for a person walking 10 hours under an operating turbine that uses a de-icing system, considering a total ice accretion of 75kg/rotor per day.

2.1 Physics of atmospheric icing

There are three types of atmospheric icing related to wind turbine: in-cloud, precipitation and frost (Boluk, 1996; Fikke et al., 2006; ISO-12494, 2001; Richert, 1996).
In-cloud icing: it happens when super-cooled water droplets hit a surface below 0°C and freeze upon impact. The droplets temperature can be as low as -30°C and they do not freeze in the air, because of their size. Accretions have different sizes, shapes and properties, depending on the number of droplets in the air (liquid water content - *LWC*) and their size (median volume diameter - *MVD*), the temperature, the wind speed, the duration, the chord length of the blade and the collection efficiency. There is a continuum of ice accretion appearance from rime at coldest temperatures to glaze at warmest.
Soft rime: thin ice with needles and flakes. Appears when temperature is well below 0°C and the MVD and LWC are small. The resulting accretion will have low density and little adhesion.
Hard rime: higher MVD and LWC will cause accretion with higher density, which is more difficult to remove.
Glaze: when a portion of the droplet does not freeze upon impact, but run back on the surface and freezes later. The resulting ice density and adhesion are strong. It is often associated with precipitation.
Precipitation: can be snow or rain. The accretion rate can be much higher than in-cloud, which causes more damage.
Freezing rain: when rain falls on a surface whose temperature is below 0°C. It often occurs during inversion. Ice density and adhesion are high when this phenomenon occurs.
Wet snow: when snow is slightly liquid at air temperature between 0 and -3°C, it sticks to the surface. It is easy to remove at first, but can be difficult if it freezes on the surface.
Frost: appears when water vapour solidifies directly on a cool surface. It often occurs during low winds. Frost adhesion may be strong.

2.2 Ice accretion measurement

The estimation of ice accretion on solid surfaces can be done using direct measurement, indirect measurement or numerical modelling. Direct methods are based on the detection of some change of physical property caused by ice accretion. These include mass, reflective properties, electrical or thermal conductivity, dielectric coefficient and inductance. Indirect methods are based on detecting weather conditions that lead to icing, such as humidity, temperature and wind speed or detecting the effects of icing, such as a reduction in power production. Then, an empirical or deterministic model is used to determine when icing occurs to evaluate the LWC and MVD, (Homola et al., 2006). Meteorological numerical prediction models have the capability to determine with some accuracy the severity and duration of icing events.

2.3 Icing evaluation during site sssessment

When the financial benefits of a blade heating system are evaluated, parameters like the icing duration and severity, as well as potential wind resources, need to be known (Laakso et al., 2003a). How cost-effective is this operation is depending on the available wind energy during the icing period and on the severity of icing. This analysis requires knowledge of both the meteorological conditions leading to ice accretion and the turbine's geometry and operating conditions. The meteorological parameters needed for ice prevention design are well known and draw on over 50 years of experimental investigation in the aeronautic field. They are mainly the liquid water content (LWC), water droplet diameter (MVD), pressure, temperature, and the horizontal distribution of these variables. One reason for the scarce use of these parameters is that they are difficult or expensive to measure. Quantitative data are hardly ever available for assessing either icing frequency or icing severity at a given site (Battisti et al., 2005a). Measurement of the icing duration is so difficult that, most of the time, it has to be estimated empirically (Kimura et al., 2000). In some conditions, it has been observed that measurements as close as 1km from a specific site may not give reliable information (Laakso et al., 2003a). It is thus recommended to measure icing events directly at the planned implementation site (Fikke et al., 2006). A significant difficulty stems from the fact that ice measurement should be done at the same height as the top blade tip (Homola et al., 2006). This section presents the most frequent methods and models to evaluate icing.

Ice sensors: The ISO-12494 norm proposes an ice mass measurement method using an ice collector that consists of a 30mm diameter cylinder, with a minimum length of 0.5m that slowly rotates around a vertical axis (ISO-12494, 2001). This technique provides poor information indicative of icing risks for wind turbines. In fact, wind velocity and the dimension of the cylinder are far out of the range of relative velocities and leading edge diameters of wind turbine blades (Battisti et al., 2005a). Other sensors, using different approaches, such as longitudinal wire waves, vibrating probes or optics exist, but they are mostly used during the operation phase. The main reason these sensors are not commonly used on remote met masts is because of their high costs and their energy demands. Also, different sensors have been tested thoroughly (Fikke et al., 2005; Tammelin et al., 2005) and none of them has perfect reliability and accuracy. The different ice sensors generated significant differences in the recording of icing events and the different sensors did not always indicate icing synchronously (Tammelin et al., 2005). Ice sensors also underestimate icing because of the heating cycle (Laakso et al., 2003b). It has been proposed to improve the predictions by using two detectors, one heated for intensity and one unheated for duration (Tammelin et al., 2005).

Double anemometry and vane: An icing event is assumed when the difference between the wind speeds measured by heated and unheated anemometers exceeds a certain limit. Tallhaug suggests a limit of 20% at wind speeds above 2 m/s (Tallhaug, 2003). Laakso (Laakso et al., 2003b) considers that a value of ± 5% is more appropriate. A larger difference, e.g. ± 10%, would have resulted in only few data sets to indicate icing, which represents too little icing time compared to the ice amounts that can be seen on photos. Craig suggests a method that uses three anemometers: one permanently heated, one unheated and one heated when a 15% difference is observed between the speeds of the heated and unheated anemometers (Craig and Craig, 1996). Equipping the measurement mast with one properly heated and one unheated anemometer to estimate wind resource measurements is cheap and advisable. This arrangement gives an overall picture of ice climate. These methods, in

addition to relative humidity measurements, may give a fairly good idea of the time that ice is likely to affect wind turbines operation (Laakso et al., 2005).
Tallhaug (Tallhaug, 2003) proposes to use an unheated wind vane and assumes that an icing event occurs when the unheated wind vane has zero standard deviation calculated from 6 following 10- minute averages and at temperatures below zero. He showed that there is a very good correlation between zero standard deviation on the unheated direction sensor and the double anemometry indication.
A disadvantage of the double anemometry method lies in the fact that the anemometers are lower than the tip of the blades, where there is more icing (Tammelin et al., 2005). Also, the method works well at relatively mild temperatures (around 0°C). Otherwise, the unheated anemometer may stay frozen for longer periods (Craig and Craig, 1996). On the other hand, the double anemometry can measure the persistency of icing (Laakso et al., 2003b). Results showed that it is possible to estimate the total period during which an icing event will affect the operation of the WT (comments from (Laakso et al., 2003b) on (Craig and Craig, 1996)), by studying its correlation with the period during which ice affects the unheated anemometer. This duration is obviously longer than the actual icing period.
Another disadvantage is that low temperatures were found to cause negative errors for unheated anemometers (i.e. the difference between heated and unheated anemometers did not result from icing (Laakso et al., 2003b)). It is difficult to determine when icing occurs only from wind speed measurement as the unheated anemometer shows both higher and lower wind speeds. Higher values of unheated anemometers have also been observed during snowfall (Seifert, 2003). Obviously, double anemometry will give no indication of icing at zero wind. At very low wind speed, the difference between heated and unheated anemometers can be explained by their different inertial characteristics rather than actual icing.
Relative humidity and dew point: Because relative humidity is high during in-cloud icing, the detection of high humidity (>95%) combined with low temperature (<0°C) is used to detect icing. A dew point detector that has been designed for subzero operation could provide valuable information, because in practice air temperature is at frost point nearly all the time when in-cloud icing occurs (Laakso et al., 2005b). Measuring the relative humidity with temperature is used much more often than the dew point measurement. However, this method has not detected icing events during the same period as ice detectors (Tammelin et al., 2005). It has been shown that 33% of all ice detector indications at 84m level occurred when relative humidity was lower than 95% (Laakso et al., 2003b). The predictability of icing events using conditions of relative humidity of more than 95-98% with temperature less than 0°C is weak. On many occasions, the temperature was below 0°C with relative humidity higher than 95% and no icing was observed (Tammelin et al., 2005).
Visibility and cloud base: In-cloud icing occurs when a structure is surrounded by a cloud at a temperature below 0°C with a minimal wind speed of around 2m/s. The cloud can be detected using the cloud base height or the horizontal visibility. Clouds are classified using qualitative quotes or visibility distance to estimate the liquid water content (LWC), which directly affects the intensity of the in-cloud icing. This is done using airport observation, a pyranometer, video monitoring or automatic sensors. Airport observation provides cloud base heights and a cloudiness index based on the observation of the cloud density, on a scale from 1 to 8. When the index is higher than 6/8 and the cloud base height is lower than the wind turbine, icing is detected or the index can be used as a ratio for accretion intensity (Tallhaug, 2003). The pyranometer measures the solar radiation intensity. It has been observed that when the intensity is lower than $300W/m^2$, ice is detected (Kimura et al.,

2000). Video monitoring can measure horizontal visibility using painted poles at distances from 50 m up to 300m from met mast (Dobesch et al., 2003). The use of painted poles gives a numerical criterion of the cloud density (Tammelin et al., 2005). Finally, automatic detection can be done using radar and microwave radiometers which can directly detect the LWC (Battisti et al., 2005a). Unfortunately, this method is too expensive and not commonly used during site assessment.

Acquiring information, such as time series of cloud base heights from the nearest airport, and comparing them with the measured data is also advisable. This method and double anemometry are likely to give a fairly good idea of the time period that ice is likely to affect the operation of the wind turbines (Laakso et al., 2005b). However, cloud height observed at the nearby weather station does not necessarily represent accurately onsite cloud conditions and it is preferable to measure the horizontal visibility on the specific site to estimate in-cloud icing. Even under these conditions, the method underestimates the real ice mass (Tammelin et al., 2005).

Aeronautics considers that icing events occur for specific values of meteorological variables: wind speed V>0, temperature T<0, liquid water content LWC>0 and median volumetric droplet diameter MVD. When such limits are simultaneously exceeded, ice will form. The direct icing duration (icing time) is defined as the minimum duration of the single event with a contemporary occurrence of LWC>0, T<0 and V>0. The intensity of the phenomenon can only be determined if turbine's geometrical features and operating conditions are given. The most important specific wind turbine parameters are the relative wind velocity, the dimension and surface properties of the airfoils subjected to icing. Correlations between LWC and MVD with base cloud heights (CH) are valid for each specific site condition. On the other hand, the range of altitudes at which icing can occur is condition specific. Thus, extrapolating from aeronautical data produces uncertain results. For mountainous sites, significant influence of topography complicates the issue (Battisti et al., 2005a).

An alternative to this technique is the creation of an ice map. Several papers have been published with the objective of creating an icing map that would directly give the number of icing days with respect to the location. In Europe, this map was introduced by Makkonen and Ahti (Makkonen and Ahti, 1995), using cloud height (CH<200m), wind speed (V>0) and temperature (T<0) as conditions for ice accretion. They also found that the severity of rime ice is strongly related to the terrain roughness. The altitude relative to the sea level (ASL) does not significantly affect the calculated ice loads but the elevation in relation to the local terrain configuration (topography and roughness) does. Later, it has been found that this formula strongly overestimated icing, probably because of wrong wind speed measurement during icing events. A corrected equation was therefore proposed (Tammelin and Säntti, 1996). Later, comments on this equation mentioned that wind speed and temperature should be taken at a 200m height above ground. However, it remains that this formula can only provide a rough estimation of the amount of rime accretion (Tammelin and Säntti, 1998). The use of heated anemometers under similar conditions results in a smaller quantity of ice (Tammelin et al., 2000). Finally, Dobesh adds the visibility factor (Dobesch et al., 2003). He considers that icing occurs when the following conditions are met:

$$\begin{cases} T_{200m} < 0^{o}C \\ CH < 200m \text{ or } Visibility < 300m \\ V_{200m} > 0 \end{cases} \tag{1}$$

The introduction of horizontal visibility conditions takes into account that frequently no cloud height is observed during foggy weather conditions. The commonly used value of 1,000m for "visibility" in meteorology is not useful to predict icing. The visibility should be less than 300 meters to observe icing using the LWC. Dobesh concluded that the effect of solar radiation on icing duration does not improve the estimation of icing event maps at the site scale (Dobesch et al., 2003). Indeed, during winter, especially at higher latitudes, the intensity of solar radiation is too weak to enhance significant melting processes at low temperatures. Also, it is very difficult to get accurate data because the radiation network is very sparse and the use of analytical models is quite uncertain for the time span of one to several hours during the day.

Furthermore, wind turbines react differently to icing (Tammelin et al., 2005). A tool for estimating the number of icing days and icing intensity at a given site is still missing. Due to local topography, great variations in icing severity and intensity are observed within short distances. Therefore, icing maps cannot be interpreted as exact and must be used in connection with local topographical information and, if possible, measurement statistics (Laakso et al., 2003a).

Models: Physical mesoscale models (MM5, MC2 and others), generally used in regional weather prediction, can be used to predict upcoming icing events or the likelihood of such events for specific projects or time frames. More sophisticated empirical or statistical models consider additional parameters such as temperature (air, object, wet-bulb and dew point), wind direction, wind speed, cloud height, cloud cover, the humidity profile, precipitation, regional topography, local topography, object size, shape and material composite and solar radiation. These models can now provide information about the amount and rate of icing instead of just the frequency of icing events (Laakso et al., 2003a).

Other methods: Visual detection uses video filming of guy wires during icing events. Ice accumulation models are in reasonable agreement to the ice thickness observed on guy wires by an onsite web camera. Icefall due to wire vibration occurs and must be accounted for in the analysis. Predictions can be improved using onsite temperature and wind speed measurements or water droplet density information from a combined analysis of onsite visibility records and cloud base observations from the airport (Harstveit et al., 2005). A rain detector with a temperature sensor can be used to detect freezing rain (Tammelin et al., 2005). However, Laakso reports a case of a rain detector that did not indicate icing even though the maintenance staff of the turbines reported freezing precipitation (Laakso et al., 2003b). Finally, ice detection based on damage analysis, such as break down of meteorological masts or power lines due to buckling or possible resonance caused by additional masses, should be an exception, but can be an additional indicator for sites where heavy icing is not expected (Seifert, 2003).

2.4 Icing evaluation during wind turbine operation

In order to use active anti-icing - de-icing system (ADIS), a reliable instrument to observe rime accretion must be available (Tammelin and Säntti, 1994). Proper and fast identification of icing events is crucial because if the heating does not start as soon as icing starts, immediate production losses follow. A reduction of power production of about 5-15% can be seen before heat is switched on (Peltola et al., 1996). Optimized ice prevention systems of wind turbines include technologies allowing proper wind speed measurements in icing conditions and reliable ice detection (Makkonen et al., 2001). The performance of a blade heating system is highly dependent on the performance of a controlling ice detector. In some cases, several hours passed between the moment when icing could first be observed on video and the ice detector indicated icing. During this time, power production decreased considerably (Peltola et al., 2003). It has been observed that blade heating equipment receives more power than needed when there is a problem with ice detectors (Laakso et al., 2003a).

These methods for blade de-icing worked effectively, but the ice sensors used in the control systems could not reliably detect the onset of icing. A reliable icing detector is important to correctly activate the de-icing system. Detection of icing on wind turbines has different requirements than detection of icing on aircraft or for meteorological purposes (Homola et al., 2006). The ice detection must be made directly on the site (Fikke et al., 2006). Homola published a complete review of ice detection methods for operating turbines in 2006 (Homola et al., 2006). Three basic requirements for ice detection on wind turbines have been identified:

Sensor position on blade tip: Based on the icing model, the rate of ice accretion is directly related to the relative velocity of the super cooled water droplets and it is at the blade tip that the highest velocity occurs. Blade tips can experience icing due to low clouds even when the nacelle is ice-free. At Pori (Finland), measurements showed that the number of in-cloud icing periods at 84m was six times the number of in-cloud icing periods at 62m (using the relative humidity measurement method). The outer ends of the blades sweep a larger volume and collect water or ice from the entire volume. This becomes more important as the blade length increases. Because sensors are positioned on the tip of the blade near the lightning protection, it is difficult to access them in the event of failure. Also, problems associated with mounting a sensor on the blade's flexing material must be taken into account. Alternatively, the sensor can be a wireless unit for retrofitting of existing wind turbines.

High sensitivity to detect small accretion: De-icing by heating of the blades requires a much higher heating power if the airflow over the blade changes from laminar flow to turbulent flow (due to the increased heat removal of the turbulent layer). High safety risk from ice cast is already present with the accumulation of 1-2 cm of ice on the leading edge. Power production from the wind turbine is already reduced with the formation of surface roughness, with corresponding losses of income.

Ability to detect ice over a large area: Ice accumulation does not always occur at the same place on the blade. The location varies depending on the mechanisms of ice accumulation. Glaze icing can occur over large areas of the blade, with water running back and freezing away from the leading edge. Rime icing generally occurs on the leading edge, around the stagnation point, but the exact location can vary depending on the angle of attack. Also, accumulated ice can be shed from the blades. This can result in some areas of the blade having little or no ice while other areas have large accumulations. These phenomena indicate that wind turbine ice sensors should be able to detect ice at more than one or two points.

Using these criteria, 29 different methods of detection are presented in Homola's review. Here, only those that are most common in the literature are presented.

Ice detectors: This is the same technology as the one presented in the assessment section. It is also the most common method used to control the ADIS. One way to improve it could be the use of two detectors, one heated for intensity and the other unheated for duration (Tammelin et al., 2005). With slight icing, several hours can pass before icing is detected. Ice build up on blades more quickly because of their high velocity compared to static ice detectors (Laakso et al., 2003b).

Power curve errors: Comparison with normal operation power curve is used along with temperature and air pressure measurements. This analysis can be done by dividing the whole into sectors to consider the effects of wakes, topography and obstacles. A power drop of 50% is recommended as a reference for ice detection for stall regulated wind turbines. This method has the advantage of detecting icing where others fail. However, it was not possible to obtain accurate indication of ice accretion by comparing calculated and actual production power afterwards. Better results should be obtained with non-stop monitoring (Tammelin et al., 2005).

In another experiment, the operation of the de-icing system (warm air) was controlled with ice detectors and also by comparing the power output of the turbine with the expected performance that is calculated on the basis of anemometer indications (Laakso and Peltola, 2005). Power curves should be used as a safety check and always be implemented, but the difference between actual production and normal operation power curve may be explained by other factors than icing (Homola et al., 2006).

Multiple anemometry: This is the same technique as the one presented in the assessment section. The highest mounting place for an anemometer is the nacelle roof. Unfortunately, that level seems too low for detecting all in-cloud icing events at the blade tip level. In addition to this, anemometers should be seated with special care to avoid false alarms caused by the wind turbine wake effect (Marjaniemi et al., 2000).

Video monitoring: A rather good instrument for detecting ice at the rotor blades seems to be a web-cam in the hub, positioned where the pressure side of an iced rotor blade can be seen. For checking the blades' surface, in order to compare ice detection with other instruments or to check for ice accretion before a complete restart of the turbine after icing events, the web cam provides appropriate information. The disadvantage of this method is that visual observation must be continuous and requires good visibility at night, which is expensive (Seifert, 2003).

This method may be suitable for short periods of time. Although various sensors have been tested, the recording of conditions at the wind turbine using this method has not yet been conclusive for continuous ice detection, for several reasons. First, in arctic regions, there is little light during the winter. Artificial lighting is therefore needed, which can have negative visual impacts if it is in the visible spectrum. . The second and perhaps most important reason is the lack of suitable automated image analysis tools. Considering that image analysis is a fast-growing field, this type of system could become viable in the near future (Homola et al., 2006).

Vibration and noise: De-icing system operation (warm air) can be controlled with vibration sensors (Laakso and Peltola, 2005). These sensors are connected to the control system and in case of higher than normal vibrations, the turbine shuts down and blade heating starts. This method cannot detect icing during stall operation (Tammelin et al., 2005).

Another detection method for small amounts of ice accretion is the increase of aerodynamic noise coming from the rotor blades. A slim layer of ice at the leading edge increases noise and shifts the frequency to higher levels. The disturbed aerodynamic flow results in fully turbulent boundary layer from the leading edge, which produces a higher noise at a frequency that can be heard clearly (Seifert, 2003). Again, this method is not able to detect icing during stall operation (Tammelin et al., 2005). While measurement of the change in sound frequency seems to be a good indication of ice accretion, this method requires further investigation to determine how background noise and varying wind speeds affect data (Homola et al., 2006).

2.5 Recommendations

Different ice detection methods give different results. The main reasons for this are related to the fact that two different icing mechanisms are concerned: in-cloud icing due to super cooled droplets in low level clouds and freezing precipitation icing due to rain drops. Cumulative and transient effects may be significant and affect the accuracy of the predictions and, therefore, no method is accurate and reliable for all situations (Marjaniemi et al., 2000).

During Assessment Phase

To correctly evaluate icing events during the assessment phase, it is recommended to take measurements during at least one year using an ice detector, heated/unheated anemometers with heated boom, dew point and visibility detectors. It is possible for each detection method to give different information on frequency and duration of icing events. None of the ice detection methods compared here proved superior to others (Tammelin et al., 2005). Hence, the simultaneous indication of icing from at least two different sources improves prediction reliability. An ice detector is strongly recommended in conjunction with usual meteorological site measurements. Different ice detecting methods are suited to different climates and for different purposes. Different devices are needed to detect the persistency of icing and the actual icing time, the two variables that are needed to estimate ice climate. It is required to determine ice accretion period in order to determine the necessary heating energy, while the persistency of icing is necessary to determine the overall ice induced production losses. Since ice detectors are expensive for the assessment, installing one properly heated and one unheated anemometer on the measurement mast to estimate wind resource is cheap and advisable. This arrangement gives an overall picture of ice climate. Acquiring information such as time series of cloud base height from the nearest airport and comparing it with the measured data is also advisable. These two methods are likely to give a fairly good idea of the time that ice is likely to affect the operation of wind turbines. A dew point detector designed for subzero operation could also provide valuable information, because in practice, air temperature is at frost point nearly all the time when in-cloud icing occurs (Laakso et al., 2005).

During Wind Turbine Operation

A major review has been done by Homola (Homola et al., 2006) and among the 29 ice sensors, no satisfactory performance report in all conditions was found. A lot of methods, including ice collecting cylinders, dew point and temperature and double anemometry were rejected because they are mounted on the nacelle of the turbine and have limited applicability. In the event that they are modified, such that they can be mounted on the blade tip, they can be suitable and in any case they can be applicable in the absence of a blade-mounted sensor. Ice detection methods that are best suited for sensors mounted near the blade tips are infrared spectroscopy through optic fibre cables, flexible resonating diaphragms, ultrasound from inside the blade and a capacitance, inductance or impedance-based sensor. These methods were selected because they can directly measure some properties of the ice itself, they are sensitive to very thin layers of ice and they can be constructed with slight weight or no electronics near the blade tip. Among the ice detection methods used on aircraft, the ones based on capacitance and ultrasounds from within the blade seem the most suitable for wind turbines. The methods based on the resonant frequency of probes have not been shown to work successfully, perhaps due to either mounting on the nacelle or lower relative droplet velocities resulting in lower collection efficiencies than on aircraft.

3. Icing mitigation systems

Icing mitigation systems result from two main strategies: anti-icing and de-icing systems (*ADIS*). Anti-icing prevents ice to accrete on the object while de-icing removes the ice layer from the surface. Both strategies can also be divided into two methods: passive and active. Passive methods take advantage of the physical properties of the blade surface to eliminate or prevent ice, while active methods use external systems and require an energy supply that is either thermal, chemical or pneumatic (Dalili et al., 2009). Most of the strategies from the aerospace industry can be transferred to the wind energy sector, although, some scaling has to be done to adjust parameters (wind speed, chord length, airfoil) (Richert, 1996).

ADIS benefit and costs: No mass produced commercial ADIS currently exist. Although, no ADIS, passive or active, are totally effective to prevent initial and subsequent icing, during icing events some systems have proved to maintain power output, minimize dynamic impact of icing and, perhaps most importantly, avoid ice hazard close to habitable sites.

Consequently, because most ADIS are based on heating, wind turbines need more power to operate. The extra power will be added to the consumption of the nacelle cold climate package. Also, additional maintenance should be planned (Laakso et al., 2005). Tammelin et al. (Tammelin et al., 2005) concluded that the total energy consumed by WT in cold climate lies between 3 and 8% while the energy used by the anti-icing system alone is less than 3% of the total energy produced. Early studies evaluated the power consumption of ADIS electrical-heating to be 25% of the nominal power output (Makkonen and Autti, 1991), but more recent studies have lower estimates to 6 to 12% for electrical anti-icing, of 100-220kW turbines (Laakso et al., 2005) and 10 to 15% for warm air ADIS (Battisti et al., 2006). Other studies reported even lower levels, Seifert (Seifert, 2003) estimating that the heating elements consume 2% of nominal power output. In global values, the de-icing consumption translates in a 1 to 4% loss of annual energy production, depending on icing severity (Peltola et al., 2003). An investment of 5% of the cost of a 600kW turbine has been estimated for the purchase and installation of ADIS. Cost percentage decreases as turbine size increases (Laakso and Peltola, 2005). Depending on icing severity on the site (i.e. power losses related to icing) and the price of electricity, ADIS' payback time will vary from 1 to 18 years. For a site with medium icing severity, with an average of 30 icing days per year, the payback time should be less than 5 years (Tammelin et al., 2005).

Moreover, similar weather conditions may produce different icing events depending on the size of the turbine,the control strategy (stall or pitch regulated) and the operation regime (angle of attack of the blades, turning speed, etc.) (Battisti et al., 2005a). This underlines the importance of a good onsite icing evaluation for sizing ADIS.

ADIS control strategies: Most icing prevention methods are active heating systems that need a control strategy. The simplest strategy, if icing events are rare, is to continue to operate or to stop the turbine. In the second case, the turbine may be restarted automatically or after visual observation. In harsh conditions, it is recommended to use ADIS (Seifert, 2003). Another simple strategy for anti-icing system is to turn the power on all the time, but this increases energy consumption (Fortin, 2009). Usually, the ADIS basic control includes an ice detection method that turn on the system when ice is detected. If it is a heating system, more power can be delivered as the blade rotates faster, because the cooling intensity becomes higher while speed increase. Surface temperature is also a good indicator to adjust power output, avoiding overheating (Marjaniemi et al., 2000).

Anti-icing requires much more energy than de-icing because of the continuous heating. In theory, the surface temperature of the blade must be kept above 0°C whenever there is icing.

Moreover, when ice melts on the heated elements, water can run back after the element and freeze again. To avoid this, the water must evaporate, which implies for the heated element temperature to be at least 100°C (Fortin, 2009). For de-icing systems, however, the power that is required to remove accretions already formed through rapid heating far exceeds the power required for anti-icing (Laakso et al., 2005). Icing identification must be fast to avoid power losses.

To lower energy consumption, the blade span can be divided in separately controlled sections (Maissan, 2001). Because of low relative wind speed and smaller contribution to power production, the first two-thirds of the blade is less important to de-ice, and therefore requires less energy. Setting up a de-icing system only on the last third of the blade would enable to decrease equipment and energy costs while maintaining 90% of the aerodynamic performance of the clean blade with only 30% of the length de-iced (Hochart et al., 2008). The tip, however, must be as clean as possible. The use of heating resistance allows a more efficient energy distribution on the blade, which is particularly difficult to achieve with a hot air system (Mayer et al., 2007). For mechanical systems, a minimal ice layer thickness is required, and thin layers formed on the leading edge are harder to remove. It has been proposed to combine a heated element on the leading edge with a mechanical system elsewhere (Fortin, 2009).

3.1 Passive Anti-Icing and De-Icing Systems (ADIS)

The passive anti-icing systems are: special coating, black paint or chemicals. The characteristics, advantages and inconvenient of each system are presented here.

Special coating: In theory, ice-phobic coatings prevent ice from sticking to the surface because of their anti-adherent property, while super-hydrophobic coatings do not allow water to remain on the surface because of repulsive features. Reducing the shear forces between the ice and the surface will also reduce sensitivity to dirt and bugs (Seifert, 2003). Currently, most manufacturers use epoxy or polyester matrix composites reinforced with glass and/or carbon fibres, although polyester and glass fibres remain the material of choice due to their lower cost. Current research is heading towards nanocomposite coatings, polymers reinforced by minute, nanometre-scale particles. These nano-composites create very high contact angles with water (Dalili et al., 2009). A combination of coatings and active ADIS should be considered for preventing ice accretion (Kimura et al., 2003).

The advantages of the special coating are: low cost, no special lightning protection required, easy blade maintenance and a protection of the whole surface (Seifert, 2003). The reduction of ice adhesion combined to a blade heating system should lower energy consumption. The most effective coatings were found to reduce the adhesion of ice to about half of what is observed on uncoated aluminum (Anderson and Reich, 1997). The adhesion strength of accreted ice is reduced when blades are treated with the appropriate coating (water-repellent agent) (Kimura et al., 2003). A single application of coating material could provide a multipurpose solution that may reduce the frequency of unscheduled shutdowns and maintenance issues (Dalili et al., 2009).

However, the icing prevention on wind turbine blades by coating alone is not realistic. Several materials have been lab and field-tested but no adequate solution has been found. Icing occurred even on coated surfaces, regardless of the temperature (Kimura et al., 2003). None of the coatings was found to be truly ice-phobic (Anderson and Reich, 1997). Other disadvantages are the ice throw, the large accretion during severe icing and the unsymmetrical accretion leading to instability (Seifert, 2003). After a short period, the coating becomes porous and loses its ability to repel ice (Tammelin et al., 2000). Detailed information for most of these coatings remains proprietary. Consequently, most data sheets and chemical compositions are classified and inaccessible (Dalili et al., 2009).

Black paint: Black paint allows blade heating during daylight and is used with an ice-phobic coating. When tested in Yukon (Canada), this method showed immediate and noticeable improvement in performance (Maissan, 2001). This method may be sufficient in sites where icing is slight, infrequent and where icing periods are followed by temperatures above 0°C or in areas with high winter solar intensity at lower altitudes (Laakso et al., 2003a). Most of the time, it is not sufficient to prevent icing (Laakso et al., 2005). Temperature of the blade's surface may affect the properties of glass-fibre reinforced plastics, as they are sensitive to high temperature (Seifert, 2003). However, another study shows that black blades do not overheat in the summer (Weis and Maissan, 2003).

Chemicals: When applied on blade surface, chemicals lower the water's freezing (Patreau et al., 1998). It is mostly used during aircraft take-off. It is a pollutant and it needs special application and a lot of maintenance (Patreau et al., 1998). It cannot remain on the surface of the blade for a long period (Tammelin et al., 2000).

The passive de-icing systems are: *flexible blades* and *active pitching*. *Flexible blades* are flexible enough to crack the ice loose. Blade flexing is known to help shed the ice but very few information is available on the subject (Dalili et al., 2009). The *active pitching* is a semi-active method that uses start/stop cycles to orient iced blades into the sun (Laakso et al., 2005). The method may work in slight icing environments but it has not been scientifically validated and may damage wind turbines (Laakso et al., 2005).

3.2 Active Anti-Icing and De-Icing Systems (ADIS)

Anti-Icing Systems: The active anti-icing systems are used to prevent the ice accretion on blade surface and are based on resistive, air layer or microwave heating. Heating resistance and warm air can be used in anti-icing mode to prevent icing. The blade temperature should be kept around 0°C to prevent icing. The advantage is that no ice accumulates on blades. Blade can be kept at -5°C, instead of 0°C, in good condition. This way, 33% of power can be saved which represents 2.3% of winter production (Mayer, 2007). The inconvenient is that it requires a lot of energy. If it is used to prevent runback at 100°C, it is close to the softening point of some epoxies and resins (although thermosetting plastics that are designed for higher operating temperatures are available). The continuous operating temperature should be less than 50°C with current blade materials (Laakso and Peltola, 2005). The air layer consists in an air flow originating inside the blade and pushed through rows of small holes near the blades' leading and trailing edges in order to generate a layer of clean and, if necessary, heated air, directly around the blade surface (Dalili et al., 2009). This method would deflect the majority of water droplets in the air and would melt the few droplets that managed to hit the surface but very few information is available for its application (Dalili et al., 2009). Microwave heating consists in heating the blade's material with microwaves to prevent ice formation. The objective is to maintain the blade surface at a temperature slightly above 0°C, in order to save up some energy that will be used for defrosting. It is recommended to cover the surface of the blade with a material that reflects microwaves (metallic material such as wire mesh) and apply paint to improve the final surface. Another method consists in heating the blades when they pass in front of the tower by fixing an emitter on the tower. (Mayer, 2007). It has been tested at the LM Glassfiber workshop on a

LM19.1 blade with a 6kW power, a frequency of 2.45GHz and an emitted power less than 0.01W/m2 but is still to be implemented commercially (Mansson, 2004).

De-Icing Systems: The active de-icing systems are used to eliminate the ice accreted on the blade using a heating resistance, hot air, flexible pneumatic boots and electro impulsive/expulsive devices.

Heating resistance: The electrical heating uses an electrical resistance embedded inside the membrane or laminated on the surface (Laakso et al., 2005). The idea is to create a water film between the ice and the surface. Once this film created, centrifugal forces will throw the ice away (Battisti et al., 2006). Electrically heated foils can be heating wires or carbon fibres (Seifert, 2003). Heating elements cover the leading edge area of the blade. The ice detector and blade surface temperature are used to control the operation of the heating system. Additional temperature sensors are installed to protect the blade from permanent damage induced by over-heating. Heating foil can be applied to most turbines (Tammelin et al., 2005). For the Finnish JE-System, the estimated heating power to keep the total blade area rime and ice free is around 1.2kW/m (Tammelin and Säntti, 1994). Most recent results have proved to be about 0.5kW/m, which represents 5% of the wind turbine rated power (Marjaniemi and Peltola, 1998). A system of 15kW per blade has been used for a 600kW wind turbine, corresponding to 1-4% of annual production, depending on climate conditions (Laakso and Peltola, 2005). A system installed on a 1.8 MW turbine will needs 82kW per blade or 14% of power output at 8m/s (Mayer, 2007). Another system was tested using about 3.4kW per blades on small Bonus 150kW turbines (Pinard and Maissan, 2003). The minimum time to keep the heating on, after the icing event has passed, is usually about 15-30min (Peltola et al., 1996). Heating demand is almost linearly dependent on the temperature difference between the air and the blade surface (Marjaniemi and Peltola, 1998). More energy is needed to de-ice the tip's leading edge than the hub's (3.5 to 3.9 times more). More energy is also needed to de-ice the tip's trailing edge than the hub's (2.6 to 2.9 times more) and to de-ice the lower surface rather than the upper (1.3 to 1.5 times more) (Mayer et al., 2007). This simple method has been used successfully in the aerospace industry for many years. It has also been also used in the wind industry since 1990 (Dalili et al., 2009). JE Finnish's equipment is the most used and is installed on 18 wind turbines (Laakso and Peltola, 2005; Makkonen et al., 2001). The needed heating energy during rime accretion is quite small considering the profitability of wind energy production (Tammelin and Säntti, 1994). Heating power seems to be adequate except in the case of super cooled rain (Marjaniemi and Peltola, 1998). Thermal efficiency is close to 100% because of direct heating (Battisti et al., 2005a). Energy demand does not increase with blade size (Laakso and Peltola, 2005). As an inconvenient, there are many commercially available products but none are mass produced (Dalili et al., 2009). The technology is still at the prototype level because of the limited market (Laakso et al., 2005). If one heater fails, it will cause major imbalance on the whole system (Maissan, 2001). In some extreme icing cases, blade heating power was found to be insufficient (Peltola et al., 2003). Icing of the run-back water at the edges of the heating elements occurs quite often. When the running water from the heating element area reaches a cold blade surface, it re-freezes and forms a barrier at the edges of the heating element. The edge barriers may grow towards the leading edge as "horns" without a contact with the heating element. This could explain why in some blade icing cases, the thermostat of the ice prevention system indicates a temperature higher than 0°C on the surface of the heating element during icing (Makkonen et al., 2001). Heating elements can attract lightning but lightning protection is efficient and no damage was detected in the ice prevention system studied by Marjaniemi (Marjaniemi et al., 2000).

Hot air: The second method consists in blowing warm air into the rotor blade at standstill with special tubes (Seifert, 2003). Blowers located in the root of each blade or inside the hub produce the hot air. The heat is transferred through the blade shell in order to keep the blade free of ice (Laakso and Peltola, 2005). Again, the idea is to develop a water film between the ice and the surface. Once developed, it allows centrifugal forces to get rid of the ice (Battisti et al., 2006), but heating is also possible during parking (Laakso and Peltola, 2005). An air circuit is created inside the blade by dividing the internal volume in two parts. Hot air is injected in one part, which sends the cold air to the heating system on the other section (Mayer, 2007). Using a closed circuit, heating power is reduced significantly compared to an open cycle where air needs to be heated to the desired temperature starting from the outside temperature. Efficiency can be improved by using waste heat from the machinery (Peltola et al., 2003). A prototype is installed on an Enercon turbine in Switzerland and consists in a 7kW hot air blower in each rotor blade for a 850kW wind turbine. It consumes 1% of the total electricity production (Horbaty, 2005). Relatively low temperatures of the warm-air (80–120 °C) are suitable for the de-icing process, allowing lower temperatures (60–80 °C) of the blade surface, compared with the anti-icing practice (Battisti and Fedrizzi, 2007). The leading edge surface and the blade's aerodynamics are not affected. The system has no negative effect on the lightning protection system (Seifert, 2003). It works well in milder climates where icing occurs mainly at temperatures close to 0°C (Laakso et al., 2005). De-icing systems have a substantial advantage over anti-icing systems in terms of energy consumption: the energy consumption ratio is 50% for all simulations (Battisti et al., 2006). One inconvenient of the method is that it uses a lot of power at high wind speed and low temperature. Also, glass-fibre reinforced plastics (GRP) material is a good insulator and, as blades increase in size and thickness, more heat needs to be pushed and transferred trough the surface and to the tip of the blade (Seifert, 2003). The maximum operating temperatures of composites must be considered (Laakso et al., 2005). As this system works once the ice is accreted, there is a safety hazard related to ice projection. Thermal efficiency is low (about 30%) (Battisti et al., 2005a). The thermal efficiency of closed loop hot air based system will remain rather poor, because large mass of material has to be heated prior to attending the blade surface. Also, the heat source, often a hot air blower, is located typically at the blade root while the highest heat fluxes are needed at the tip of the blade (Laakso and Peltola, 2005).

Flexible pneumatic boots inflate to break ice. In the normal non-inflated state, tubes lay flat and are attached to the airfoil surface on which the de-icer is bonded. After the build up of generally 6 to 13 mm of ice on the surface of the airfoil, de-icers are inflated with compressed air. The inflation cycle lasts for a few seconds to achieve optimal ice shed and prevent additional ice formation on the inflated surface. After the ice has cracked, its bond to the surface is broken and it is removed through centrifugal and aerodynamic forces. De-icers are then allowed to deflate. Vacuum is then applied to ensure that there is no lifting of the surface on the low-pressure side of the airfoil (Botura and Fisher, 2003). Goodrich has tested this method in laboratory. Three 6 by 1m de-icers where tested on a simulated 1.5MW wind turbine rotor blade. De-icers for wind turbine applications have equivalent ice shedding and residual ice performance as conventional aircraft de-icers. Working at higher pressures for wind turbine applications, tests indicated satisfactory icing shedding on glaze ice at temperatures above -10°C and residual ice at temperatures between -10 and -20°C.

During in-field operation, residual ice is reduced due to blade vibration and centrifugal forces (Botura and Fisher, 2003). The system is installed on many aircrafts and has low energy consumption (Mayer, 2007). However, the method has yet to be field-tested for wind turbine application. The test is currently on hold pending agreement with a suitable wind turbine manufacturer or operator (Botura and Fisher, 2003). It may disturb the aerodynamics by increasing drag and cause more noise. Ice expulsion is a potential problem. During the 20 years of operation, it will require intensive maintenance, which may be expensive. High centrifugal loads at the outer radius of the pneumatic system will inflate itself or has to be divided in short sections (Seifert, 2003).

Electro impulsive/expulsive devices: This consists in very rapid electromagnetically induced vibration pulses in cycles that flex a metal abrasion shield and crack the ice (Dalili et al., 2009). A spiral coil is placed near the surface of the blade. When current is applied to the coil, a magnetic field is created between the coil and the thickness of the blade. The result is a rapid movement of the surface and the expulsion of the accumulated ice (Mayer, 2007). The method has been recently certified for use on Raytheon's Premier I business jet (Dalili et al., 2009). It is used by Hydro-Quebec for transmission lines and Goodrich is currently developing this method for aeronautical applications. The system is efficient, environmentally friendly, has low energy consumption, causes no interference with Hertz transmission and is easily automated (Mayer, 2007). In the mean time, it is a new technology that has not yet been tested on wind turbines. Ice expulsion is a potential problem (Mayer, 2007).

3.3 Synthesis and conclusion of existing methods for evaluation and mitigation of ice accretion on wind turbines (literature review)

Considering the current available technology, the following recommendations can be made for wind turbines exposed to icing and for the use of ADIS.

Icing assessment with multiple anemometry and relative humidity: double anemometry is a proven way to estimate onsite icing. As opposed to icing sensors, anemometers are cheap and have low energy consumption, which is a great advantage for remote site met masts. Triple anemometry seems a promising way to measure the severity and the duration of icing (Craig and Craig, 1996). Relative humidity or dew point detectors are also a cheap way to detect clouds and can identify icing for temperatures below 0°C. As this method is not reliable on its own (Tammelin et al., 2005), combining it with multiple anemometry seems ideal for assessment. Another way to detect clouds is video monitoring, but this method has yet to be automated.

Icing detection by ice sensors and power curve check during operation: ice sensor methods are currently the only proven way to directly measure icing during operation. It is also the most used method for current ADIS. Unfortunately, this method has several disadvantages. The most important one is that measurements are made at the nacelle level (Dalili et al., 2009). Combining this method with power curve checks can improve accuracy. Methods currently being developed, including capacitance and infrared stereoscopy, will be able to measure icing on several blade points (mainly close to the tip) with great precision (Dalili et al., 2009).

Passive method of special coating with active heating elements: This is the only method currently available and it has been tested for more than 20 years. This method is simple and its efficiency is close to 100% because it involves direct heating of the blades. It requires a large amount of energy that can be reduced through different strategies. First, a better control strategy that properly uses de-icing instead of anti-icing. Second, as detection methods improve, heating will be more efficiently started. Third, a combination with special coating will reduce adhesion of ice and run-backs. New developments in special coating will help reduce the energy demand. The warm air method will be more difficult to use with larger blades. Commercially produced anti-icing or de-icing systems have not yet been proven reliable and there have been reports on damage of prototype heating systems. Therefore, some manufacturers prefer using special coatings of the blade's surface instead of heating systems (Seifert, 2003).

4. Experimental analysis of wind turbine icing and optimization of electro-thermal de-icing

The wind farm near Murdochville, Quebec, is a good example of the severe effects of cold climate on wind turbines. The farm has 60 Vestas 1.8 MW turbines and is located between 850 and 950 m altitude. During the 2004-05 winter and spring, the meteorological station operated at 610 m altitude by the Wind Energy TechnoCentre, located near the wind park of Murdochville, recorded 13 icing events (Fortin et al. 2005a). Among these 13 events, five were considered severe and a hazard for the wind farm. Two events out of the five were selected for wind-tunnel simulation to study their effects on the Vestas-V80 wind turbine, through a quantitative study of ice-accretion shape, lift reduction and drag increase. The two icing events selected for the simulations were in-fog icing conditions as shown in Table 1. They were characterized by their liquid water content (LWC), median volume diameter (MVD) of the super cooled droplets, air speed (V_∞), air temperature (T_∞), and duration of the event (t):

Event	LWC (g/m^3)	MVD (μm)	V_∞ (m/s)	T_∞ (°C)	T (min)
1	0.218	38.3	8.8	-1.4	360
2	0.242	40.5	4.2	-5.7	264

Table 1. Characteristics of measured icing events used for wind-tunnel simulation of in-fog icing (Fortin et al. 2005a)

The tests were carried out in the AMIL (Anti-icing Materials International Laboratory) icing wind tunnel (IWT) at the Université du Québec à Chicoutimi (Figure 1). It is a refrigerated closed loop wind tunnel 4.5 m wide and 9.5 m long. The test section is 0.6 m high, 1.5 m long and 0.5 m wide. The working temperature range is -30 °C to +25 °C. The maximum wind speed is 70 m/s.

In-fog icing is produced using an oscillating spray-nozzle assembly located upwind from the convergent. The spray nozzles are set to produce water droplets with a diameter of 27.6 μm. The lift and drag forces are measured using an aerodynamic scale made up of two aluminum arms linked together by a bearing. A load cell is placed at the end of each arm to record the lift and drag forces on blade airfoil in the test section.

Generally, the shapes of ice deposits used in wind-tunnel aerodynamic simulations are measured directly on blades during icing events, or calculated by ice-accretion simulation software. An artificial deposit is then moulded and glued along the blade profile to simulate the 2D runoff on an iced blade profile. Seifert and Richert (Seifert and Richert, 1997) presented experimental measurements of lift and drag on a blade airfoil, the leading edge of which was covered with artificial ice deposits shaped from actual deposits collected from a small, horizontal-axis wind turbine during different icing periods. Jasinski (Jasinski et al., 1997) made the same measurements, but used artificial ice shapes created with the LEWICE

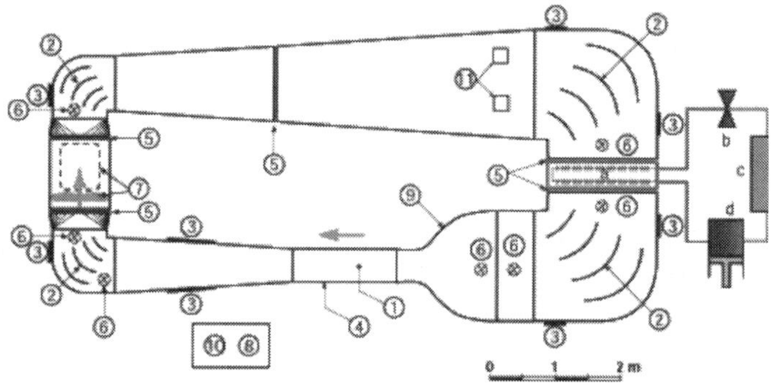

1	Test section	9	Contraction cone
2	Corner vanes	10	Control console
3	Access doors	11	Traps
4	Access panels	a	evaporator
5	Thermal expansion joints	b	Expansion valve
6	Spray nozzle ramp	c	Condenser
7	Fan and Motor	d	Compressor
8	Motor control panel		

Fig. 1. The AMIL Refrigerated Wind Tunnel (Hochart et al. 2008)

ice-accretion simulation software at NASA. The special feature of the experiments described here (Hochart et al. 2008) resides in the way the ice deposits on the blade airfoil was obtained by simulating in a wind tunnel the meteorological and operating conditions of the wind turbine during in-fog icing. The effects of ice accretion were determined in two phases: one phase of ice-accretion when the shape has been determined and a second phase to determine the aerodynamic characteristics (lift and drag) of the iced airfoil. A load calculation based on the blade element theory [Burton et al. 2001] was used to estimate the effects of icing on the driving and bending forces, as well as torque. The resulting data were used as a basis to determine the power loss and the best position for the heating-element of a de-icing system.

A second analysis was done to establish the de-icing parameters in order to optimize the heating process and minimize electric energy consumption. The calculation of the power used for the de-icing is based on the evaluation of the convective heat transfer on the airfoil surface. The experimental study quantifies the power consumption for the whole icing event as well as the evolution of the surface temperature and heating.

The Vestas-V80 wind-turbine blade uses a NACA 63 XXX airfoil between the blade tip and its centre, and a FFA W3 XXX airfoil between the centre of the blade and the hub (Anonymous, 2004). Because the exact blade airfoil configuration was unknown, a 0.2 m (chord) x 0.5 m (width) NACA 63 415 airfoil was chosen for testing. The model for the analysis of ice accretion shape was cut from a block of 6061-T6 aluminum, has a 200 μm surface finish and was horizontally mounted, suction side upwards, in the test section (Figure 3a). The blade section used for the analysis and optimisation of de-icing system is made with fibreglass tissue as close as possible to that of the real blades. Considering its size, the fibreglass tissue layers of the blade section follows the orientation [±45°/0°/±45°] (McKittrick et al., 2001). Consequently, the thickness of fibreglass is approximately 1.96 mm along the airfoil. It is equipped with 12 resistant heating elements and instrumented with 12 thermocouples (Figure 3b).

(a) (b)

Fig. 2. (a) Blade Sections for Ice Accretion and (b) De-Icing Analysis (based on NACA 63-415 airfoil) (Hochart et al., 2008, Mayer et al. 2007)

5. Experimental evaluation of icing effect on the wind turbine performance

5.1 Test cases

To determine the effect of ice accretion at different span positions across the blade, the cinematic conditions have been simulated at three radial positions, 12 m, 23.5 m and 35 m, of the 40 m blade. Each simulation included two major parameters, the relative wind speed (V_{rel}) and the angle of attack (α). As shown in Figure 3, these parameters were calculated from the wind speed at the rotor disc entrance (V_{vent}), the tangential speed (V_{tang}) and the pitch angle (φ).The relative wind speed was:

$$V_{rel} = \sqrt{V_{vent}^2 + V_{\tan g}^2} \tag{2}$$

and the angle of attack (α) was:

$$\alpha = \arctan\left(\frac{V_{vent}}{V_{\text{tang}}}\right) - \varphi \tag{3}$$

The wind speed at the rotor disc entrance (V_{vent}) was calculated using the actuator disc concept (Burton et al., 2001),

$$V_{vent} = V_{\infty}(1 - a) \tag{4}$$

and the tangential speed (V_{tang}) of the blade section was derived from the rotor disk theory (Burton et al., 2001):

$$V_{tang} = \omega r(1 + a') \quad (5)$$

The axial induction factor, a, was assumed to be 1/3. This is an optimal value for the wind turbine power coefficient (C_p), according to the actuator disc concept. The radial induction factor was assumed to be very small ($a' << 1$) and tip corrections were not included. The twist angle was calculated for an optimal lift to drag ratio along the blade with a free stream speed (V_∞) of 8 m/s. These assumptions, as explained in the blade element theory (Burton et al., 2001), are usually good approximations for fairly well designed wind turbines in normal conditions (without ice). Therefore, they were considered as acceptable to the aim of this work, which is not to accurately calculate air flow or aerodynamic forces along the blade but only to emphasize the difference between iced and non iced situations.

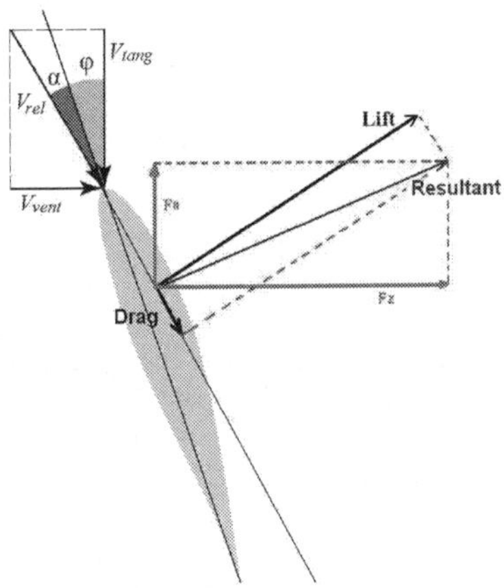

Fig. 3. Cinematic of the blade section (speed and angle of attack)

The meteorological conditions for the two in-fog icing conditions selected were scaled down to wind-tunnel dimensions. The method described by Anderson (Anderson, 2004) was used. The fixed variables for scaling were the model chord, 0.2 m, and the median volume diameter (MVD) of the water droplets, 27.6 μm. The imposed variable was the air speed in the wind tunnel, which corresponds to the relative air speed at the radial position tested. The free variables were the liquid water content, air temperature, and duration of the event. The simulation parameters for the six tests are shown in Table 2. They are the radial position (r), angle of attack (α), liquid water content, median volume diameter of the supercooled water droplets, relative air speed (V_{rel}), experimental Reynolds numbers (Re), wind-tunnel temperature (T_∞), and duration of the event (t).

The liquid water content (LWC) was calibrated using the rotating cylinder method (Stallabrass, 1978), which consists in accreting ice on a rotating cylinder of 5 cm diameter during one hour. The spray nozzles were adjusted to yield, at a given speed, the desired liquid water content. The experimental method for the simulations consisted in positioning the blade airfoil (Figure 2a) at the desired angle of attack; setting the speed, temperature, and liquid water content in the test section; accreting ice on the airfoil for a specified duration; measuring the lift and drag coefficients; weighing the blade profile to determine the mass of accreted ice; and draw the ice shape at the centre of the blade section. Each simulation was repeated once to ensure conformity of results.

Test	Fog	R (m)	α (°)	LWC (g/m^3)	MVD (μm)	V_{rel} (m/s)	Re	$T\infty$ (°C)	T (min)
1	1	11.9	13	0.37	27.6	19.9	2.65×10^{05}	-1.4	14.8
2	1	23.4	13	0.48	27.6	38.0	5.07×10^{05}	-1.4	15.1
3	1	34.8	13	0.48	27.6	56.0	7.47×10^{05}	-1.4	24.8
4	2	11.8	3	0.37	27.6	18.7	2.49×10^{05}	-5.7	10.6
5	2	23.3	7	0.48	27.6	36.7	4.89×10^{05}	-5.7	11.8
6	2	35.0	9	0.48	27.6	55.0	7.33×10^{05}	-5.7	19.6

Table 2. Wind-tunnel simulation parameters

5.2 Results

The results of the six simulations for ice mass, ice-deposit shape, lift reduction and drag increase are described in this section.

In-fog icing event 1

Tests 1 to 3 simulated the effects of in-fog icing event 1 on three positions of a Vestas 1.8 MW wind-turbine blade. The icing event characteristics were as follows: LWC of 0.218 g/m^3; temperature of -1.4 °C; wind speed of 8.8 m/s; duration of 6 hrs. For this wind speed, the angle of attack was calculated to 13° for all simulations. Simulations 1, 2, and 3 correspond to 11.9 m, 23.4 m, and 34.8 m span positions from the hub, respectively.

Figure 5a shows the masses and shapes of the ice deposits for simulations 1 to 3 of in-fog icing 1. For the three simulations, the deposits on the blade were glaze, a transparent ice of high-density (917 kg/m^3) characteristic of wet-regime accretions. A fraction of the water striking the leading edge of the blade profile froze upon impact while the rest ran along the pressure surface and, at very high speeds, along the suction surface as well. All or some of the running water may freeze on the pressure and suction surfaces of the blade airfoil.

Figure 5b shows the lift coefficient reduction and the drag coefficient increase for wet-regime simulations 1, 2 and 3. The lift coefficients measured on the iced profiles were 0.697, 0.685 and 0.553 for the simulations corresponding to radial position 11.9 m, 23.4 m and 34.8 m respectively. The drag coefficients measured for the same simulations were 0.068, 0.090, and 0.195 respectively.

The unfrozen water flowed to the trailing edge where some of it froze and the rest sprayed off into the air. Moreover, because of the sharp angle of attack, some droplets struck the pressure surface, thus increasing the water flow. In the ice accretion simulation near the hub (Figure 6a), the glaze on the leading edge followed the contour of the blade profile. In the ice accretion simulation near the middle of the blade (Figure 6b and Figure 6c), the glaze on the leading edge and that on the pressure surface followed the contour of the blade profile. In the ice accretion simulation near the blade tip (Figure 6d), the glaze on the leading edge was horn shaped, on the pressure side followed the contour of the blade profile, while on the suction side formed rivulets. The glaze on both sides of the airfoil was the result of runoff water that froze nearly completely for the simulation near the hub, and partially for the simulations near the mid and tip positions. For these last two positions, a fraction of the runoff water froze on the trailing edge. The quantities of captured water and glaze increased with an increase in the relative air velocity seen by the blade section. The ice masses experimentally accreted on the blade section in the tunnel were 48 g, 130 g and 354 g for the simulations corresponding to radial positions 11.9 m, 23.4 m and 34.8 m respectively.

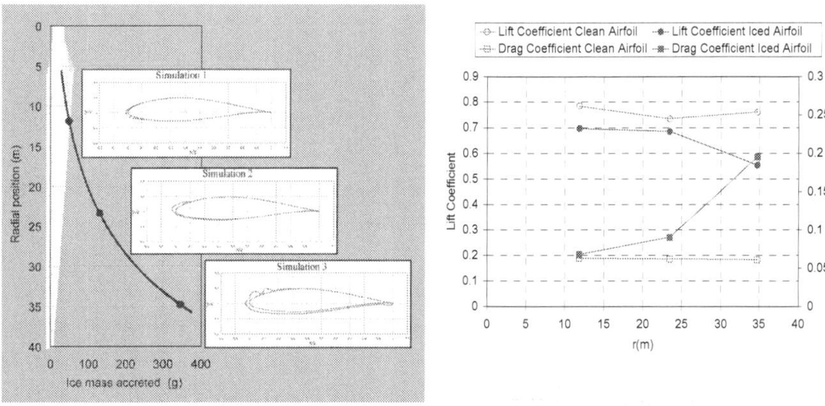

Fig. 5. (a) Masses and shapes of ice deposits for icing event 1 and (b) Lift and drag coefficients for icing event 1

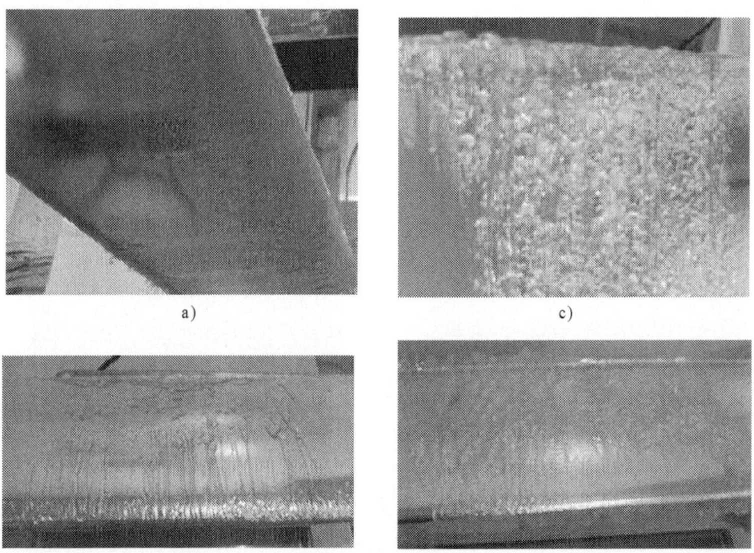

Fig. 6. Iced blade shape at the end of the simulations, a) simulation 1, view from below, b) simulation 2, view from above, c) simulation 2, view from below, d) simulation 3, view from below

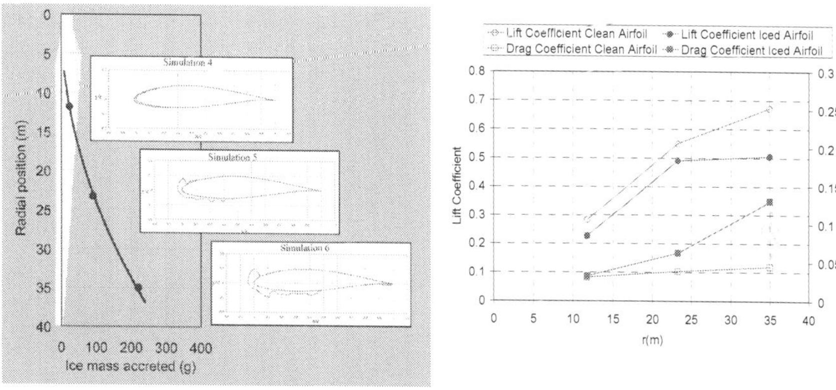

Fig. 7. (a): Masses and shapes of ice deposits for icing event 2 and (b): Lift and drag coefficients for icing event 2

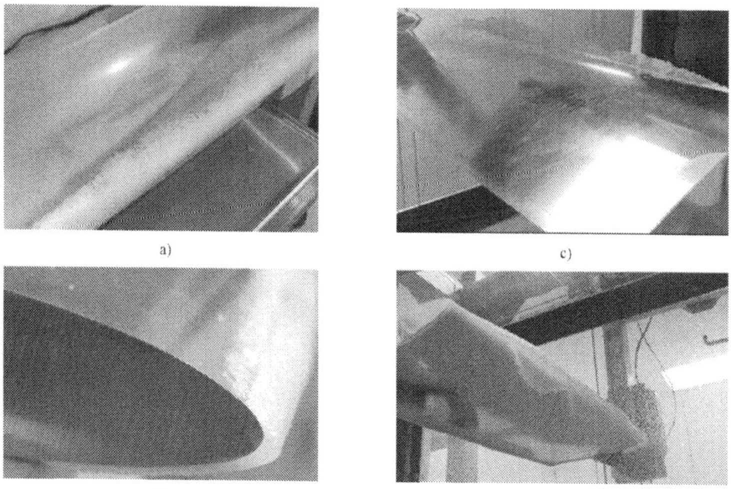

Fig. 8. Iced blade profiles at the end of the simulations, a) simulation 4, view from below, b) simulation 5, side view and from below, c) simulation 6, view from above, d) simulation 6, view from below

All the water striking the leading edge and the blade profile froze upon impact. For the simulation corresponding to the cross section closest to the hub (Figure 8a), the rime on the leading edge, pressure surface, and suction surface partially followed the contour of the blade profile and formed slight protrusions. For the cross section near the middle of the blade (Figure 8b) and closest to the tip (Figure 8c and Figure 8d), the rime on the leading edge had a double-horn shape, and that on the pressure and suction side partially followed the contour of the blade profile and exhibited protrusions. The rime was oriented in the direction of the water droplets incidence angle, creating zones of shadow with little accretion, leading to the formation of protrusions. The quantity of accreted rime increased with an increase in the relative air velocity seen by the blade section, due to the proportional increase of captured water, as follows: 24 g, 91 g and 220 g for the simulations at radial positions 11.8 m, 23.3 m and 35 m respectively.

5.3 Analysis

The dry-regime simulations (icing event 2) were easier to carry out than those in wet regime (icing event 1) because they have better reproducibility. Each simulation was repeated once. Tables 3, 4 and 5 show the mean values and the standard deviations of ice mass, lift coefficient and drag coefficient for the two simulations carried for each regime. Standard deviations are based on the two average values measured during the experiments and not on the signals in time.

Icing Event	Wet Regime (Event 1)			Dry Regime (Event 2)		
Radial Position(m)	11.9	23.4	34.8	11.8	23.3	35.0
Average mass of ice (g)	48	130	354	24	91	220
Standard Deviation(g)	0.25	9.25	4.5	1.75	0.25	5.5
Standard Deviation (%)	0.52	7.07	1.27	7.36	0.28	2.5

Table 3. Average values and standard deviations of ice mass

Icing Event	Wet Regime (Event 1)			Dry Regime (Event 2)		
Radial Position(m)	11.9	23.4	34.8	11.8	23.3	35.0
Average Lift Coefficient	0.697	0.685	0.553	0.227	0.491	0.226
Standard Deviation	0.021	0.011	0.088	0.004	0.012	0.016
Standard Deviation (%)	3.04	1.54	15.9	1.87	2.39	3.2

Table 4. Average values and standard deviations of lift coefficient

Icing Event	Wet Regime (Event 1)			Dry Regime (Event 2)		
Radial Position(m)	11.9	23.4	34.8	11.8	23.3	35.0
Average Drag Coefficient	0.068	0.09	0.195	0.033	0.063	0.13
Standard Deviation	0.01	0.017	0	0.0026	0.0017	0.009
Standard Deviation (%)	14.7	18.4	0	7.9	2.8	6.9

Table 5. Average values and standard deviations of drag coefficient

As shown in Figures 5a and 7a, in both wet (icing event 1) and dry (icing event 2) regimes, because of local cinematic conditions, the ice mass accreted on the airfoil increases as the cross section moves from the hub to the blade tip. In order to show dimensionless results (Table 6), the accreted ice masses on the experimental blade airfoil have been revaluated for the six simulations considering a standard 1 m (chord) x 1 m (width) NACA 63 415 airfoil and the six ice shapes of Figure 5 and Figure 8. For a real blade, the chord length decreases with the radial position from the hub to the blade tip. Considering this size variation, it is in the median zone of the full size blade that the largest quantity of ice accretes, as shown in Figure 9. The total mass of accreted ice is estimated to 709 kg for event 1 (11% of the blade initial mass, which is 6500 kg) and 434 kg for event 2 (6.7% of the blade initial mass).

Icing Event	Wet Regime (Event 1)			Dry Regime (Event 2)		
Radial Position(m)	11.9	23.4	34.8	11.8	23.3	35.0
Average mass of ice (g)	2400	6500	17700	1200	4550	11000

Table 6. Average values of ice mass, dimensionless 1 m (chord) x 1 m (width) profile

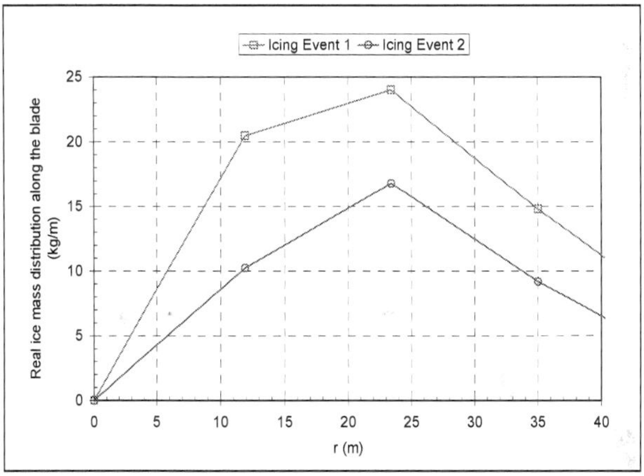

Fig. 9. Mass of ice accumulated along the full size rotor blade

In both dry and wet regimes, the lift and drag coefficients are more affected as we move from the hub to the blade tip. In Figures 7 and 10 it is illustrated how the lift coefficient decreases and drag coefficient increases with the radial position on the blade. The drag coefficient variation with radius follows approximately a power law. Especially between the middle and the blade tip, drag coefficient increased considerably and, combined with lift decreases, lead to a significant reduction of lift to drag ratio. In wet regime (icing event 1), we estimated that drag coefficient increased 7.7 % at 11.9m, 45.7 % at 23.4 m and 220 % at 34.8 m, according to the test results corresponding to the respective positions on the real blade. Using the same assumptions, the lift coefficient decreased 11.2 % at 11.9 m, 6.8% at 23.4 m and 27.2 % at 34.8 m. The drag coefficient increase at the blade tip (40 m) is estimated to 365% and the lift coefficient reduction to 40 %. In dry regime (icing event 2), drag coefficient increased 5.5 % at 11.8 m, 61.3 % at 23.3 m and 190 % at 35.0 m, according to the test results corresponding to the respective positions on the real blade. Lift coefficient decreased 19.8 % at 11.8 m, 10.7 % at 23.3 m and 24.8 % at 34.8 m. The drag coefficient increase at the blade tip (40 m) was estimated to 250 % and the lift coefficient reduction to 37%.

In order to assess the effect of ice on the aerodynamic forces on the full size rotor, a load calculation based on the blade element theory (Burton et al, 2001) has been used. The orthoradial force component, which generates rotor torque, is called driving force and noted as F_θ. The force component perpendicular to F_θ, noted as F_Z, is oriented in the direction of the rotor axis and serves to estimate the bending force applied to the blade. The formulas for dF_θ and dF_z are:

$$dF_\theta(r) = \left(\tfrac{1}{2}\rho\, c(r)\, \sqrt{V_{vent}^2 + r^2\omega^2}\right)\left(C_L(r)V_{vent} - C_D(r)\;\omega r\right)dr \quad (5)$$

$$dF_z(r) = \left(\tfrac{1}{2}\rho\, c(r)\sqrt{V_{vent}^2 + r^2\omega^2}\right)\left(C_L(r)\omega r + C_D(r)V_{vent}\right) dr \qquad (6)$$

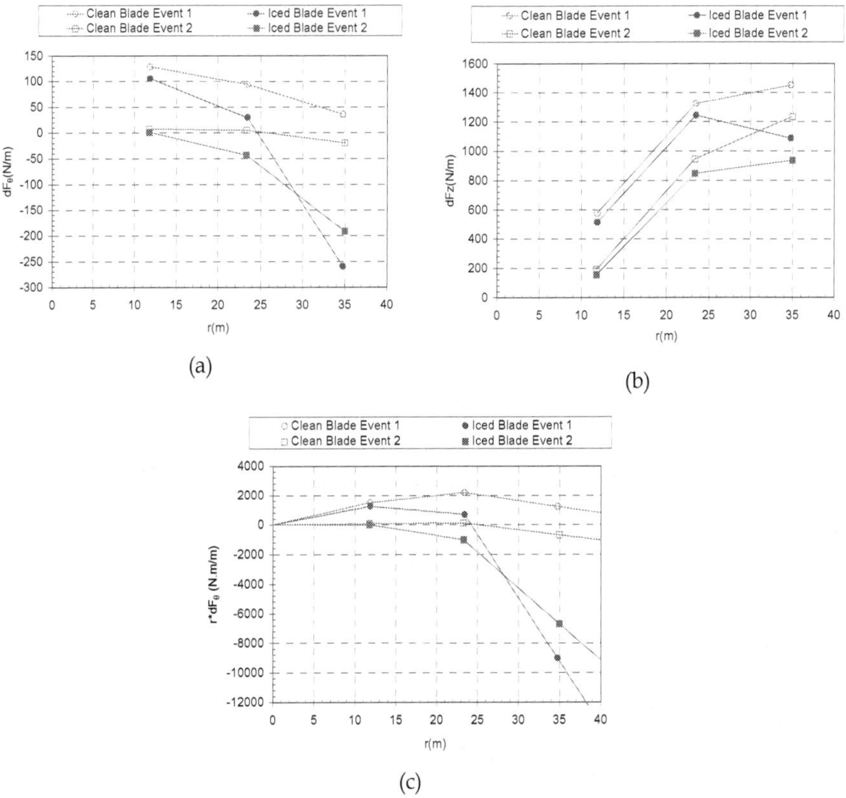

Fig. 10. (a) Distribution along the full size blade of the driving force per length unit dF_θ; (b) Distribution along the full size blade of the bending force per length unit dF_z and (c) Distribution along the full size blade of the torque per length unit $r*dF_\theta$

Here, *r is* the radial position in m, ρ the air density in kg/m³, c the blade chord in m, ωr the tangential speed in m/s, and V_{vent} the wind speed in m/s at the rotor disc entrance. The driving (dF_θ) and bending force (dF_Z) variation along the blade span are shown on Figure 10a and 10b. Figure 10c shows the torque distribution ($r \times dF_\theta$) linearly interpolated over the entire blade length in order to estimate the total torque. During both wet and dry accretion regimes, the driving and bending forces acting on the blade decrease, leading to a drastic torque reduction. In both cases, the drag force becomes so large compared to lift, that a negative torque occurs leading to rotor deceleration and possible stop. Torque reduction is more significant on the outer third of the blade so that the efficiency of a de-icing system would be increased in that region.

In Figure 11 we illustrate the variation of the total torque produced by the blade with the length of the de-icing system. The de-icing system is installed over a given length starting from the blade tip and the lift and drag coefficients of the clean airfoil are used where the de-icing system is operational. We notice again that the most efficient zone to be de-iced is near the blade tip as approximately 90% of the torque penalty compared to the clean blade is recuperated with only 30% length de-iced.

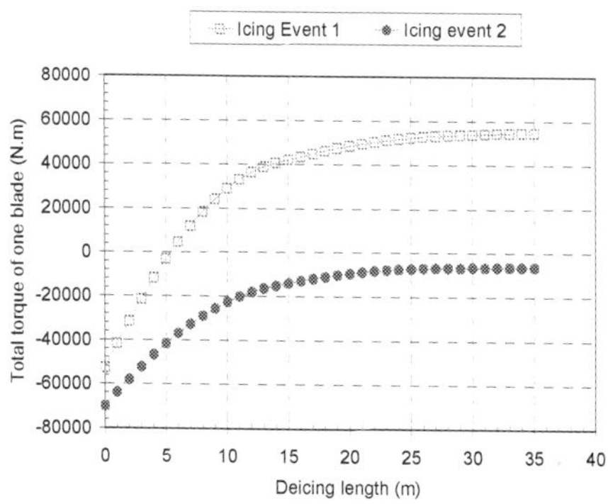

Fig. 11. Blade torque as a function of de-icing length, starting from the blade tip

5.4 Conclusion

The study provide the experimental assessment of the impact of glaze (icing event 1, wet regime) and rime (icing event 2, dry regime) on a wind-turbine blade. The liquid water content (LWC) for glaze accretion was 0.218 kg/m³, at -1.4 °C and 8.8 m/s wind speed, while the LWC for rime accretion was 0.242 kg/m³, at -5.7 °C and 4.2 m/s wind speed. In wet-regime (icing event 1), the angles of attack along the blade were 13° in average and glaze formed mostly at the leading edge and on the pressure side. Some ice accreted by runoff on the trailing edge for cinematic conditions corresponding to the blade airfoils located at the centre and the tip. In dry-regime (icing event 2), the angles of attack were below 9° and rime accreted mostly on the leading edge and partially on the pressure side for cinematic conditions of the blade airfoils located between the middle and the tip. The rime accreted on the leading edge was horn shaped, which considerably increased the surface roughness. The total mass of accumulated glaze on the blade was estimated to 709 kg (11% of the blade initial mass, which is 6500 kg) and the total mass of accumulated rime was estimated to 434 kg (6.7% of the blade initial mass). When glaze or rime accreted on the blade profile, lift decreased and drag increased. In both dry and wet regimes the lift reduction varied only slightly on the first two thirds of the blade, 9 %, but increased to 25 % on the last third, near the tip. The lift reduction at the blade tip was estimated at 40 % for both events. Drag increased along the blade following approximately a power law. The increase at the blade tip was in the order of 365 % for glaze and 250 % for rime. The amount by which lift decreased or drag increased depended on the quality, shape, and position of the ice. Finally, based on blade element model estimations, for both icing conditions the lift reduction and drag increase lead to a decrease in the bending and driving forces, and consequently a decrease in torque. The drag force becomes so important compared to lift that the torque is negative, resulting in rotor deceleration and stop. Torque reduction is more significant on the outer third of the blade. Setting up a de-icing system on the last third of the blade only, would enable to decrease equipment and heating energy costs while maintaining 90% of the aerodynamic performance of the clean blade.

6. Experimental analysis and optimisation of an electro-thermal de-icing system

6.1 Scaling

Based on the results of the previous study on the effects of ice accretion on wind turbine performance, a second experimental study has been conducted to determine what would be the optimum positioning and functioning parameters for an electro-thermal de-icing system. The blade section is a NACA 63-415 airfoil, 0.5 m long with a 0.2 m chord. It is equipped with 12 resistant heating elements and instrumented with 12 thermocouples (Figure 2b). The meteorological parameters used for the experimental analysis of the de-icing system are presented In Table 7. These values have been scaled for simulation in the IWT (Figure 1). The main difference with the observed values is the colder temperature used in the IWT simulation which guarantees a rime ice along the blade. Rime ice is the most frequent during icing events in wind farms and it is the major cause of production losses.

LWC (g/m^3)	MVD (μm)	U_∞ (m/s)	T_{amb} (°C)	T (min)
0.218	38.3	8	-10	360

Table 7. Characteristics of the icing event for the tests (full scale)

The blade geometry used for the testing corresponds to a Vestas V80 turbine of 1.8 MW. This type of wind turbine is presently installed where the meteorological data has been gathered. To simulate the icing event in the IWT, the icing conditions and the blade geometry need to be scaled. The wind speed selected for simulation is 8 m/s and corresponds to a configuration for which the angle of attack is 12° along most of the blade. The blade section is made of fibreglass and uses a NACA 63 415 airfoil. For the airfoil, 12° is the ideal angle of attack in terms of aerodynamic performance. All the tests were performed for this wind speed that corresponds to the maximal power coefficient of the wind turbine. Three different span positions on the real blade have been modelled. For the experimental value of the LWC in Table 1, the wind velocity in the IWT test section is limited to 40 m/s. Consequently the maximum span location that can be simulated is limited to 23 m. The relative velocity (V_{rel}) expressed in Equation (2) and simulated in the icing wind tunnel (IWT) is the combination of the wind velocity (V_{vent}) and the tangential velocity (V_{tang}) due to the rotation of the blade (Figure 3). The relation between the span position r and the relative velocity V_{rel} is obtained using the rotor disc theory (Burton et al. 2001). The wind speed that crosses the rotor plane (V_{vent}) is expressed in Equation (4) as a function of the free stream speed (V_∞) and the axial flow induction factor (a). The tangential velocity (V_{tang}) at the span location r is expressed in Equation (5) as a function of the rotational speed (ω), the span position r and the tangential flow induction factor (a'). Using the optimum value of the axial flow induction factors (a=1/3) the tangential flow induction factor becomes:

$$a' = \frac{U_\infty^2 a(1-a)}{\omega^2 r^2} \tag{8}$$

The relation between the span position and the relative wind speed simulated in the IWT is expressed as:

$$r = \frac{\omega\sqrt{U_{rel}^2 - U_w^2} + \sqrt{(U_{rel}^2 - U_w^2)\omega^2 - 4\omega^2 \cdot \frac{2}{9} U_\infty^2}}{2\omega^2} \tag{9}$$

Three different tests are performed considering a relative wind speed of 20m/s, 30m/s and 40m/s respectively. According to these calculations, Table 8 presents the corresponding span positions and the blade chord lengths for the selected IWT test section speeds. The chord length corresponds to the dimensions of the Vestas V80 wind turbine of 1.8 MW.

N° test	U_∞ (m/s)	U_{rel} (m/s)	r (m)	c (m)
1	8	20	10.6	2.7
2	8	30	16.6	2.3
3	8	40	22.5	1.8

Table 8. Corresponding real chord and span positions for the tests performed

According to the icing conditions and the geometry, the real characteristics are scaled using a scaling method developed by Anderson (Anderson, 2004). The Anderson scaling method gives excellent results for rime ice scaling. The experimental conditions for the three wind tunnel tests, corresponding to the span positions in Table 8, are presented in Table 9. The ice accretion in the wind tunnel is performed both with and without de-icing system working.

N° test	LWC (g/m^3)	MVD (μm)	α (°)	c (m)	U_{rel} (m/s)	T_{amb} (°C)	T (min)
1	0.4	26.7	12	0.2	20	-10	14.3
2	0.4	26.7	12	0.2	30	-10	16.9
3	0.4	26.7	12	0.2	40	-10	21.7

Table 9. Test conditions set for the IWT (each test number corresponds to the equivalent span position in Table 8)

6.2 Heating elements

The heating elements are Kapton flexible heaters with a wattage density of 10 W/po^2 (1.55 W/cm^2). They are distributed on the airfoil as shown in Figure 12. The blade section is built in two parts in order to be able to open the airfoil if a technical problem with the thermocouples or the heating elements appears. The heating elements 0 and 3 are on the upper section and the heating elements 1 and 2 are on the lower section of the airfoil.

For each heating element, a thermocouple is placed between the surface and the heater (external thermocouple, referred as Th_e) and another one is placed inside and in front of the airfoil (internal thermocouple, referred as Th_i).

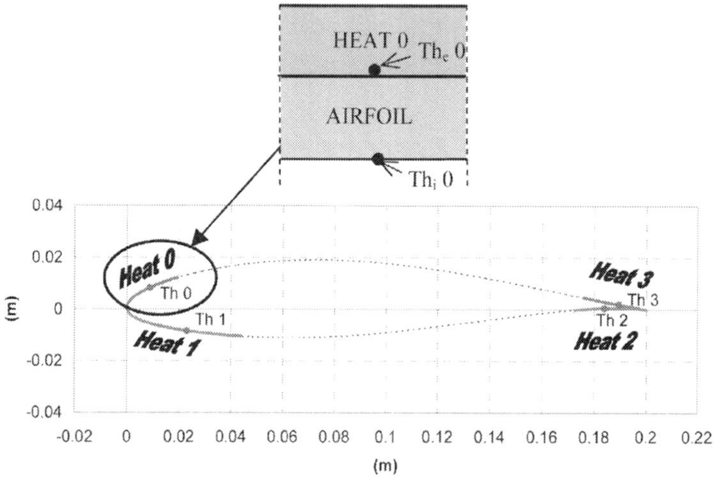

Fig. 12. Location of thermocouples and heating elements

6.3 Heating control

A computer controls the electric power sent to each heating element based on the temperature measured by the external thermocouples. The power consumed and the surface temperature of the system are recorded during the test. This allows studying the system's response depending on the icing condition applied. A comprehensive scheme of the de-icing system is presented in Figure 13.

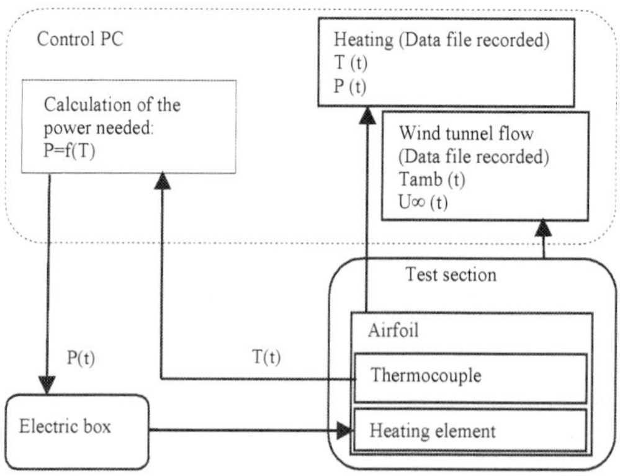

Fig. 13. Control Scheme of the De-Icing Control System

To optimize the power supplied to the heating elements, the amount of heat is related to the convective energy transfer during ice accretion at the blade surface. The other forms of energy transfer during ice accretion, which are adiabatic heating, conduction, evaporation and radiation, are all considered negligible compared to convection. The adiabatic heating is

low compared to the convection when the Mach number is less than 0.3. The conduction is negligible compared to the convection when the Biot number is high. Due to the low air temperature, little evaporation occurs during ice accretion and the radiation is negligible due to the clouds coverage during an icing event.
The power of the heating elements used on the airfoil's surface is expressed in Equation (10) and depends on the convective heat transfer coefficient h_1, the ambient, target and surface temperatures T_{amb}, T_c, T_h and the heating element area A:

$$P(t) = h_1 A(T_{amb} - T_c) + h_1 A(T_c - T_h) \tag{10}$$

The first right-hand-side term represents an approximation of the power needed for the de-icing and the second one represent the correction, considering the gap between the target and measured surface temperatures. The «heating coefficient» h_1 is approximated with the convective heat transfer coefficient, evaluated using a flat plate hypothesis given by Equation (11):

$$h_1 = \rho_{air}.c_{p_{air}}.U_{rel}.St \tag{11}$$

where the Stanton number is:

$$St = \frac{1}{2}.c_f \tag{12}$$

The friction coefficient is evaluated (Equation 13) for a turbulent flow because heating elements are located in the turbulent zone of the airfoil:

$$c_f = \frac{0.058}{\mathrm{Re}^{0.2}} \tag{13}$$

Finally the Reynolds number is:

$$\mathrm{Re} = \frac{\rho_{air}.U_{rel}.c}{\mu_{air}} \tag{13}$$

The target temperatures T_c for each test has been chosen so that we obtain a complete de-icing of all the heating elements surfaces during each test. The values of the different parameters used to calculate the power for each de-icing test are given in Table 10. With the increase of the wind speed (see each test parameters in Tables 8 and 9) the «heating coefficient» h_1 varies from 0.64 for Test 1 to 1.12 for Test 3. Also, the target temperature required for a complete de-icing of the heating elements 0 and 1 is different from the test 1 to 3.

N° test	h_1	T_c (°C) (Heat 0 and 1)	T_c (°C) (Heat 2 and 3)
1	0.64	5	1
2	0.88	10	1
3	1.12	15	1

Table 10. Test conditions set for the heating control

6.4 Results and analysis

De-Icing Results

The shapes and locations of the ice accretion on the surface are analysed in Figures 14 to 16 and illustrate the effectiveness of de-icing the wind turbine blade for the three test cases which parameters are indicated in Tables 8 to 10.

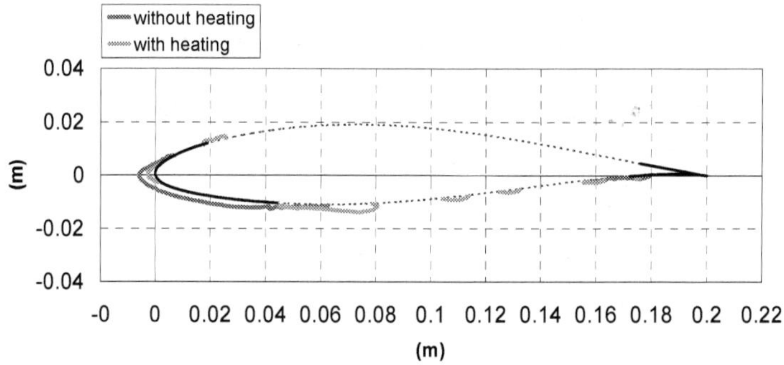

Fig. 14. Ice shape for Test 1 corresponding to r=10.6 m span position

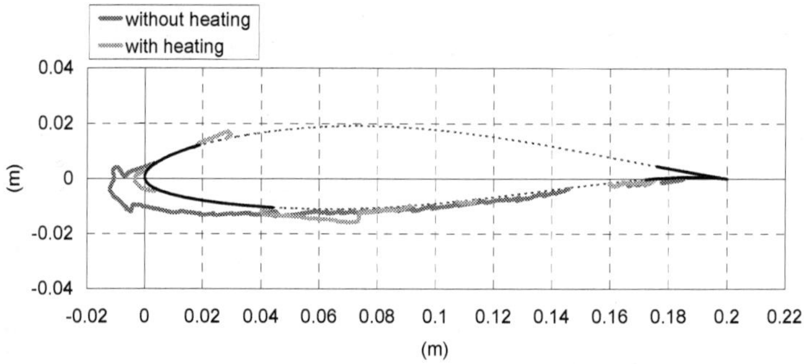

Fig. 15. Ice shape for Test 2 corresponding to r=16.6 m span position

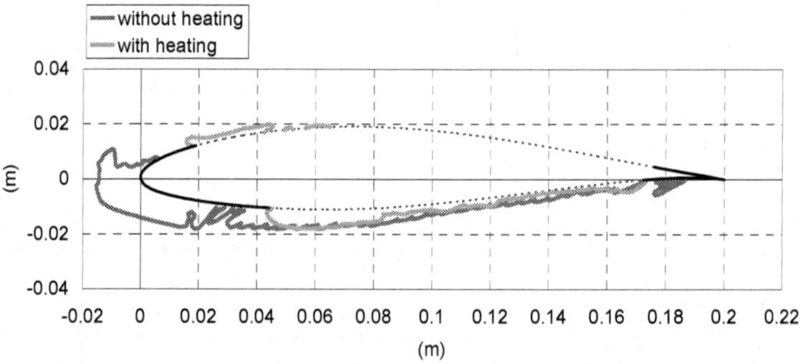

Fig. 16. Ice shape for Test 3 corresponding to r=22.5 m span position

Rime ice with milky appearance and characterized by a low density accretes on the airfoil during the tests without heating. This ice accretion is characteristic of a dry regime. On the other hand, the ice observed during the de-icing tests is transparent, with a high density, characteristic of glaze ice formed by runback water. Stream wise ice is observed at the leading edge for Test 1, corresponding to the span position nearest to the hub (10.6 m). For Tests 2 and 3, corresponding to a span position near the middle of the blade (16.6 m and 22.5 m), a protuberance with high roughness appears at the leading edge. Ice covers most of the lower surface of the airfoil at 16.6 m and covers completely the lower surface at 22.5 m.
For de-icing tests 1 and 2, corresponding to positions nearest to the hub (10.6m and 16.6 m) ice still accretes at the leading edge, probably because the heating elements number 0 and 1 are not exactly end-to-end. The liquid water at the leading edge runs back and refreezes at the end of the heating element to form a ridge. This is the worse inconvenient of the de-icing system. Leading edge is almost ice-free after de-icing. As the airfoil aerodynamic characteristics are more sensitive to the leading edge geometric modifications, aerodynamic performances should be better than for the iced airfoil without heating, but still worse than for a clean airfoil. This point needs more investigations to evaluate the real gain of the de icing.

Temperature Records

Figure 17 (a,b) illustrate temperature variation for icing tests without heating, while Figure 18 (a,b) illustrates results of the de-icing tests, with the de-icing system working. The temperatures measured by the thermocouples are recorded every 0.2 second during the whole test. These temperatures follow the icing wind tunnel (IWT) temperature variations as shown in Figure 17a. The IWT temperature variation is ±1 °C due to thermal inertia. The thermocouples have quick variations under short periods (26 s). This phenomenon is more significant for thermocouple 1 which is positioned in the zone of impacting droplets because of the high angle of attack of the blade. These quick temperature variations are partly due to the water spray ramp oscillations. The nozzles are fixed horizontally on the ramp, which moves up and down to insure a homogenous LWC in the test section. The ramp speed is about 0.08 m/s. When the ramp is in front of the airfoil, the surface collects more droplets and the temperature increases because of the liberation of latent heat of solidification. Thermocouple 0 is still warmer than thermocouples 2 and 3 for each icing test because of the smaller convection in this zone due to the high angle of attack. During the icing tests, the internal temperature is more stable than the external one, as shown in Figure 17b by thermocouples 1 and 3 in icing test 2. The temperature measured by thermocouple 1 decreases during the test. This is more obvious for tests 2 and 3 because of the important ice thickness collected at leading edge and due to ice insulation properties (Figures 15 and 17b).
For the de-icing tests, the surface temperature is strongly dependent on the heating. The recorded external temperatures follow the quick power variations (Figure 18a). The internal temperature measured by thermocouple 1 is less stable than that of the other thermocouples during the de-icing tests. This thermocouple is the most exposed to the impinging water droplets and the system has a short response time (Figure 18b). The external thermocouple 2 runs smoothly compared to the others (Fig. 11) while the internal thermocouple 2 temperature is more sensitive to the IWT temperature's variation than the other internal ones. This behavior is observed for all tests and is probably due to the small thickness of the airfoil in this area or to the contact between the thermocouple and a cold surface that increases the impact of the IWT.

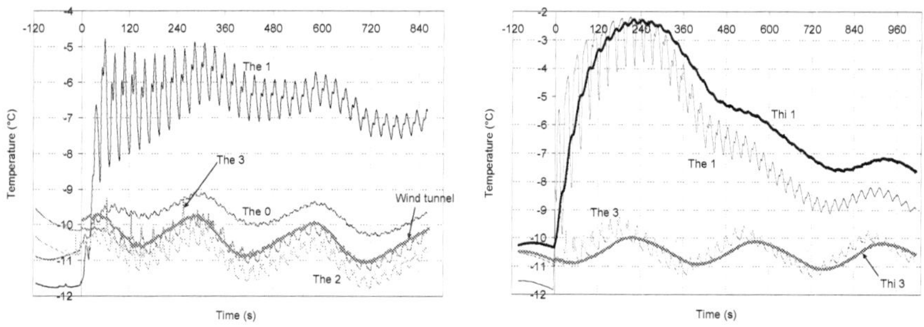

(a) (b)

Fig. 17. (a): External and IWT temperature evolution for icing test 1 and (b): Temperature evolution of the internal and external thermocouples 1 and 3 for icing test 2

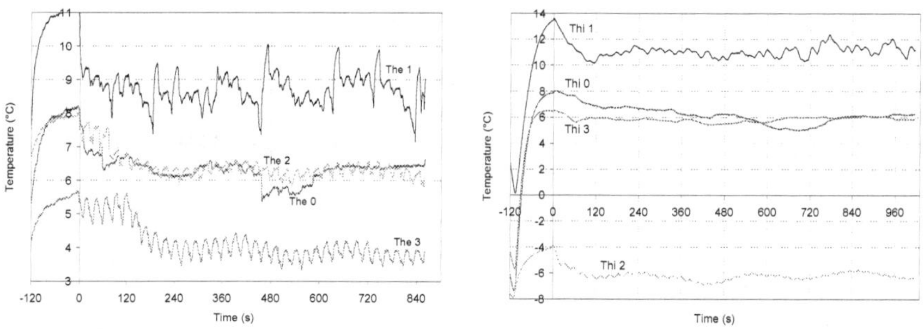

(a) (b)

Fig. 18. (a): External temperature evolution for de-icing test 1 and (b): Internal temperature evolution for de-icing test 2

Power Consumption

The power is calculated and recorded every 0.2 second during the whole test. The power consumption varies from one test to another under the same experimental conditions because of the IWT temperature variations; these variations are lower than 2%. Due to the thermocouple's short response time, the quick power variations are instantaneously measured by the thermocouples and, as the temperature is then used for the heating control, this leads to rapid power variations. As an example, Figure 19a shows the variation of the power consumption for each element for de-icing test 1 which illustrates these fast variations. As for the temperature variation (Figure 18a), the power is less stable for heater 1 which is the most exposed to water droplets. Figure 19b shows the power consumed by every heating element for all tests parameters detailed in Tables 8 to 10 and corresponding to three different span positions. The power required for the de-icing increases with the radius of the span position. This is due to the increase of relative wind speed that has an important impact on the convective heat transfer of the airfoil. The system needs 3.5 to 3.9

more power near the tip position (22.5 m) than near the hub position (10.6 m) for the leading edge and 2.6 to 2.9 more power for the trailing edge. It needs 1.3 to 1.5 more power for the lower surface than for the upper surface. It is to be noted that heating element 3 is practically not useful for the test conditions selected in this study because of the high angle of attack and icing conditions. This heating element required more power than the heating element 2 due to the high convective heat exchange between the surface and the air.

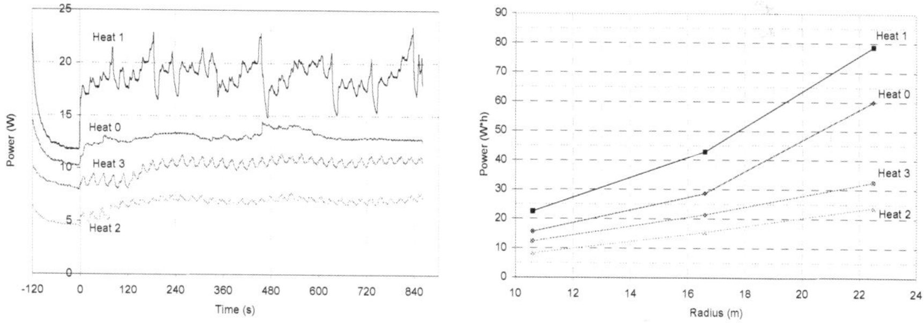

(a) (b)

Fig. 19. (a): Power consumption evolution of the heating elements during de-icing test 1 and (b): Heating elements power consumption at the 3 span positions

6.5 Conclusions

The design of the de-icing system is based on the different partial heating of four areas on the airfoil near the leading and the trailing edge and is subject to experimental limitations of IWT facilities. However the results are very encouraging. The de-icing is very efficient at keeping the leading edge free of ice, which is very important for the aerodynamic properties of the blade even if ridge ice forms at the end of the heating element. The airfoil instrumentation provided numerous results contributing to a better understanding of the process and further optimization of de-icing. As the time step of 0.2 s chosen for the power calculation lead to rapid variations of surface temperature, power calculation should be adjusted based on an average surface temperature computed over a longer period (1s or more). The results show that more energy is needed to de-ice the leading edge at the tip than at the hub (3.5 to 3.9 more), to de-ice the trailing edge at the tip than at the hub (2.6 to 2.9 more) and to de-ice the lower surface than the upper surface (1.3 to 1.5 more). It will be useful to confirm these wind tunnel results with full scale tests. The power control method leads to an optimization of the de-icing depending on the span position but the target temperatures need to be adjusted to obtain a clean airfoil.

7. Acknowledgments

This work represents a synthesis of over 5 years of collaboration between the Anti-icing Materials International Laboratory at the Université du Québec à Chicoutimi (Professors Jean Perron and Guy Fortin) and the Wind Energy Research Laboratory from the Université du Québec à Rimouski. Financial support from NSERC (Natural Sciences and Engineering

Research Council of Canada) and FQRNT (Fonds Québécois de Recherche sur la Nature et les Technologies) is kindly acknowledged. The author acknowledges also the contributions of graduate students, Olivier Parent, Clément Hochart and Christine Mayer.

8. References

Anderson, D.N. and Reich, A.D., 1997. Tests of the Performance of Coatings for Low Ice Adhesion. In: NASA (Editor), Aerospace Science Meeting and Exhibit, Reno, USA, pp. 14.

Anderson, D.N., 2004. *Manual of Scaling Methods*, NASA/CR-2004-212875, Ohio Aerospace Institute, Brook Park, Ohio, March 2004

Anonymous, 2004. *V80-1.8MW Pitch regulated wind turbine with OptiSlip and OptiTip*, General Specification, Technical document, Vestas, February 2004

Battisti, L., Baggio, P. and Fedrizzi, R., 2006. Warm-Air Intermittent De-Icing System for Wind Turbines. Wind Engineering, 30(5): 361-374.

Battisti, L., Brighenti, A., Dal Savio, S. and Dell'Anna, S., 2005a. Evaluation of Anti-Icing Energy and Power Requirement for Wind Turbine Rotors in Cold Climates, BOREAS VII. FMI, Saariselkä, Findland, pp. 13.

Battisti, L. and Fedrizzi, R., 2007. 2D Numerical Simulation of a Wind Turbine De-Icing System, Using Cycled Heating. Wind Engineering, 31(1): 33-42.

Battisti, L., Fedrizzi, R., Dell'Anna, S. and Rialti, M., 2005b. Ice Risk Assessment for Wind Turbine Rotors Equipped with De-Icing Systems, BOREAS VII. FMI, Saariselkä, Findland, pp. 11.

Boluk, Y., 1996. Adhesion of Freezing Precipitates to Aircraft Surfaces. Transport Canada, pp. 44.

Botta, G., Cavaliere, M. and Holttinen, H., 1998. Ice Accretion at Acqua Spruzza and its Effects on Wind Turbine Operation and Loss of Energy Production, BOREAS IV. FMI, Hetta, Finland, pp. 77-86.

Botura, G. and Fisher, K., 2003. Development of Ice Protection System for Wind Turbine Applications, BOREAS VI. FMI, Pyhätunturi, Finland, pp. 16.

Burton, T., Sharpe, S., Jenkins, N., Bossanyi, E., 2001. *Wind Energy Handbook,* John Wiley & Sons, LTD, 2001

Craig, D. and Craig, D., 1996. An Investigation of Icing Events on Haeckel Hill, BOREAS III. FMI, Saariselkä, Finland, pp. 169-193.

Dalili, N., Edrisy, A. and Carriveau, R., 2009. A Review of Surface Engineering Issues Critical to Wind Turbine Performance. Renewable and Sustainable Energy Reviews(13): 428-438.

Dobesch, H., Zach, S. and Tran, H.V., 2003. A New Map of Icing Potentials in Europe - Problems and Results, BOREAS VI. FMI, Pyhätunturi, Finland, pp. 9.

Fikke, S. et al., 2006. COST-727, Atmospheric Icing on Structures: 2006, Measurements and data collection on icing: State of the Art. MeteoSwiss, 75: 110.

Fikke, S.M., Säntti, K. and Laakso, T., 2005. Detectors for Atmospheric Icing, BOREAS VII. FMI, Saariselkä, Findland, pp. 8.

Fortin, G., Perron, J., Ilinca, A., 2005a. *A Study of Icing Events at Murdochville: Conclusions for the Wind Power Industry,* International Symposium "Wind Energy in Remote Regions", Magdalen's Island, October 2005

Fortin, G., Perron, J. and Ilinca, A., 2005b. Behaviour and Modeling of Cup Anemometers under Icing Conditions, IWAIS XI, Montréal, Canada, pp. 6.

Fortin, G., 2009. Thermodynamique de la Glace Atmosphérique. UQAC, pp. 9.

Gillenwater, D., 2008. Pertes de Puissance Associées aux Phénomènes Givrants sur une Éolienne Installée en Climat Nordique. Master Thesis, ÉTS, Montréal, Canada, 162 pp.

Harstveit, K., Tallhaug, L. and Fidje, A., 2005. Ice Accumulation Observed by Use of Web Camera and Modelled from Meteorological Parameters, BOREAS VII. FMI, Saariselkä, Findland, pp. 10.

Hochart, C., Fortin, G., Perron, J. and Ilinca, A., 2008. Wind Turbine Performance under Icing Conditions. Wind Energy(11): 319-333.

Homola, M.C., Nicklasson, P.J. and Sundsbo, P.A., 2006. Ice Sensors for Wind Turbines. Cold Regions Science and Technology(46): 125-131.

Horbaty, R., 2005. Wind Energy in Cold Climates - The Swiss Experience, BOREAS VII. FMI, Saariselkä, Findland, pp. 10.

ISO-12494, 2001. Atmospheric Icing of Structures. ISO copyrigth office, Geneva, Switzerland, pp. 56.

Jasinski, W.J., Noe, S.C., Selig, M.S. and Bragg, M.B., 1997. Wind Turbine Performance Under Icing Conditions, Aerospace Sciences Meeting & Exhibit. AIAA, Reno, USA, pp. 8.

Kimura, S., Sato, T. and Kosugi, K., 2003. The Effect of Anti-Icing Paint on the Adhension Force of Ice Accretion on a Wind Turbine Blade, BOREAS VI. FMI, Pyhätunturi, Finland, pp. 9.

Kimura, S., Tammelin, B. and Säntti, K., 2000. Estimation of Reduction of Power due to Icing from the Exixting Meteorological Data, BOREAS V. FMI, Levi, Finland, pp. 12.

Laakso, T. et al., 2003a. State-of-the-art of Wind Energy in Cold Climates. IEA Wind Annex XIX: 53.

Laakso, T., Peltola, E., Antikainen, P. and Peuranen, S., 2003b. Comparison of Ice Sensors for Wind Turbines, BOREAS VI. FMI, Pyhätunturi, Finland, pp. 11.

Laakso, T. and Peltola, E., 2005a. Review on Blade Heating Technology and Future Prospects, BOREAS VII. FMI, Saariselkä, Finland, pp. 12.

Laakso, T. et al., 2005b. Wind Energy Projects in Cold Climates. International Energy Agency: 36.

Maissan, J.F., 2001. Wind Power Development in Sub-Arctic Conditions with Severe Rime Icing, Circumpolar Climate Change Summit and Exposition, Whitehorse, Canada, pp. 18.

Makkonen, L. and Ahti, K., 1995. Climatic Mapping of Ice Loads Based on Airport Weather Observations. Atmospheric Research, 36: 185-193.

Makkonen, L. and Autti, M., 1991. The Effects of Icing on Wind Turbines, EWEC, Amsterdam, Netherlands, pp. 575-580.

Makkonen, L., Laakso, T., Marjaniemi, M. and Finstad, K.J., 2001. Modelling and Prevention of Ice Accretion on Wind Turbines. Wind Engineering, 25(1): 3-21.

Mansson, J., 2004. Why De-Icing of Wind Turbine Blades?, Global Windpower, Chicago, USA, pp. 12.

Marjaniemi, M. et al., 2000. Wind Turbines in Light Icing Conditions - Experience of the Pori 8 MW Wind Farm, BOREAS V. FMI, Levi, Finland, pp. 13.

Marjaniemi, M. and Peltola, E., 1998. Blade Heating Element Design and Practical Experiences, BOREAS IV. FMI, Hetta, Finland, pp. 197-209.

Mayer, C., 2007. Système Électrothermique de Dégivrage pour une Pale d'Éolienne. Master Thesis, UQAR, Rimouski, Canada, 193 pp

Mayer, C., Ilinca, A., Fortin, G. and Perron, J., 2007. Wind Tunnel Study of Electro-Thermal De-Icing of Wind Turbine Blades. International Journal of Offshore and Polar Engineering, 17(3): 182-188.

McKittrick, L R, Cairns, D S, Mandell, J, Combs, D C, Rabern, D A, Van Luchene, R D (2001). *Analysis of a Composite Blade Design for the AOC 15/50 Wind Turbine Using a Finite Element Model*, Sandia National Laboratories.

Patreau, V., Morency, F. and Paraschivoiu, I., 1998. Analysis of Thermal De-Icing System for Horizontal Axis Wind Turbine Blade, BOREAS IV. FMI, Hetta, Finland, pp. 222-235.

Peltola, E., Laakso, T., Antikainen, P. and Peuranen, S., 2003. Prevention of Icing Effects, BOREAS VI. FMI, Pyhätunturi, Finland, pp. 9.

Peltola, E., Marjaniemi, M., Kaas, J. and Aarnio, E., 1996. Pyhätunturi Operational Experiences, BOREAS III. FMI, Saariselkä, Finland, pp. 131-144.

Pinard, J.-P. and Maissan, J.F., 2003. Experience from Use of Heated Wind Sensors and Rime Ice Detectors over the Past 12 Years, BOREAS VI. FMI, Pyhätunturi, Finland, pp. 7.

Richert, F., 1996. Is Rotorcraft Icing Knowledge Transferable to Wind Turbines?, BOREAS III. FMI, Saariselkä, Finland, pp. 366-380.

Seifert, H., Richert, F., 1997. *Aerodynamics of iced airfoils and their influence on loads and power production*, Germany, 1997.

Seifert, H., 2003. Technical Requirements for Rotor Blades Operating in Cold Climate, BOREAS VI. FMI, Pyhätunturi, Finland, pp. 13.

Stallabrass, J.R., 1978. *An Appraisal of the single Rotating Cylinder Method of Liquid Water Content Measurement*, Canadian Research Council Report, LTR-L T-92, November 1978.

Tallhaug, L., 2003. Calculation of Potential Ice Risk in Norway, BOREAS VI. FMI, Pyhätunturi, Finland, pp. 8.

Tammelin, B. et al., 2000. Wind Energy Production in Cold Climate (WECO). 41, FMI, Helsinki, Finland.

Tammelin, B. and Säntti, K., 1994. Effect of Rime Accretion on Wind Energy Production in the Top Areas of Fells, BOREAS II. FMI, Pyhätunturi,Finland, pp. 265-275.

Tammelin, B. and Säntti, K., 1996. Estimation of Rime Accretion at High Altitudes - Preliminary Results, BOREAS III. FMI, Saariselkä, Finland, pp. 194-210.

Tammelin, B. and Säntti, K., 1998. Icing in Europe, BOREAS IV. FMI, Hetta, Finland, pp. 125-132.

Tammelin, B. et al., 2005. Wind Turbines in Icing Environment: Improvement of Tools for Siting, Certification and Operation - NEW ICETOOLS. FMI: 127.

Weis, T.M. and Maissan, J., 2003. The Effects of Black Blades on Surface Temperatures for Wind Turbines, Pembina Institute.

9

An Experimental Study of the Shapes of Rotor for Horizontal-Axis Small Wind Turbines

Yoshifumi Nishizawa
Collaborative Research Center, Ashikaga Institute of Technology
Japan

1. Introduction

Aggravation of the problems of both the environment and the limited fossil energy at present has enabled the wind energy, the limitless source of energy without environmental pollution, to draw a good deal of public attention.

The large-scale wind turbines with the grid connection system operated in wind farms are being used for the power generation in many advanced nations. On the other hand, the small-scale wind turbines have been used as the independent source of electricity regardless of the places. In recent years, these small wind turbines of a diameter under 2m are employed for street lights, park lights, environmental monuments, power sources for emergencies and teaching materials. Because such small wind turbines are usually installed in urban areas, the noise generate from rotor, the failures due to the sudden changes of the wind direction and wind speed which are peculiar to those areas became particularly noticeable these days with increase of the installation.

The rotors for small wind turbines designed to have low tip speed ratio are increasing in number because of their reduced noise. The development of generators and controllers suitable for low rotational speed is largely accelerated these years. While there are many papers on the small wind turbine rotors with high tip speed ratio adjusted for conventional generators, a number of the paper systematically dealing the rotor shapes applied for low tip speed ratio is very small.

In this study, two types of experimental models are prepared to compare the performance at low tip speed. One has a rotor with tapered blades designed by the combined blade element and momentum theory, and the other has a rotor with inversely tapered blades in which the calculated chord lengths are applied in an opposite way.

2. Experimental models

The experimental models were designed according to the combined blade element and momentum theory. The following of equations shows the expressions used for the model design.

Local tip speed ratio: $$\lambda_{rd} = \lambda_d \frac{r}{R} \qquad (1)$$

Inflow angle: $$\theta = \frac{2}{3}\tan^{-1}\frac{1}{\lambda_{rd}} \tag{2}$$

Blade pitch angle: $$\beta = \theta - \alpha \tag{3}$$

Blade chord: $$C = \frac{8\pi r}{BC}(1-\cos\phi) \tag{4}$$

B : Number of blades, C : Chord length[m], R : Rotor radius[m], r : Local rotor radius[m], λ_{rd} : Local design tipp speed ratio, λ_d : Design tip speed ratio, θ : Inflow angle, α : Design angle of attack, β : Blade pitch angle

Rotor diameter is 0.6[m] and aerofoils of the blades are all Clark Y. Figure 1 shows the characteristics of Lift on Clark Y. For each tapered and inversely tapered type, we prepared five different shapes of blades. For typical three blade-shapes out of the five of both types, blade numbers are varied from two to six as shown in Table 1 and Table 2, which shows the design values of each model corresponding to the figure 2 to figure 6.

Nominal designations of blades in the table are defined by the authors to distinguish the taper-types and shapes of the blades as follows. "2" in the last digit stands for the normally tapered-blade and "4", the inversely tapered one. "2" to "4" in the first digit stand for the different tip speed ratio in a five-bladed rotor where the bigger the number, the smaller the chord length. The blades with "5" in the first digit are used in a three-bladed rotor and "6", in a five-bladed one. The 5xx and 6xx blades are designed to have the same tip speed ratio, λ=3.4.

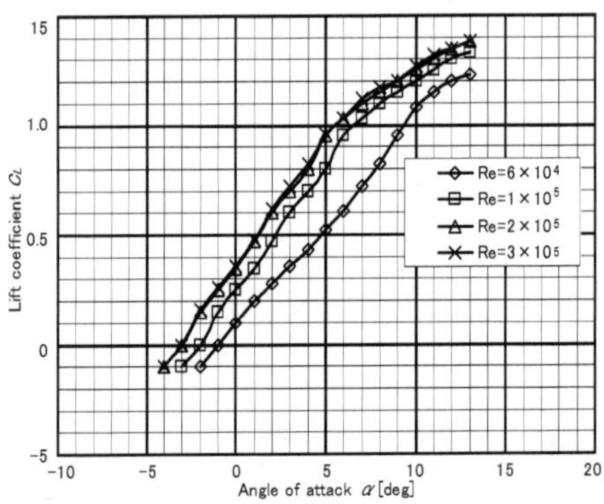

Fig. 1. Characteristics of Lift at Clark Y

Rotor diameter[mm]	600
Aerofoil	CLARK Y
Design lift coefficient	1.1
Design angle of attack[°]	8
Linearization point 1[%]	75
Linearization point 2[%]	95

Table 1. Preliminary design values for experimental models

Blade type No.	202 (Tapered) 、204 (Inversely tapered)				
Number of blades	2	3	4	5	6
Design tip speed ratio	3.4	2.72	2.29	2	1.78
Blade pitch angle at 80% from root[°]	5.6	8.6	11.2	13.4	15.5
Tip chord length of tapered type[mm]	64.4				
Tip chord length of inversely tapered type[mm]	113.3				
Blade type No.	502 (Tapered) 、504 (Inversely tapered)				
Number of blades		3			
Design tip speed ratio		3.4			
Blade pitch angle at 80% from root[°]		5.6			
Tip chord length of tapered type[mm]	40.9				
Tip chord length of inversely tapered type[mm]	84.0				
Blade type No.	302 (Tapered) 、304 (Inversely tapered)				
Number of blades	2	3	4	5	6
Design tip speed ratio	5	4	3.4	3	2.72
Blade pitch angle at 80% from root[°]	1.4	3.7	5.6	7.2	8.6
Tip chord length of tapered type[mm]	22.5				
Tip chord length of inversely tapered type[mm]	38.0				
Blade type No.	602 (Tapered) 、604 (Inversely tapered)				
Number of blades				5	
Design tip speed ratio				3.4	
Blade pitch angle at 80% from root[°]				5.6	
Tip chord length of tapered type[mm]	24.6				
Tip chord length of inversely tapered type[mm]	50.4				
Blade type No.	402 (Tapered) 、404 (Inversely tapered)				
Number of blades	2	3	4	5	6
Design tip speed ratio	6.4	5.2	4.5	4	3.4
Blade pitch angle at 80% from root[°]	-0.6	1.1	2.4	3.7	4.9
Tip chord length of tapered type[mm]	18.0				
Tip chord length of inversely tapered type[mm]	38.0				

Table 2. Design values of each blade for experimental models

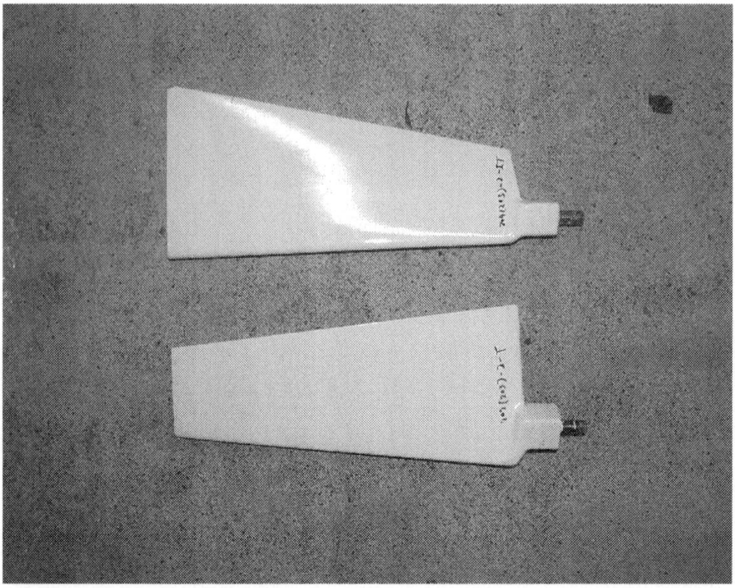

Fig. 2. Experimental blades of 202, 204.

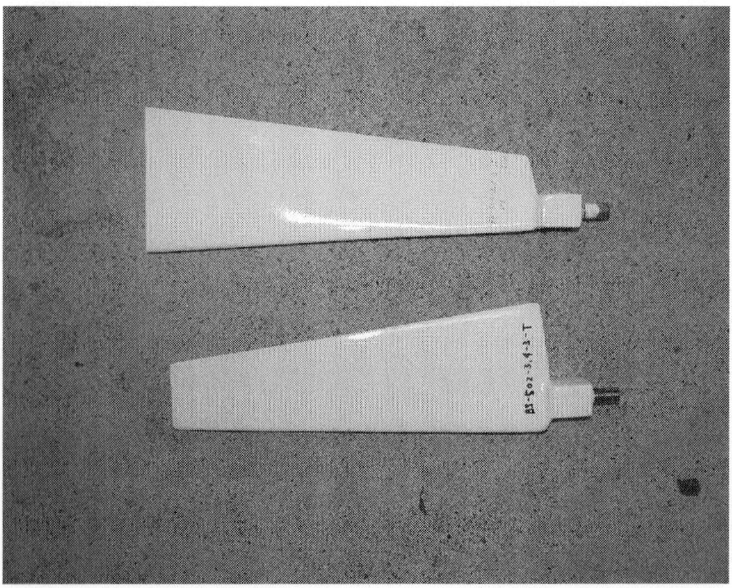

Fig. 3. Experimental blades of 502, 504

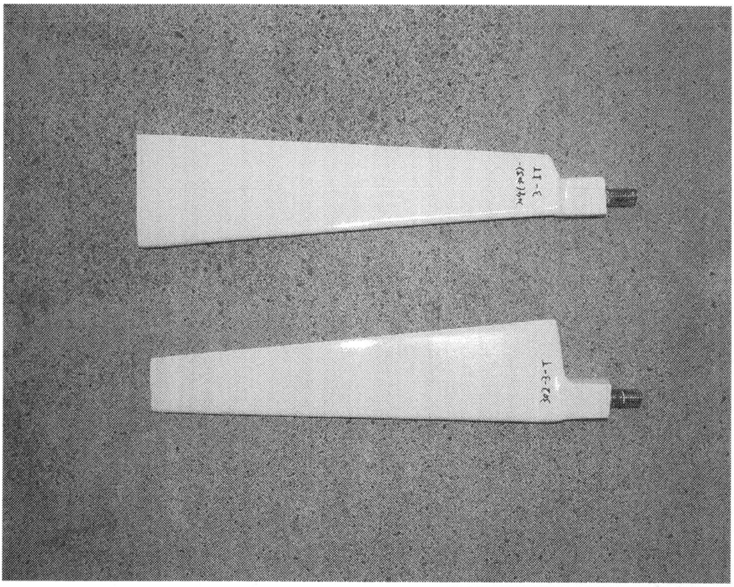

Fig. 4. Experimental blades of 302, 304

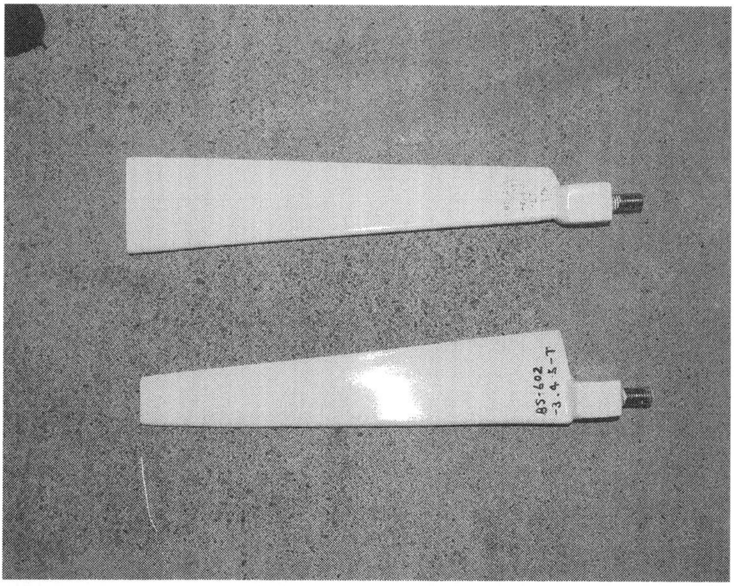

Fig. 5. Experimental blades of 602, 604

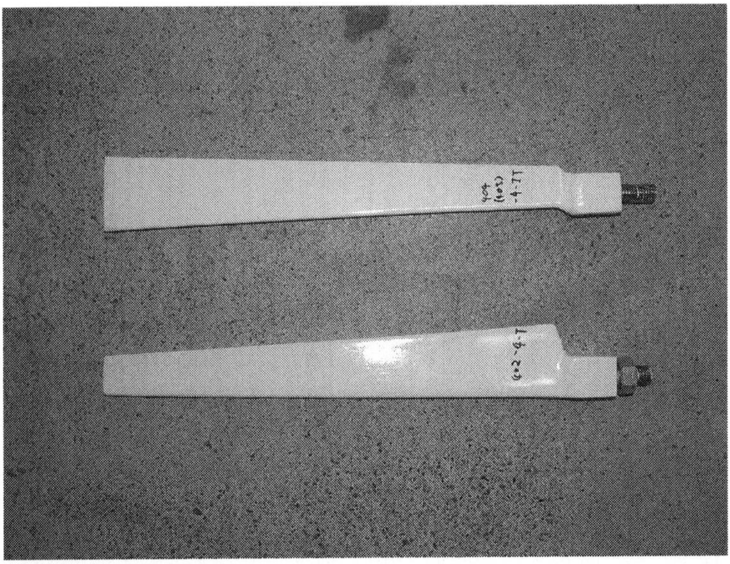

Fig. 6. Experimental blades of 402, 404

3. Experiments

The wind tunnel used for our experiment is of the Effel type with an exit of 1.05m times 1.05m. The wind speed is adjustable between 2m/s and 20m/s.
An induction motor is employed as load in the experiment and the synchronized frequency is controlled by an inverter. For arbitrary determined frequencies, corresponding torque and rotational speed are measured by a torque converter and a revolution counter. From the results of the experiment the power coefficient Cp and the tip speed ratio are calculated to obtain the power characteristics for different blade types. Fig.7 shows the schematic of the experimental apparatus.
In the experiment of 202, 204, 302, 304, 402 and 404 blades, the wind speed in the wind tunnel is set at 4, 6, 8, 10 and 12m/s and the torque corresponding to the revolution is measured by gradually increasing the load.
The same procedure is followed for the blades of 5xx and 6xx, but they have the fixed blade numbers of three and five, respectively.

4. Results and consideration

4.1 Comparison among the blades with identical design tip speed ratio

As table2, If we look at the different number of blades with same tip speed ratio, say 3.4, we can see that the shorter the chord length, the larger the number of blades.
Torque coefficient of the tapered blades, power coefficient of the tapered blades, and the power coefficient of the inversely tapered blades at wind speed of 10m/s and the design tip speed ratio λ=3.4 are shown in the Fig.8,9 and 10, respectively to study the contribution of the blade numbers.

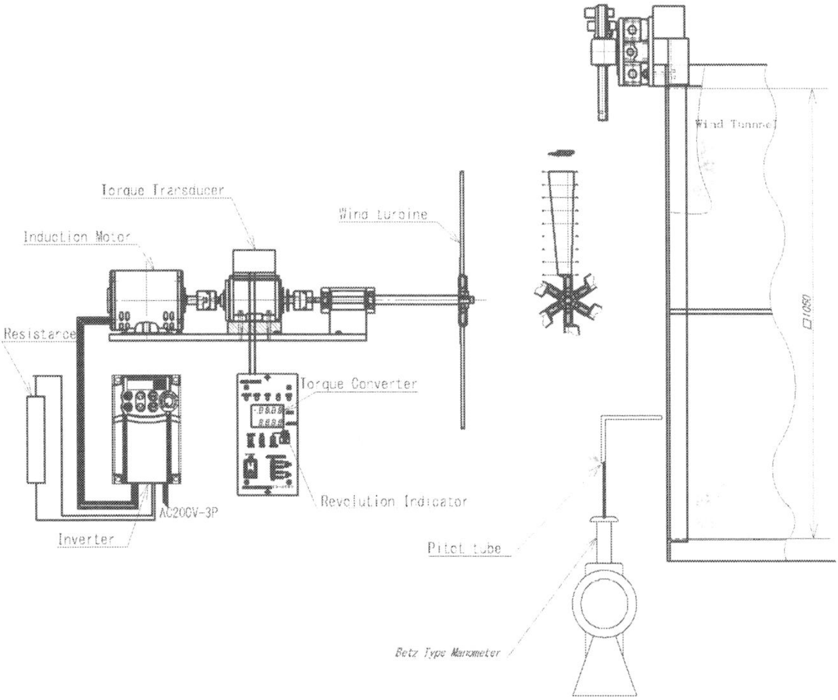

Fig. 7. Layout of the experimental apparatus

The results for the normally tapered blades show that the highest torque coefficient is obtained for the two bladed rotor and the maximum torque coefficient decreases with the increase of blade-number. However, tip speed ratio corresponding to each maximum torque coefficient increases with the number of blades.

The maximum power coefficients are over 0.38 for 2- to 5-bladed rotor while that for 6-bladed rotor is under 0.3. The highest power coeffient, Cpmax=0.39 is obtained for the 3-bladed rotor when λ=4.06.

However, six bladed rotor has lower values than other bladed type. Table 3 shows the calculated result of Reynolds number of these rotors at 80% position of span. The Reynolds number of the six bladed rotor was lower than the other rotors.

	Rev.[rpm]	Chord length	Radius[mm]	Tip speed[m/s]	Re. Number
2 bladed	1225	75.3	300	38.48451	1.9E+05
3 bladed	1293	50.5	300	40.620793	1.3E+05
4 bladed	1327	38	300	41.688935	1.0E+05
5 bladed	1411	30.3	300	44.327872	8.6E+04
6 bladed	1427	22.5	300	44.830527	6.5E+04

Table 3. Calculated result of Reynolds number (Tapered blades)

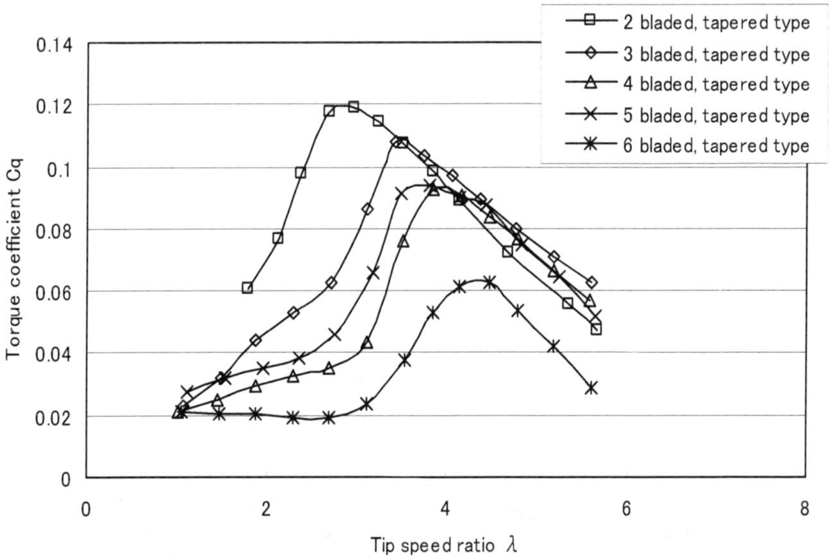

Fig. 8. Torque coefficient of taped types (λ =3.4)

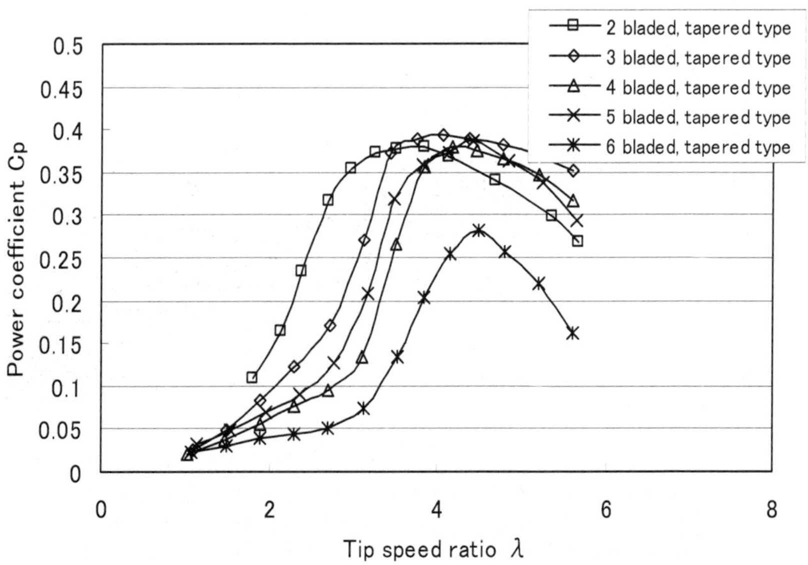

Fig. 9. Power coefficient of taped types (λ=3.4)

However, while Reynolds number of the five bladed rotor is as low as 8.6 times ten to the fourth power, power coefficient is not low. From these results, we found that the border of the superiority and inferiority of power coefficient of tapered type corresponds to the Reynolds number of 6.5 to 8.6 times ten to the fourth power.

For the inversely tapered blades, Cp for the two bladed rotor reached 0.38 when λ=3.72. However, power coefficients of other rotors were lower than that of 2bladed rotor.

Table 4 shows the calculated result of Reynolds number of these rotors at 80% position of span. The impaired power coefficient of three bladed rotor the Reynolds number was two times ten to the fifth power.

From these results, We found that the border of the superiority and inferiority of power coefficient of inversely tapered type did not correspond to Reynolds number only.

As the result of performance comparison among the different blades with identical design tip speed ratio, we found that 3bladed tapered rotor was most efficient. Moreover, we found that power coefficient did not differ between tapered and inversely tapered rotor with the longest chord length.

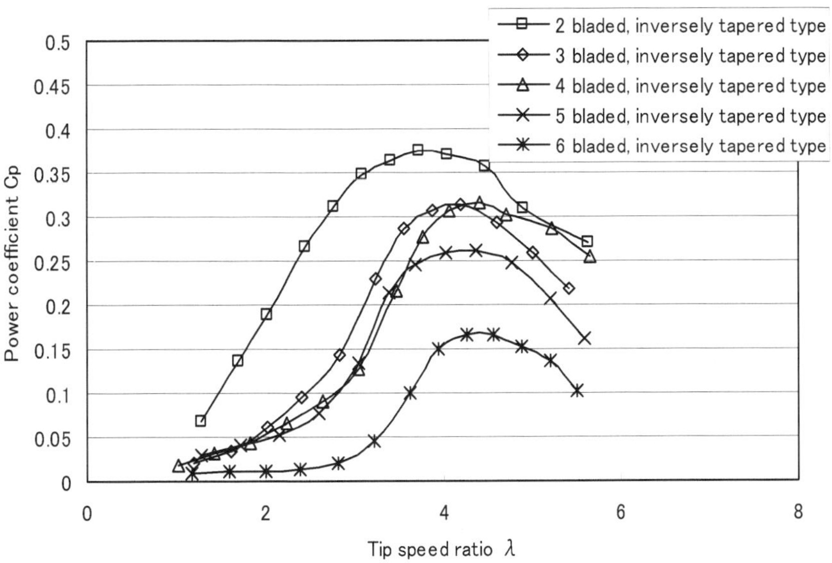

Fig. 10. Power coefficient of inversely taped types (λ=3.4)

	Rev.[rpm]	Chord length	Radius[mm]	Tip speed[m/s]	Re.Number
2 bladed	1186	102.5	300	37.259289	2.4E+05
3 bladed	1336	74.5	300	41.971678	2.0E+05
4 bladed	1407	55.2	300	44.202209	1.6E+05
5 bladed	1421	43.6	300	44.642032	1.2E+05
6 bladed	1452	33.6	300	45.615925	9.8E+04

Table 4. Calculated result of Reynolds number (Inversely tapered blades)

4.2 Characteristics of each blade in a rotor with different blade-number

The Fig.11 through 13 shows the characteristics of power coefficient for tapered blades, while the Fig.14 through 16 show those for inversely tapered blades.

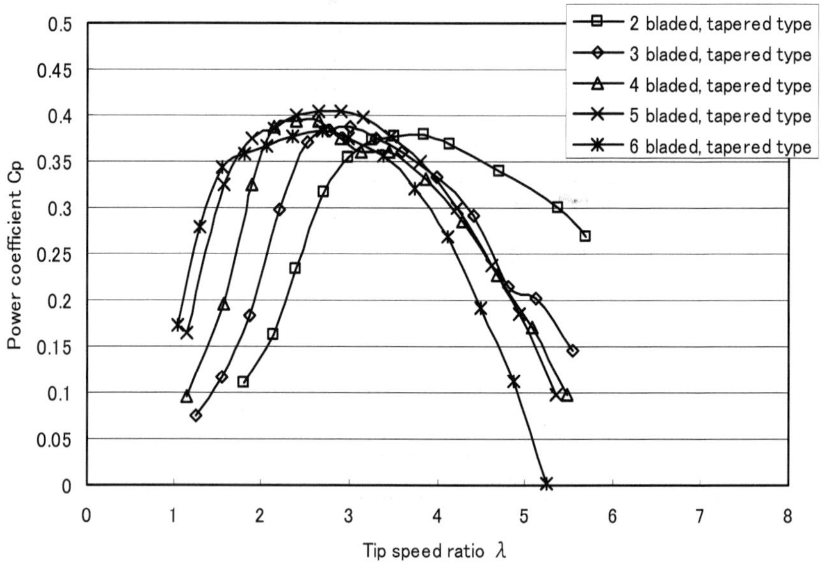

Fig. 11. Power coefficient of 202 (Tapered type)

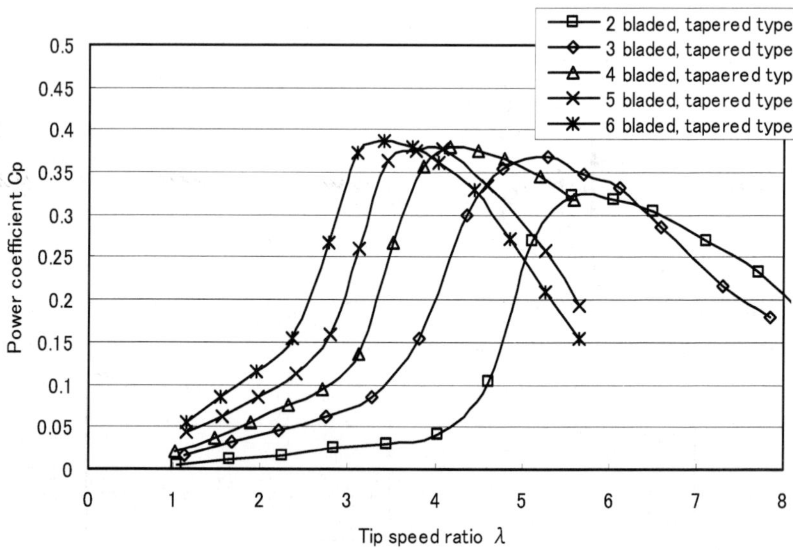

Fig. 12. Power coefficient of 302 (Tapered type)

For both tapered and inversely tapered types, the maximum power coefficient is obtained for the largest chord length. With regard to the number of the blades, five bladed rotor has highest maximum power coefficient.
In addition, the inversely tapered blades show an obvious increase of power coefficient with the number of blades than normally tapered blades.

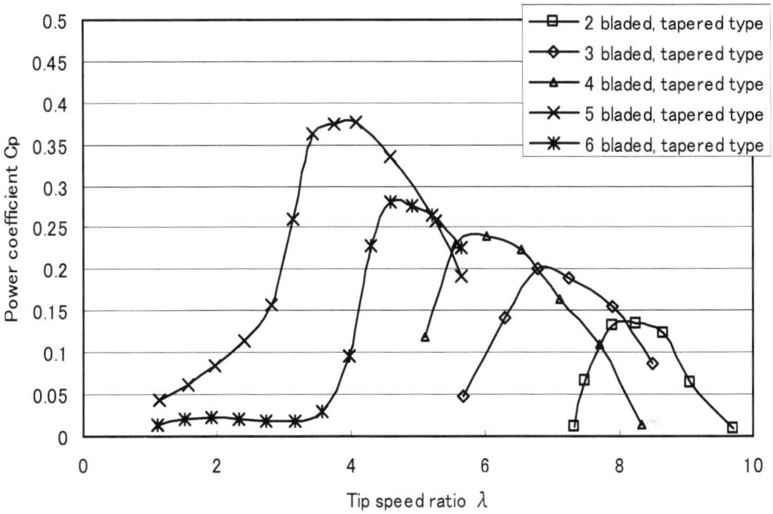

Fig. 13. Power coefficient of 402 (Tapered type)

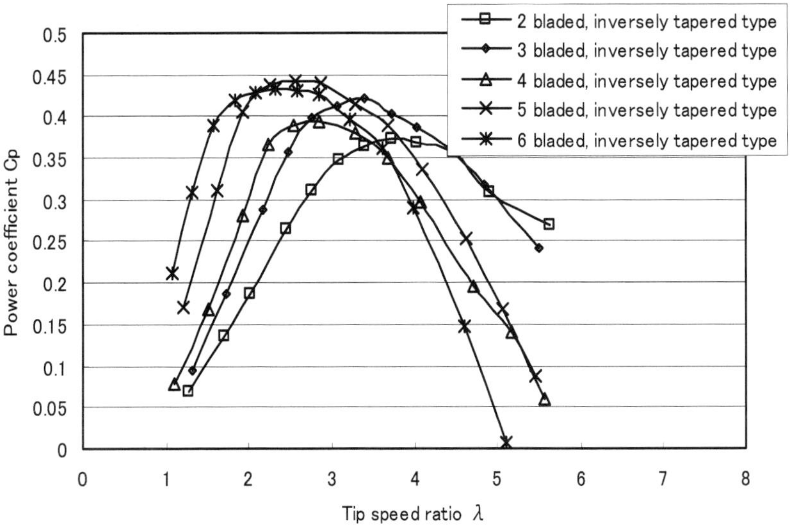

Fig. 14. Power coefficient of 204 (Inversely tapered type)

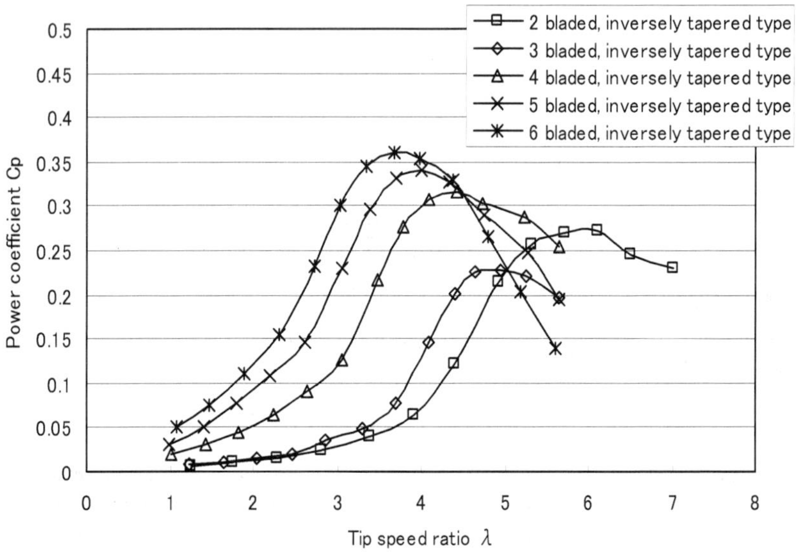

Fig. 15. Power coefficient of 304 (Inversely tapered type)

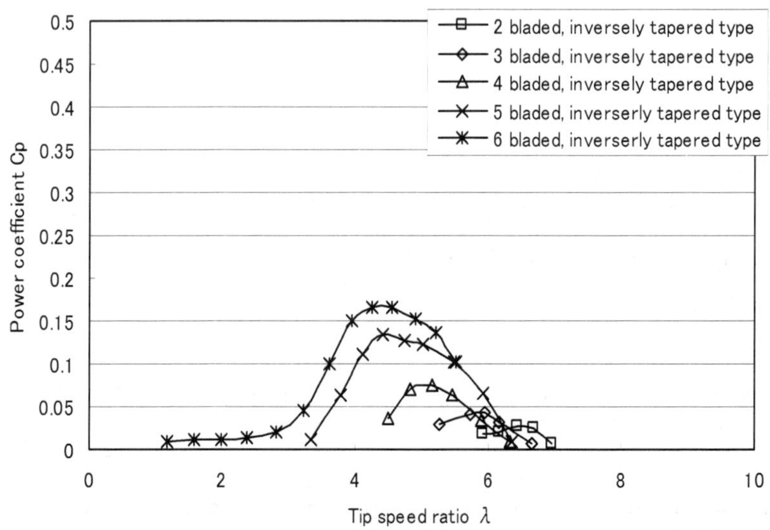

Fig. 16. Power coefficient of 404 (Inversely tapered type)

From these results, we found that the performance of a rotor designed at low tip speed ratio (about λ=2) can be improved by increasing the number of blades and by adopting the inversely tapered blades.

5. Visualization around rotors

Our study is obvious that the power coefficient of inversely tapered type was higher than tapered type.

Then, the visualization tests were conducted and the behavior of the air flow were analyzed.

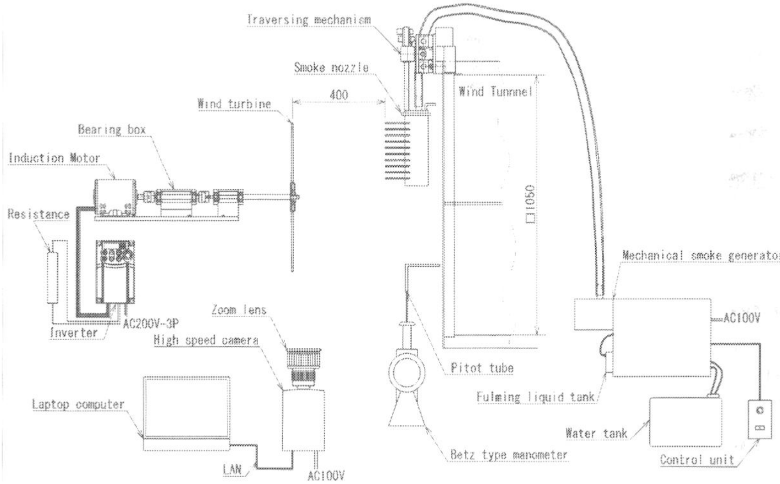

*Luminous source: Halogen lamp

Fig. 17. Visualization for experimental apparatus

5.1 Experiments

Figure 17 shows the experimental apparatus.

In this experiment, visualization test of the air flow around the rotor was conduced using the smoke of oil, high-speed camera and PIV analysis.

The authors took photos at the rotational speed correspond to the maximum power coefficient for the wind speed of 6, 8, and 10[m/s].

Figure 18 shows the grid of PIV analysis.

5.2 Results and consideration

The figure 19 and 20 shows the characteristics of Visualization for tapered blades, while the figure 21 and 22 show those for inversely tapered blades at 10m/s.

From the visualization test, the wind speed of the wake for tapered type is 5~8[m/s], on the contrary, the inversely tapered type is 3~6[m/s].

These results are presumed to be the latter type is effectively converting the air flow than the former type.

6. Conclusion

As the result of performance comparison among the blades with different design tip speed ratio, number of blades, and taper types, we obtained following conclusions.

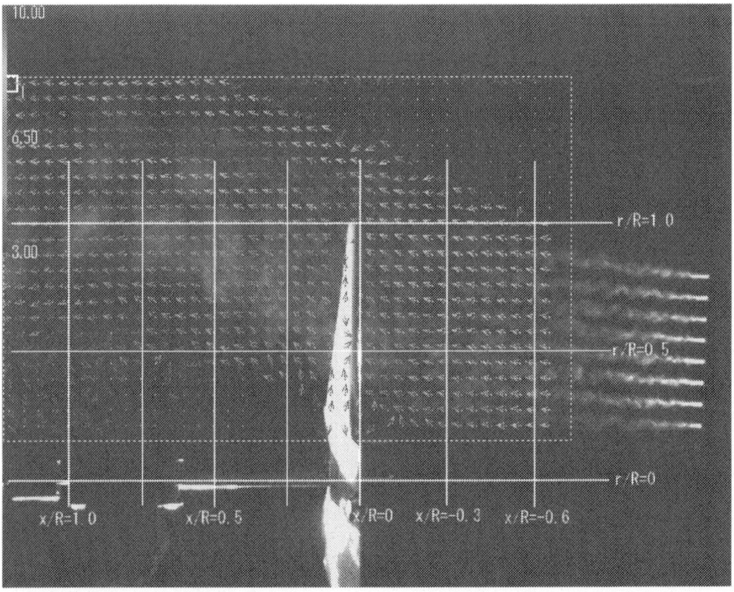

Fig. 18. Grid of PIV analysis

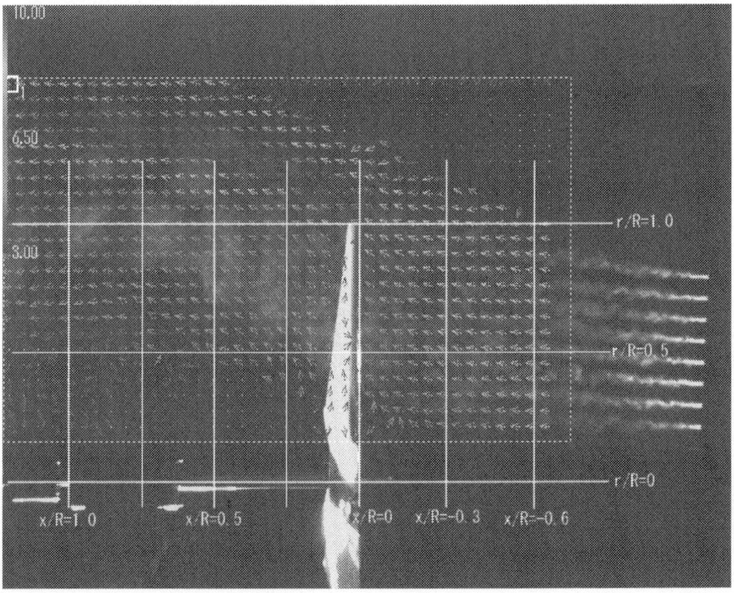

Fig. 19. Visualization analysis of vector line around tapered type rotor

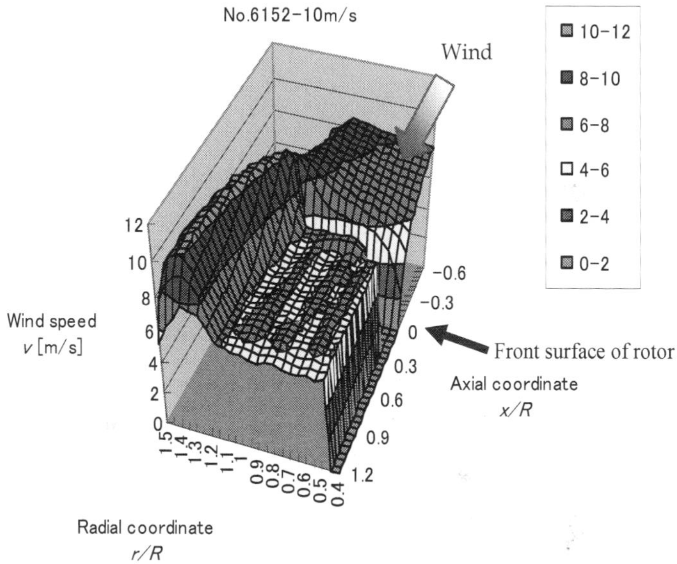

Fig. 20. Wind speed distribution around tapered type rotor

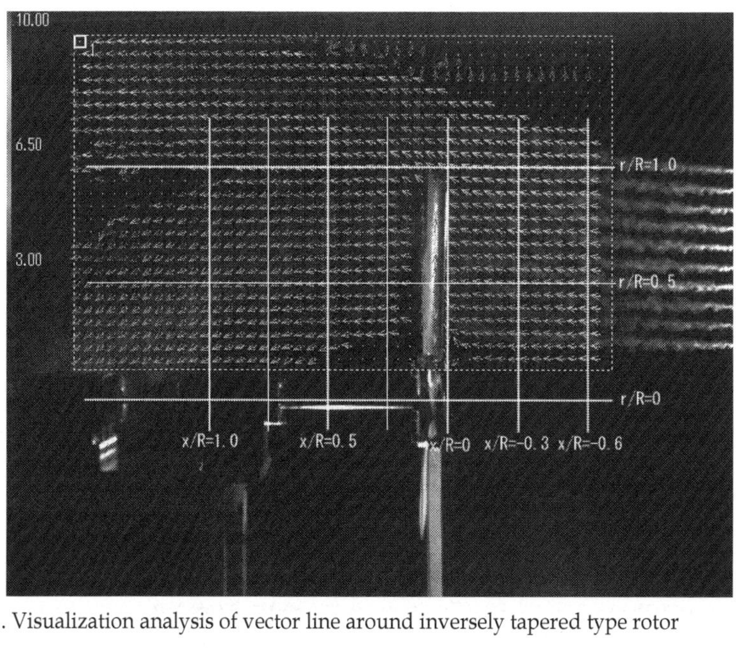

Fig. 21. Visualization analysis of vector line around inversely tapered type rotor

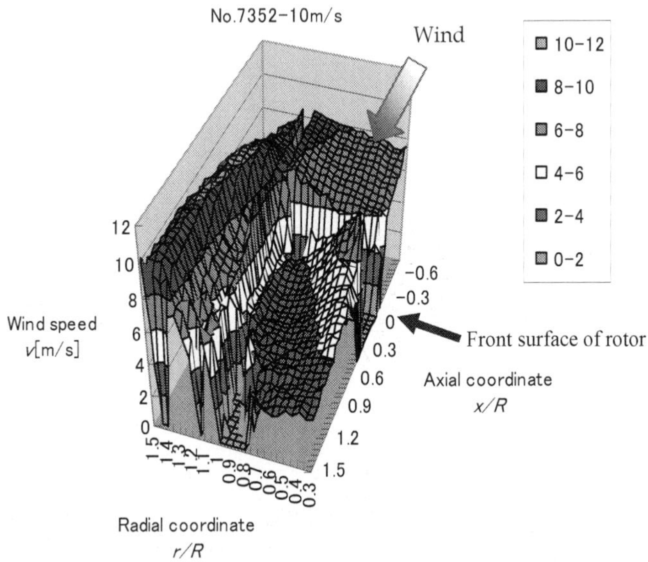

Fig. 22. Visualization analysis of vector line around inversely tapered type rotor

1. The border of the superiority and inferiority of power coefficient of tapered type corresponds to the Reynolds number of 6.5 to $8.6{\times}10^4$.
2. The border of the superiority and inferiority of power coefficient of inversely tapered type did not correspond to the Reynolds number only.
3. As the result of performance comparison among the blades with identical design tip speed ratio, we found that 3bladed tapered rotor was most efficient. In addition, the power coefficient did not differ between tapered and inversely tapered rotor for the longest chord length.
4. As the result of performance comparison between the blades with the longest chord length in a rotor with different blade-number, we found that 5 bladed tapered and inversely tapered rotor was most efficient. Moreover, power coefficient of inversely tapered rotor is larger than tapered type.

7. References

Y. Nishizawa, M. Suzuki, H. Taniguchi and I. Ushiyama, An Experimental Study of the Shapes of Blade for a Horizontal - Axis Small Wind TUribnes "Optimal Shape for Low Design Tip Speed of Rotor, JSME-B, Vol.75, No.751, (2009), pp547-549

H. Tokuyama, I. Ushiyama and K. Seki, Experimental Determination of Optimum Design Configuration for Micro Wind Turbines at Low Wind Speeds, Journal of Wind Engineering, (2000), pp.65-70

E.H. Lysen, Introduction to Wind Energy, SWD Publication, (1982), pp.65-73

C.A. Lyon, A.P. Broeren, P. Giguere, Ashok Gopalarathnam and Michael S. Selig, Summary of Low-Speed Airfoil Data, Volume 3, Soar Tech Publications, (1997), pp.80-87

T. Sano, Introduction to PIV, VSJ-PIV-S1, (1998), pp.17-21

10

Selection, Design and Construction of Offshore Wind Turbine Foundations

Sanjeev Malhotra, PE, GE
Parsons Brinckerhoff, Inc.
United States of America

1. Introduction

In the past twenty five years, European nations have led the way in the development of offshore wind farms. However, development in offshore wind energy is picking speed in other continents as well. More recently, there has been explosive growth in investment in the clean energy sector, with onshore and offshore wind power taking by far the largest share of that investment. About 50 billion US dollars were invested each year since 2007. Although economic crises may have impeded investment in 2010. In the last few years nearly 30 to 40 percent of all new installed power generation capacity in Europe and the United States is attributed to wind energy. The European Wind Energy Association estimates that between 20 GW and 40 GW of offshore wind energy capacity will be operating in the European Union by 2020. The US Department of Energy predicts that 50 GW of installed offshore wind energy will be developed in the next 20 years (NWTC, 2006). This means at least US$100 billion of capital investment with about US$50 billion going to offshore design and construction contracts.

In the United States, offshore wind power development has not been a focus area because there is great potential for wind power on land. However, high quality onshore wind resources are mostly located in the Midwest and Central United States while the demand centers are located along the coasts, thereby making the cost of transmission high. On the northeast coast of the United States, offshore development is an attractive alternative because electricity costs are high and transmission line construction from the mid-west faces many obstacles. Higher quality wind resources, proximity to coastal population centers, potential for reducing land use, aesthetic concerns, and ease of transportation and installation are a few of the compelling reasons why power companies are turning their attention to offshore development. Offshore turbines are being made larger to economize in the foundation and power collection costs. As the technology for wind turbines improves, the industry has developed wind turbines with rotor diameters as large as 150 m and power ratings of over 7.5 MW to 10 MW. As increasing number of wind farms are being planned 15 to 50 km from shore in water depths of over 50 m, the combination of water depth, the increasing wind tower heights and rotor blade diameters create loads that complicate the foundation design and consequently place a greater burden on the engineer to develop more innovative and cost-effective foundations and support structures. Moreover, offshore foundations are exposed to additional loads such as ocean currents, storm wave loading, ice loads and potential ship impact loads. All of these factors pose significant challenges in the design and construction of wind turbine support structures and foundations. This chapter summarizes current practices in selecting and designing such foundations.

2. Background

2.1 Wind turbine farm layout

Primary components of a typical offshore wind farm include several wind turbines located in the water, connected by a series of cables to an offshore transformer station which in turn is connected by an undersea cable to an onshore transformer station linked to the existing power grid (Figure 1). The wind turbines are usually spaced laterally at several (4 to 8) times the rotor diameter and staggered so as to minimize wake effects. Placing turbines closer reduces the quantity of electric cable required but it increases turbulence and wake effects thereby reducing power generation. Therefore, laying out wind turbine farms includes minimizing the length of cabling required yet maximizing power generation so as to optimize costs per unit of power produced.

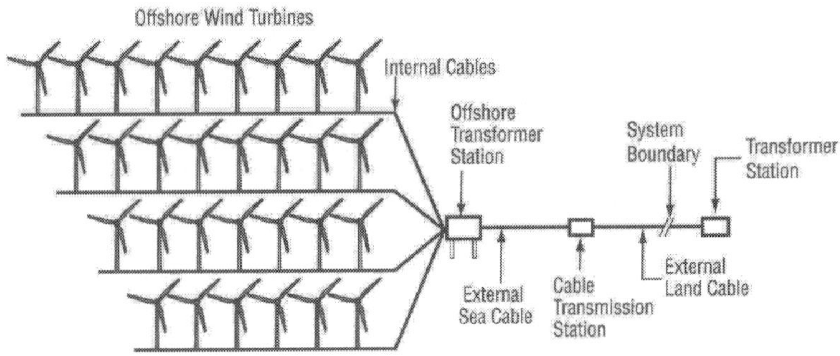

Fig. 1. Wind Farm Components and their Layout, (Malhotra, 2007c)

2.2 Wind turbine components

The components of a wind turbine system (Figure 2) include the foundation, the support structure, the transition piece, the tower, the rotor blades and the nacelle. The foundation system and support structure, used to keep the turbine in its proper position while being exposed to the forces of nature such as wind and sea waves, can be made using a variety of materials such as reinforced concrete or steel. Support structures connect the transition piece or tower to the foundation at seabed level. In some cases, the foundations serve as support structures as well by extending from the seabed level to above the water level and connecting directly to the transition piece or tower. The transition piece connects the tower to the support structure or foundation. The transition piece also provides a means to correct any misalignment of the foundation that may have occurred during installation. The towers are made of steel plate rolled into conical subsections that are cut and rolled into the right shape, and then welded together. The nacelles contain the key electro-mechanical components of the wind turbine, including the gearbox and generator. The rotor blades are made of fiberglass mats impregnated with polyester or carbon fiber composites. The power cable from each turbine is inserted in a "J" shaped plastic tube which carries the cable to the cable trench in the seabed.

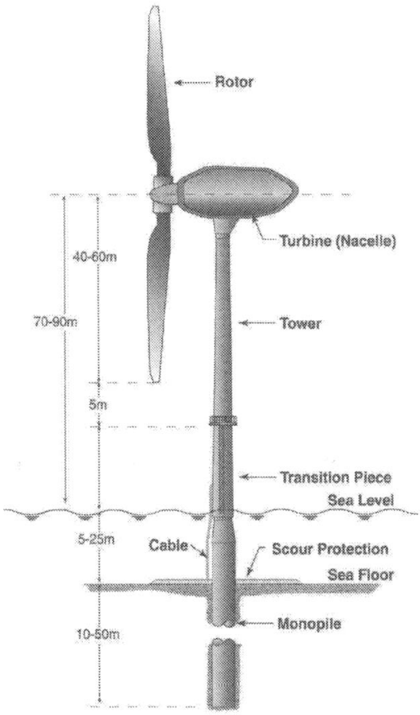

Fig. 2. Wind Turbine System Components (Malhotra, 2007c)

2.3 Wind turbine operation

As wind flows through a turbine it forces the rotor blades to rotate, transforming kinetic energy of the wind to mechanical energy of the rotating turbine. The rotation of the turbine drives a shaft which through a gear box drives a power generator which generates current through the principal of electromagnetic induction. The shaft, gearbox and generator are located in the nacelle. The nacelle is able to revolve about a vertical axis so as to optimally direct the turbine to face the prevailing wind. The electric current thus generated is converted to a higher voltage via a transformer at the base of the tower. The power that can be harnessed from the wind is proportional to the cube of wind speed up to a theoretical maximum of about 59 percent. However, today's wind turbines convert only a fraction of the available wind power to electricity and are shut down beyond a certain wind speed because of structural limitations and concern for wear and tear. So far, it is considered cost optimal to start power regulation at 10-min wind speed of 9-10 m/s, have full regulation at mean wind speeds above 14-15 m/s and shut-down or idle mode at 25 m/s. Power regulation is the ability of a system to provide near constant voltage over a wide range of load conditions. To minimize fluctuation and to control the power flow, the pitch of the blades of offshore wind turbines is regulated. At lower wind speeds, variable rotor speed regulation is used to smooth out power output. The yaw of the turbine is also varied every 30-sec to 60-sec, to maximize operating efficiency which creates gyroscopic loads. The pitching and yawing creates non-linear aerodynamics and hysteresis which have to be modeled in turbine response calculations.

2.4 Wind turbine foundation performance requirements

Deformation tolerances are usually specified by the wind turbine manufacturer and are based on the requirements for the operation of the wind turbine. Typically, these tolerances include a maximum allowable rotation at pile head after installation, and also a maximum accumulated permanent rotation resulting from cyclic loading over the design life. For an onshore wind turbine, the maximum allowable tilt at pile head after installation is typically between 0.003 to 0.008 radian (0.2 degrees to 0.45 degrees). A somewhat larger tilt 0.009 (0.5 degrees) may be allowed for offshore wind turbines. Any permanent tilt related to construction tolerances must be subtracted from these specified tolerances. Typical values of construction tolerances range from 0.003 to 0.0044 radians (0.20 degrees to 0.25 degrees). Allowable rotation of the support structure/foundation during operation is generally defined in terms of rotational stiffness which typically ranges between 25 GNm/radian to 30 GNm/radian (Vestas, 2007).

2.5 Foundation dynamics

Foundation dynamics is an important consideration in the design of an offshore wind turbine. As the offshore wind turbine rotates, the blades travel past the tower creating vibrations to which the offshore wind turbine is sensitive. It has been shown that when a three bladed rotor encounters a turbulent eddy it resists peak forces at frequencies of 1P and 3P, where P is the blade passing frequency. For a typical variable speed turbine, the blade passing frequency is between an approximate range of 0.18 Hz and 0.26 Hz, and rotation frequency, which is between about 0.54 Hz and 0.78 Hz. Meanwhile, cyclic loading from sea waves typically occurs at a frequency between 0.04 Hz and 0.34 Hz (Gaythwaite, 1990). Therefore, to avoid resonance, the offshore wind turbine (turbine, tower, support structure and foundation) have to be designed with a natural frequency that is different from the rotor frequencies as well as wave frequencies as shown in Figure 3.

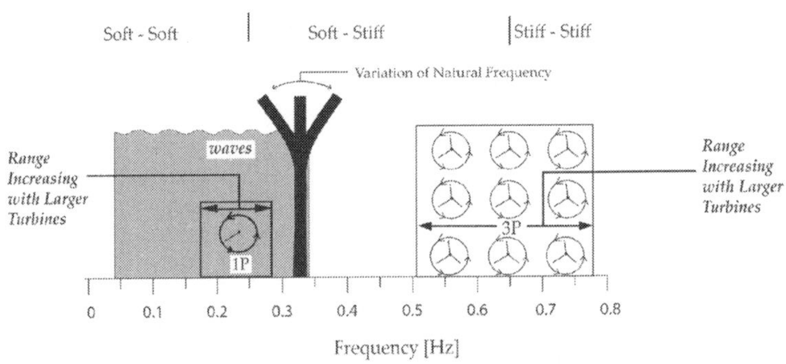

Fig. 3. Typical ranges for frequencies for waves, rotors, blade passing and structure (Malhotra, 2009).

Larger turbine diameters will require taller towers and heavier nacelles. The range of natural rotational frequencies 1P and 3P will also increase linearly with the blade diameter.

Since the natural frequency of the tower system is inversely proportional to the height of tower squared, the frequency of the higher towers will decrease rapidly and will fall in the region of wave frequencies, thereby imposing even greater demands on the design of the foundation and support structure. Accordingly, the support structure and foundation system would need to be made relatively stiff. A stiffer foundation would require more materials and therefore cost more than a flexible foundation.

3. Design process

The design process involves an initial site selection followed by an assessment of external conditions, selection of wind turbine size, subsurface investigation, assessment of geo-hazards, foundation and support structure selection, developing design load cases, and performing geotechnical and structural analyses. A flow diagram for the design process of a typical offshore wind turbine is shown in Figure 4. For achieving economies of scale, wind

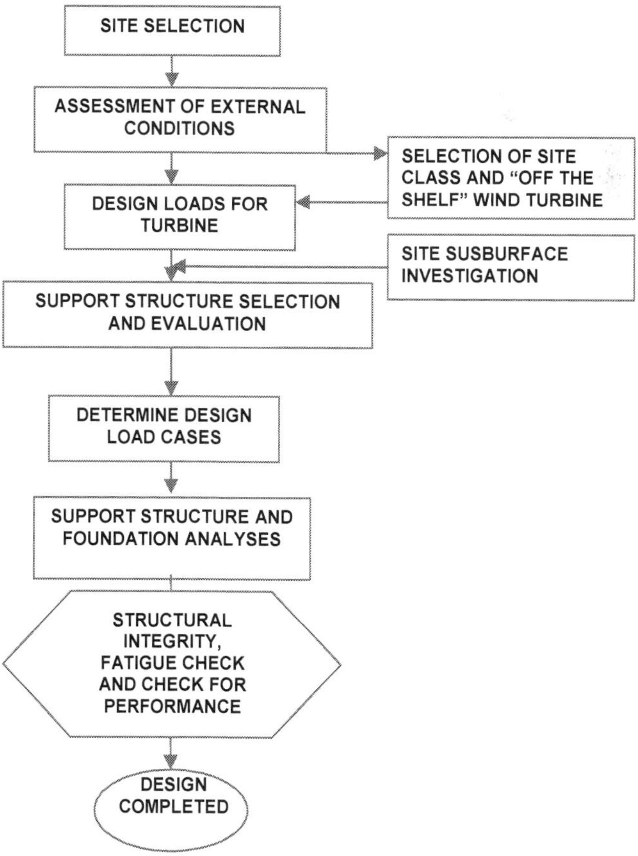

Fig. 4. Design Process for a typical offshore wind turbine (Malhotra, 2007c)

turbines are generally mass produced and available in four predefined classes based on wind speed. Consequently, the designer simply selects one of the predefined turbine classes that may apply to the wind farm site. Because the water depth, seabed conditions, sea state statistics (wave heights and current velocities), ice climate etc., may vary widely between sites, the use of a generic support structure concept is not feasible. Therefore, the tower, substructure and foundation, are designed for site specific conditions. The foundation system is selected based on several factors such as the level of design loads, depth of water at the site, the site geology and potential impact to the marine environment. As larger, customized wind turbines are developed, they will require an integrated analytical model of the turbine, support structure and foundation system and rigorous analyses with site specific wind and wave regimes.

3.1 Site selection

Besides favourable wind conditions, factors that govern selection of a wind farm site include site availability, visibility and distance from shore, proximity to power demand sites, proximity to local electricity distribution companies, potential impact to existing shipping routes and dredged channels, interference with telecom installations, buried under-sea cables and gas lines, distance from local airports to avoid potential interference with aircraft flight paths and interference with bird flight paths.

An offshore wind farm faces numerous challenges in all phases. During early development an environmental impact study phase requires extensive public involvement, while the permitting process is time consuming and requires ample input from various stakeholders, such as fishermen, local communities, aviation authorities, the Coast Guard authorities, the Corps of Engineers and others. A proactive approach with early community involvement generally helps the process. During this time perhaps by focusing on works that are more visible to the community such as onshore substations and cable routes the developer may be able to achieve progress. Locating the wind array farther from shore obviously will reduce visual impact. Obtaining suitable connections to the power grid and early collaborations with various suppliers of the wind turbine and cable systems are crucial for the successful project design and implementation. An early identification and evaluation of potential grid connection points to develop various substation locations and cable routes is essential for gaining public approval. From an electrical engineering standpoint, the compatibility between the wind farm export power cables and the grid require careful evaluation with respect to grid code compliance and system interface.

3.2 Assessment of external conditions

Following initial site selection, the developer makes an assessment of external conditions such as the level of existing wind conditions, water depth, currents, tides, wave conditions, and ice loading, the site geology and associated geo-hazards, such as sea-floor mudslides, scour and seismic hazards.

3.2.1 Design loads

Since wind loading is the dominant loading on an offshore wind turbine structure, it results in dynamics characteristics that are different from the wave and current loading that dominates the design of foundations for typical oil and gas installations. The loading on wind turbine foundations is characterized by relatively small vertical loading and larger horizontal and moment loads which are also dynamic. The design loads are classified into permanent, variable and environmental loads.

3.2.2 Permanent loads

Permanent loads include the mass of the structure in air, including the mass of grout and ballast, equipment, or attachments which are permanently mounted onto the access platform and hydrostatic forces on the various members below the waterline. These forces include buoyancy also. Permanent loads from typical offshore wind turbines are presented in Table 1.

	Typical 3.0 MW Turbine 80 m Hub Height	Typical 3.6 MW Turbine 80 m Hub Height	Typical 5 MW Turbine 90 m Hub Height	Future 7.5 MW Turbine 100 m Hub Height
Tower	156 ton	178 ton	347 ton	~550 ton
Nacelle	68 ton	70 ton	240 ton	~300 ton
Rotor	40 ton	40 ton	110 ton	~180 ton

Table 1. Permanent Loads from a Typical Offshore Wind Turbine (Various Sources)

3.2.3 Variable loads

Variable loads are loads that may vary in magnitude, position and direction during the period under consideration. These include personnel, crane operational loads, ship impacts from service vessels, loads from fendering, access ladders, platforms and variable ballast and also actuation loads. Actuation loads result from the operation of the wind turbine. These include torque control from the generator, yaw and pitch actuator loads and mechanical braking loads. In addition to the above, gravity loads on the rotor blades, centrifugal and Coriolis forces, and gyroscopic forces due to yawing must be included in design. Loads that arise during fabrication and installation of the wind turbine or its components also classify as variable loads. During fabrication, erection lifts of various structural components generate lifting forces, while in the installation phase forces are generated during load out, transportation to the site, launching and upending, as well as during lifts related to installation. The necessary data for computation of all operating loads are provided by the operator and the equipment manufacturers. The data need to be critically evaluated by the designer. Forces generated during operations are often dynamic or impulsive in nature and must be treated as such. For vessel mooring, design forces are computed for the largest ship likely to approach at operational speeds. Generally, permanent and variable loads can be quantified with some certainty.

3.2.4 Environmental loading

Environmental loads depend on the site climate and include loads from wind, wave, ice, currents and earthquakes and have a greater degree of uncertainty associated with them (Figure 5). These loads are time dependent, covering a wide range of time periods ranging from a fraction of a second to several hours. These loads act on the wind tower through different load combinations and directions under different design conditions and are then resolved into an axial force, horizontal base shear, an overturning moment and torsional moment to be resisted by the foundation.

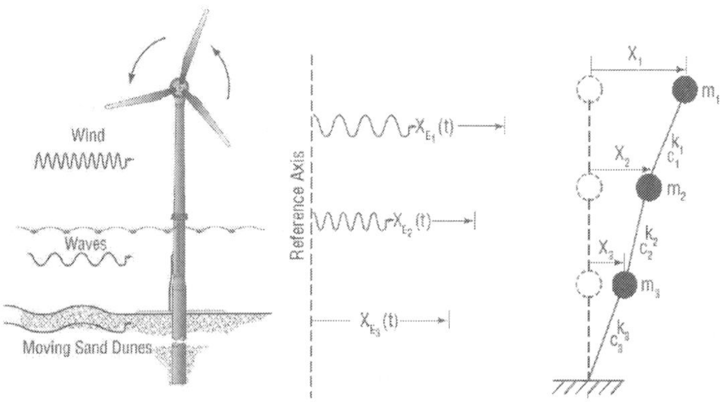

Fig. 5. Loads from wind, waves, currents and moving sand dunes (Malhotra, 2007c, 2009).

Wind Loading. Site specific wind data collected over sufficiently long periods are usually required to develop the wind speed statistics to be used as the basis of design. The design wind is represented by a mean wind speed, a standard deviation and a probability distribution for each of these parameters. Wind speed data are height dependent. To develop a design wind speed profile, a logarithmic or an exponential wind speed profile is often used. In areas where hurricanes are known to occur the annual maximum wind speed should be based on hurricane data.

Hydrodynamic Loads. Site specific measured wave data collected over long continuous periods are preferable. When site specific wave data are unavailable, data from adjacent sites must be transformed to account for possible differences due to water depths and different seabed topographies. Because waves are caused by winds, the wave data and wind data should correlate. However, extreme waves may not occur in the same direction as an extreme wind. Therefore, the directionality of the waves and wind should be recorded.

Loads from Currents. Tidal and wind generated currents such as those caused by storm surge have to be included in the design. In shallower waters usually a significant component of the hydrodynamic load is from currents.

Ice Loads. In areas where ice is expected to develop or where ice may drift ice loads have to be considered in design. The relevant data for sea ice conditions include the concentration and distribution of ice, the type of ice, mechanical properties of ice, velocity and direction of drifting ice, and thickness of ice.

Seismic Loads. For wind turbines to be located in seismic areas, a site response spectrum is usually developed for horizontal and vertical directions. For the analyses, the wind turbine is represented by a lumped mass at the top of the tower and it includes the mass of the nacelle, the rotors and part of the tower. Buckling analyses of the tower are conducted with the loads from the vertical ground acceleration.

3.2.5 Environmental loading conditions in the United States

Across the globe, foundation designers for offshore wind farms will face varied environmental conditions. For example, in the United States alone, environmental loading conditions include hurricanes in the southeastern United States and the Gulf of Mexico, and Northeast storms along the east coast from Maine to Virginia, and floating freshwater ice in the Great Lakes region. Hurricanes are large, revolving tropical cyclones which form well defined spirals with a distinct low pressure center and can be as large as 1000 km in diameter, traveling at a velocity of up to 11 m/s. Wind data for a number of hurricanes that made landfall in the US over a 50 year period are available. However, measured wave data from hurricanes are however quite limited and simplified methods are often employed to estimate design load parameters for wave loads.

Northeast winter storms are generated in the winter at higher latitudes with colder air at their core and do not have a well defined spiral and are often much larger in diameter that hurricanes. Even though these storms produce winds with lower velocities than hurricanes, their larger diameter can develop bigger high energy waves. Approximately 30 northeast storms occur in the northern portion of the Atlantic coast every year. Therefore, these storms must be considered in the determination of the wind turbine design loads.

In most European waters, sea ice is not a common phenomenon. It mostly occurs in the Barents Sea, northern and western parts of the Norwegian Sea, and inland waters such as the Baltic and Skagerak. Moreover, most offshore wind turbines have been installed in saline water either in the North Sea or the Baltic Sea. Meanwhile, the Great Lakes region of the US consists of large bodies of fresh water and is more susceptible to the formation of floating ice than are salt seas. Floating fresh water ice is harder than salt water ice and should be considered in the design of support structures for turbines in certain locations. Data on environmental conditions obtained from various projects at various locations are summarized in Table 2.

	North Sea	*Baltic Sea*	*Coastline United Kingdom*	*US East Coast*	*Gulf of Mexico*	*US West Coast*
			Wind			
50 yr Extreme 10 min Mean Wind (m/s)	50	54.5	38-46	46	45	33
50 yr Extreme 5 sec Gust (m/s)	60	65	46 - 55	55	54	40
			Waves and Currents			
50-yr Max. Wave Height (m)	22.3	5	8-12	24.2	19.9	17
Related Wave Period (sec)	14.5	13.0	14.0	13.7- 14.5	13	15 -17
50-yr Tidal Current Surface Velocity (m/s)	1.71	1.7	3.0	2.06	1.03	1.0
50-yr Storm Surge Current Surface Velocity (m/s)	0.43	0.30	0.7	0.5	0.25	0.25

Table 2. Comparison of Environmental Loading Conditions (Various Sources)

3.2.6 Application of available design standards

The lack of available guidelines for offshore wind turbine structures in the United States drives the designers of support structures for offshore wind turbines to look at the established design practice for conventional fixed offshore platforms as outlined in guidelines prepared by the American Petroleum Institute (API), of Washington, D.C. However, designers must first recognize the differences in the two types of structures and how they respond to applied dynamic loads.

The assessment of the dynamic response of offshore wind turbines differs from that of offshore oil and gas platforms and also onshore wind turbines. Offshore platforms are designed using static or quasi-static response calculations for external design loads, whereas, offshore wind turbines are driven by a combination of wind, wave and current loading in a non-linear dynamic analyses. The natural frequency of the offshore wind turbine is somewhere between the wave and rotor frequencies. On the other hand fixed platforms for the offshore oil industry are usually designed to have natural frequencies well above the wave frequencies. Unlike common practice in the offshore platforms, frequency domain analysis of dynamic response is seldom used for offshore wind turbines. The non-linear behavior of aerodynamic loading of the rotor, time domain simulations are generally required for an accurate assessment of both fatigue and ultimate limits states. Since the operating state of the wind turbine varies along with variable wind conditions, a number of load cases need to be analyzed. Compared with onshore wind turbines, wave and current climate cause a large extension of the number of load cases. Moreover, the influence of the foundation and support structure on the overall dynamic behavior is much larger compared to that of an onshore wind turbine.

Extreme wave loads generally govern the design of conventional fixed offshore platforms with wind loads contributing a mere 10 percent to the total load. Therefore, existing offshore standards emphasize wave loading but pay little attention to the combination with wind loads. In contrast, the design of offshore wind turbines is generally governed by extreme wind, wave and current loads, with almost equal weight being given to wind and wave loads depending on the site location. In addition, given the highly flexible response of the wind turbine structure, fatigue loads are critical.

So far, a key assumption in the design of wind turbines in Europe is that the turbines must be able to withstand extreme events with a return period of 50 years, whereas, the oil and gas industry structures are designed to withstand 100 year events. Therefore, the resulting reliability for offshore wind turbines and conventional fixed offshore platforms is understandably different. Extending the use of design loads obtained from API in the design of support structure and foundation will result in a higher degree of conservatism for the foundation design than for the wind turbine and consequently lead to higher construction costs. For the design of wind turbines a 10-minute average wind speed is considered long enough to cover all high frequency fluctuations of the wind speed and short enough to have statistical stable values. This is significantly different from offshore platform design where 1-hour average values are used.

For the design of offshore structures in the United States three exposure category levels corresponding to the consequence of failure are considered (API, 1993). Consequences would include environmental impact, danger to human life or economic loss. The failure of manned facilities or those with oil and gas storage facilities are considered of high consequence. Failure of platforms that may be manned but are evacuated during storms or do not have oil and gas storage is considered to be of medium consequence. Structures that are never manned and have low consequence of failure fall in the low consequence category. For the Gulf of Mexico, associated with each of these categories are a minimum wave height and period, wind speed and current speed to be used for design.

Offshore wind turbines are generally unmanned in storm situations so that the risk of human injury is low compared to typical manned offshore structures. Moreover, economic consequences of collapse and the related environmental pollution are low. For now, offshore wind turbines are likely to fall in the low consequence category. But as they become more integrated into the power grid and supply more power to the grid, the consequences of their failure are likely to increase.

API-RP2A (1993) guidelines suggest that the recurrence interval for the oceanic design criteria should be several times the design life of the offshore platform. Typical offshore platforms have a design life of about 20 years and are designed using 100 year return period oceanic criteria. However, for offshore wind turbine foundation design, a 50 year recurrence period is being used in Europe and appears appropriate for the United States as well.

Ultimate load cases may result from different environmental conditions (wind, wave, current, ice) and system operating conditions or installation procedures. Per the DNV (2004), for offshore wind turbine foundation design, a recurrence interval of 50 years is considered for extreme environmental conditions. For installation, operation and normal environmental conditions a recurrence interval of 1 year is considered. In the past few years, the International Electrotechnical Commission, of Geneva, Switzerland, and the Det Norske Veritas, a classification organization that has its headquarters in Oslo, Norway, have developed guidelines for offshore wind turbines, guidelines that for the interim are being used for projects in the United States.

4. Typical support structures

Support structures for offshore wind towers can be categorized by their configuration and method of installation into six basic types, described below.

Gravity Structures As the name implies, gravity structures resist the overturning loads solely by means of its own gravity. These are typically used at sites where installation of piles in the underlying seabed is difficult, such as on a hard rock ledge or on competent soil sites in relatively shallow waters. Gravity caissons are typically concrete shell structures. These structures are cost-effective when the environmental loads are relatively low, and the dead load is significant, or when additional ballast can be provided at a reasonable cost.

Monopiles This is a simple design in which the wind tower, made up of steel pipe, is supported by the monopile, either directly or through a transition piece. The monopile consists of a large diameter steel pipe pile of up to 6 m in diameter with wall thicknesses as much as 150 mm. Depending on the subsurface conditions, the pile is typically driven into the seabed by either large impact or vibratory hammers, or the piles are grouted into sockets drilled into rock. Compared to the gravity base foundation, the monopile has minimal and localized environmental impact. By far, the monopile is the most commonly used foundation for offshore wind turbines.

Guyed Monopile Towers The limitation of excessive deflection of a monopile in deeper waters is overcome by stabilizing the monopile with tensioned guy wires.

Tripods Where guyed towers are not feasible, tripods can be used to limit the deflections of the wind towers. The pre-fabricated frame is triangular in plan view and consists of steel pipe members connecting each corner. A jacket leg is installed at each corner which is diagonally and horizontally braced to a transition piece in the center. The tripod braced frame and the piles are constructed onshore and transported by barge to the site. Another construction advantage of these types of foundations is that they do not require any seabed preparation. In 2009, tripods were installed in 30 m of water for the Alpha Ventus wind farm located 45 km from the island of Borkum, Germany.

Braced Lattice Frames A modification of the tripod frame, the lattice frame has more structural members. The jacket consists of a 3-leg or 4-leg structure made of steel pipes interconnected with bracing to provide the required stiffness. Braced lattice frames have been used in deep water installations offshore of Scotland and are being planned for wind farms offshore of New Jersey.

Floating Tension Leg Platforms Floating structures are partially submerged by means of tensioned vertical anchor legs. The submerged part of the structure helps dampen the motion of the system. Installation is simple because the structure can be floated to the site and connected to anchor piles or suction caissons. The structure can be subsequently lowered by use of ballast tanks, tension systems, or both. The entire structure can be disconnected from the anchor piles and floated back to shore for major maintenance or repair of the wind turbine. Another version of the floating foundation requires merely a counterweight lowered to the seabed, in effect anchoring the floating platform. Several concepts for floating foundations are in the testing stage, with at least one in the demonstration phase in the Adriatic Sea off the south coast of Italy.

The following factors should be considered when selecting support structures for offshore wind turbines:

- Required dynamic response in the given water depth;
- Constructability and logistics of installation, including contractor experience and availability of equipment;
- Costs of fabrication and availability of steel and other materials; and
- Environmental effects.

The required dynamic response of the overall system in the given water depth is the main consideration. Because the dynamic response of a typical wind turbine depends on the stiffness of the support structure, which in turn is inversely proportional to its free standing height (or water depth) to the third power, one can use the water depth as a main factor for selecting the support structure in initial design. In 2004 and 2005, the author surveyed nearly 40 wind farms in Europe to obtain data on such details as water depths, distance from shore, soil conditions, and types of foundations and support structures employed. Water depth was correlated with support structure type to create Figure 6. The author believes that each of these support structures can be installed in even greater water depths with innovation and improved designs.

5. Typical foundations

Foundations anchor the support structures to the seabed, and typically fall into the six types described herein.

Gravity Caissons This type of foundation has been used for several offshore wind farms in Europe. For economical fabrication of gravity caissons one requires a shipyard or a drydock near the site (Figure 7) which allows the massive foundation structures to be floated out to

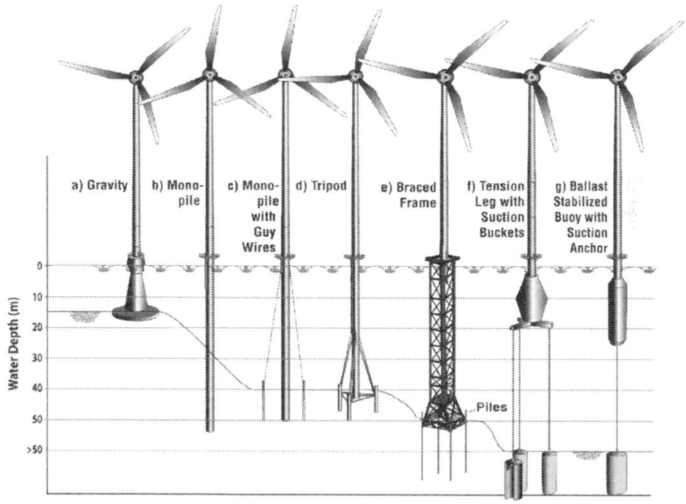

Fig. 6. Various types of support structures and their applicable water depth (Malhotra, 2007b, c).

Fig. 7. Gravity Base Foundation being constructed for Nysted Offshore Wind Farm at Rødsand, Denmark. *(Courtesy of Bob Bittner, Ben C. Gerwick, Inc.)*

the site and sunk. Site preparation and placement required for gravity caissons typically involves dredging several meters of generally loose, soft seabed sediment and replacement with compacted, crushed stone to prepare a level bed for the gravity caisson to rest on.

Special screeds and accurate surveying is required to accomplish this task. Installation of these structures is relatively time consuming. For example, approximately 29 days were needed to complete four gravity foundations at the Nysted wind farm constructed in Denmark in 2003.

Driven Pipe Pile The driven steel pipe pile option is an efficient foundation solution in deep waters. The typical method of offshore and near-shore installation of piled structures is to float the structure (monopile, tripod or braced frame) into position and then to drive the piles into the seabed using hydraulic hammers (Figure 8). The handling of the piles requires the use of a crane of sufficient capacity, preferably a floating crane vessel. Use of open-ended driven pipe piles allows the sea bottom sediment to be encased inside the pipe thus minimizing disturbance. The noise generated during pile driving in the marine environment might cause a short term adverse impact to aquatic life. Since the number of piles is typically few and spread apart, these adverse impacts are only short term and relatively minor. Installation times for driven monopiles are relatively short. For example, individual monopiles constructed in 2004 as part of the Scroby Sands wind farm in Norfolk, United Kingdom, required less than 24 hours to install. Although available offshore pile-driving hammers with a rated energy of 3,000 kJ or more are capable of installing piles with diameters as large as 4.5 m, newer, higher-capacity models with adaptors for even larger piles are being developed. Pile driveability evaluations and hammer selection are crucial parts of the process.

Fig. 8. Menck Pile driving hammer atop a steel pipe pile at Kentish Flats Offshore Wind Farm, UK (Courtesy: Elsam)

Post-Grouted Closed-end Pile in Predrilled Hole In this design, a closed-ended steel pipe pile is placed into a predrilled hole and then grouted in place. This option (Figure 9) is often used for offshore pile foundations less than 5 m in diameter and offers significant advantages over the cast-in-place drilled shaft option, including advance fabrication of the pile, better quality control, and much shorter construction time on the water. This option requires a specially fabricated large diameter reverse circulation drill. It also requires handling and placement of a long, large diameter pile, with considerable weight. Closed-end piles can be floated to the site and lowered into the drill hole by slowly filling them with water. Installation times for drilled and post-grouted monopiles are relatively long, averaging about 50 hours per monopile.

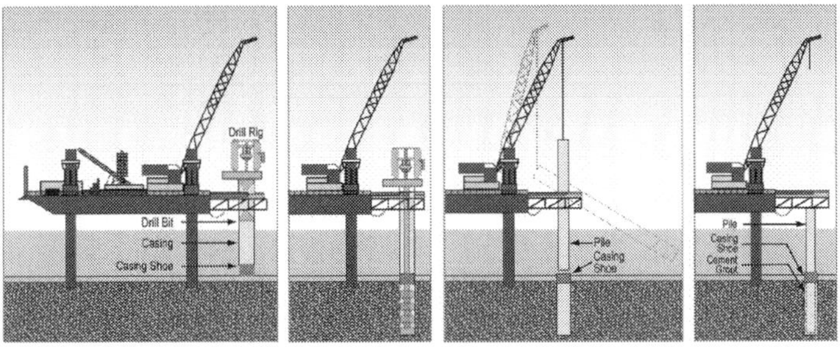

Fig. 9. Typical Installation Sequence for a post grouted closed end pipe pile in predrilled hole (Malhotra, 2007c)

Drilled Shafts The installation of bored, cast-in-place concrete pile requires driving a relatively thin (25 mm) walled casing through the soft sediment to the underlying denser material (if necessary to establish a seal), then drilling through and below the casing to the required base elevation. Bending resistance is provided by a heavy reinforcing cage utilizing high strength, large diameter bars, with double ring, where necessary. The casing provides excavation support, guides the drilling tool, contains the fluid concrete, and serves as sacrificial corrosion protection. This approach requires a large, specially fabricated reverse circulation drill. The use of drilled shafts for offshore wind turbine foundations suffers from several disadvantages, including the need for placement of reinforcement, as well as the need to transport and place large quantities of concrete offshore. The logistics associated with offshore concreting and reinforcement placement make drilled shafts uneconomical for offshore wind turbines.

Composite "Drive-Drill-Drive" Pile This procedure requires an adaptation of existing drilling and piling techniques and involves a combination of drive-drill-drive sequence to achieve the design depth. Installation times for monopiles using this composite sequence are relatively long. For example, the construction of monopiles as part of the North Hoyle wind farm in the United Kingdom in 2003 required approximately 70 to 90 hours on average.

Suction Anchor Suction anchors consist of a steel canister with an open bottom and closed top. Like piles, suction anchors (Figure 10) are cylindrical in shape but have larger diameters (10 m to 15 m) and subsequently shallower penetration depths. They are installed by sinking into the seabed and then pumping the water out of the pile using a submersible pump (Figure 11). Pumping the water creates a pressure difference across the sealed top, resulting in a downward hydrostatic force on the pile top. The hydrostatic pressure thus developed pushes the anchor to the design depth. Once the design depth is achieved, the pumps are disconnected and retrieved. Installing suction caissons is relatively time consuming, as evidenced during the construction of the Hornes Rev II wind farm in Denmark in 2008. Approximately 32 hours were needed to complete installation of a single suction caisson for a meteorological mast, of which 10 hours involved penetration. Sunction anchors resist tension loads by relying on the weight of the soil encased by the steel bucket along with side friction on the walls and hydrostatic pressure. The stability of the system is ensured because

Fig. 10. Suction Anchors for an offshore platform being transported to site in the Gulf of Mexico (Courtesy Prof Aubeny, TAMU).

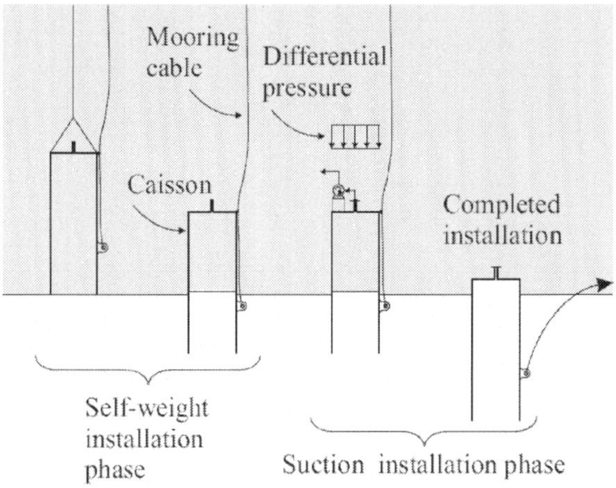

Fig. 11. Installation stages of a Suction Anchor

there is not enough time for the bucket to be pulled out of the soil during a wave passage. As the bucket is pulled up, a cavity is formed between the soil surface and the bottom of the bucket which creates a suction pressure that resists the uplift loads. These foundations carry compression loads by side friction and end bearing. Suction anchors are expected to be particularly suitable for foundations in soft cohesive sediments. These foundations cannot be used in rock, in gravel or in dense sand. Suction anchors are cheaper to install since they do not require underwater pile drivers. At the end of a wind turbine's life, a suction anchor may be removed completely from the seabed, unlike piled foundations. This provides room for recycling and reuse.

Foundation selection considerations for offshore wind turbines include:

- Soil Conditions that facilitate installation and performance,
- Driveability for driven piles and penetrability for suction anchors,
- Constructability and logistics of installation, including Contractor experience and availability of equipment, and
- Costs of fabrication, availability of steel and other materials.
- Environmental impact considerations.

Soil conditions at a project site will generally drive the method of installation and constructability aspects. Driven monopiles are most adaptable to a variety of soil conditions. They are currently the most commonly used foundation for offshore wind turbine projects. Their construction procedure can be modified to suit the site conditions encountered. For example, in the presence of cobbles and boulders, or very dense sands the Contractor may use a sequence of drilling and driving to achieve the required design depth for the monopile. In cohesive till and in soft rock, drilled shafts or post-grouted closed end pipe in drilled hole may be most suitable. Gravity base foundations will be feasible in shallow waters, where competent bearing stratum, such as a rock ledge or glacial till is available at shallow depth. Suction caissons will be geotechnically feasible in soft clay strata and medium dense sands. The final selection of the foundation may be driven by other factors such as environmental impact, costs of construction, availability of equipment and contractor preference.

6. Environmental impact of foundation installation

The type of foundation selected will also have an impact on the environment. If drilled shafts are selected as the foundation then the issue of disposing the excavated material will need to be addressed. The larger areas required for gravity caissons also pose significant disturbance to the seabed environment. To limit the area of dredging required for the gravity base foundation, some form of ground improvement can be performed. The use of various available ground improvement techniques for such purposes should be further examined. Use of open-ended driven pipe piles allows the sea bottom sediment to be encased inside the pipe thus minimizing disturbance.

Airborne Noise: During construction of offshore farms, airborne noise from construction work (vessels, ramming, pile driving, etc.) will likely affect birds and marine mammals, but as the construction operations are of limited duration, the effects are expected only to be temporary. However, sensitive time periods like breeding or nursery periods should be avoided if the construction site is placed near important biological areas - which may be in conflict with the intentions of the developers to establish offshore wind farms when stormy weather is least probable.

Underwater Construction Noise: During construction, underwater noise from construction vessels and drilling or piling equipment may have a detrimental effect on marine mammals, fish and benthos. These effects are especially evident when driving monopiles. Noise from pile driving can either cause behavioral changes, injury or mortality in fish when they are very close to the source and exposed to either sufficiently prolonged durations of noise or elevated pressure levels. Accurately analyzing and addressing these effects is somewhat complicated. The sound pressure levels and acoustic particle motion produced from pile installation can vary depending on pile type, pile size, soil conditions, and type of hammer. Furthermore, the diversity in fish anatomy, hearing sensitivity, and behavior, as well as the acoustic nature of the environment itself for example, water depth, bathymetry, tides, further complicates the issue of how fish are affected by pile driving noise and how severe those effects may be. Experiences from a variety of marine projects in the US and offshore wind farm projects in Sweden indicate that barotrauma from pile driving noise results either in mortality or trauma in fish, resulting in loss of consciousness and drifting on the water surface as if they were dead. However, the effect is considered temporary. In the case of fish larvae, noise from construction work at sensitive periods may result in a very high fish mortality rate. Accordingly, construction work during larvae season should be avoided. Generally, peak sound pressure levels of more than about 160 dB at a reference pressure of 1 μPa are considered harmful to aquatic life and marine mammals (Elmer et al, 2007).

Available approaches for mitigating noise related to pile driving include prolonging hammer impact, using an air-bubble curtain or bubble tree, using an isolation casing with foam coating, or using a vibratory hammer. Prolonging hammer impact results in lower velocity amplitudes and frequencies, lowering overall noise levels. A bubble curtain involves pumping air into a network of perforated pipes surrounding the pile. As the air escapes the perforations, it forms an almost continuous curtain of bubbles around the pile, preventing the sound waves from being transmitted into the surrounding water. A foam-coated isolation casing works in a similar manner. Vibratory hammers operating between 20 and 40 Hz generate sounds that are 15 to 20 decibels lower than those generated by impact driving. Although vibratory hammers are effective within a limited range of soil conditions, they are easily adaptable to pile diameters of as much as 6 m.

Underwater Operational Noise: During operation, noise from offshore turbines can be transmitted into the water in two ways: the noise either enters the water via the air as airborne sound, or the noise is transmitted into the water from tower and foundation as structural noise. The frequency and level of underwater noise is thereby to a certain degree determined by the way the tower is constructed and by the choice of foundation type and material (monopile/steel - or caisson type/concrete - foundation). Underwater noise from offshore wind turbines must of course exceed the level of underwater background noise (ambient noise, especially from ships) in order to have any impacts on marine fauna.

Measurements from offshore farms Vindeby in Denmark (caisson foundation type) and Bockstigen in Sweden (monopile) indicate that underwater noise is primarily a result of the structural noise from tower and foundation (Bach et al., 2000). When the results were scaled up, based on measurements from a 2MW onshore wind turbine, it was concluded that the underwater noise might be audible to marine mammals within a radius of 20 m from the foundation. Generally, it is believed that for frequencies above 1 kHz, the underwater noise from offshore turbines will not exceed the ambient noise, whereas it is expected that for frequencies below 1 kHz, noise from turbines will have a higher level than the background noise. Only measurements and impact studies after the construction can reveal if underwater noise will really affect marine mammals. The impact on fish from low frequency sounds (infrasound, below 20 Hz) is uncertain. The effects from noise and electromagnetic fields on fish communities living at the seabed are still a subject of further study.

7. Foundation design considerations

7.1 Geotechnical Investigation

Since foundation construction costs can balloon from unanticipated subsurface conditions, such as paleochannels, the presence of boulders or foreign objects such as shipwrecks, abandoned anchors, pipelines, or lost casings from prior explorations, it is an important risk management strategy to perform a thorough bathymetric survey and a geotechnical investigation program. With the purpose of defining the soil stratification and investigating the seabed for unidentified objects, a geotechnical investigation program should be performed in at least three phases: a desk top geological study, followed by a bathymetric and geophysical survey, and then by Cone Penetration Test (CPT) soundings, and geotechnical borings. A wind farm site would extend over a substantial area and there may be significant variation in subsurface conditions across the site. The generally limited number of geotechnical borings may not reveal the true variation in subsurface conditions across the site. In this regard, a geologic study and geophysical methods become invaluable in alerting the designer to the range of variation in subsurface conditions across the site. The wind turbine array may need to be modified based on the subsurface conditions (Figure 12).

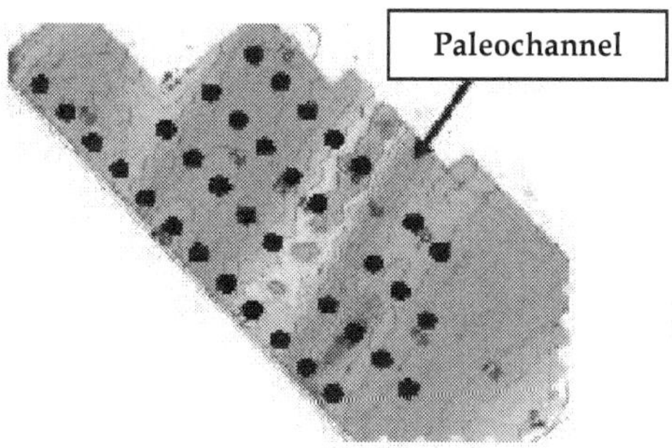

Fig. 12. Paleochannel filled with soft mud at a wind farm site requiring a change in the layout.

Geophysical methods usually based on seismic and sonar methods, including side-scan sonar and sub-bottom profilers, can be used to obtain information on seabed stratigraphy and topography over the proposed site along with information on any foreign objects. With the purpose of confirming the results of the geophysical profiling, cone soundings should be performed in an array across the site. These should be supplemented by several borings located strategically across the study area. Proper surveying methods and locating systems must be employed to ensure that the bathymetry, geophysical methods, geotechnical soundings and borings, and the actual installation are all controlled to the same positioning relative to each other. To determine the geotechnical engineering characteristics of the soils, a program of laboratory soil testing should be performed.

The geotechnical investigations should extend throughout the depth and areal extent of soils that affect or be affected by the installation of the planned structure and foundation. The number of borings will depend on the foundation type selected and the anticipated geologic conditions. In general, one boring per turbine foundation is recommended. Since the main expense of performing an offshore drilling program is the mobilization of the drill barge, performing fewer borings may not be as cost-effective. Moreover, obtaining geotechnical information at foundation location goes a long way in reducing risk. If pile foundations are selected, the borings and soundings should extend at least 5 m to 10 m beyond the preliminary estimated depth of pile foundations. If a gravity structure is selected, the borings should extend to a depth such that stress increase in the soil and the resulting strain caused by the gravity structure is negligible. Moreover, multiple borings may be required for each gravity structure to account for variation in subsurface conditions. All in all, a geotechnical engineer experienced in offshore investigations should be retained for planning and executing such investigations. He (or she) should be able to modify the initially proposed plan based on site conditions encountered. The program should be equipped to switch between soil sampling, rotary coring, and insitu testing, where needed. For cable routing studies, relatively shallow cone soundings or vibro-core borings may be performed to obtain information for the cable trench.

7.2 Local geology and geologic hazards

Geologic hazards that have the potential for causing failures of offshore engineered manmade structures include slope failures, fault ruptures, and adverse soil conditions. Such hazards can be triggered by external events, including earthquakes and surface storms. For instance, Hurricane Ivan, Katrina and Rita destroyed 118 oil drilling platforms and caused substantial damage to underwater pipelines by triggering slope failures. Pipelines and undersea cables face several geohazards including:

- Submarine slope failures including rotational or translational failures and sheet flows,
- Sea floor rupture associated with seismic activity, and
- Rough sea floor where the cable or pipeline may have to bridge over rock outcrops and channels.

Seafloor Mudslides. If the seafloor at the proposed site of the wind farm is even mildly sloping, it could be susceptible to localized or areal sea-wave or earthquake induced mudslides or submarine slides that could adversely impact the foundations. Numerous submarine slides have been documented on the coasts of continental United States. Identifying slope stability hazards in the marine environment is a challenging and important task. Marine slope failures can occur on slopes with gradients as low as 2 to 6 degrees. But their most notable feature is their large aerial extent. For example, the Sigsbee Escarpment in the Gulf of Mexico and the Storegga Slide in the Norwegian Sea are among the largest documented submarine slides in the world. A submarine slide can either result in

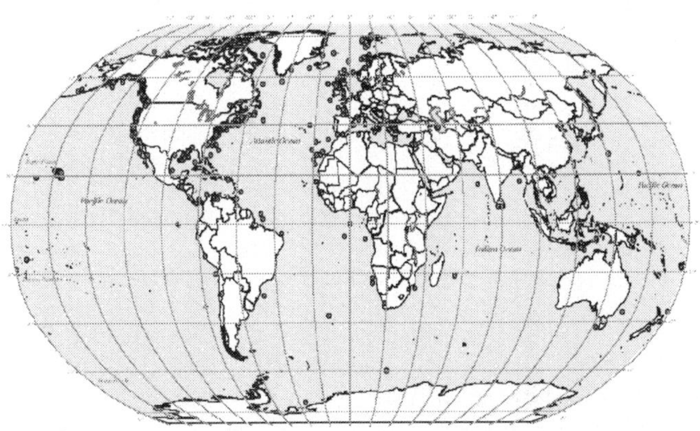

Fig. 13. Documented Submarine slope failures in the United States (Hance, 2003).

mud flow or the sliding mass will stop in a short distance. Debris flow analyses require an estimation of the depth of debris and run-out distances. Seismic stability of submarine slopes should also be evaluated. Moreover, slopes offsite could potentially impact the wind farm foundations. Therefore, these issues should be identified early in the process since they can significantly influence site selection. Where identified, the potential loading from seafloor mudslides should be incorporated in foundation design. Seafloor mudslides can potentially overstress the foundations by lateral loads and damage buried cables.

Scour Considerations. Removal and deposition of seabed soils caused by waves and currents is an ongoing geologic process that may be further exacerbated when the natural flow regime is interrupted by man-made structures. For an offshore wind farm different types of scour need to be considered: local scour; occurring immediately around the foundation, global scour; occurring to a large extent as a shallow scoured basin due to effects of multiple foundations; and overall seafloor movements, including sand waves, ridges and shoals.

The consequence of scour is either reduction of lateral and vertical support for foundations, or imposition of additional forces resulting in settlement or overstressing of the foundations. Scour will have an impact on the length of foundation embedment, the natural frequency of the wind turbine system, and the J-tube. Since the offshore wind turbine foundation primarily provides lateral support, and for monopiles the lateral load resistance is offered by soils primarily in the upper 5 to 8 pile diameters, the effect of scour on monopile design is rather pronounced. Moreover, changes in seabed level conditions whether by scour or by sand waves will result in a decreased or increased lateral pile stiffness and natural frequency. In design, the most probable combination of maximum scour depth and maximum extreme loads must be used for foundation analyses. As noted earlier, the power cable from each turbine is inserted in a "J" shaped plastic tube which carries the cable to the cable trench. In the event of scour, the cable would be freely spanning over the scour hole. Since the function of the cable is crucial to the whole operation, any damage to the cable whether by scour or other factors must be prevented. Scour protection measures are presented in a later section. Fatigue, which spans the lifetime of the structure, is particularly sensitive to any changes in the dynamic stiffness of the foundation. The effect of scour on fatigue response is discussed in a later section.

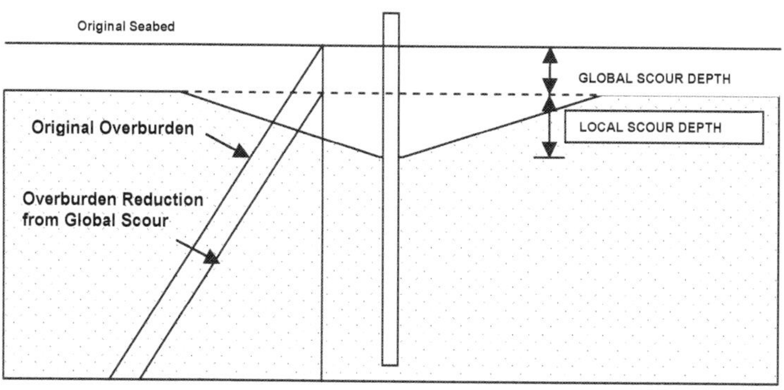

Fig. 14. Overburden stress reduction due to scour (Malhotra, 2007c)

Seismic Considerations. The level of seismicity of the site where the wind turbine structure is to be installed should be assessed either through available data or through detailed investigations. If the area is determined to be seismically active and the wind turbine will be affected by such activity, then site specific response spectra and design criteria should be developed. Depending on the location of the site, seismic hazards of liquefaction, seismic settlement, lateral spreading and earthquake loads could adversely impact the wind tower foundations and should be addressed in the design. The potential for seismic induced sea waves also known as tsunamis should also be assessed. Generally, earthquake loads seldom drive the design for wind turbines. For conceptual level studies seismic ground accelerations may be obtained from prior studies or relevant seismic hazard map.

Liquefaction Potential. Soil liquefaction, defined as a significant reduction in soil strength and stiffness as a result of increase in pore pressure during dynamic loading, is a major cause of damage to built structures during earthquakes. Typically, the hazard from liquefaction occurs in four ways, including: a) bearing failure, b) settlement, c) localized differential lateral movements, and d) ground loss or highly localized subsidence associated with expulsion of material such as "sand boils". Usually, for soil liquefaction to occur offshore, two conditions must exist: including a) presence of loose, sandy soils or silty soils of low plasticity, and, b) a source of sudden or rapid loading, typically associated with earthquakes. Generally, soil conditions found at offshore wind farm sites would be susceptible to liquefaction should an earthquake occur. Therefore, liquefaction potential at all wind farm sites should be performed in the early phase of design development.

Lateral Spread Potential. Lateral spreading occurs primarily by horizontal displacement of soils at the seabed due to liquefaction of underlying granular deposits. The degradation in the undrained shear resistance arising from liquefaction may lead to limited lateral spreads induced by earthquake inertial loading. Such spreads can occur on gently sloping ground or where nearby drainage or stream channels can lead to static shear biases on essentially horizontal ground. The determination of lateral spread potential and an assessment of its likely magnitude ought to be addressed as a part of the hazard assessment process for an offshore wind farm site. In the event that there is a potential for lateral spreading at a wind farm site, it should be incorporated in the design of the foundation. For instance, if pile foundations are employed, passive forces from the moving soil mass should be applied to the pile foundation and checked for moment and shear capacity and related deflection (Figure 15). Alternately, the expected soil displacement may be applied to the pile model to evaluate the related shear and moment developed in the pile.

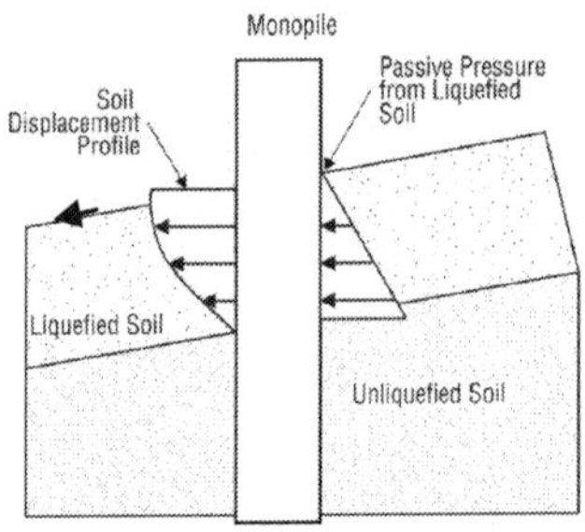

Fig. 15. Forces on Monopile Foundation from lateral spreading of liquefied seabed soils (Malhotra, 2007c).

8. Foundation analyses

Gravity Base Foundations: In gravity base foundations, the passage of waves produces dynamic horizontal forces concentrated on the top of the structure. At the same time, pressure fluctuations of the passing crest and trough reduces the apparent weight of the gravity structure and creates a mode for vertical vibration. With or without earthquake forces, the eccentricity of the wave forces results in coupled rocking and sliding modes of vibrations. During a major storm, the simultaneous occurrence of vertical, sliding and rocking loads would create high stress concentrations at the edges of the foundation resulting in excess pore pressures and possible localized yielding of foundation soils below the foundation edge opposing the storm. Following the passage of the storm, the excess pore pressures would dissipate causing settlement and related differential settlement possibly resulting in tilting of the foundation. Therefore, for gravity foundations, uplift, overturning, sliding, bearing capacity, lateral displacement and settlement are potential failure modes

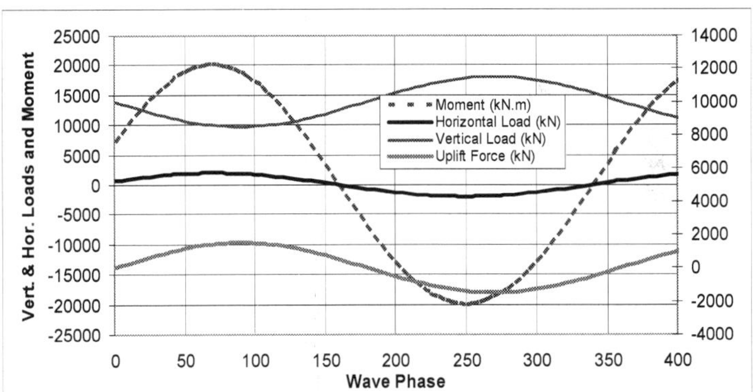

(a)

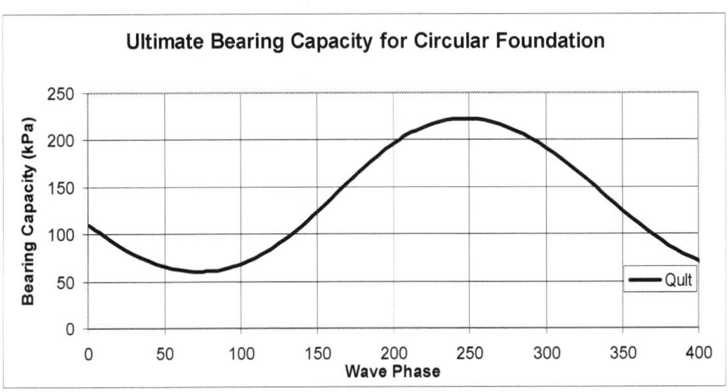

(b)

Fig. 16. (a and b): Bearing capacity of Gravity Base with varying load levels.

and require evaluation. Immediate, primary and secondary consolidation settlement, permanent horizontal displacements and resulting differential settlements have to be evaluated. Differential settlements should incorporate the lateral variation of the soil conditions, unsymmetrical weight distributions, predominant directional loading and seismically induced settlement. The required sliding resistance determines the minimum weight of the system, based on Coulomb's relation for frictional material. It has been found that heave force on the gravity base is a more dominant factor leading to instability by sliding than the overturning moment from the aerodynamic loads (Henderson et. al., 2003). Heave force can be calculated using Bernoulli's theorem for the undisturbed wave kinematics (instead of the more complicated diffraction theory). Vertical bearing capacity is computed using bearing capacity theory developed by Prandtl, Brinch Hansen and Terzaghi. In bearing capacity calculations, the inclined and eccentric loads on the gravity

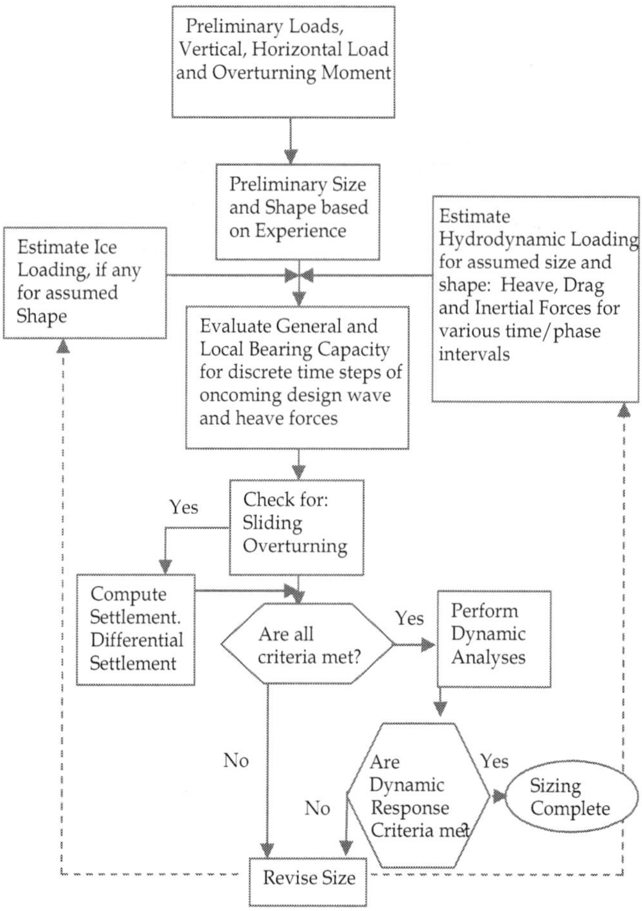

Fig. 17. Design process for an offshore Gravity Base Structure (Malhotra, 2007c)

base can create a severe reduction in the effective area and allowable bearing capacity, therefore reduction factors for inclined and eccentric loads have to be applied. Because of its sensitivity to the ratio of the vertical and lateral loading, an evaluation of bearing capacity and hydrodynamic heave and lateral forces is required for many phases of the oncoming design wave (Figure 16, a and b). The design of the gravity base foundation is iterative in the sense that the large contribution from hydrodynamic loading on the gravity base depends on the shape and size of the gravity structure, which is in turn is governed by its load bearing function (Figure 17).

Dynamic analyses of gravity base foundations require consideration of soil-structure interaction effects. For homogeneous soil conditions, the soils can be modeled as an elastic half space with an equivalent shear modulus G. Foundation stiffness coefficients can be determined based on elastic theory. These stiffnesses can be used in structural analyses for wind and wave loading on the turbine and its support structure. To account for the strain dependency of shear modulus and internal soil damping, a range of shear modulii and damping should be considered in developing foundation springs. In the case where heave from wave passage and excessive overturning moments lead to formation of a gap at the base of the gravity foundation, these procedures overestimate the foundation stiffness. Gravity base structures are generally massive and constitute a majority of the weight of the wind tower system. Therefore, the fundamental period of vibration of the gravity base should be small compared to the rest of the wind tower system. Since the modes of vibration of the wind tower and the gravity base are different, an uncoupled analysis would be expected to provide reasonable estimates of foundation behavior (Figure 18).

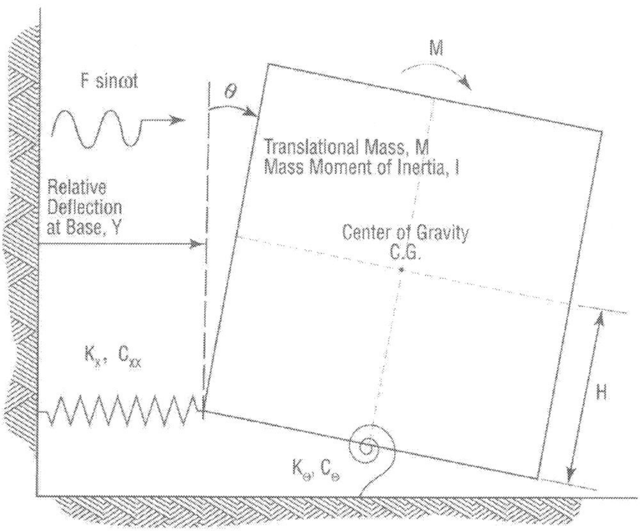

Fig. 18. Analytical Model of Gravity base showing Uncoupled Translational and Rotational Springs (Malhotra, 2007c).

Deep Foundations Geotechnical design methods for driven pile foundations focus on three aspects. 1) Axial load carrying capacity in compression and tension, 2) Response under cyclic lateral loads, and 3) installation. The main objective of the design is to select a foundation size that can develop the required axial capacity, perform adequately under lateral loads without excessive deflection and rotation at mudline, and withstand the installation stresses needed to penetrate the pile without buckling the walls. A typical driven pile design will require the prediction of: 1) Penetration resistance with depth, 2) Depth to Fixity under lateral loads, 3) Minimum required penetration depth for fixity and axial capacity considerations, 4) Driveability evaluation.

Bearing capacity and lateral load deformation performance of the pile is evaluated using finite element procedures or other available procedures using the p-y approach. For a detailed treatment of axial pile capacity for pipe piles, the reader is referred to Malhotra (2002).

Dynamic analyses of pile foundations can be performed in a variety of ways (Figure 19). The most common methods are: a) A finite difference model of a beam supported on non-linear 'p-y', 't-z' and 'Q-z' curves to represent the pile-soil interaction, b) Coupled foundation stiffness method, c) Uncoupled stiffness model, d) Equivalent Cantilever Method with an effective depth to fixity. Soil-pile-structure interaction under such cyclic loading is a complex process involving several simultaneously occurring phenomena such as cyclic degradation, pore water pressure generation, and gap-slap-scour near the pile head. All these factors may occur simultaneously to modify the axial and lateral stiffness and response of the pile foundations. A detailed discussion of these phenomena is provided in Malhotra (2004).

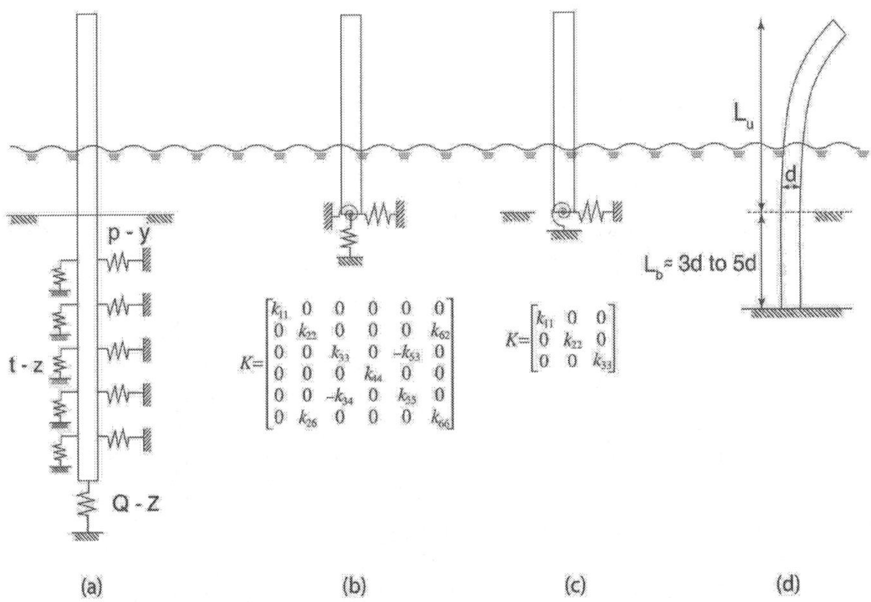

Fig. 19. Available analytical models for deep foundations (Malhotra, 2007c)

Pile Driveability: Driveability evaluations are a crucial part of the design process. One-dimensional wave equation analyses have been widely used to evaluate pile driveability and hammer performance. The purpose of wave equation analyses is to predict the behavior of the pile during installation for specific site conditions and driving equipment. In the United States, wave equation analyses are typically performed using computer program GRLWEAP developed by Goble, Rausche and Likens (GRL) (1997) to assess the range of pile hammer energies required to drive the piles. The major purpose of this study is to 1) estimate the feasibility of driving the piles to design depth with the selected hammer, 2) to evaluate pile tensile and compressive stresses and pile driving resistances for the selected range in hammer energies, and 3) to develop preliminary driving criteria. A case study on driveability of very large diameter piles can be found in Malhotra 2007c.

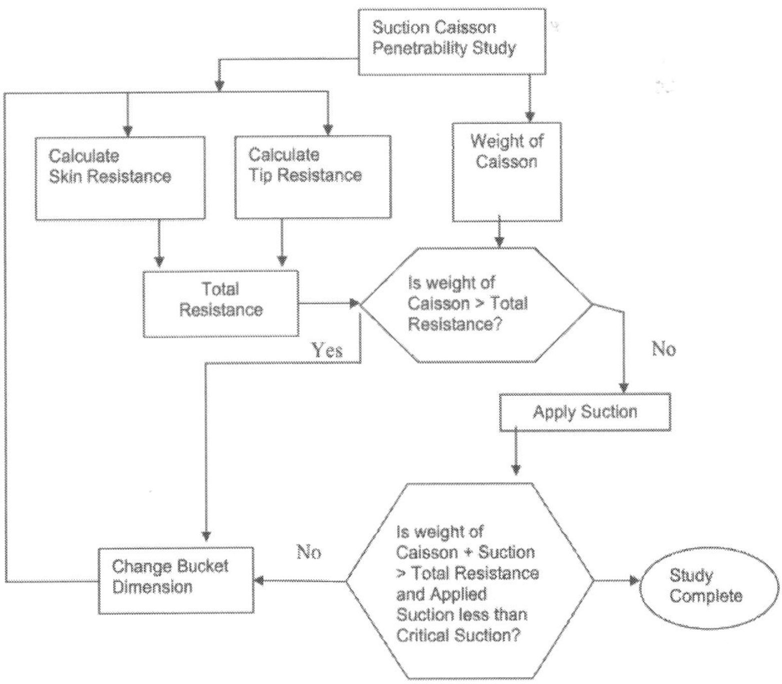

Fig. 20. Typical flow diagram in the design of suction caissons, (Malhotra, 2007c).

Suction Caissons Geotechnical design methods for suction caissons focus on two aspects. 1) installation and retrieval, and 2) uplift capacity. The main objective of the design is to select a caisson size that can develop the required uplift capacity and withstand the installation suction needed to penetrate the caisson without buckling the walls or failing the soil plug. A typical suction caisson design will require the prediction of: 1) Penetration resistance with depth, 2) Required suction with depth, 3) Optimum location of pad eye for lateral loads (if any), 4) Self-weight penetration depth, and, 5) Maximum penetration depth. Once the caisson is sized, a caisson penetrability study is performed. This study is performed to ensure that the soil resistance offered is less than the driving force on the caisson. The driving force on the caisson consists of the buoyant weight of the caisson and the applied suction. It is necessary to ensure that critical suction is not exceeded. A detailed discussion of the design process with design examples (Figure 20) are provided in Malhotra (2007c).

9. Fatigue

Fatigue is the process of gradual damage done to materials when subjected to continually changing stresses. Due to these repeated stress changes, the material slowly deteriorates, initiating cracks which will eventually propagate and lead to eventual failure. Offshore wind turbines are by default subjected to loads varying in time from wind as well as waves. This means that the stresses on the support structure will also vary continuously making them prone to fatigue. Tubular steel structures such as monopiles, braced lattice frames and tripods in deep water and exposed to wave loading are particularly susceptible to fatigue. To be able to take fatigue into account in the design process, an empirical design method for the design of steel structures is commonly used. Fatigue evaluation during design involves comparing the intended design life of the structure with its predicted fatigue life as limited by "hot-spot" stresses, i.e. areas of high local stress reversals. Fatigue, which spans the lifetime of the structure, is particularly sensitive to any changes in the dynamic stiffness of the foundation. Small increases in stress levels will increase fatigue damage by the power of 4. Minor amounts of scour can result in relatively minor changes in foundation stiffness and consequent cyclic stresses, yet result in major changes in fatigue response. It can be shown that for a typical monopile, lowering the natural frequency of the foundation by a small amount (5 to 8 percent), will have a dramatic effect on fatigue (almost 100 percent) (Malhotra, 2007c). At lower natural frequency of the pile more wave energy will create a resonant response of the wind turbine and increase fatigue. Therefore, great emphasis is placed on foundation selection and foundation dynamics.

10. Other considerations

Corrosion Protection: Since the most intensive corrosion typically occurs in the splash zone, the transition piece is usually provided with a heavy duty protective coating. To protect the underwater part of the transition piece and piles against corrosion, sacrificial cathodic protection (in addition to the shop-applied coating) is used.
Scour Potential: A site specific study should be performed to assess potential scouring of the seafloor. Typical scour protection measures include scour mats made up of buoyant

polypropylene fronds and polyester webbing which is anchored securely to the seabed or crushed rock mattresses.

Marine Growth: The plant and animal life on the site causes marine growth on structural components in the water and in the splash zone. The potential for marine growth should be addressed, since it adds weight to the structural components of the system and may increase hydrodynamic loads. Special paints and coatings are available that prevent marine growth.

Impact from Ship Collision: Accidents between ships and wind farms can result in damage to the wind farm, the ship and the local environment in a variety of ways. Although the probability of occurrence of ship collisions should be relatively low, the consequences could be serious. Accordingly, mitigating measures are warranted. Mitigation techniques involve a two pronged approach. Reduce frequency of collision and reduce consequences of collision. An optimum approach would be one which balances between the two approaches. Collision risks can be reduced by passive measures, such as proper marking of the wind farm and the individual wind turbines using classical techniques as marking lights, painting, buoys or by active measures such radar based ship detection in combination with emergency towing capabilities. Collision damage can be reduced by classical fendering techniques designed based on the existing experience with pier, ship, tug and buoy fendering and specifically developed for offshore wind farm applications. Damage reduction solutions for offshore wind farms will be different from classic applications: for individual monopile wind turbines since one has to balance on the one hand the need to absorb without damage the impact of small vessels and on the other hand the requirement of quick total collapse of the monopile with a structural fuse such as a plastic hinge mechanism forming at or below the sea-bottom when the colliding ship is an larger oil or chemical tanker in order to avoid damage to the tanker.

11. Construction considerations

Transportation: Monopiles, tripods, braced frames and suction caissons are usually transported to the site on barges. In some cases, for the larger diameter monopiles, the ends are capped and sealed and they are floated to the site. Gravity caissons either are floated to the site, and then filled with ballast or transported by barge (Figure 21). The turbine is usually transported in one piece.

Erection: A jack-up rig serves as a fairly stable platform for carrying out the operation of installing the piles, the tower, nacelle and rotors. However, its inherent stability brings with it a lack of maneuverability during installation of the tower. Ship-shaped vessels and flat bottom barges are most commonly available but provide the least stability for construction work. Ship-shaped vessels with rotating cranes offer the optimum mix of stability and maneuverability in carrying out the construction work (Figure 22).

Offshore electric cabling: Undersea power transmission cables are deployed using an environmentally sensitive process called "hydro-plowing" in which the cables are buried 2-3 m under the ocean floor. This process uses high-powered jets to fluidize a pathway in the sea floor. The cables are then laid in the pathway and are buried as the sediment settles around them. This technique is minimally invasive and quickly returns the sea bed to its original form.

(a) Monopile being floated out to site.

(b) Gravity Bases being lowered from barge

Fig. 21. Transportation of Wind Turbine Components *(Courtesy of Bob Bittner, Ben C. Gerwick, Inc.)*

Fig. 22. Wind Turbine being erected with a crane from a special ship at Kentish Flats Offshore Wind Farm, UK *(Courtesy: Elsam)*

12. Construction operations planning

Significant cost and schedule savings can be achieved by evaluating each step of the construction process and assuring that it is indeed necessary and being performed in the most efficient manner. Minor changes in design can often lead to significant changes in the construction process, such as reducing the number of construction steps or removing a construction step altogether. Therefore, designers should be familiar with the construction processes involved and be able to assess that would consider a constructability review as well as schedule and cost analysis.

The wind farm at Scroby Sands in Norfolk, United Kingdom is an example of how minor design changes can lead significant savings in construction schedule and costs. The designers decided to make minor modifications to the monopile by welding a flange to which the wind tower could be bolted thereby getting rid of the transition piece and the expensive grouting used to connect it to the monopile altogether.

The Nysted wind farm at Rodsand in Denmark involved fabricating the gravity bases on the barge itself at a casting yard off of Poland's Baltic Coast. The significantly lower labor cost along with the more efficient construction process led to considerable cost savings.
The gravity bases used at Middlegrunden Wind Farm were designed using composite construction, i.e., internal steel can was encased in reinforced concrete and filled with ballast to create the gravity base. Composite construction saved substantial construction time and led to cost savings.
Similarly, means and methods of transportation and erection can also be streamlined by minimizing the number of operations required to perform the work. For example, the monopiles and the wind tower can all be placed on the same jack-up barge before towing to the site, thereby reducing the need for another ship.

13. Maintenance considerations

Maintenance cost is also consideration in the overall wind farm economics and should be considered in the early phases of the development project. The operation of wind turbines at sea is much more challenging than onshore operations. Offshore maintenance and operations involve more costs, since the sea conditions are more irregular and extreme, a higher level of specialist expertise is required, stricter security requirements must be met and longer travel times are involved. Maintenance is categorized into preventive and corrective maintenance. Preventive maintenance comprises planned turbine maintenance according to a fixed maintenance schedule at intervals of 4 to 6 months. Corrective maintenance is required when a malfunction has occurred. In some cases, it is possible to remedy the problem remotely. In other cases, a visit to the turbine is required. The costs of maintaining the turbines can add up over their design life and should be considered in the initial financial rate of return analyses. Key factors to be considered when estimating costs for operational maintenance include: Distance from port, availability of boats, weather and insurance premiums.

14. Conclusions

Increasing wind turbine and tower sizes and installations in deeper waters have clearly demonstrated a need for more innovative and cost-effective foundations. There is room for improvement in all areas of wind farm development; in design, through the innovative use of composite materials, support structures and foundations; and in construction processes, through improvements in installation techniques, fabrication, and transportation. Six basic types of offshore wind turbine support structures and foundations have been described, along with their advantages and drawbacks. General considerations in the selection, design, analyses and construction of these structures have been presented. This paper will assist wind farm developers and engineers in identifying various issues that are likely to arise in the development phase of an offshore wind farm.

15. Acknowledgements

The author wishes to thank Dr. George Munfakh of Parsons Brinckerhoff for promoting innovation in the geo-industry; and encouraging this work, and for his review and valuable suggestions. The author is grateful to his employer Parsons Brinckerhoff for the opportunity to work on design studies for various offshore wind farms first in the United Kingdom and now in the United States. The author is grateful to Dr. Aubeny at TAMU, Mr. Bob Bittner, then of Ben C. Gerwick, Inc., and Elsam, for various construction photographs, and to Mr. Pedro Silva for creating various illustrations. Opinions expressed in this paper are solely of the writer and are not necessarily consistent with the policy or opinions of Parsons Brinckerhoff, Inc.

16. References

[1] American Petroleum Institute, "Recommended Practice for Planning, Design and Constructing Fixed Offshore Platforms - Working Stress Design, API RP2A-WSD," Twentieth Edition, July 1993.

[2] American Petroleum Institute, "Recommended Practice for Planning, Design and Constructing Fixed Offshore Platforms - Load and Resistance Factor Design, API RP2A-LRFD," First Edition, July 1993.

[3] Bach, S., Teilmann, J. & Henriksen, O.D. 2000. VVM-redegørelse for havmølleparker ved Rødsand. Teknisk rapport vedrørende marsvin. Rapport til SEAS. 41 pp.

[4] Elmer, W. J. Gerasch, T. Neumann, J. Gabriel, K. Betke, M. Schultz-von Glahn, (2007), "Measurement and Reduction of Offshore Wind Turbine Construction Noise," *DEWI Magazine Nr. 30, February 2007.*

[5] Gaythwaite, J.H., (1990) "Design of Marine Facilities for the Berthing, Mooring, and Repair of Vessels," *Van Nostrand Reinhold Company ISBN-13: 978-0442229009*, June 1990.

[6] Hance, J. J., (2003) "Submarine Slope Stability," *Project Report for the Minerals Management Service Under the MMS/OTRC Coop. Research Agreement 1435-01-99-CA-31003, Task Order 18217*, MMS Project 421, August 2003.

[7] Henderson, Andrew, M.B. Zaaijer, T.R. Camp, "Hydrodynamic Loading on Offshore Wind Turbines," (2003) *Proceedings Offshore Wind Energy in Mediterranean and Other European Seas (OWEMES) Conference 2003*, Naples, Sept 2003.

[8] Malhotra Sanjeev, (2009), "Design Considerations for Offshore Wind Turbine Foundations in the United States," *International Society of Offshore Polar Engineers*, Osaka, Japan, June 24-26, 2009.

[9] Malhotra Sanjeev, (2007c), "Selection, Design and Construction Guidelines for Offshore Wind Turbine Foundations," PB Research & Innovation Report, Oct., 2007.

[10] Malhotra Sanjeev, (2007b), "Design and Construction Considerations for Offshore Wind Turbine Foundations," *26th International Conference on Offshore Mechanics and Arctic Engineering*, San Diego, California, June 10-15, 2007.

[11] Malhotra Sanjeev, (2007a), "Performance of Open-Ended Pipe Piles in Cretaceous Soils," ASCE Geo-Institute's Geo-Denver 2007 Conference, Denver, Feb. 2007.

[12] Malhotra, Sanjeev, (2004), "Soil Pile Structure Interaction during Earthquakes," Proceedings of the GeoInstitute's Conference on Geotechnical Engineering for Transportation Infrastructure, GeoTrans04, Los Angeles, 2004.

[13] Malhotra, Sanjeev, (2002), "Axial Capacity of Open Ended Steel Pipe Piles in Sand," Proceedings of the International Congress on Deep Foundations, ASCE, Orlando, Florida, 2002.

[14] National Wind Technology Center, (2006), Golden, Colorado, http://www.nrel.gov/wind/windpact/ Jan., 2006.

[15] U.S. Department of Energy, (2004) "Wind Power Today and Tomorrow," March 2004.

[16] Vestas Wind Systems, "Personal Communication with Vestas, March 2007.

Part 2

Wind Turbine Controls

11

Control System Design

Yoonsu Nam
Department of Mechanical Engineering, Kangwon National University 192-1 Hyoja-2 dong, Chunchon, Kangwon 200-701 Korea

1. Introduction

A wind turbine control system is a complex and critical element in a wind turbine. It is responsible for the autonomous, reliable, and safe operation of the machine in all wind conditions. Two levels of control operations are required. One is supervisory control and the other is dynamic feedback control of blade pitch and generator torque for maximizing power production and minimizing mechanical loads on the wind turbine. The supervisory control system is one operating system of the wind turbine and has the following functions:

- operational state (stand-by, start-up, power production, shutdown) transition control
- control of subsystems (cooling, heating, hydraulics, etc.)
- diagnostics, alarm management
- data logging, statistics, post-processing
- user access (management), visualization.

A supervisory controller monitors the turbine and wind condition in order to decide when to start and shut down the wind turbine. Fig. 1 shows a schematic of the operational state transition logic which is implemented in supervisory control. A wind turbine operator can start and shut down turbine operation through a SCADA (supervisory control and data acquisition) system as shown in Fig. 1. The SCADA system can communicate with the supervisory controller in order to control and monitor the wind turbine.

The main topic of this chapter is the design of a control algorithm for the dynamic feedback controller which manages the blade pitch, the generator torque, and the yaw system. Most multi-MW wind turbines are equipped with variable speed and variable pitch (VSVP) technology (Leithead[b] & Connor, 2000; Bianchi et al., 2007; Muller et al., 2002; Boukhezzar et al., 2007). In the below rated wind speed conditions, the rotor speed varies with wind speed, while the pitch is fixed in order to maximize the energy capture from the wind. However, in the above rated wind speed conditions, the pitch is varied, while the rotor speed is fixed, in order for the machine to produce the rated power. All the analysis and design issues covered in this chapter for the dynamic feedback controller using VSVP technology target an upwind type horizontal axis multi-MW wind turbine having 3 blades. Yawing control is not dealt with here because of its simple on-off control logic. All the control algorithms covered in this chapter are based on classical control theory (Franklin et al., 2006; Dorf & Bishop, 2007).

This chapter is composed of 3 sections. Section 2 begins with a mathematical description of the wind, which is not only the source of energy but also a disturbing input to the wind

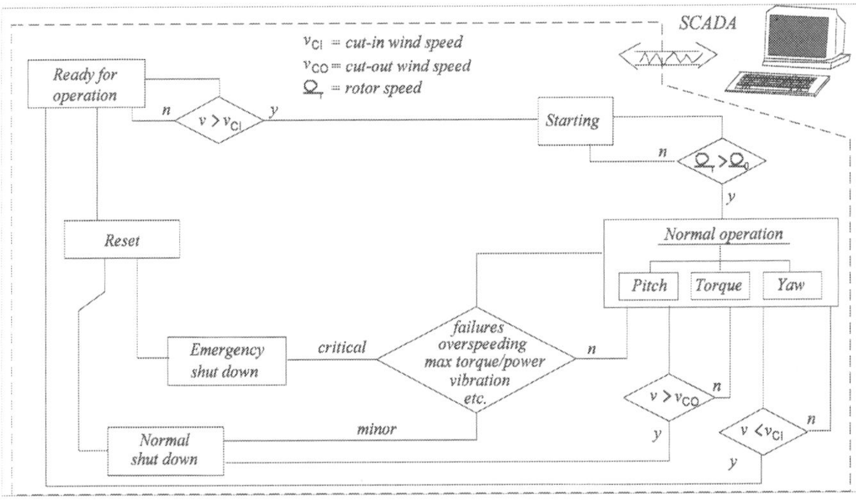

Fig. 1. Schematic of a wind turbine control system

turbine. The wind turbine control strategy and structure based on VSVP technology are explained. Simplified dynamic models of the pitch actuator and generator, which are elements of the wind turbine control system, are followed. Section 3 covers dynamic modeling and steady state characteristics of the wind turbine, based on a drive train model. Variables comprising rotor speed, wind speed, and pitch angle completely specify an operating condition of a wind turbine. How the dynamic characteristics vary with different operating conditions is analyzed in this section. Finally, Section 4 presents the main topic of this chapter, i.e. wind turbine control system design. It starts with the control system design requirements which are determined from the wind characteristics and the aeroelastic properties of the wind turbine structure. A methodology on how to set PI controller gains is introduced, considering gain scheduling and the integrator anti-windup problem. The section includes a feedforward pitch control system design using a wind speed estimator to enhance the performance of the output power regulation. This section concludes with the introduction of individual pitch control for mechanical load alleviation of the blades.

2. Wind turbine control system

2.1 Wind

Wind is highly variable. To accurately predict the wind ahead of time is almost impossible. Statistical measures such as mean wind speed and turbulence intensity are frequently used. Turbulence intensity is given by

$$I = \frac{\sigma}{\overline{v}} \tag{1}$$

where σ is the standard deviation of the wind speed and $\overline{v}$ is the mean wind speed, usually defined for 10 minutes of wind data. Fig. 2 shows two different winds, even though these have the same mean wind speed and turbulence intensity. One further statistical property, i.e. autocorrelation, is necessary to discern the wind more specifically. The autocorrelation is defined as

$$\phi_{xx}(\tau) = E\{x(t)x(t+\tau)\} = \lim_{T\to\infty}\left\{\frac{1}{2T}\int_{-T}^{T} x(t)x(t+\tau)dt\right\} \tag{2}$$

where $x(t)$ is a de-trended time series. Therefore, $\varphi_{xx}(0)=\sigma_x^2$. As signal frequency increases and time-lag τ gets larger, the autocorrelation becomes smaller. The power spectral density, which is the Fourier transform of the autocorrelation, is defined as

$$\Phi_{xx}(\omega) = \int_{-\infty}^{\infty} \phi_{xx}(\tau)e^{-j\omega\tau}d\tau\ . \tag{3}$$

A sample of the autocorrelation and power spectral density for two different time series is shown in Fig. 3.

Van der Hoven observed at Brookhaven, New York, in 1957 that there were distinct periodicities in wind, as shown in Fig. 4 (van der Hoven, 1957). Three peaks, namely synoptic, diurnal, and turbulent peaks, are clear in this plot. On a short time scale of less than 2 hours, turbulent wind has most energy in the wind spectrum. The power spectral density of turbulent wind can be modeled as

$$\Phi_{vv}(\omega) = \frac{4\sigma_u^2 L_u / \overline{v}}{\left(1+70.8\left(\omega L_u/(2\pi\overline{v})\right)^2\right)^{5/6}} \tag{4}$$

where σ_u is the standard deviation for a turbulent wind and L_u is the length scale (Burton et al., 2001). The length scale is a site-specific parameter and depends on the surface roughness a and height z. It is given by

$$L_u = 280(z/z_i)^{0.35}\ . \tag{5}$$

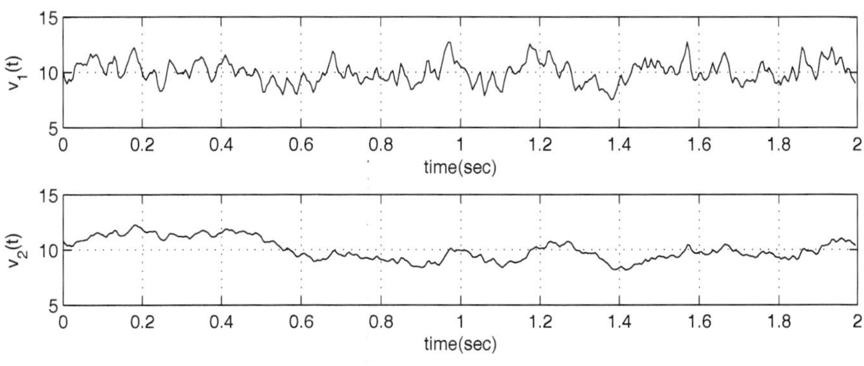

Fig. 2. Two different winds with the same mean wind speed and turbulence intensity

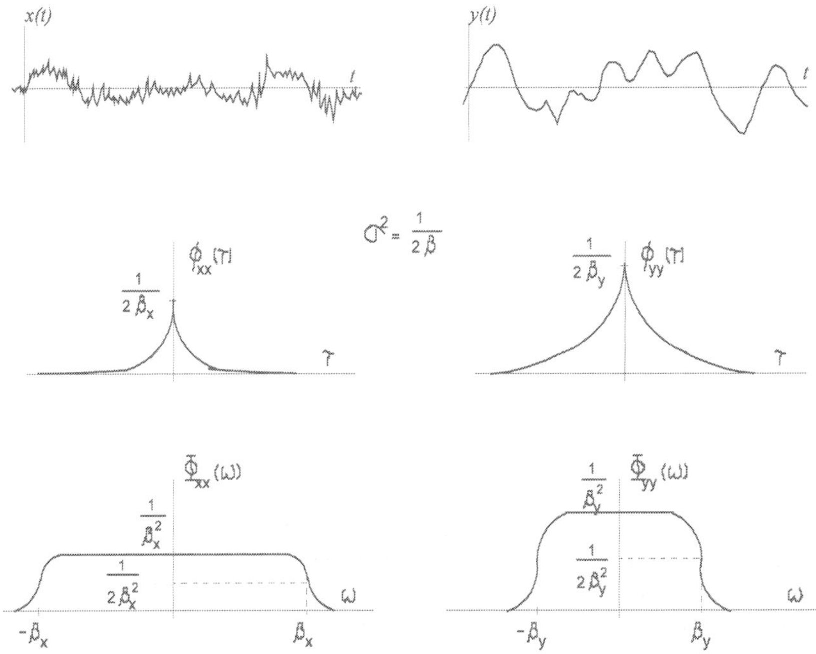

Fig. 3. Autocorrelation and power spectral density

Some representative values of a are given in Table 1 with the value of z_i. The turbulence becomes isotropic for the condition of $L_u \geq 280$ m at an altitude of $z>z_i$. The above expression of Eq. (4) is known as the von Karman spectrum in the longitudinal direction. The von Karman expressions in the lateral and vertical directions can be found in the literature (Burton et al., 2001).

2.2 Control system strategies and structure

The mechanical power of an air mass which has a flow rate of dm/dt with a constant speed of v is given by

$$P = \frac{d}{dt}(E) = \frac{d}{dt}\left(\frac{1}{2}mv^2\right) = \frac{1}{2}\frac{dm}{dt}v^2 = \frac{1}{2}\rho A v^3 \qquad (6)$$

where ρ is the air density and A is the cross sectional area of the air mass. Only a portion of the wind power given by Eq. (6) is converted to electric power by a wind turbine. The efficiency of the power conversion by a wind turbine depends on the aerodynamic design and operational status of the wind turbine. Usually, the power generated by the wind turbine is represented by

$$P = C_P\left(\frac{1}{2}\rho A v^3\right) \qquad (7)$$

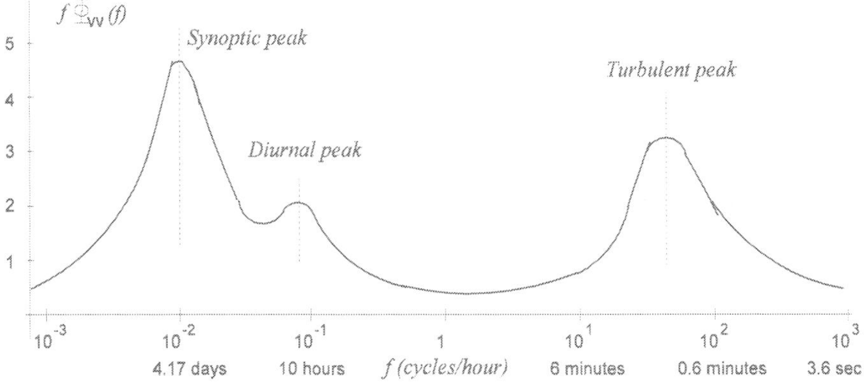

Fig. 4. van der Hoven wind spectrum

Type of terrain	Roughness length α (m)	z_i=1000$a^{0.18}$ (m)
Cities, forests	0.7	937.8
Suburbs, wooded countryside	0.3	805.2
Villages, countryside with tress and hedges	0.1	660.7
Open farmland, few trees and buildings	0.03	532.0
Flat grassy plains	0.01	436.5
Flat desert, rough sea	0.001	288.4

Table 1. Surface roughness (Burton et al., 2001)

where Cp represents the efficiency of wind power conversion and is called the power coefficient. The ideal maximum value of Cp is 16/27= 0.593, which is known as the Betz limit (Manwell et al., 2009).

As shown in Fig. 5, the power coefficient, Cp is a function of pitch angle β and tip speed ratio λ which is defined as

$$\lambda = \frac{R\Omega_r}{v} \tag{8}$$

where R is the rotor radius and Ω_r is the rotor speed of the wind turbine. Fig. 5 is a sample plot of Cp for a multi-MW wind turbine. The curve with dots shows the variation of Cp with λ for a fixed pitch angle of β_0. As the pitch angle is away from β_0, the value of Cp becomes smaller. Therefore, Cp has the maximum with the condition of $\lambda=\lambda_0$ and $\beta=\beta_0$. In order for a wind turbine to extract the maximum energy from the wind, the wind turbine should be operated with the max-Cp condition. That is, the wind turbine should be controlled to maintain the fixed tip speed ratio of $\lambda=\lambda_0$ with the fixed pitch of $\beta=\beta_0$ in spite of varying wind speed. Referring to Eq. (8), there ought to be a proportional relationship between the wind speed v and the rotor speed Ω_r to keep the tip speed ratio at constant value of λ_0.

Fig. 6 represents a power curve which consists of three operational regions. Region I is max-Cp, Region II is a transition, and Region III is a power regulation region.

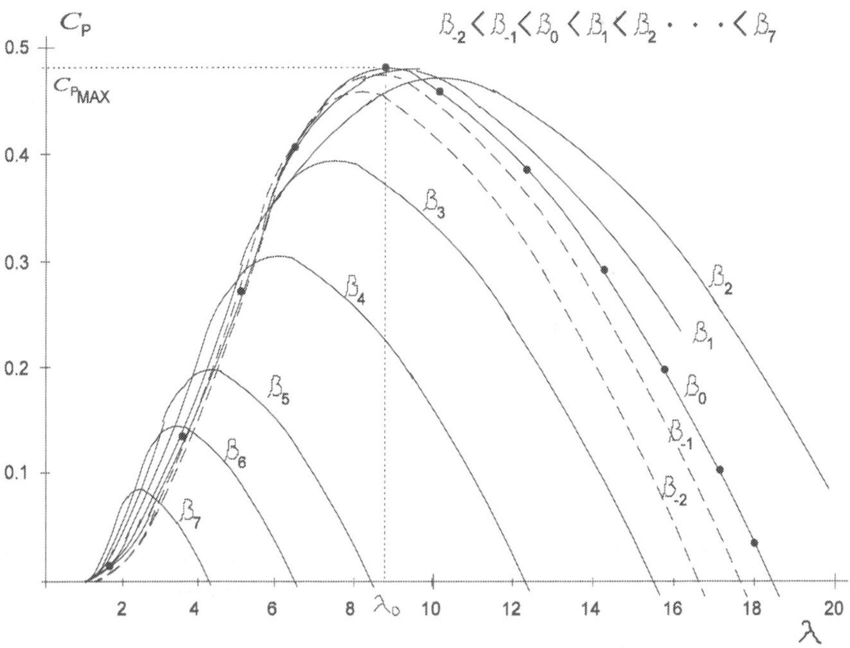

Fig. 5. Sample plot of Cp as a function of λ and β

- Region I: The wind turbine is operated in max-Cp. The blade pitch angle is fixed at β_0 and the rotor speed is varied so as to maintain the tip speed ratio constant (λ_0). Therefore, the rotor speed is changed so as to be proportional to the wind speed by controlling the generator reaction torque. In the max-Cp region, the generator torque control is active only, while the blade pitch is fixed at β_0.
- Region II: This is a transition region between the other two regions, that is the max-Cp (Region II) and power regulation region (Region III). Several requirements, such as a smooth transition between the two regions, a blade-tip noise limit, minimal output power fluctuations, etc., are important in defining control strategies for this region.
- Region III: This is the above rated wind speed region, where wind turbine power is regulated at the rated power. Therefore, rotor speed and generator reaction torque are maintained at their rated values. In this region, the value of Cp has to be controlled so as to be inversely proportional to v^3 to regulate the output power to the rated value. This is easily found by noting Eq. (1). In this region, the blade pitch control plays a major role in this task.

A control system structure for a wind turbine is shown schematically in Fig. 7. There are two feedback loops. One is the pitch angle control loop and the other is the generator torque control loop. Below the rated wind speed region, i.e. in Regions I and II, the blade pitch angle is fixed at β_0 and the generator torque is controlled by a prescheduled look-up table (see Section 3.2). The most common types of generator for a multi-MW wind turbine are a doubly fed induction generator (DFIG) (Soter & Wegerer, 2007) and a permanent magnet

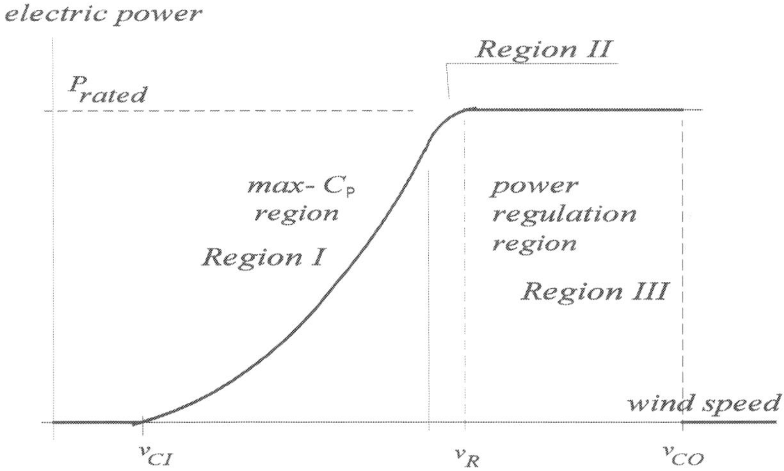

Fig. 6. Power curve

synchronous generator (PMSG) (Haque et al., 2010). These electric machines are complicated mechanical and electric devices including AC-DC-AC power converters. For the purposes of control system design, however, it is suffiicient to use a simple model of generator dynamics:

$$\frac{T_g(s)}{T_g^C(s)} = \frac{\omega_{ng}^2}{s^2 + 2\varsigma_{ng}\omega_{ng}s + \omega_{ng}^2} \tag{9}$$

where $T_g{}^C$ is a generator torque command, ω_{ng} (~ 40 r/s) is a natural frequency of the generator dynamics and ζ_{ng} (~ 0.7) is a damping ratio (van der Hooft et al., 2003). Blade pitch angle is actuated by an electric motor or hydraulic actuator which can be modeled as

$$\frac{\beta(s)}{\beta^C(s)} = \frac{1}{1 + \tau_p s} \tag{10}$$

where β^C is a pitch angle demand and τ_p (~ 0.04 r/s) is a time constant of the pitch actuator. It is necessary and important for a realistic simulation to include saturation in actuator travel and its rate as depicted in Fig. 8 (Bianchi et al., 2007). In general, the pitch ranges from -3° to 90° and a maximum pitch rates of ±8°/s are typical values for a multi-MW wind turbine.

Power curve tracking and mechanical load alleviation are two main objectives of a wind turbine control system. For a turbulent wind, the wind turbine control system should not only control generation of electric power as specified in the power curve but also maintain structural loads of blades, drive train, and tower as small as possible. In the below rated wind speed region (max-Cp region), the generator torque control should be fast enough to follow the variation of turbulent wind. Generally, this requirement is not an issue because the electric system is much faster than the fluctuation of the turbulent wind. In the above rated wind speed region (power regulation region), the rotor speed should be maintained at

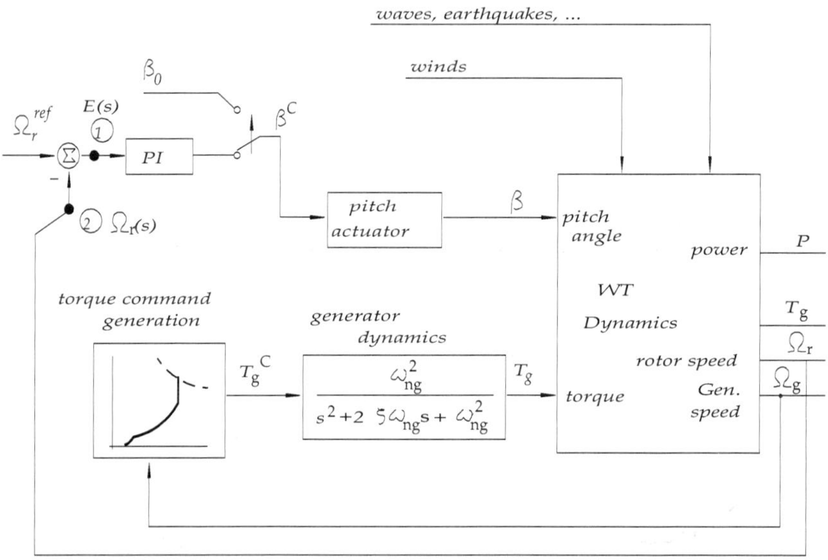

Fig. 7. Wind turbine control system structure

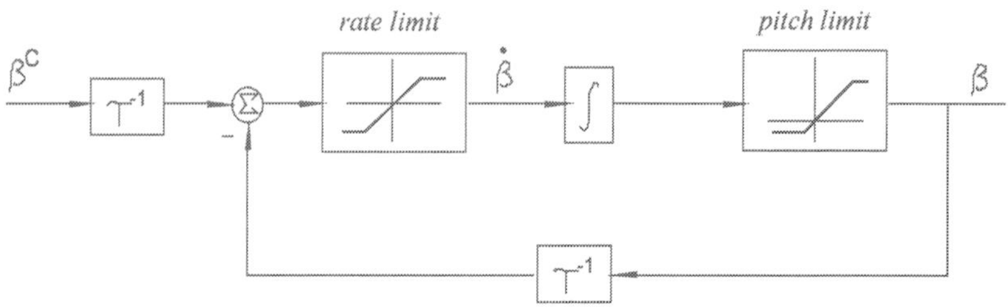

Fig. 8. Pitch actuator model

its rated speed by the blade pitch control, irrespective of wind speed fluctuation. The design of pitch control loop affects the mechanical loads of blades and tower as well as the performance of the wind turbine. Combined control of torque and pitch or the application of feedforward control (see Section 4.3) is a promising alternative for enhancing the power regulation performance. The alleviation of mechanical loads by the individual blade pitch control is discussed in Section 4.4.

3. Dynamic model and steady state operation

3.1 Drive train model and generator torque scheduling

A wind turbine is a complicated mechanical structure which consists of rotating blades, shafts, gearbox, electric machine, i.e. generator, and tower. Sophisticated design codes are necessary for predicting a wind turbine's performance and structural responses in a turbulent wind field. However, the simple drive train model of Fig. 9 is sufficient for control system design (Leithead[a] & Connor, 2000). The parameters referred to in Fig. 9 are summarized in Table 2. The aerodynamic torque developed by the rotor blades can be obtained using Eq. (7) and Eq. (8) as follows

$$T_a = \frac{P}{\Omega_r} = \frac{1}{2}\rho\pi R^2 \frac{C_P(\lambda,\beta)}{\Omega_r} v^3 = \frac{1}{2}\rho\pi R^3 \frac{C_P(\lambda,\beta)}{\lambda} v^2 = \frac{1}{2}\rho\pi R^3 C_Q(\lambda,\beta) v^2 \quad (11)$$

where $C_Q=C_p/\lambda$ is the torque constant. The torque of Eq. (11) is counteracted by the generator torque. Therefore, the governing equations of motion for a drive train model are

$$\begin{aligned} J_r \frac{d\Omega_r}{dt} &= T_a - k_S(\theta_r - \frac{1}{N}\theta_g) - c_S(\Omega_r - \frac{1}{N}\Omega_g) - B_r\Omega_r \\ J_g \frac{d\Omega_g}{dt} &= \frac{k_S}{N}(\theta_r - \frac{1}{N}\theta_g) + \frac{c_S}{N}(\Omega_r - \frac{1}{N}\Omega_g) - B_g\Omega_g - T_g \end{aligned} . \quad (12)$$

It is useful to understand the physical meaning of Fig. 10 which shows the relationship between rotor speed (Ω_r) and torque on a high speed shaft ($(T_a)_{HSS}$). The several mountain-shaped curves in this figure represent the aerodynamic torque on a high speed shaft for different wind speeds and rotor speeds at a fixed pitch β_o. These are easily calculated using Eq. (11) and power coefficient data from Fig. 5 for any specific wind turbine. On this plot, the max-Cp operational condition is shown as a dashed line, which satisfies the quadratic relation:

$$\begin{aligned} (T_a)_{HSS} &= \frac{1}{2N}\rho\pi R^3 \left(\frac{C_{P\max}}{\lambda_o}\right) v^2 = \frac{1}{2N}\rho\pi R^3 \left(\frac{C_{P\max}}{\lambda_o}\right)\left(\frac{\Omega_r R}{\lambda_o}\right)^2 \\ &= \frac{1}{2N}\rho\pi R^5 \frac{C_{P\max}}{\lambda_o^3}(\Omega_r)^2 = k_{op}\Omega_r^2 \end{aligned} . \quad (13)$$

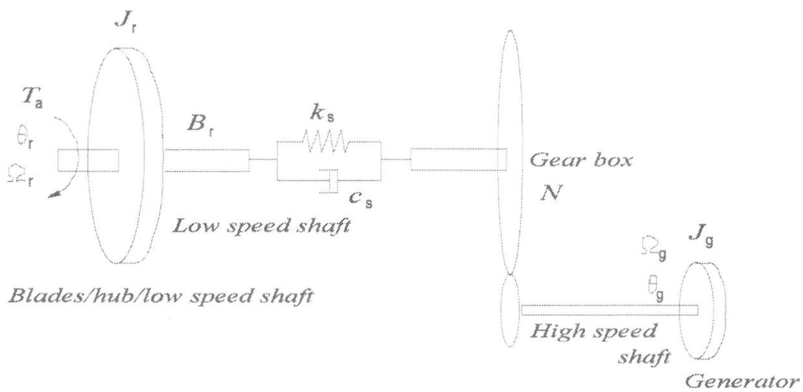

Fig. 9. Drive train model

Symbol	Description	unit
J_r	Inertia of three blades, hub and low speed shaft	Kgm^2
J_g	Inertia of generator	Kgm^2
B_r	Damping of low speed shaft	Nm/s
B_g	Damping of high speed shaft	Nm/s
k_s	Torsional stiffness of drive train axis	N
c_s	Torsional damping of drive train axis	Nm/s
N	Gear ratio	-
T_g	Generator reaction torque	Nm
Ω_g	Generator speed	r/s

Table 2. Parameters for the drive train model of Fig. 9

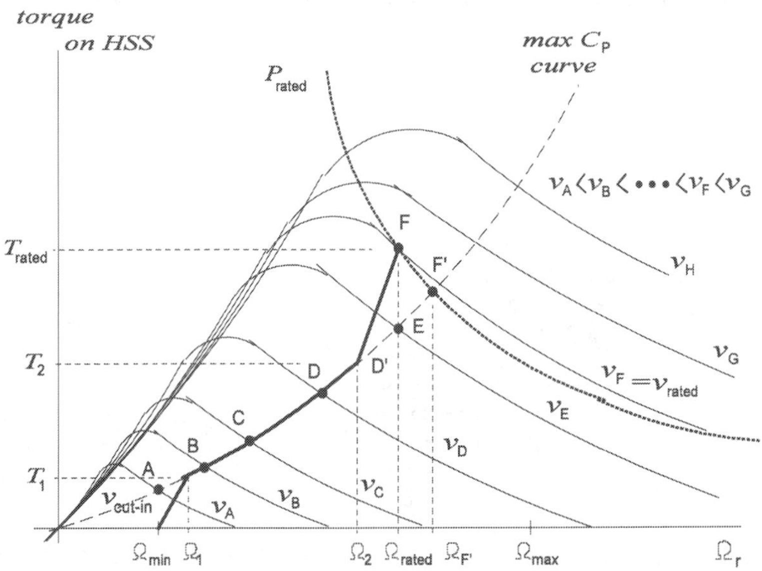

Fig. 10. Characteristic chart for torque on a high speed shaft and rotor speed

In the below rated wind speed region, a wind turbine is to be operated with the max-Cp condition to extract maximum energy from the wind. This means that the wind turbine should be operated at the point B for a steady wind speed v_B, the point C for a wind speed v_C, and so on in Fig. 10. For steady state operation, the aerodynamic torque of Eq. (13) should be counteracted by the generator reaction torque plus the mechanical losses from viscous friction, i.e. $B_r\Omega_r/N$ and $B_g\Omega_g$. Considering only the maximum energy capture, a torque schedule of A-B-C-D-E-F′ for a variable rotor speed is the optimal. However, the rated rotor speed might not be allowed to be as large as $\Omega_{F'}$ because of the noise problem. If the tip speed ($R\Omega_r$) of a rotor is over around 75 m/s (Leloudas et al., 2007), then noise from the rotor blades could be critical for on-shore operation. Therefore, as the size of a wind turbine becomes larger, the rated rotor speed becomes smaller. Because of this constraint, the toque schedule for most multi-MW wind turbines has the shape of either A-B-C-D-E-F or A-B-C-D′-F. Wind turbines using a permanent magnet synchronous generator (PMSG) often have the torque schedule of A-B-C-D-E-F. In this case, the generator torque control of Fig. 7 using a look-up table is not appropriate because of the vertical section E-F. A PI controller with the max-Cp curve as the lower limit can be applied (Bossanyi, 2000).

3.2 Aerodynamic nonlinearity and stability

The nonlinearity of a drive train model comes from the aerodynamic torque of Eq. (11), which is a nonlinear function of three variables, (Ω_r, v, β). A single set of these variables defines a steady state operating condition of a wind turbine. The aerodynamic torque can be linearized for an operating condition of (Ω_{ro}, v_o, β_o) as follows:

$$
\begin{aligned}
T_a &= \frac{1}{2}\rho\pi R^3 C_Q(\lambda,\beta)v^2 = T_a(\Omega_r, v, \beta) \\
&\simeq T_a(\Omega_{r0}, v_0, \beta_0) + \left(\left.\frac{\partial T_a}{\partial \Omega_r}\right|_{(\Omega_{r0}, v_0, \beta_0)}\right)\delta\Omega_r + \left(\left.\frac{\partial T_a}{\partial v}\right|_{(\Omega_{r0}, v_0, \beta_0)}\right)\delta v + \left(\left.\frac{\partial T_a}{\partial \beta}\right|_{(\Omega_{r0}, v_0, \beta_0)}\right)\delta\beta \\
&= T_a(\Omega_{r0}, v_0, \beta_0) + B_\Omega\delta\Omega_r + B_v\delta v + k_\beta\delta\beta
\end{aligned}
\tag{14}
$$

where $\delta\Omega_r = \Omega_r - \Omega_{ro}$, $\delta v = v - v_o$, $\delta\beta = \beta - \beta_o$.

Note that the sign of B_Ω is related with the stability of the wind turbine. The operating condition of (Ω_{ro}, v_o, β_o) where the B_Ω value is positive is unstable. This is clear on substituting the linearized aerodynamic torque of Eq. (14) into Eq. (12). Therefore, if a wind turbine is operating on the left side hill (positive slope, i.e. positive B_Ω region, which is also known as the stall region) of the mountain-shaped curve of Fig. 10, this means that the wind turbine is naturally (open loop) unstable. The coefficient B_v denotes just the gain of aerodynamic torque for a wind speed increase. The coefficient k_β represents the effectiveness of pitching to the aerodynamic torque. Fig. 11 shows a sample plot of these three coefficients as a function of wind speed for a multi-MW wind turbine. This plot is easily obtained using a linearizing tool, Matlab/Simulink© with Eq. (11). The line marked with 'x' shows B_Ω variation with wind speed in Nm/rpm. B_v data are shown with the symbol '+' in Nm/(m/s). The effectiveness of pitch angle on aerodynamic torque, i.e. k_β, is represented by the line with '◊' in Nm/deg. The values of k_β are zero in the low wind speed region, which means that the wind turbine is operating at the top of the Cp-curve, i.e. max-Cp (see Fig. 5). It gradually becomes negative because a blade pitching to feathering position decreases the aerodynamic torque. Note that the magnitudes of k_β in the rated wind speed region (12 m/s) are relatively small compared to those at high wind speed. Because of this property, gain scheduling of the pitch loop controller is required (see Section 4.2).

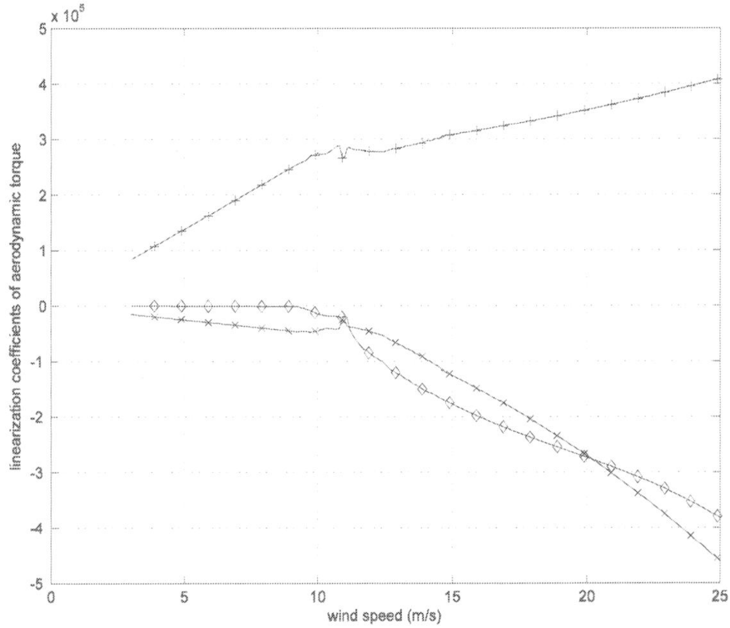

Fig. 11. Variation of B_Ω , B_v , and k_β with steady wind speeds for a multi-MW WT

3.3 Steady state operation

For a steady wind speed, a wind turbine should also be in steady state operation, i.e. with constant rotor speed and pitch angle. Therefore, a set of three variables, (Ω_r, v, β) defines a steady state operation condition of a wind turbine. How to determine these sets of variables is the topic of this section. In steady state operation, the dynamic equations of motion of Eq. (12) are combined to a nonlinear algebraic equation:

$$\frac{T_a}{N}-\frac{B_r\Omega_r}{N}-B_g\Omega_g-T_g=\frac{1}{2N}\rho\pi R^3C_Q(\lambda,\beta)v^2-\frac{B_r\Omega_r}{N}-NB_g\Omega_r-T_g=0. \qquad (15)$$

Assuming that generator torque scheduling is completed as explained in Section 3.1 (see Fig. 10), generator torque T_g would be a function of rotor speed Ω_r. Therefore, a set of three variables, (Ω_r, v, β) constitutes the above nonlinear equation. To find one set of variables, (Ω_r, v, β) for a given wind speed v, one further relationship between these variables is needed, apart from Eq. (15). Fortunately, depending on the wind speed region, either pitch angle or rotor speed is fixed as explained in Section 2.2.

In the below rated wind speed region, blade pitch angle is fixed at β_0. Therefore, only one variable, which is the rotor speed, is unknown and can be determined by Eq. (15). However, an analytic solution is not possible, because the equation includes terms having numeric

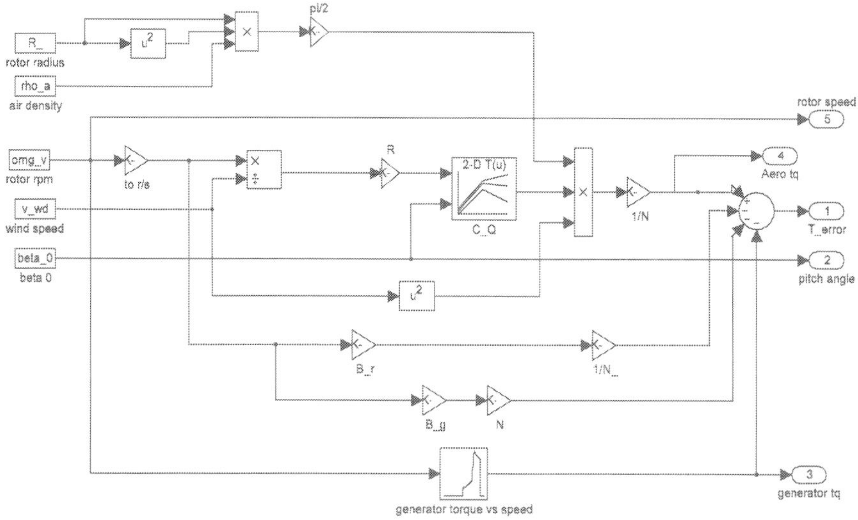

Fig. 12. Simulink model of Eq. (15) in the below rated wind speed region

data for C_Q and T_g. A numerical method using an optimization algorithm can be applied to solve this problem. Fig. 12 shows a Matlab/Simulink© model of Eq. (15). The #4 output ('Aero tq') in this figure corresponds to the first term of Eq. (15). The other blocks below this represent the remaining terms of Eq. (15). Therefore, the #1 output ('T_error') is the total sum of terms in left side of Eq. (15). An optimization algorithm which minimizes the magnitude of 'T_error' can be applied to find an appropriate rotor speed ('omg_v' in Fig. 12) for a fixed wind speed ('v_wd') and a fixed pitch angle ('beta_0'). By iterating the above procedure for wind speeds in the whole below rated region, an appropriate rotor speed schedule similar to Fig. 13 can be sought out. Exactly the same algorithm as the above is applied to find a pitch angle variation in the above rated wind speed region, where the rotor speed is fixed at rated speed. Fig. 13 shows full sets of three variables, (Ω_r, v, β), which are obtained using the above algorithms. The trajectory in this figure defines the steady state operating point for each wind speed from the cut-in to the cut-out wind speed envelope.

Fig. 14 provides some additional insights on the steady state operations of Fig. 13. Note how the power coefficient, C_P, varies with changes in wind speed, pitch angle, and rotor speed. A torque schedule similar to the one shown as a thick solid line in Fig. 10 is applied in this analysis. As the wind speed increases from zero to $v_{D'}$ in Fig. 10, the wind turbine starts to rotate and then reaches and stays for a while at the max-C_P operational state. Because of the torque schedule of Fig. 10, the magnitude of Cp decreases in a transition region from the max-C_P value and goes toward zero in the above rated wind speed region, being inversely proportional to the third power of wind speed as explained in Section 2.2. Note also how C_P varies with the pitch angle. In this figure, try to identify the matching rotor speeds, Ω_{min}, Ω_1, Ω_2, and Ω_{rated}, of Fig. 10. The final plot of Fig. 14 shows the variation of tip speed ratio, λ, as a function of wind speed.

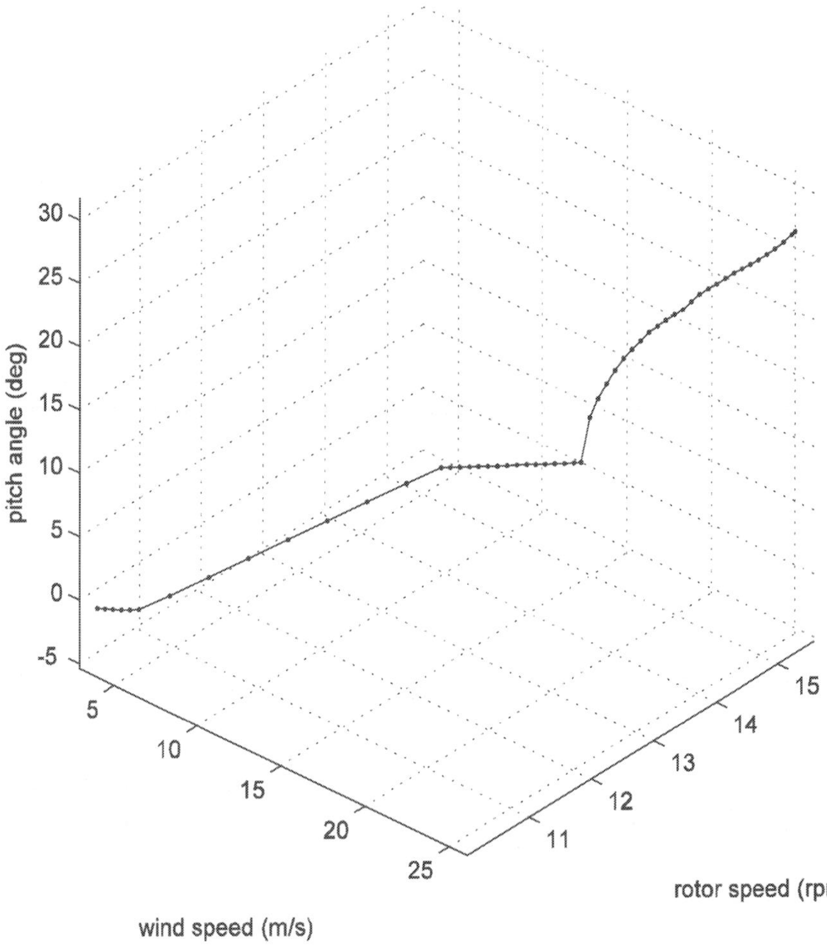

Fig. 13. Locus of operating point variation with wind speed

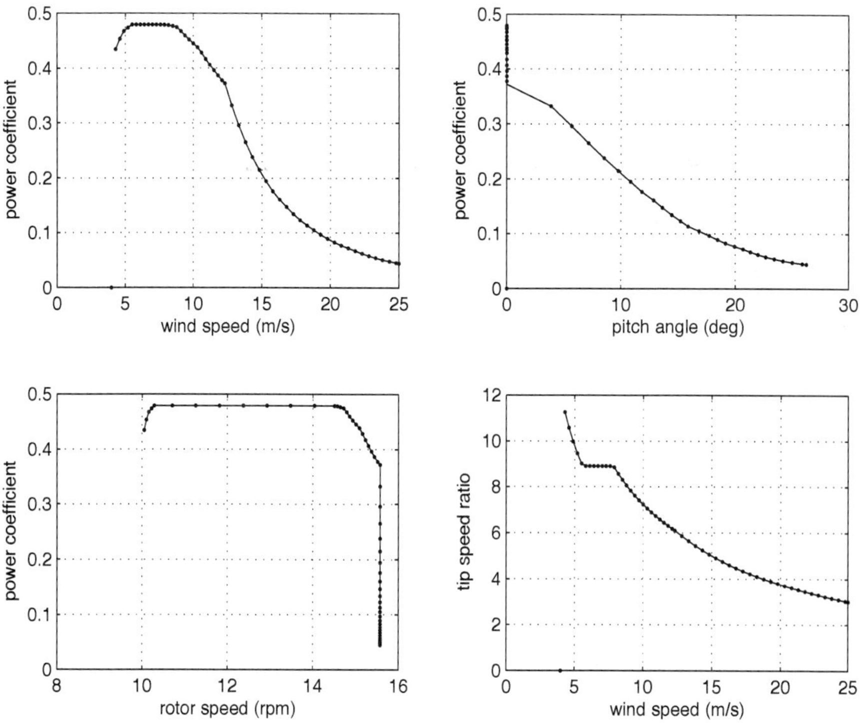

Fig. 14. Variation of C_P and λ with changes in wind speed, pitch angle, and rotor speed

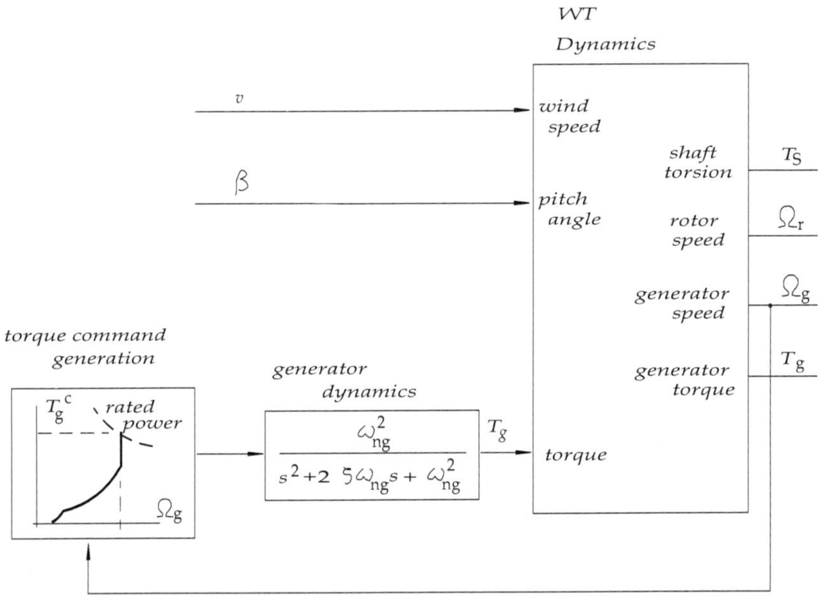

Fig. 15. Schematic open pitch loop structure of wind turbine

3.4 Dynamic characteristic change with varying wind speed

A wind turbine should be maintained to operate on a locus of Fig. 13 for varying winds if it produces electric power as specified in the power curve. Then, would the dynamic characteristics of the wind turbine be the same for all operational points of this locus? If different, by how much would they differ? It is important to understand these characteristics well for a successful pitch control system design, which will be covered in the next section. Fig. 15 shows a schematic open pitch loop structure. Generator torque control is implemented by high speed switching power electronics. Therefore, it has much faster dynamics than a pitch control loop. In Fig. 15, the generator torque control system is modeled as a second order system of Eq. (9) and controlled as specified with the torque-rotor speed schedule table. The 'WT Dynamics' block of Fig. 15 can be represented with the drive train model of Eq. (12), which is easily programmed with Matlab/Simulink©. A linearized model for each operating point, $(\Omega_{ro}, v_o, \beta_o)$, on the locus of Fig. 13 can be found as follows:

$$\begin{aligned} \dot{\mathrm{x}} &= A\mathrm{x} + B\mathrm{u} \\ \mathbf{y} &= C\mathrm{x} + D\mathrm{u} \end{aligned} \tag{16}$$

where $\mathrm{x} = \begin{bmatrix} x_1 & x_2 & x_3 & x_4 & x_5 & x_6 \end{bmatrix}^T = \begin{bmatrix} \delta\Theta_g & \delta\Theta_r & \delta\Omega_r & \delta\Omega_g & \delta T_g & \delta(dT_g/dt) \end{bmatrix}^T$

$\mathrm{u} = \begin{bmatrix} u_1 & u_2 \end{bmatrix}^T = \begin{bmatrix} \delta v & \delta\beta \end{bmatrix}^T$

$\mathbf{y} = \begin{bmatrix} y_1 & y_2 & y_3 & y_4 \end{bmatrix}^T = \begin{bmatrix} \delta T & \delta\Omega_r & \delta\Omega_g & \delta T_g \end{bmatrix}^T.$

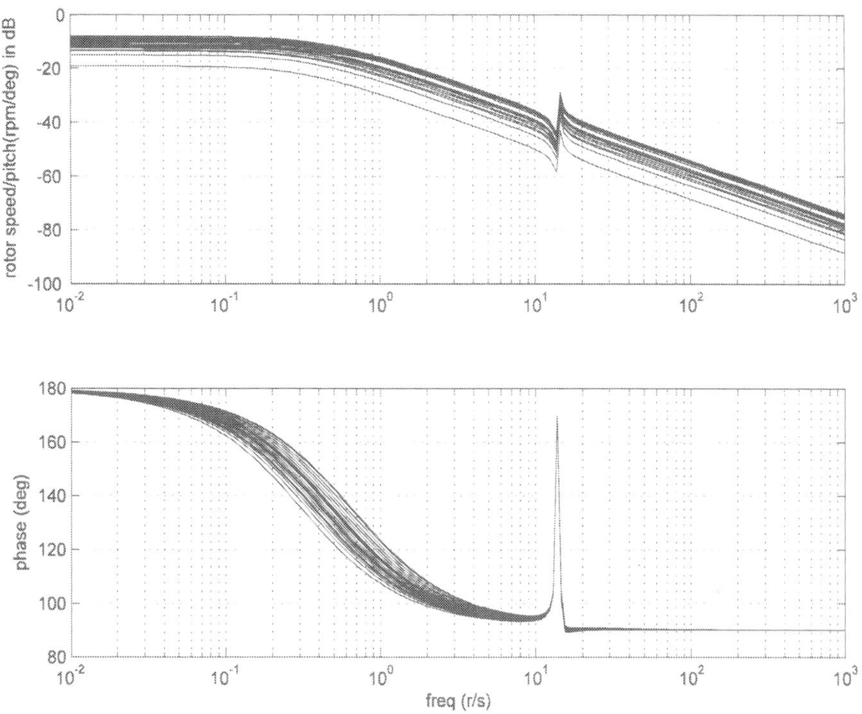

Fig. 16. Frequency response of $G_{22}(s)=\delta\Omega_r(s)/\delta\beta(s)$ for operating points in the above rated wind speed region

Θ_g and Θ_r in the above equation are the rotational displacement of the generator and rotor. Note that the linear model of Eq. (16) is meaningful only in the vicinity of (Ω_{ro}, v_o, β_o).

A transfer function of rotor speed for the pitch angle input, $G_{22}(s)=\delta\Omega_r(s)/\delta\beta(s)$ can be obtained from the linear model of Eq. (16). This transfer function is important in the pitch controller design. A sample of frequency response of this transfer function for a multi-MW wind turbine is shown in Fig. 16. Frequency responses only for the above rated wind speed region are displayed, because a pitch control is active only in this region. Overall, it behaves like a first order system but has some variations in DC gain and low frequency pole location with different operating conditions. The difference of DC gain for each operating point comes from the pitch effectiveness variation with wind speed. As already shown in Fig. 11, the pitch effectiveness, k_β (the plot with '◊' in Fig. 11), becomes larger with an increase of wind speed. Therefore, the frequency responses having larger DC gain in Fig. 16 correspond to those at high wind speed operating points. The peaks at around 16 r/s represent the torsional vibration mode of the drive train. As mentioned in the above, the dynamics of a wind turbine is similar to a first order dynamic system. Because of the huge moment of inertia of three blades for a multi-MW machine, it usually takes more than several seconds to reach steady state operation for abrupt changes in wind speed or pitch angle. Fig. 17 shows changes in dominant pole (i.e. pole of the first order system) locations with different

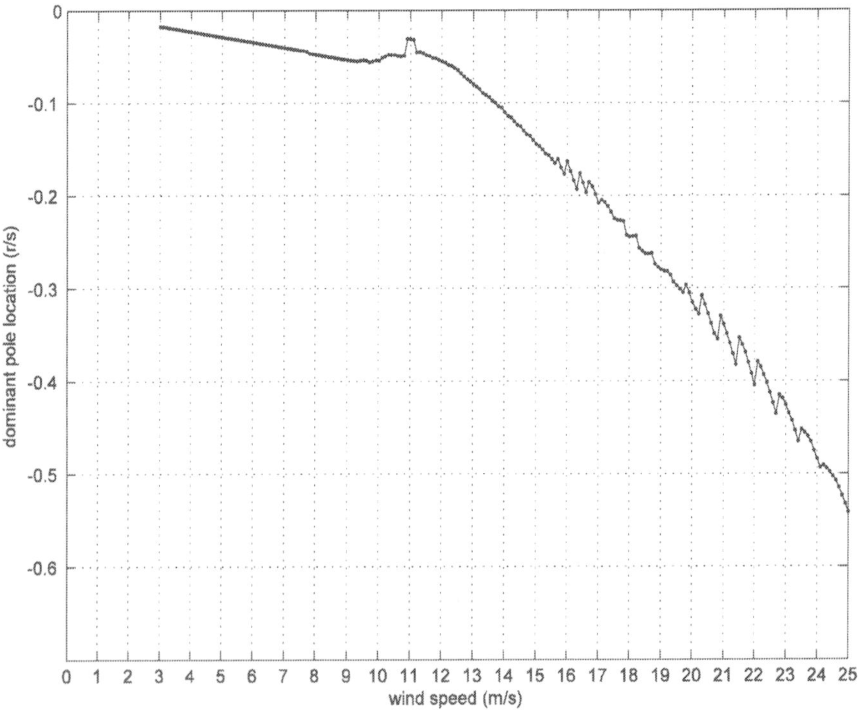

Fig. 17. Variation of dominant pole locations with wind speed for a multi-MW wind turbine

operating conditions. A wind turbine having the operating locus of Fig. 13 has stable but very slow dynamics, especially in the low wind speed region. In designing a pitch controller at around the rated wind speed region, the characteristics of slow dynamics and low DC gain should be considered.

4. Control system design

4.1 Control system design requirements

A control law structure for power curve tracking is introduced in Fig. 7. It consists of two feedback loops. One is the generator torque control loop, which is covered in Section 3, and the other is the pitch control loop. As mentioned earlier, depending on wind speed, there are two control regimes. In the below rated wind speed region, the pitch angle is fixed at β_0 and the generator torque is controlled to maintain max-Cp operation in the face of turbulence. But in the above rated region pitch control is active to regulate the rotational speed of the rotor to the rated rpm while maintaining generator torque at the rated value. Therefore, the electric power of a wind turbine is regulated as rated in this region. The design of the pitch control loop of Fig. 7 is a matter of selecting suitable PI gains to make the control system satisfy some design criteria. How to set the pitch control system bandwidth is one of the design criteria.

The bandwidth of a wind turbine control system should be fast enough to extract the wind power in a turbulent wind spectrum. Assuming that Eq. (4) in Section 2.1 can be approximated as

$$\Phi_{vv}(\omega)=\frac{4\sigma_u^2 L_u/\bar{v}}{\left(1+70.8\left(\omega L_u/(2\pi\bar{v})\right)^2\right)^{5/6}}\approx\frac{4\sigma_u^2 L_u/\bar{v}}{1+70.8\left(\omega L_u/(2\pi\bar{v})\right)^2}, \tag{17}$$

a turbulent wind could be modelled using a first order Markov process (Gelb, 1974). The power spectral density of the output signal of the first order Markov process, $y(t)$, for the input, $x(t)$, i.e. white noise, is given as

$$\Phi_{yy}(\omega)=\left|G(j\omega)\right|^2\Phi_{xx}(\omega)=\frac{k^2}{\omega^2+\beta_1^2}=\frac{(k/\beta_1)^2}{1+(\omega/\beta_1)^2} \tag{18}$$

where $G(s)=k/(s+\beta_1)$ is a first order low pass filter system. Comparing Eq. (18) with Eq. (17), one can notice that a turbulent wind can be generated by filtering a white noise with a first order low pass filter which has a cut-off frequency of

$$\beta_1=\frac{2\pi\bar{v}}{\sqrt{70.8}L_u}. \tag{19}$$

Therefore, a design criterion for the bandwidth of a wind turbine system is that it should be larger than β_1, which has values in the range of 0.0196 ~ 0.148 r/s, depending on the type of terrain (see Table 1, Eq. (3), and Eq. (4): 0.148 r/s for a mean wind speed of 25 m/s in a flat desert or rough sea). However, the pitch control loop bandwidth cannot be set too high because of non-minimum phase (NMP) zero dynamics of the wind turbine. Fig. 18 shows the frequency response of the rotor rpm for the pitch demand. This plot is obtained from a linearized aeroelastic model of a multi-MW wind turbine at a wind speed of 13 m/s. An abrupt phase change of 360 degrees at around 2 r/s implies the existence of NMP zero dynamics of

$$\frac{s^2-2\varsigma\omega_z s+\omega_z^2}{s^2+2\varsigma\omega_z s+\omega_z^2} \tag{20}$$

in the transfer function of $G_{22}(s)=\delta\Omega_r(s)/\delta\beta(s)$. This NMP zero dynamics is related with the first mode of tower fore-aft motion (Dominguez & Leithead, 2006). This is a common characteristic of most multi-MW wind turbines. It is well known that a NMP zero near to the origin in the s-plane sets a limit for the crossover frequency (~bandwidth) of the loop gain transfer function. Therefore, the frequency of the NMP zero coming from the tower fore-aft motion determines an upper bound of the pitch control loop bandwidth.

4.2 Pitch controller design and gain scheduling

There are two design parameters, i.e. k_p and k_I/k_p, for the pitch control system structure of Fig. 7. As discussed in the former section, these parameters are to be selected such as to meet the crossover frequency requirement. The loop gain transfer function from the point ① to the point ② in Fig. 7 is the most important in the pitch controller design and has the following form:

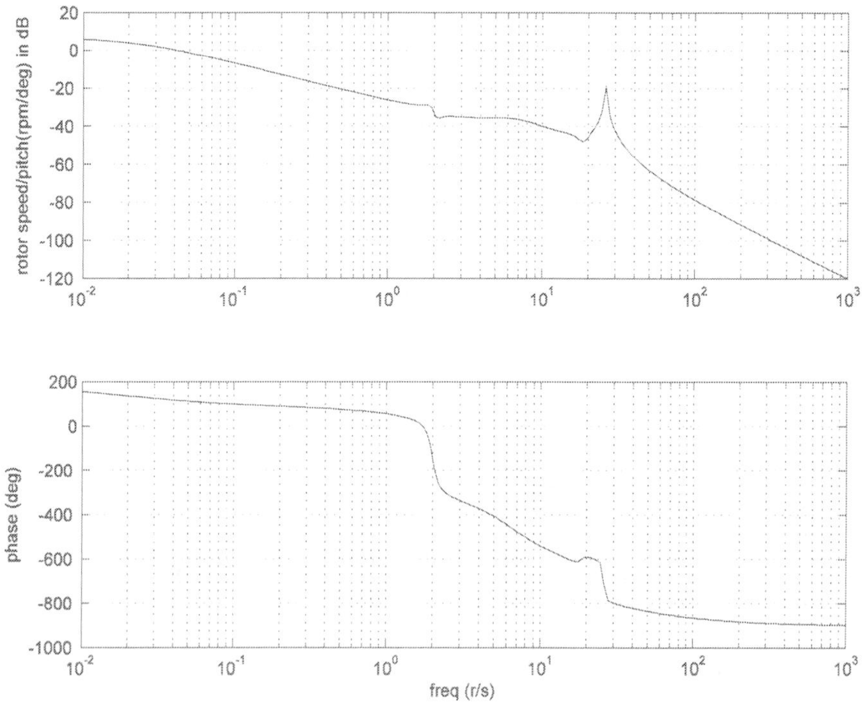

Fig. 18. Frequency response of rotor speed for the pitch demand ($G_{22}(s)=\delta\Omega_r(s)/\delta\beta(s)$) from an aeroelastic model of a multi-MW wind turbine at 13 m/s

$$L(s) = G_{22}(s)\left(\frac{1}{1+\tau_p s}\right)\left(k_p + \frac{k_I}{s}\right) = G_{22}(s)\left(\frac{1}{1+\tau_p s}\right)\left(\frac{k_p(s + k_I / k_p)}{s}\right). \tag{21}$$

Fig. 19 shows the frequency response of the pitch loop gain transfer function at a wind speed of 22.8 m/s for a multi-MW wind turbine, which is obtained using the linearized drive train model of Eq. (12). A crossover frequency of 1 r/s and phase margin of 90° are achieved for the selection of k_p=-5.844 *(deg/rpm)* and k_I/k_p=0.55 *(1/s)*. Increasing (decreasing) the magnitude of the proportional gain, k_p, from 5.844 results in a higher (lower) crossover frequency than 1 r/s. Then, how does the parameter, k_I/k_p affect the pitch loop design? As explained in Section 3.4 (see Fig. 16), the dynamics of $G_{22}(s)=\delta\Omega_r(s)/\delta\beta(s)$ can be approximated as a first order transfer function, the pole of which varies with the wind speed as shown in Fig. 17 for a multi-MW machine. By referring to Fig. 20, which is a root-locus plot for the pitch control loop, the question of how to set the parameter k_I/k_p is answered. Depending on the selection of k_I/k_p, the shape of the root-locus differs greatly. If this parameter is chosen to be smaller than the magnitude of the open loop pole in Fig. 17 for the design wind speed, it would be difficult to achieve the requirement on the crossover frequency of 1 r/s, even when applying very high proportional gain.

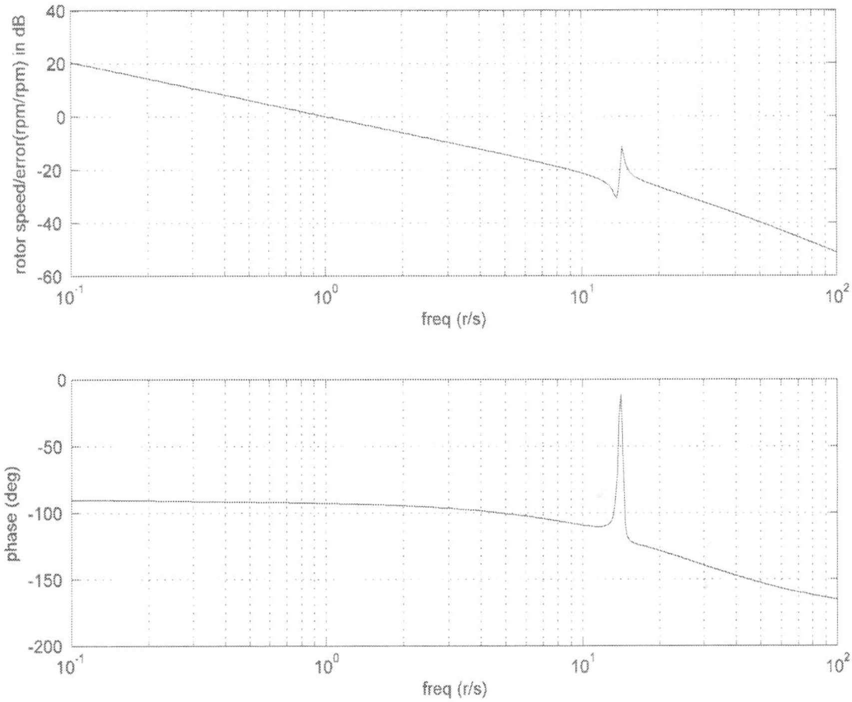

Fig. 19. Frequency response of pitch loop gain transfer function at 22.8 m/s

Successful completion of a pitch loop design for a certain design point (for example, 22.8 m/s in the above) does not guarantee the same level of design for any other design point, because the pitch effectiveness, k_β, varies with the wind speed (see the plot with '◊' in Fig. 11). As shown in this figure, the pitch effectiveness in the rated wind speed region is the lowest, which matches the frequency response of $G_{22}(s)=\delta\Omega_r(s)/\delta\beta(s)$ having the lowest DC gain in Fig. 16. If the same pitch controller gains as those for 22.8 m/s are used in the rated wind speed region, the crossover frequency of the pitch loop would be so low that there might be a large excursion of rotor speed from rated. A gain scheduling technique can be applied to compensate the DC gain variation with the wind speed. The PI controller in Eq. (21) multiplied by a scheduled gain, $k_G(\beta)$, is given as

$$k(s) = \frac{\beta^C(s)}{E(s)} = k_G(\beta)\left(k_P + \frac{k_I}{s}\right) = k_G(\beta)\cdot\frac{k_p(s + k_I / k_p)}{s}. \tag{22}$$

It is common to schedule the gain, k_G, as a function of pitch angle, β as in Eq. (22), because the wind speed is not only difficult to estimate but also too high frequency for a gain scheduling operation. A sample of scheduled gain, $k_G(\beta)$, for a multi-MW wind turbine is shown in Fig. 21. These gains are determined from the plot of the pitch effectiveness, k_β, with the pitch angle, which is similar to the plot with '◊' in Fig. 11. However, too much

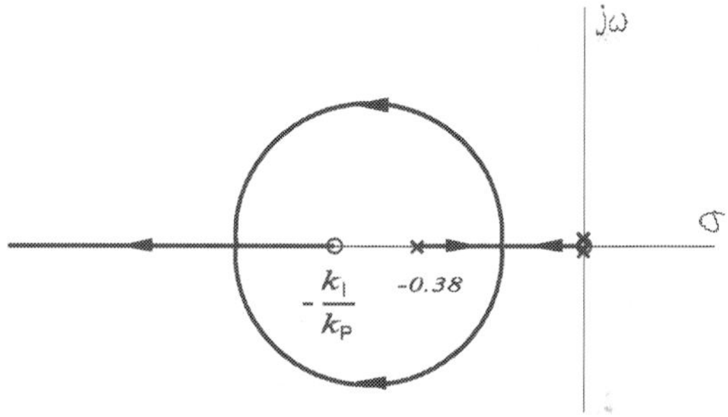

(A) Root locus when $\frac{k_I}{k_P} > 0.38$

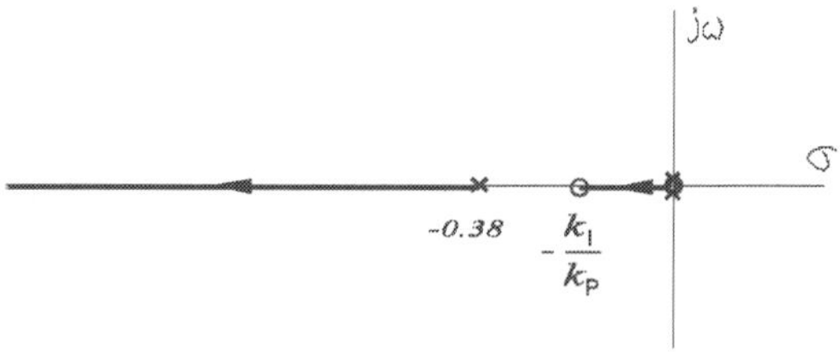

(B) Root locus when $\frac{k_I}{k_P} < 0.38$

Fig. 20. Change of root locus of the pitch control loop depending on k_I/k_p selection

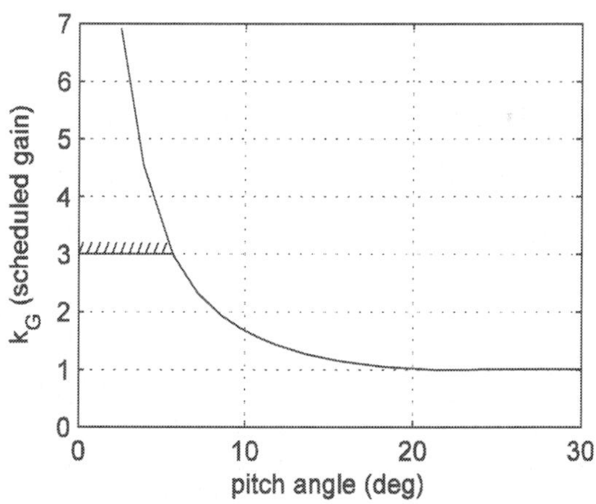

Fig. 21. Gain scheduling as a function of pitch angle

scheduled gain in the rated wind speed region might result in large mechanical loads on the blades and tower. Therefore, it would be reasonable to limit the scheduled gain to a certain value as depicted in Fig. 21. The appropriate limit should be determined through a full nonlinear simulation covering aeroelastic behaviour of structural loads.

As pointed out in Section 2.2 (see Fig. 8), the pitch actuator has saturation in speed and displacement. If integral control is used with the actuator having travel limits, a well-known integrator windup problem arises. This might cause too large an excursion in rotor rpm from the rated value or might make the pitch system unstable. By preventing integral action when the pitch is at the limit, which is called integrator anti-windup, this problem can be solved. Discrete implementation of an anti-windup PI controller is shown in Fig. 22, where the approximation of integral action is made by the following relations. For the sampling interval T, the output of the PI controller of Eq. (22) at the k-th sampling time, $\beta^C(kT)$, is given by

$$\beta^C(kT) = k_G(\beta)\left(k_P e(kT) + k_I \int_0^{kT} e(\tau)d\tau\right). \tag{23}$$

The output of the PI controller at $t=(k-1)T$, $\beta^C((k-1)T)$, can be obtained in a similar way. Then, the following relation holds:

$$\Delta\beta_k^C = \beta^C(kT) - \beta^C((k-1)T) \simeq k_G(\beta)\left\{k_P\{e(kT) - e((k-1)T)\} + k_I T\frac{e(kT)+e((k-1)T)}{2}\right\}. \tag{24}$$

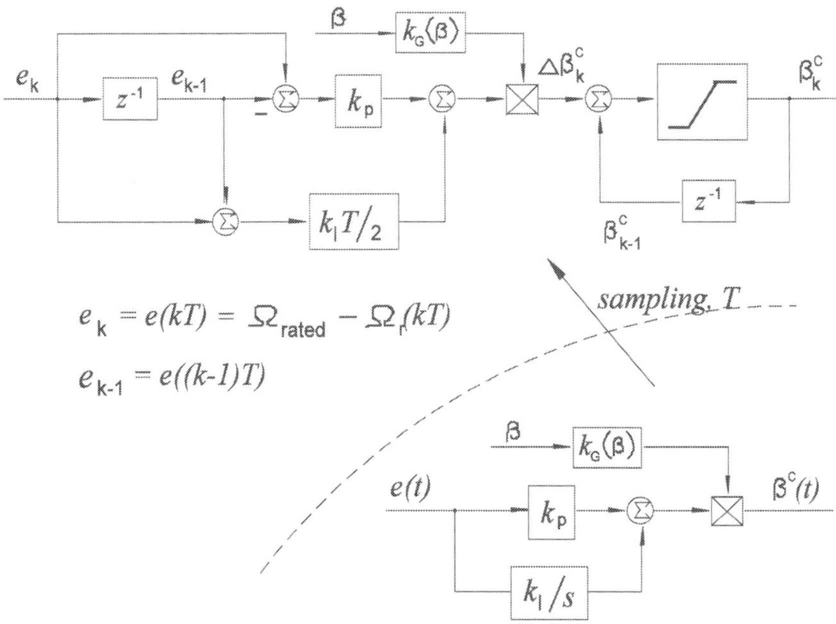

Fig. 22. Discrete PI controller with the integrator anti-windup

Fig. 22 is just a graphical representation of Eq. (24) with the integral anti-windup function, which is implemented with a limiter.

This section concludes by introducing simulation results which were obtained by applying the above pitch control system for a mean wind speed of 15 m/s with a turbulence intensity of 17.5 %. These are shown in Fig. 23 and Fig. 24 and were obtained using GH Bladed© (Bossanyi, 2009) for a 2 MW wind turbine. A generator torque schedule similar to A-B-C-D-E-F of Fig. 10 is applied in both simulations. A gain scheduling technique is applied, in which the maximum gain is limited to 7. Too much gain scheduling in the rated wind speed (small pitch) region might result in large fluctuations in rotor speed and an increase in mechanical loads on the wind turbine structure. The results in Fig. 23 are for a pitch loop design of 0.8 r/s crossover frequency, while those in Fig. 24 are for 1.7 r/s. The first window in these figures represents the wind speed in m/s and the second shows the rotor speed in rpm. After these, the pitch angle (deg), generator torque (kNm), and output power (kW) are shown. The final 4 windows in these figures show the structural response of the blade and tower. The first two plots show the bending moments (MNm) at the root of the #1 blade in the in-plane and out-of-plane direction (see Section 4.4 for the definition of these directions). The next two plots show the root bending moments (MNm) of the tower in the side-side and fore-aft direction. As the pitch loop bandwidth (~crossover frequency) is designed to be higher, the rotor speed fluctuations over the rated rpm become much smaller (compare the second plots of Fig. 23 and Fig. 24). This is the anticipated consequence because the high bandwidth pitch control system tries to manage the rpm errors tightly compared to the low bandwidth system. Noting the relationship (power)=(generator torque)x(rotor speed), compare also the responses of output power in the fifth plot of the figure. While the difference of performance response such as rotor speed or power is very comparable in Fig. 23 and Fig. 24, it is difficult to differentiate between the structural response such as blade or tower bending moment in these figures. The damage equivalent loads (DEL), M_{eq}, can be a quantitative measure which shows differences between the two mechanical load responses and is given by

$$M_{eq} = \left(\frac{\sum_k (M_k)^m n_k}{n_{tot}} \right)^{1/m} \quad (25)$$

where n_k is the number of cycles in mechanical load range M_k and n_{tot} is the total number of cycles in a mechanical load signal. m in this equation is the material specific number, for example m=3.5 for a steel tower structure and m=10 for a fiber glass blade (Bossanyi, 2008). The larger the DEL value, the more prone to end up in fatigue failure. The statistical data of performance response and DEL data of structural load response for the two cases of different crossover frequencies are summarized in Table 3. A remarkable performance improvement is expected for the pitch control system design with the 1.7 r/s crossover frequency compared to the system with the frequency of 0.8 r/s. As the rotor fluctuates more around the rated rpm, the DEL of the blade bending moment in the out-of-plane direction and the tower root bending moment in the fore-aft direction becomes larger. However, the effects of these on the in-plane blade loads and the side-side tower loads seem to be minor.

4.3 Wind speed estimation and feedforward pitch control

The dynamic response of a multi-MW wind turbine to the wind is slow. Assuming that the crossover frequency of a pitch control loop is around 1 r/s, it would take more than 2

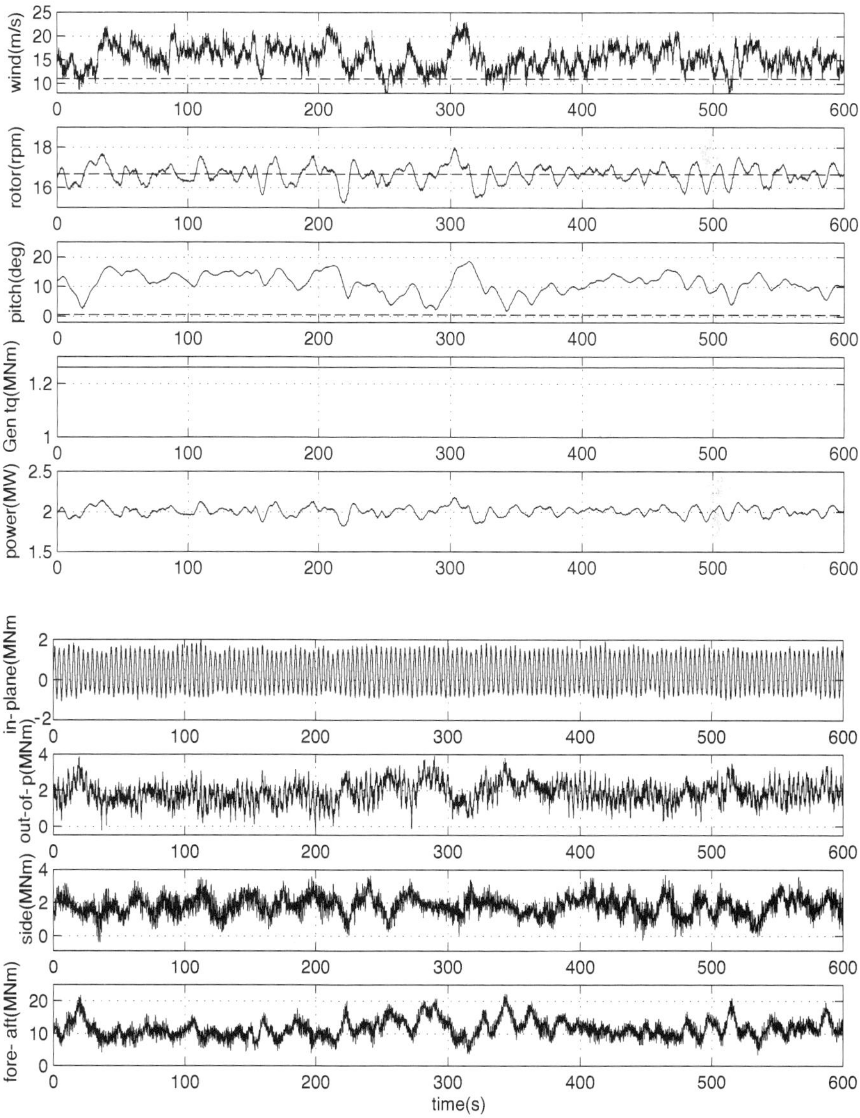

Fig. 23. Performance and structural response in time domain for a turbulent wind when the crossover frequency of the pitch loop is 0.8 r/s

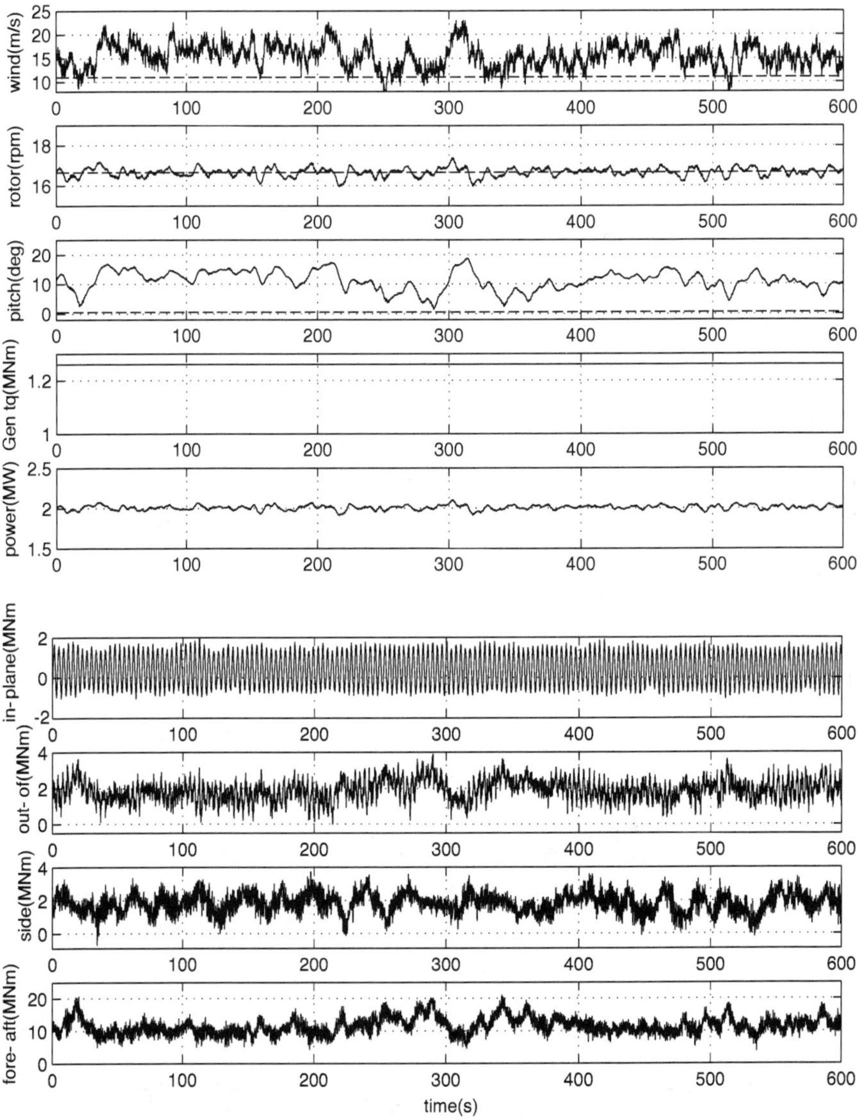

Fig. 24. Performance and structural response in time domain for a turbulent wind when the crossover frequency of the pitch loop is 1.7 r/s

seconds to reach a steady state for a step change in wind speed. A pitch control system will respond in the right way only after some error in rotor speed for the change of wind speed is developed. If the pitch actuation system is commanded with the wind speed information instead of the rotor speed error, then the fluctuation of rotor speed can be decreased by anticipating pitch action.

Feedforward pitch control using wind speed estimation improves the performance of power output regulation for turbulent wind operation. Fig. 25 shows a schematic structure of a feedforward pitch control. It consists of an aerodynamic torque estimator, a 3-dimensional look-up table which outputs the estimated wind speed, and two feedforward gains. The aerodynamic torque of Eq. (11), which is rewritten as Eq. (26), is the basic relation for the wind speed estimation:

$$T_a = \frac{1}{2}\rho\pi R^3 \frac{C_P(\lambda,\beta)}{\lambda} v^2 = \frac{1}{2}\rho\pi R^3 C_Q(\lambda,\beta) v^2 . \qquad (26)$$

The above equation has 4 variables, namely the aerodynamic torque T_a, rotor speed Ω_r (see Eq. (8)), pitch angle β, and wind speed v. Among these variables, the rotor speed and pitch angle are measurable. Therefore, if estimation of the aerodynamic torque is possible, then the wind speed can be pre-calculated and presented as a 3-dimensional look-up table as depicted in Fig. 25. In high wind speed ranges up to cut-out, the relationship between the

Crossover Frequency, w_c (r/s)	Performance data				Structural loads data (DEL)			
	Rotor (rpm)		Power (kW)		Blade (MNm)		Tower (MNm)	
	mean	std.	mean	std.	in-plane	out-of-plane	side-side	fore-aft
w_c = 0.8 (A)	16.657	0.441	2007.8	58.17	2.119	2.334	1.457	5.582
w_c = 1.7 (B)	16.659	0.215	2008.1	28.32	2.124	2.203	1.452	5.276
(B-A)/A(%)	0.01	-51.3	0.01	-51.3	0.24	-5.56	-0.34	-5.48

Table 3. Comparison of performance and structural response data for different crossover frequencies

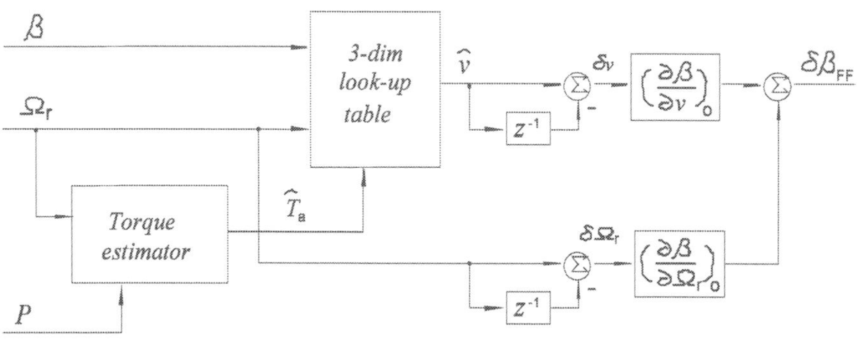

Fig. 25. Schematic diagram of the feedforward pitch control

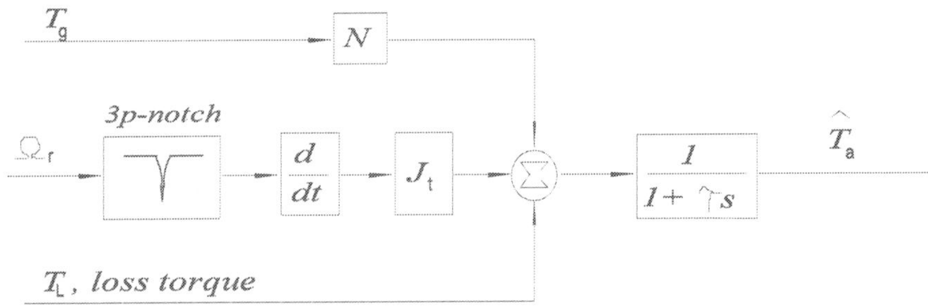

Fig. 26. Aerodynamic torque estimator based on a simple drive train model

aerodynamic torque and the wind speed in Eq. (26) is one-to-one where the wind turbine is operating at a large pitch angle. However, this relation is broken on operating near to the rated wind speed region where the pitch is small, because of the concaveness of $C_Q(\lambda,\beta)$ in Eq. (26). In this region, there could be two solutions for the wind speed for a certain aerodynamic torque. In this case, the smaller wind speed, which is near to the rated wind speed, is the normal solution, while the larger one corresponds to the wind speed at stall condition.

There are several ways of estimating the aerodynamic torque. The simplest one uses the drive train model of Eq. (12). The two equations in Eq. (12) are combined to give

$$J_t \frac{d\Omega_r}{dt} = T_a - NT_g - T_L \tag{27}$$

where $J_t = J_r + N^2 J_g$ and T_L is a mechanical loss torque. All the viscous frictional terms in Eq. (12), such as $B_r\Omega_r$ are combined in the mechanical loss. A schematic of a torque estimator using this equation is shown in Fig. 26. A low pass filter is used to suppress the high frequency noise generated from the differentiation of measured rotor speed. Filtering out the 3p (1p=rotational frequency) or drive train resonant frequency component in the rotor speed signal might be necessary (van der Hooft & van Engelen, 2003). The use of a dynamic estimator such as a Kalman filter is another option. Augmenting the unknown input, T_a, in the state vector, Eq. (27) becomes

$$\begin{aligned} \dot{x} &= \begin{Bmatrix} \dot{\Omega}_r \\ \dot{T}_a \end{Bmatrix} = \begin{bmatrix} 0 & 1/J_t \\ 0 & 0 \end{bmatrix} \begin{Bmatrix} \Omega_r \\ T_a \end{Bmatrix} + \begin{pmatrix} -N/J_t \\ 0 \end{pmatrix} T_g + \begin{pmatrix} -1/J_t \\ 0 \end{pmatrix} T_L + \begin{pmatrix} -N/J_t \\ 1 \end{pmatrix} w_g \\ y &= [1 \quad 0] \begin{Bmatrix} \Omega_r \\ T_a \end{Bmatrix} + v \end{aligned} \tag{28}$$

where w_g and v are input process and sensor noise. The state estimator can be designed by combining Eq. (28) with a measurement update term:

$$\dot{\hat{x}} = \begin{Bmatrix} \dot{\hat{\Omega}}_r \\ \dot{\hat{T}}_a \end{Bmatrix} = \begin{bmatrix} 0 & 1/J_t \\ 0 & 0 \end{bmatrix} \begin{Bmatrix} \hat{\Omega}_r \\ \hat{T}_a \end{Bmatrix} + \begin{pmatrix} -N/J_t \\ 0 \end{pmatrix} T_g + \begin{pmatrix} -1/J_t \\ 0 \end{pmatrix} T_L + L\left(\Omega_r - \hat{\Omega}_r\right) \tag{29}$$

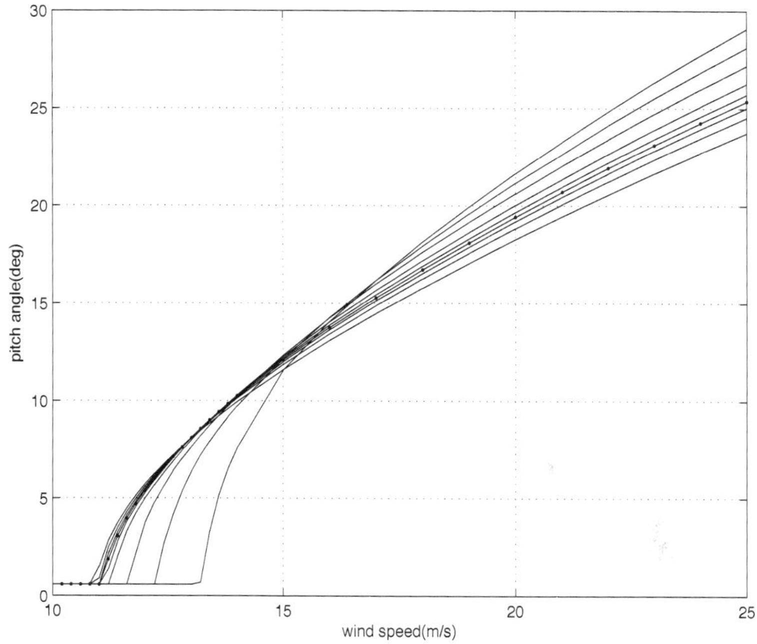

Fig. 27. Family of a set of parameters, $(\Omega_{ro}, v_o, \beta_o)$, which produce the rated power

where the variables with hat are estimated and L is the estimator gain. The estimated torque and two other measurable variables, i.e. rotor speed and pitch angle, are input to a 3-dimensional look-up table, and then the pre-calculated wind speed can be obtained. Note that the estimated wind speed from the 3-dimensional table is not real. It is so called effective wind speed, which is the spatial average of the wind field over the rotor plane with the wind stream being unaffected by the wind turbine (Ostergaard et al., 2007).

The purpose of feedforward pitch control is to minimize rotor speed fluctuation so that good quality of power regulation is achieved. The wind turbine would generate the rated power P_{rated}, for a set of parameters, $(\Omega_{ro}, v_o, \beta_o)$, which can be sought by solving the equation:

$$P = \frac{1}{2}\pi R^2 C_P(\lambda, \beta) v^3 = P_{rated} = constant\,. \tag{30}$$

Fig. 27 shows a family of these parameters as a function of rotor speed from 13.28 rpm to 18.26 rpm with a step of 0.83 rpm for a multi-MW wind turbine. The line with dots represents steady state operating conditions producing rated power, i.e. a set of parameters, $(\Omega_{ro}, v_o, \beta_o)$, at the rated rotor speed. The line with the highest pitch angle at wind speeds above 17 m/s corresponds to steady state operating conditions at a rotor speed of 13.28 rpm, and the line at the bottom is for 18.26 rpm. The relationship between a set of parameters, $(\Omega_{r0}, v_0, \beta_0)$, in Fig. 27 can be expressed as

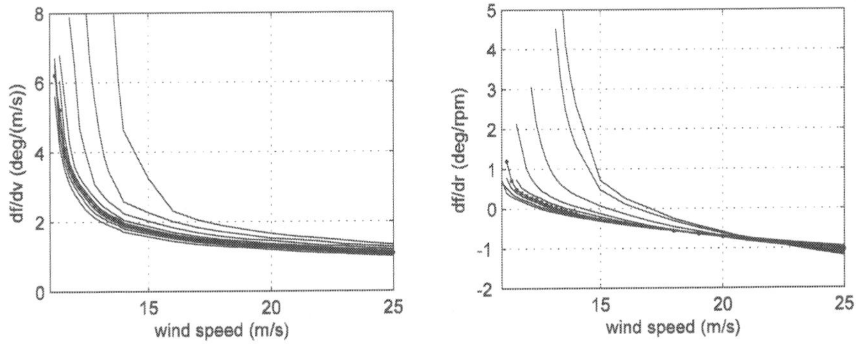

Fig. 28. A sample of $(\partial f/\partial v)_o$ and $(\partial f/\partial \Omega_r)_o$ as a function of wind speed and rotor rpm for a multi-MW wind turbine

$$\beta_0 = f(v_0, \Omega_{r_0}) \, . \tag{31}$$

The above equation means that the pitch angle should be maintained at β_0 for a wind speed of v_0 and a rotor speed of Ω_{r0} in order for a wind turbine to output a rated power. Assuming a sudden change of the wind speed from v_0 to $v_0+\delta v$ and the rotor speed from Ω_{r0} to $\Omega_{r0}+\delta\Omega_r$, the above equation can be approximated as

$$\begin{aligned} \beta_0 + \delta\beta &\simeq f(v_0, \Omega_{r_0}) + \left(\frac{\partial f}{\partial v}\right)_o \delta v + \left(\frac{\partial f}{\partial \Omega_r}\right)_o \delta\Omega_r \\ \Rightarrow \quad \delta\beta &= \left(\frac{\partial f}{\partial v}\right)_o \delta v + \left(\frac{\partial f}{\partial \Omega_r}\right)_o \delta\Omega_r \end{aligned} \, , \tag{32}$$

where $\delta\beta$ is the amount of pitch angle to compensate the combined variations of wind speed and rotor speed. This amount is $\delta\beta_{FF}$ in Fig. 25, which is the feedforward command to the pitch control system. Fig. 28 shows a sample of $(\partial f/\partial v)_o$ and $(\partial f/\partial \Omega_r)_o$ as a function of wind speed for several rotor speeds. The same set of rotor rpm as used in Fig. 27 is applied in this calculation. The line with dots in this figure represents the variation of $(\partial f/\partial v)_o$ and $(\partial f/\partial \Omega_r)_o$ with wind speed for the rated rotor speed. The lines at the top in both plots correspond to the variation of these gains for a rotor speed of 13.28 rpm. Note that the signs of $(\partial f/\partial v)_o$ and $(\partial f/\partial \Omega_r)_o$ at most wind speeds above rated are opposite. This results from the contrary aerodynamic effects of the wind and rotor speed on the rotating blades. An increase in the wind speed makes the angle of attack of the rotor blade larger, while an increase in the rotor speed makes it smaller (Manwell et al., 2009).

The simulation results for a 2 MW wind turbine using the above feedforward algorithm are summarized in Fig. 29. In this simulation, a torque schedule of the A-B-C-D-E-F shape in Fig. 10 is applied with the PI pitch controller. The first plot of this figure is the hub height wind speed, of which the mean and turbulence intensity are 16 m/s and 18%. The second is the estimated wind speed. The straight line in these plots represents the rated wind speed of the wind turbine, which is 11 m/s. The estimation was based on the Kalman filter of Eq. (29). In this calculation, 5% of the rated rpm and 8% of the rated torque are assumed as

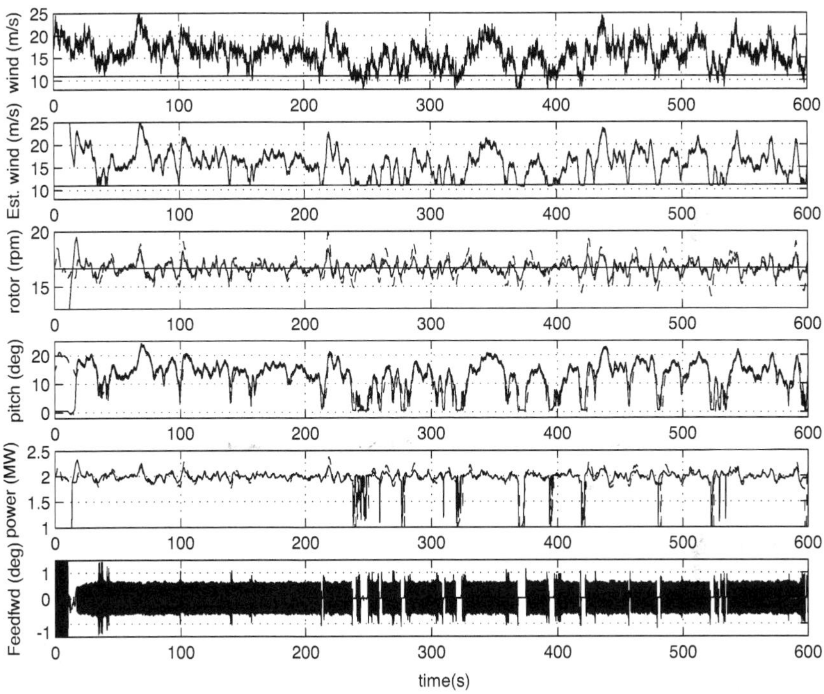

Fig. 29. Simulation results of the feedforward control for the mean wind speed of 16 m/s and 18% turbulence intensity

Performance data		Feedforward on	Feedforward off
Rotor speed (rpm)	Mean	16.6720	16.6408
	Standard deviation	0.4956	0.8971
Power (MW)	Mean	1.9920	1.9959
	Standard deviation	0.0619	0.1031

Table 4. Rotor speed and power data for the feedforward control on and off

measurement and process noise, respectively. The estimation errors up to around 20 seconds are due to the mismatched initial conditions in the estimator. The estimated wind speed tracks well and looks like a low-pass filtered wind speed because of the estimator dynamics. Note that there is a lower bound in the wind speed estimation, which is the rated wind speed. The third plot of Fig. 29 is the rotor rpm. The solid line shows the rotor speed response when the feedforward control is active, while the dashed line shows the response when it is off. As shown in this plot and summarized in Table 4, the benefit of applying the feedforward control is clear. The fourth plot is the pitch response. The solid line shows the response when the feedforwad control is active. A more high-frequency pitching action can be noticed in the response, compared to the pitch response when the feedforward is off (dashed

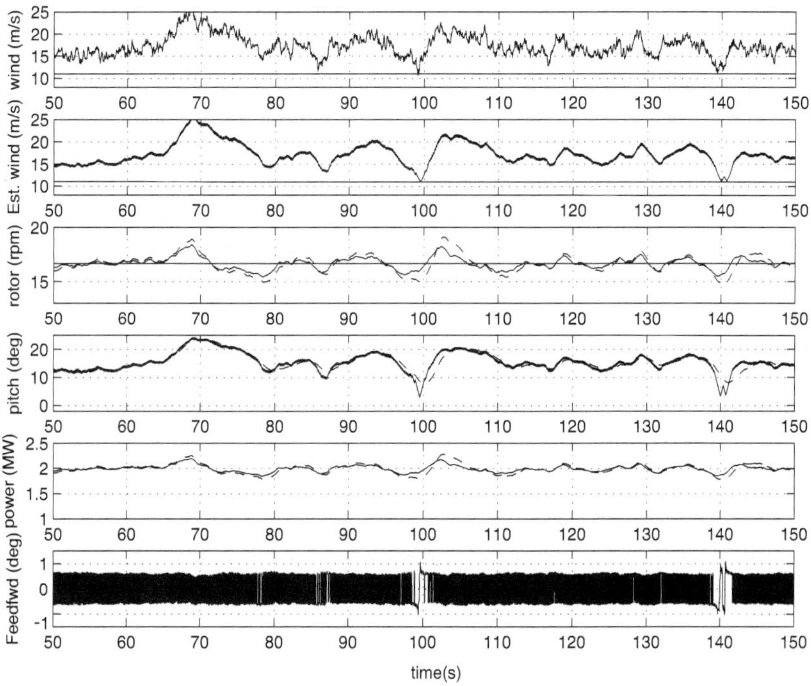

Fig. 30. Simulation results of the feedforward control for the time period from 50 to 150 seconds (re-plot of Fig. 29)

plot). Also note a time-ahead pitch motion in order to compensate the wind speed variation, compared to that of the baseline pitch control action, i.e. the pitch response when the feedforward is off. The fifth plot is the output power, which shows the performance of the feedforward control. Sudden drops of power in the response are due to the vertical torque schedule at the rated rpm (see the section E-F of Fig. 10). When the statistical data on power in Table 4 were calculated, the data during the power dip were not included. This is fair data processing, because the feedforward is not active for operating conditions in the below rated wind speed region. The final plot of this figure is the pitch demand by the feedforward control. Note that even where the magnitude of the pitch demand is not very large, there is distinct advantage in applying the feedforward control. The above mentioned points can also be seen in Fig. 30, which is a re-plot of Fig. 29 for a time period from 50 to 150 seconds.

4.4 Individual pitch control

The modern wind turbine is a huge mechanical structure. For example, the rotor diameter of even a 3 MW machine is over 90 m. In a turbulent wind condition, the wind speed variation of this machine experienced through a full rotation could easily be over 10 m/s, which imparts asymmetric aerodynamic forces to the rotating blades. During wind turbine operation, there are many sources of asymmetric forces on a blade such as

- vertical and horizontal wind shear

- turbulent winds
- yawing misalignment
- gravitational effect due to tilt angle, tower shadowing, etc.

These forces are periodic with rotational angle, i.e. azimuth angle except those due to turbulence. Theoretically, periodic forces can be reduced by cyclic pitch control using azimuth angle information. However, because of stochastic mechanical loadings by turbulent winds, it is not sufficient to cover the problem of load alleviation using only cyclic pitch control.

Prior to discussion of individual pitch control, we will define mechanical loads in two coordinate systems, namely the blade coordinate system and hub coordinate system of Fig. 31 (Bossanyi, 2009). The system on the right side of Fig. 31 is the blade coordinate system, which is attached to the blade. The ZB-axis is along the blade pitch axis, the XB-axis points towards the tower, and the YB-axis completes a right-handed coordinate system. The plane which is made by the ZB-axis during full rotational motion is the rotor in-plane. The rotor out-of-plane is clear from this definition. Note that the force and moment relation between the in-plane/out-of-plane and flapwise/edgewise direction at each blade section is given by

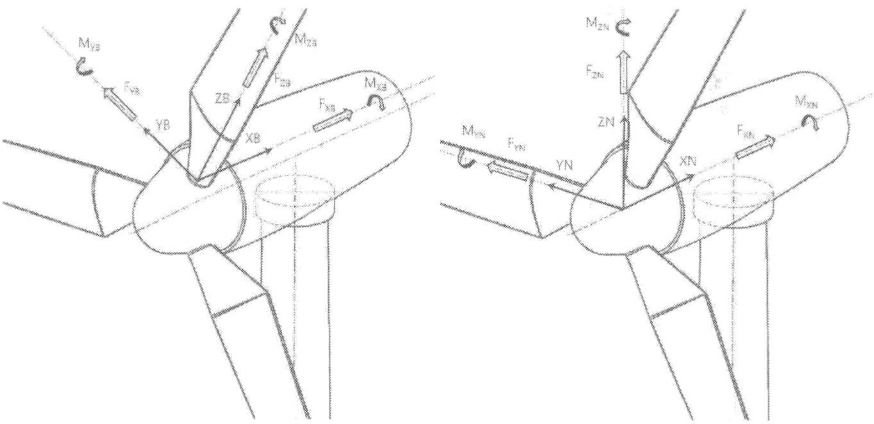

Fig. 31. Hub and blade coordinate system

$$\begin{Bmatrix} M_{XB} \\ M_{YB} \end{Bmatrix} = \begin{bmatrix} \sin(\theta_T + \beta) & \cos(\theta_T + \beta) \\ \cos(\theta_T + \beta) & -\sin(\theta_T + \beta) \end{bmatrix} \begin{Bmatrix} M_{FLAP} \\ M_{EDGE} \end{Bmatrix} \tag{33}$$

where θ_T and β represent the twist and pitch angle at the blade section, respectively. M_{XB} and M_{YB} are the bending moments at any blade section in the in-plane and out-of-plane direction, while M_{FLAP} and M_{EDGE} are the bending moments at the same blade section in the flapwise and edgewise direction. Note that strain gauge measurements on the blade root are usually made along the flapwise and edgewise directions.

The hub coordinate system is shown on the left side of Fig. 31 and is a fixed reference system. The XN-axis coincides with the shaft axis pointing towards the tower, the ZN-axis is in a vertically upward direction but inclined with a tilt angle, and the YN-axis completes a right-handed coordinate system. The out-of-plane bending moment of each blade at the root section, i.e. M_{YB1}, M_{YB2}, M_{YB3}, can be transformed to a tilting moment, M_{YN}, and a yawing moment, M_{ZN}, in the hub coordinate system, that is

$$\begin{Bmatrix} M_{YN} \\ M_{ZN} \end{Bmatrix} = \begin{bmatrix} \cos\psi_1 & \cos\psi_2 & \cos\psi_3 \\ \sin\psi_1 & \sin\psi_2 & \sin\psi_3 \end{bmatrix} \begin{Bmatrix} M_{YB1} \\ M_{YB2} \\ M_{YB3} \end{Bmatrix} \tag{34}$$

where ψ_i is the azimuth angle of the i-th blade and is defined as zero when the blade is in the upright position. The above equation shows how the blade loads developed in a rotating reference frame are transferred to the tower in a fixed reference frame. The transform of Eq. (34) is similar to the inverse Coleman transform (van Engelen et al., 2007) or the d-q axis transform for electric machinery (Bossanyi, 2003; Krause et al., 2002), which is just a change of variables from a rotating to a fixed reference frame, that is

$$\begin{Bmatrix} M_{F1} \\ M_{F2} \\ M_{F3} \end{Bmatrix} = \begin{bmatrix} 1/3 & 1/3 & 1/3 \\ 2/3\cos\psi_1 & 2/3\cos\psi_2 & 2/3\cos\psi_3 \\ 2/3\sin\psi_1 & 2/3\sin\psi_2 & 2/3\sin\psi_3 \end{bmatrix} \begin{Bmatrix} M_{YB1} \\ M_{YB2} \\ M_{YB3} \end{Bmatrix} \tag{35}$$

where M_{F1}, M_{F2}, and M_{F3} are variables in a fixed reference frame. M_{F1} is the average of three variables M_{YB1}, M_{YB2}, M_{YB3}. Note that M_{F2} and M_{F3} are proportional to a tilting moment, M_{YN}, and a yawing moment, M_{ZN}, in the hub coordinate system. The inverse of the transform of Eq. (35) is the Coleman transform, which is given by

$$\begin{Bmatrix} M_{YB1} \\ M_{YB2} \\ M_{YB3} \end{Bmatrix} = \begin{bmatrix} 1 & \cos\psi_1 & \sin\psi_1 \\ 1 & \cos\psi_2 & \sin\psi_2 \\ 1 & \cos\psi_3 & \sin\psi_3 \end{bmatrix} \begin{Bmatrix} M_{F1} \\ M_{F2} \\ M_{F3} \end{Bmatrix}. \tag{36}$$

The above relation illuminates the core idea of the individual pitch control (IPC) algorithm. If M_{F2} and M_{F3} can be maintained close to zero, then only small amounts of oscillatory out-of-plane motion of the blade result. This is self-evident from Eq. (36). Fig. 32 is a block diagram representation of the above IPC algorithm. As shown in this figure, the first step of the IPC is to transform the bending moment of each blade in the out-of-plane direction to fixed frame quantities, i.e. M_{F2} and M_{F3}. The next step is the application of appropriate control laws which make M_{F2} and M_{F3} zero. The final step is to convert the amounts of pitch demand calculated in the fixed frame to quantities of individual pitch demand by using the Coleman transform.

A schematic structure of a wind turbine control system with the IPC algorithm is depicted in Fig. 33. As mentioned in the above, the core of the individual pitch control (IPC) is how to minimize the magnitudes of M_{F2} and M_{F3}. The same pitch and torque control structure as that explained in Section 2.2 (compare Fig. 33 with Fig. 7) is applied. Two control loops, i.e. the power curve tracking control loop (pitch and torque control loop) and the IPC loop, are not coupled, so the independent design of each loop is possible. Reduction of mechanical loads on the blades is a critical issue for a large wind turbine, because the amplitude of mechanical loads is directly related with the life cycle of the blades. The efficiency of the IPC algorithm depends on how the control laws are programmed. Various IPC control laws and simulation results can be found in the references (Selvam, 2007; Bossanyi & Wright, 2009; van Engelen, 2007).

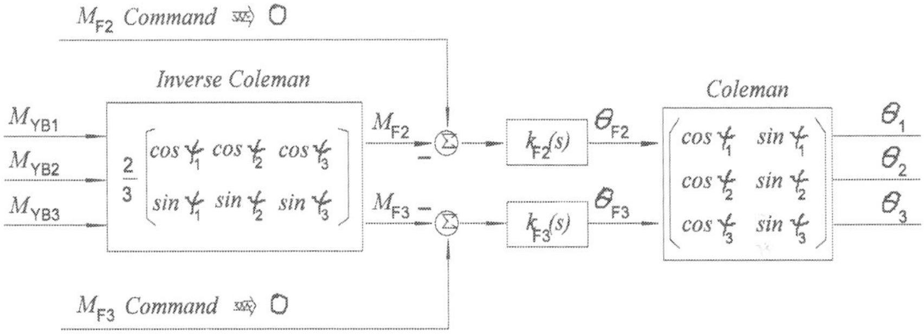

Fig. 32. Schematic of IPC algorithm

Because of the diverse sources of cyclic loads such as wind shear, tower shadowing, etc., most mechanical loads on blades consist of a 1p (1 per revolution, i.e. rotational frequency) component and its harmonics as shown in the upper part of Fig. 34. Due to averaging through a full rotation of the rotor, however, only the 3p and its harmonics of mechanical loads are left and transferred to the hub, drive train, and generator. Noting that the azimuth angle of the i-th blade can be represented as $\psi_i=\omega t+\psi_{i0}$ in Eq. (36), where ω is the 1p in the unit of r/s, the relationship of the mechanical loads between a rotating reference frame and a fixed reference frame can be understood. As depicted in this figure, the 0p and 1p motions in a rotating frame affect only the 0p motion in a fixed frame. The IPC algorithm explained in Fig. 32 is based on the 1p motion control, which is effective only in decreasing the 1p component magnitude of the blade mechanical loads, in other words, DC mechanical loads in the non-rotating frame, such as the yaw bearing moments. In order to alleviate the 3p component of non-rotating mechanical loads, the 2p IPC should be applied. The 2p IPC has the same control structure as that of the 1p IPC in Fig. 32, but the arguments in sine and cosine functions should be doubled (Bossanyi & Wright, 2009).

The simulation results of applying the IPC algorithm to a 3 MW wind turbine are summarized in Fig. 35, Fig. 36, and Fig. 37. The IPC simulation was performed in the GH Bladed© (Bossanyi, 2008) environment. Simple proportional controllers for $k_{F2(s)}$ and $k_{F3(s)}$ are used in this IPC simulation. The wind speed shown in these figures is the hub height wind speed, which has a mean of 20 m/s and 10% turbulence intensity. Fig. 35 contains the performance-related time responses, which are the rotor speed (rpm), the #1 pitch angle (deg), the generator torque (kNm), and the power (MW). The dashed lines in these figures represent the time domain responses, which are obtained when the IPC is off. As expected, there is almost no interaction between the power curve tracking control and IPC loop. The dashed plot in the third window of Fig. 35 is the #1 pitch response when the IPC is off, while the solid plot is the response when it is on. The 1p pitching is dominant in the solid plot, which results from the IPC action to minimize the blade bending moment in the out-of-plane direction. The effect of the IPC on the blade load alleviation is shown in Fig. 36. The first window is the wind speed, which is the same wind as that in Fig. 35. The three consecutive plots after this are the bending moments of #1, #2, and #3 blade at the root section in the out-of-plane direction. Again, the dashed lines are the responses when the IPC is off. Note how the peak-to-peak magnitude of the out-of-plane bending moment of each blade is changed with the application of the IPC. As summarized in Table 5, there are more than 20% DEL decreases of the blade moments in the out-of-plane direction for all blades.

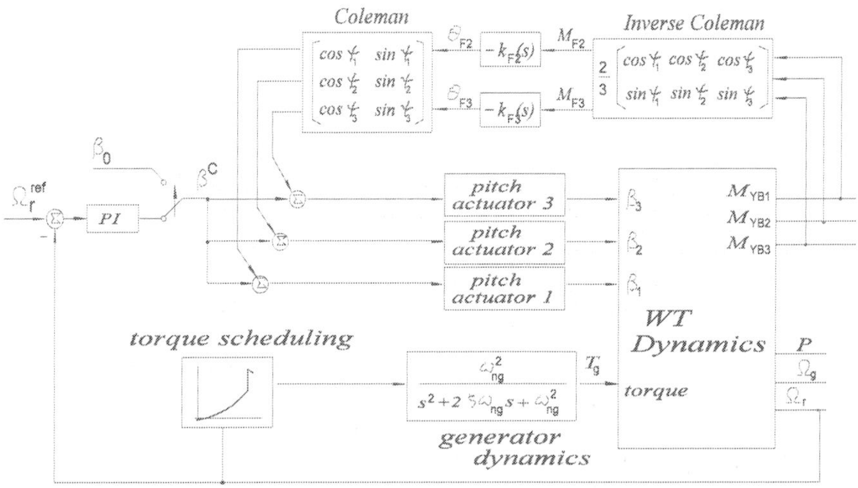

Fig. 33. Schematic of collective pitch control system with IPC algorithm

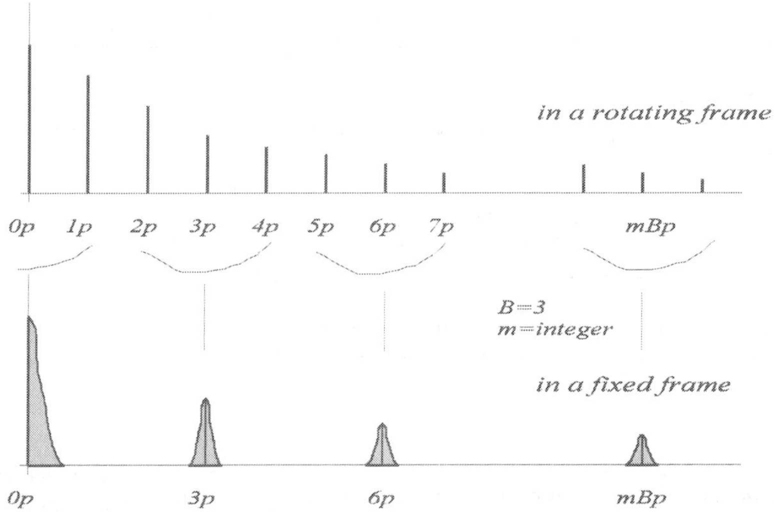

Fig. 34. Amplitudes of harmonic components for the mechanical loads in a rotating and fixed frame

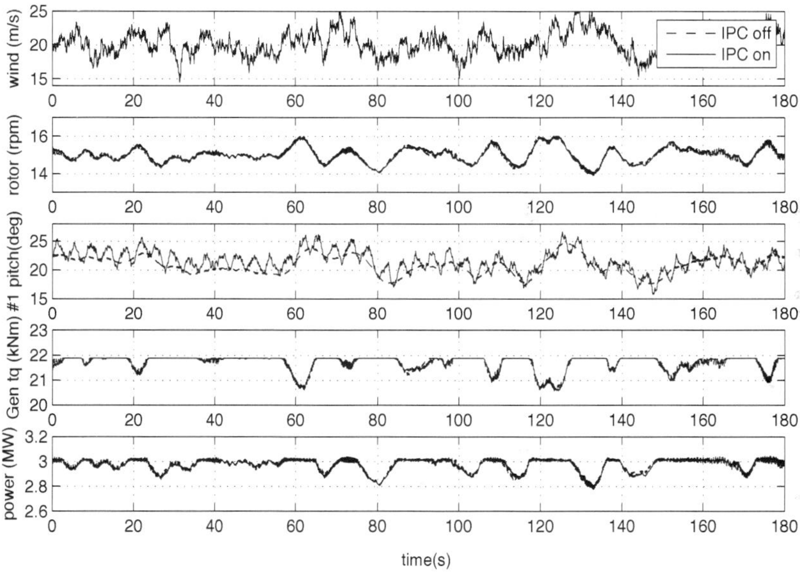

Fig. 35. Time domain responses of performance related data when the IPC is on (solid lines) and off (dashed lines)

The final three plots of Fig. 36 are M_{F1}, M_{F2}, and M_{F3}. Due to the averaging of three blade moments, M_{F1} comes to have the 3p component signal. Note how much the mean levels of M_{F2} and M_{F3} move toward zero when the IPC is on. As mentioned earlier, the core of the IPC algorithm is minimization of M_{F2} and M_{F3} (see Fig. 32). Depending on how $k_{F2(s)}$ and $k_{F3(s)}$ of Fig. 32 are designed, the performance of the IPC, i.e. how large the mean values of M_{F2} and M_{F3} are, would be determined. Even for the application of a simple proportional IPC controller, the above simulation results look promising. The tower root bending moments in the side-side and fore-aft direction are shown in the second and third windows of Fig. 37. The peak-to-peak magnitudes of the tower mechanical loads are increased when the IPC is on. Especially, the tower loads in the side-side direction deteriorate more than those in the fore-aft (see Table 5). The fourth window of Fig. 37 is the yawing moment of the yaw bearing. Note the similarity between this plot and the M_{F3} response of Fig. 36. The final three plots of Fig. 37 are the blade pitch angles of all three blades. The dashed line shows the collective pitch response when the IPC is off.

5. Conclusions

Control system design of wind turbines is a complex task which has to consider the various aspects of performance and safety issues of wind turbine operation. The role of the control system becomes more critical in a multi-MW wind turbine design. This chapter covers all the areas of control system design. Starting from a simple drive train model, how a PI pitch controller is designed and what are the critical design problems are explained. Some interesting themes of wind speed estimation, feedforward pitch control, and individual pitch control system design are included, together with numeric simulation results.

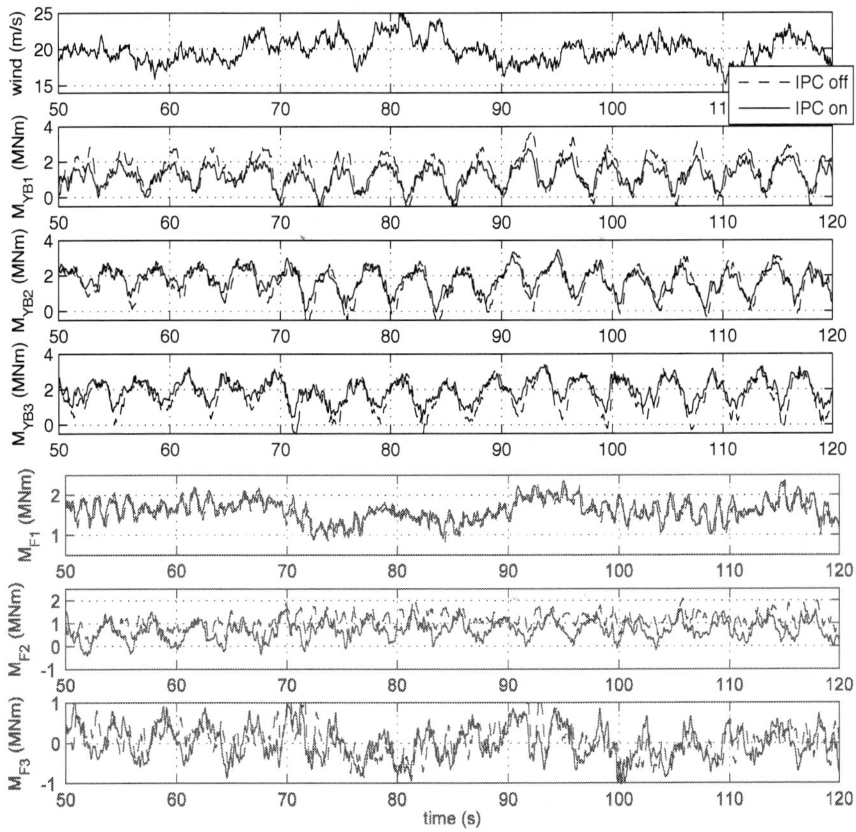

Fig. 36. Time domain responses of blade mechanical loads when the IPC is on (solid lines) and off (dashed lines)

	Blade loads in the out-of-plane direction (MNm)			Tower loads (MNm)	
	#1 blade	#2 blade	#3 blade	Side-side	Fore-aft
IPC off (A)	2.906	2.906	2.906	9.212	12.66
IPC on (B)	2.097	2.275	2.184	11.32	13.14
(B-A)/A (%)	-27.87	-21.71	-24.84	22.8	3.79

Table 5. Damage equivalent loads (DEL) of out-of-plane blade bending moments and tower root moments with IPC on and off

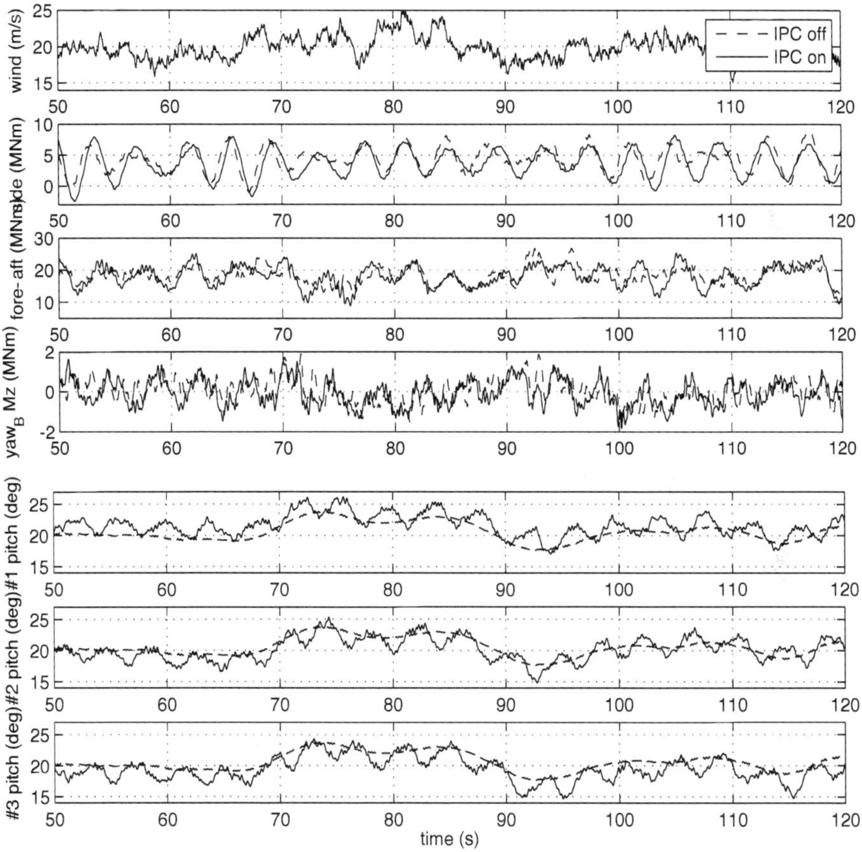

Fig. 37. Time domain responses of tower loads, yaw bearing loads, and blade pitches when the IPC is on (solid lines) and off (dashed lines)

6. References

Bianchi, F.; Battisa, H. & Mantz, R. (2007). Wind turbine control systems, principles, modelling and gain scheduling design, Springer, ISBN: 1846284929

Bossanyi, E. (2009). GH Bladed user manual (version 3.81), Garrad Hassan and Partners, 282-BR-010

Bossanyi, E. (2008). GH Bladed theory manual (version 3.81), Garrad Hassan and Partners, 282-BR-009

Bossanyi, E. & Wright, A. (2009). Field testing of individual pitch control on the NREL CART-2 wind turbine, EWEC2009-European Wind Energy Conference & Exhibition

Bossanyi, E. (2003). Individual blade pitch control for load reduction. Wind Energy, No. 6, pp. 119-128, ISSN: 1099-1824

Bossanyi, E. (2000). The design of closed loop controllers for wind turbines. Wind Energy, No. 3, pp. 149-163, ISSN: 1099-1824

Boukhezzar, B.; Lupu, L., Siguerdidjane, H. & Hand, M. (2007). Multivariable control strategy for variable speed variable pitch wind turbines. Renewable Energy, Vol. 32, pp. 1273-1287, ISSN: 0960-1481

Burton, T.; Sharpe, D., Jenkins, N. & Bossanyi, E. (2001). Wind energy handbook, A John Wiley and Sons, ISBN: 9780471489979

Dominguez, S. & Leithead, W. (2006). Size related performance limitations on wind turbine control performance, International Control Conference (ICC 2006), FB4-181, Glasgow

Dorf, R. & Bishop, R. (2007). Modern control systems, Prentice Hall, ISBN: 0132270285.

Franklin, G.; Powell, J. & Emami-Naeini, A. (2006). Feedback control of dynamic system, Prentice Hall, ISBN: 0-13-149930-0

Gelb, A. (1974). Applied optimal estimation, The MIT Press, ISBN: 10- 0262570483

Haque, M.; Negnevitsky, M. & Muttaqi, K. (2010). A novel control strategy for a variable-speed wind turbine with a permanent-magnet synchronous generator. IEEE Transactions on Industry Applications, Vol. 46, Issue 1, pp. 331-339, ISSN: 0093-9994

Krause, P.; Wasynczuk, O. & Sudhoff, S. (2002). Analysis of electric machinery and drive systems, A John Wiley and Sons, ISBN: 0-471-14326-X

Leithead[a], W. & Connor, B. (2000). Control of variable speed wind turbines: dynamic models. International Journal of Control, Vol. 73, No. 13, pp. 1173-1188, ISSN: 0020-7179

Leithead[b], W. & Connor, B. (2000). Control of variable speed wind turbines: design task. International Journal of Control, Vol. 73, No. 13, pp. 1189-1212, ISSN: 0020-7179

Leloudas, G.; Zhu, W. Sorensen, J., Shen, W. & Hjort, S. (2007). Prediction and reduction of noise for a 2.3MW wind turbine. Journal of Physics: Conference series, Vol. 75, pp. 1-9, ISSN: 1742-6588

Manwell, J.; McGowan, J. & Rogers, A. (2009). Wind energy explained, Theory, design and application, A John Wiley and Sons, ISBN: 9780470015001

Muller, S.; Deicke, M. & de Donker, R. (2002). Doubly fed induction generator systems for wind turbines. IEEE Industry Applications Magazine, May-June 2002, pp. 26-33, ISSN: 1077-2618

Ostergaard, K; Brath, P. & Stoustrup, J. (2007). Estimation of effective wind speed. Journal of Physics: Conference series, Vol. 75, pp. 1-9, doi: 10.1088/1742-596/75/1 /012082

Selvam, I. (2007). Individual pitch control for large scale wind turbines, ECN-E-07-053

Soter, S. & Wegerer, R. (2007). Development of induction machines in wind power technology, IEMDC 2007-IEEE International Electrical Machines and Drives Conference, pp. 1490-1495

van Engelen, T. (2007). Control design based on aero-hydro-servo-elastic linear models fromTURBU(ECN), ECN-M-07-054

van Engelen, T.; Markou, H., Buhl, T. & Marrant, B. (2007). Morphological study of aeroelastic control for wind turbines, ECN-E-06-056

van der Hooft, E.; Schaak, P., & van Engelen, T. (2003). Wind turbine control algorithm, ECN-C-03-111

van der Hooft, E. & van Engelen, T. (2003). Feedforward control of estimated wind speed, ECN-C-03-137

van der Hoven (1957). Power spectrum of horizontal wind speed spectrum in the frequency range from 0.0007 to 900 cycles per hour. Journal of Meteorology, Vol. 14, pp. 160-164

12

Using Genetic Algorithm to Obtain Optimal Controllers for the DFIG Converters to Enhance Power System Operational Security

João P. A. Vieira, Marcus V. A. Nunes and Ubiratan H. Bezerra
Federal University of Pará
Brazil

1. Introduction

The wide spread use of Doubly Fed Induction Generator technology (DFIG) is mostly due to the characteristics of its static converters that permit a variable speed operation with very reduced nominal power requirement which is approximately 25% of the generator power capacity specification [1-2]. With the increasing number of DFIGs being connected to the electrical grids some instability effects are arising which may worsen the power system security [3]. Time domain simulation techniques have being employed to evaluate the DFIG dynamic behavior and how it is dynamically impacting the power system security [4-7]. These studies are offering a better comprehension of the DFIG intrinsic dynamics when it is connected in electrical power grids, and these studies may be very useful when projecting controllers for this generator.

Recently the particle swarm optimization technique (PSO) has being used to adjust the DFIG controllers gains with the objective of reducing the rotor current and also improving the small-signal stability [8-9]. In both references [8] and [9] the results obtained with the application of the PSO technique are compared with those obtained with a trial and error approach. The Bacteria Foraging optimization method has also being applied to the gain adjustment of the controller that is responsible to improve the damping of the DFIG oscillation modes [10]. However, the characteristics of exhibiting a robust damping and also improving the stability margin may not be guaranteed simultaneously when a change in the operation condition occurs.

This chapter presents a multi-objective optimization methodology based on genetic algorithm (AG) to obtain the controller gains of the rotor side DFIG converter with the main objective of improving the DFIG ride-through capability, the voltage control and the overall power system stability margin.

The methodology in this chapter, combines the GA with the pole placement method, since the gains obtained by the pole placement technique are chosen to compose one of the individuals of the initial population of the GA, in order that the optimal design procedure may start from a good initial solution, which may improve the convergence of the evolutionary procedure towards better solutions.

2. DFIG model

For power system stability studies, the generator may be modeled as an equivalent voltage source behind transient impedance [11]. The differential equations of the stator and rotor circuits of the induction generator with stator current and equivalent voltage behind transient impedance as state variables can be given in a *d-q* reference frame rotating at synchronous speed. For adequately representing the DFIG dynamics involved in the controllers design the fourth order model of the induction generator is used as presented in [12].

$$\frac{d\overline{e}_d'}{dt} = -\frac{1}{T_o'}\cdot\left[\overline{e}_d' - \left(\overline{X} - \overline{X}'\right)\overline{i}_{qs}\right] + s\omega_s\overline{e}_q' - \omega_s\frac{\overline{L}_m}{\overline{L}_{rr}}\overline{v}_{qr}$$

$$\frac{d\overline{e}_q'}{dt} = -\frac{1}{T_o'}\cdot\left[\overline{e}_q' + \left(\overline{X} - \overline{X}'\right)\overline{i}_{ds}\right] - s\omega_s\overline{e}_d' + \omega_s\frac{\overline{L}_m}{\overline{L}_{rr}}\overline{v}_{dr}$$

$$\frac{\overline{X}'}{\omega_s}\frac{d\overline{i}_{ds}}{dt} = -\overline{v}_{ds} - \left[\overline{R}_s + \frac{\left(\overline{X} - \overline{X}'\right)}{\omega_s T_o'}\right]\overline{i}_{ds} + \overline{X}'\overline{i}_{qs} + (1-s)\overline{e}_d' - \frac{1}{T_o'\omega_s}\overline{e}_q' + \frac{\overline{L}_m}{\overline{L}_{rr}}\overline{v}_{dr}$$

$$\frac{\overline{X}'}{\omega_s}\frac{d\overline{i}_{qs}}{dt} = -\overline{v}_{qs} - \left[\overline{R}_s + \frac{\left(\overline{X} - \overline{X}'\right)}{\omega_s T_o'}\right]\overline{i}_{qs} - \overline{X}'\overline{i}_{ds} + (1-s)\overline{e}_q' + \frac{1}{T_o'\omega_s}\overline{e}_d' + \frac{\overline{L}_m}{\overline{L}_{rr}}\overline{v}_{qr}$$

Where X´ and X are respectively the transient reactance and the open circuit reactance; e_d' and e_q' are respectively the d-axis and q-axis components of the internal voltage; T_o' is the open circuit time constant in seconds.
To represent the electrical and mechanical interaction between the electrical generator and wind turbine in transient stability studies, the two masses model is presented, according to [3]:

$$\frac{d\overline{\omega}_t}{dt} = \frac{1}{2H_t}\left(\overline{T}_m - K_s\delta_s\right) - \overline{D}_t\overline{\omega}_t$$

$$\frac{d\delta_s}{dt} = \omega_o\left(\overline{\omega}_t - \overline{\omega}_g\right)$$

$$\frac{d\overline{\omega}_g}{dt} = \frac{1}{2H_g}\left(K_s\delta_s - \overline{T}_e - \overline{D}_g\overline{\omega}_g\right)$$

Where K_s is the axis stiffness in p.u./rad-elect; $\overline{D}_t$ and $\overline{D}_g$ are respectively the wind rotor and the electrical rotor damping coefficients; ω_o is the electrical system angular velocity in rad-elect/sec; and H_t and H_g are the wind turbine and the electrical generator inertia constants respectively, in seconds.

3. The DFIG converters models

The model of the converter system includes the representation of the rotor-side converter, the grid-side converter and the dc link and the converter control. The rotor-side and grid-side converters are modeled as voltage sources. The control system for the rotor-side converter was implemented so that the field-oriented current control loop is used to control the rotor current, in which the q-axis component is responsible for the rotor speed control, and the d-axis component controls the terminal voltage, as shown in Fig. 1.

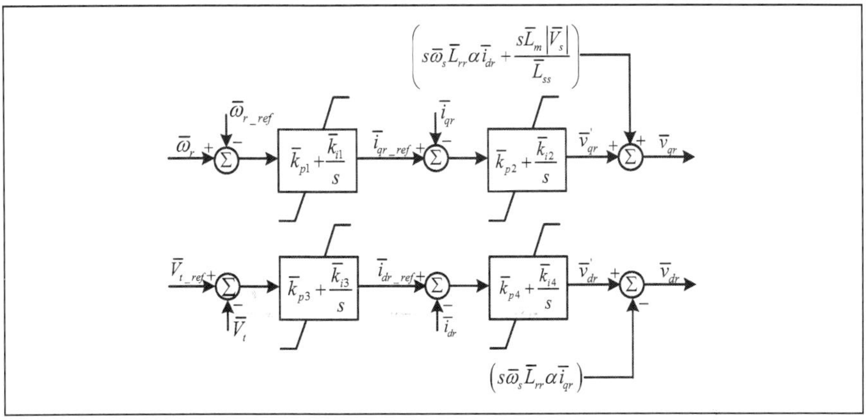

Fig. 1. Rotor-Side Converter Control Scheme

5. GA-Multi objective optimal control

The problem of adjusting the controller's gains of the rotor-side DFIG converter, considering a specific operating point, may be formulated as a multi-objective optimization problem. The objectives to be optimized are the absolute errors between the rotor reference currents, which are established by the PI controllers, and the rotor measured currents along the q and d axis respectively and the magnitude of the rotor voltage.

By considering the DFIG vector control formulation as presented in [13], it can be shown that the q and d components of the rotor current are very effective in controlling both the DFIG stator active power and the terminal voltage respectively. This way, improving the rotor current dynamic response (which may be obtained by minimizing the error between the rotor reference and measured currents) may reflect also in a better dynamic performance for the DFIG stator active power and terminal voltage. Besides that, the minimization of an additional term in the objective function that will be responsible for obtaining optimized responses for the magnitude of the rotor voltage may improve the dynamic behavior of other variables which are controlled by the grid side converter, as for example the rotor active power which is a function of the rotor voltage, as well as the dc-link voltage, and the current and reactive power of the grid-side converter.

This way the global objective is to improve the DFIG dynamic behavior after the occurrence of faults in the electrical network enhancing the ride-through capability, voltage control, and also increasing the small-signal and transient stability margins of the power system. A measure that indicates if a good adjustment for the parameters of the rotor side converter has been achieved is given by the fitness function which is composed by the weighted sum of three objectives which will be minimized by the genetic algorithm optimization procedure:

$$\overline{F} = \int_0^{tsim} \left(\omega_1 \left| \overline{i}_{dr\_ref} - \overline{i}_{dr} \right| + \omega_2 \left| \overline{i}_{qr\_ref} - \overline{i}_{qr} \right| + \omega_3 \sqrt{\overline{v}_{dr}^2 + \overline{v}_{qr}^2} \right) dt$$

Where ω_1 , ω_2 and ω_3 are weight factors.

The gains obtained by the pole placement technique as described in [14], form one of the individuals of the GA initial population which may improve the convergence of the GA once the evolutionary process is started with a good initial solution.

6. Electrical network

The electrical network used for the simulation studies is a real power system belonging to the COSERN electric power utility that operates in the northeast region of Brazil, in the state of Rio Grande do Norte. In this study, the wind park to be connected is considered as a dynamic equivalent, represented by an equivalent wind generator of 20 MW and 960 V. The wind park must be connected to the distribution electrical grid by 0.96 kV/69 kV transformers.

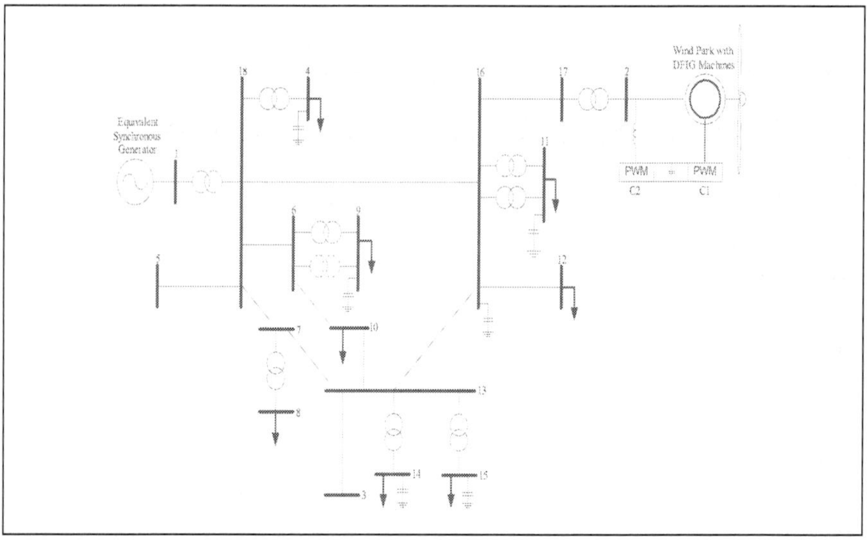

Fig. 2. Electrical Network

7. Simulations and results

Firstly, it will be presented the gains obtained for the PI rotor-side controller using the GA optimal design technique. In this optimization procedure a three-phase short circuit was applied at t=0.1s for 100 ms at bus 2. The simulation time was 4 s and it was considered the base operational condition for the electrical network as shown in Fig. 2, without the "crow-bar" protection arrangement.

The gains obtained by the pole placement project and by the GA project are presented in tables 1 and 2, respectively. It may be noticed that the switching frequency used for the CA-

CC-CA converter system was 2 kHz [15], which is a key parameter for the adjustment of the static converter controls in DFIG generators. The objective function weight factors ω_1, ω_2 and ω_3 were set equal to 1.

K_{P1}	K_{I1}	K_{P2}	K_{I2}	K_{P3}	K_{I3}	K_{P4}	K_{I4}
-0.27	-0.016	0.4	0	0.006	0.004	0.405	0

Table 1. Poles Placement Gains Adjustments for the PI Controllers of Rotor-Side Converter

K_{P1}	K_{I1}	K_{P2}	K_{I2}	K_{P3}	K_{I3}	K_{P4}	K_{I4}
-0.87	-0.016	0.45	7.9	0.19	0.004	0.36	0.06

Table 2. GA Gains Adjustments for the PI Controllers of Rotor-Side Converter

To evaluate the performance and robustness of the proposed GA optimization methodology, as well as the effectiveness of the crow-bar protection scheme, three case studies are presented: a) base case load as informed by the electrical utility; b) 20% load reduction in all load buses with respect to the base case; c) 20% load increase in all load buses with respect to the base case. In the results presented in this chapter, the optimal design refers to the results obtained by the GA optimization procedure, and formal design refers to the results obtained by the pole placement techniques.

Case a) A three phase short circuit lasting for 100 ms is applied at t1 = 1s, at the end of line 18-16, near bus 16. The fault is cleared by the protection scheme and the electrical system changes to a new operational point disconnecting transmission line 18-16.

In Fig. 3 it is shown the transient behavior of the DFIG rotor current. It can be observed that the rotor current limit specified for the rotor-side converter, which is approximately 0.406 p.u., is exceeded right after starting the fault which implies in activating the crow-bar protection, at t2 = 1.0016 s, by the insertion of external resistances in the DFIG rotor. The inserted resistances reduce significantly the rotor current until the fault is cleared at t3 = 1.1 s.
It must be emphasized that during the fault period the rotor-side converter remains connected to the DFIG once the rotor current is flowing through the external resistances and not through the converter itself. Immediately after the fault is cleared the crow-bar protection is deactivated and simultaneously the DFIG returns to normal operation, activating again the rotor-side converter controllers.
But when the fault is cleared the rotor current oscillates again as can be seen in Fig. 4. In this case the projected PI controllers, by either pole placement technique or by GA technique, present a good performance in damping the oscillation without the need of activating the crow-bar protection scheme again.
However, it is noticed in Fig. 4 that when using the optimal gains of the GA projected PI controller the rotor current presents a better time response when compared with the pole placement projected PI controller. This improvement is evident in the second oscillation when the current overshoot is higher for the pole placement projected controller, reaching values above 0.3 p.u., as compared with the response obtained by the GA PI controller. Besides that, the GA PI controller reduced more significantly the oscillation after t = 2 s, with respect to the pole placement PI controller.

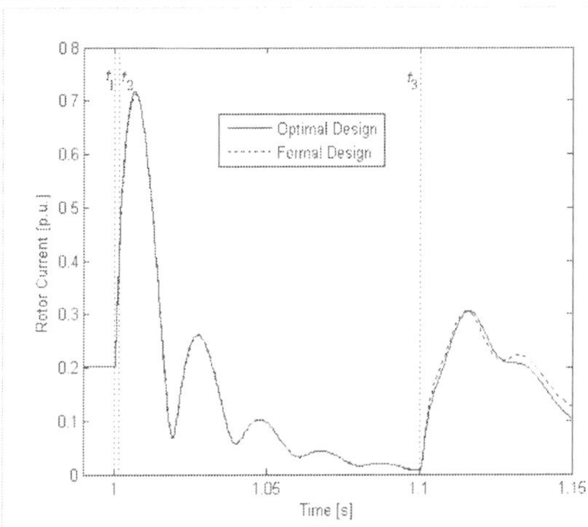

Fig. 3. DFIG Rotor Current

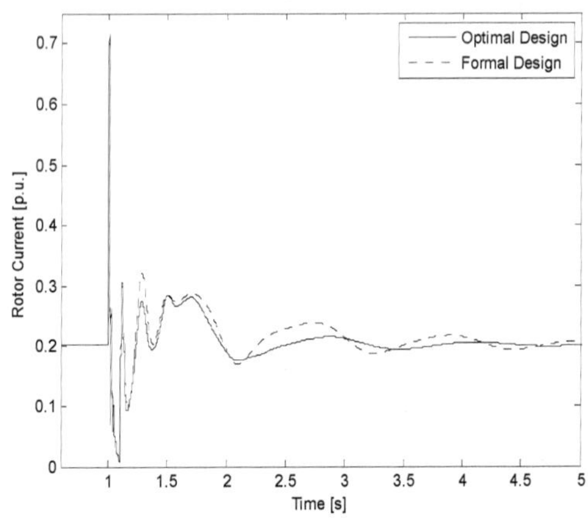

Fig. 4. DFIG Rotor Current

It is shown in Fig. 5 the DFIG rotor voltage. It is observed that the adopted crow-bar protection strategy was efficient, once the rotor voltage oscillation does not exceed the maximum allowed limit value which is specified by the rotor-side converter and is equal to 0.3 p.u. It is noticed also that during the fault the rotor voltage is obtained by the applied voltage to the external resistances of the crow-bar protection scheme, which is equal to the rotor-side converter voltage.

After the fault is cleared, both PI controllers, adjusted by pole placement and by GA techniques, have presented a good performance when submitted to voltage sags. As

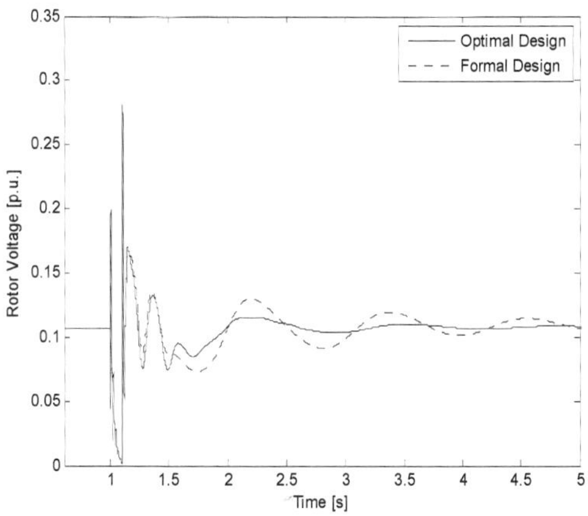

Fig. 5. DFIG Rotor Voltage

observed previously for the current behavior, the optimal PI controller design also reduced the rotor voltage oscillation after t = 2 s, as compared to the PI controller designed by the pole placement technique.

The DC link voltage time responses are shown in Fig. 6, and it can be seen that the response that corresponds to the PI controller projected by the GA technique presents oscillation with lower overshoot and higher damping as compared to the response obtained by the PI controller which gains were adjusted by the pole placement procedure. This is an important aspect to consider since the DC link voltage is one of the variables that may activate the crow-bar protection scheme.

The time response of the DFIG terminal voltage is presented in Fig. 7. It may be observed that by using the GA procedure to project the PI controller, it is obtained for the DFIG terminal voltage a less oscillatory response containing lower overshoot after the fault is cleared, when compared with the PI controller projected by the pole placement technique.

These results are very relevant as much as high voltage values for the wind generator buses may disconnect the DFIG machines by the overvoltage protection scheme. The grid operators in some European countries, for example, are including this recent requisite, known as High Voltage Ride-Through [16], to be attended by wind parks to be connected to the grid.

Besides that, the problem of poorly damped oscillations in distributed generation systems may affect significantly the power quality for the consumers. This happens because such oscillations directly influence the magnitude and frequency of the voltage waveform in load buses.

In Fig. 8 it is presented the plot of the DFIG stator active power. It can be observed a less oscillatory response after the fault is cleared when using the PI controller designed by the GA procedure. The proposed optimization procedure improves the behavior of variables that are decoupled by the vector control strategy employed for the DFIG, namely the terminal voltage (or reactive power) and active power (or rotor speed) as shown in Figs. 7

and 8 respectively. This way it is justified the methodology of improving the transient behavior of the d and q axis components of the rotor current because this improvement has as consequence a better transient behavior for the terminal voltage (or reactive power) and active power (or rotor speed).

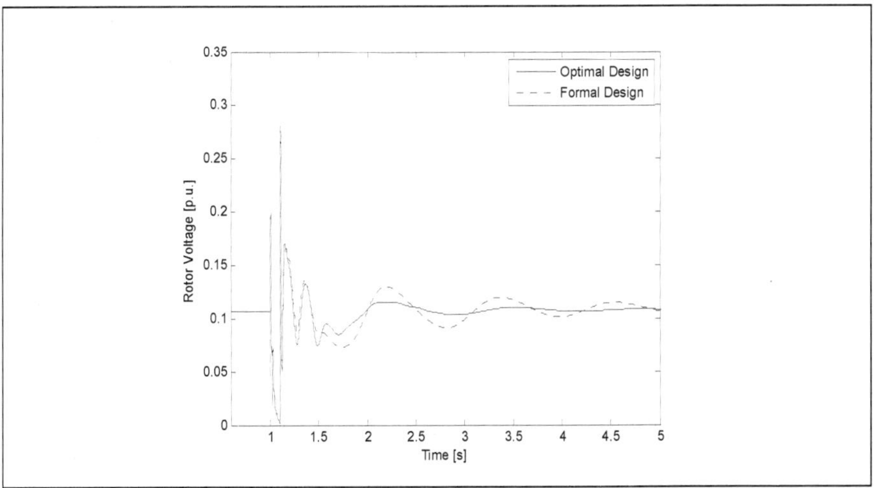

Fig. 6. DC-Link Voltage

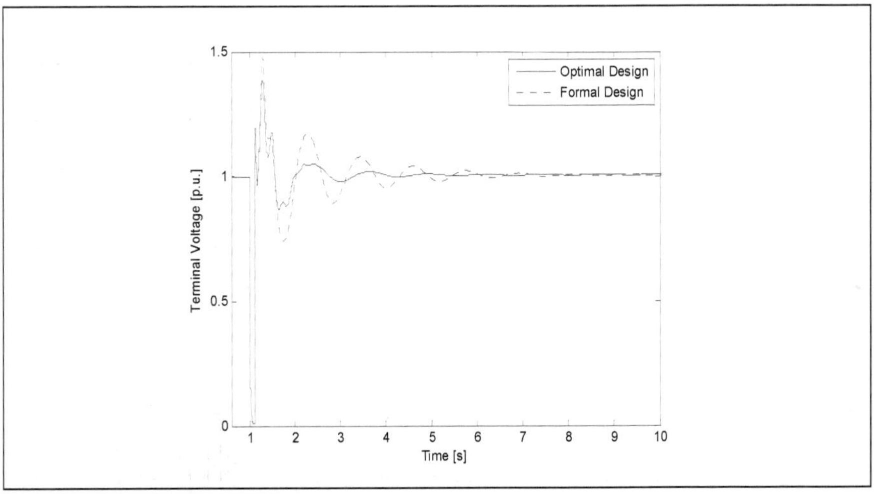

Fig. 7. DFIG Terminal Voltage

Fig. 9 presents the grid-side converter reactive power transient response. It is evident that when the PI controller projected by the GA procedure is used the transient response is less oscillatory presenting a better overall performance. The behavior presented by the grid-side converter reactive power, as well as the DC link voltage (which are variables controlled by the grid-side converter) demonstrates the effectivity of the GA optimization procedure in improving the grid-side and rotor-side converters overall performance, although the optimal gain adjustment GA procedure was applied only to the rotor-side controller.

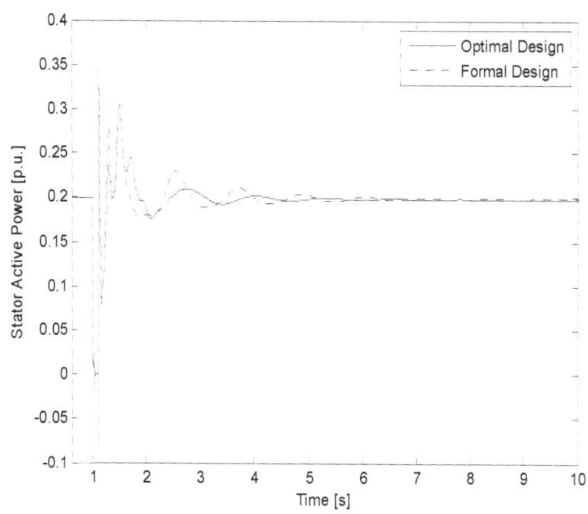

Fig. 8. DFIG Stator Active Power

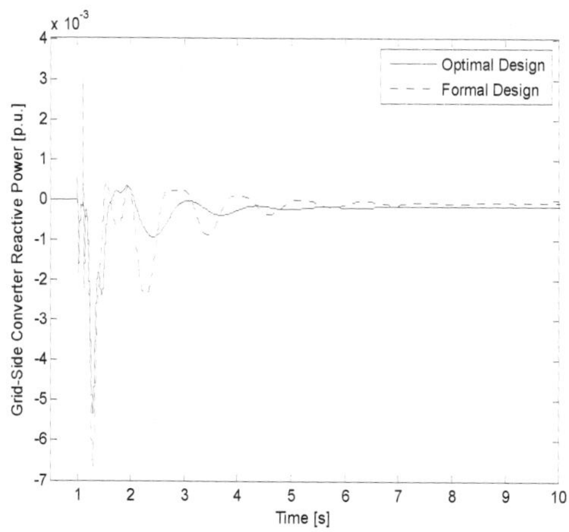

Fig. 9. DFIG Grid-Side Converter Reactive Power

Fig. 10 presents the rotor angle transient response of the equivalent synchronous generator connected at bus 1 of the Açu electrical system. It is evident that the synchronous generator rotor angle time response is more oscillatory when the PI controller designed by the pole placement procedure is used. In this case the risk of small signal instability is more evident. On the other side, when using the PI controller designed by the GA technique, the low frequency oscillation is reduced which improve the small signal stability margin.

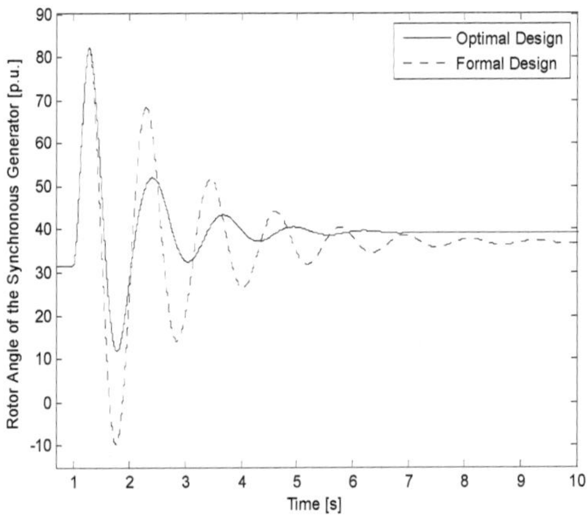

Fig. 10. Rotor Angle of the Synchronous Generator

This way, the proposed GA optimization process to obtain the gains of the DFIG rotor-side converter, besides contributing to a better characteristic of terminal voltage recovery, and ride-though the fault capability, it also improved considerably the system damping characteristic reducing the magnitude of the electromechanical oscillation, without the need of a power system stabilizer (PSS) in the equivalent synchronous generator.

It is worth mentioning that the objective of damping the electromechanical oscillations is not directly included in the GA fitness function. However, the DFIG capacity to introduce damping in the synchronous generator oscillations can be reinforced by an appropriate adjustment of the rotor angle δ, and of the DFIG rotor flux $\overline{\lambda}_r$, which are accomplished by the quadrature rotor current component $\overline{i}_{qr}$, that is used in the proposed vector control adopted here, to control the DFIG rotor speed or the active power.

Case b) 20% load reduction in all buses. A three phase short circuit lasting for 100 ms at bus 10 is applied.

The time responses of the DFIG variables in this case study are very similar to those presented in Case a. These results presented in Figs. 11 to 15 demonstrate the better performance exhibited by the PI controllers designed by the GA approach, demonstrating robustness and effectiveness when the system operation point is changed. It is observed in Fig. 15 that the rotor angle of the synchronous generator presents smaller low frequency oscillations and a larger transient stability margin, when the PI controller projected by the GA approach is used.

In this case, the proposed optimal solution contributes: to enhance the DFIG capacity to withstand voltage sags events; to improve voltage control; to increase transient and small signal stability margins, contributing, this way, to improve the overall system security.

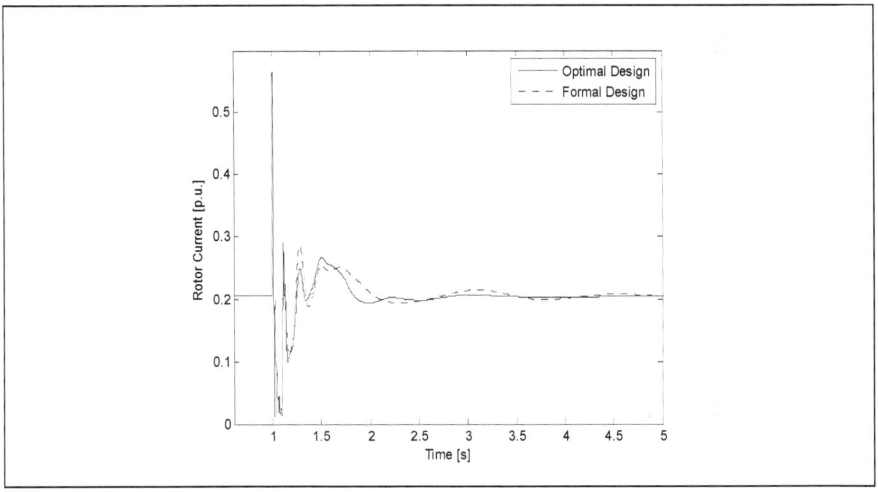

Fig. 11. DFIG Rotor Current

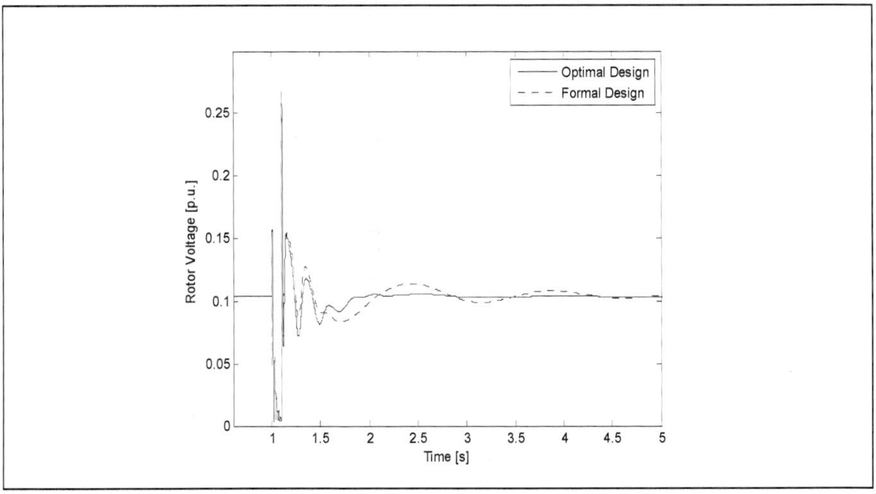

Fig. 12. DFIG Rotor Voltage

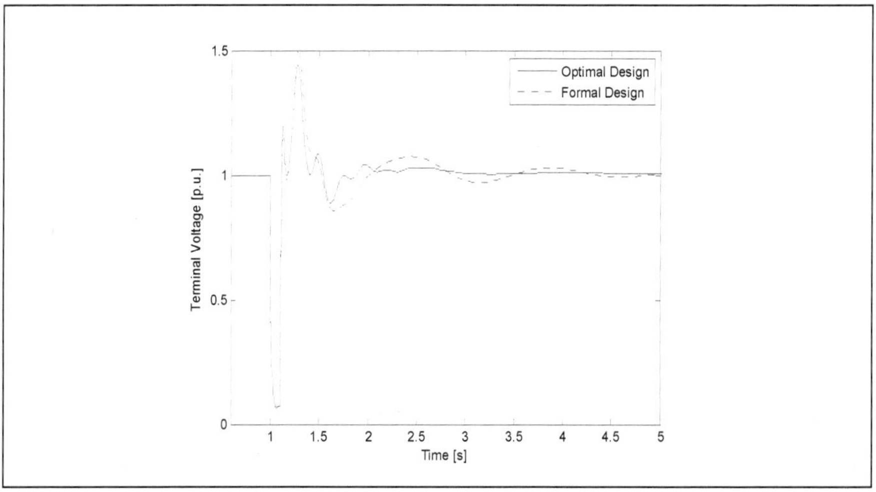

Fig. 13. DFIG Terminal Voltage

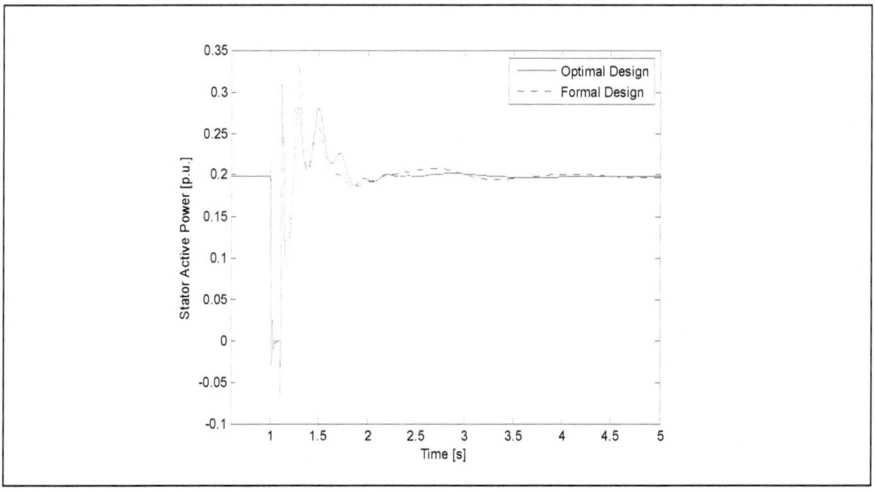

Fig. 14. DFIG Stator Active Power

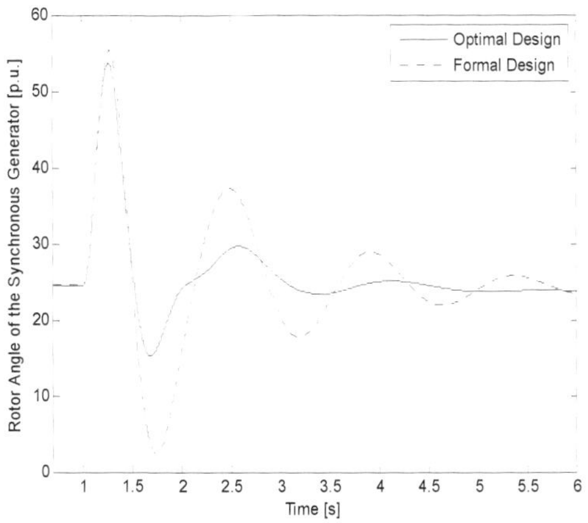

Fig. 15. Rotor Angle of the Synchronous Generator

Case c) 20% load increase in all buses. A three phase short circuit lasting for 100 ms at bus 6 is applied.

The DFIG rotor current and rotor voltage time responses are shown in Figs. 16 and 17. It can be seen that immediately after the short circuit is applied the rotor current limit is exceeded, activating the crow-bar protection scheme. After the short circuit is cleared the crow-bar protection is deactivated, and simultaneously the DFIG generators return to normal operation, with the activation of the rotor-side PI controllers.

Just after the fault is cleared both controllers, namely that designed by pole placement and the other by GA, succeeded in maintaining the wind park connected to the grid avoiding the activation of the crow-bar protection, although the PI controller designed by GA procedure was more effective in reducing the rotor voltage and current oscillations in this time period.

However, in approximately t = 2.5 s it is observed that in the case of using the PI controller projected by the pole placement technique, the converter specified current limit is exceeded again, which activates the crow-bar protection, for a period of 100 ms, which is the transition time imposed by the crow-bar logic.

After this transition time the crow-bar is deactivated and immediately the PI controllers start to function again. However, it may be observed that in the case of using the PI controller designed by the pole placement technique, it was not possible to introduce sufficient damping in the current oscillation and the system became instable.

It is worth noting that the proposed crow-bar protection logic does not allow the activation of the protection scheme for more than two times in a short time period. Besides that, the activation of the crow-bar scheme makes the DFIG machine to operate as a conventional induction machine, lacking the advantage of using the converter control actions.

In Fig. 18 it is presented the DFIG terminal voltage response. When using the PI controller designed by the GA procedure it can be seen an improvement in the terminal voltage control, besides presenting smaller low frequency oscillations, after the fault is cleared. On the other side, when using the PI controller designed by the pole placement technique it was not able to recover the terminal voltage, as can be seen in Fig. 18.

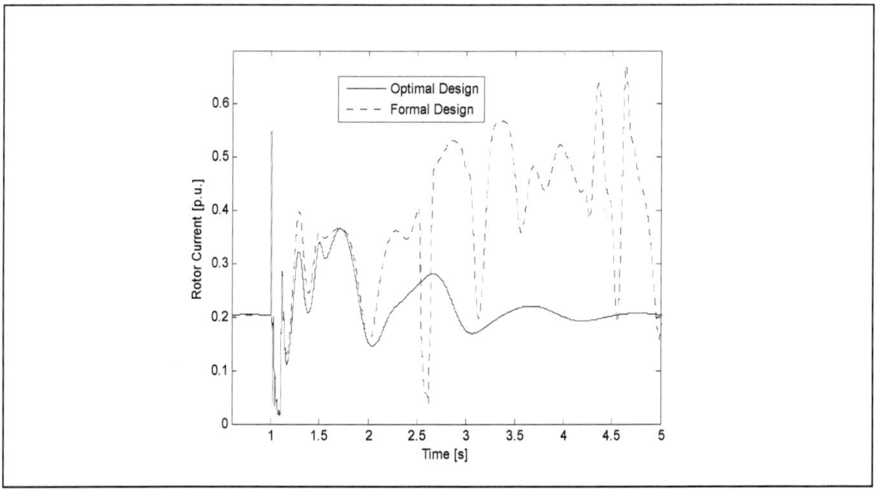

Fig. 16. DFIG Rotor Current

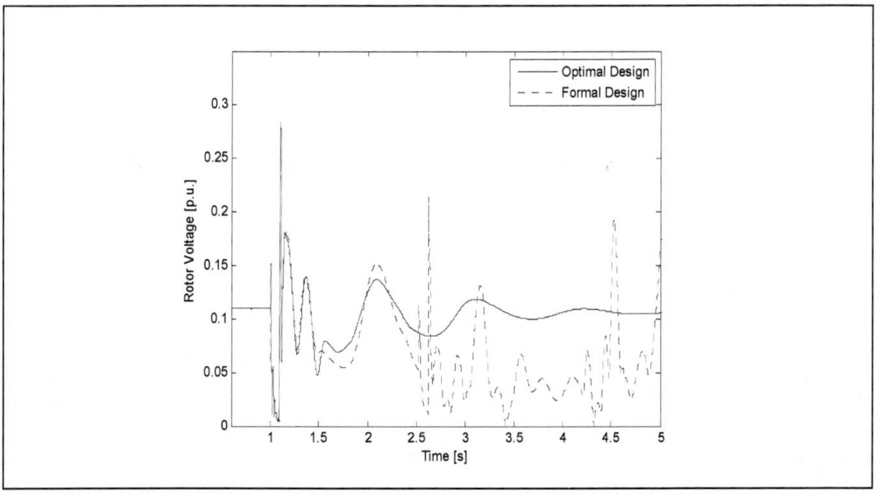

Fig. 17. DFIG Rotor Voltage

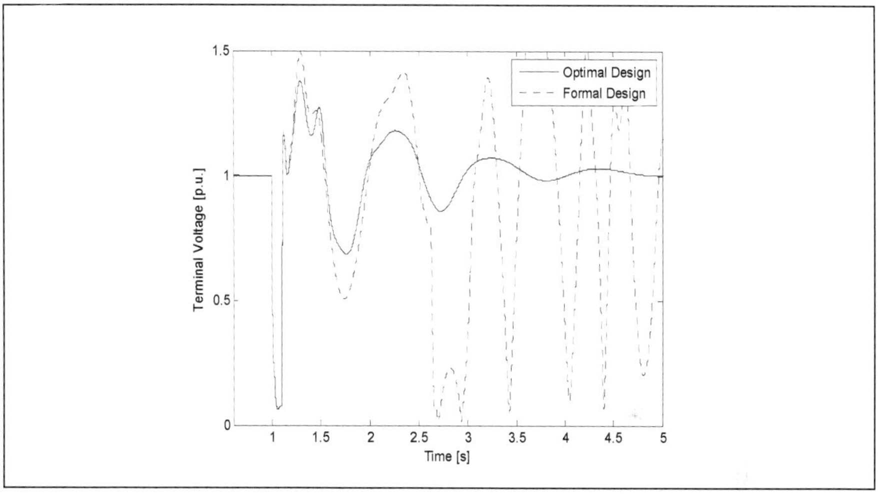

Fig. 18. DFIG Terminal Voltage

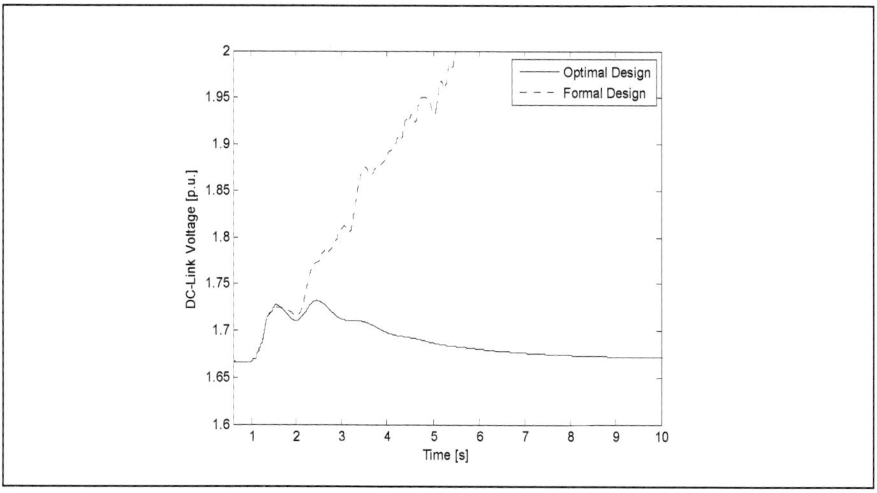

Fig. 19. DC-Link Voltage

The DC link voltage is shown in Fig.19. When using the controller designed by the pole placement technique it is observed a power unbalance between the grid-side converter and the DFIG rotor, which energy is stored continuously in the capacitor, resulting in increasing voltage and the DC link voltage becomes instable. This behavior is not observed when the PI controller designed by the GA procedure is used, maintaining the DC link voltage stable.

In Fig. 20 it is presented the equivalent synchronous generator rotor angle time response. It can be seen that the synchronous generator looses synchronism in the case the PI controller designed by the pole placement technique is used. The same does not happen when using the PI controller designed by the GA procedure, which maintain the synchronous generator synchronism, besides improving the small signals stability margin.

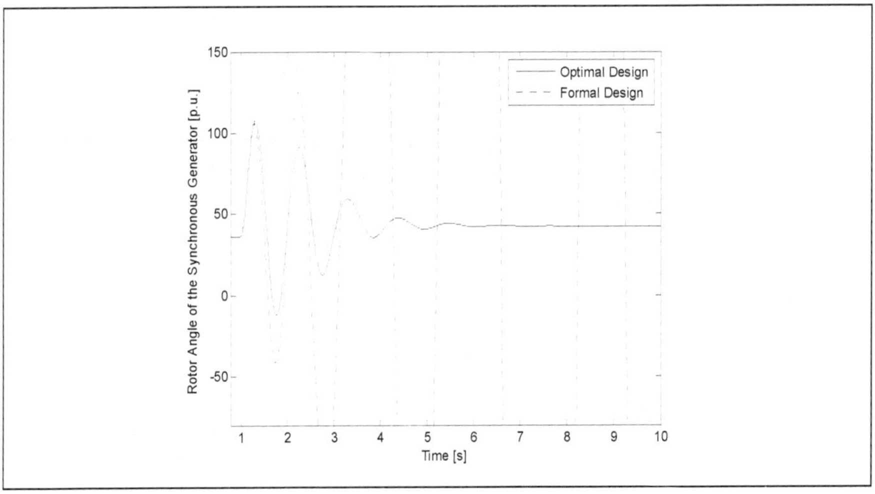

Fig. 20. Rotor Angle of the Synchronous Generator

7. Conclusion

This chapter presented a design procedure based on genetic algorithms combined with the formal pole placement methodology to obtain optimal gains for the PI controllers used in the control loop of the DFIG rotor-side converter in order to increase the ride-through capability and the overall stability margin of the power system. The effectiveness of this proposed approach was assessed for the DFIG-based plants using a real electrical network, in three different operational conditions, and the results obtained confirmed the effectiveness of the proposed control design procedure.

8. References

Jenkins N.; Ekanayake J. B.; Holdsworth L. & Wu X. (2003). Dynamic Modeling of Doubly Fed Induction Generator Wind Turbines. *IEEE Transactions* on *Power Systems*. Vol. 18, (May 2003). pp. 803–809,

Muller S.; Deicke M.; & De Doncker R. W. (2002). Doubly Fed Induction Generator Systems for Wind Turbines. *IEEE Industry Applications Magazine*. Vol. 8, (May/June 2002) pp. 26–33,

Akhmatov V. (2003). Analysis of Dynamic Behaviour of Electric Power Systems with Large Amount of Wind Power. Ph.D. Dissertation. Tech. University of Denmark, Lyngby, Denmark.

Slootweg J. G. (2003). Wind Power Modelling and Impact on Power Systems Dynamics. Ph.D. dissertation, Delft Univ. Technol, Delft, The Netherlands.

Nunes M. V. A.; Lopes J. A. P.; Zurn H. H.; Bezerra U. H. & Almeida R. G. (2004). Influence of the Variable-Speed Wind Generators in Transient Stability Margin of the Conventional Generators Integrated in Electrical Grids. *IEEE Transactions on Energy Conversion*. Vol. 19, No. 4, pp. 692–701,

Mei F. & Pal B. (2007). Modal Analysis of Grid-Connected Doubly Fed Induction Generators. *IEEE Transactions on Energy Conversion*, Vol. 22, No. 3, pp. 728–736.

Miao Z.; Fan L.; Osborne D. S. & Yuvarajan. (2009). Control of DFIG-based Wind Generation to Improve Interarea Oscillation Damping. *IEEE Transactions on Energy Conversion*. Vol. 24, No. 2. pp.415-422.

Qiao W.; Venayagamoorthy G. K. & Harley R.G. (2006). Design of Optimal PI Controllers for Doubly Fed Induction Generators Driven by Wind Turbines Using Particle Swarm Optimization. *Proc. Int. Joint Conf. on Neural Network*, Canada, pp. 1982-1987.

Wu F.; Zhang X. P.; Godfrey K. & Ju P. (2007). Small Signal Stability Analysis and Optimal Control of a Wind Turbine with Doubly Fed Induction Generator. *IET Generation, Transmission & Distribution*. Vol. 1, No. 5. pp. 751-760.

Mishra Y.; Mishra S.; Tripathy M.; Senroy N. & Dong Z.Y. (2009). Improving Stability of a DFIG based Wind Power System with Tuned Damping Controller. *IEEE Transactions on Energy Conversion*.

Kundur P.. (1994). *Power System Stability and Control*. New York. McGraw-Hill.

Mei F. & Pal B. C. (2008). Modelling of Doubly-fed Induction Generator for Power System Stability Study. *Proc. of IEEE PES General Meeting*, pp. 1-8.

Pena R.; Clare J. C. & Asher G. M. (1996). Doubly Fed Induction Generator Using Back-to-Back PWM Converters and its Application to Variable Speed Wind-Energy Generation. *Proc. Inst. Elect. Eng., Elect. Power Applicat.*, Vol. 143, No. 3.

Vieira J. P. A.; Nunes M. V. A.; Bezerra U. H. & Nascimento A. C. (2009). Designing Optimal Controllers for Doubly Fed Induction Generators Using Genetic Algorithm. *IET Generation, Transmission & Distribution*, Vol. 3, No. 5, pp. 472-484.

Xu L. (2008). Coodinated Control of DFIG's Rotor and Grid-Side Converters During Network Unbalance. *IEEE Transactions on Power Elect.* Vol. 23, pp.1041-1049.

Feltes C.; Engelhardt S.; Kretschamann J.; Fortmann J.; Koch F. & Erlich I. (2008). High Voltage Ride-Through of DFIG-based Wind Turbines. *IEEE PES General Meeting. Pittsburgh,* USA.

13

Intelligent Approach to MPPT Control Strategy for Variable-Speed Wind Turbine Generation System

Whei-Min Lin and Chih-Ming Hong
Department of Electrical Engineering,
National Sun Yat-Sen University
Kaohsiung 80424
Taiwan, R.O.C.

1. Introduction

Recently, wind generation systems are attracting great attentions as clean and safe renewable power sources. Wind generation can be operated by constant speed and variable speed operations using power electronic converters. Variable speed generation is attractive because of its characteristic to achieve maximum efficiency at all wind velocities (Pena et al. 2000; Senjyu et al. 2006; Sakamoto et al. 2006; Ramtharan et al. 2007; Fernandez et al. 2008), the improvement in energy production, and the reduction of the flicker problem. In the variable-speed generation system, the wind turbine can be operated at the maximum power operating point for various wind speeds by adjusting the shaft speed. These characteristics are advantages of variable-speed wind energy conversion systems (WECS). In order to achieve the maximum power control, some control schemes have been studied.

A variable speed wind power generation system (WPGS) needs a power electronic converter and inverter, to convert variable-frequency, variable-voltage power into constant-frequency constant-voltage, to regulate the output power of the WPGS. Traditionally a gearbox is used to couple a low speed wind turbine rotor with a high speed generator in a WPGS. Great efforts have been placed on the use of a low speed direct-drive generator to eliminate the gearbox. Many of the generators of research interest and for practical use in wind generation are induction machines with wound-rotor or cage-type rotor (Simoes et al. 1997; Li et al. 2005; Karrari et al. 2005; Wang & Chang 2004). Recently, the interest in PM synchronous generators is increasing. High-performance variable-speed generation including high efficiency and high controllability is expected by using a permanent magnet synchronous (PMSG) for a wind generation system.

Previous research has focused on three types of maximum wind power extraction methods, namely tip speed ratio (TSR) control, power signal feedback (PSF) control and hill-climb searching (HCS) control. TSR control regulates the wind turbine rotor speed to maintain an optimal TSR. PSF control requires the knowledge of the wind turbine's maximum power curve, and tracks this curve through its control mechanisms. Among previously developed wind turbine maximum power point tracking (MPPT) strategies, the TSR direction control method is limited by the difficulty in wind speed and turbine speed measurements

(Thiringer & Linders 1993; Chedid et al. 1999; Tanaka & Toumiya 1997; Morimoto et al. 2005; Koutroulis & Kalaitzakis 2006). Many MPPT strategies were then proposed to eliminate the measurements by making use of the wind turbine maximum power curve, but the knowledge of the turbine's characteristics is required. HCS control has been proposed to continuously search for the peak output power of the wind turbine. In comparison, the HCS MPPT is popular due to its simplicity and independence of system characteristics. In this paper, a Wilcoxon radial basis function network (WRBFN)-based with HCS MPPT strategy is proposed for PMSG wind turbine generator (WTG). The proposed control structure, WRBFN with modified particle swarm optimization (MPSO) algorithm is forces the system to reach its equilibriums quickly where the turbine inertia effect is minimized. HCS can be fast and effective in spite of the variations in wind speeds and the presence of turbine inertia.

Intelligent control approaches such as neural network and fuzzy system do not require mathematical models and have the ability to approximate nonlinear systems. Therefore, there were many researchers using intelligent control approaches to represent complex plants and construct advanced controllers. Moreover, the locally tuned and overlapped receptive field is a well-known structure that has been studied in regions of cerebral cortex, visual cortex, and so on (Jang & Sun 1997). Based on the biological receptive fields, the RBFN that employs local receptive fields to perform function mappings was proposed in (Jang & Sun 1993). Furthermore, the RBFN has a similar feature to the fuzzy system. First, the output value is calculated using the weighted sum method. Second, the number of nodes in the hidden layer of the RBFN is the same as the number of if-then rules in the fuzzy system. Finally, the receptive field functions of the RBFN are similar to the membership functions of the premise part in the fuzzy system. Therefore, the RBFN is very useful to be applied to control the dynamic systems (Seshagiri & Khail 2000).

2. Analysis of wind generation system

2.1 Wind turbine characteristics and modeling

In order to capture the maximal wind energy, it is necessary to install the power electronic devices between the WTG and the grid where the frequency is constant. The input of a wind turbine is the wind and the output is the mechanical power turning the generator rotor (Li et al. 2005; Karrari et al. 2005; Wang & Chang 2004). For a variable speed wind turbine, the output mechanical power available from a wind turbine could be expressed as

$$P_m = \frac{1}{2}\rho A C_p(\lambda,\beta)V_\omega^3 \tag{1}$$

where ρ and A are air density and the area swept by blades, respectively. V_ω is the wind velocity (m/s), and C_p is called the power coefficient, and is given as a nonlinear function of the tip speed ratio (TSR) λ defined by

$$\lambda = \frac{\omega_r r}{V_\omega} \tag{2}$$

where r is wind turbine blade tip radius, and ω_r is the turbine speed. C_p is the function of the λ and the blade pitch angle β, general defined as follows:

$$C_p = 0.73\left(\frac{151}{\lambda_i} - 0.58\beta - 0.002\beta^{2.14} - 13.2\right)e^{\frac{-18.4}{\lambda_i}}$$
$$\lambda_i = \frac{1}{\frac{1}{\lambda - 0.02\beta} - \frac{0.003}{\beta^3 + 1}} \tag{3}$$

By using (3), the typical C_P versus λ curve is shown in Fig. 1. In a wind turbine, there is an optimum value of TSR λ_{opt} that leads to maximum power coefficient $C_{p\max}$. When TSR in (2) is adjusted to its optimum value $\lambda_{opt} = 6.9$ with the power coefficient reaching $C_{p\max} = 0.4412$, the control objective of the maximum power extraction is achieved.

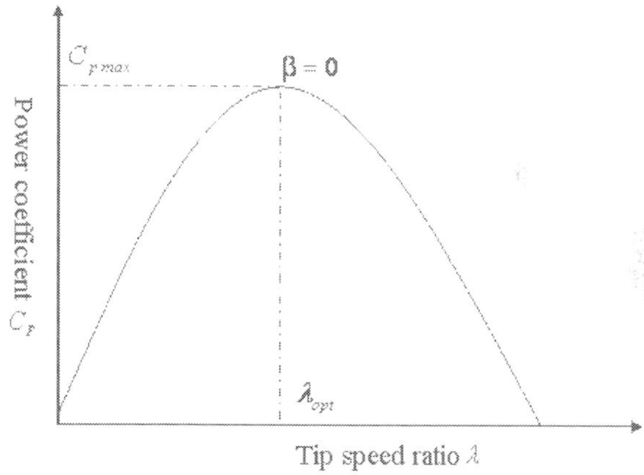

Fig. 1. Typical C_P versus λ curve

2.2 PMSG

The wind generator is a three-phase PMSG, where the mechanical torque (T_m) and electrical torque (T_e) can be expressed as

$$T_m = \frac{P_m}{\omega_r} \tag{4}$$

$$T_e = \frac{P_e}{\omega_e} = \frac{2}{P}\frac{P_e}{\omega_r} \tag{5}$$

In general, the mechanical dynamic equation of a PMSG is given by

$$J\frac{d\omega_r}{dt} = T_m - (P/2)T_e \tag{6}$$

where ω_e and P are electrical angular frequency, and the number of poles. J is the inertia moment of WTG.

2.3 Wind turbine emulation

The emulation of the wind turbine is implemented by a dc motor drive with torque control. In the prototype, a 1.5kW, 1980rpm dc motor was used. A computer program reads the wind input file obtained with various test conditions, and calculates the wind turbine torque by taking into account wind velocity, turbine rotational speed, and the wind turbine power coefficient curve. The control algorithms for turbine emulation are implemented in a control board dSPACE DS1102. This board is a commercial system designed for rapid prototyping of real control algorithms; it is based on the Texas Instruments TMS320C32 floating-point DSP. The DS1102 board is hosted by a personal computer.

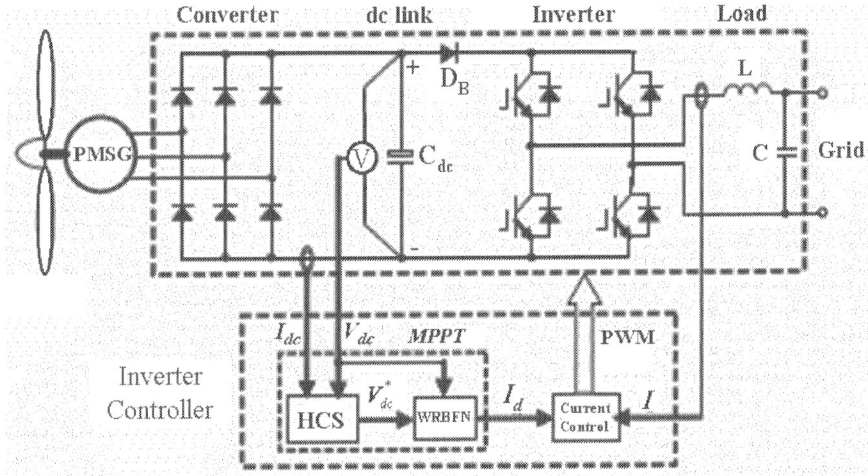

Fig. 2. PMSG WT generation system

3. HCS control method

3.1 System configuration

Fig. 2 presents the block diagram of the WT generation system in our research, where a PMSG is driven by a WT to feed the extracted power from wind resources to the grid through a single-phase inverter. A variable speed WPGS needs a power electronic converter and inverter, to convert variable-frequency, variable-voltage AC power from a generator to DC and then into constant-frequency constant-voltage power. In the dc-link of the inverter, a blocking diode D_B is used to improve the power delivering capability as well as to guarantee that the dc-link voltage transfers to the output voltage. An inverter controller is designed to deal with two aspects, the MPPT control for power maximization and the the current control for output PWM to inverter. The dc-link voltage and current, V_{dc} and I_{dc} are sampled to provide the power ($P_{dc} = V_{dc} \cdot I_{dc}$) input to the controller, and V_{dc} reference signal V_{dc}^* is updated in real time using an HCS method so as to lead the system to its optimal operation point. On the other hand, a WRBFN controller is designed to force V_{dc} to follow V_{dc}^* by adjusting the load current reference for the inverter current controller.

3.2 Optimal DC-link voltage (V_{dc}) search

From the C_P - λ characteristics, the turbine mechanical power P_m can be shown as a function of V_{dc}, and an optimal V_{dc} exists for the maximum P_m output, provided that a PMSG is employed in the generation system. Fig. 3 shows a group of P_m - V_{dc} curves and the corresponding maximum power curve formed with various optimal operating points, where wind speeds $u1 < u2 < u3 < u4$. In order to extract maximum power from wind, the optimal V_{dc} is searched in real time using the HCS method. With HCS, if the previous increment of V_{dc}^* is followed by an increase of P_m, then the search of V_{dc}^* continues in the same direction; otherwise, the search reverses its direction. An example can be seen in Fig. 3, where the wind change from $u3 \rightarrow u4 \rightarrow u2$ with the search of V_{dc} from $A \rightarrow B \rightarrow C \rightarrow D \rightarrow E$. The increment of P_m is approximated by that of P_{dc}, and the search is executed at dynamic equilibrium operation points where P_{dc} is approximately equal to P_m and the effect of turbine inertia J can be minimized. In dynamic states, V_{dc}^* will be held and the WRBFN will adjust the load current in real time to drive the system to its equilibrium point as soon as possible. Fig. 3 illustrates the searching process $(ABCDE)$ for the maximum power points when the wind speed varies.

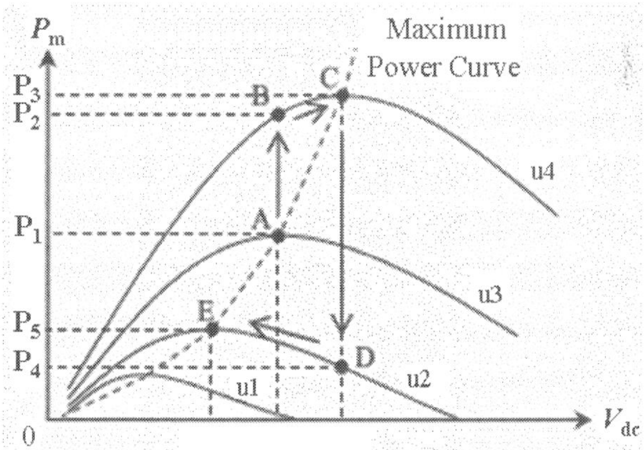

Fig. 3. Principle of HCS control method.

4. The proposed intelligent MPPT control algorithm

4.1 Wilcoxon radial basis function network

The linear Wilcoxon regressor is quite robust against outliers (Hogg et al. 2005), which motivates the design of wilcoxon neural networks. A three-layer neural network shown in Fig. 4 is adopted to implement the proposed WRBFN. The WRBFN with MPSO controller is used, and the control law I_d is generated from the WRBFN The WRBFN input is $x_1^{(1)}$ and $x_2^{(1)}$ of the first layer, where $x_1^{(1)} = V_{dc} - V_{dc}^* = e$ and $x_2^{(1)} = \dot{e}$ in this study. In the proposed WRBFN, the units in the input, hidden, and output layers are two, nine and one, respectively. The signal propagation and the basic function in each layer can be found (Lin & George Lee 1996).

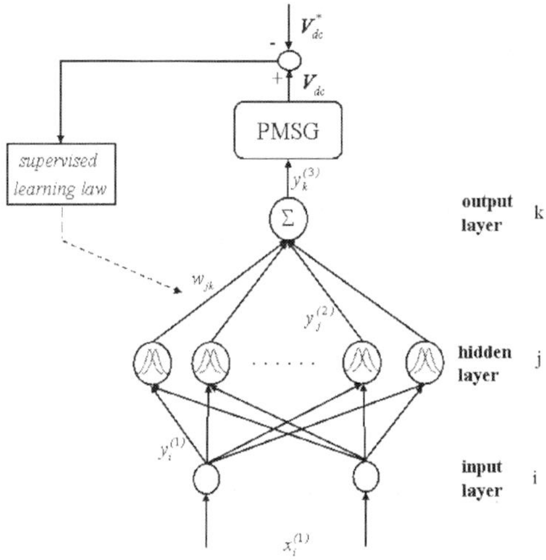

Fig. 4. Architecture of the WRBFN

Layer 1 : input layer

The nodes at this layer are used to directly transmit the numerical inputs to the next layer. That is, for the *i*th node of layer 1, the net input and output are defined by

$$\begin{aligned} net_i^{(1)} &= x_i^{(1)}(N) \\ y_i^{(1)}(N) &= f_i^{(1)}\left(net_i^{(1)}(N)\right) = net_i^{(1)}(N) \qquad i = 1,2 \end{aligned} \tag{7}$$

Layer 2 : hidden layer

At this layer, every node performs a Gaussian basis function. The Gaussian basis function, a particular example of radial basic functions, is used here as a membership function. Then

$$\begin{aligned} net_j^{(2)}(N) &= -\sum_{i=1}^{n}(x_i^{(1)} - c_{ij})^2 / v_{ij} \\ y_j^{(2)}(N) &= f_j^{(2)}\left(net_j^{(2)}(N)\right) = \exp\left(net_j^{(2)}(N)\right) \qquad j = 1,...,9 \end{aligned} \tag{8}$$

where $c_j = \left[c_{1j} \quad c_{2j} \quad \cdots \quad c_{ij}\right]^T$ and v_{ij} denote respectively, the mean and the standard deviation (STD) of the Gaussian basis function.

Layer 3 : output layer

The single node k in this layer is denoted by Σ, which computes the overall output as the summation of all incoming signals by

$$\begin{aligned} net_k^{(3)} &= \sum_{j-1}^{m} w_{jk} y_j^{(2)}(N), \\ y_k^{(3)}(N) &= f_k^{(3)}(net_k^{(3)}(N)) = net_k^{(3)}(N) = I_d \end{aligned} \tag{9}$$

where the connection weight w_{jk} is the connective weight between the hidden node, and the output layer k.

4.2 The network training and learning process

Once the WRBFN has been initialized, a supervised learning law is used to train this system. The basis of this algorithm is gradient descent. The derivation is the same as that of the back-propagation algorithm. It is employed to adjust the parameters w_{jk}, c_{ij}, v_{ij} of the WRBFN by using the training patterns. By recursive application of the chain rule, the error term for each layer is first calculated. The adaptation of weights to the corresponding layer is then given. The purpose of supervised learning is to minimize the error function E expressed as

$$E = \frac{1}{2}\left(V_{dc} - V_{dc}^{*}\right)^{2} = \frac{1}{2}e_{L}^{2} \tag{10}$$

where V_{dc}^{*} and V_{dc} represent the dc-link voltage reference and actual dc-link voltage feedback of the generator.

<u>Layer 3 : update weight</u> w_{jk}

At this layer, the error term to be propagated is given by

$$\delta_k = -\frac{\partial E}{\partial net_k^{(3)}} = \left[-\frac{\partial E}{\partial y_k^{(3)}}\frac{\partial y_k^{(3)}}{\partial net_k^{(4)}}\right] \tag{11}$$

Then the weight w_{jk} is adjusted by the amount

$$\Delta w_{jk} = -\frac{\partial E}{\partial w_{jk}} = \left[-\frac{\partial E}{\partial y_k^{(3)}}\frac{\partial y_k^{(3)}}{\partial net_k^{(3)}}\right]\left(\frac{\partial net_k^{(3)}}{\partial w_{jk}}\right) = \delta_k y_j^{(2)} \tag{12}$$

We have

$$w_{jk}(N+1) = w_{jk}(N) + \eta_w \Delta w_{jk}(N) \tag{13}$$

where η_w is the learning rate for adjusting the parameter w_{jk}.

<u>Layer 2 : update</u> c_{ij} and v_{ij}

The multiplication operation is done in this layer. The adaptive rule for c_{ij} is

$$\Delta c_{ij} = -\frac{\partial E}{\partial c_{ij}} = \left[-\frac{\partial E}{\partial net_k^{(3)}}\frac{\partial net_k^{(3)}}{\partial y_j^{(2)}}\frac{\partial y_j^{(2)}}{\partial c_{ij}}\right] = \delta_k w_{jk} y_j^{(2)} \frac{2\left(x_i^{(1)} - c_{ij}\right)}{v_{ij}} \tag{14}$$

and the adaptive rule for v_{ij} is

$$\Delta v_{ij} = -\frac{\partial E}{\partial v_{ij}} = \left[-\frac{\partial E}{\partial net_k^{(3)}}\frac{\partial net_k^{(3)}}{\partial y_j^{(2)}}\frac{\partial y_j^{(2)}}{\partial v_{ij}}\right] = \delta_k w_{jk} \frac{2\left(x_i^{(1)} - c_{ij}\right)^{2}}{\left(v_{ij}\right)^{2}} \tag{15}$$

We have

$$\begin{aligned} c_{ij}(k+1) &= c_{ij}(k) + \eta_m \Delta c_{ij} \\ v_{ij}(k+1) &= v_{ij}(k) + \eta_\sigma \Delta v_{ij} \end{aligned} \tag{16}$$

where η_m and η_σ are the learning rates for adjusting the parameters c_{ij} and v_{ij}, respectively. The exact calculation of the jacobian of the system, which is contained in $\partial E / \partial y_k^{(3)}$, cannot be determined due to the uncertainty of the PMSG dynamic. To overcome this problem and to increase the on-line learning ability of the connective weights, the delta adaptation law (Lin & George Lee 1996) is implemented as follows to solve the difficulty

$$\delta_k \equiv e_L + \dot{e}_L$$

The learning rates η_w, η_m, η_σ are adjusted by MPSO as stated below.

5. WRBFN learning rates adjustment using MPSO

PSO is a population-based optimization method first proposed by Kennedy and Eberhart. PSO technique finds the optimal solution using a population of particles. Each particle represents a candidate solution to the problem. PSO is basically developed through simulation of bird flocking in two-dimensional space (Esmin et al. 2005).

Step 1: Define basic conditions

In the first step of MPSO, one should determine the parameters that need to be optimized and give them minimum and maximum ranges. The number of groups, population size of each group, and initial radius of each *gbest* are also assumed in this step.

Step 2: Initialize random swarm location and velocity

To begin, initial location $R_i^d(N)$ and velocities $v_i^d(N)$ of all particles are generated randomly in whole search space. Moreover, the population size is set to $P = 15$ and the dimension of the particle is set to $d = 3$ in this study. The generation particles are $R_i^d = [R_i^1, R_i^2, R_i^3]$, where R_i^1, R_i^2, R_i^3 are the RBFN learning rates, respectively. The initial *pbest* of a particle is set by its current position. Then, *gbest* of a group is selected among the *pbests* in the group.
The random generation of $R_i^d(N)$ initial value ranged as

$$R_i^d \sim U[\eta_{\min}^d, \eta_{\max}^d]$$

where $\eta_{\min}, \eta_{\max}$ are the lower and upper bound of the learning rates.

Step 3:Update velocity

In the classical PSO algorithm, the velocity of a particle was determined according to the relative location from *pbest* and *gbest.*
During each iteration, every particle in the swarm is updated using (17) and (18). Two pseudorandom sequences $r_1 \sim U(0,1)$ and $r_2 \sim U(0,1)$ are used to produce the stochastic nature of the algorithm. For dimensions *d*, let R_i^d, $Pbest_i^d$,and v_i^d be the current position, current personal best position. The velocity update step is

$$v_i^d(N+1) = wv_i^d(N) + c_1 \cdot r_1 \cdot (Pbest_i^d - R_i^d(N)) + c_2 \cdot r_2 \cdot (Gbest^d - R_i^d(N)) \quad (17)$$

Step 4: Update position

The new velocity is then added to the current position of the particle to obtain its next position

$$R_i^d(N+1) = R_i^d(N) + v_i^d(N+1) \quad i = 1,\ldots,P \quad (18)$$

Step 5: Update *pbests*

If the current position of a particle is located within the analysis space and does not intrude territory of other *gbests*, the objective function of the particle is evaluated. If the current fitness is better than the old *pbest* value, *pbest* is replaced by the current position. The fitness value of each particle is calculated by

$$FIT = \frac{1}{0.1 + abs(V_{dc} - V_{dc}^*)} \quad (19)$$

Step 6: Update *gbests*

In the conventional PSO, *gbest* is replaced by the best *pbest* among the particles. However, when such a strategy is applied to multimodal function optimization, some *gbests* of different groups can be overlapped. To maintain fast convergence rate of PSO, *gbest* of the group could be selected among the $Pbest_i^d = [Pbest_1^d, Pbest_2^d, \ldots Pbest_p^d]$ having high fitness value.

Step 7: Repeat and check convergence

Steps 3-6 are repeated until all particles are gathered around the *gbest* of each group, or a maximum iteration number is encountered. The final $Gbest_i^d$ is the optimal learning rate $(\eta_w, \eta_m, \eta_\sigma)$ of RBFN.
The inertia weight w in (17) is used to control the convergence behavior of the PSO. Small w results in rapid convergence usually on a suboptimal position, while a large value may cause divergence. In this paper, the inertia weight w is set according to the equation that

$$w = w_{\max} - \frac{w_{\max} - w_{\min}}{iter_{\max}} \cdot iter \quad (20)$$

where $iter_{\max}$ is maximum number of iterations, and *iter* is the current iteration number.

6. Experimental results

The WRBFN with MPSO performance is compared with two baseline controllers: the fuzzy (Chen et al. 2000) and the proportional-integral (PI) controller. The WRBFN with MPSO, fuzzy and PI methods were tested through experimental. The obtained performance with the different controllers are shown in Fig. 5 to Fig. 7, and summarized in Table 1. Various cases were conducted, the wind profile is simulation with a $5msec$ sampling time the wind profile is assumed a volatile sinusoidal wave. The conventional PI type controller is widely used in the industry due to its simple control structure, ease of design and inexpensive cost. The average power of PI is compared with that of WRBFN with MPSO algorithm and Fuzzy-Based algorithm.

The WTG system used for the experimental has the following parameters:

1. Wind turbine parameters:

$$P_m = 750W\ ;\ 3.75A\ ;\ 3000r/\min;\ \rho = 1.25kg/m^3;\ r = 0.5m;\ J = 1.32\times10^{-3}Nm\sec^2$$

2. Generator parameters:

$$R = 1.47\Omega\,;L_d = L_q = 5.33mH\ ;\ L_{md} = 4.8mH\ ;\ I_{fd} = 46.75A\,;\ K_t = 0.6732Nm/A$$

6.1 PI algorithm for V_{dc} control

The WRBFN with MPSO algorithm replaced by PI algorithm is shown in Fig. 2. Fig. 5 illustrates the experimental result for PI control. The average power is 205W for the same period. It can be found that TSR is always round 6.9 and C_p is 0.4412. The verification of maximum power tracking control is shown in Fig. 5(a). The dc-link voltage tracking response is shown in Fig. 5(b). Fig. 5(c) and 5(d) shows power coefficient C_p and TSR λ.

6.2 Fuzzy-based algorithm for V_{dc} control

The WRBFN with MPSO algorithm replace by Fuzzy-Based algorithm as shown in Fig. 2. A fuzzy logic control (FLC) algorithm is characterized by "IF-THEN" rules. The algorithm is suitable for wind turbine control with complex nonlinear models and parameters variation. The input variables of Fuzzy-Based MPPT are dc-link power tracking error and the difference of dc-link power tracking error.

Fig. 6 shows that 217W (an increase of 5.36% compared with that of PI control) is obtained by the Fuzzy-Based algorithm during the 50 sec. It can be found that λ and C_p are close to the optimal values of 6.9 and 0.4412, respectively. The wind speed profiles of turbine power P_m and dc-link power P_{dc} are also shown in Fig. 6(a). The dc-link voltage tracking response is shown in Fig. 6(b). Fig. 6(c) and 6(d) are shows that power coefficient C_p and TSR λ.

6.3 WRBFN with MPSO algorithm for V_{dc} Control

WRBFN with MPSO algorithm control is considered and the experimental result is shown in Fig. 7. The verification of maximum power tracking control is shown in Fig. 7(a), where the wind speed profiles of turbine power P_m and dc-link power P_{dc} are also shown. The dc-link voltage tracking response is shown in Fig. 7(b). Fig. 7(c) shows power coefficient C_p which is close to its maximum value during the whole wind speed profile, same for λ of Fig. 7(d). The efficiency of the maximum power extraction can be clearly observed as the power coefficient is fixed at the optimum value $C_p = 0.4412$ and $\lambda = 6.9$. The average power is 224W. Compared with that from the PI control method, it increases by 9.27%.

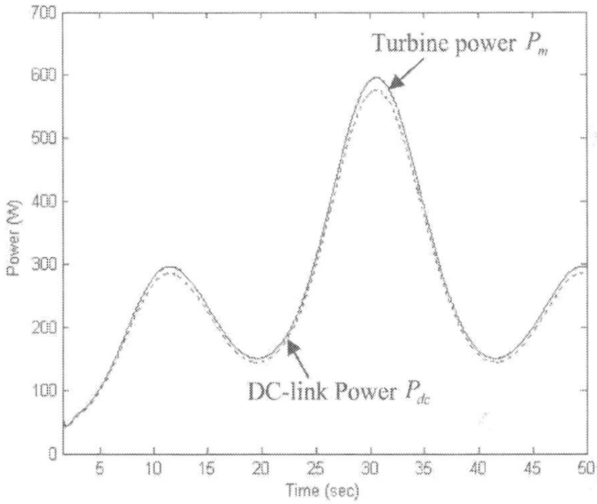
Turbine power P_m
DC-link Power P_{dc}
Power (W)
Time (sec)

(a)

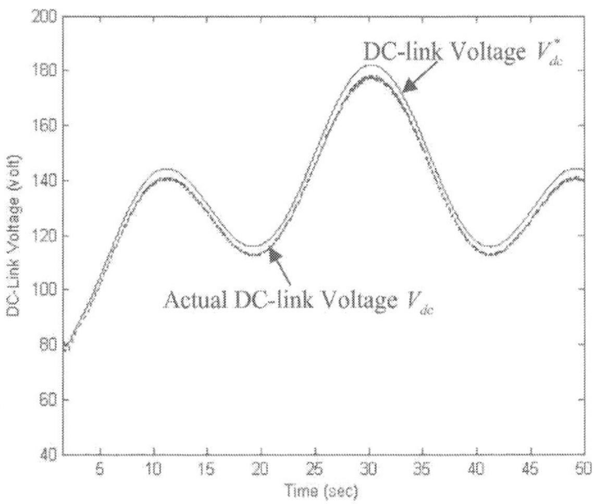
DC-link Voltage V_{dc}^*
Actual DC-link Voltage V_{dc}
DC-Link Voltage (volt)
Time (sec)

(b)

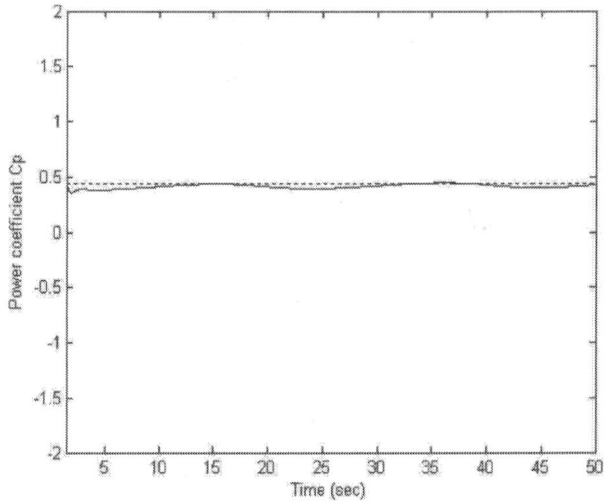

(c)

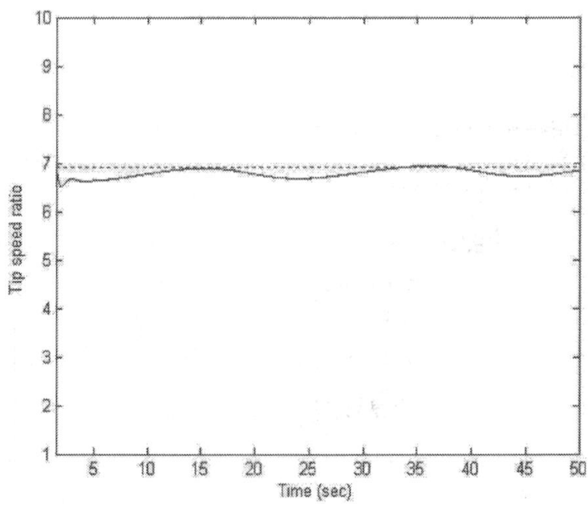

(d)

Fig. 5. Experimental results of the wind speed profile : (a)The maximum power tracking control signal. (b)The dc-link voltage tracking response. (c)Power coefficient C_p. (d)Tip-speed ratio λ .

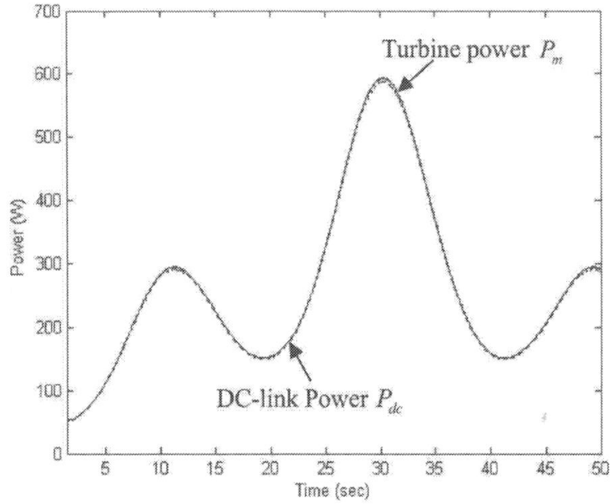
Turbine power P_m
DC-link Power P_{dc}
Power (W)
Time (sec)

(a)

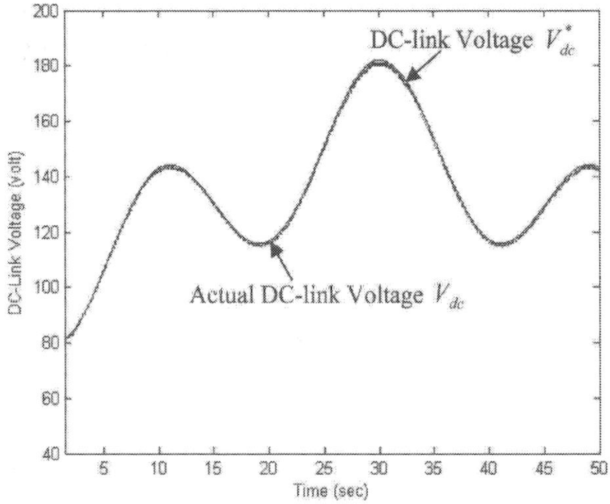
DC-link Voltage V_{dc}^*
Actual DC-link Voltage V_{dc}
DC-Link Voltage (volt)
Time (sec)

(b)

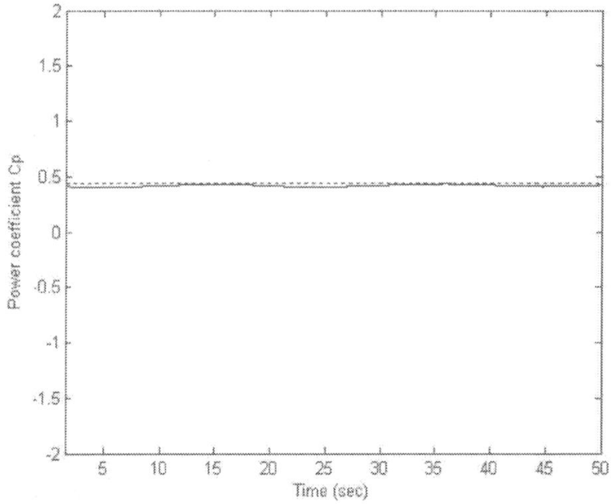

(c)

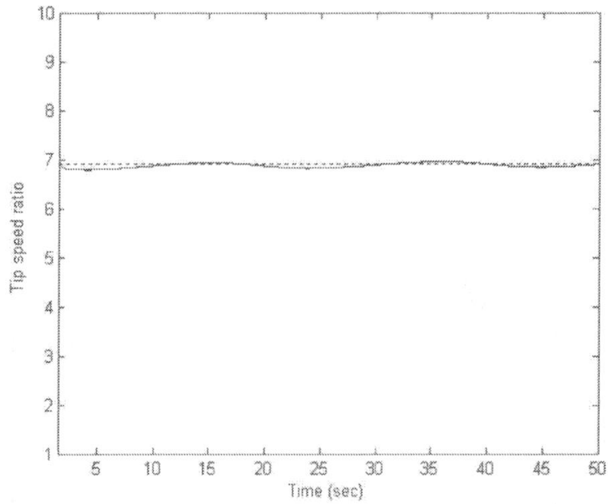

(d)

Fig. 6. Experimental results of the wind speed profile : (a)The maximum power tracking control signal. (b)The dc-link voltage tracking response. (c)Power coefficient C_p. (d)Tip-speed ratio λ .

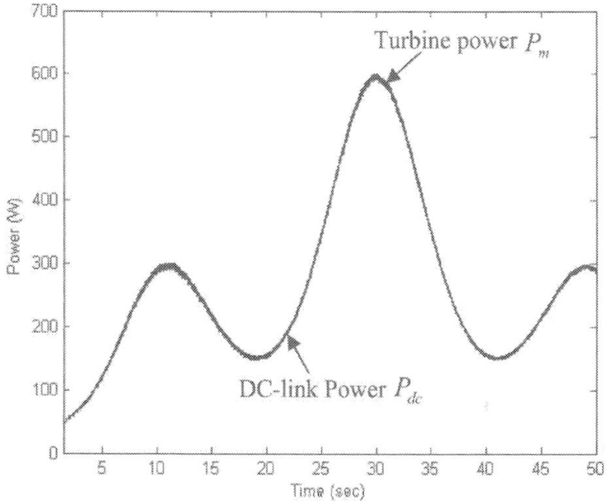
Turbine power P_m
DC-link Power P_{dc}
Power (W)
Time (sec)
700
600
500
400
300
200
100
0
5
10
15
20
25
30
35
40
45
50

(a)

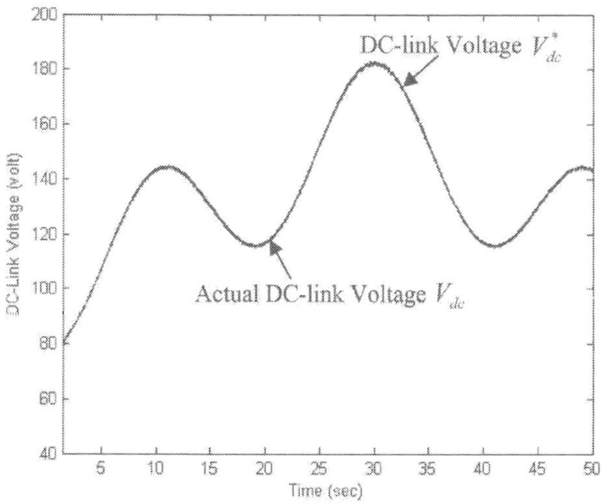
DC-link Voltage V_{dc}^*
Actual DC-link Voltage V_{dc}
DC-Link Voltage (volt)
Time (sec)
200
180
160
140
120
100
80
60
40
5
10
15
20
25
30
35
40
45
50

(b)

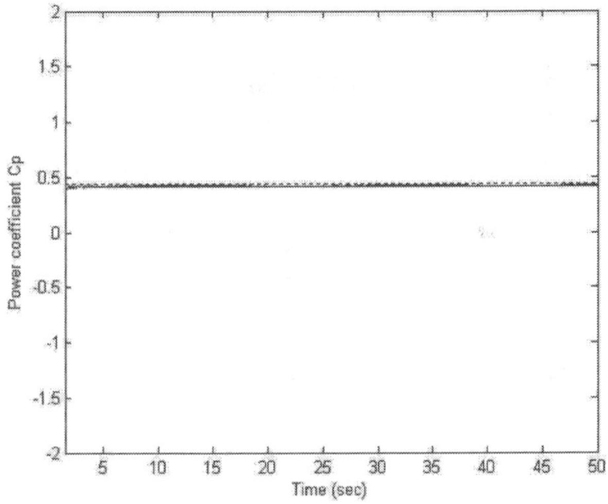

(c)

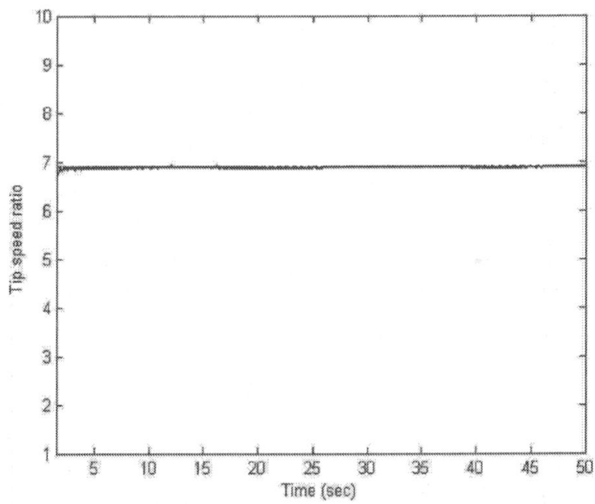

(d)

Fig. 7. Experimental results of the wind speed profile : (a)The maximum power tracking control signal. (b)The dc-link voltage tracking response. (c)Power coefficient C_p. (d)Tip-speed ratio λ .

Control Characteristic / Controller Type	Average Power (P_{dc}) (W)	Increasing power percentage (%)	Max. error of power coefficient (C_p) (%)	Max. error of DC-link power (%)	Max. error of DC-link voltage (%)
WRBFN with MPSO algorithm method	224	9.27	1.79	0.32	0.074
Fuzzy-Based algorithm method	216	5.36	9.95	1.4	0.9
PI method	205	----	18.4	2.58	2.25

Table 1. Performance for various control methods

From the performance comparison for various methods above experimental results, we can see that MPPT is important for either high or low wind speeds, as shown in Table 1. Table 1 shows the average power, maximum error of power coefficient, maximum error of dc-link power and percentage of power increase from each control method. On the other hand, the maximum error of the power coefficient is around 23% in [9], and the maximum power deviation is about 7% in [14]. The proposed method in comparison with other methods [9,14] has better performance.

7. Conclusion

This paper focuses on the development of maximum wind power extraction algorithms for inverter-based variable speed WPGS. The HCS method is proposed in this paper for maximum power searching with various turbine inertia. Without a need for measurements of wind speed and turbine rotor speed, HCS is simple to implement. When exciting the system with a real wind profile, the system is able to track maximum power using generated power as input. The proposed system has been implemented, with a commercial PMSG and a dc drive to emulate the wind turbine behavior. The process is running in a dSPACE board that includes a TMS320C32 floating-point DSP. Experimental results show the appropriate behavior of the system.
Three MPPT control algorithms are proposed in this paper, without any wind speed sensor. It is found that the PI method can operate near the optimal *Cp*. However, the PI-type controller may not provide perfect control performance if the controlled plant is highly nonlinear or the desired trajectory is varied with higher frequency. The proposed output maximization control of WRBFN can maintain the system stability and reach the desired performance even with parameter uncertainties.

8. References

Pena, R. S., Cardenas, R. J. , Asher G. M. & Clare J. C. (2000). Vector controlled indction machines for stand-alone wind energy applications, IEEE Industry Application Annu. Meeting, Vol.3, pp.1409-1415.

Senjyu, T., Sakamoto, R., Urasaki, N., Funabashi, T. & Sekine, H. (2006). Output Power Leveling of Wind Farm Using Pitch Angle Control with Fuzzy Neural Network. IEEE Power Engineering Society General Meeting, page(s):8pp.

Sakamoto, R., Senjyu, T., Sakamoto, R., Kaneko, T., Urasaki, N., Takagi, T., Sugimoto, S. & Sekine, H. (2006). Output Power Leveling of Wind Turbine Generator by Pitch Angle Control Using H_∞ Control, The IEEE PSCE Conf., pp. 2044-2049.

Ramtharan, G., Ekanayake, J. B. & Jenkins, N. (2007). Frequency support from doubly fed induction generator wind turbines, IET Renewable Power Generation, Vol. 1, No. 1, pp. 3-9.

Fernandez, L. M., Garcia, C. A. & Jurado, F. (2008). Comparative study on the performance of control systems for doubly fed induction generator (DFIG) wind turbines operating with power regulation, Energy, Vol. 33, Issue. 9, pp. 1438-1452.

Simoes, M. G., Bose, B. K. & Spiegel, R. J. (1997). Fuzzy logic-based intelligent control of a variable speed cage machine wind generation system, IEEE Transactions on Power Electronics, Vol. 12, No. 1, pp. 87-95.

Li, H., Shi, K. L. & McLaren, P. G. (2005). Neural-Network-Based Sensorless Maximum Wind Energy Capture with Compensated Power Coefficient, IEEE Transactions Industry Application, Vol. 41, No. 6, pp. 548-1556.

Karrari, M., Rosehart, W. & Malik, O. P. (2005). Comprehensive Control Strategy for a variable Speed Cage Machine Wind Generation Unit, IEEE Transactions on Energy Conversion, Vol. 20, No. 2, pp. 415-423.

Wang, Q. & Chang, L. (2004). An intelligent maximum power extraction algorithm for inverter-based variable speed wind turbine systems, IEEE Transactions Power Electronics, Vol. 19, No. 5, pp. 1242-1249.

Thiringer, T. & Linders J. (1993). Control by variable rotor speed of a fixed-pitch wind turbine operating in a wide speed range, IEEE Transactions Energy Conversion, Vol. EC-8, pp. 520-526.

Chedid, R., Mrad, F. & Basma, M. (1999). Intelligent control of a class of wind energy conversion systems, IEEE Transactions on Energy Conversion, Vol. EC-14, pp. 1597-1604.

Tanaka, T. & Toumiya, T. (1997). Output control by Hill-Climbing method for a small wind power generating system, Renewable Energy, Vol. 12, Issue. 4, pp. 387-400.

Morimoto, S., Nakamura, T., Sanada, M. & Takeda, Y. (2005). Sensorless Output Maximization Control for Variable-Speed Wind Generation System Using IPMSG, IEEE Transactions on Industry Applications, Vol.41, No.1, pp. 60-67.

Koutroulis, E. & Kalaitzakis, K. (2006). Design of a Maximum Power Tracking System for Wind-Energy-Conversion Applications, IEEE Transactions on Industrial Electronics, Vol. 53, No. 2, pp. 486-494.

Jang, J. S. R. & Sun, C. T. (1997). Neuro-Fuzzy and Soft Computing: A computational approach to learning and machine intelligenc. Upper Saddle River, NJ: Prentice-Hall.

Jang, J. S. R. & Sun, C. T. (1993). Function equivalence between radial basis function networks and fuzzy inference systems, IEEE Transactions on Neural Networks, Vol. 4, No. 1, pp. 156-159.

Seshagiri, S. & Khail, H. K. (2000). Output feedback control of nonlinear systems using RBF Function equivalence between radial basis function neural networks, IEEE Transactions Neural Networks, Vol. 11, No. 1, pp. 69-79.

Hogg, R. V., McKean, J. W. & Craig, A. T. (2005). Introduction to Mathematical Statistics. Sixth Edition, New Jersey: Prentice-Hall.

Lin, C. T. & George Lee, C. S. (1996). Neural Fuzzy Systems. Upper Saddle River, NJ:Prentice-Hall, Inc..

Esmin, A. A., Torres, G. L. & Souza, C. Z. (2005). A Hybrid Particle Swarm Optimization Applied to Loss Power Minimization, IEEE Transaction on Power Systems, Vol. 20, No. 2, pp. 859-866.

Chen, Z., Gomez, S. A. & McCormick, M. (2000). A fuzzy logic controlled power electronic system for variable speed wind energy conversion systems, Eighth International Conference on Power Electronics and Variable Speed Drives, pp. 114-119.

14

A Simple Prediction Model for PCC Voltage Variation Due to Active Power Fluctuation for a Grid Connected Wind Turbine

Sang-Jin Kim and Se-Jin Seong
Junma Engineering Co., Ltd. Dept. of Electrical Engineering,
Chungnam National University
Korea

1. Introduction

A great deal of attention is focused on the source of new renewable energy under the present circumstance that serious problems occur all over the industry and the economy due to the high cost of energy according to the high price of oil. For these reasons, wind energy has been installed all over the world and is the fastest growing energy source in the world. In cases of micro grid areas and islands that are isolated from the main power supply and are provided by electric power through diesel engines, the COE(Cost of Energy) is much higher due to shipping expenses of materials for the generation of electric power. It is, therefore, important to have a big interest in a hybrid generation system that makes use of the rich source of wind by connecting wind turbines with the established power system in parallel(Song et al., 2003; Slootweg & Kling, 2003).
A wind turbine has output characteristics in which the output power follows the cube of the blowing wind speed which is an uncontrollable element apart from other generating sources. This means that the output power of a wind turbine cannot be controlled. When it connects with a weak grid system of a diesel generation system, it can cause potential unstable elements that can bring about various power problems. Thus, it is necessary that the possible problems according to wind speed be predicted and analyzed when a wind turbine is connected with the existing generating source(Jeong et al., 2007). The purpose of this study is, therefore, to examine voltage variation among several power quality problems that appear when wind speed increases; that is, the output increases at the connecting point between wind turbine and diesel generating system.
It takes actual islands as a model to inquire into what elements other than output variation of wind turbine affect voltage variation at the PCC (Point of Common Coupling) at which a wind turbine is connected with the existing power system. This then tries to predict the amount of voltage variation in advance based on rated capacity of wind turbine to be established and short circuit capacity at the connecting point. It is, ultimately, expected that this study would contribute to predicting power quality including voltage variation when the small-sized wind power generating system is installed and operated in domestic islands, cutting down the cost of power production in islands by ensuring security and reliability of carrying out projects, and improving the penetration ratio of new renewable energy source.

2. System modeling

In order to carry out this study, a domestic island was chosen to operate on by building a small-sized electric power network, which was sampled. The electric power network and the connected wind turbine were modeled briefly and the amount of voltage variation through the simulation based on PSCAD/EMTDC was verified. The Sapsido system as a simplified model formed with power impedance and load impedance was suggested and the amount of voltage variation calculated. Also, the elements which affect voltage variation at the Point of Common Coupling (PCC) of wind turbine through actual measurement by comparing three results was confirmed(KEPC, 2005; IEEE 1547, 2003; E-On Netz GMBH, 2006).

2.1 Equivalent iron loss resistance

This paper made a model by simplifying the Sapsido system located in Boryeong, Chungcheongnamdo to develop the model that a small-sized wind turbine is connected to a system. Sapsido is a small island with an area of 3.78km^2 located in Boryeong, Chungcheongnamdo. The formation of the overall power supply line in Sapsido, which is an isolated system, is shown in Fig. 1 by obtaining data at the spot. The total 4 diesel generators are being operated and supply power in Sapsido has a 900[kW] facility capacity. Electric power produced at a power plant is supplied to every kind of load through transformers and lines within the island, and a 10[kW] wind turbine is installed and connected to a low voltage line on the seashore.

In order to make the Sapsido system as a model, the power supply line at Fig. 1 is simplified as that of Fig. 2 in which a diesel generator, transformer and loads are indicated in the diagram of Fig. 1. Also, Fig. 2 is formed by regarding the point that a 10[kW] wind turbine is connected to a power supply system as PCC (Point of Common Coupling). Load is divided into Load 1, which is the total one at the front of the wind turbine, and Load 2, which is one at the village on the seashore for the simple modeling.

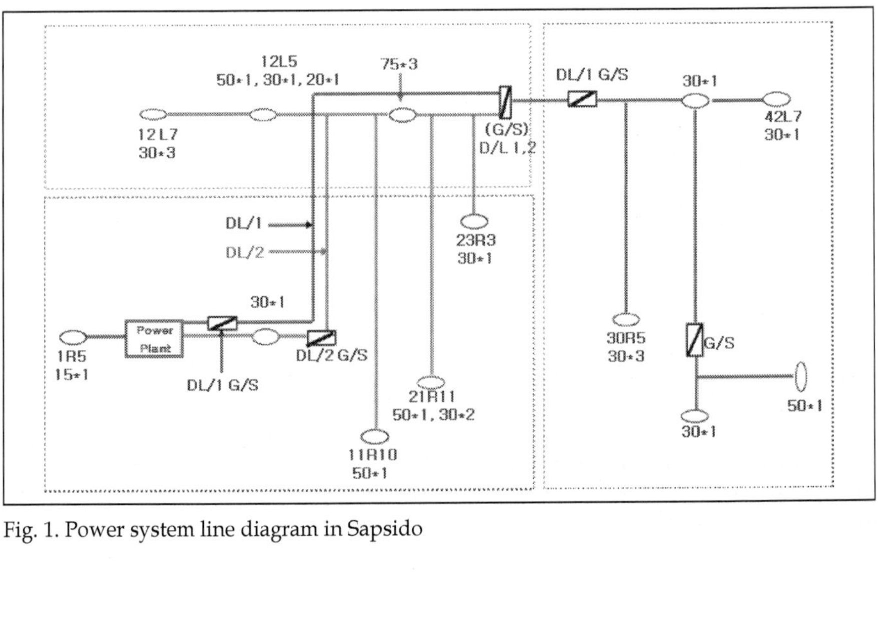

Fig. 1. Power system line diagram in Sapsido

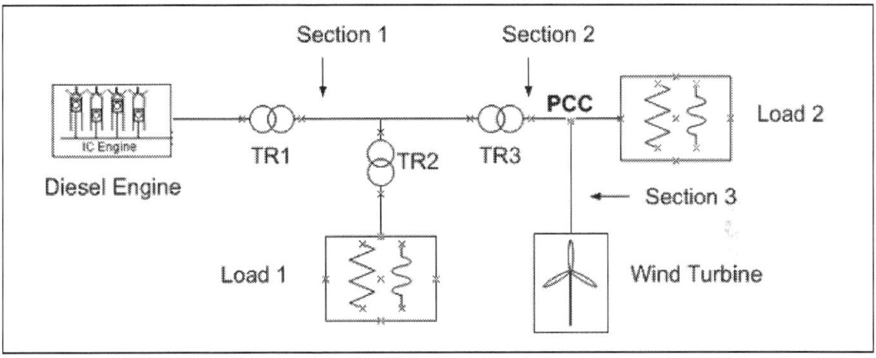

Fig. 2. Outline of power supply system in Sapsido

A more detailed explanation of the components is shown in Table 1. All contents of Table 1, except for the generator, are applied in modeling and only one 300[kVA] generator is modeled for convenience of modeling.

Impedance of the line attains its scale value at the point indicated as the section in Fig. 2, just as indicated in Table 2 according to kinds and length of cable. It applies an almost similar value even though there is little difference according to each manufacturing company.

Classification	Explanation	Reference
Diesel Generator	Sapsido independent generator (4 generators total)	150[kVA] 2set 300[kVA] 2set
TR1	380/6600[V] booster transformer	750[kVA]
TR2	6600/380[V] drop transformer	120[kVA]
TR3	6600/380[V] drop transformer	15[kVA]
Load 1	Load of whole island except for 2 houses in the village on the seashore	Maximum load 120[kW], 50[kVAR]
Load 2	Single phase load of 2houses in the village on the seashore	10[kW], 92~93% power factor
Wind Turbine	BERGEY, BWT-10240	Single phase 10[kW]

Table 1. Power system parameter in Sapsido

Section	Sort	Distance [m]	Scale line Resistance[Ω]
Section 1	ACSR-AW/OC 8[mm^2]	93	0.0484
Section 2	OW 22[mm^2]	93, 139	0.0849, 0.1273
Section 3	OW 22[mm^2]	89	0.0849

Table 2. Data of power supply line in Sapsido

2.2 PSCAD modeling of Sapsido power system full model

This paper developed a simulation model of the power supply system in Sapsido by using PSCAD/EMTDC, and emulated output variation of wind turbine with the model(Rajendiran et al.). A simulation model was made up based on the power supply system in Sapsido in Fig. 2, and the exciter and the governor of diesel generator at a power plant were assumed as ideal ones(Bialasiewics et al., 1998).

A grid connected inverter that adopts the up-to-date Digital Signal Processor (DSP) takes the role of connecting the output of variable DC voltage of decentralized power from not only a wind turbine but also a solar and fuel cell with the power system containing constant voltage and frequency. If it is presumed that the loss of an inverter operated by PWM switching method is very small and that the AC power that is almost the same as the input DC power will be generated, which will finally reveal output current variation.

Fig. 3 shows the inner structure of the installed grid connected wind turbine system and the simplified equivalent modeling of the grid connected inverter(Kim & Song, 2005). Fig. 4 is designed to generate a current with the same phase as the system voltage by adopting a Phase Locked Loop (PLL) on a grid connected inverter because the power production synchronized with the phase of a power system voltage should be essential in case of grid connection of an AC generating power.

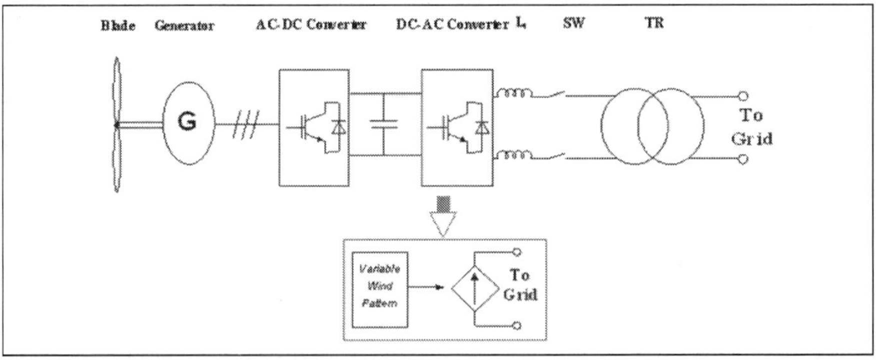

Fig. 3. Wind turbine & Grid connected inverter model

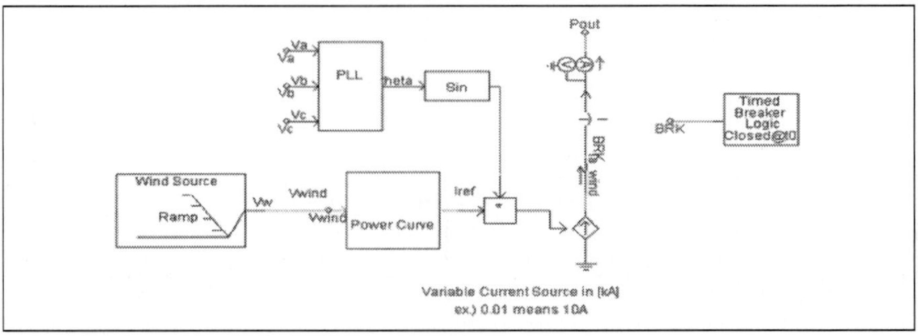

Fig. 4. Simulation model of wind turbine

Fig. 5 is a graph which indicates the variation of power output according to Excel-S wind speed. This paper takes the characteristics of Excel-S which is a 10[kW] model of two models(www.windturbinewarehouse.com).

The simulation model in Fig. 4 takes on a kind of sinusoidal wave current without the modeling of separate equipment for the simplified modeling based on the model of an Excel-S wind turbine. Wind speed input is fixed to generate 10[kW] at 16[m/s] according to the characteristics in Fig. 5.

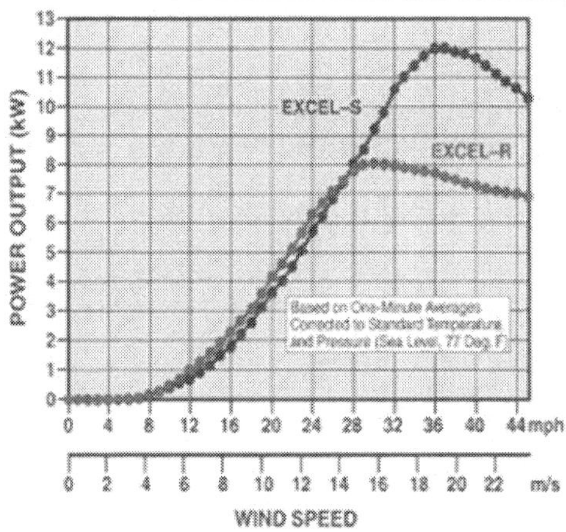

(a)

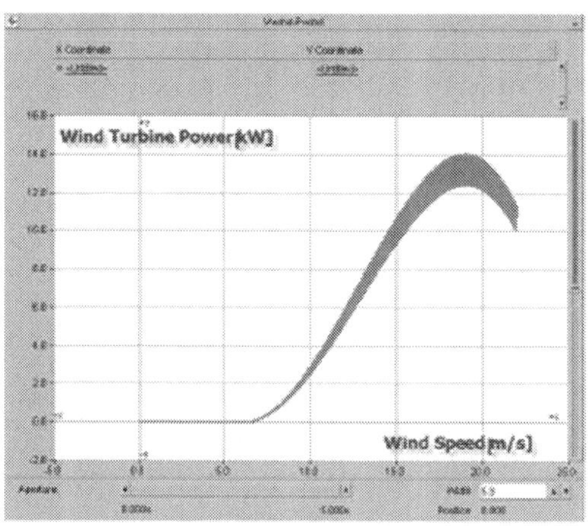

(b)

Fig. 5. The compare of characteristic curves

(a) The characteristic curve of wind turbine output

(b) The output characteristics of simulation model of wind turbine

Fig. 5 (b) shows the output characteristics of the simulation model of the developed wind turbine per wind speed. We can see that it accords with the characteristics in Fig. 5 (a).
Fig. 6 shows the overall simulation model in Sapsido. All parameters follow to the actual ones, and a grid connected inverter is operating as the control system of output current in proportion to the variation of output power(Sim et al., 2007).

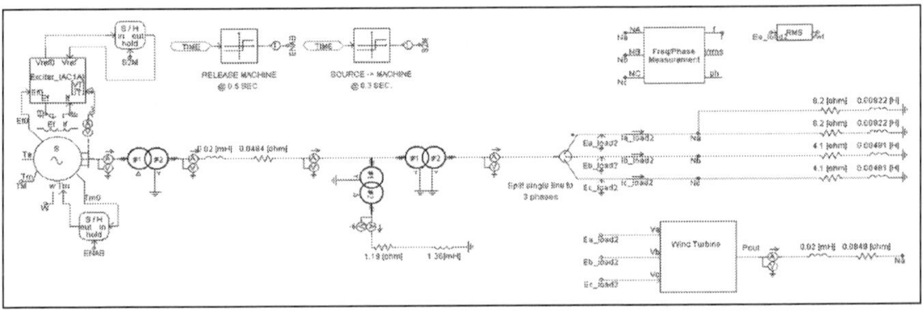

Fig. 6. The simulation formation of power supply system in Sapsido

In simulation, it is observed by emulating the variation of wind speed to investigate the influences in case of grid connected operation from the perspective of voltage variation. If the output of a wind turbine increases under the condition of a constant load, the power plant can reduce as much as the output amount. A graph in Fig. 7 can be gained when the RMS voltage at PCC is observed through the simulation model according to the output of the wind turbine. The output of the wind turbine shows the voltage variation from 6[kW] to about 6-7[V].

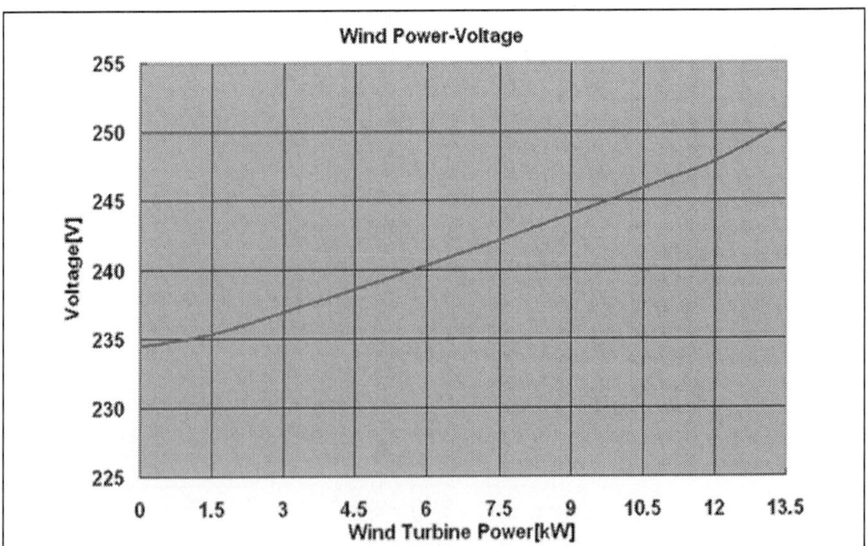

Fig. 7. The simulation of PCC voltage variation in case of wind speed change

2.3 Simplified prediction model for voltage variation study

The overall impedance of the power system in Sapsido can be indicated as shown in Fig. 8 to analyze not the simulation as the above executed, but the actual voltage variation.

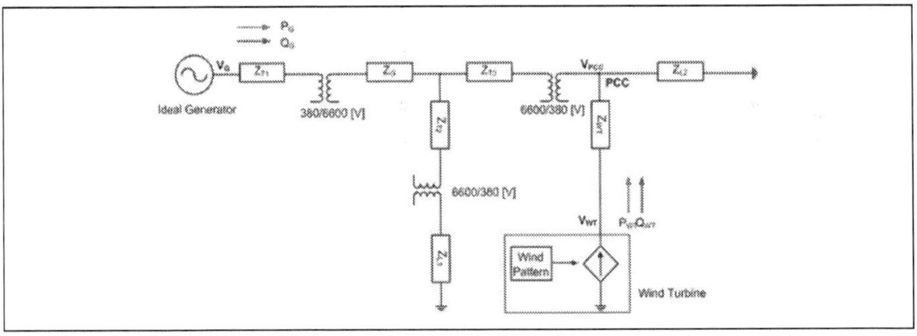

Fig. 8. The overall impedance of Sapsido system

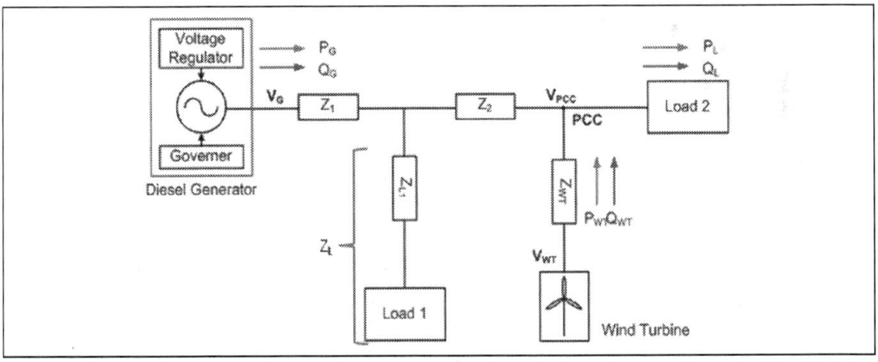

Fig. 9. The simplified impedance formation of power system in Sapsido

System impedance in Sapsido as shown in Fig. 8 can be more simplified as shown in Fig. 9, and indicated as shown in Fig. 10 if the Load1 impedance is much higher than the line impedances Z_1 and Z_2.

The total equivalent line impedance Z_G can be expressed as (1).

$$Z_G \approx Z_1 + Z_2 \tag{1}$$

In Fig. 10, the equivalent load impedance Z_{LL} includes the line impedance Z_L to the load and the load impedance Z_{Load} as shown in (2). When load current I_L is fixed, output current I_G is changed as shown in (3) if output current I_{WT} of the wind turbine is changed. It can be shown as (4) and (5) if it is indicated in terms of power.

$$Z_L + Z_{Load} = Z_{LL} \tag{2}$$

$$I_G + I_{WT} = I_L \tag{3}$$

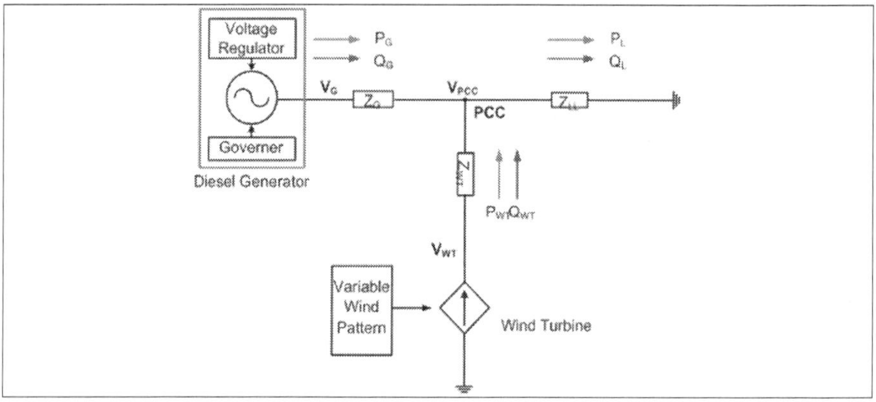

Fig. 10. The simplified model connected with wind turbine

$$P_G + P_{WT} = P_L \tag{4}$$

$$Q_G + Q_{WT} = Q_L \tag{5}$$

The voltage V_{PCC} at the grid connected point can be shown as (6). PCC voltage is composed of the first term by the existing internal voltage source V_G and the second one by current source IWT of the wind turbine from the principle of superposition shown in Fig. 11.

$$V_{PCC} = \left(\frac{Z_{LL}}{Z_{LL} + Z_G} \right) (V_G + Z_G I_{WT}) \tag{6}$$

Fig. 11(a) shows the case that wind turbine is not working (I_{WT}=0), and PCC voltage is determined as (7) by impedance distribution between power side and load one. The PCC voltage without wind power generation(I_{WT}=0) can be considered as nominal voltage V_{PCC0} in (7). The nominal voltage is determined by source voltage V_G and the impedance distribution. From (6) and (7) the PCC voltage variation, $\triangle V_{PCC}$ due to the wind power is proportional to I_{WT} and the equivalent effective impedance Z_{effect} as (8).

These results can also be explained using the principle of superposition in Fig. 11. Z_{effect} corresponds to parallel impedance of Z_G and Z_{LL} as shown in (9). There is one thing to watch out here that line impedance Z_{WT} connected to current source I_{WT} in series does not affect PCC voltage variation; instead, it affects port voltage V_{WT} of the wind turbine according to the change of current I_{WT} as shown in (10).

$$V_{PCC0} = \left(\frac{Z_{LL}}{Z_{LL} + Z_G} \right) V_G \tag{7}$$

$$\Delta V_{pCC} = Z_{effect} \cdot I_{WT} \tag{8}$$

$$Z_{effect} = \frac{Z_{LL} \cdot Z_G}{Z_{LL} + Z_G} \tag{9}$$

$$V_{WT} = V_{pCC} + Z_{WT} \cdot I_{WT} \tag{10}$$

Here,
V_G : output voltage of internal generator
I_{WT} : output current of wind turbine
Z_{Load} : equivalent load impedance
Z_G : equivalent circuit impedance between internal generator and PCC
Z_L : equivalent line impedance between PCC and load
Z_{WT} : line impedance between PCC and wind turbine

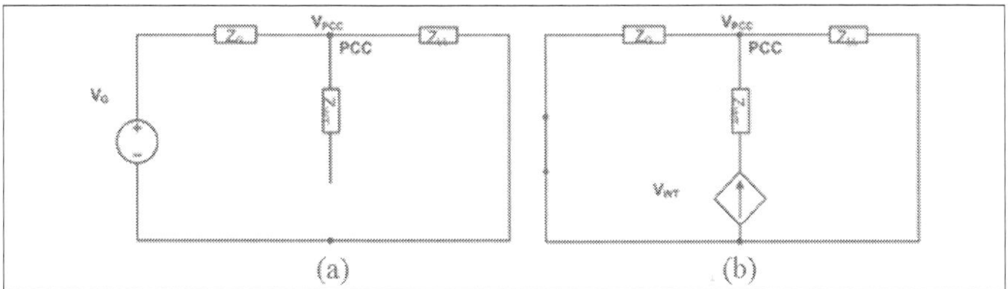

a) The case that wind turbine is not working (I_{WT}=0)
b) The analysis of voltage variation port by I_{WT} in case of operating wind turbine

Fig. 11 The analysis of equivalent circuit of the connected model with wind turbine

In order to verify the analysis, the analysis is done by converting impedance in Fig. 8 into the actual impedance parameter such as in Table 3.

Classification	Explanation	Reference
Ideal Generator	Sapsido power plant (3 phase)	380[V]
Z_{T1}	750[kVA] transformer impedance (380[V] scale value)	$0.013 \angle 71.57$
Z_{T2}	120[kVA] transformer impedance (380[V] scale value)	$0.076 \angle 71.57$
Z_{T3}	15[kVA] transformer impedance (380[V] scale value)	$0.611 \angle 71.57$
Z_G	Line impedance of high voltage side (380[V] scale value)	$0.0002 \angle 7.13$
Z_{L1}	Load impedance of whole island except for 2 houses in the village on the seashore	$1.35 \angle 25.73$
Z_{L2}	Load impedance of 2 houses in the village on the seashore	$4.45 \angle 23$
Z_{WT}	Line impedance of wind turbine side	$0.085 \angle 5.05$

Table 3. The actual impedance parameter

The simplified impedance shown in Fig. 10 is indicated in Table 4 below.

Classification	Explanation	Value
Ideal Generator	Sapsido power plant (3 phase)	380[V]
Z_G	Scale line impedance	$0.625 \angle 72.37$
Z_{LL}	Scale load impedance	$4.19 \angle 22.24$
Z_{WT}	Line impedance of wind turbine side	$0.085 \angle 5.05$

Table 4. Scale impedance parameter

$$\therefore Z_{effect} = \frac{Z_{LL} \cdot Z_G}{Z_{LL} + Z_G} = 0.57 \angle 66.4° \tag{11}$$

$$\therefore \Delta V_{pCC} = 0.57 \angle 66.4° \times I_{WT} \tag{12}$$

The final equation on voltage variation occurring at the actual PCC with data in Table 4 is shown in (11) and (12).

2.4 The actual measurement

Data is measured by installing a power measuring instrument at the PCC step within the actual system in Sapsido to enhance reliability in this study. Fig. 12 shows the location where measurement is carried out. Valid/invalid power graph of the wind turbine according to time based on this measured data is shown in Fig. 13.

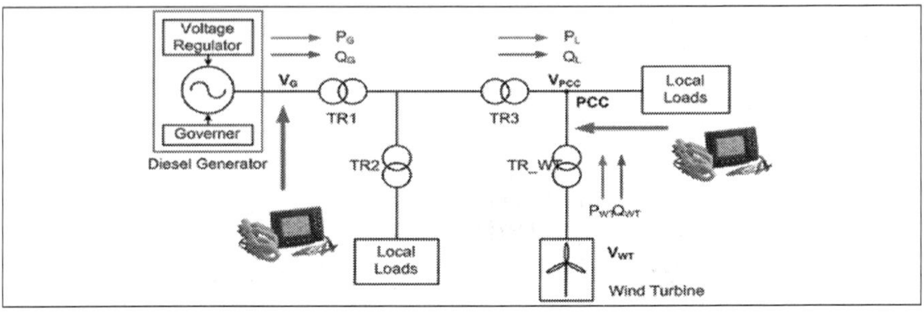

Fig. 12. Measuring location of power in Sapsido

Fig. 14 shows voltage variation amount for the output. The measured dots show the tendency of increase proportional to the power output and are represented as a solid line. Dots inside the circle, near zero output power, show the voltage variation in normal operating conditions at the PCC even without wind power generation. An approximate 7[V] voltage variation is shown in case of producing about 5.5[kW].

Table 5 shows the comparative results of the voltage variation using two simulation models and one experimental measurement.

At first, the voltage variation of 6.4V results from the PSCAD simulation model which has full modeling of the wind turbine and complex line impedance parameters. The second result of 6.5V comes from the proposed simplified prediction model of PCC voltage using (8). The last result of 6.9V PCC voltage deviation is a measured value as shown in Fig. 14 during the field test.

From this comparison the result of the proposed prediction model highly matches the simulation and measurement results.

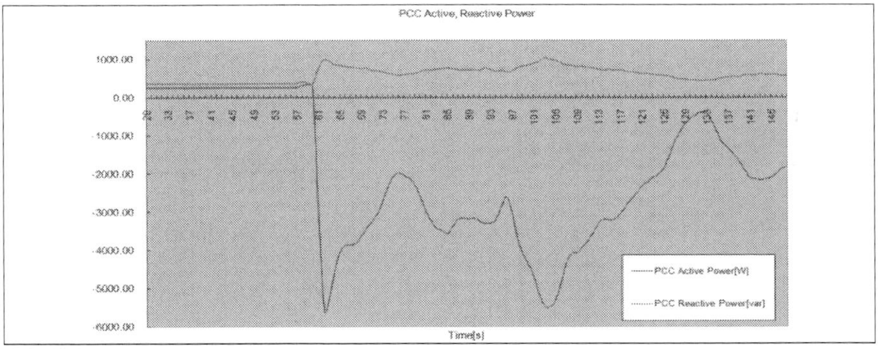

Fig. 13. PCC valid· invalid power in case of wind speed change

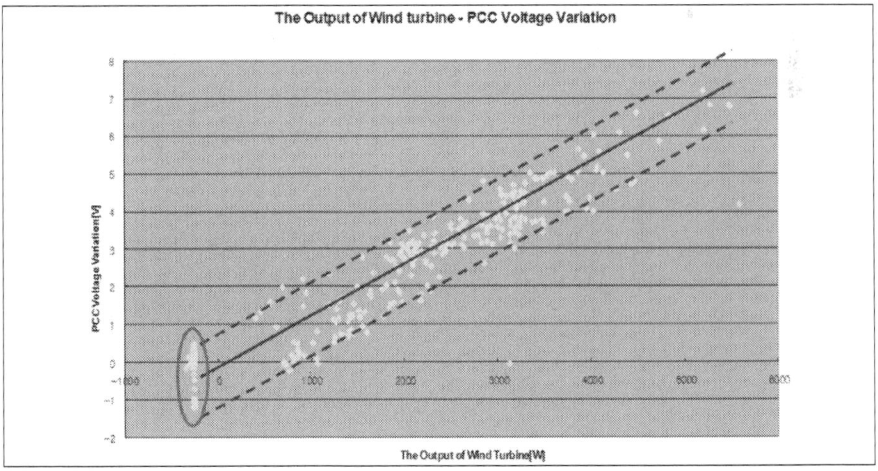

Fig. 14. Measured value of PCC voltage variation amount for the output of wind turbine

Classification	Amount of voltage variation in case of 5.5[kW] generation (V)
PSCAD Model	6.4
Simplified prediction model	6.5
Results of actual measurement	6.9

Table 5. The comparative table of PCC voltage variation amount in case of active power fluctuation of wind turbine

3. Conclusion

This study analyzed voltage variation according to the active power fluctuation of a wind turbine connected with the existing generating facilities in parallel in a small-sized isolated system, e.g. an island. The prediction of the voltage variation amount at the connected point is possible under the consideration of the change of installation location of a wind turbine and load fluctuation in the future. The variations of voltage drop happen according to the fluctuation of generated wind amount, and voltage change (rising) in receptive value inevitably. The simulation model was developed by using PSCAD/EMTDC to analyze voltage variation of the simplified model in the small-sized isolated system; and suggested the method to analytically obtain voltage variation amount through the analysis of a lot more generalized equivalent circuit. It was confirmed that the suggested method, simulation and the results of actual measurement matched in the error range of about less than 7%. It is considered that the study on the additional control method of a wind turbine to make the penetration ratio of wind plant increase in comparison to load capacity is necessary along with the development of a much more precise simulation model by taking account of the uniqueness of the excessive condition of generators in the future.

4. References

Bialasiewicz, J; Muljadi, E. & Drouilhet, S. & Nix, G. (1998) Modular Simulation of a Hybrid Power System with Diesel and Wind Turbine Generation, Windpower '98 Bakersfield, CA

E-ON Netz Gmbh, (2006) *Grid Code-High and Extra High Voltage*

IEEE 1547 (2003) *IEEE Standard for Interconnecting Distributed Resources with Electric Power Systems*

Jeong, W.; Yoon, K. & Kim, S. & Lee, H. (2007) The analysis of voltage variation according to grid connection at wind plant in Yangyang, *International Spring Academy Conference of Korea Wind Energy, papers*, 35-40

Kim, J. & Song, H. (2005) The development of PSCAD/EMTDC simulation model of variable speed wind plant system by permanent magnet synchronous generator, *Society of Power & Electronics, papers*, 610-617

Korea Electric Power Corporation (2005) *Technology standard connected with the distributed power supply system*

Rajendiran, K.; Keerthipala, W. & Nayar, C. PSCAD/EMTDC Based Simulation of a Wind-Diesel Conversion Scheme

Sim, M.; Palle, B. & Chakraborty, S. & Uriarte, C. (2007) *Electrical Model Development and Validation for Distributed Resources*, Subcontract Report, NREL

Slootweg, J. & Kling, W. (2003) Is the Answer Blowing in the Wind?, *IEEE Power & Energy Magazine*, No., 6, 26-33

Song, S.; Kim, I. & Han, B. (2003) Electric characteristics of wind turbine and interaction in case of grid connection, *Power & Electronics Magazine, Ed.*, Vol., 8, No., 6, 21-27

www.windturbinewarehouse.com/pdgs/bergey/BergeyExcel-S 10kW Turbine Specs.03.pdf

15

Markovian Approaches to Model Wind Speed of a Site and Power Availability of a Wind Turbine

Alfredo Testa, Roberto Langella and Teresa Manco
Second University of Naples
Italy

1. Introduction

This chapter is devoted to markovian approaches to model first the wind speed in a given site starting from experimental results and, then, power availability of a wind turbine installed in assigned wind conditions.

The chapter organization is as follows: section 2 is devoted to wind data and their treatment; section 3 presents two wind models with different accuracy; section 4 describes produced power models for wind turbines, first in hypotheses of ideal failure-free turbine and, then, accounting for a real behaviour; section 5 is devoted to model analytical solutions; section 6 reports the results of numerical examples developed for a very simple case-study; section 7 describes some possible model applications; section 8 concludes the chapter; section 9 is constituted of an appendix on the basic Markov model concepts.

A significant part of the chapter content derives from references (Manco & Testa, 2007 and to other references of the chapters' authors reported in Section 10).

2. Wind data and their treatment

The power generated from a wind turbine depends on the site specific wind speed, which randomly fluctuates with time. Therefore, wind power studies require accurate models to forecast wind speed variations for wind farm locations of interest (Billinton et al., 2006).

Historical data for a wind farm site in Sardinia for a measurement campaign of one year are considered. The data are collected every twenty minutes.

In Fig. 1 the wind speed profile versus the sample number is represented. Being the duration of a single measurement exactly of twenty minutes, the sample number coincides with the number of twenty minutes interval elapsed from the beginning to the end of the observation year.

The wind speed range of interest can be divided into classes equally spaced. The wind classes, $W_{(.)}$, with the respective speed ranges, the number of samples observed for each class, $N_{(.)}$, and the consequent estimated probability of each class frequency, $f_{(.)}$, are reported in Table 1. N_T is the total number of samples considered, so an estimation of $f_{(.)}$ has been obtained as:

$$\hat{f}_{(\cdot)} = \frac{N_{(\cdot)}}{N_T}\,. \qquad (1)$$

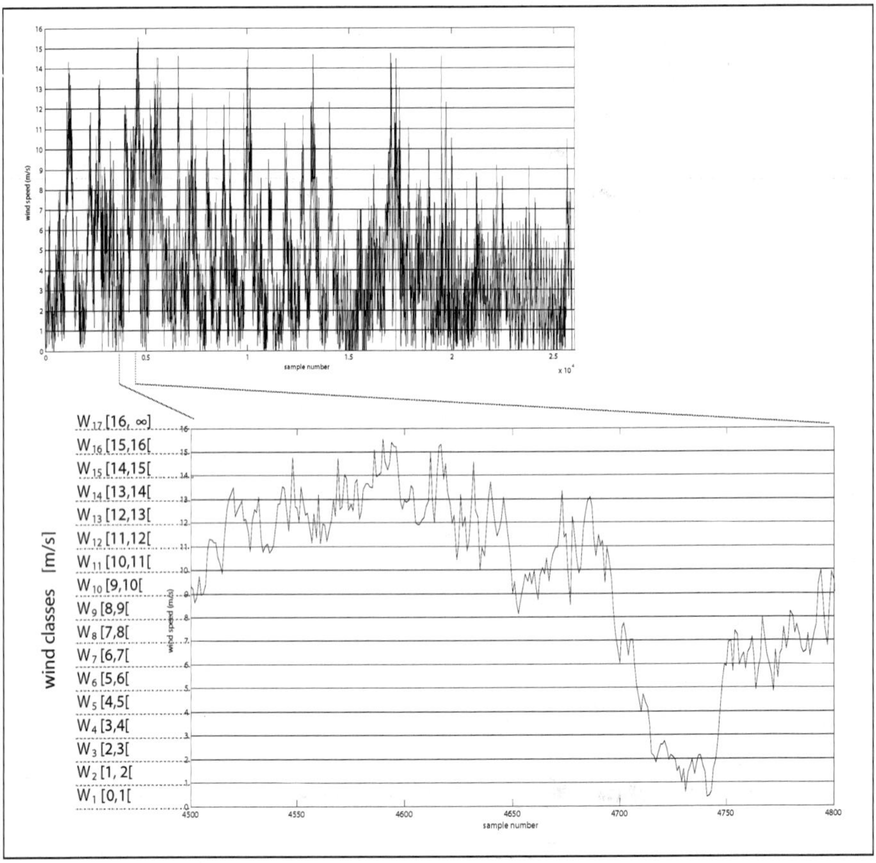

Fig. 1. Wind speed values versus the number of samples and wind classification.

Wind Class	Speed range (m/s)	Samples in the class, $N_{(\cdot)}$	Class frequency, $f_{(\cdot)}$
W_1	[0,1[	N_1	f_1
W_2	[1,2[	N_2	f_2
W_3	[2,3[	N_3	f_3
W_4	[3,4[	N_4	f_4
...	...	...	...
W_N	[W_{max}, ∞[	N_N	f_N

Table 1. Wind classification according to wind profile.

In Fig. 2 the pdf of the wind speed is depicted, and compared with a Weibull distribution characterized by the same mean value. Weibull distribution parameters assume the following values: α=0.3407, β=1.4052 (Allan & Billinton, 1992).
It is possible to observe that the Weibull distribution correctly estimates the high speed wind pdf values while the lower values are overestimated for some classes and underestimated for some other.

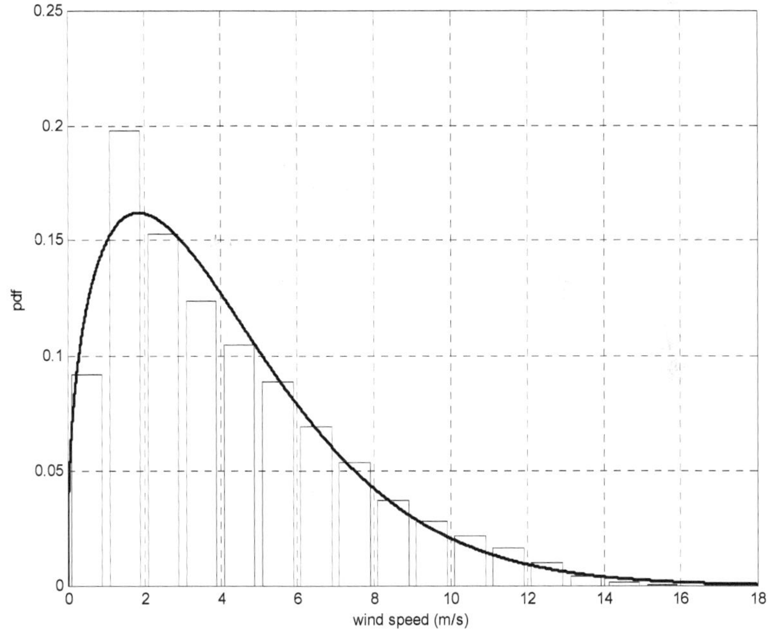

Fig. 2. Wind speed pdf versus wind speed classes; the histogram represents the distribution observed while the black line represents Weibull distribution.

3. Wind models

The wind speed range of interest is divided into N classes that are not equally spaced, because of the nonlinear characteristic of the turbines, as shown in next Section 4. In the first class all the values of wind speed under the cut in (w_{min}) and over the cut off (w_{max}) of the wind turbine are considered.
The wind classes, $W_{(\cdot)}$, with the respective speed ranges, the number of samples observed for each class, $N_{(\cdot)}$, and the consequent estimated probability of each class, $p_{(\cdot)}$, are reported in Table 2. N_T is the total number of samples considered, so an estimation of $p_{(\cdot)}$ has been obtained as:

$$p_{(\cdot)} = \frac{N_{(\cdot)}}{N_T}. \tag{2}$$

Wind class	Speed range (m/s)	Samples in the class, $N_{(\cdot)}$	Class probability, $p_{(\cdot)}$
W_1	$[0, w_{min}[\cup [w_{max} + \infty[$	N_1	p_1
W_2	$[w_{min}, w_2[$	N_2	p_2
W_3	$[w_2, w_3[$	N_3	p_3
W_4	$[w_3, w_4[$	N_4	p_4
...	...	...	...
W_{N-1}	$[w_{N-1}, w_N[$	N_{N-1}	p_{N-1}

Table 2. Wind classification according to wind profile.

Two models of different accuracy are presented in the following subsections.

3.1 Models accounting for transitions among all states

In Table 3 the number of transitions n_{ij}, from the i-th wind class to the j-th wind class, are reported observed passing from each sample to the subsequent sample. The rows represent the i-th state, the columns the j-th state. The i-i transitions stay for a permanence in the i-th state.

i\j	1	2	3	4	...	N
1	n_{11}	n_{12}	n_{13}	n_{14}	...	n_{1N}
2	n_{21}	n_{22}	n_{23}	n_{24}	...	n_{2N}
3	n_{31}	n_{32}	n_{33}	n_{34}	...	n_{3N}
4	n_{41}	n_{42}	n_{43}	n_{44}	...	n_{4N}
...	...	...	...	...	...	...
N	n_{N1}	n_{N2}	n_{N3}	N_{N4}	...	n_{NN}

Table 3. Number of transitions observed in the wind sample.

In Fig. 3 the Markovian model of the wind is shown. Each state corresponds to a class, with a proper wind speed. The model is constituted by n states. Each state is represented with explicit reference to the level of wind speed.

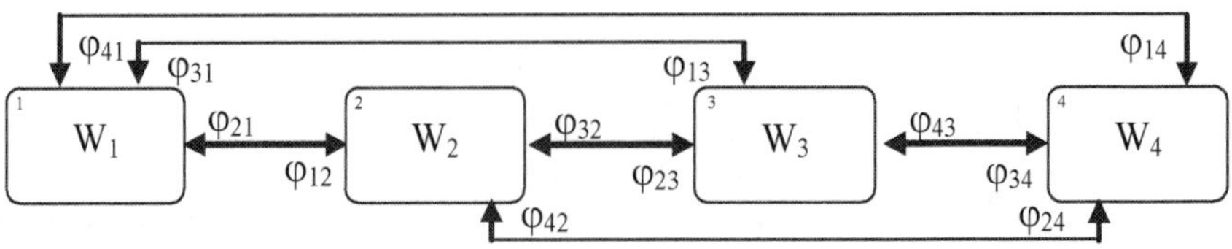

Fig. 3. Proposed Markovian model of the wind, N=4.

The time spent in each of the four states is exponentially distributed, so the process can be viewed as a homogeneous Markov process. φ_{ij} represents the transition rate between the state i and the state j. The transition rates among different states (different wind classes) and the permanence in the i-th state can be calculated from the measurement campaign considered by these formulas:

$$\varphi_{ij_{i\neq j}} = \frac{n_{ij} / N_T}{N_i / N_T} \cdot \frac{1}{\Delta t} = \frac{n_{ij}}{N_i} \text{ [events/hour]}, \tag{3}$$

$$\varphi_{ii} = 1 - \sum \varphi_{ij_{i\neq j}} , \tag{4}$$

being Δt in the field measurement considered equal to equal to the measurement interval. In Table 4 the transition rates φ_{ij} obtained by formulas (3) and (4) are reported. The rows represent the *i* state, the columns the *j* state.

φ_{ij}	1	2	3	4	...	N
1	φ_{11}	φ_{12}	φ_{13}	φ_{14}	...	φ_{1N}
2	φ_{21}	φ_{22}	φ_{23}	φ_{24}	...	φ_{2N}
3	φ_{31}	φ_{32}	φ_{33}	φ_{34}	...	φ_{3N}
4	φ_{41}	φ_{42}	φ_{43}	φ_{44}	...	φ_{4N}
...	...	...	...	...	...	...
N	φ_{N1}	φ_{N2}	φ_{N3}	φ_{N4}	...	φ_{NN}

Table 4. Transitions rates.

3.2 Models accounting for transitions only among contiguous states

It is possible to observe that the model presented in (Allan & Castro Sayas, 1996) considers that wind speeds do not increase or decrease instantaneously, but change continuously, albeit over very short periods of time. For this reason, the transitions among non adjacent wind speed states are considered as not possible. To reproduce this condition the transitions among non adjacent classes observed in the measurement campaign are summed to the transitions of the adjacent classes:

$$\varphi'_{i,i-1} = \sum_{j>i} \varphi_{ij} , \tag{5}$$

$$\varphi'_{i,i-1} = \sum_{j<i} \varphi_{ij} , \tag{6}$$

$$\varphi'_{i,j\ j\neq i,i-1,i-1} = 0 , \tag{7}$$

$$\varphi'_{ii} = 1 - \sum \varphi'_{ij_{i\neq j}} . \tag{8}$$

In Fig. 4 the markovian model of the wind, obtained under these assumptions for N=4 is reported, and its transition rates are shown in Table 5.

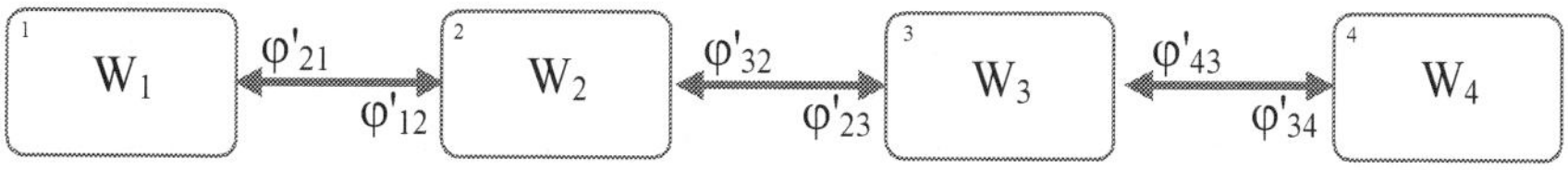

Fig. 4. Markovian model of the wind, N=4, according to (Allan & Castro Sayas, 1996).

φ'_{ij}	1	2	3	4	...	N
1	φ'_{11}	φ'_{12}	*0*	*0*	...	*0*
2	φ'_{21}	φ'_{22}	φ'_{23}	-	...	*0*
3	*0*	φ'_{32}	φ'_{33}	φ'_{34}	...	*0*
4	*0*	-	φ'_{43}	φ'_{44}	...	*0*
...	...	...	...	...	...	...
N	*0*	*0*	*0*	*0*	...	φ'_{NN}

Table 5. Transitions rates.

4. Power models

The wind turbine is characterized by its operational curve. An example of wind turbine power versus wind speed is reported in Fig. 5.

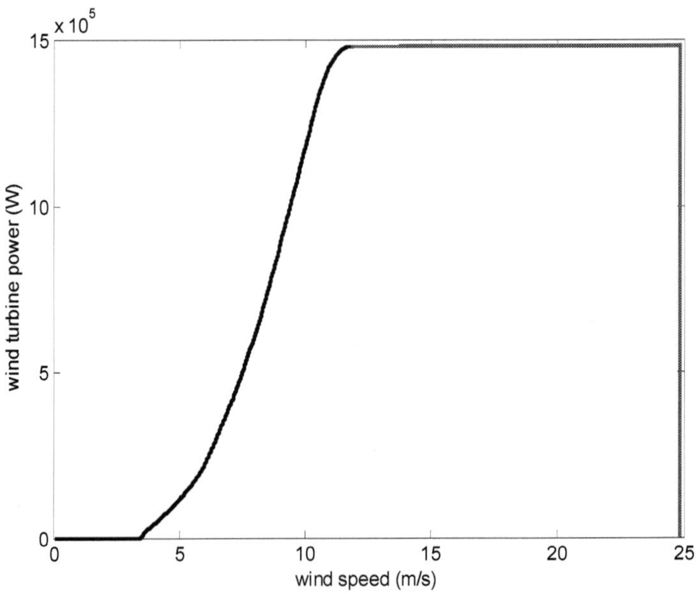

Fig. 5. Power of the wind turbine on the wind.

The typical operational data of a turbine are reported in Table 6.

Cut-in wind speed (m/s)	Nominal wind speed (m/s)	Stop wind speed (m/s)	Failure rate (event/year)	Repair rate (event/hour)
w_{min}	w_n	w_{max}	λ_{WT}	μ_{WT}

Table 6. Wind Turbine Operational Data.

4.1 Model for a failure free turbine

Starting from the wind data of Fig. 1 and utilizing the turbine characteristic depicted in Fig.5 the power profile of Fig. 6 is obtained.

As already described for wind speed values, the power range of interest is divided into classes. In the first class all the conditions of wind speed that give the 0 class value power ($w<w_{min}$ and $w\geq w_{max}$) are concentrated; while in the final class all the conditions that give maximum power ($w_{N-1}\leq w<w_{max}$) are concentrated.

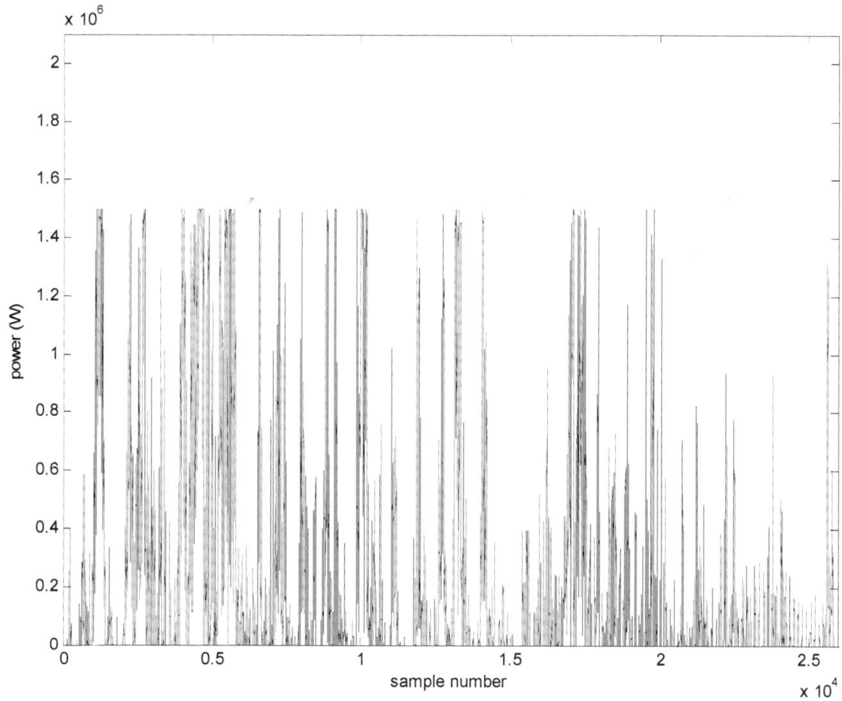

Fig. 6. Power values versus the number of samples.

The typical results are summarized in Table 7, where for each power class P_{Wi} the corresponding power range, the number of samples and the corresponding class frequency are reported.

As concerns the number of transitions observed from different power states, a Table similar to Table 4, obtained in the case of wind speed data, has to be calculated. It is not reported here for the sake of brevity. It is utilized to evaluate the transition rates among different states (different power classes).

Of course, power models of different accuracy can be considered corresponding to the two wind models described in Fig. 3 and Fig. 4 of section 3.

Class symbol	Power range (kW)	Samples in the class, $N_{(\cdot)}$	Class frequency, $f_{(\cdot)}$
P_{W1}	0	N_1	f_1
P_{W2}	]0,P2]	N_2	f_2
P_{W3}	]P2,P3]	N_3	f_3
P_{W4}	]P3,P4]	N_4	f_4
...	...	...	...
P_{WN}	P_N	N_N	f_N

Table 7. Power classification.

4.2 Model for a wind turbine affected by failures

A very simple two states wind turbine model is considered. It is depicted in Fig. 7, where λ_{WT} and μ_{WT} represent the failure rate and the repair rate, respectively.

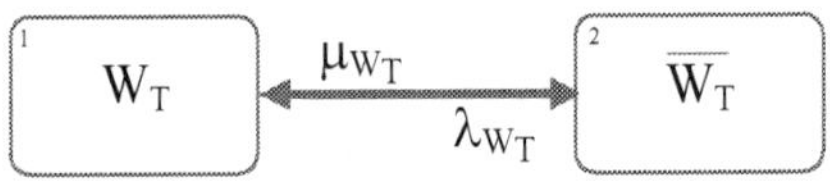

Fig. 7. Two state model for a Wind Turbine

The direct integration between the wind model of Fig. 3 and the considered two states wind turbine model gives the model depicted in Fig. 8.

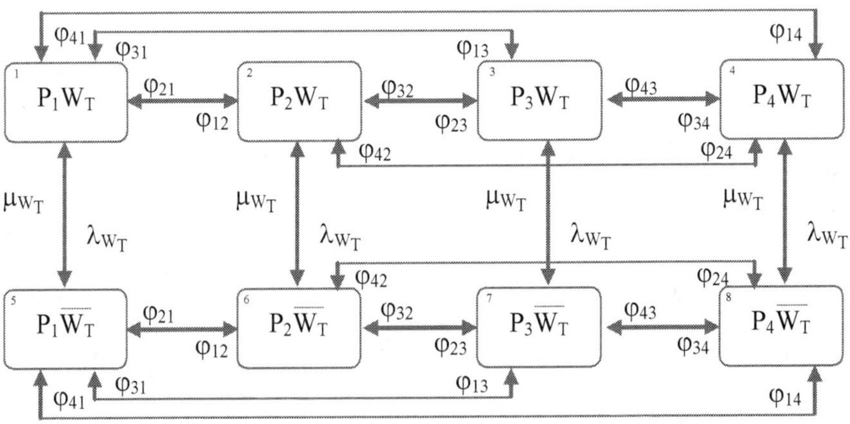

Fig. 8. Markovian model of a two states wind turbine in presence of the wind modelled as in Fig. 3.

It is interesting to consider the possibility of reducing states 5-8 of Fig. 8 into a unique equivalent state 5 as depicted in Fig. 9, where the coefficients α_i introduced to weight the repair rates are the steady state probabilities of Fig. 3 model, that can be calculated according to the formulas reported in next section 5. The only problem that seems to arise is

that one related to the presence of possible correlation effects deriving from the repair time of the turbine and the duration of cyclic effects in the wind speed, related for instance to the different hours of the day. A very long repair time seems a good guaranty of negligible correlation effects, while repair durations of the order of 24 hours or less could excite night-day correlation effects.
Models of Fig. 8 and Fig. 9 refer to the wind model of Fig. 3; it is possible and easy to obtain models also starting from the wind model of Fig. 4.

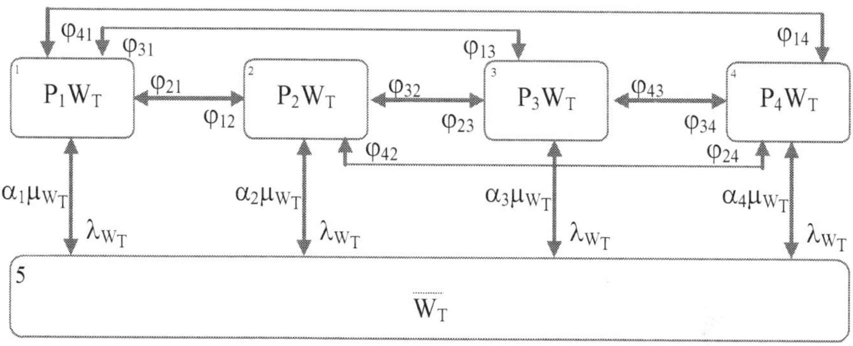

Fig. 9. Markovian simplified model of a two states wind turbine in presence of the wind modelled as in Fig. 3.

5. Model solution

Typically for the kind of problems related to both wind or produced power models, steady-state solutions are of prevalent or exclusive interest. The models have to be solved in terms of steady-state probabilities, $\alpha(\cdot)$, frequencies, $f(\cdot)$, and durations, $d(\cdot)$, of each state of interest, as shown in Table 8.

State	Steady-state Probability $\alpha(\cdot)$	frequency $f(\cdot)$	Duration $d(\cdot)$
1	α_1	f_1	d_1
2	α_2	f_2	d_2
3	α_3	f_3	d_3
4	α_4	f_4	d_4
...	...	...	...
N	α_N	f_N	d_N

Table 8. Results.

The state probabilities can be calculated by considering the following balance equations between input and output frequencies for each system state:

$$\alpha_1 \cdot \sum_{\substack{j=1 \\ j \neq 1}}^{N} \varphi_{1j} = \sum_{\substack{j=1 \\ j \neq 1}}^{N} \alpha_j \cdot \varphi_{j1} \,,$$

$$\alpha_2 \cdot \sum_{\substack{j=1 \\ j \neq 2}}^{N} \varphi_{2j} = \sum_{\substack{j=1 \\ j \neq 2}}^{N} \alpha_j \cdot \varphi_{j2} \,, \tag{9}$$

....

....

$$\alpha_N \cdot \sum_{\substack{j=1 \\ j \neq N}}^{N} \varphi_{Nj} = \sum_{\substack{j=1 \\ j \neq N}}^{N} \alpha_j \cdot \varphi_{jN} \,.$$

This system can be seen as a linear system of N equations of unknowns α_1, α_2, ..., α_N in which one of the equations is linearly dependent from the others. By eliminating the last equation, and adding to the steady-state probability congruency equation:

$$\sum_{i=1}^{N} \alpha_i = 1 \,, \tag{10}$$

to the remaining N-1 equations linearly independent, a linear system of N equations of N unknown is obtained.
Once solved the previous system, also the frequencies can be calculated as:

$$f_i = \alpha_i \cdot \sum_{\substack{j=1 \\ j \neq i}}^{N} \varphi_{ij} \qquad i = 1, 2, \ldots, N. \tag{11}$$

The mean state durations, d_1, d_2...... d_N can be obtained directly from the transition rates as:

$$d_i = 1 / \sum_{\substack{j=1 \\ j \neq i}}^{N} \varphi_{ij} \qquad i = 1, 2, \ldots, N. \tag{12}$$

6. Numerical experiments

In Fig. 10 the wind speed profile considered versus the sample number is represented. Being the duration of a single measurement exactly of one hour, the sample number coincides with the number of hours elapsed from the beginning of the observation interval.
The wind classes, $W_{(\cdot)}$, with the respective speed ranges, the number of samples observed for each class, $N_{(\cdot)}$, and the consequent estimated probability of each class, $p_{(\cdot)}$, are reported in Table 9. $N_T = 722$ is the total number of samples considered.

6.1 Wind modelling

In Table 10 the number of transitions nij, from the i-th wind class to the j-th wind class for the classes of Table 9, observed passing from each sample to the subsequent sample, are reported. The fifth wind class has not be considered because of the extremely low probability of occurrence (0 samples during the observation period).

Wind class	Speed range (m/s)	Samples in the class, $N_{(\cdot)}$	Class probability, $p_{(\cdot)}$
W_1	$[0, 4[\cup[25,+\infty[$	41	0.0568
W_2	[4,7[	106	0,1468
W_3	[7,10[	416	0,5762
W_4	[10,15[	159	0,2202
W_5	[15,25[	0	0

Table 9. Wind classification according to wind profile.

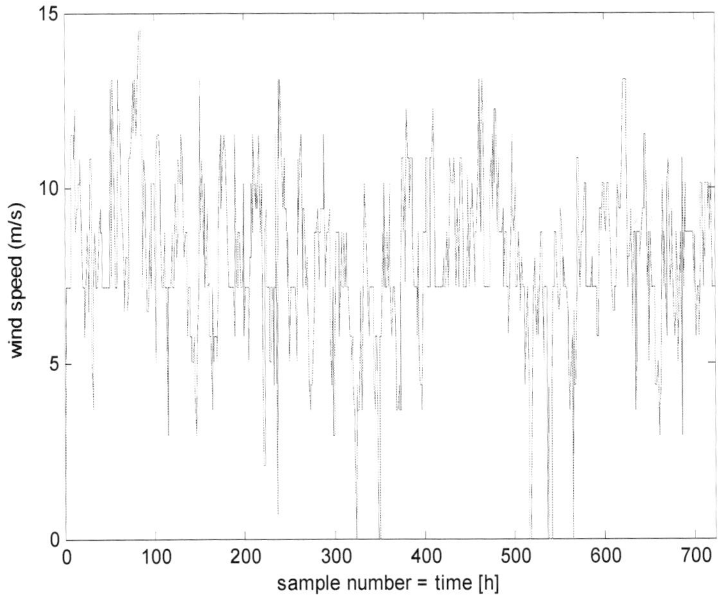

Fig. 10. Wind speed values versus the number of samples which coincides with the number of hours elapsed from the beginning of the observation interval.

φ_{ij}	1	2	3	4
1	17	15	7	2
2	15	44	41	6
3	7	45	308	56
4	1	2	61	95

Table 10. Transitions observed.

In Table 11 and Table 12 the transitions rates evaluated from Table 10 data, for the model of Fig. 3 and the model of Fig. 4, are reported respectively.

$\varphi_{i\backslash j}$	1	2	3	4
1	0,4146	0,3659	0,1707	0,0488
2	0,1415	0,4151	0,3868	0,0566
3	0,0168	0,1082	0,7404	0,1346
4	0,0063	0,0126	0,3836	0,5975

Table 11. Transitions rates for the model of Fig. 3.

$\varphi'_{i\backslash j}$	1	2	3	4
1	0,4146	0,5854	0	0
2	0,1415	0,4151	0,4434	0
3	0	0,1250	0,7404	0,1346
4	0	0	0,4025	0,5975

Table 12. Transitions rates for the model of Fig. 4.

In Table 13 and Table 14 the results of the models described in Fig. 3 and Fig. 4 are reported, respectively.

By comparing Table 14 and Table 13 results it is easy to observe that the estimations of the model of Fig. 4 are sometimes optimistic (state 1: under evaluation of probability and frequency; state 3: over evaluation of probability and frequency), and sometimes pessimistic (state 4). Further comments are added in Section 6.2.

State	Steady state probability	Frequency (event/h)	Duration (h)
1	0,0543	0,0318	1,7083
2	0,1458	0,0853	1,7097
3	0,5792	0,1504	3,8519
4	0,2208	0,0889	2,4844

Table 13. Results of the wind model of Fig. 3.

State	Steady state probability	Frequency (event/h)	Duration (h)
1	0,0405	0,0237	1,7083
2	0,1674	0,0979	1,7097
3	0,5936	0,1541	3,8519
4	0,1985	0,0799	2,4844

Table 14. Results of the wind model of Fig. 4.

6.2 Failure-free turbine modelling

The wind turbine considered is a highly reliable turbine of 850 kW, of a leading producer of high-tech wind power systems. This turbine is ideal for populated and remote areas, with compact dimensions that make it easy to transport overland. It uses pitch technology to optimise the output under medium to high wind conditions and it is available in a wide range of tower heights from 40-86 m. Its power curve is reported in Fig. 11.

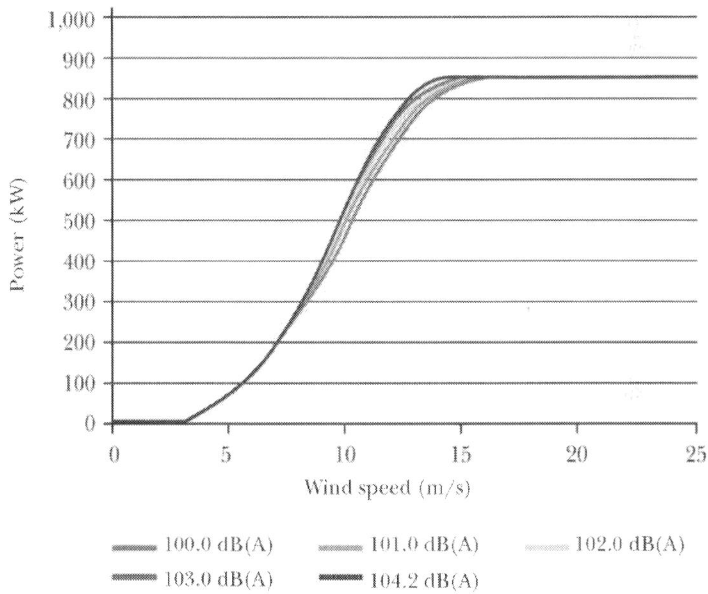

Fig. 11. Power of the wind turbine on the wind speed at different sound levels.

The operational data of the wind turbine, in particular referred to in this paper, are reported in Table 15.

Cut-in wind speed (m/s)	Nominal wind speed (m/s)	Stop wind speed (m/s)	Failure rate (event/year)	Repair rate (event/hour)
4	16	25	1	0.0069

Table 15. Wind Turbine Operational Data.

In Table 16 the power classes, the corresponding wind classes, the minimum (P_m), the maximum (P_M) and the mean (P_{mean}) values of the electric power, obtained for each class on the basis of the power curve represented in Fig.11, are reported.

For an ideal failure free turbine, the models corresponding to those of the wind in Fig. 3 and Fig. 4 can be considered, because the fifth power class has not be represented due to the extremely low probability of occurrence (0 samples during the observation period). For the particular choice of wind and power classes effected, the results are the same reported in Table 13 and Table 14.

Power class	Wind Class	P_M(KW)	P_M(kW)	P_{mean}(kW)
P_1	W_1	0	0	0
P_2	W_2	0	200	100
P_3	W_3	200	500	350
P_4	W_4	500	850	675
P_5	W_5	850	850	850

Table 16. Power classes.

By comparing Table 14 and Table 13 it is easy to observe that the estimations of the model of Fig. 4 are sometimes optimistic (state 1: underevaluation of probability and frequency; state 3: overevaluation of probability and frequency), and sometimes pessimistic (state 4) or depending on the minimum acceptable level of electric power available (state 2 characterized by a minimum non zero power). A further comparison can be made by considering the mean power evaluated as:

$$[P] = \sum \alpha_i \cdot p_i \ . \qquad (13)$$

The model of Fig. 3 gives the value of 366 kW, while the model of Fig.4 gives the value of 358 kW; so, from this point of view the second model is pessimistic (about 2%). On the contrary, in the cases in which the no power condition, state 1, is a major concern (isolated systems), the second model makes a largely optimistic estimation of the probability and frequency (-25%), with the same state duration.

Assuming that the model of Fig.3 is the most accurate, because it does not introduce approximations about possible state transitions, the conclusion is that the simplification introduced by the model of Fig. 4 may result in great inaccuracies of unpredictable sign.

6.3 Power availability modelling

In particular, the parameters considered for the turbine failure rate and repair rate are those reported in Table 15, and derived from (ABB AG et al., 2006). The direct integration between the wind model of Fig. 3 and the considered two states wind turbine model gives the eight states model depicted in Fig. 8.

The results of the Markovian model of Fig. 8 are reported in Table 17.

State	Steady state Probability	Frequency (event/h)	Duration (h)
1	0,0534	0,0313	1,7080
2	0,1434	0,0839	1,7093
3	0,5698	0,1480	3,8502
4	0,2172	0,0875	2,4837
5	0,0009	0,0005	1,6883
6	0,0024	0,0014	1,6896
7	0,0094	0,0025	3,7515
8	0,0036	0,0015	2,4422

Table 17. Results of the model in Fig. 8.

Then states 5-8 of Fig. 8 have been reduced into a unique equivalent state 5 as depicted in Fig. 9.
The results of the model in Fig. 9 are reported in Table 18. It is possible to observe that states 1-4 have the same state probabilities, frequencies and durations of those in Fig.11. So, these states are unaffected by the simplification operated in Fig. 9.

State	Steady state probability	Frequency (event/h)	Duration (h)
1	0,0534	0,0313	1,7080
2	0,1434	0,0839	1,7093
3	0,5698	0,1480	3,8502
4	0,2172	0,0875	2,4837
5	0,0162	0,0001	144

Table 18. Results of model in Fig. 9.

The model of Fig. 9 is more compact and more useful than the model of Fig. 8 because allows making simple considerations on the reserve dimensioning in absence of wind or in case of the wind turbine failure.

7. Applications

Some of the possible applications of the Markov Model presented in the previous sections are here reported.

7.1 Energy production estimation

Simple formulas allow to evaluate the expected energy production of a wind turbine in one year.
One possibility is based on the use of steady-state probabilities:

$$E_{year} = 8760 \cdot \sum_{i=1}^{N} P_i \alpha_i = 8760 \cdot [P] \text{ [kWh].} \tag{14}$$

Equivalently, it is possible to use frequencies and durations:

$$E_{year} = \sum_{i=1}^{N} f_i^{'} P_i d_i \text{ [kWh],} \tag{15}$$

with

$$f_i^{'} = 8760 \cdot f_i \text{ [event/year].} \tag{16}$$

7.2 Hybrid wind-diesel stand alone systems

In (Carpentiero et al., 2008), the approaches previously described are utilized to size a hybrid wind-diesel stand alone system. Starting from an opportune probabilistic load model and from the wind turbine model, the energy produced by the wind turbine and used by the load for the different levels of the load can be calculated.

Then, it is possible to evaluate the energy that diesel generators have to supply, when the power produced by the turbine is not available or not enough to satisfy the load.
Relevant quantities that can be evaluated are:

- the maximum energy that could be produced in a year by the turbine, independently from the load demand;
- the energy that could be produced in a year by the turbine which is not utilized by the load.

7.3 Energy reserve sizing

It can be interesting to compensate the power unavailability of a wind turbine generator not only in stand-alone systems for obvious reasons, but also in grid-connected systems of the future for particular availability needs (Manco & Testa, 2007). The origin of the power unavailability is a consequence of the reduced or excessive wind speed (state 1 of the model in Fig. 9) or of the turbine failure (state 5 of Fig.9). Of course, in the reality, these two conditions are characterized by different duration mean values and different probability distribution functions. By sure in the framework of maintenance phenomena representation, the lognormal distribution is one of the most popular together with more complex multimodal distributions.
In (Manco & Testa, 2007) the models of some possible distributions for repair time are considered.

7.4 Very short term wind forecast

Markov Chain models are generally used for the generation of synthetic wind speed and wind power time series. In (Carpinone et al., 2010) they are here used to develop a probabilistic forecasting method that allows to provide estimates of future wind power generation, not only as point forecasts, but also as estimate of the probability distributions associated to the point forecasts.
Some first results of the application of Markov Chain of first and second order are obtained and compared with the persistent model results. The second order model allows to improve the forecast performance by reducing the prediction error.

7.5 Further applications

Interesting references for other applications are (Allan & Billinton, 1988), (Bagen et al., 2003), (Filios et al., 2006), (Islam &Liu, 2006), (Kim & Singh, 1988), (Lago-Gonzalez & Singh, 1985), (Carpentiero et al., 2010), (Langella et al., 2010).

8. Conclusion

Markovian approaches to model the energy production and power availability of a wind turbine have been described. The wind speed time variability is taken into account by means of the wind speed classification in a discrete reduced number of contiguous classes, each corresponding to an opportune range of values, starting from field measurement data. The duration of each class is statistically treated so to preserve information about its duration and the transition rates into all the other classes. Two different accuracy models have been described.
Then, the wind turbine availability is taken into account by means of a simple model. Simple numerical experiments have been presented.

9. Appendix

The Markov approach is based on the following hypotheses:

- the prediction of the future states of a system, based on the present state alone, does not differ from that formulated on the basis of the whole history of the system (Markov property);
- when this transition probability does not depend on the age of the system (time), the Markov process is called homogeneous.

In a homogeneous Markov process, the time between successive transitions has an exponential distribution.

In some applications, the modeling of the failure-repair process is based on the following hypotheses:

- the successive time-to-failure values are independent and identically distributed random variables;
- the successive time-to-repair (or recovery time) values are independent and identically distributed random variables.

In a case in which both the time-to-failure and time-to-repair values are exponentially distributed, the failure-repair process can be viewed as a two state homogeneous Markov process. When the random variables time-to-failure and/or time-to-repair cannot be assumed to be exponentially distributed, extensions of the homogeneous Markov process have to be adopted.

10. References

ABB AG, Hanson, J., Mannheim, Osterholt, A., Underbrink, A., & Zimmermann, W., (2006). *Reliability Calculations for the Grid Connection of an Offshore Wind Farm,* 9th International Conference on Probabilistic Methods Applied to Power Systems KTH, Stockholm, Sweden - June 11-15, 2006.

Allan, R.N., & Billinton, R, (1988), *Reliability assessment of large electric power systems,* Kluwer Academic, Boston.

Allan, R.N., & Billinton, R, (1992), *Reliability evaluation of engineering systems,* 2nd edition, Plenum Press.

Allan, R.N., & Castro Sayas, F., (1996). *Generation availability assessment of wind farms,* Vol. 143, No. 5, (September 1996), IEE Proceeding-General Transmission Distribution, pp. 507-518.

Bagen, Billinton, R., & Cui, Y., (2003). *Reliability evaluation of small stand-alone wind energy conversion systems using a time series simulation model,* IEE Proc.-Gener. Transm. Distrib., Vol. 150 No. 1. January 2003, pp. 96-100.

Billinton, R., Hu, P., & Karki, R. (2006). *A Simplified Wind Power Generation Model for Reliability Evaluation,* Vol. 21, No. 2, (June 2006), IEEE Transaction on Energy Conversion, pp. 533-540.

Carpentiero, V., Langella, R., Manco, T., & Testa, A. (2008). *A Markovian Approach to Size a Hybrid Wind-Diesel Stand Alone System,* proc. of the 10th International Conference on Probabilistic Methods Applied to Power Systems, Puertorico, USA, May 2008.

Carpentiero, V., Langella, R., & Testa, A., (2010). *Hybrid Wind-Diesel Stand-Alone System Sizing Accounting for Fuel Price Uncertainty,* 11th International Conference on Probabilistic Methods Applied to Power Systems, Singapore, June 2010.

Carpinone A., Langella, R., & Testa, A. (2010). *Very Short-term Probabilistic Wind Power Forecasting based on Markov Chain Models,* 11th International Conference on Probabilistic Methods Applied to Power Systems, Singapore, June 2010.

Filios, A., Kaldellis, J.K., & Kondili, E., (2006). *Sizing a hybrid wind-diesel stand-alone system on the basis of minimum long-term electricity production cost,* Applied Energy, pp. 1384-1403.

Islam, S., & Liu, X., (2006). *Reliability Evaluation of a Wind-Diesel Hybrid Power System with Battery Bank Using Discrete Wind Speed Frame Analysis,* 9th International Conference on Probabilistic Methods Applied to Power Systems KTH, Stockholm, Sweden - June 11-15, 2006.

Kim, Y., & Singh, C., (1988). *An efficient technique for reliability analysis of power systems including time dependent sources, IEEE Trans. Power Syst., 1988, 3, (3), pp. 1090-1096.*

Lago-Gonzalez, A., & Singh, C., (1985). *Reliability modeling of generation systems including unconventional energy sources, IEEE Trans., 1985, PAS-104, (5), pp. 1049-1056.*

Langella, R., Manco, T., & Testa, A., (2010). *Unifying Supply Reliability and Voltage Quality in the Representation of an Electrical System Node,* Vol. 25, No. 2, (2010), IEEE Transactions on Power Delivery, pp. 1172-1181.

Manco, T., & Testa, A., (2007). *A Markovian Approach to Model Power Availability of a Wind Turbine*, proc. of IEEE Power Tech 2007, Lausanne, Switzerland, July 1-5, 2007.

16

Modelling and Control Design of Pitch-Controlled Variable Speed Wind Turbines

Marcelo Gustavo Molina and Pedro Enrique Mercado
CONICET, Instituto de Energía Eléctrica, Universidad Nacional de San Juan
Argentina

1. Introduction

In the past decade, many problems related to energy factors (oil crisis), ecological aspects (climatic change), electric demand (significant growth) and financial/regulatory restrictions of wholesale markets have arisen worldwide. These difficulties, far from finding effective solutions, are continuously increasing, which suggests the need of technological alternatives to assure their solution. Under these circumstances, distributed or dispersed generation (DG) arises as the technological alternative with the ability of giving an effective solution to such difficulties. Distributed generation consists of generating electricity as near as possible of the consumption site, in fact like it was made in the beginnings of the electric industry, but now incorporating the advantages of the modern technology. Here it is consolidated the idea of using clean non-conventional technologies of generation that use renewable energy sources (RESs) that do not cause environmental pollution, such as wind, photovoltaic, wave, hydraulic, and more sophisticated systems based on hydrogen. The main advantages of using RESs as DG systems are the elimination of harmful emissions and inexhaustible resources of the primary energy (Heier, 2006).

Medium to large grid-connected wind turbine generators (WTGs) are particularly becoming today the most important and fastest growing form of electricity generation among the renewable technologies. They attract interest as one of the most cost-effective ways to generate electricity from RESs (Guerrero, et al., 2010). Indeed, this RES technology started in the eighties with a few tens of kW power capacity to date with multi-MW size wind turbines that are being installed. This trend is expected to be increased in the near future, sustained by the cost competitiveness of wind power technology and the development of new power electronics technologies, new circuit topologies and control strategies. These profits include the strong support provided by governments of different countries, as investment subsidies and incentives that impact directly on the commercial acceptance of wind turbine generators.

The growing number of distributed generators, particularly based on wind power systems, brings new challenges to the operation and management of the power distribution system, especially when the intermittent energy source constitutes a significant part of the total system capacity (Rahman, 2003). Under this scenario, the power electronic technology plays an important role in the integration of DG into the electrical grid since the DG system is subject to requirements related not only to the RES itself but also to its effects on the power system operation. The use of power electronic converters enables wind turbines to operate at variable (or adjustable) speed, and thus permits to provide more effective power capture than the fixed speed counterparts (Timbus et al., 2009). In fact, variable speed wind turbines have demonstrated to capture 8-15% more energy than constant speed machines. In variable speed operation, a control system designed to extract maximum power from the wind turbine and to provide constant grid voltage and frequency is required. As well as becoming larger, wind turbine designs were progressing from fixed speed, stall-controlled and with drive trains with gear boxes to become pitch controlled, variable speed and with or without gearboxes.

Among variable speed wind turbine generators, direct-in-line systems and doubly-fed induction generator (DFIG) systems have increasingly drawn more interests to wind turbine manufactures due to their advantages over other variable speed wind turbines and currently have the most significant potential of growth. Direct-in-line systems consists of a direct-driven (without gearbox) permanent magnet synchronous generator (PMSG) grid-connected via a full-scale power converter, while DFIG systems are built with a common induction generator with slip ring and a partial-scale converter connected to the rotor windings. Both modern pitch-controlled variable speed wind turbines technologies are emerging as the preferred technologies and have become the dominating type of yearly installed wind turbines in recent years.

2. Wind energy development

Worldwide, the development of wind energy is experiencing dramatic growth. During the last decade, the installed wind energy capacity has grown rapidly. According to the Global Wind Energy Council (GWEC), 15 197 MW of wind power capacity have been installed in 2006 in more than 40 countries, an increase of 32% over 2005. The installation of the total global wind energy capacity was increased to 74 223 MW by the end of 2006 from 59 091 MW of 2005. In terms of economic value, the wind energy sector has presently become one of the important players in the energy markets, with the total value of new generating equipment installed in 2006 reaching $ 23 billion (approximately € 18 billion) (Global wind energy council, 2006).

Europe continues to be the world major player in the installation of wind power systems. In 2006, the country having the highest total installed capacity was Germany with 20 621 MW. Spain and the United States are in second and third place, each with a little more than 11 603 MW installed. India is in fourth place, and Denmark ranks fifth. Asia experienced the strongest increase in installed capacity outside of Europe, with an addition of 3679 MW, taking the total capacity over 10 600 MW, about half that of Germany. The Chinese market was increased by the country's new Renewable Energy Law. China has more than doubled its total installed capacity by installing 1347 MW of wind energy in 2006, a 70% increase over 2005. This brings China up to 2604 MW of capacity, making it the sixth largest market worldwide. It is expected that more than 40 GW will be installed by 2020; this may become China the third major power supply by that year. Growth in African and Middle Eastern market also picked up in 2006, with 172MW of new installed capacity, mainly in Egypt, Morocco, and Iran, bringing the total up to 441 MW, a 63% growth (Blaabjerg & Chen, 2006). The European Wind Energy Association (EWEA) has set a target to satisfy 23% European electricity needs with wind by 2030. The exponential growth of the wind industry reflects the increasing demand for clean, safe and domestic energy and can be attributed to government policies associated with the environmental concerns, and research and development of innovative cost-reducing technologies.

The large scale development of wind power results in the wind turbines/farms becoming a significant part of the generation capacity in some area, which requires that the power system treats the wind turbines/farms like a power source, not only an energy source. The wind power penetration would result in variations of load flows in the interconnected systems, as well as re-dispatch of conventional power plants, which may cause a reduction of reserve power capacity (Slootweg & Kling, 2003). Some actions become necessary to accommodate large scale wind power penetration. For example, the electric grid may need an expansion for bulk electricity transmission from offshore wind farms to load centers, and it may require reinforcement of existing power lines or construction of new power lines, installation of Flexible AC Transmission system (FACTs) devices, etc.

3. Modern wind power systems

The discovery of electricity generated using wind power dates back to the end of last century and has encountered many ups and downs in its more than 100 year history. In the beginning, the primary motivation for essentially all the researches on wind power generation was to reinforce the mechanization of agriculture through locally-made electricity generation. Nevertheless, with the electrification of industrialized countries, the role of wind power drastically reduced, as it could not compete with the fossil fuel-fired power stations. This conventional generation showed to be by far more competitive in providing electric power on a large scale than any other renewable one.

Lack of fossil fuels during World War I and soon afterward during World War II created a consciousness of the great dependence on fossil fuels and gave a renewed attention to renewable energies and particularly to wind power. Although this concern did not extend long. The prices for electricity generated via wind power were still not competitive and politically nuclear power gained more attention and hence more research and development funds. It took two oil crises in the 1970s with supply problems and price fluctuations on fossil fuels before wind power once again was placed on the agenda. And they were these issues confronting many countries in the seventies which started a new stage for wind power and motivated the development of a global industry which today is characterized by relatively few but very large wind turbine manufacturers.

The beginning of modern wind turbine development was in 1957, marked by the Danish engineer Johannes Juul and his pioneer work at a power utility (SEAS at Gedser coast in the Southern part of Denmark). His R&D effort formed the basis for the design of a modern AC wind turbine - the well-known Gedser machine which was successfully installed in 1959. With its 200kW capacity, the Gedser wind turbine was the largest of its kind in the world at that time and it was in operation for 11 years without maintenance. The robust Gedser wind turbine was a technological innovation as it became the hall mark of modern design of wind turbines with three wings, tip brakes, self-regulating and an asynchronous motor as generator. Foreign engineers named the Gedser wind turbine as 'The Danish Concept'.

Since then, the main aerodynamic concept has been this horizontal axis, three-bladed, upwind wind turbine connected to a three-phase electric grid, although many other different concepts have been developed and tested over the world with dissimilar results. An example of other concepts is the vertical axis wind turbine design by Darrieus, which provides a different mix of design tradeoffs from the conventional horizontal-axis wind turbine. The vertical orientation accepts wind from any direction with no need for adjustments, and the heavy generator and gearbox equipment can rest on the ground instead of on top of the tower.

The aim of wind turbine systems development is to continuously increase output power, as depicted in Fig. 1. Since the rated output power of production-type units reached 200 kW various decades ago, by 1999 the average output power of new installations climbed to 600 kW. Today, the manufactured turbines for onshore applications are specified to deliver 2-3 MW output power. In this sense, the world's first wind park with novel "multi-mega power class" 7 MW wind turbines was manufactured by the German wind turbine producer Enercon (11 E-126 units) and put into partial operation in Estinnes, Belgium, in 2010 (to be completed by July 2012). The key objective of this 77 MW pilot project is to introduce a new power class of large-scale wind energy converters (7 MW WECs) into the market with potential to significantly contribute to higher market penetration levels for wind electricity, especially in Europe. On the other hand, sea-based wind farms are likely to mean bigger turbines than on land, with models that produce up to three times the power of standard on-shore models. Series production of offshore wind turbines can reach to date up to 5 MW or more, being the largest onshore wind turbine presently under development a 10 MW unit. At least four companies are working on the development of this "giant power class" 10 MW turbine for sea-based applications, namely American Superconductors (U.S.), Wind Power (U.K.), Clipper Windpower (U.K.) and Sway (Norway). Even more, it is likely that in the near future, power rating of wind turbines will increase further, especially for large-scale offshore floating wind turbine applications.

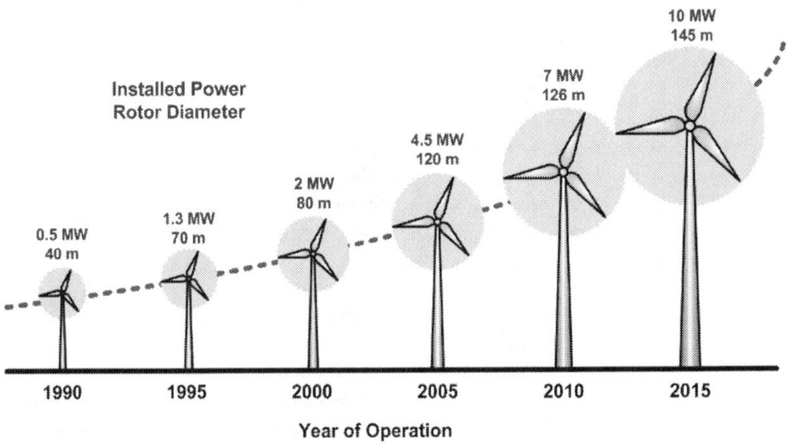

Fig. 1. Size evolution of wind turbines over time

4. Wind energy conversion

A wind turbine is a rotary engine that captures power from a fluid flow (the wind) using aerodynamically designed blades and convert it into useful mechanical power. The available power depends on the wind speed but it is important to be able to control and limit the power at higher wind speeds so as to avoid the damage of the unit. The power limitation may be done by some of the three following methods, namely stall control (the blade position is fixed but stall of the wind appears along the blade at higher wind speed), active stall (the blade angle is adjusted in order to create stall along the blades) or pitch control (the blades are turned out of the wind at higher wind speed). Essentially, three types of typical wind generator systems are the most widely spread. The first type is a constant-speed wind turbine system with a standard squirrel-cage induction generator (SCIG) directly connected to the grid. The second type is a variable speed wind turbine system with a doubly fed induction generator (DFIG). The power electronic converter feeding the rotor winding has a power rating of approximately up to 30% of the rated power; the stator winding of the DFIG is directly connected to the grid. The third type is a variable speed wind turbine with full-rated power electronic conversion system and a synchronous generator or a SCIG. A multi-stage gearbox is usually used with the first two types of generators. Synchronous generators, including permanent magnet synchronous generator (PMSG), may be direct driven though a low-ratio gear box system; one or two stage gearbox, becomes an interesting option. Fig. 2 summarized the major parts included in the mechanical and electrical power conversion of a typical wind turbine system (Chen & Blaabjerg, 2009).

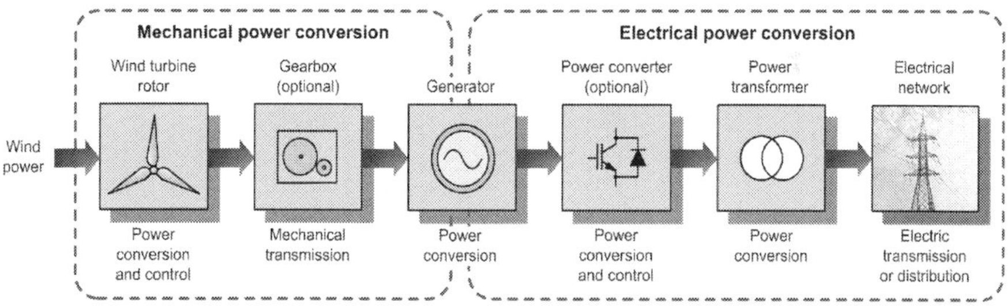

Fig. 2. General description of a wind turbine system

The appropriate voltage level is related to the generated power level. A modern wind turbine is often equipped with a transformer stepping up the generator terminal voltage, usually a voltage below 1 kV (E.g. 575 or 690 V), to a medium voltage around 20-30 kV, for the local electrical connection within a wind farm (distribution level). If the wind farm is large and the distance to the electrical grid is long, a transformer may be used to further step up the medium voltage in the wind farm to a high voltage at transmission level. For instance, for large onshore wind farms at hundreds of MWs, high voltage overhead lines above 100 kV are usually employed. For offshore wind farms with a long distance transmission to an onshore grid, the power generated is transferred by submarine cables buried in the sea bed. The cables between the turbines are linked to a transformer substation, which, at most cases, is placed offshore near the wind farm due to the long distance to shore (more than 5 km from the shore).

Basically, a wind energy conversion system consists of a turbine tower which carries the nacelle, and the wind turbine rotor, consisting of rotor blades and hub. Most modern wind turbines are horizontal-axis wind turbines (HAWTs) with three rotor blades usually placed upwind of the tower and the nacelle, as illustrated in Fig. 3. On the outside, the nacelle is usually equipped with anemometers and a wind vane to measure the wind speed and direction, as well as with aviation lights. The nacelle contains the key components of the wind turbine, i.e. the gearbox, mechanical brake, electrical generator, control systems, yaw drive, etc. The wind turbines are not only installed dispersedly on land, but also combined as wind farms (or parks) with capacities of hundreds MWs which are comparable with modern power generator units. Consequently, their performance could significantly affect power system operation and control (Hansen, et al. 2004).

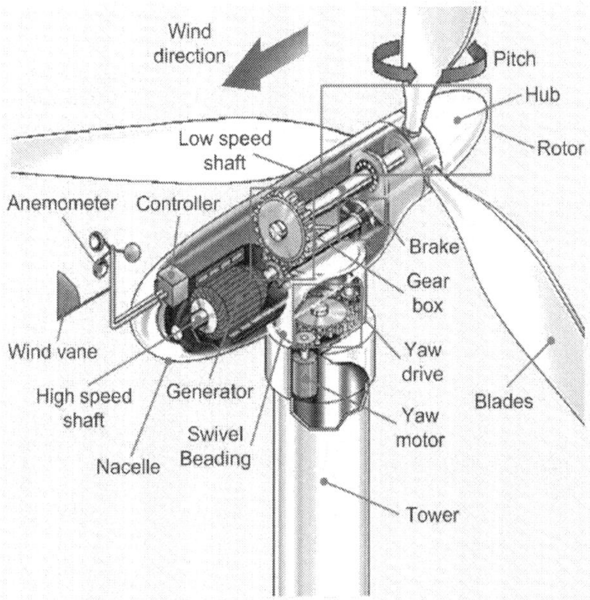

Fig. 3. Major components of a typical horizontal axis, three-bladed, upwind wind turbine

5. Wind turbine concepts

Wind turbines can either be designed to operate at fixed speed (actually within a speed range about 1%) or at variable speed. Many low-power wind turbines built to-date were constructed according to the so-called "Danish concept" that was very popular in the eighties, in which wind energy is transformed into electrical energy using a simple squirrel-cage induction machine directly connected to a three-phase power grid (Qiao et al., 2007). The rotor of the wind turbine is coupled to the generator shaft with a fixed-ratio gearbox. At any given operating point, this turbine has to be operated basically at constant speed. On the other hand, modern high-power wind turbines in the 2-10 MW range are mainly based on variable speed operation with blade pitch angle control obtained mainly by means of power electronic equipment, although variable generator rotor resistance can also be used. These wind turbines can be mostly developed using either a direct-in-line system built with a direct-driven (without gearbox) PMSG grid-connected via a full-scale power converter, or a doubly-fed induction generator (DFIG) system that consists of a DFIG with a partial-scale power converter connected to the rotor windings. Based on these concepts, the most commonly applied wind turbine designs can be classified into four wind turbine concepts, as described below.

5.1 Type A – Fixed speed wind turbine

This topology corresponds to the constant or fixed speed controlled wind turbine, with asynchronous squirrel cage induction generator (SCIG) directly connected to the electric grid using a step up power transformer, as depicted in Fig. 4. Since the squirrel cage induction generator always draws reactive power from the AC network, this concept requires a reactive power compensator, such as a capacitor bank, in order to reduce or even eliminate the reactive power demand from these turbine generators to the grid. It is typically achieved by continuously switching capacitor banks (5-25 steps) according to the active power generated. A smoother grid connection occurs by including a soft starter. Regardless the power control principle in a fixed speed machine, the wind fluctuations are converted into mechanical fluctuations and further into electrical power fluctuations. These can cause voltage fluctuations at the point of common coupling (PCC) of the wind turbine to the electric grid when the network is weak. Because of these voltage fluctuations, the fixed speed wind turbine draws fluctuating reactive power from the utility grid (in the case of no use of capacitor bank), which increases both the voltage fluctuations and the line losses. Fixed speed systems have the advantage of simplicity and low cost; however, the main drawbacks of this concept include the inability of supporting speed control, the requirement of a stiff grid (fixed voltage and frequency), and the necessity of a robust mechanical structure in order to support the high mechanical stress caused by wind gusts.

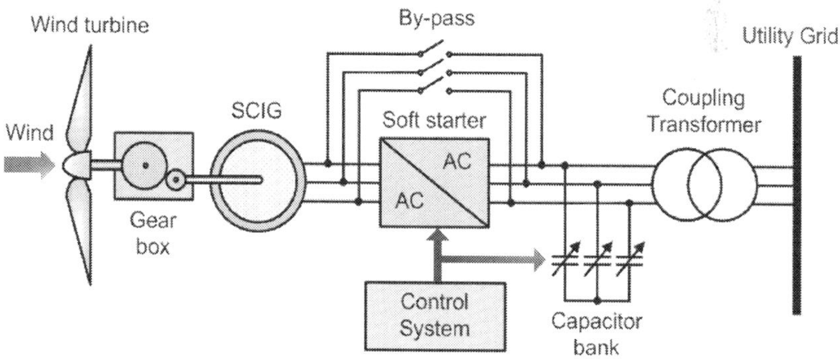

Fig. 4. Type A wind turbine concept: Fixed speed wind turbine directly connected to the electric grid via a squirrel cage induction generator

5.2 Type B – Partial variable speed wind turbine with variable rotor resistance

This topology corresponds to the partial variable speed controlled wind turbine with variable generator rotor resistance, aka OptiSlip by the Danish manufacturer Vestas™ Wind Systems (Krüger & Andresen, 2001), as presented in Fig. 5. It uses a wound rotor induction generator (WRIG) and has been used since the mid-nineties. In this case, the configuration is analogous to the type A wind turbine concept, with a generator directly connected to the electric grid. However, the rotor windings of the generator are connected in series with a controlled resistance, whose size defines the range of the variable speed (typically 0-10% above the synchronous speed). A capacitor bank performs reactive power compensation and a smoother grid connection is obtained by including a soft starter. The distinctive feature of this concept is that it has a variable additional rotor resistance, which is changed by an optical controlled converter mounted on the rotor shaft. Thus the rotor resistance is controllable, but eliminates the need for costly slip rings, which needs brushes and maintenance, through the optically coupling patented system. By varying the rotor resistance, the slip and thus the power output of the wind turbine can be controlled. The dynamic speed control range varies with the size of the variable rotor resistance and commonly reaches up to 10% above the synchronous speed. The energy coming from the external power conversion unit is dumped as heat loss.

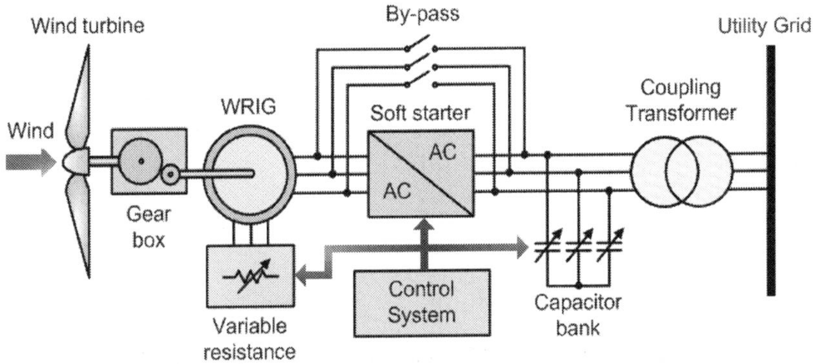

Fig. 5. Type B wind turbine concept: Partial variable speed wind turbine directly connected to the electric grid via a wound rotor induction generator with variable rotor resistance

5.2 Type C – Variable speed wind turbine with partial-scale power converter

This concept, aka doubly-fed induction generator (DFIG), corresponds to a variable speed controlled wind turbine with a wound rotor induction generator (WRIG) and a partial-scale power converter (rated approximately at 30% of nominal generated power) on the rotor circuit (Muller et al, 2002). The use of power electronic converters enables wind turbines to operate at variable (or adjustable) speed, and thus permits to provide more effective power capture than the fixed-speed counterparts. In addition, other significant advantages using variable speed systems include a decrease in mechanical loss, which makes possible lighter mechanical designs, and a more controllable power output (less dependent on wind variations), cost-effectiveness, simple pitch control, improved power quality and system efficiency, reduced acoustic noise, and island-operation capability. As shown in Fig. 6, the rotor stator is directly connected to the electric grid, while a partial-scale power converter controls the rotor frequency and consequently the rotor speed. The partial-scale power converter is composed of a back-to-back four-quadrant AC/DC/AC converter design based on insulated gate bipolar transistors (IGBTs), whose power rating defines the speed range (typically around ±30% of the synchronous speed). Moreover, this converter allows controlling the reactive power compensation and a smooth grid connection. The control range of the rotor speed is larger than that of the type C concept (Vestas´s OptiSlip). Even more, it captures the energy which is burned off in the controllable rotor resistance of the type C concept; this allows enhancing the efficiency of the overall system. The smaller power converter makes this concept attractive from an economical point of view. However, its main drawbacks are the use of slip rings, which needs brushes and maintenance, and the complex protection schemes in the case of grid faults.

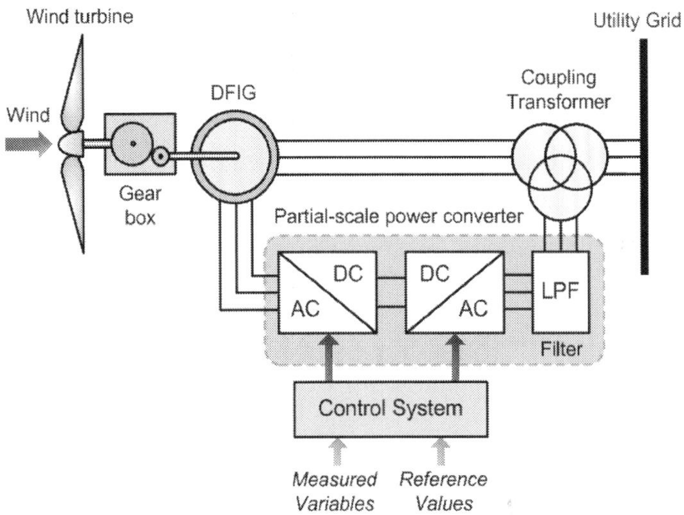

Fig. 6. Type C wind turbine concept: Variable speed wind turbine directly connected to the electric grid via a doubly-fed induction generator controlled with a partial-scale power converter

5.3 Type D – Direct-in-line variable speed wind turbine with full-scale power converter

This configuration corresponds to the direct-in-line full variable speed controlled wind turbine, with the generator connected to the electric grid through a full-scale power converter. A synchronous generator is used to produce variable frequency AC power. The power converter connected in series (or in-line) with the wind turbine generator transforms this variable frequency AC power into fixed-frequency AC power. This power converter also allows controlling the reactive power compensation locally generated, and a smooth grid connection for the entire speed range. The generator can be electrically excited (wound rotor synchronous generator, WRSG) or permanent magnet excited type (permanent magnet synchronous generator, PMSG). Recently, due to the development in power electronics technology, the squirrel-cage induction generator (SCIG) has also started to be used for this concept. As illustrated in Fig. 7, the generator stator is connected to the grid through a full-scale power converter, which is composed of a back-to-back four-quadrant AC/DC/AC converter design based on insulated gate bipolar transistors (IGBTs). Some full variable speed wind turbine systems have no gearbox (shown in dotted lines in Fig. 7) and use a direct driven multi-pole generator.

Direct-in-line variable speed wind turbines have several drawbacks respect to the former variable speed DFIG concepts, which mainly include the power converter and output filter ratings at about 1 p.u. of the total system power. This feature reduces the efficiency of the overall system and therefore results in a more expensive device. However, as the full scale power converter decouples entirely the wind turbine generator from the utility grid, grid codes such as fault ride through and grid support are easier to be accomplished, as required from modern applications. In addition, since a direct-in-line system can operate at low speeds, the gearbox can be omitted (direct-driven). Consequently, a gearless construction

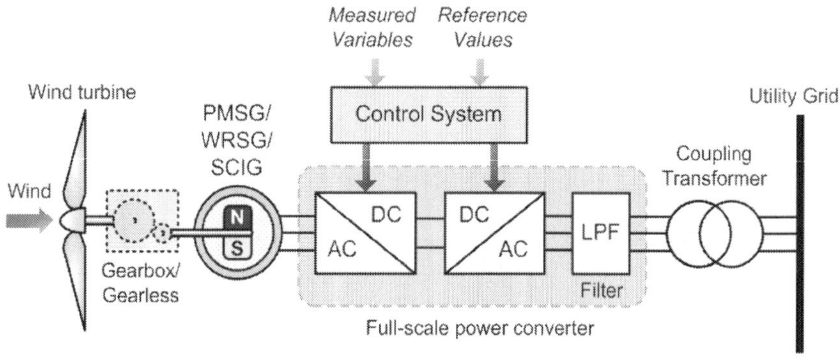

Fig. 7. Type D wind turbine concept: Direct-in-line variable speed wind turbine connected to the electric grid through a full-scale power converter

represents an efficient and robust solution that is beneficial, especially for offshore applications, where low maintenance requirements are essential. Moreover, using a permanent magnet synchronous generator, the DC excitation system is eliminated and allows reducing weight, losses, costs, and maintenance requirements (no slip rings are required). Even more, due to the intensified grid codes around the world, direct-driven PMSG wind turbine systems could be favored in the future compared to DFIG wind turbine concepts (Li et al., 2009).

6. Modelling of a direct-driven PMSG wind turbine system

This section presents the mathematical model for each component of the individual direct-in-line wind turbine system, including the wind turbine, the mechanical shaft system, the generator and the power electronic interface with the electric utility grid. The modeling approach of the proposed wind turbine system is based on the general structure presented in Fig. 7. The wind turbine generator considered in this work employs a direct-driven (without gearbox) PMSG directly coupled to the wind turbine and connected to the electric grid through the power conditioning system (PCS). The stator windings of the PMSG are directly connected to the PCS composed of a full-scale power converter built using a back-to-back AC/DC/AC power converter topology which includes a three-phase rectifier bridge (AC/DC conversion), a DC/DC converter and a grid-side converter with an intermediate DC link (DC/AC conversion) (Blaabjerg et al., 2004).

6.1 Wind turbine

The wind turbine employed in this work is a classic three-bladed horizontal-axis (main shaft) wind turbine design. This turbine was implemented and characterized using a laboratory-scale 0.5 kW rated power (at 12.5 m/s) prototype. Since the turbine corresponds to a small-scale one, no active blade pitch control is implemented and instead a self-regulation (passive stall control) through blades twisting is employed.
The proposed model is based on the steady-state aerodynamic power characteristics of the wind turbine. The output mechanical power available from a variable speed wind turbine can be expressed through the following algebraic relation (Freris, 1990; Ackermann, 2005).

$$P_m = \frac{1}{2} \rho \, A \, v^3 \, C_p(\lambda, \beta) \,, \tag{1}$$

where:

ρ: air density (typically 1.225 kg/m³ at sea level with standard conditions, i.e. temperature of 15 °C and atmospheric pressure of 101.325 kPa)

A: area swept by the rotor blades

v: wind speed

C_p: so-called power coefficient (aka coefficient of performance) of the wind turbine.

The power coefficient C_p is a nonlinear function of the blade pitch angle β and the tip-speed ratio λ as given by equation (2).

$$\lambda = \left(\frac{R \, \omega_m}{v} \right) , \tag{2}$$

with:

R: radius of the turbine blades

ω_m: angular speed of the turbine rotor

As can be derived from equation (1), the power coefficient C_p is given in terms of the blade pitch angle β and the tip-speed ratio λ. Since the proposed wind turbine can operate over a wide range of rotor speeds, the assumption of linear torque versus speed characteristic (at a given wind speed and blade pitch angle) cannot be used and thus the aerodynamic system results very complex to be analytically determined. Consequently, numerical approximations have been developed in order to calculate the mechanical power characteristic of the wind turbine and a bi-dimensional characteristic function of C_p has been used (Raiambal & Chellamuthu, 2002) and validated in the laboratory.

$$C_p(\lambda, \beta) = \frac{1}{2} \left(\frac{98}{\lambda_i} - 0.4\beta - 5 \right) \exp\left(\frac{-16.5}{\lambda_i} \right) , \tag{3}$$

with:

$$\lambda_i = \left[\frac{1}{(\lambda + 0.089)} - \frac{0.035}{(\beta^3 + 1)} \right]^{-1} \tag{4}$$

The characteristic function C_p vs. λ, for various values of the pitch angle β, is shown in Fig. 8. The maximum value of C_p, that is C_{pmax}=0.47, is achieved for β=0° and for λ=6.75. This particular value λ_{opt} results in the point of optimal efficiency where the maximum power is captured from wind by the wind turbine.

Fig. 9 illustrates the mechanical power versus the rotating speed of the proposed wind power system with no blade pitch angle control (β=0°) at various wind speeds. As can be derived, there exists a good agreement between the results obtained with both scaled laboratory prototype and the corresponding model. It can be also observed that, for each wind speed, there exists a specific point in the wind generator power characteristic, aka maximum power point (MPP), where the output power is maximized. Thus, the control of the wind turbine power results in a variable speed operation aiming at tracking the MPP for

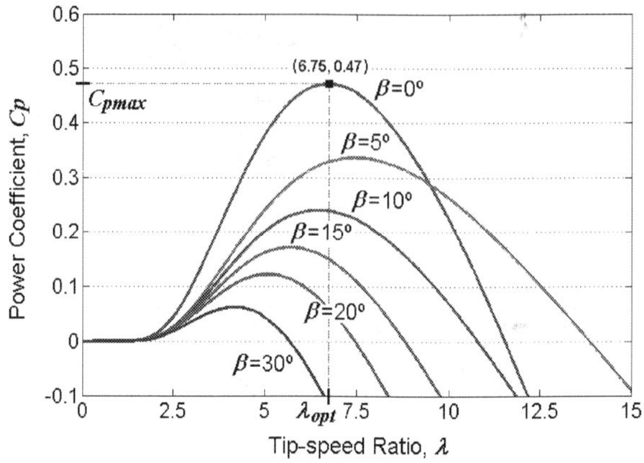

Fig. 8. Characteristics function C_p vs. λ, at various pitch angle values for the studied wind turbine generator

the particular operating conditions such that the maximum power can be continuously extracted from the wind (MPP tracking control or MPPT).

6.2 Mechanical shaft system

The mechanical shaft system of the WTG can be usually represented either by a two-mass system or by a single-mass lumped-parameter system. In the case of direct-in-line variable speed wind power systems, because the wind turbine is connected to the electric grid through a full-scale converter, the shaft properties are hardly reflected at the grid connection due to the decoupling effect of the power conditioning system. In this way, the turbine rotor

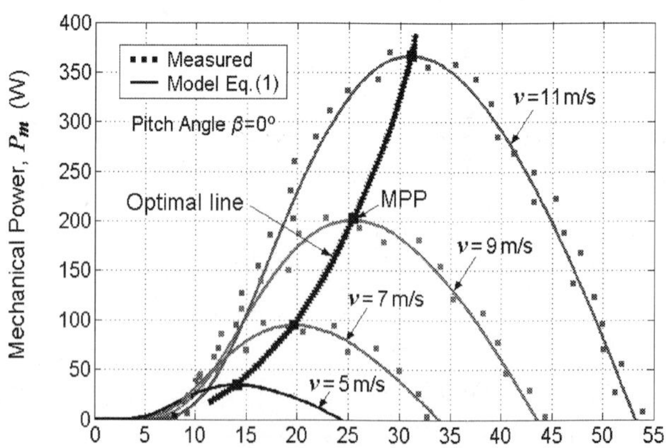

Fig. 9. Mechanical power versus rotor speed curves measurements and simulations at various wind speeds for the studied wind turbine generator

is modeled as a lumped mass and the shaft dynamics is neglected. Even more, since multi-pole PMSGs have increasingly been used in variable speed generators, in such a way that they can operate at low speeds, the gearbox can be omitted. Consequently, a direct-driven (gearless) construction represents an efficient and robust solution that is beneficial.
The wind turbine rotor dynamics can be modeled as:

$$T_e = T_l + B_f \omega_m + J_c \frac{d\omega_m}{dt} \tag{5}$$

where:

T_e: electromagnetic torque of the electric machine
T_l: load torque
B_f: viscous friction coefficient
J_c: combined inertia moment of the WTG rotor and PMSG
ω_m: rotor mechanical speed, which is related to the rotor angular speed of the electric machine ω_s, through:

$$\omega_m = \frac{\omega_s}{p_p} \tag{6}$$

with p_p being the number of pole-pairs of the generator.

6.3 Permanent magnet synchronous generator

The permanent magnet synchronous machine can be electrically described in steady-state using a simple equivalent circuit with an armature equation including back electromotive forces (emfs). This model assumes that saturation is neglected, the induced emfs are sinusoidal, the Eddy currents and hysteresis losses are negligible, and that there are no field current dynamics. In this way, voltage equations for the PMSG are given by:

$$\begin{bmatrix} u_{am} \\ u_{bm} \\ u_{cm} \end{bmatrix} - \begin{bmatrix} u_a \\ u_b \\ u_c \end{bmatrix} = (\mathrm{R_m} + s\mathrm{L}) \begin{bmatrix} i_{am} \\ i_{bm} \\ i_{cm} \end{bmatrix} \tag{7}$$

where:

$$\mathrm{R_m} = \begin{bmatrix} R_m & 0 & 0 \\ 0 & R_m & 0 \\ 0 & 0 & R_m \end{bmatrix}, \qquad \mathrm{L} = \begin{bmatrix} L_{aa} & L_{ab} & L_{ac} \\ L_{ab} & L_{bb} & L_{bc} \\ L_{ac} & L_{bc} & L_{cc} \end{bmatrix}, \tag{8}$$

being:

s: Laplace variable, with $s = d/dt$ for t > 0 (Heaviside operator p also used)
u_{im} (i=a,b,c): stator phase voltages in a-b-c coordinates
u_i: back emfs in a-b-c coordinates
i_{im}: stator currents in a-b-c coordinates
L_{ij}: stator winding inductances, including self and mutual ones (combinations of i and j=a, b, c). It is considered symmetry for mutual inductances, so that L_{ij}=L_{ji}

The terminal voltages applied from the machine-side converter to the stator, u_{im} and the back emfs, u_i are balanced three-phase voltages, being the later defined as follows:

$$u_i = \omega_s \Psi_{mi} \tag{9}$$

with:

Ψ_{mi}: permanent-magnet flux linkage in *a-b-c* coordinates

Since there is no functional equation for instantaneous reactive power in the *a-b-c* reference frame, it is useful to apply a transformation to the synchronous-rotating orthogonal *d-q* set aligned with the rotor flux, to equations (7) and (8) in order to analyze the electric machine. This is performed by applying Park's transformation and defining the *q*-axis to be always coincident with the instantaneous stator magneto-motive forces (mmfs), which rotate at the same angular speed as that of the rotor (yielding u_q equals $|u|$, while u_d is null). This is beneficial because any AC signals that spins at ω_s become DC quantities in the rotor *d-q* frame. Then, by neglecting the zero sequence components, equations (10) and (11) are derived.

$$\begin{bmatrix} u_{dm} \\ u_{qm} \end{bmatrix} - \begin{bmatrix} u_d \\ u_q \end{bmatrix} = (\mathrm{R_m} + s\mathrm{L'_s}) \begin{bmatrix} i_{dm} \\ i_{qm} \end{bmatrix} + \begin{bmatrix} -\omega_s & 0 \\ 0 & \omega_s \end{bmatrix} \mathrm{L'_s} \begin{bmatrix} i_{qm} \\ i_{dm} \end{bmatrix} \tag{10}$$

where:

$$\mathrm{R_m} = \begin{bmatrix} R_m & 0 \\ 0 & R_m \end{bmatrix}, \qquad \mathrm{L'_s} = \begin{bmatrix} L_d & 0 \\ 0 & L_q \end{bmatrix}, \qquad u_d = \omega_s \Psi_{qm}, \qquad u_q = \omega_s \Psi_{dm} \tag{11}$$

Flux Linkages in the *d-q* frame can be expressed in terms of the stator currents, inductances, and the flux linkage due to the permanent magnets of the rotor linking the stator, Ψ_m as:

$$\Psi_{dm} = L_d i_{dm} + \Psi_m \tag{12}$$

$$\Psi_{qm} = L_q i_{qm} \tag{13}$$

By rewriting equation (10), the following state equation is obtained:

$$s \begin{bmatrix} i_{dm} \\ i_{qm} \end{bmatrix} = \begin{bmatrix} \dfrac{-R_m}{L_d} & \omega_s \\ -\omega_s & \dfrac{-R_m}{L_q} \end{bmatrix} \begin{bmatrix} i_{dm} \\ i_{qm} \end{bmatrix} + \begin{bmatrix} \dfrac{u_{dm}}{L_d} \\ \dfrac{u_{qm} - |u|}{L_q} \end{bmatrix} \tag{14}$$

being $|u| = \omega_s \Psi_m$

In the rotor *d-q* frame, the active and reactive power flows are calculated as follows:

$$p = \frac{3}{2}(v_{dm} i_{dm} + v_{qm} i_{qm}) \tag{15}$$

$$q = \frac{3}{2}(v_{dm} i_{qm} - v_{qm} i_{dm}) \tag{16}$$

The developed electromagnetic torque of the electric machine takes the following convenient form:

$$T_e = \frac{3}{2} p_p \left[\psi_m i_{qm} + (L_d - L_q)\, i_{dm} i_{qm} \right] \tag{17}$$

For a non-salient-pole machine, as the considered in this analysis, the stator winding direct and quadrature inductances L_d and L_q, are approximately equal. Indeed this application uses a surface mount permanent magnet synchronous machine (SPMSM) which has zero saliency. This means that the direct-axis current i_{dm} does not contribute to the electrical torque T_e. The key concept is to keep null the direct current i_{dm} using appropriate transformation synchronization in order to obtain maximal torque with minimum current i_{qm} (Li et al., 2009).

6.4 Power conditioning system

The power conditioning system (PCS) used for connecting renewable energy sources to the distribution utility grid requires generation of high quality electric power, being at the same time flexible, efficient and reliable (Mohan et al., 1995). Fig. 10 shows the detailed model of a modern direct-in-line variable-speed direct-driven PMSG wind turbine connected to the utility distribution grid, which is composed of a back-to-back AC/DC/AC power converter that fulfills all the requirements mentioned above (Carrasco et al., 2006).

Since the permanent magnet synchronous generator produces an output voltage with variable amplitude and frequency, additional conditioning is required to meet the amplitude and frequency requirements of the roughly stiff utility grid. A three-phase uncontrolled full-wave rectifier bridge is proposed here for performing the AC/DC conversion. This device has the benefit of being simple, robust, cheap, and needs no control system. On the other hand, a three-phase three-level DC/AC voltage source inverter (VSI) using IGBTs is employed for connecting to the grid. As the power rating of the inverter is lesser than a few MWs, the output voltage control of the VSI can be achieved through pulse width modulation (PWM) techniques. This power inverter topology is proposed above other ones because generates a more sinusoidal output voltage waveform than conventional structures without increasing the switching frequency. In this way, the harmonic performance of the inverter is improved, also obtaining better efficiency and reliability respect to the conventional two-level inverter topology. The connection to the utility grid is made through a step-up transformer and a low pass filter in order to reduce the perturbation on the distribution system from high-frequency switching harmonics generated by the PWM control.

As the VSI needs a fixed DC link in order to allow a decoupled control of both active and reactive power exchange with the electric grid, an interface in the DC side of the VSI is required. For this purpose, an intermediate DC/DC converter in a boost topology is used, linking the output of the full-wave rectifier bridge to the DC bus of the power inverter. Only one power switching device is used in the DC/DC converter, resulting in a low cost and simple control. This two-stage AC/DC energy conversion system offers an additional degree of freedom in the operation of the system when compared with conventional one-stage configurations, permitting to pursue various control objectives simultaneously with the WTG system operation.

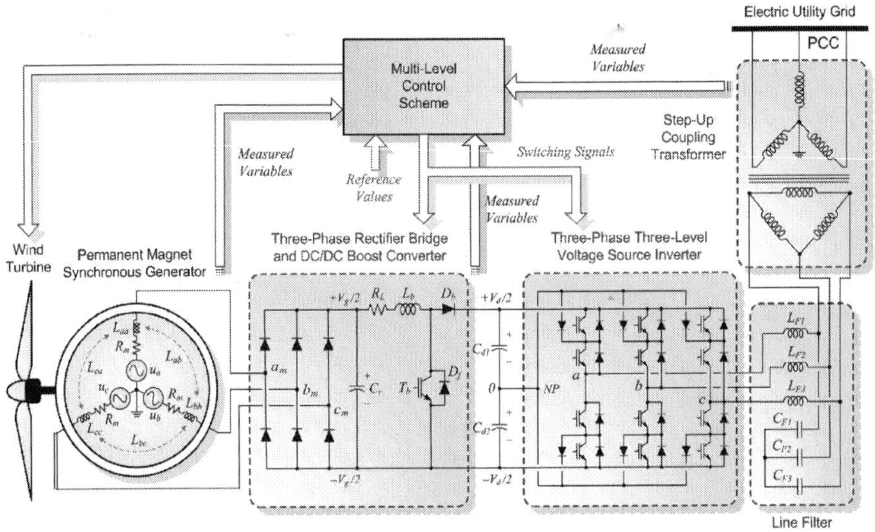

Fig. 10. Detailed model of a modern variable-speed direct-driven PMSG wind turbine connected to the utility distribution grid

6.4.1 DC/DC Converter

The standard unidirectional topology of the DC/DC boost converter (also known as step-up converter or chopper) consist of a switching-mode power device containing basically two semiconductor switches (a rectifier diode and a power transistor with its corresponding anti-parallel diode) and two energy storage devices (an inductor and a smoothing capacitor) for producing an output DC voltage at a level greater than its input dc voltage. This converter acts as an interface between the full-wave rectifier bridge and the VSI, by employing pulse-width modulation (PWM) control techniques.

Operation of the DC/DC converter in the continuous (current) conduction mode (CCM), i.e. the current flowing continuously in the inductor during the entire switching cycle, makes simple the development of the state-space model because only two switch states are possible during a switching cycle, namely, (*i*) the power switch T_b is *on* and the diode D_b is *off*; or (*ii*) T_b is *off* and D_b is *on*.

In steady-state CCM operation and neglecting parasitic components, the state-space equation that describes the dynamics of the DC/DC boost converter is given by equation (18) (Molina & Mercado, 2008).

$$\begin{bmatrix} sI_L \\ sV_d \end{bmatrix} = \begin{bmatrix} 0 & -\frac{1-S_{dc}}{L} \\ -\frac{1-S_{dc}}{C} & 0 \end{bmatrix} \begin{bmatrix} I_L \\ V_d \end{bmatrix} + \begin{bmatrix} \frac{1}{L} & 0 \\ 0 & -\frac{1}{C} \end{bmatrix} \begin{bmatrix} V_g \\ I_d \end{bmatrix}, \tag{18}$$

where:

I_L: chopper input current (inductor current).

V_g: chopper input voltage, the same as the three-phase rectifier output voltage.
V_d: chopper output voltage, coinciding with the inverter DC bus voltage.
i_d: chopper output current.
S_{dc}: switching function of the boost DC/DC converter.

The switching function S_{dc} of the power converter is a two-levelled waveform characterizing the signal that drives the power switch T_b of the DC/DC boost converter, defined as follows:

$$S_{dc} = \begin{cases} 0, \text{for the switch } T_b \textit{ off} \\ 1, \text{for the switch } T_b \textit{ on} \end{cases} \tag{19}$$

If the switching frequency of the power switches is significantly higher than the natural frequencies of the DC/DC converter, this discontinuous model can be approximated by a continuous state-space averaged (SSA) model, where a new variable D is introduced. In the [0, 1] interval, D is a continuous function and represents the modulation index of the DC/DC converter, defined as the ratio of time during which the power switch T_b is turned-on to the period of one complete switching cycle, T_s. This variable is used for replacing the switching function in equation (18), yielding the following SSA expression:

$$\begin{bmatrix} sI_L \\ sV_d \end{bmatrix} = \begin{bmatrix} 0 & -\frac{1-D}{L} \\ -\frac{1-D}{C} & 0 \end{bmatrix} \begin{bmatrix} I_L \\ V_d \end{bmatrix} + \begin{bmatrix} \frac{1}{L} & 0 \\ 0 & -\frac{1}{C} \end{bmatrix} \begin{bmatrix} V_g \\ I_d \end{bmatrix} \tag{20}$$

Since, in steady-state conditions the inductor current variation during both, on and off times of T_b are essentially equal, so there is not net change of the inductor current from cycle to cycle, and assuming a constant Dc output voltage of the boost converter, the steady-state input-to-output voltage conversion relationship of the boost converter is easily derived from equation (20) by setting the inductor current derivative at zero, yielding equation (21):

$$V_d = \frac{V_g}{(1-D)} \tag{21}$$

In the same way, the relationship between the average inductor current I_L and the DC/DC converter output current I_d in the CCM can be derived, as follows:

$$I_d = (1-D)\, I_L \tag{22}$$

6.4.2 Voltage source inverter

The three-phase three-level voltage source inverter proposed corresponds to a DC/AC switching power inverter using IGBTs operated through sinusoidal PWM (Molina & Mercado, 2006). The VSI structure proposed is designed to make use of a three-level twelve pulse pole structure, also called neutral point clamped (NPC), instead of a standard two-level six pulse inverter structure (Rodriguez et al., 2002, Soto & Green, 2002). This three-level VSI topology generates a more smoothly sinusoidal output voltage waveform than conventional two-level structures without increasing the switching frequency and effectively doubles the power rating of the VSI for a given semiconductor device. Moreover, the three level pole attempts to address some limitations of the standard two-level by offering an additional flexibility of a level in the output voltage, which can be controlled in duration, either to vary the fundamental output voltage or to assist in the output waveform construction. This extra feature is used here to assist in the output waveform structure. In this way, the harmonic performance of the inverter is improved, also obtaining better efficiency and reliability. The output line voltage waveforms of a three-level VSI connected to a 380 V utility system are shown in Fig. 11. It is to be noted that in steady-state the VSI generates at its output terminals a switched line voltage waveform with high harmonics content, reaching the voltage total harmonic distortion (VTHD) almost 45% when unloaded. At the output terminals of the low pass sine wave filters proposed, the VTHD is reduced to as low as 1%, decreasing this quantity to even a half at the coupling transformer secondary output terminals (PCC). In this way, the quality of the voltage waveforms introduced by the PWM control to the power utility is improved and the requirements of IEEE Standard 519-1992 relative to power quality (VTHD limit in 5%) are entirely fulfilled (Bollen, 2000).

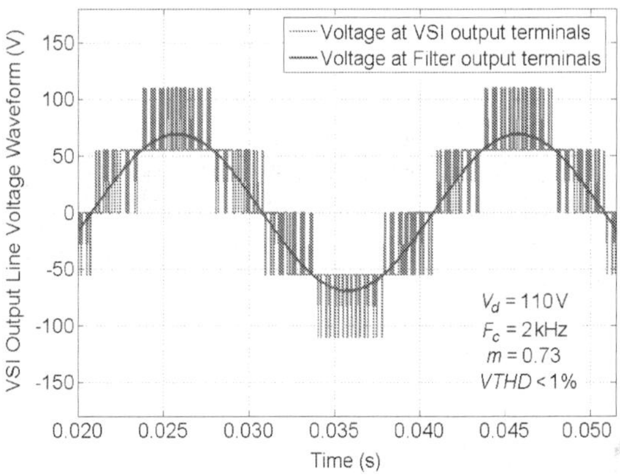

Fig. 11. Three-level NPC voltage source inverter output line voltage waveforms

The mathematical equations describing and representing the operation of the voltage source inverter can be derived from the detailed model shown in Fig. 10 by taking into account some assumptions respect to its operating conditions. For this purpose, a simplified equivalent VSI connected to the electric system is considered, also referred to as an averaged model, which assumes the inverter operation under balanced conditions as ideal, i.e. the voltage source inverter is seen as an ideal sinusoidal voltage source operating at fundamental frequency. This consideration is valid since, as shown in Fig. 11, the high-frequency harmonics produced by the inverter as result of the sinusoidal PWM control techniques are mostly filtered by the low pass sine wave filters and the net instantaneous output voltages at the point of common coupling resembles three sinusoidal waveforms phase-shifted 120° between each other.

This ideal inverter is shunt-connected to the network at the PCC through an equivalent inductance L_s, accounting for the leakage of the step-up coupling transformer and an equivalent series resistance R_s, representing the transformers winding resistance and VSI semiconductors conduction losses. The magnetizing inductance of the step-up transformer can also be taken into consideration through a mutual equivalent inductance M. In the DC side, the equivalent capacitance of the two DC bus capacitors, C_{d1} and C_{d2} (C_{d1}=C_{d2}), is described through C_d=$C_{d1}/2$=$C_{d2}/2$ whereas the switching losses of the VSI and power losses in the DC capacitors are considered by a parallel resistance R_p. As a result, the dynamics equations governing the instantaneous values of the three-phase output voltages in the AC side of the VSI and the current exchanged with the utility grid can be directly derived by applying Kirchhoff's voltage law (KVL) as follows:

$$\begin{bmatrix} v_{inv_a} \\ v_{inv_b} \\ v_{inv_c} \end{bmatrix} - \begin{bmatrix} v_a \\ v_b \\ v_c \end{bmatrix} = \left(\mathrm{R_s} + s\mathrm{L_s}\right) \begin{bmatrix} i_a \\ i_b \\ i_c \end{bmatrix}, \tag{23}$$

where:

$$\mathrm{R_s} = \begin{bmatrix} R_s & 0 & 0 \\ 0 & R_s & 0 \\ 0 & 0 & R_s \end{bmatrix}, \quad \mathrm{L_s} = \begin{bmatrix} L_s & M & M \\ M & L_s & M \\ M & M & L_s \end{bmatrix} \tag{24}$$

Under the assumption that the system has no zero sequence components (operation under balanced conditions), all currents and voltages can be uniquely transformed into the synchronous-rotating orthogonal two-axes reference frame, in which each vector is described by means of its d and q components, instead of its three a, b, c components. Thus, the new coordinate system is defined with the d-axis always coincident with the instantaneous voltage vector, as described in Fig. 12. By defining the d-axis to be always coincident with the instantaneous voltage vector v, yields v_d equals $|v|$, while v_q is null. Consequently, the d-axis current component contributes to the instantaneous active power and the q-axis current component represents the instantaneous reactive power. This operation permits to develop a simpler and more accurate dynamic model of the inverter.
By applying Park's transformation (Krause, 1992) stated by equation (25), equations (23) and (24) can be transformed into the synchronous rotating d-q reference frame as follows (equation (26)):

$$\mathrm{K_s} = \frac{2}{3} \begin{bmatrix} \cos\theta & \cos\left(\theta - \frac{2\pi}{3}\right) & \cos\left(\theta + \frac{2\pi}{3}\right) \\ -\sin\theta & -\sin\left(\theta - \frac{2\pi}{3}\right) & -\sin\left(\theta + \frac{2\pi}{3}\right) \\ \frac{1}{2} & \frac{1}{2} & \frac{1}{2} \end{bmatrix}, \tag{25}$$

with:

$\theta = \int_0^t \omega(\xi)d\xi + \theta(0)$: angle between the d-axis and the reference phase axis, and ξ: integration variable

ω: synchronous angular speed of the network voltage at the fundamental system frequency f (50 Hz throughout this chapter).

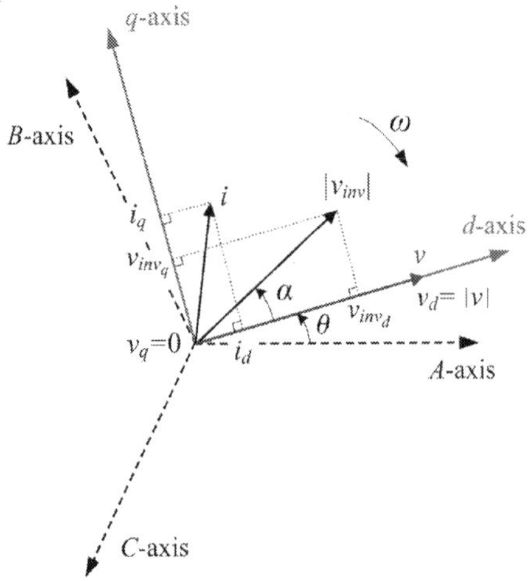

Fig. 12. Voltage source inverter vectors in the synchronous rotating *d-q* reference frame

Thus,

$$\begin{bmatrix} v_{inv_d} - v_d \\ v_{inv_q} - v_q \\ v_{inv_0} - v_0 \end{bmatrix} = \mathrm{K_s} \begin{bmatrix} v_{inv_a} - v_a \\ v_{inv_b} - v_b \\ v_{inv_c} - v_c \end{bmatrix}, \qquad \begin{bmatrix} i_d \\ i_q \\ i_0 \end{bmatrix} = \mathrm{K_s} \begin{bmatrix} i_a \\ i_b \\ i_c \end{bmatrix} \tag{26}$$

Then, by neglecting the zero sequence components, equations (27) and (28) are derived.

$$\begin{bmatrix} v_{inv_d} \\ v_{inv_q} \end{bmatrix} - \begin{bmatrix} v_d \\ v_q \end{bmatrix} = (\mathrm{R_s} + s\mathrm{L'_s}) \begin{bmatrix} i_d \\ i_q \end{bmatrix} + \begin{bmatrix} -\omega & 0 \\ 0 & \omega \end{bmatrix} \mathrm{L'_s} \begin{bmatrix} i_q \\ i_d \end{bmatrix}, \tag{27}$$

where:

$$\mathrm{R_s} = \begin{bmatrix} R_s & 0 \\ 0 & R_s \end{bmatrix}, \qquad \mathrm{L'_s} = \begin{bmatrix} L'_s & 0 \\ 0 & L'_s \end{bmatrix} = \begin{bmatrix} L_s - M & 0 \\ 0 & L_s - M \end{bmatrix} \tag{28}$$

It is to be noted that the coupling of phases *a-b-c* through the term M in matrix $\mathrm{L_s}$ (equation (24)), was fully eliminated in the *d-q* reference frame when the VSI transformers are magnetically symmetric, as is usually the case. This decoupling of phases in the synchronous-rotating system allows simplifying the control system design.
By rewriting equation (27), the following state equation can be obtained:

$$s \begin{bmatrix} i_d \\ i_q \end{bmatrix} = \begin{bmatrix} \dfrac{-R_s}{L'_s} & \omega \\ -\omega & \dfrac{-R_s}{L'_s} \end{bmatrix} \begin{bmatrix} i_d \\ i_q \end{bmatrix} + \frac{1}{L'_s} \begin{bmatrix} v_{inv_d} - |v| \\ v_{inv_q} \end{bmatrix} \tag{29}$$

A further major issue of the *d-q* transformation is its frequency dependence (ω). In this way, with appropriate synchronization to the network (through angle θ), the control variables in steady state are transformed into DC quantities. This feature is quite useful to develop an efficient decoupled control system of the two current components. Although the model is fundamental frequency-dependent, the instantaneous variables in the *d-q* reference frame contain all the information concerning the three-phase variables, including steady-state unbalance, harmonic waveform distortions and transient components.

The relation between the DC side voltage V_d and the generated AC voltage v_{inv} can be described through the average switching function matrix in the *d-q* reference frame $\mathbf{S}_{\mathbf{av},dq}$ of the proposed inverter, as given by equation (30). This relation assumes that the DC capacitors voltages are balanced and equal to $V_d/2$.

$$\begin{bmatrix} v_{inv_d} \\ v_{inv_q} \end{bmatrix} = \mathbf{S}_{\mathbf{av},\,dq}\, V_d\,, \tag{30}$$

and the average switching function matrix in *d-q* coordinates is computed as:

$$\mathbf{S}_{\mathbf{av},\,dq} = \begin{bmatrix} S_{av,d} \\ S_{av,q} \end{bmatrix} = \frac{1}{2} m_i\, a \begin{bmatrix} \cos\alpha \\ \sin\alpha \end{bmatrix}, \tag{31}$$

being,

m_i: modulation index of the voltage source inverter, $m_i \in [0, 1]$.

α: phase-shift of the inverter output voltage from the reference position,

$a = \dfrac{\sqrt{3}}{\sqrt{2}}\dfrac{n_2}{n_1}$: turns ratio of the step-up Δ-Y coupling transformer,

The AC power exchanged by the inverter is related to the DC bus power on an instantaneous basis in such a way that a power balance must exist between the input and the output of the inverter. In this way, the AC power should be equal to the sum of the DC resistance (R_p) power, representing losses (IGBTs switching and DC capacitors) and to the charging rate of the DC equivalent capacitor (C_d) (neglecting the wind generator action):

$$P_{AC} = P_{DC} \tag{32}$$

$$\frac{3}{2}\left(v_{inv_d} i_d + v_{inv_q} i_q\right) = -\frac{C_d}{2} V_d s V_d - \frac{V_d^2}{R_p} \tag{33}$$

Essentially, equations (23) through (33) can be summarized in the state-space as described by equation (34). This continuous state-space averaged mathematical model describes the steady-state dynamics of the ideal voltage source inverter in the *d-q* reference frame, and will be subsequently used as a basis for designing the middle level control scheme to be proposed. As reported by Acha et al. (2002), modelling of static inverters by using a synchronous-rotating orthogonal *d-q* reference frame offer higher accuracy than employing stationary coordinates. Moreover, this operation allows designing a simpler control system than using *a-b-c* or α-β.

$$s\begin{bmatrix} i_d \\ i_q \\ V_d \end{bmatrix} = \begin{bmatrix} \frac{-R_s}{L'_s} & \omega & \frac{S_{av,d}}{2L'_s} \\ -\omega & \frac{-R_s}{L'_s} & \frac{S_{av,q}}{2L'_s} \\ -\frac{3}{2C_d}S_{av,d} & -\frac{3}{2C_d}S_{av,q} & -\frac{2}{R_pC_d} \end{bmatrix} \begin{bmatrix} i_d \\ i_q \\ V_d \end{bmatrix} - \begin{bmatrix} \frac{|v|}{L'_s} \\ 0 \\ 0 \end{bmatrix} \tag{34}$$

7. Control strategy of the direct-driven PMSG wind turbine system

The proposed hierarchical three-level control scheme for the grid-connected direct-in-line wind turbine system is depicted in Fig. 13. This control system consists of three distinct blocks, namely the external, middle and internal level. Its design is based on concepts of instantaneous power on the synchronous-rotating *d-q* reference frame. This structure has the goal of rapidly and simultaneously controlling the active and reactive power provided by the wind turbine system (Molina & Mercado, 2009).

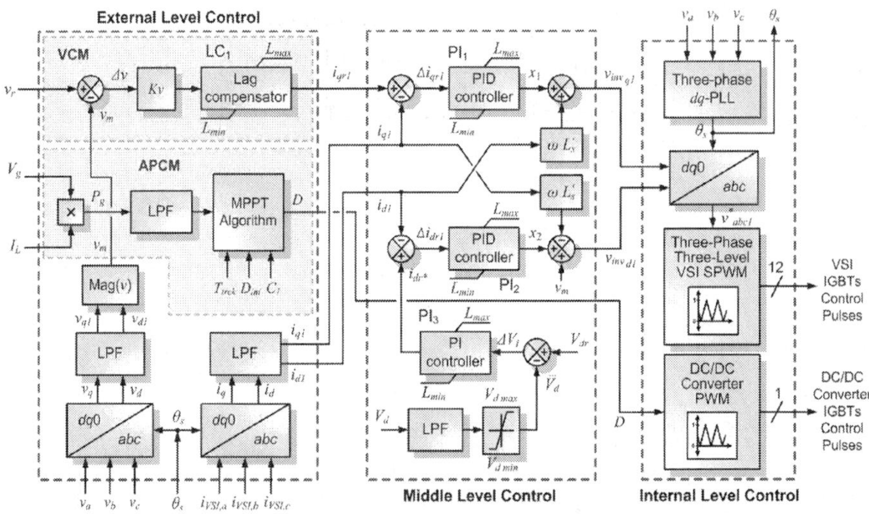

Fig. 13. Multi-level control scheme for the proposed three-phase grid-connected wind energy conversion system

7.1 External level control

The external level control, which is outlined in Fig. 13 (left side) in a simplified form, is responsible for determining the active and reactive power exchange between the WECS system and the utility grid. This control strategy is designed for performing two major control objectives: the voltage control mode (VCM) with only reactive power compensation capabilities and the active power control mode (APCM) for dynamic active power exchange with the AC network. To this aim, the instantaneous voltage at the PCC is computed by employing a synchronous-rotating reference frame. In consequence, by applying Park's transformation, the instantaneous values of the three-phase AC bus voltages are transformed into d-q components, v_d and v_q respectively, and then filtered to extract the fundamental components, v_{d1} and v_{q1}. As formerly described, the *d-axis* was defined always coincident with the instantaneous voltage vector v, then v_{d1} results in steady-state equal to $|v|$ while v_{q1} is null. Consequently, the d-axis current component of the VSI contributes to the instantaneous active power p while the q-axis current component represents the instantaneous reactive power q, as stated in equations (35) and (36). Thus, to achieve a decoupled active and reactive power control, it is required to provide a decoupled control strategy for i_{d1} and i_{q1} (Timbus et al., 2009).

$$p = \frac{3}{2}(v_{d1}i_{d1} + v_{q1}i_{q1}) = \frac{3}{2}|v|\, i_{d1}\,, \tag{35}$$

$$q = \frac{3}{2}(v_{d1}i_{q1} - v_{q1}i_{d1}) = \frac{3}{2}|v|\, i_{q1}\,, \tag{36}$$

In this way, only v_d is used for computing the resultant current reference signals required for the desired SMES output active and reactive powers. Independent limiters are use for restrict both the power and current signals before setting the references i_{dr1} and i_{qr1}. Additionally, the instantaneous actual output currents of the wind turbine system, i_{d1} and i_{q1}, are computed for use in the middle level control. In all cases, the signals are filtered by using second-order low-pass filters to obtain the fundamental components employed by the control system.

The standard control loop of the external level is the VCM and consists in controlling (supporting and regulating) the voltage at the PCC through the modulation of the reactive component of the inverter output current, i_{q1}. This control mode has proved a very good performance in conventional reactive power static controllers. The design of this control loop in the rotating frame is simpler than using stationary frame techniques, and employs a standard proportional-integral (PI) compensator including an anti-windup system to enhance the dynamic performance of the VCM system. This control mode compares the reference voltage set by the operator with the actual measured value in order to eliminate the steady-state voltage offset via the PI compensator. A voltage regulation droop (typically 5%) R_d is included in order to allow the terminal voltage of the WTG to vary in proportion with the compensating reactive current. Thus, the PI controller with droop characteristics becomes a simple phase-lag compensator (LC_1), resulting in a stable fast response compensator. This feature is particularly significant in cases that more high-speed voltage compensators are operating in the area. This characteristic is comparable to the one included in generators´ voltage regulators.

The main purpose of a grid-connected wind turbine system is to transfer the maximum wind generator power into the electric system. In this way, the APCM aims at matching the active power to be injected into the electric grid with the maximum instant power generated by the wind turbine generator. This objective is fulfilled by using the output power signal P_g as an input for the maximum power point tracker (MPPT), which will be subsequently described. Maximum power point tracking means that the wind turbine is always supposed to be operated at maximum output voltage/current rating. From equations (3) and (4), the optimal rotational speed ω_{opt} of the wind turbine rotor for a given wind speed can be used to obtain the maximum turbine efficiency $\eta_{h\max}$ and then the maximum mechanical output power of the turbine. Unfortunately, measuring the wind flowing in the wind turbine rotor is difficult and increases complexity and costs to the DG application, especially for small generating systems; so that to avoid using this measurement for determining the optimal rotor speed, an indirect approach can be implemented.

The wind turbine power is directly controlled by the DC/DC boost converter, while the generator speed in critical conditions is regulated by the pitch angle of the turbine blades. The pitch angle controller is only active in high wind speeds. In these circumstances, the rotor speed can no longer be controlled by increasing the generated power, as this would lead to overloading the generator and/or the converter. To prevent the rotor speed from becoming too high, which would result in mechanical damage, the blade pitch angle is changed in order to reduce the power coefficient C_p. At partial load, the pitch angle is kept constant to its optimal value, while the control of the electrical system via the chopper assures variable speed operation of the WTG.

The proposed MPPT strategy is based on directly adjusting the DC/DC converter duty cycle according to the result of the comparison of successive WTG output power measurements (Datta & Ranganathan, 2003). The control algorithm uses a "Perturbation and Observation" (P&O) iterative method that proves to be efficient in tracking the MPP of the WECS for a wide range of wind speeds. The algorithm, which was widely used for photovoltaic solar systems with good results (Molina et al., 2007), has a simple structure and requires few measured variables, as depicted in Fig. 14.

The WECS MPPT algorithm operates by constantly perturbing, i.e. increasing or decreasing, the rectified output voltage $V_g(k)$ of the WTG and thus controlling the rotational speed of the turbine rotor via the DC/DC boost converter duty cycle D and comparing the actual output power $P_g(k)$ with the previous perturbation sample $P_g(k\text{-}1)$. If the power is increasing, the perturbation will continue in the same direction in the following cycle so that the rotor speed will be increased, otherwise the perturbation direction will be inverted. This means that the WTG output voltage is perturbed every MPPT iteration cycle k at sample intervals T_{trck}. Therefore, when the optimal rotational speed of the rotor ω_{opt} for a specific wind speed is reached, the P&O algorithm will have tracked the MPP and then will settle at this point but with small oscillations. This allows driving the turbine automatically into the operating point with the highest aerodynamic efficiency and consequently leads to optimal energy capture using this controller. Above rated wind speed the pitch angle is increased to limit the absorbed aerodynamic power and the speed is controlled to its rated value ω_{lim}.

7.2 Middle level control

The middle level control makes the expected output, i.e. positive sequence components of i_d and i_q, to dynamically track the reference values set by the external level. The middle level control design, which is depicted in Fig. 13 (middle side), is based on a linearization of the state-space averaged model of the VSI in d-q coordinates, described in equation (34). Inspection of this equation shows a cross-coupling of both components of the inverter output current through ω. Therefore, in order to fully decouple the control of i_d and i_q, appropriate control signals have to be generated. To this aim, it is proposed the use of two control signals x_1 and x_2, which are derived from assumption of zero derivatives of currents ($s\, i_d$ and $s\, i_q$) in the upper part (AC side) of equation (34). This condition is assured by employing conventional PI controllers with proper feedback of the inverter actual output current components, as shown in Fig. 13. Thus, i_d and i_q respond in steady-state to x_1 and x_2 respectively with no crosscoupling, as derived from equation (37). As can be noticed, with the introduction of these new variables this control approach allows to obtain a quite effective decoupled control with the VSI model (AC side) reduced to first-order functions.

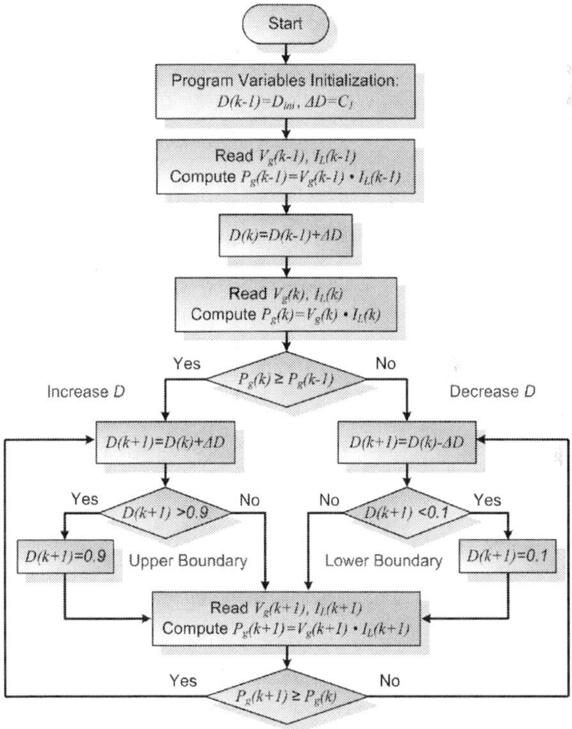

Fig. 14. Flowchart for the P&O MPPT algorithm

$$s\begin{bmatrix} i_d \\ i_q \end{bmatrix} = \begin{bmatrix} \dfrac{-R_s}{L'_s} & 0 \\ 0 & \dfrac{-R_s}{L'_s} \end{bmatrix} \begin{bmatrix} i_d \\ i_q \end{bmatrix} - \begin{bmatrix} x_1 \\ x_2 \end{bmatrix} \tag{37}$$

From equation (34), it can be seen the additional coupling resulting from the DC capacitors voltage V_d, as much in the DC side (lower part) as in the AC side (upper part). This difficulty demands to maintain the DC bus voltage as constant as possible, in order to decrease the influence of the dynamics of V_d. The solution to this problem is obtained by using another PI compensator which allows eliminating the steady-state voltage variations at the DC bus, by forcing the instantaneous balance of power between the DC and the AC sides of the inverter through the contribution of a corrective signal i_{dr^*}, and thus by the modulation of the duty cycle D of the DC/DC chopper.

7.3 Internal level control

The internal level provides dynamic control of input signals for the DC/DC and DC/AC converters. This level is responsible for generating the switching control signals for the twelve valves of the three-level VSI, according to the control mode (SPWM) and types of valves (IGBTs) used and for the single valve (IGBT) of the boost two-level DC/DC converter. Fig. 13 (right side) shows a basic scheme of the internal level control of the WTG unit. This level is mainly composed of a line synchronization module and a firing pulses generator for both the VSI and the chopper. A phase locked loop (PLL) is used for synchronizing through the phase θ_s, the pulses generated for the three-phase inverter. The phase signal is derived from the positive sequence components of the AC voltage vector measured at the PCC of the inverter. In the case of the sinusoidal PWM pulses generator block, the controller of the VSI generates pulses for the carrier-based three-phase PWM inverter using three-level topology. In the case of the DC/DC converter firing pulses generator block, the PWM modulator is built using a standard two-level PWM generator.

8. Digital simulation results

In order to investigate the effectiveness of the proposed models and control algorithms, digital simulations were performed using SimPowerSystems of MATLAB/Simulink (The MathWorks Inc., 2010). For validation of both control strategies, i.e. APCM and VCM of the wind power system, two sets of simulations were employed.

Simulations depicted in Fig. 15 show the case with only active power exchange with the utility grid, i.e. with just the APCM activated at all times, for the studied 0.5 kW WTG connected to a 380V/50Hz weak feeder. The incident wind flowing at the WTG rotor blades is forced to vary quickly in steps every 1s in the manner described in Fig. 15(a). This wind speed variation produces proportional changes in the maximum power that can be drawn from the WTG (MPP actual power shown in red dashed lines). As can also be observed in blue solid lines, the P&O maximum power tracking method proves to be accurate in following the MPP of the WTG with a settling time of almost 0.35s. The trade-off between fast MPP tracking and power error in selecting the appropriate size of the perturbation step can notably be optimized in efficiency. As can be noted in Fig. 15(b), all the active power generated by the WTG is injected into the electric grid trough the PCS, except losses, with small delays in the dynamic response (blue dashed lines). It can also be seen the case with fixed voltage control of the rectified voltage V_d, i.e. with no MPPT control and consequently with near constant rotor speed operation (green dotted lines). In this case, the power injected into the electric grid is much lesser than with MPPT, about up to 30% in some cases. Eventually, no reactive power is exchanged with the electric grid since the VCM is not activated (shown in red solid lines). In this way, as can be observed in Fig. 15(c), the instantaneous voltage at the point of coupling to the ac grid is maintained almost invariant at about 0.99 p.u. (per unit of 380 V base line-line voltage). It is also verified a very low transient coupling between the active and reactive (null in this case) powers exchanged by the grid-connected WTG due to the proposed full decoupled current control strategy in *d-q* coordinates.

Simulations of Fig. 16 show the case with both, active and reactive power exchange with the utility grid, i.e. the APCM is activated all the time while the VCM is activated at t=0.5 s. The WTG is now subjected to the same previous profile of wind speed variations, as described in Fig. 16(a). As can be seen, the maximum power for each wind speed condition is rapidly and

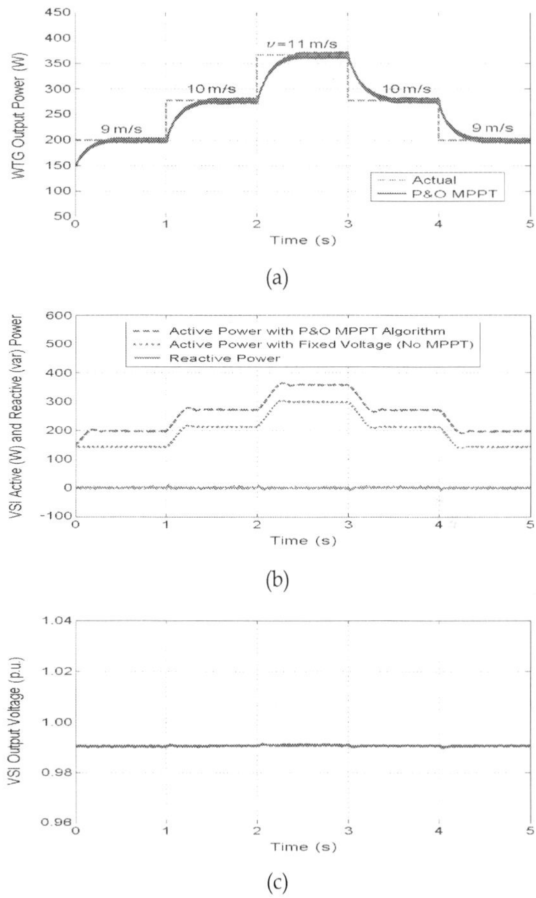

Fig. 15. Simulation results for active power exchange with the utility distribution grid (APCM). (a) PMSG generator output power. (b) VSI output active and reactive power. (c) PCC terminal voltage.

accurately drawn by the P&O MPPT method in the same way as in the previous case study, as depicted in blue solid lines in Fig. 16(b). This power is injected into the electric grid (blue dashed lines), except losses. These losses are increased with the injection of reactive power, causing a slightly lower exchange of active power than the previous case studied with both controls of the inverter (with and without MPPT). As shown in Fig. 16(c), the rapid injection of almost 320 var of reactive capacitive power into the electric utility system (red solid lines) when the VCM is activated aims at controlling the magnitude of the instantaneous voltage at the PCC around the reference voltage of 1 p.u. (or 380 V). Here it is also verified a very low transient coupling between the active and reactive powers injected into the AC grid due to the full decoupled current control strategy in the *d-q* reference frame.

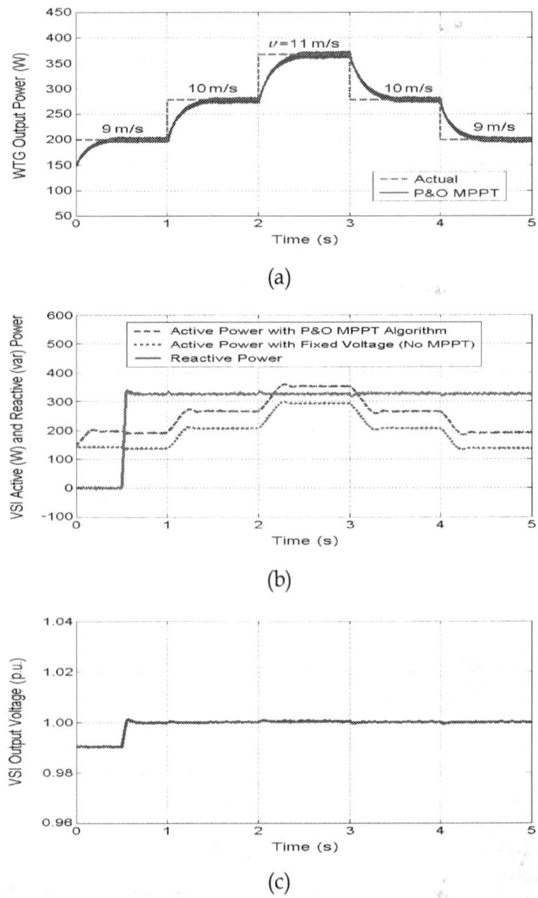

Fig. 16. Simulation results for simultaneous active and reactive power exchange with the utility distribution grid (APCM and VCM). (a) PMSG generator output power. (b) VSI output active and reactive power. (c) PCC terminal voltage.

9. Conclusion

This chapter has provided an overall perspective of modern wind power systems, including a discussion of major wind turbine concepts and technologies. More specifically, of the various wind turbine designs, pitch-controlled variable speed wind turbines controlled by means of power electronic converters have been considered. Among them, direct-in-line wind turbines with full-scale power converter and using direct-driven permanent magnet synchronous generators have increasingly drawn more interests to wind turbine manufactures due to its advantages over the other variable-speed wind turbines. Based on this issue, major operating characteristics of these devices have been thoroughly analyzed and a three-phase grid-connected wind turbine system, incorporating a maximum power point tracker for dynamic active power generation has been presented. Moreover, a simplified state-space averaged mathematical model of the wind turbine system has been provided. An efficient power conditioning system of the selected wind turbine design and a new three-level control scheme by using concepts of instantaneous power in the synchronous-rotating *d-q* reference frame in order to simultaneously and independently control active and reactive power flow in the distribution network level have been proposed. Dynamic system simulation studies in the MATLAB/Simulink environment has been used in order to demonstrate the effectiveness of the proposed multi-level control approaches in *d-q* coordinates and the full detailed models presented. The fast response of power electronic devices and the enhanced performance of the proposed control techniques allow taking full advantage of the wind turbine generator.

10. Acknowledgments

The author wishes to thank CONICET (Argentinean National Research Council for Science and Technology), IEE/UNSJ (Institute of Electrical Energy at the National University of San Juan) and ANPCyT (National Agency for Scientific and Technological Promotion) under grant FONCYT PICT 2006 - Cod. No. 1790, for the financial support of this work.

11. References

Ackermann, T. (2005). *Wind Power in Power Systems*, John Wiley & Songs, 1st Ed., United Kingdom.

Acha, E.; Agelidis, V.; Anaya-Lara, O. & Miller, T. (2002). *Power Electronic Control in Electrical Systems*, Newness, 1st ed., United Kingdom.

Blaabjerg, F.; Chen, Z & Kjaer, S. B. (2004). Power Electronics as Efficient Interface in Dispersed Power Generation Systems. *IEEE Transactions on Power Electronics*, Vol. 19, No. 5, pp. 1184–1194.

Blaabjerg, F. & Chen, Z. (2006). *Power Electronics for Modern Wind Turbines*, Morgan & Claypool Publishers, 1st Ed., Seattle, WA, USA.

Bollen, M. H. J. (2000). *Understanding Power Quality Problems*. IEEE Press, Piscataway, New Jersey, USA.

Carrasco J. M.; Garcia-Franquelo, L.; Bialasiewicz, J. T.; Galván, E; Portillo-Guisado, R. C.; Martín-Prats, M. A.; León, J. I. & Moreno-Alfonso, N. (2006). Power Electronic Systems for the Grid Integration of Renewable Energy Sources: A Survey. *IEEE Trans. on Industrial Electronics*, Vol. 53, No. 4, pp. 1002-1016.

Chen, Z. & Blaabjerg, F. (2009). Wind Farm - A Power Source in Future Power Systems. *Renewable and Sustainable Energy Reviews*, Vol. 13, No. 6-7, pp. 1288–1300.

Datta R. & Ranganathan, V. T. (2003). A Method of Tracking the Peak Power Points for a Variable Speed Wind Energy Conversion System", *IEEE Transactions on Energy Conversion*, Vol.18, pp.163-168.

Freris, L. L. (1990). *Wind Energy Conversion Systems*, Prentice-Hall, 1st Ed., New Jersey, USA.

Global wind energy council (2006) Global Wind Energy Markets Continue to Boom - 2006 another record year, Available from http://www.gwec.net/ [June, 2007].

Guerrero, J. M.; Blaabjerg,F.; Zhelev, T.; Hemmes, K.; Monmasson, E.; Jemei, S.; Comech, M. P.; Granadino, R. & Frau, J. I. (2010). Distributed Generation: Toward a New Energy Paradigm. *IEEE Industrial Electronics Magazine*, Vol. 4, No. 1, pp. 52-64, March 2010.

Hansen, A. D.; Iov, F.; Blaabjerg, F. & Hansen L. H. (2004). Review of contemporary wind turbine concepts and their market penetration, Journal of Wind Engineering, Vol. 28, No. 3, pp. 247-263.

Heier, S. (2006). *Grid Integration of Wind Energy Conversion Systems*, John Wiley & Sons, 1st ed., United Kingdom.

Krause, P.C. (1992). *Analysis of Electric Machinery*, Mc Graw-Hill, New York, USA.

Krüger T. & Andresen, B. (2001) Vestas OptiSpeed - Advanced control strategy for variable speed wind turbines. *Proceedings of European Wind Energy Conference*, pp. 983-86, July 2-6, Copenhagen, Denmark.

Li, S.; Haskew, T. A.; Muljadi, E. & Serrentino, C. (2009). Characteristic Study of Vector-Controlled Direct-Driven Permanent Magnet Synchronous Generator In Wind Power Generation. *Electric Power Components and Systems*, Vol. 37, No. 10, pp. 1162-1179.

Mohan, N.; Undeland, T & Robbins, W. (1995). *Power Electronics: Converters, Applications and Design*, John Wiley & Sons, 1st Ed., New York.

Molina M. G. & Mercado P. E. (2006). Control Design and Simulation of DSTATCOM with Energy Storage for Power Quality Improvements, *Proceedings of IEEE/PES Transmission and Distribution Conference Latin America*, Caracas, Venezuela, Aug. 2006.

Molina M. G. & Mercado P. E. (2008). A New Control Strategy of Variable Speed Wind Turbine Generator for Three-Phase Grid-Connected Applications, *Proceedings of IEEE/PES Transmission and Distribution LA*, Bogotá, Colombia.

Molina M. G. & Mercado P. E. (2009). Control of Tie-Line Power Flow of Microgrid Including Wind Generation by DSTATCOM-SMES Controller. *Proceedings of IEEE Energy Conversion Congress and Exposition*, San José-CA, USA, Sept. 2009, pp. 2014-2021.

Molina, M. G.; Pontoriero, D. H. & Mercado, P. E. (2007). An Efficient Maximum-Power-Point-Tracking Controller for Grid-Connected Photo-Voltaic Energy Conversion System. *Brazilian Journal of Power Electronics*, Vol.12, No.2, pp. 147–154.

Muller, S.; Deicke, M. & De Doncker, R. W. (2002). Doubly Fed Induction Generator Systems for Wind Turbines. *IEEE Industry Applications Magazine*, Vol. 8, No. 3, pp. 26-33.

Qiao, W.; Harley, R. G. & Venayagamoorthy, G. K. (2007). Dynamic Modeling of Wind Farms With Fixed-Speed Wind Turbine Generators, *Proceedings of IEEE PES 2007 General Meeting*, June 24-8, Tampa, USA.

Raiambal, K. & Chellamuthu, C. (2002). Modeling and Simulation of Grid Connected Wind Electric Generating System, *Proc. IEEE TENCON*, pp.1847–1852.

Rahman, S. (2003). Going Green: The Growth of Renewable Energy, *IEEE Power and Energy Magazine*, Vol. 1, No. 6, pp. 16-18.

Rodríguez, J.; Lai, J. S. & Peng, F.Z. (2002). Multilevel Inverters: A Survey of Topologies, Controls, and Applications. *IEEE Transactions on Industrial Electronics*, Vol. 49, No. 4, pp. 724-738.

Soto, D. & Green, T. C. (2002). A Comparison of High-Power Converter Topologies for the Implementation of FACTS Controllers. *IEEE Trans. on Industrial Electronics*, Vol. 49, No. 5, pp. 1072-1080.

Slootweg J. G. & Kling, W. L. (2003). The Impact of Large Scale Wind Power Generation on Power System Oscillations, *Electric Power Systems Research*, Vol. 67, No. 1, pp. 9-20.

The MathWorks Inc. (2010). *SimPowerSystems for Use with Simulink: User's Guide*, R2010a, Available from http://www.mathworks.com [July, 2010].

Timbus, A.; Liserre, M.; Teodorescu, R.; Rodriguez, P. & Blaabjerg, F. (2009). Evaluation of Current Controllers For Distributed Power Generation Systems, IEEE Transactions on Power Electronics, Vol. 24, No. 3, pp. 654-664, March 2009.

17

Wind Park Layout Design Using Combinatorial Optimization

Ivan Mustakerov and Daniela Borissova
Institute of Information and Communication Technologies of Bulgarian Academy of Sciences Bulgaria

1. Introduction

The energy sector has essential influence on climate change and atmospheric pollution. Wind energy, being a clean and renewable energy, can greatly contribute to decreasing of the air pollution negative impacts. Generally speaking, the production of renewable energy from wind can have a positive socioeconomic benefit - it not only help to reduce the climate changing but also support meeting of the long-term world economical goals. Taking that into account, many countries are encouraging the building of industrial wind parks. The building of a wind park is an expensive and complex task involving a wide range of engineering and scientific knowledge. The design of a wind park can have profound implications for its future profitability. Sustainable wind park development has to be done in an ecological way which means evaluating of the all possible positive and negative influences on the environment. The development of new wind energy projects requires also a significant consideration of land use issues. One of the most important factors in selecting a wind energy site is the availability of proper wind resources. The wind itself is a variable source of energy. The ability of a wind turbine to extract power from varying wind is a function of three main factors - the wind power availability, the power curve of the generator, and the ability of the turbine to respond to wind fluctuations (Üstüntaşa & Şahin, 2008). Wind turbines are available in various sizes and power output. They are designed to operate over a range of wind speeds (3-25 m/s) and can be erected singly by an individual property owner or grouped together to form a wind farm (wind park) connected to a public grid (Rodman & Meentemeyer, 2006). The common challenge for the wind park designer is to maximize the energy capture within the given restrictions (White et al., 1997; Kusiak & Song, 2010). As there is pressure to build more compact wind farms to optimize land utilization, an determination of array losses for very close turbine spacing is required (Smith et al., 2006). The investigations on wind energy using essentially cover four principal topics (Ettoumi et al., 2008). The first one deals with the sensors and instrumentation used for wind measurements. The second one examines the evaluation of wind energy potential for a given region using various statistical approaches. The third is focused on the design and characterization of wind energy turbines. The fourth is the design and development of wind parks. The results of investigations in those areas are used by wind park planners to develop cost-effective wind parks.

The investigations discussed here concern the problems associated with the forth topic - design of the wind park layout, including choice of the turbines type, number and their placement in the wind park area. How to choose the number and the type of the turbines to install depends on a variety of factors - wind conditions, terrain, investments costs, power output, environmental influence, etc. More powerful turbine is usually preferred to the less powerful one since both the cost of a turbine and the energy it generates is usually proportional to its nominal power. The placement of wind turbines on the wind park site (i.e. wind park layout) is affected by several factors which have to be taken into account - the number of turbines, wind direction, wake interactions between wind turbines, land availability (area and shape), etc. For the goal a combinatorial design model for defining wind turbines type, number and placement is proposed. It is used for formulation of mixed-integer nonlinear discrete combinatorial optimization tasks satisfying different design requirements and restrictions. The tasks solutions results define different optimal wind park layout designs. The solutions results can be used also to evaluate the impact of alternative wind park layout schemes on the investment costs and wind park power output.

2. Mathematical methods in wind park design

An interdisciplinary branch of applied mathematics and formal science that uses mathematical modelling methods and algorithms to arrive at optimal or near optimal solutions to complex practical problems is known as *operations research.* Operations research helps the management to achieve its goals using scientific methods and can be used in particular for wind park design decisions. It is often concerned with optimizing of some objectives (maximum of profit, performance, etc. or minimum of loss, risk, cost, etc.) at limited resources. The majority of real-world optimization problems are multiobjective by nature - they have more then one and usually conflicting objectives that must be satisfied simultaneously. Instead of aiming at a single solution finding, the multiobjective optimization methods try to produce a set of good trade-off solutions (Pareto-optimal solutions) from which the decision maker could select. Nevertheless, there exist some practical problems where the single criterion optimization would be able to get an optimal solution with less calculation difficulties. One of the questions that should be answered when using optimization methods for the wind park design is the effectiveness and advisability of single or multicriteria optimization application.

The general wind park design goal is maximizing electrical energy production while minimizing the costs. Up to day different approaches to the wind park design are developed. At the early stage a suitable wind park site is identified by its geographical location and based on long-term wind records for annual wind speed variation (White et al., 1997). The determining of the wind turbines number in the designed wind park is a decision making parameter that should take into consideration the limited wind park geographical area. The separation distance between the turbines is another parameter to be considered when defining the wind park layout and depends on the needed recovery of wind energy behind the neighbouring turbines (Grady et al., 2005; Sørensen, 2006). The number and separation distances between turbines are used to define the wind park layout. The wind park layout design is usually performed in a heuristic trial and-error iterative approach taking into account all social, environmental and technical constrains (Mora et al., 2007). The intuitive spacing scheme results in sparse wind parks layouts and in inefficiently using of the wind energy potential of the site (Ammara et al., 2002). The production of industrial quantities of electricity requires getting as much as possible electrical power output, i.e. the wind park with as many powerful wind turbines as it is possible. The best solution should define the wind turbines number, type and their placement in the given geographical area. Profit-to-cost or profit-to-area ratios could be used as performance metrics of investment (Kongnam et al., 2009). Several methods have been applied in order to try to optimize wind park layout. Genetic algorithms are used for defining of optimal wind turbines placement. A square shape of the wind park is subdivided into a 10 x 10 grid and cells to install turbines are determined so as to minimize the cost per unit energy (Mosetti et al., 1994; Grady et al., 2005; Emami & Noghreh, 2010). The effectiveness of genetic algorithms is demonstrated for solving the wind park micro-siting optimization problem. However, due to the binary coding method of the genetic algorithms, turbines could only be installed in the center of selected cells. Another application of genetic algorithm is also used to maximize an economic function, which is related to turbine parameters and locations also in a square grid (Mora et al., 2007). A generic model for optimizing in-land wind farm layout considers circular wind park radius and turbine distance constraints. The model maximizes the energy production by placing wind turbines in such a way that the wake loss is minimized (Kusiak & Song, 2010). The uncertainties of wind speed distribution and power–speed characteristics are considered in mixed-integer nonlinear programming approach used to determine the optimum wind park electricity generation. It takes into account the profit-to-cost and profit-to-area ratios to evaluate the investment alternatives (Kongnam et al., 2009). Optimal placement and arrangement of wind turbines in a wind park while maximizing energy production and minimizing installation cost, is done by mean of Monte Carlo simulation (Marmidis et al., 2008). Variations of the vertex packing problem are used to determine the optimal positions of wind turbines within a wind farm. The maximum power generation is sought considering constraints for the number of turbines, turbine proximity, and turbine interference (Donovan, 2005). The characteristic of wind turbine generation is represented by an analytical expression, which is used to solve distributed generation sizing and placement problem while minimizing the real power loss in the system. It is used in a methodology for finding the optimal size and location for connecting wind type distributed generation in primary distribution systems (Mahat et al., 2006).

As could be seen from the survey, the wind park layout design is an important and complex problem. Different mathematical methods and approaches are developed to get mathematically reasoned (optimal) solutions in contrast to the experience based heuristic approaches. Using of mathematical methods has also the advantage of getting of preliminary theoretical estimation about the designed wind park characteristics and satisfaction of the given requirements. The preliminary evaluation is important because it decreases the risks of taking ineffective or even wrong wind park design decisions.

2.1 Generalized discrete combinatorial optimization description

The wind park design has common specifics of the engineering systems design. One of the possible approaches to these problems is using of discrete combinatorial optimization modelling (Mustakerov & Borissova, 2007). That approach could be shortly described by assuming of the existence of finite discrete sets of values for some design variables *a, b,…, c*:

$$a = \begin{pmatrix} a_1 \\ a_2 \\ \dots \\ a_i \end{pmatrix}, b = \begin{pmatrix} b_1 \\ b_2 \\ \dots \\ b_j \end{pmatrix}, \dots, c = \begin{pmatrix} c_1 \\ c_2 \\ \dots \\ c_k \end{pmatrix} \tag{1}$$

The values of the design variables define some discrete sets of system parameters:

$$
\begin{aligned}
&\forall i : P_{a_i} = \{P_{a_i}^1, P_{a_i}^2, \ldots, P_{a_i}^r\} \\
&\forall j : P_{b_j} = \{P_{b_j}^1, P_{b_j}^2, \ldots, P_{b_j}^s\} \\
&\ldots \\
&\forall k : P_{c_k} = \{P_{c_k}^1, P_{c_k}^2, \ldots, P_{c_k}^t\}
\end{aligned} \tag{2}
$$

The choice of particular elements of *a, b,…, c* is formalized as choice of the system parameters by means of binary integer decision variables x, y, …, z:

$$
\begin{aligned}
&\forall r : P_a^r = \sum_i P_{a_i}^r x_i \\
&\forall s : P_b^s = \sum_j P_{b_j}^s y_j \\
&\ldots \\
&\forall t : P_c^t = \sum_k P_{c_k}^t z_k
\end{aligned} \tag{3}
$$

where:

$$
x = \begin{pmatrix} x_1 \\ x_2 \\ \ldots \\ x_i \end{pmatrix}, y = \begin{pmatrix} y_1 \\ y_2 \\ \ldots \\ y_j \end{pmatrix}, \ldots, z = \begin{pmatrix} z_1 \\ z_2 \\ \ldots \\ z_k \end{pmatrix} \tag{4}
$$

are subject to restrictions for single element choice:

$$
\begin{aligned}
&\sum_i x_i = 1, x_i \in \{0,1\} \\
&\sum_j y_j = 1, y_j \in \{0,1\} \\
&\ldots \\
&\sum_k z_k = 1, z_k \in \{0,1\}
\end{aligned} \tag{5}
$$

The real engineering design problems are characterized by the existing of some technological and user requirements between the design variables described as functions:

$$
\begin{aligned}
&g_1 = f_1(a, b, c, \ldots) \\
&g_2 = f_2(a, b, c, \ldots) \\
&\ldots \\
&g_m = f_m(a, b, c, \ldots)
\end{aligned} \tag{6}
$$

and restrictions for system parameters:

$$
L_a^r \le P_a^r \le U_a^r \tag{7}
$$

The combinatorial optimization design approach needs defining of some objective functions:

$$\begin{aligned} f_\phi &= \phi(P_a^r, P_b^s, ..., P_c^t) \\ f_\varphi &= \varphi(P_a^r, P_b^s, ..., P_c^t) \\ &\dots \\ f_\gamma &= \gamma(P_a^r, P_b^s, ..., P_c^t) \end{aligned} \tag{8}$$

They are used to formulate single criterion

$$\max(f_\phi + f_\varphi + ... + f_\gamma) \tag{9}$$

or multicriteria:

$$\max(f_\phi, f_\varphi, ..., f_\gamma) \tag{10}$$

formulations of discrete combinatorial optimization tasks whose solutions define optimal or Pareto-optimal design elements combinations.

2.2 Discrete combinatorial optimization for wind park layout design

The described generalized discrete combinatorial optimization approach is applied to wind park layout design. In most cases, the wind park layout design should conform to a preliminary chosen geographical area with known parameters - wind conditions and area shape and size. The design goal is to determine the optimal wind turbines type, number and placement to get maximal power output while minimizing the investment costs and considering different technological and functional requirements and restrictions. One of the requirements concerns the spacing between turbines in a wind park. If the wind strikes a next turbine before the wind speed has been restored from striking the previous turbine, the energy production from the second turbine will be decreased, the so called *wake effect*. Spacing the turbines further apart will produce more power, but at the expense of more land, more roads, and more electrical wire. The separation distance between the turbines depends on the needed recovery of wind energy behind the neighbouring turbines (Grady et al., 2005; Sørensen, 2006). The spacing of a cluster of wind turbines in a wind park depends on the terrain, the wind direction and speed, and on the turbines' size. There exist some recommendations for the turbines separation distances depending on the wind directions and rotor diameters sizes. Two typical cases of wind directions are assumed - uniform and predominant wind direction - Fig. 1.

The uniform wind direction indicates equal distances between turbines in rows and columns of approximately 5 rotor diameters (Grady et al., 2005). In the case of predominant wind direction, the recommended turbines spacing is 8 to 12 rotor diameters in rows apart in the windward direction and 1.5 to 3 (Grady et al., 2005; Marmidis et al., 2008; Donovan, 2005) or 2 to 4 rotor diameters (Johnson, 2006) in the crosswind direction. Spacing the turbines further apart will decrease the wake influence and presumably will capture more wind power but at the expense of land wasting. The concept of operating wind turbines as closely as it is possible has the potential advantages in reducing capital costs associated with design and construction, while improving wind turbine system reliability and operating efficiency (Ransom et al., 2010).

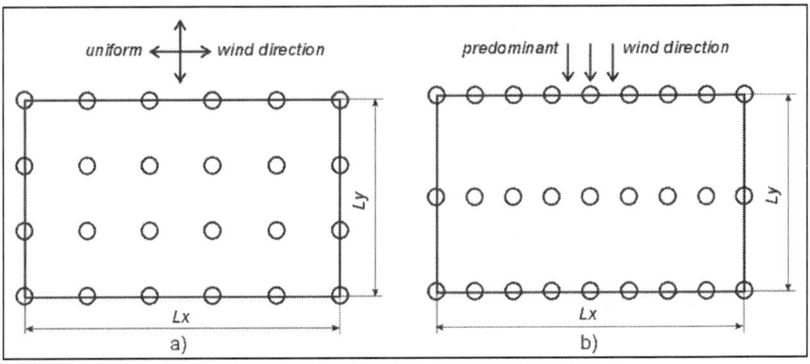

Fig. 1. Wind turbines placements patterns: (a) uniform and (b) predominant wind direction

One of the essential parameters to be considered in the wind park layout design modeling is the overall turbines number. Using the patterns shown in Fig. 1, the total number of turbines N within park area can be defined as multiplication of rows and columns turbines numbers N_{row} and N_{col}:

$$N = N_{row}N_{col} \tag{11}$$

The number of wind turbines in row and column N_{row} and N_{col} for rectangular area with dimensions L_x and L_y and separation distances between turbines SD_x and SD_y can be defined as (Johnson, 2006):

$$N_{row} = \frac{L_x}{SD_x} + 1 \tag{12}$$

$$N_{col} = \frac{L_y}{SD_y} + 1 \tag{13}$$

The existing recommendations about the turbines separation distances are taken into account by introducing of coefficients k_{row} and k_{col} (Mustakerov & Borissova, 2010). By using k_{row} and k_{col}, the separation distances can be expressed by the turbine rotor diameter D as:

$$SD_x = k_{row}\, D \tag{14}$$

$$SD_y = k_{col}\, D \tag{15}$$

Substituting (14) and (15) into (12) and (13) the row and column turbines numbers can be defined as functions of turbines rotor diameter:

$$N_{row} = \frac{L_x}{k_{row}D} + 1, \text{integer} \tag{16}$$

$$N_{col} = \frac{L_y}{k_{col}D} + 1, \text{integer} \tag{17}$$

To increase the flexibility of the modeling, the coefficients k_{row} and k_{col} are considered to be continuous variables limited by upper and lower values (Mustakerov & Borissova, 2010):

$$k_{row}^{\min} \le k_{row} \le k_{row}^{\max} \tag{18}$$

$$k_{col}^{\min} \le k_{col} \le k_{col}^{\max} \tag{19}$$

As a result of technological advances, there exist a number of different types and models of wind turbines that can be used in the wind parks design. The different wind turbines have different technical characteristics reflecting on power production and price. Because of the fact that purchasing of the wind turbines is one of the most essential investment for the wind park developing, the choice of optimal wind turbines' type and number is a critical step for wind park design. The choice of identical turbines for particular wind park developing is assumed to be an advantage from the technological and maintenance point of view. The wind turbines type choice is modeled by using of binary integer variables $x_i \in \{0, 1\}$, complying with restriction:

$$\sum_i x_i = 1 \tag{20}$$

The binary integer variables and the integrality requirement on the turbines number transform the model into discrete mixed integer programming problem. It should be pointed out that most of engineering design optimization problems are mixed integer-discrete-continuous, non-linear programming problems. It is quite often in practice of wind park design to solve them as continuous problems and to use the nearest available discrete value (Johnson, 2006). The result in this case is not optimal or even unfeasible if equality types of restrictions are existing (Lampinen & Zelinka, 1999; Taha, 2006; Shapiro, 1979). The existence of integrality requirement increases the difficulty of finding a solution but could be not avoided if a true optimal solution is needed.

The developed combinatorial wind park layout design approach is based on simultaneous choice of multiple wind turbine parameters considered as design variables (Mustakerov & Borissova, 2010). For example:

- an important parameter influencing the overall wind park electrical energy output is the wind turbine rated power. The turbine choice among the set of m available different turbines with rated power P_{wt}^i is formalized as:

$$P_{wt} = \sum_i^m x_i P_{wt}^i \tag{21}$$

- the rotor diameter D is another of the important wind turbine parameters defining the turbine's effectiveness. It defines also the separation distance between turbines and indirectly the number of turbines in a given limited wind park area. The choice of particular wind turbine rotor diameter among the given set of m different types of turbines with known rotor diameters can be done analogically to (21) as:

$$D = \sum_i^m x_i D_{wt}^i \tag{22}$$

- the wind speed in the wind park geographical location influences the turbine's choice by considering of their nominal wind speed parameter S^i_{wt} :

$$S = \sum_i^m x_i S^i_{wt} \tag{23}$$

- the height of the wind turbine tower H is another parameter that can be considered in the turbines choice (the wind speed increases with the height and the turbine effectiveness increases respectively):

$$H = \sum_i^m x_i H^i_{wt} \tag{24}$$

Using the same technique other essential wind turbine parameters (electrical output voltage, frequency, noise, etc.) could be considered in the wind turbine type choice.
Because the wind turbines parameters are taken as design variables some requirements about their values can be taken into account by introducing of lower and upper limits:

$$P^{\min}_{wt} \le P_{wt} \le P^{\max}_{wt} \tag{25}$$

$$D_{\min} \le D \le D_{\max} \tag{26}$$

$$S_{\min} \le S \le S_{\max} \tag{27}$$

$$H_{\min} \le H \le H_{\max} \tag{28}$$

The turbine's optimal choice should be done accordingly to some optimization criteria. A widely accepted criterion for wind park design is the minimum cost per unit energy produced ratio.

$$\left(\frac{costs}{P}\right) \tag{29}$$

where the wind park costs per year is defined as function of the turbines number N (Grady et al., 2005; Mosetti et al., 1994; Marmidis et al., 2008):

$$costs = N\left(\frac{2}{3} + \frac{1}{3}\exp(-0.00174N^2)\right) \tag{30}$$

and total power extracted by all of the turbines in the wind park is (Grady et al., 2005; Johnson, 2006)

$$P = NP_{wt} \tag{31}$$

where P is the wind park total power output, N is number of wind turbines and P_{wt} is the single wind turbine power production. The single wind turbine power P_{wt} depends on the wind conditions (wind speed, direction, intensity and probability) and on turbine's type –

rotor and tower size, rated power, thrust coefficient, etc. (Kongnam et al., 2009; Mosetti et al., 1994; Marmidis et al., 2008). The wind park electrical energy production for a year can be estimated by using some practically determined nominal power utilization coefficient μ(Lepa et al., 2008) of the total wind park installed power. In the proposed modelling approach the wind park power output per year is estimated by the number of the hours over the year h_y, nominal power utilization coefficient μ, number of wind park turbines N and single turbine power rating P_{wt} as (Mustakerov & Borissova, 2010):

$$P = h_y \eta N P_{wt} \tag{32}$$

The advantage of the described modeling approach is the possibility to introduce different criteria, relations and restrictions to satisfy more precisely different requirements about the designed wind park layout.

2.3 Optimization task formulation

The proposed modeling approach is used to formulate a single criterion mixed-integer nonlinear optimization task:

$$\min\left(\frac{costs}{P}\right) \tag{33}$$

subject to:

$$costs = N\left(\frac{2}{3}+\frac{1}{3}\exp(-0.00174N^2)\right) \tag{34}$$

$$P = h_y \eta N P_{wt} \tag{35}$$

$$N = N_{row} N_{col}\text{, integer} \tag{36}$$

$$N_{row} = \frac{L_x}{k_{row}D}+1\text{, integer} \tag{37}$$

$$N_{col} = \frac{L_y}{k_{col}D}+1\text{, integer} \tag{38}$$

$$k_{row}^{\min} \le k_{row} \le k_{row}^{\max}\text{, } k_{row} > 0 \tag{39}$$

$$k_{col}^{\min} \le k_{col} \le k_{col}^{\max}\text{, } k_{col} > 0 \tag{40}$$

$$P_{wt} = \sum_i^m x_i P_{wt}^i \tag{41}$$

$$D = \sum_i^m x_i D_{wt}^i \tag{42}$$

$$H = \sum_{i}^{m} x_i H_{wt}^{i} \tag{43}$$

$$\sum_{i}^{m} x_i = 1, x \in \{0,1\} \tag{44}$$

The objective function defines the profitability of the wind park design by minimization of the investments while maximize the wind park electrical power generation. Thus, the solution of the formulated optimization task (33) - (44) will be a compromise between the overall wind park investments costs and wind turbines energy output. Both depend on the type of wind turbines, on their number and placement within park area. On the other hand, the number of the wind turbines depends of the size of the wind park and wind direction. The wind park energy output is function of the number of installed turbines, their rated power and wind conditions. The wind conditions for the particular wind park are taken into account by using the statistically defined nominal power utilization coefficient μ of the total wind park installed power.

The solution of the formulated optimization task (33) - (44) define the optimal type and number of wind turbines for a given wind park area. It also defines the values of the coefficients k_{row}, k_{col} and the corresponding separation distances between turbines. The turbines number and their separation distances define the optimal wind park layout design. The calculated values of the costs and wind park electrical power output give some preliminary theoretical estimation of the wind park design and its effectiveness.

3. Numerical testing

The proposed combinatorial optimization approach to wind park layout design is tested numerically by using formulation (33) - (44) and real data of commercial wind turbines shown in Table 1 (Wind-energy-market, 2010; Enercon, 2010; Vestas, 2010). Each row of Table 1 represents a particular turbine type and each column represents the corresponding parameter value to be used in the numerical testing. The number of turbines types and the number of their parameters define the number of the optimization model variables and restrictions. It should be noted, that the turbines and parameters shown in Table 1 are just a sample used to illustrate the practical applicability of the proposed wind park design approach. Real design problem would include much more design items leading to large scale optimization tasks formulations.

Two basic wind conditions are considered - uniform wind direction and known predominant wind direction. Changing the layout of a wind park will presumably result in different investment costs and park power output. That is why the different layouts have to be compared in order to find the design that minimizes the investments and supplies the expected level of power output. For the goal, a number of different optimization tasks formulations defining different layouts are tested numerically by means of LINGO v. 11 optimization solver.

The values of the coefficients h_y and μ used to calculate expected wind park power output per year are taken to be h_y = 8760 hours over the year (365 days x 24 hours) and μ= 0.3 (30% nominal turbines power utilization).

i	Wind turbine		Rated power output, P^i_{wt}, MW	Rotor diameter, D^i_{wt}, m	Tower height, H^i_{wt}, m
	Type	Model			
1	1.1	enercon's E-33	0.330	33.4	37
2	1.2				44
3	1.3				50
4	2.1	enercon's E-48	0.800	48.0	50
5	2.2				60
6	2.3				76
7	3.1	enercon's E-53	0.800	52.9	60
8	3.2				73
9	4.1	Vestas V52	0.850	52.0	49
10	4.2				65
11	4.3				74
12	5.1	enercon's E-44	0.900	44.0	45
13	5.2				55
14	6.1	Vestas V82	1.650	82.0	78
15	7.1	Vestas V80	2.000	80.0	60
16	7.2				78
17	7.3				100
18	8.1	enercon's E-82	2.000	82.0	78
19	8.2				138
20	9.1	enercon's E-70	2.300	71.0	58
21	9.2				113
22	10.1	Vestas V90	3.000	90.0	95
23	10.2				105
24	10.3				125

Table 1. Commercial wind turbines parameters

3.1 Uniform wind direction case

Eight mixed-integer discrete nonlinear optimization tasks defining wind park layout design for uniform wind direction case are solved. Four of them consider square wind park shape with dimensions L_x = 2 km, L_y = 2 km and the other four - rectangular wind park shape with the same area (L_x = 4 km, L_y = 1 km).

The task (33) - (44) for square wind park has four different formulations as follows:

- **Task 1a -** (39) and (40) use boundary values $k_{row}^{\min} = k_{col}^{\min} = 4.8$ and $k_{row}^{\max} = k_{col}^{\max} = 5.2$ for the turbines separation coefficients following the recommendations for the uniform wind direction case.
- **Task 1b** differs from **Task 1a** by adding of the restriction for the wind turbine rotor diameter $D < 70$ m. That additional restriction shifts the choice of wind turbines toward to smaller wind turbines types.
- **Task 1c** and **Task 1d** are equivalent to the tasks 1a and 1b but have relaxed boundaries for the coefficients $k_{row}^{\min} = k_{col}^{\min} = 4.5$ and $k_{row}^{\max} = k_{col}^{\max} = 5.5$ in (39) and (40) to illustrate the influence of the separation distances on the tasks solution.

The tasks solutions (Table 2) define optimum wind turbines placement illustrated on Fig. 2.

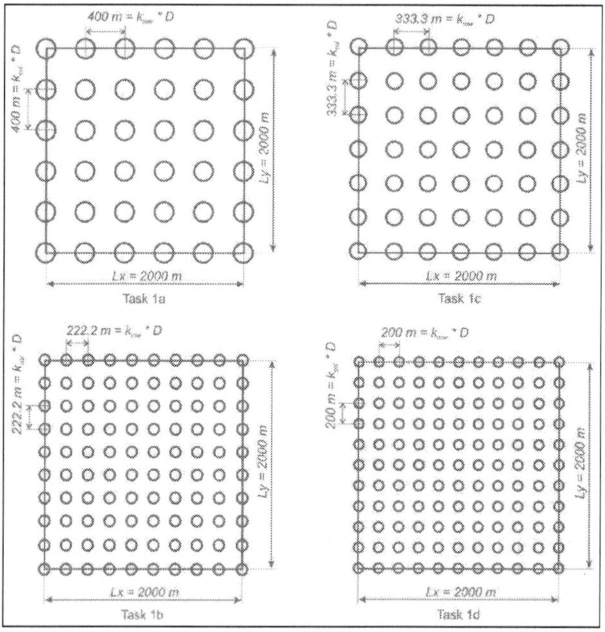

Fig. 2. Wind turbines placement for uniform wind direction and square wind park area

Task	Chosen type of wind turbine	Turbines number	Total installed power, MW	Separation coefficients	Separation distances, m	Cost per unit energy
Task 1a $4.8 < k_{row} < 5.2$ $4.8 < k_{col} < 5.2$	8.1 enercon's E-82 P_{wt} = 2 MW, D = 82 m, H = 78 m	N = 36 N_x = 6 N_y = 6	72	k_{row} = 4.88 k_{col} = 4.88	SD_x = 400 SD_y = 400	0.000133
Task 1b $4.8 < k_{row} < 5.2$ $4.8 < k_{col} < 5.2$ $D < 70$	5.1 enercon's E-44 P_{wt} = 0.9 MW, D = 44 m, H = 45 m	N = 100 N_x = 10 N_y = 10	90	k_{row} = 5.05 k_{col} = 5.05	SD_x = 222.2 SD_y = 222.2	0.000282
Task 1c $4.5 < k_{row} < 5.5$ $4.5 < k_{col} < 5.5$	9.1 enercon's E-70 P_{wt} = 2.3 MW, D = 71 m, H = 58 m	N = 49 N_x = 7 N_y = 7	112.7	k_{row} = 4.7 k_{col} = 4.7	SD_x = 333.3 SD_y = 333.3	0.000111
Task 1d $4.5 < k_{row} < 5.5$ $4.5 < k_{col} < 5.5$ $D < 70$	5.1 enercon's E-44 P_{wt} = 0.9 MW, D = 44 m, H = 45 m	N = 121 N_x = 11 N_y = 11	108.9	k_{row} = 4.54 k_{col} = 4.54	SD_x = 200 SD_y = 200	0.000282

Table 2. Optimal tasks solutions for uniform wind direction (L_x = 2000 m, L_y = 2000 m)

It is interesting to evaluate the influence of the wind park geometrical shape on the wind turbines choice and placement for uniform wind direction case. The described above optimization tasks are solved for the equivalent wind park area but with rectangular shape (Lx = 4 km and Ly = 1 km) as Task 1a-1, Task 1b-1, Task 1c-1 and Task 1d-1.
The solution results define optimum wind turbines placement illustrated on Fig. 3 and the corresponding solution data are shown in Table 3.

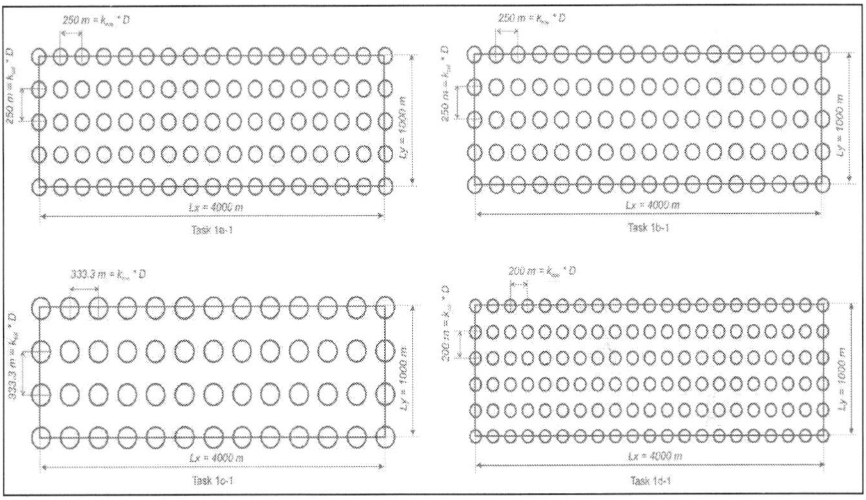

Fig. 3. Wind turbines placement for uniform wind direction and rectangular wind park area

Task	Chosen type of wind turbine	Turbines number	Total installed power, MW	Separation coefficients	Separation distances, m	Cost per unit energy
Task 1a-1 $4.8 < k_{row} < 5.2$ $4.8 < k_{col} < 5.2$	4.3 Vestas V52 $P_{wt} = 0.85$ MW, $D = 52$ m, $H = 74$ m	$N = 85$ $N_x = 17$ $N_y = 5$	72.25	$k_{row} = 4.8$ $k_{col} = 4.8$	$SD_x = 250$ $SD_y = 250$	0.000298
Task 1b-1 $4.8 < k_{row} < 5.2$ $4.8 < k_{col} < 5.2$ $D < 70$	4.1 Vestas V52 $P_{wt} = 0.85$ MW, $D = 52$ m, $H = 49$ m	$N = 85$ $N_x = 17$ $N_y = 5$	72.25	$k_{row} = 4.8$ $k_{col} = 4.8$	$SD_x = 250$ $SD_y = 250$	0.000298
Task 1c-1 $4.5 < k_{row} < 5.5$ $4.5 < k_{col} < 5.5$	9.1 enercon's E-70 $P_{wt} = 2.3$ MW, $D = 71$ m, $H = 58$ m	$N = 52$ $N_x = 13$ $N_y = 4$	119.6	$k_{row} = 4.69$ $k_{col} = 4.69$	$SD_x = 333.3$ $SD_y = 333.3$	0.000111
Task 1d-1 $4.5 < k_{row} < 5.5$ $4.5 < k_{col} < 5.5$ $D < 70$	5.1 enercon's E-44 $P_{wt} = 0.9$ MW, $D = 44$ m, $H = 45$ m	$N = 126$ $N_x = 21$ $N_y = 6$	113.4	$k_{row} = 4.54$ $k_{col} = 4.54$	$SD_x = 200$ $SD_y = 200$	0.000282

Table 3. Optimal tasks solutions for uniform wind direction ($L_x = 4000$ m, $L_y = 1000$ m)

The tasks 1a-1, 1b-1, 1c-1 and 1d-1 are also solved for changed wind park orientation i.e. for L_x = 1 km and L_y = 4 km. The solutions results show that changing of the wind park rectangular shape orientation for the uniform wind direction case does not affect the turbines choices but only interchanges the values of N_{row} and N_{col}.

3.2 Predominant wind direction case

The predominant wind direction case is characterized by a statistically defined prevailing wind direction that blows most frequently across a particular geographical region. The **Task 1a** and **Task 1b** are reformulated as **Task 2a** and **Task 2b**, considering the predominant wind direction by using different separation distances coefficients limits in (39) and (40) accordingly to the recommended values as: $k_{row}^{\min} = 1.5$, $k_{row}^{\max} = 3$, $k_{col}^{\min} = 9$, $k_{col}^{\max} = 11$. The tasks 2a and 2b are reformulated as **Task 2c** and **Task 2d** by using relaxed separation distances coefficients limits $k_{row}^{\min} = 2$, $k_{row}^{\max} = 4$, $k_{col}^{\min} = 8$, $k_{col}^{\max} = 12$. The corresponding optimum wind turbines placement is shown on Fig. 4 accordingly to the solutions results in Table 4.

Task	Chosen type of wind turbine	Turbines number, N	Total installed power, MW	Separation coefficients,	Separation distances, m	Cost per unit energy
Task 2a $1.5 < k_{row} < 3$ $9 < k_{col} < 11$	9.1 enercon's E-70 P_{wt} = 2.3 MW, D = 71 m, H = 58 m	$N = 76$ $N_x = 19$ $N_y = 4$	174.8	$k_{row} = 1.56$ $k_{col} = 9.39$	$SD_x = 111.1$ $SD_y = 666.6$	0.000110
Task 2b $1.5 < k_{row} < 3$ $9 < k_{col} < 11$ $D < 70$	5.1 enercon's E-44 P_{wt} = 0.9 MW, D = 44 m, H = 45 m	$N = 138$ $N_x = 23$ $N_y = 6$	124.2	$k_{row} = 2.07$ $k_{col} = 9.09$	$SD_x = 91$ $SD_y = 400$	0.000282
Task 2c $2 < k_{row} < 4$ $8 < k_{col} < 12$	10.2 Vestas V90 P_{wt} = 3 MW, D = 90 m, H = 105 m	$N = 36$ $N_x = 12$ $N_y = 3$	108	$k_{row} = 2.01$ $k_{col} = 11.11$	$SD_x = 181.8$ $SD_y = 1000$	0.000089
Task 2d $2 < k_{row} < 4$ $8 < k_{col} < 12$ $D < 70$	5.1 enercon's E-44 P_{wt} = 0.9 MW, D = 44 m, H = 45 m	$N = 132$ $N_x = 22$ $N_y = 6$	118.8	$k_{row} = 2.16$ $k_{col} = 9.09$	$SD_x = 95.2$ $SD_y = 400$	0.000282

Table 4. Optimal tasks solutions for predominant wind direction (L_x = 2000 m, L_y = 2000 m)

It could be expected that in the case of predominant wind direction the geometrical shape of the wind park influence the choice of wind turbines type and number. To investigate that the optimization tasks 2a, 2b, 2c, 2d are solved as **Task 2a-1, Task 2b-1, Task 2c-1, Task 2d-1** for rectangular wind park shape with dimensions L_x = 4 km and L_y = 1 km. Their solution results are shown in Table 5 and optimum wind turbines placement - on Fig. 5.

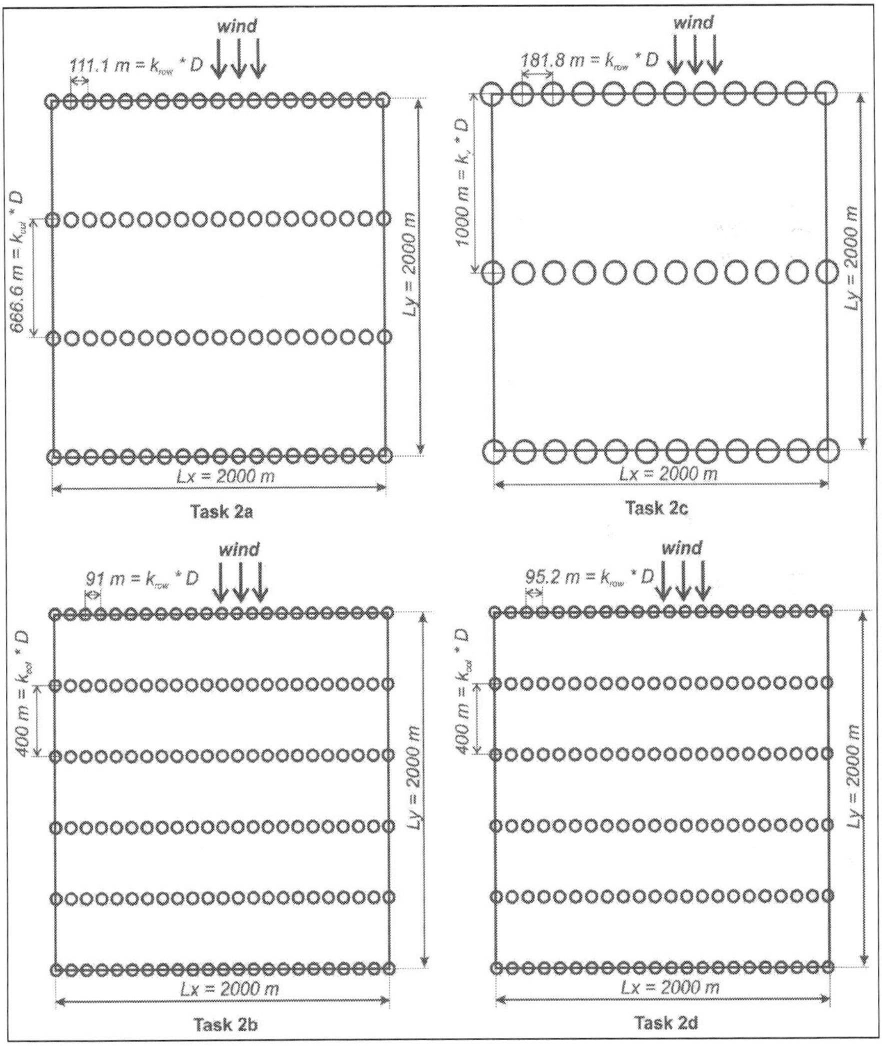

Fig. 4. Wind turbines placement for predominant wind direction and square wind park area

To investigate the influence of the rectangular wind park orientation toward predominant wind direction the tasks 2a-1, 2b-1, 2c-1 and 2d-1 are reformulated as **Task 2a-2**, **Task 2b-2**, **Task 2c-2** and **Task 2d-2** with interchanged wind park dimensions as L_x = 1 km and L_y = 4 km, i.e. for changed wind park orientation toward wind direction. The corresponding solutions results and wind turbines placements are shown in Table 6 and on Fig. 6 respectively.

Task	Chosen type of wind turbine	Turbines number, *N*	Total installed power, MW	Separation coefficients,	Separation distances, m	Cost per unit energy
Task 2a-1 $1.5 < k_{row} < 3$ $9 < k_{col} < 11$	4.3 Vestas V52 P_{wt} = 0.85 MW, D = 52 m, H = 74 m	N = 156 N_x = 52 N_y = 3	132.6	k_{row} = 1.51 k_{col} = 9.61	SD_x = 78.43 SD_y = 500	0.000298
Task 2b-1 $1.5 < k_{row} < 3$ $9 < k_{col} < 11$ $D < 70$	4.1 Vestas V52 P_{wt} = 0.85 MW, D = 52 m, H = 49 m	N = 156 N_x = 53 N_y = 3	132.6	k_{row} = 1.51 k_{col} = 9.61	SD_x = 78.4 SD_y = 500	0.000298
Task 2c-1 $2 < k_{row} < 4$ $8 < k_{col} < 12$	10.2 Vestas V90 P_{wt} = 3 MW, D = 90 m, H = 105 m	N = 46 N_x = 23 N_y = 2	138	k_{row} = 2.02 k_{col} = 11.11	SD_x = 181.8 SD_y = 1000	0.000086
Task 2d-1 $2 < k_{row} < 4$ $8 < k_{col} < 12$ $D < 70$	5.1 enercon's E-44 P_{wt} = 0.9 MW, D = 44 m, H = 45 m	N = 132 N_x = 44 N_y = 3	118.8	k_{row} = 2.11 k_{col} = 11.36	SD_x = 93 SD_y = 500	0.000282

Table 5. Optimal solutions for predominant wind direction and rectangular wind park area (L_x = 4 km, L_y = 1 km)

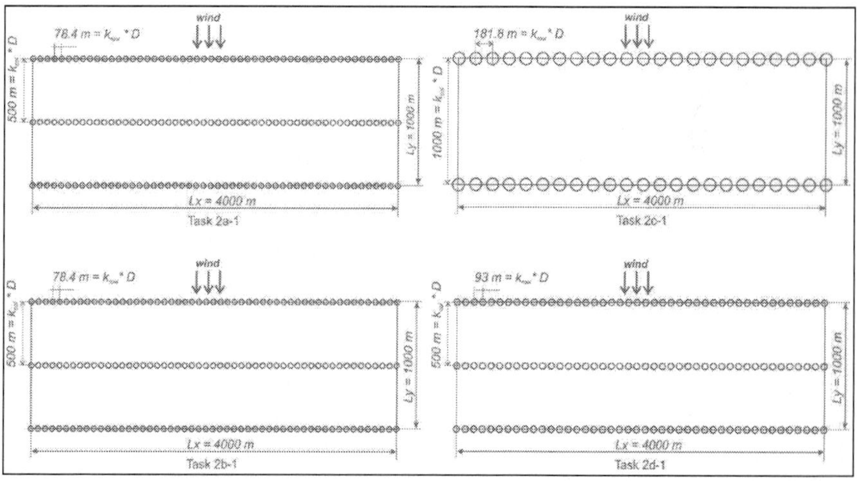

Fig. 5. Wind turbines placement for predominant wind direction toward the long side of rectangular wind park shape

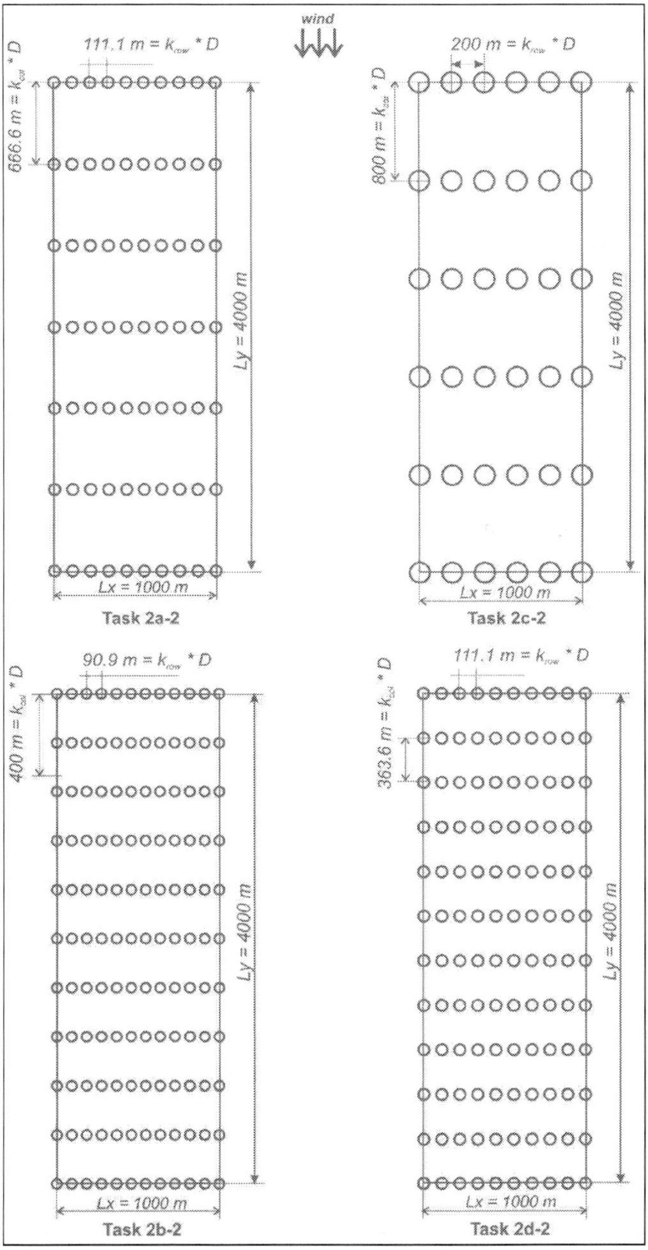

Fig. 6. Wind turbines placement for predominant wind direction toward the short side of rectangular wind park shape

Task	Chosen type of wind turbine	Turbines number, N	Total installed power, MW	Separation coefficients,	Separation distances, m	Cost per unit energy
Task 2a-2 $1.5 < k_{row} < 3$ $9 < k_{col} < 11$	9.1 enercon's E-70 $P_{wt} = 2.3$ MW, $D = 71$ m, $H = 58$ m	$N = 70$ $N_x = 10$ $N_y = 7$	161	$k_{row} = 1.56$ $k_{col} = 9.39$	$SD_x = 111.1$ $SD_y = 666.6$	0.000110
Task 2b-2 $1.5 < k_{row} < 3$ $9 < k_{col} < 11$ $D < 70$	5.1 enercon's E-44 $P_{wt} = 0.9$ MW, $D = 44$ m, $H = 45$ m	$N = 132$ $N_x = 12$ $N_y = 11$	118.8	$k_{row} = 2.07$ $k_{col} = 9.09$	$SD_x = 90.9$ $SD_y = 400$	0.000282
Task 2c-2 $2 < k_{row} < 4$ $8 < k_{col} < 12$	10.2 Vestas V90 $P_{wt} = 3$ MW, D = 90 m, H = 105 m	$N = 36$ $N_x = 6$ $N_y = 6$	108	$k_{row} = 2.22$ $k_{col} = 8.88$	$SD_x = 200$ $SD_y = 800$	0.000089
Task 2d-2 $2 < k_{row} < 4$ $8 < k_{col} < 12$ $D < 70$	5.1 enercon's E-44 $P_{wt} = 0.9$ MW, $D = 44$ m, $H = 45$ m	$N = 120$ $N_x = 10$ $N_y = 12$	108	$k_{row} = 2.52$ $k_{col} = 8.26$	$SD_x = 111.1$ $SD_y = 363.6$	0.000282

Table 6. Optimal solutions for predominant wind direction and rectangular wind park area ($L_x = 1$ km, $L_y = 4$ km)

4. The numerical testing analysis and discussions

The first obvious result of the tasks solutions is that the different tasks formulations with different restrictions define different turbines' type and number choices. The increasing of the feasible space for the separation coefficients k_{row} and k_{col} by defining wider limits of their values also define different choices as it is illustrated by the solutions of **Task 1a** compared to **Task 1c, Task 1a-1** to **Task 1c-1, Task 2a** to **Task 2c, Task 2a-1** to **Task 2c-1** and **Task 2a-2** to **Task 2c-2**. Generally speaking, in the uniform wind direction case the larger feasible interval of the separation coefficients k_{row} and k_{col} define better choices in respect to the installed power and cost per unit of energy **(Task 1a** and **Task 1c, Task 1a-1** and **Task 1c-1)**. For the predominant wind direction case such tendency is not observed **(Task 2a** and **Task 2c, Task 2a-1** and **Task 2c-1, Task 2a-2** and **Task 2c-2)**. Changing of the wind park square shape to rectangular shape influences the turbines type and number choices in most uniform and predominant wind direction cases - **Task 1a** and **Task 1a-1, Task 1c** and **Task1c-1, Task 2a** and **Task 2a-1** and **Task 2a-2, Task 2c** and **Task 2c-1** and **Task 2c-2**. Some tasks solutions define the same turbine's type choice but different turbines number. As it is expected, the changing of the wind park rectangular shape orientation for the uniform wind direction case does not affect the solutions choices but only interchanges values of N_{row} with N_{col} respectively. For the predominant wind direction case the turbines choice depends on the rectangular shape orientation toward the wind direction (**Task 2a-1** and **Task 2a-2**, Task **2b-1** and **Task 2b-2, Task 2c-1** and **Task 2c-2, Task 2d-1** and **Task 2d-2**). In that case again some tasks solutions define the same turbines type choice but different turbines number (**Task 2c-1** and **Task 2c-2, Task 2d-1** and **Task 2d-2**). The solution results show that the wind park shape and orientation should be taken into account when formulating the optimization tasks to get optimal wind park layout design. The turbines rotor diameter restriction narrows the turbines type choice and in some cases results in equivalent turbines type (**Task 1b** and **Task 1d, Task 2b** and **Task 2d, Task 2b-2** and **Task 2d-2**) but with different turbines number. The wind turbines choice optimality is judged by the used optimization criterion – the value of costs per unit energy. The comparison of the those values in last columns of tables 2, 3, 4, 5 and 6 shows that using of small number of big wind turbines (with bigger rated power and rotor diameter) is more economically justified to that optimization criterion then using of large number of smaller wind turbines. The best costs per unit of energy values are obtained for **Task 1c** (P_{wt} = 2.3 MW, D = 82 m), **Task 1c-1** (P_{wt} = 2.3 MW, D = 71 m), **Task 2a** (P_{wt} = 2.3 MW, D = 71 m), **Task 2c** (P_{wt} = 3 MW, D = 90 m), **Task 2c-1** (P_{wt} = 3 MW, D = 90 m), **Task 2a-2** (P_{wt} = 2.3 MW, D = 71 m) and **Task 2c-2** (P_{wt} = 3 MW, D = 90 m). Nevertheless it is important to note that the size of wind turbines is closely related to the wind conditions of the given wind park area. That means the optimal wind turbines choice should take into account the wind conditions of the particular area by introducing of the proper relations and restrictions. The tasks **1b, 1d, 1b-1, 1d-1, 2b, 2d, 2b-1, 2d-1, 2b-2** and **2d-2** illustrate that by adding the restriction for the size of the turbine rotor diameter which defines the choice of smaller turbines.

The formulated optimization tasks are based on the sample data from Table 1. They are used to illustrate numerically the applicability of the proposed wind park layout design approach. The tasks solution times are about a few seconds on PC with Intel Core i3 CPU/2.93 GHz and 3 GB RAM. Larger sets of turbines and parameters together with additional restrictions can be taken into consideration to get a better conformance to the particular wind park design requirements. That will reflect in increasing of the optimization tasks solutions space but also will increase the size of the tasks and eventually their computational difficulties.

The widely used in wind park design cost per unit energy ratio optimization criterion is one of the possible optimization criteria. It could be refined by using of statistical data for the wind conditions of particular geographical area and calculating of more precise value for the nominal power utilization coefficient μ. Other analytical formulas for wind park power output and other optimization criteria including different economical, technological and ecological requirements could be used.

5. Conclusions

The wind park layout design is important for ensuring of the expected wind power capture without increasing of the project costs. A combinatorial optimization modeling approach for wind park layout design considering the wind park area shape, size, orientation and different requirements and restrictions is proposed. It is used for formulation of single criterion mixed-integer nonlinear optimization tasks. The relation of investment costs and power output as function of wind turbines number and type is used as optimization criterion. The solutions of formulated optimization tasks give the optimal wind turbines type and number for given wind park area. It is assumed that from technological point of view it is better to have all wind park turbines of the same type. The wind turbines number is defined on the basis of given wind park area size and turbines' spacing recommendations. Two basic wind directions cases are presumed - uniform and predominant wind directions for square and rectangular wind park, both with equivalent area. The turbines' spacing is modeled by introducing of variable separation distances coefficients within given lower and upper limits. The values of the separation distance coefficients and turbines number define the wind park turbines placement taking into account the wind direction.

The developed wind park design approach was tested numerically by solving of different real data optimization tasks formulations involving different separation coefficients limits and wind park shapes and orientations. The wind park orientation for the uniform wind direction does not change significantly the layout design - it only exchanges the numbers of turbines in rows and columns for the case of rectangular wind park. For the predominant wind direction case and rectangular shape of the wind park the turbines choice and placement depends on the park orientation toward the wind direction. The optimization tasks solutions show that the different practical requirements and restrictions define in general, different turbines choices. The solution results confirm the practical conclusion that using of big size turbines is more profitable than large numbers of small size turbines. The numerical testing shows the applicability of the developed optimization approach to wind park layout design. It could be used in the design of both onshore and offshore wind farms design by introducing of specific requirements as variables relations and restrictions. The proposed approach can be programmed as module in computer aided design system. It could be used also as a computer simulation tool for different wind park design conditions thus assisting mathematically reasoned decision making as contradiction to the heuristic approaches.

6. References

Ammara, I.; Leclerc, Ch. & Masson, Ch. (2002). A viscous three-dimensional differential/actuator-disk method for the aerodynamic analysis of wind farms. *Journal of Solar Energy Engineering*, Vol. 124, No. 4, pp. 345–356, ISSN: 0199-6231.

Donovan, S. (2005). Wind farm optimization, *Proceedings of the 40th Annual ORSNZ Conference*, pp. 196-205, 02-03.12.2005, New Zealand, Wellington.

Emami, A. & Noghreh P. (2010). New approach on optimization in placement of wind turbines within wind farm by genetic algorithms. *Renewable Energy*, Vol. 35, No 7, pp. 1559-1564, ISSN: 0960-1481.

Ettoumi, F. Y.; Adanea, Abd El Hamid; Benzaouia, M. L. & Bouzergui, N. (2008). Comparative simulation of wind park design and siting in Algeria. *Renewable Energy*, Vol. 33, pp. 2333–2338, ISSN: 0960-1481.

Grady, S.A; Hussainia, M.Y & Abdullah, M.M. (2005). Placement of wind turbines using genetic algorithms. *Renewable Energy*, Vol. 30, No 2, pp. 259–270, ISSN: 0960-1481.

Johnson, G. L. (2006). *Wind energy systems*. Electronic edition, Manhattan.

Kongnam, C.; Nuchprayoon, S.; Premrudeepreechacharn, S. & Uatrongjit, S. (2009). Decision analysis on generation capacity of a wind park. *Renewable and Sustainable Energy Reviews*, Vol. 13, No 8, pp. 2126–2133, ISSN: 1364-0321.

Kusiak, A. & Song, Zhe. (2010). Design of wind farm layout for maximum wind energy capture. *Renewable Energy*, Vol. 35, pp. 685-694, Elsevier, ISSN: 0960-1481.

Lampinen, J. & Zelinka, I. (1999). Mixed Integer-Discrete-Continuous Optimization by Differential Evolution, Part 1: the optimization method. *Proceedings of MENDEL'99, 5th International Mendel Conference on Soft Computing*, pp. 71-76, ISBN 80-214-1131-7, June 9-12, 1999, Brno.

Lepa, J.; Kokin, E.; Annuk, A.; Põder, V. & Jürjenson, K. (2008). Wind power stations performance analysis and power output prognosis. *Proceedings of Engineering for Rural Development*, pp. 92-96, Latvia, May 29-30, 2008, Jelgava.

Mahat, P.; Ongsakul, W. & Mithulananthan, N. (2006). Optimal placement of wind turbine DG in primary distribution systems for real loss reduction. *Proceedings of Energy for Sustainable Development: Prospects and Issues for Asia*, pp. 1-6, Thailand: ESD2006; March 1-3, 2006, World Academy of Science, Engineering and Technology, Phuket.

Marmidis, G.; Lazarou, S. & Pyrgioti, E. (2008). Optimal placement of wind turbines in a wind park using Monte Carlo simulation. *Renewable Energy*, Vol. 33, No 7, pp. 1455-1460, ISSN: 0960-1481.

Mora, J. C.; Barón, J. M. C.; Santos, J. M. R. & Payán, M. B. (2007). An evolutive algorithm for wind farm optimal design. *Neurocomputing*, Vol. 70, Issue 16-18, pp. 2651-2658, ISSN: 0925-2312.

Mosetti, G.; Poloni, C. & Diviacco B. (1994). Optimization of wind turbine positioning in large wind farms by means of a genetic algorithm. *Journal of Wind Engineering and Industrial Aerodynamics*; Vol. 51, No 1, pp. 105-116, ISSN: 0167-6105.

Mustakerov, I. & Borissova, D. (2007). Technical systems design by combinatorial optimization choice of elements on the example of night vision devices design. *Comptes rendus de l'Academie bulgare des Sciences*, Tom 60, No 4, pp. 373-380, ISSN 1310-1331.

Mustakerov, I. & Borissova, D. (2010). Wind turbines type and number choice using combinatorial optimization. *Renewable Energy*, Vol. 35, No 9, pp. 1887-1894, Elsevier, ISSN: 0960-1481.

Ransom, D.; Moore, J. J. & Heronemus-Pate, M. (2010). Performance of Wind Turbines in a Closely Spaced Array. *RenewableEnergyWorld.com*, May, http://www.renewableenergyworld.com/rea/news/article/2010/05/performance-of-wind-turbines-in-a-closely-spaced-array

Rodman, L. C. & Meentemeyer, R. K. (2006). A geographic analysis of wind turbine placement in Northern California. *Energy Policy*, Vol. 34, pp. 2137-2149, ISSN: 0301-4215.

Shapiro, J. F. (1979). *Mathematical Programming: Structures and Algorithms*. John Wiley & Sons Inc. ISBN-10: 0471778869, 406 p.

Smith, G.; Schlez, W.; Liddell, A.; Neubert, A. & Peña, A. (2006). Advanced wake model for very closely spaced turbines. pp. 1-9, *European Wind Energy Conference*, 27-Feb-2Mar, 2006, Greece, Athens.

Sørensen, J. D. (2006). Optimal reliability-based design of offshore wind turbine parks, *Proceedings of the 2nd International Forum on Engineering Decision Making conference*, pp. 1-12, April 2006, Canada, Lake Louise.

Taha, H. A. (2006). *Operations Research: An Introduction*. 8th. Edition, Prentice Hall

Üstüntaşa, T. & Şahin, A. (2008). Wind turbine power curve estimation based on cluster center fuzzy logic modeling. *Journal of Wind Engineering and Industrial Aerodynamics*. Vol. 96, Issue 5, pp. 611-620, ISSN: 0167-6105.

White, G.; Garrad, A.; Tindal, A. (1997). Integrated design methodology for wind farms. *IPENZ Transactions*, Vol. 24, No. 1, pp. 62–8.

Wind-energy-market, http://www.wind-energy-market.com/ [accessed June 2010].

Wind Turbines - Enercon, http://www.enercon.de/ [accessed June 2010].

Wind Turbines - Vestas, http://www.vestas.com/ [accessed June 2010].

18

Genetic Optimal Micrositing of Wind Farms by Equilateral-Triangle Mesh

Jun Wang[1], Xiaolan Li[2] and Xing Zhang[3]
[1]*Tongji University*
[2,3]*Tsinghua University*
China

1. Introduction

The micrositing problem designs the layout, i.e. the number of turbines and specific location of each one, for a given farm based on the information of the weather, terrain and landscape of the farm. It aims to capture the wind energy of a farm more effectively while satisfying the constrains on economical, social and environmental issues. The micrositing process, a challenging subject involving fluid dynamics and decision making, plays a crucial role in wind farm planning (Conover & Davis, 1994).

In engineering practice, designers usually calculate the wind flow of a given wind farm by commercial software, and empirically determine the installation positions of turbines based on the flow field. The flow field usually does not include the influence of turbines on the deflexion of the original air flow, i.e. wake effects. However, as the wake effects are complicated and strongly coupled, they play a crucial role in wind farm micrositing. In academia, the micrositing problem with the consideration of turbine wake effects were studied for relative flat terrains. Patel (1999) suggested that wind turbines should be placed in rows $8 \sim 12$ rotor diameters apart in the windward direction, and $1.5 \sim 3$ rotor diameters apart in the crosswind direction. As wind profiles were not considered, the solution was still an "empirical" one. Mosetti et al. (1994), for the first time, applied genetic algorithms (GA) to solve the problem of wind farm micrositing in a systematical manner. Grady et al. (2005) improved Mosetti's work in terms of programming and computing, and obtained more reasonable results. Wan et al. (2009) introduced the Weibull function to describe the probability of wind speed distributions and employed turbine speed-power curves to estimate turbine power generation.

In Mosetti et al. (1994) and Grady et al. (2005), a square wind farm of 2km $\times$ 2km was partitioned into 10×10 squares and the turbines could only be installed in the center of suitable small squares. The square mesh (SM) simplified and reformulated the micrositing problem into a discrete-time optimization one, which could be tackled by a binary-coded GA. Although SM is the most simple and instinctive choice, it is worthwhile investigating alternative, and probably better meshing methods. In this paper, a novel equilateral-triangle mesh (ETM) is presented, which proves to be a more suitable method in terms of wind farm production and energy efficiency.

The remainder of this paper is organized as follows. Section 2 introduces the methodology of the optimal micrositing problem. Section 3 carries out computational simulations and analyzes the results. Section 4 makes concluding remarks.

2. Methodology

In this section, the mathematical model of the optimal micrositing problem is firstly formulated, the equilateral-triangle mesh is then presented, and the genetic algorithm for the problem is introduced.

2.1 Mathematical modeling

The models of wake effects, cost and optimal objective functions in Mosetti et al. (1994) and Grady et al. (2005) are used in this paper for a fair comparison. For clarity and completeness, the models are briefly repeated in the following but with some new insights.

2.1.1 Wake effects

Jensen (1983) developed a simplified wake model, which neglected the near field behind a wind turbine. As illustrated in Figure 1, the wake has an initial diameter roughly equal to the turbine diameter and spreads linearly with the downwind distance.

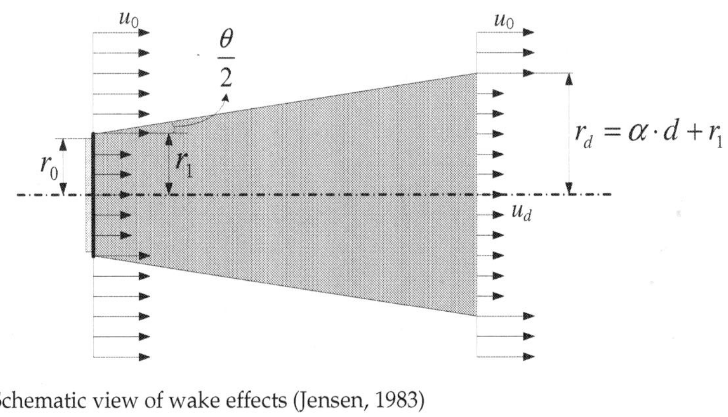

Fig. 1. Schematic view of wake effects (Jensen, 1983)

Based on the theorem of momentum balance, the wake velocity can be determined by the following expression

$$u_d = u_0 \left(1 - \frac{2a}{(1 + \alpha \cdot d/r_1)^2} \right) \tag{1}$$

where r_1 is the wake radius just behind the rotor, a is the initial velocity deficit, and the other symbols are illustrated in Figure 1. Readers can refer to Grady et al. (2005) for the details.

The entrainment constant α is empirically calculated by

$$\alpha = \frac{0.5}{\ln (h/z_0)} \tag{2}$$

where h is the turbine hub height, and z_0 is the surface roughness. The divergence angle θ, as indicated in Figure 1, is then defined as $\theta = 2 \arctan \alpha$.
The relation between roughness lengths, terrain surface characteristics and roughness classes is given in the European Wind Atlas (Troen & Petersen, 1989). The roughness length ranges from 0.0001m to 1.00m, where 0.0001m represents water areas and 1.00m represents cities. Figure 2 shows the corresponding divergence angle θ, under different roughness lengths and turbine hug heights, ranges roughly from $4° \sim 15°$.

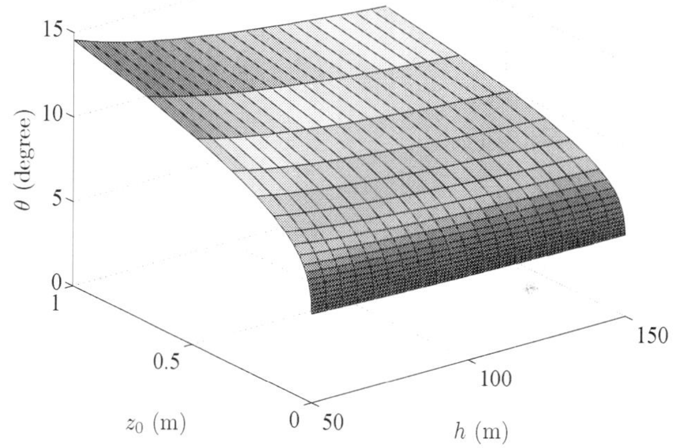

Fig. 2. Divergence angle

When multiple wakes overlap at the downstream turbines, Katic et al. (1986) assumed the kinetic energy deficit of a mixed wake to be equal to the sum of the energy deficits for each wake at the calculated position. So in the ith wind direction, the downstream wind speed $u_{[i,j]}$ at the jth turbine can be computed by (Kiranoudis et al., 2001)

$$u_{[i,j]} = u_0 - \sqrt{\sum_{k=1}^{N_{[i,j]}} \left(u_{[i,k]} - u_{[i,j,k]} \right)^2} \tag{3}$$

where u_0 is the freedom wind speed, $N_{[i,j]}$ is the number of the turbines preceding the jth one, and $u_{[i,j,k]}$ is the actual wind speed from the kth turbine at the position of the jth turbine.

2.1.2 Objective function

The optimal micrositing problem in this paper is to maximize energy production while minimizing project costs.
For a wind farm with N turbines and the wind distributed among M directions, the total output of the farm can be evaluated by (Grady et al., 2005; Kiranoudis et al., 2001)

$$\mathcal{P} = 0.3 \sum_{i=1}^{M} \left(p_i \sum_{j=1}^{N} u_{[i,j]}^3 \right) \tag{4}$$

where p_i represents the annual frequency of the ith wind direction.

Capital costs are one of the primary factors, which should be considered when determining optimum turbine spacing (Conover & Davis, 1994). The Department for Business, Enterprise and Regulatory Reform (United Kingdom) carried out a study on the cost breakdown of a wind energy investment in Europe in 2007 (Department of Trade and Industry, 2007), which claimed that turbine ex works accounted for 66% of the capital cost. And the Spanish report from Intermoney-AEE claimed that 72% of the total costs is for the turbine ex works (Intermoney-AEE, 2006). In this paper, we follow Mosetti et al. (1994) and Grady et al. (2005) and only consider the investment on the wind turbines. The total cost per year of the whole wind farm project is (Grady et al., 2005; Mosetti et al., 1994)

$$\mathcal{C} = N\left(\frac{2}{3} + \frac{1}{3}e^{-0.00174N^2}\right) \tag{5}$$

In conclusion, the objective function is to minimize the cost per unit energy, i.e. (Grady et al., 2005)

$$\min \frac{\mathcal{C}}{\mathcal{P}} \tag{6}$$

while guaranteeing the safe distance between any turbines.

2.2 Equilateral-triangle mesh

The micrositing problem defined above is a constrained optimal control one, which is rather technically challenging and computationally time-consuming due to the constraints on turbine distances. To tackle the problem, it is natural to reduce such a constrained problem into an unconstrained one.

To guarantee the minimal distance between any turbines, the most convenient way is to partition a wind farm into square cells of predefined width and to only allow turbines to be placed in the center of appropriate cells (Grady et al., 2005; Marmidis et al., 2008; Mosetti et al., 1994), as illustrated in Figure 3. The square meshing is simple and intuitive, and easy to implement. It guarantees any turbine in a farm is the same distance to adjacent ones in the same row or column if exist. However, the turbines in a diagonal direction will be unnecessarily spaced apart, i.e. the distance is magnified by $\sqrt{2}$, and therefore the wind farm is not fully exploited.

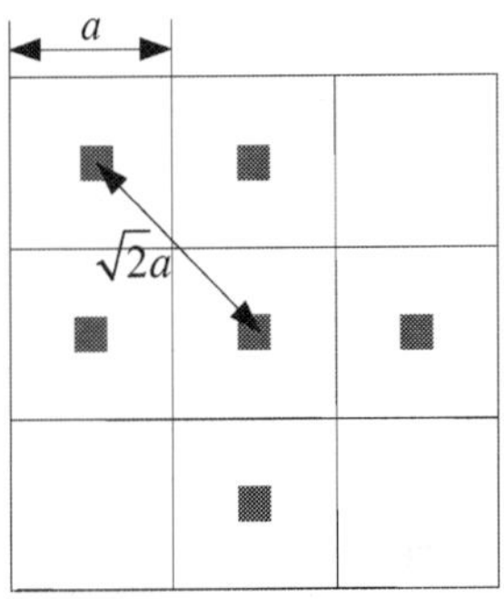

Fig. 3. An example of square meshing

An intuitive idea is to locate the wind turbines at the center of some circular cells, which are tangent to each other as illustrated in Figure 4(a). When the centers of the cells are connected, we obtain intertwined equilateral hexagons shown in Figure 4(b), seemingly a "honeycomb" mesh. If further analyzed, the hexagons can be decomposed into six equilateral triangles and the triangle vertices represent the possible positions of turbines, as shown in Figure 4(c). Therefore, the mesh is called the *equilateral-triangle mesh*.

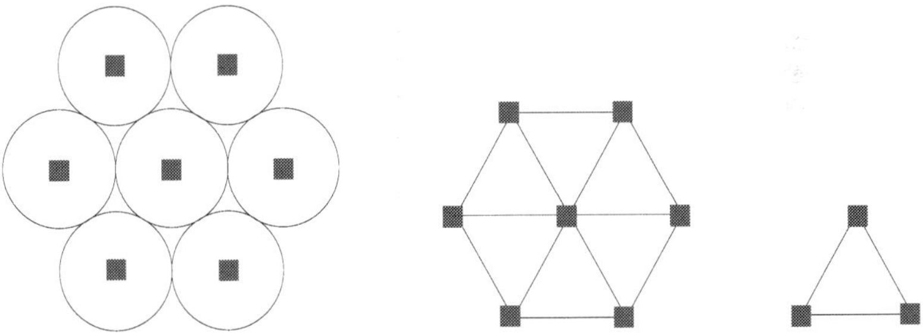

(a) Tangent circles (b) Hexagons (c) Triangle

Fig. 4. Equilateral-triangle mesh

As recommended in Troen & Petersen (1989), for a flat farm with unidirectional wind, turbines should be place about 3 $\sim$ 5 times of rotor diameter apart in columns and about 5 $\sim$ 9 times in rows. In this paper, we follow Mosetti et al. (1994), Grady et al. (2005) and Marmidis et al. (2008), and set the side length of the triangle as five times of the turbine rotor diameter.

Definition 1 (ETM orientation). *Pick up any equilateral triangle in a mesh, construct a vector from the center of the triangle to the vertex and obtain the angle ϕ (in degrees) of this vector from the north-direction vector (i.e. y-axis) clockwise, as illustrated in Figure 5. The orientation of the mesh is defined as*

$$\psi = \mod(\phi, 60^\circ)$$

where mod *stands for the modulo operation. For convenience, an ETM with an orientation angle ψ is denoted as ETM-ψ.*

Then, the orientation of the traditional SM can be similarly defined as follows.

Definition 2 (SM orientation). *Pick up any square in the mesh, and construct a vector from the center of the square towards one of its vertices. The clockwise angle from the north-direction vector towards it is ϕ (in degrees). The orientation of the traditional SM is defined as*

$$\psi = \mod(\phi, 90^\circ)$$

Under this definition, the orientation of the square meshes used in Mosetti et al. (1994), Grady et al. (2005), and Marmidis et al. (2008) were 45°, which can be denoted as ETM-45° in short.

2.3 Genetic algorithms

Due to the complexity of the optimal micrositing, genetic algorithms are introduced to solve it. Unlike the traditional calculus-based methods, GAs are robust, global, and do not require

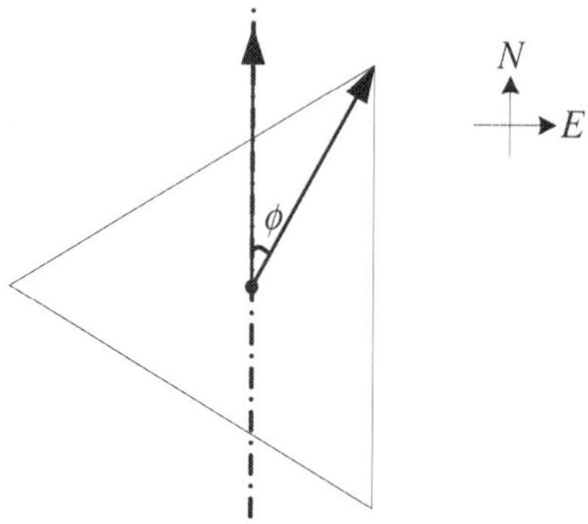

Fig. 5. Orientation of mesh

the existence of derivatives of objective functions. The basic procedures of the GA are as follows (Houck et al., 1995):

Step 1 Encode the micrositing problem into a binary string.

Step 2 Randomly generate a population representing a group of possible solutions.

Step 3 Calculate the fitness values for each individual.

Step 4 Select the individuals according to their fitness values.

Step 5 Perform crossover and mutation operations on the selected individuals to create a new generation.

Step 6 Check whether the progress is convergent, or meets the terminating condition. If not, return to Step 3.

Encoding is the first step of the GA procedures. Suppose a wind farm is a square region partitioned into equilateral-triangle cells, whose vertices represent the possible positions for placing turbines. Each bit corresponds to a vertex and all of the bits are connected serially into a binary string in a top-down left-right sequence. In the string, "1" represents that a turbine is placed on the corresponding vertex, while "0" stands for no wind turbine.
The selection, crossover and mutation are the fundamental operators of GAs. Generally, a probabilistic selection is performed based upon the individual's fitness such that the better individuals have an increased chance of being selected, and the probability is assigned to each individual based on its fitness value. The crossover takes two individuals and produces two new individuals while the mutation alters one individual to produce a single new solution (Houck et al., 1995). The crossover probability is usually between 0.6 $\sim$ 0.9, and the mutation probability between 0.01 $\sim$ 0.1 (Sivanandam & Deepa, 2008). In this paper, the crossover probability is chosen to be 0.7 through trial-and-error processes, and the mutation probability 0.05.

3. Simulation results and analyses

In this paper, the Genetic Algorithm Optimization Toolbox is utilized for simulations. The micrositing results of the ETM method are compared with the SM method employed by

Mosetti et al. (1994) and Grady et al. (2005). For a fair comparison, the same turbines are utilized, i.e. turbines with the hub height 60m, the rotor radius 20m and the thrust coefficient 0.88. The ground roughness length of the site is $z_0 = 0.3$m, and the minimal-distance between wind turbines is 200m. Note that, due to the different mesh methods, the effective region for micrositing is 1800×1800 square meters in this paper while 2000×2000 square meters in Mosetti et al. (1994), Grady et al. (2005) and Marmidis et al. (2008).
The following three cases in Grady et al. (2005) are investigated and the wind rose map of Case 3 is given in Figure 6.

- Case 1: Single-direction wind with a speed of 12m/s;
- Case 2: Multiple-direction (36 directions) wind with a speed of 12m/s;
- Case 3: Multiple-direction (36 directions) wind with typical speeds of 8, 12 and 17m/s.

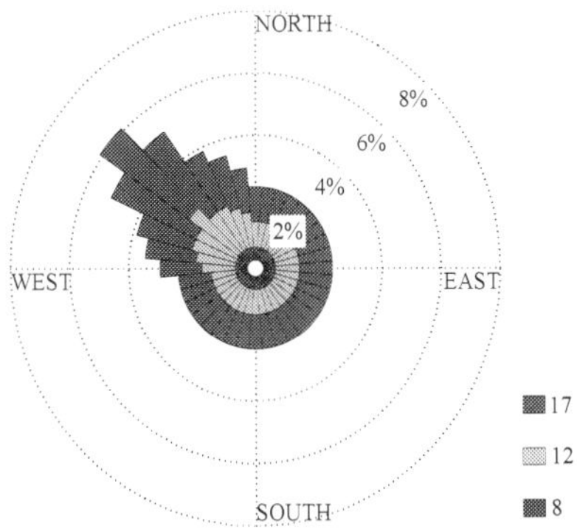

Fig. 6. Rose map of Case 3

3.1 Case 1: Single-direction & uniform-speed wind

The optimal micrositing layouts by the ETM method are presented in Figure 7(c) and Figure 7(d) while the SM-based result in Grady et al. (2005) is shown in Figure 7(a). By using the ETM-30°, turbines are roughly arranged in three evenly-spaced groups, which is similar to the layout by the SM method (Grady et al., 2005). Due to the nature of the ETM, the turbines in each group (two rows) are staggered, which is consistent with the "empirical" scheme. By using the ETM-0°, turbines are arranged into two rows in the top of the farm and the other two in the bottom. Compared to the layout by the ETM-30°, the turbines in each group are more closely placed. Note that, the ETM-0° is the same direction as the wind, while the ETM-30° is perpendicular to the wind direction. And what will happen if we chose a SM whose direction is perpendicular to the wind? The optimal layout of SM-0° is presented in Figure 7(b).
Table 1 compares the fitness values, total power output and the numbers of wind turbines for each layout. It is clear that both ETM-based schemes achieve smaller fitness values. In particular, the fitness value of the ETM-0° layout is 7.89% lower than the ETM-30°, 5.74%

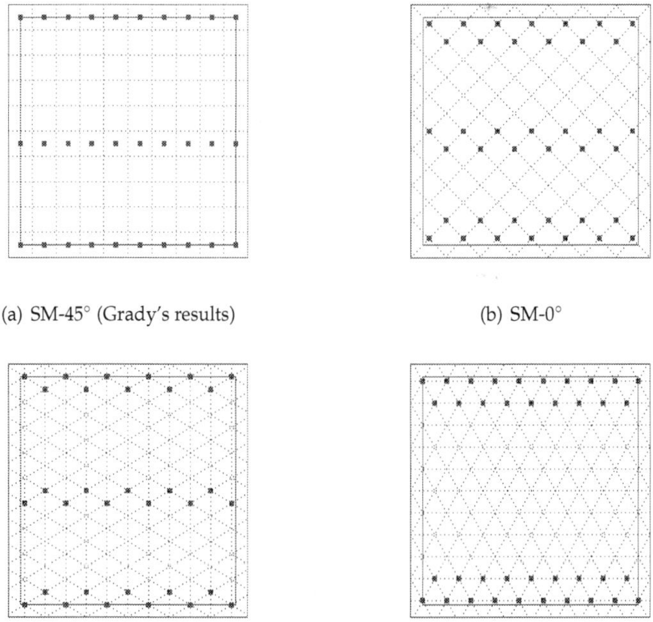

(a) SM-45° (Grady's results) (b) SM-0°

(c) ETM-30° (d) ETM-0°

Fig. 7. Optimal micrositing layouts using different meshing methods (Case 1)

Meshing Methods	Fitness ($\times 10^{-3}$)	Output (kW)	WT Numbers
SM-45°	1.5436	14310	30
SM-0°	1.4809	18180	39
ETM-30°	1.5152	15611	33
ETM-0°	1.3959	18884	38

Table 1. Results of ETM and SM-based optimal micrositing for Case 1

lower than the SM-0° and 9.57% lower than the SM-45°. So the results prove the advantages of the ETM method over the traditional SM method.

Moreover, Table 1 also shows that the fitness values of the layouts are better when the mesh orientation is along the wind direction. It indicates that the performance can be further improved if the mesh orientation is appropriately chosen. In order to study how to choose the mesh orientation, several more simulations using different orientations of the ETMs are carried out, and their results are listed in Table. 2.

It is clear that the rotationally symmetrical ETM-10° and ETM-50° gain the best fitness. The layouts of these two orientations are presented in Figure 8. This is related to the divergence angle of the wind turbines. Since the wake effects decrease as the distance downstream of the turbine increases, we would prefer to place adjacent wind turbines outside of the region of wind turbine wakes. The divergence angle of the wind turbines determine the orientation

Meshing Methods	Fitness ($\times 10^{-3}$)	Output (kW)	WT Numbers
ETM-10°	1.3727	21737	44
ETM-20°	1.3842	21556	44
ETM-40°	1.3832	22449	46
ETM-50°	1.3721	21746	44

Table 2. Results of different orientations of ETMs for Case 1

of mesh based on their geometrical relationship. From Figure 2, we can observe that the divergence angle θ ranges roughly from 4° to 15°. So the corresponding orientation angle ϕ of ETM should be better within $(\beta + \frac{\theta}{2} - 30°, \beta - \frac{\theta}{2} + 30°)$ to avoid wake effects, where β is the dominant direction of the wind. Taking into account the side length of the triangle, we generally choose ψ within

$$\text{mod } (\beta \pm 10°, 60°) \tag{7}$$

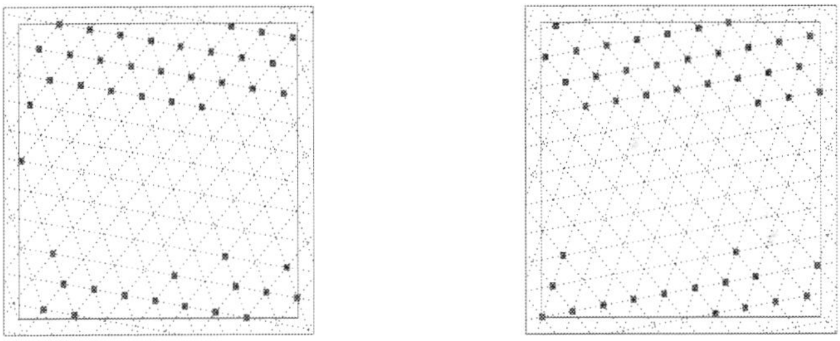

(a) ETM-10° (b) ETM-50°

Fig. 8. Optimal micrositing layouts by using ETM-10° and ETM-50° (Case 1)

3.2 Case 2: Multiple-direction & uniform-speed wind

In this case, the wind is evenly distributed in 36 directions and the wind speed in each direction is constant. Hence, the orientation of the ETM does not affect the micrositing and we choose ETM-0° in order to obtain the maximum number of mesh grid. Figure 9(b) shows the optimal layout by using the ETM method. It is clear that the layout is 6-fold rotational symmetry, which is consistent with the 36-fold rotational-symmetry rose map. The layout by the SM method, shown in Figure 9(a), is not as symmetrical as the ETM-based one, although it is evenly distributed in general.

Table 3 compares the micrositing results by both methods. The ETM-based layout produces 18256kW with 39 wind turbines and its fitness value is 5.87% lower than Grady's. The efficiency of turbines, defined as the ratio of their actual power to the rated one, is improved by 6.02%, from Grady's 85.174% to 90.299%. The results indicate that the ETM method is more suitable for a farm with even distribution of wind directions.

3.3 Case 3: Multiple-direction & multiple-speed wind

This case represents a more practical situation, where the wind is generally evenly distributed but slightly dominated in the north-west direction (about 310°) as one can observe from

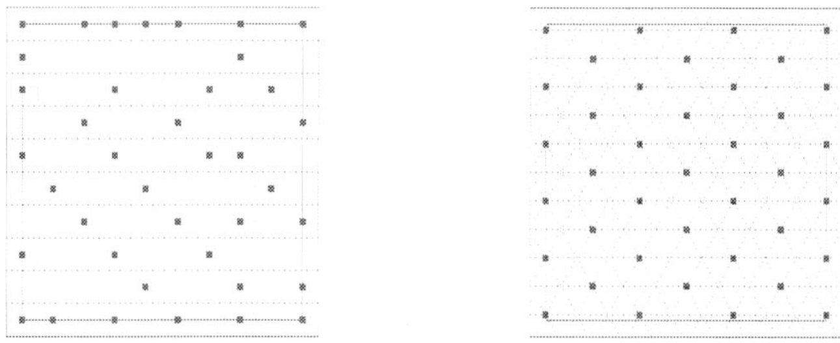

(a) Square mesh (Grady's results) (b) ETM-0°

Fig. 9. Optimal micrositing layouts using different meshing methods (Case 2)

Meshing Methods	Fitness ($\times 10^{-3}$)	Output (kW)	WT Numbers
SM-45°	1.5666	17220	39
ETM-0°	1.4746	18256	39

Table 3. Results of different orientations of ETMs for Case 2

Figure 6. We choose an ETM with an orientation 10° since $\mod(310°, 60°) = 10°$. The optimal layouts by the SM method and the ETM one are presented in Figure 10.
Table 4 compares the present study with the Grady's, and proves that all of the ETM's fitness values are better than SM's. The fitness value of ETM-0° is decreased by 4.48%, and the efficiency is increased by 4.23%. The ETM-40° uses the same number of wind turbines as Grady's, but produces more power, gains a lower fitness value and a higher efficiency. Again, the selection of the ETM orientation agrees with the "thumb of rule" given in Equation (7). The ETM method is more suitable for wind farm micrositing than the SM one.

Meshing Methods	Fitness ($\times 10^{-4}$)	Output (kW)	WT Numbers
SM-45°	8.4240[1]	31958	39
ETM-0°	8.0465	34164	40
ETM-10°	8.2490	31957	38
ETM-40°	8.2133	32779	39

[1] Note that, for Case 3, the fitness value in Grady et al. (2005) is not consistent with its fitness curve. So we re-calculate the fitness value according to Grady's layout.

Table 4. Results of different orientations of ETMs for Case 3

4. Conclusions

This paper presented a novel meshing method, i.e. the equilateral-triangle mesh, for optimal micrositing of wind farms. The ETM method, compared with the traditional square mesh, guarantees the same distance between adjacent wind turbines and matches the empirical staggered-siting style. Computational simulations consistently illustrated the advantages of

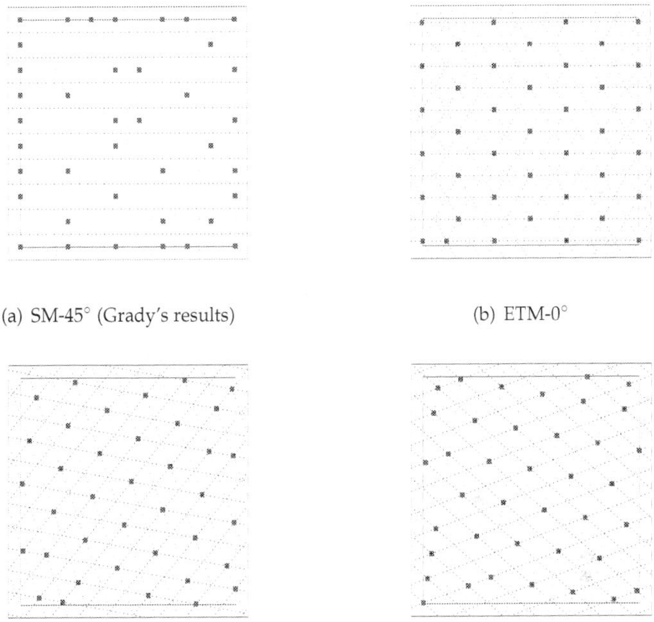

(a) SM-45° (Grady's results) (b) ETM-0°

(c) ETM-10° (d) ETM-40°

Fig. 10. Optimal micrositing layouts using different meshing methods (Case 3)

the ETM method especially when the orientation of the mesh was appropriately adjusted according to the dominant wind direction of a wind park.

5. Acknowledgments

This work was supported in part by the National High-Tech R&D Program of China (863 Program) under Grant No. 2007AA05Z426, and the Natural Science Foundation of China under Grant No. 61075064.

6. References

Conover, K. & Davis, E. (1994). Planning your first wind power project, *Technical Report 104398*, Electric Power Research Institute.

Department of Trade and Industry (2007). Impact of banding the renewables obligation — costs of electricity production, *Technical Report URN 07/948*.

Grady, S. A., Hussaini, M. Y. & Abdullah, M. M. (2005). Placement of wind turbines using genetic algorithms, *Renewable Energy* 30(2): 259–270.

Houck, C., Joines, J. & Kay, M. (1995). A genetic algorithm for function optimization: A matlab implementation, *Technical report*, North Carolina State University.

Intermoney-AEE (2006). Anaálisis y diagnóstico de la situación de la energía eólica en españa.

Jensen, N. O. (1983). A note of wind generator interaction, *Technical Report Risø-M-2411*, Risø National Laboratory.

Katic, I., Hojstrup, J. & Jensen, N. O. (1986). A simple model for cluster efficiency, *European Wind Energy Association Conference and Exhibition*, Rome, Italy, pp. 407–410.

Kiranoudis, C. T., Voros, N. G. & Maroulis, Z. B. (2001). Short-cut design of wind farms, *Energy Policy* 29(7): 567–578.

Marmidis, G., Lazarou, S. & Pyrgioti, E. (2008). Optimal placement of wind turbines in a wind park using monte carlo simulation, *Renewable Energy* 33(7): 1455–1460.

Mosetti, G., Poloni, C. & Diviaccoa, B. (1994). Optimization of wind turbine positioning in large windfarms by means of a genetic algorithm, *Journal of Wind Engineering and Industrial Aerodynamics* 51(1): 105–116.

Patel, M. R. (1999). *Wind and Power Solar Systems*, CRC Press, Boca Raton, Florida.

Sivanandam, S. N. & Deepa, S. N. (2008). *Introduction to Genetic Algorithms*, Springer, New York.

Troen, I. & Petersen, E. L. (1989). *European Wind Atlas*, Risø National Laboratory, Roskilde.

Wan, C.-Q., Wang, J., Yang, G., Li, X.-L. & Zhang, X. (2009). Optimal micro-siting of wind turbines by genetic algorithms based on improved wind and turbine models, *Proceedings of the IEEE Conference on Decision and Control*, Shanghai, P.R. China, pp. 5092–5096.

19

Wind Turbines Integration with Storage Devices: Modelling and Control Strategies

Samuele Grillo[1], Mattia Marinelli[2] and Federico Silvestro[2]

[1]*Dipartimento di Elettrotecnica – Politecnico di Milano*

[2]*Dipartimento di Ingegneria Navale ed Elettrica – Università di Genova*

Italy

1. Introduction

The electric power system is facing an evolution from the traditional concept of energy generation by few localized power plants interconnected to a meshed system to distributed medium and small scale generators (Thomas, 2009; Neural et al. 2008).
Moreover some typologies of these generators embedded into the distribution network are fed by renewable sources like wind and sunlight. Their main drawback is their hardly predictable behaviour and uncontrollable output. This means having for example maximum production during minimum demand period or excess of generation in congested parts of the electric network, thus causing bottlenecks and overvoltage situations in some critical sections of the grid (Grillo et al., 2009).
The presence of energy storage systems may allow a better management of the electric system allowing the full exploitation of renewable energy sources. Nowadays the cost per stored energy is quite high and so it might not be economically feasible to install huge amount of batteries. The size of the storage systems can considerably vary and, depending on their sizes, different tasks can be performed as shown in Fig1 (Fioravanti et al., 2009).

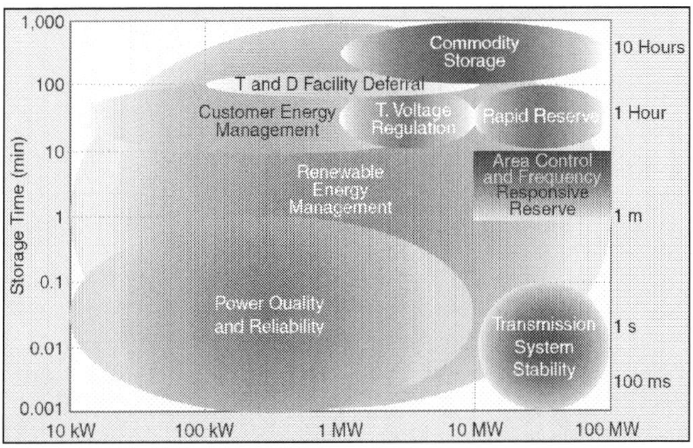

Fig. 1. Storage power requirements for electric power utility applications

Hence the possible duties range from short-term fluctuation levelling and power quality improvement to primary frequency-power regulation and, in case of large storage sizing, compliance to day-ahead generation dispatching (Oudalov et al., 2005).
The present work focuses on the development of models of wind turbines and storage systems, in Matlab-Simulink environment, for implementing integrated control strategies of the whole resulting system in order to describe the benefits that storage can provide. Hence, the idea is to control the battery charging and discharging phases in order to control the whole plant output.
The wind park is composed by four 2 MW wind turbines and a storage system of 2 MWh - 2.5 MW equipped with $Na\text{-}NiCl_2$ batteries. Both the wind turbine and the storage models have general validity and are suited for electrical studies (Di Rosa et al., 2010).
The chapter is organized as follows:

- Paragraph 2 describes the model of the wind turbine and analyzes the wind speed profiles used in the study;
- Paragraph 3 illustrates the storage model;
- Paragraph 4 analyzes the layout of the plant system and the control strategy implemented;
- Paragraph 5 describes the result of the simulations performed.
- Paragraph 6 reports the conclusion and the further developments.

2. Wind turbine model

2.1 Main assumptions

The wind turbine model is described from an electromechanical perspective, thus it provides: an analysis of the aerodynamic behaviour of the rotor including the pitch control system, the shaft dynamic and the maximum power tracking characteristic (Ackermann et al., 2005; Marinelli et al., 2009).
The wind turbine model is tuned for a 2 MW full converter direct drive equipped generator. This typology of wind turbine is characterized by the absence of the gearbox and the presence of ac/dc/ac converter sized for the whole power, as depicted in Fig. 2.

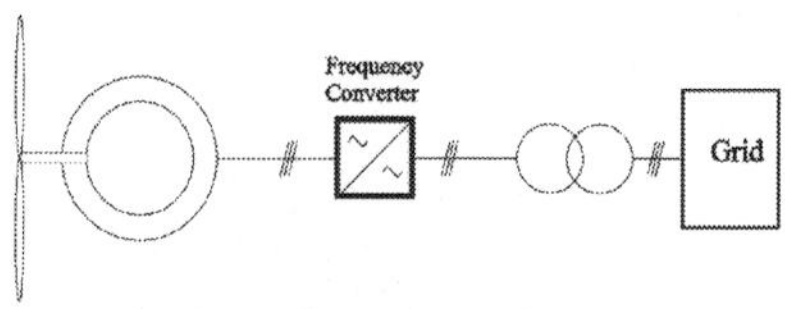

Fig. 2. Full converter direct drive wind turbine concept

Since the model is not intended to analyze dynamics faster than a fraction of second, there is no need to characterize in a detailed way the generation/conversion system, which thus it is modelled as a negative load (Achilles & Pöller, 2004). The rest of the electromechanical conversion system needs an accurate detail due to the interest in studying the possibility to reduce the output in certain conditions. There is, in fact, the need to model the delays introduced by the pitch controller and by the shaft rotational speed.
The block diagram that describes the main model components and their mutual interaction is depicted in Fig. 3. Reading the picture from left to right the first block met is the

aerodynamic one that evaluates the power harvested by the rotor that depends on wind speed, rotational speed and blade angle. This accelerating power, along with the rotational speed of the generator, enters the block that describes the shaft behaviour and allows the evaluation of the power at the end of the shaft and of the turbine rotational speed. The accelerating power at the end of the shaft is the input for the block called generator and MPT (Maximum Power Tracking) that describes the dynamic of the generator and of the MPT control characteristic, as well as the efficiency of the conversion system. At the end the electrical power produced is calculated and is the main output of the turbine model.

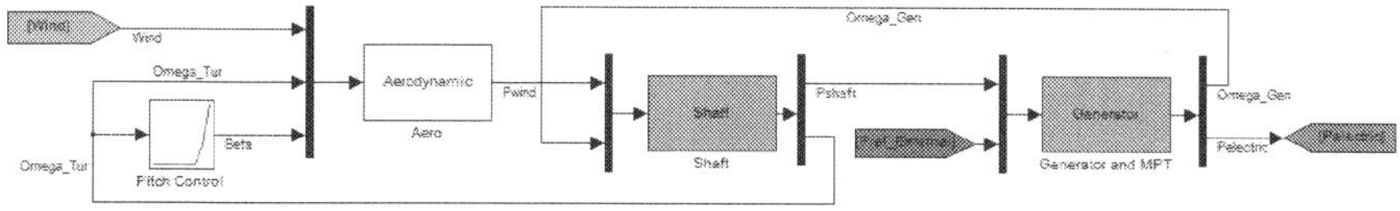

Fig. 3. Conceptual block diagram of the wind turbine

The paragraph develops as follow:

- Wind speed data analysis
- Rotor aerodynamic
- Shaft dynamic
- MPT characteristic
- Pitch controller

2.2 Wind speed data

When studying the wind turbine output, special care should be devoted to the analysis and the proper use of the wind speed data. For power system studies it is common practice to consider just one wind profile per turbine while, in reality, during their sweeping action the blades face different wind profiles: this variety is generated by the turbulence induced by the local terrain. This assumption however is commonly accepted as long as this wind is representative of the wind seen by the whole rotor and it is generally called hub wind (Sørensen et al., 2001).

Due to the interest in studying the fluctuation induced in the turbine power output it is necessary to have appropriated wind speed data or an accurate wind model. For this purpose, data related to the power outputs and to the wind speeds measured at the nacelle of 4 wind turbines belonging to the same farm are used. The data are sampled with a five seconds time step that gives accurate information on the fluctuation included in the wind. A comparison between the wind speeds, measured by the anemometer placed in the rear of the nacelle, and the electrical output power highlights that the wind measured by the anemometer is not necessary the same that is seen by the whole rotor. As shown in Fig. 4 it can be seen that there is a tight correlation between the wind measured data and power output. This correlation determines the datasheet power curve of the turbine, although being this correspondence not exact.

Due to this weak correlation, it is chosen to evaluate the wind speed starting from the output power profile by means of the static power curve of the turbine. Moreover the reduction in the power output due to the height of the installation of the farm, caused by the lower atmospheric pressure compared to sea level, is taken in account. The wind series thus

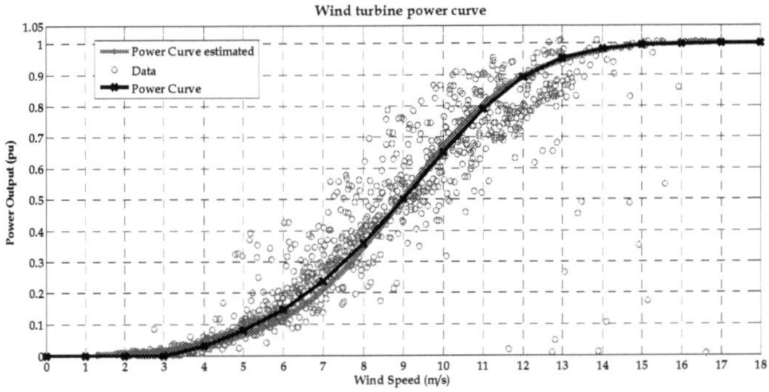

Fig. 4. Datasheet power curve (black curve) and curve estimated from the data (red curve)

created contain information related to the turbulence due to the local terrain roughness or others as for example the tower shadow effect or the wake of the surrounding machines.

Fig. 5 offers a visual comparison between two 12-hours series of wind speed: the one on the left side reports the wind measured at the nacelle, while the one on the right side shows the wind calculated from the power production.

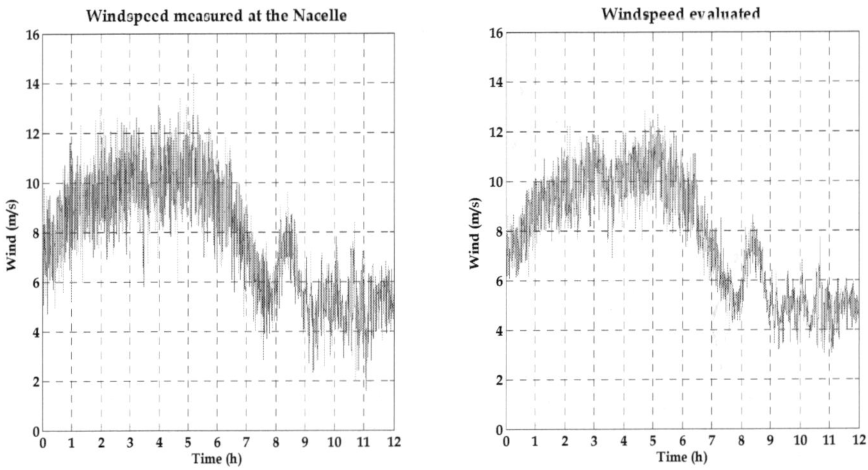

Fig. 5. 12-hours length wind speed profiles, 5-seconds sampled, measured by the nacelle anemometer (left picture) and deduced from the power output (right picture)

As it can be noticed the wind calculated is smoother because the wind measured by the nacelle anemometer has the turbulence induced by the blades themselves.

To analytically evaluate the turbulence it is common practice to introduce the turbulence intensity, generally defined as the ratio between the standard deviation and the average wind speed in a 10-minutes length wind series. In the specific case the average turbulence intensity on all the 12-hours profile, calculated as the average of all the 10-minutes measures, values 12% for the wind series measured at the nacelle and 9% for the one evaluated from the power output. Fig. 6 shows the two 10-minutes average wind speeds and the related turbulence intensity profiles.

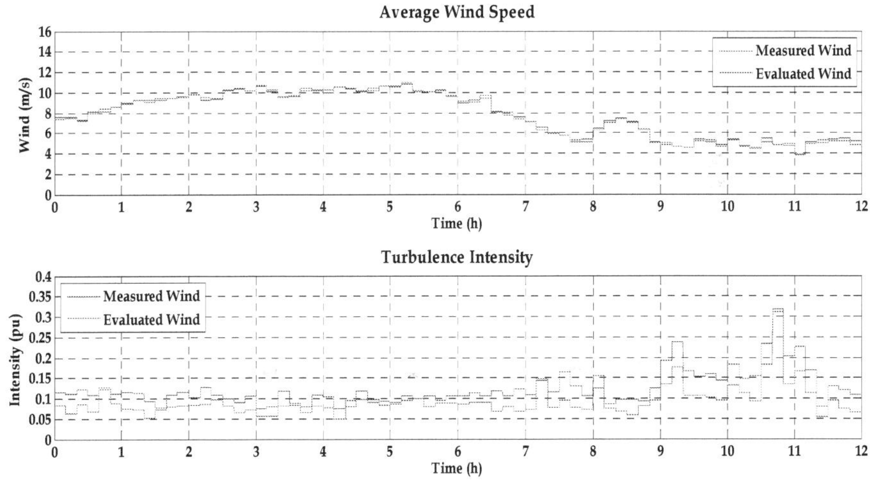

Fig. 6. Measured and Evaluated winds: 10-minutes average (first diagram) and turbulence intensity (second diagram)

In addition, the four turbines are fed by different wind speeds profiles calculated, as explain previously, from the power output of four wind turbines belonging to the same wind farm. Fig. 7 shows a window of 2-hours in order to better appreciate the correlation between the four series.

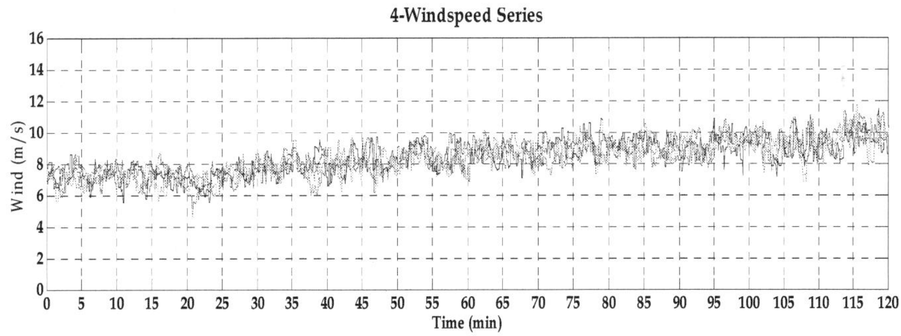

Fig. 7. Four wind speed profiles, 5-seconds sample time, 2-hours timeframe.

2.3 Aerodynamic

The power that the rotor of area, A (m^2), can extract from the wind flux depends mainly on the wind speed, U (m/s), the air density, ρ (kg/m^3), and rotor aerodynamic efficiency, c_p (pu), which is also known as power coefficient and quantifies the rate of power that the rotor extracts from the wind, P_{rotor} (W):

$$P_{rotor} = \frac{1}{2} c_p \cdot \rho \cdot A \cdot U^3 \quad (1)$$

This coefficient is function of the Tip Speed Ratio, λ (pu), and of the blade pitch angle, β (deg). The value of λ depends on the ratio between blade peripheral speed and wind speed.

$$\lambda = \omega \cdot R / U \quad (2)$$

A possible analytic representation can be expressed by the following equation:

$$c_p = c_1 * \left\{ c_6 * \lambda + \frac{\left[-c_4 - c_3 * (2,5+\beta) + c_2 * \frac{1}{\lambda + c_7 * (2,5+\beta)} - \frac{c_8}{\lambda + (2,5+\beta)^3} \right]}{\exp[c_5 * (\frac{1}{\lambda + c_7 * (2,5+\beta)} - \frac{c_8}{\lambda + (2,5+\beta)^3})]} \right\} + c_9 * \lambda \quad (3)$$

where the coefficients c_i are:

c_1	c_2	c_3	c_4	c_5	c_6	c_7	c_8	c_9
0,645	116	0,4	5	21	0,00912	0,08	0,035	0,001

Table 1. Coefficient values

The graphical representation of the power coefficient curves in function of the tip speed ratio and parameterized at different pitch angle values is depicted in Fig. 8

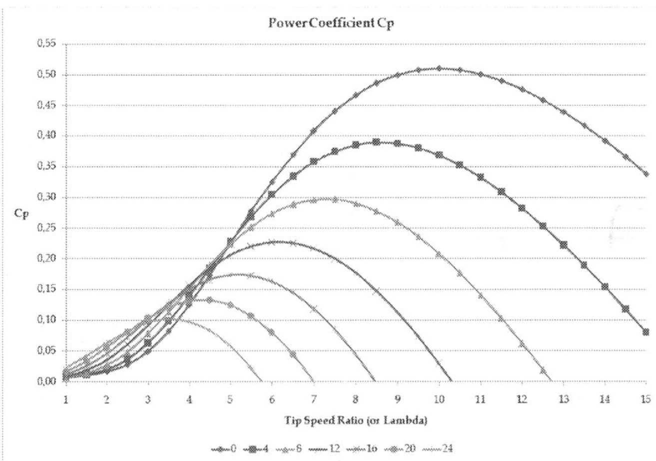

Fig. 8. Power Coefficient characteristic plotted in function of the Tip Speed Ratio (lambda) and parameterized with the pitch angle (beta).

2.4 Shaft

Wind turbines have a relatively soft shaft, and the eigenvalues of the drive train are inside the range of values normally taken in account for power system studies (0.1 ÷ 10 Hz).

Moreover the greater is the number of pole pairs of the generator the softer gets the shaft (Akhmatov, 2003). This can cause the initiation of oscillations whenever there is a sudden step of torque, due for example to wind gust or faults on the network.
In fact, whenever there is a sudden difference between mechanical and electrical torques, respectively T_{wind} and $T_{electromagnetic}$ (Nm), the two shafts, the one of the turbine and the one of the generator, have the possibility to rotate one against the other, thus there is the need to define two different rotational speeds: $\omega_{turbine}$ and $\omega_{generator}$ (rad/s). A two masses representation is therefore necessary due to the interest in evaluating the oscillations induced by the wind in power output. The differential equations that describe the dynamic of the system are reported below: starting from the response determined by the inertia of the turbine, $J_{turbine}$ (kg m²), passing through the shaft stiffness, k (Nm/rad), and damping, D (Ns/rad), values getting to the generator inertia, $J_{generator}$:

$$\begin{cases} T_{wind} - T_{shaft} = J_{tur} * \dfrac{d\omega_{tur}}{dt}; \\ \omega_{turbine0} = \omega_{initial} \end{cases} \tag{4}$$

$$\begin{cases} T_{shaft} = k * \vartheta + D * (\omega_{turbine} - \omega_{generator}); \\ \vartheta = \int (\omega_{turbine} - \omega_{generator}) * dt; \\ \vartheta_0 = {T_{shaft0}}/{k} \end{cases} \tag{5}$$

$$\begin{cases} T_{shaft} - T_{electromagnetic} = J_{gen} * \dfrac{d\omega_{gen}}{dt}; \\ \omega_{generator0} = \omega_{initial} \end{cases} \tag{6}$$

In order to help the comprehension of the shaft dynamics, a comparison with the electrical equivalent system can be done. In fact, if it is assumed that the torques behave like the currents and the rotational speeds like the voltages then the inertial effects are described by means of capacitors, the stiffness by means of inductor and the damping by means of resistance. The electrical circuit is shown in Fig. 9.

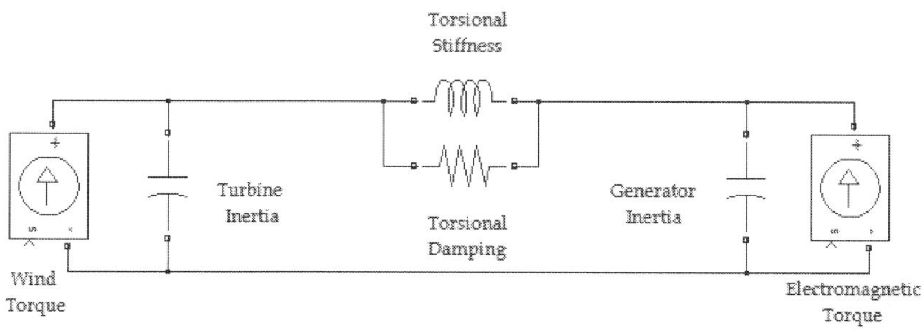

Fig. 9. Electrical equivalent of the two masses shaft.

2.5 Maximum power tracking characteristic

As already described in the aerodynamic characteristic, the maximum efficiency is available only for a small range of values of λ. Thus, since the wind is continuously changing, the Maximum Power Tracking (MPT) has to control the generator rotational speed, ω, in order to have the desired (and optimal) power output. This tracking action is realized by following the curve, shown in Fig. 10., which generates the reference power in function of the rotational speed of the turbine.

In theory it would be better to express the reference power as function of the wind speed, which, unfortunately, cannot be measured with accuracy. So, instead of using wind speed, another control variable, the rotational speed of the turbine is used (Hansen et al., 2007).

The logic behind this curve is quite simple: the control system sets a reference value of power to the generator, depending on the actual rotational speed, and, if the torque that the generator imposes on the shaft is greater than the one caught by the rotor blades, then the shaft slows down. Hence the reference power is reduced and if it is equal to the one produced by the turbine the system is steady otherwise the tracking action goes on.

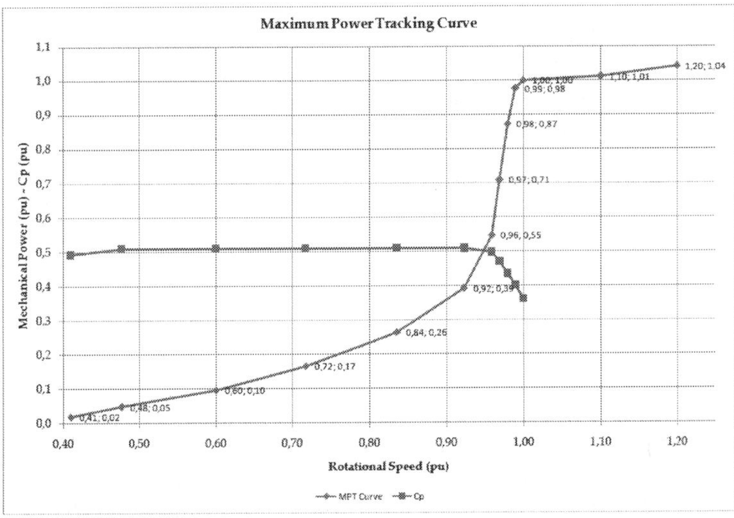

Fig. 10. Power reference (MPT characteristic) and c_p in function of the rotational speed.

This curve can be divided into three zones:

- the first one goes from the cut-in rotational speed (40% of the nominal speed) to the 90% and describes the part where the turbine is pursuing the maximum power coefficient (depicted by the red curve) and covers the wind range between 3 to 9 m/s.
- the second zone is characterized by the steep increase in the reference power and it is due to the fact that there is no more interest in collecting all the power in the wind because it is blowing close to the nominal values. Thus the power coefficient is progressively reduced. It includes the winds between 9 and the nominal one, which is equal to 12.5 m/s for the modelled turbine.
- the last one is characterized by a flat curve that sets the nominal reference power to the generators and allows the machine to go in overspeed to absorb the rapid wind speed

variations. Generally a 20% of overspeed is acceptable and a slight increase of the power output (about 4%) can help to reduce the stress on the pitch blade actuators. The wind speed covered by this area obviously ranges from the nominal wind speed to the cut-out one (25 m/s).

The c_p curve is no more represented since from there the blade angle can assume different transient values from 0° (optimal angle) to 32° and hence it is not possible to define a unique value power coefficient value.

Fig. 11 reports the influence of the reference power set by the MPT curve in the modeling block diagram. As it can be seen this reference power value can be modified by an external control signal that reduces the reference power in order to force a reduction in the turbine power output in case of request by the overall park controller. This reduction in the electromagnetic torque (the braking one) will cause an acceleration of the turbine speed that will be duty of the pitch control system to counteract.

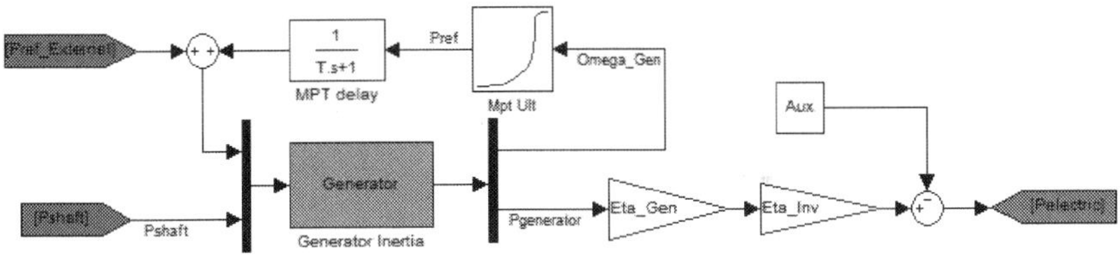

Fig. 11. MPT and generator block diagram.

2.6 Pitch angle control

The pitch control system has to reduce the aerodynamic efficiency by increasing the blade attack angle. Its control is sensible to rotational speed: if this value goes above 1 per unit, the PI (proportional-integral) control system commands the increase of the blade pitch angle. Fig. 12 shows the block diagram of the pitch control system that includes also the delay of the actuator that is realized by the integrator with unitary feedback (on the right side of the figure).

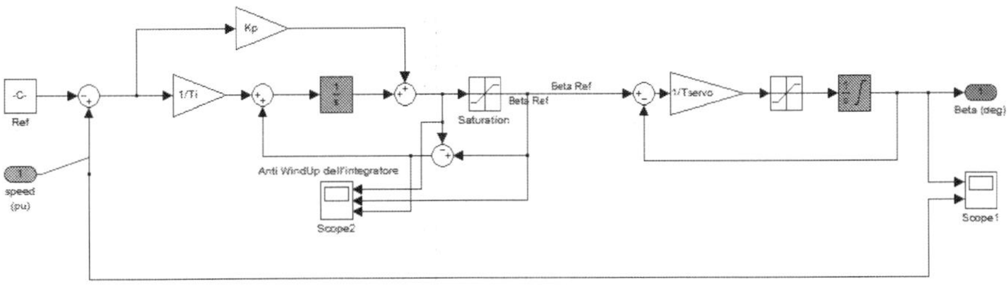

Fig. 12. Block diagram of the pitch controller.

3. Storage model

3.1 Main assumptions

The storage model proposed is suited for electrical studies and it has a general validity (Chen & Rincon Mora, 2006). The model has a nominal power of 2.5 MW and a nominal

energy of 2 MWh, it is composed by a set of 140 units, each with nominal values of 17.8 kW - 14.2 kWh; each unit is composed by 2 parallels of 108 cells connected in series. It is assumed that all the cells are perfectly balanced and thus the tasks requested to the storage system are equally divided among the 140 units. Under this assumption all the dynamics are built in the single equivalent cell; the overall storage desired size is then obtained by multiplying/dividing the cell parameters for the number of series/parallel elements.
The modelled dynamics regard the State-of-Charge (SOC) behaviour, the electrochemical conversion and the thermal characterization. The main state variables are therefore the state of charge and the temperature: all the characteristic elements of the storage system (as open circuit voltage, internal resistance and protection thresholds) present some kind of dependence from these state variables. Therefore the relationships among them are highlighted further on. An overview of the electrical equivalent of the abovementioned dynamics of the equivalent cell is proposed in Fig. 13.

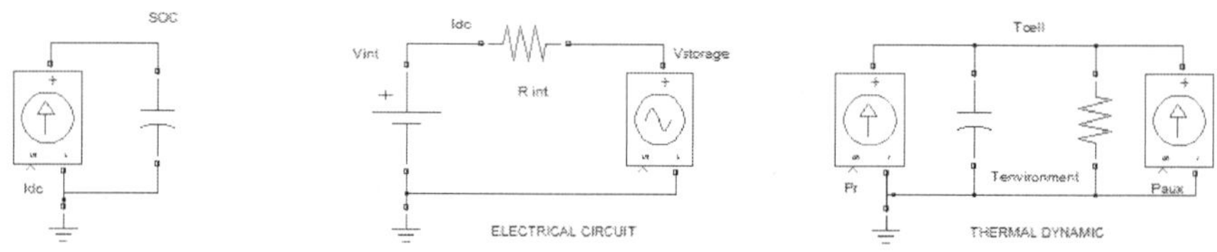

Fig. 13. Electrical equivalent of the main dynamics analysed for the description of the storage system.

The relationships of these dynamics are depicted in the conceptual block diagram shown in Fig. 14.

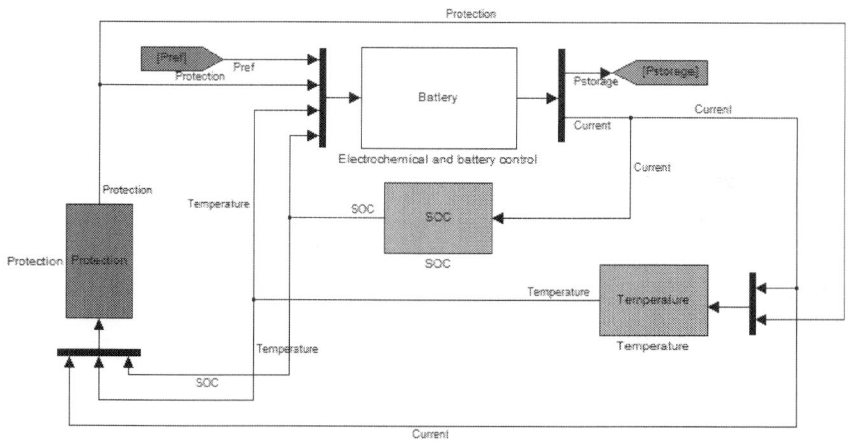

Fig. 14. Conceptual block diagram of the battery model.

The cell electrochemical conversion relations and the current controller are included in the block named Battery. The input values are the desired power that the battery must produce (or absorb, if negative), the protection signal that blocks the battery due to low or high state

of charge or temperature, the values of the SOC and of the temperature, which are needed to evaluate specific cell parameters such as the Open Circuit Voltage and the internal resistance. The output is of course the effectively produced power as well as the current flowing in the cells necessary to describe the behaviour of the SOC and the increase in the internal temperature due to Joule losses.
The paragraph develops as follow:

- State-Of-Charge
- Electrochemical dynamic
- Current control loop
- Thermal dynamic
- Protections and limitations

3.2 State-Of-Charge

The first dynamic described is related to the behaviour of the state of charge (SOC). This variable gives information about the quantity of energy still stored in the battery.
Its value is 1 when the battery is fully charged and 0 when fully discharged. Because of the nature of $Na\text{-}NiCl_2$ it is not recommended to discharge it below 0.2. The differential equation is shown further on: I_{dc} (A) is the current flowing in the battery and used by the auxiliary system; C is the nominal charge capacity, in Coulomb or Ah, of the battery.

$$\begin{cases} I_{dc} = C * {dSOC}/{dt} \\ SOC(0) = SOC_0 \end{cases} \tag{7}$$

The $Na\text{-}NiCl_2$ battery hasn't auto-discharge, so no shunt/dissipative elements are considered. In case of need to model the self discharge the previous equation is modified by the introduction of a resistive element, R, as follows:

$$\begin{cases} I_{dc} = {SOC}/{R} + C * {dSOC}/{dt} \\ SOC(0) = SOC_0 \end{cases} \tag{8}$$

3.3 Electrochemical dynamic

To evaluate the amount of energy that is stored or released by the battery the electrochemical dynamic has to be detailed. A first order model takes into account two voltage generators with a resistor in series as shown in Fig. 15. The first one, V_{oc}, generates what is commonly known as open circuit voltage and its nominal value is, for the $Na\text{-}NiCl_2$,

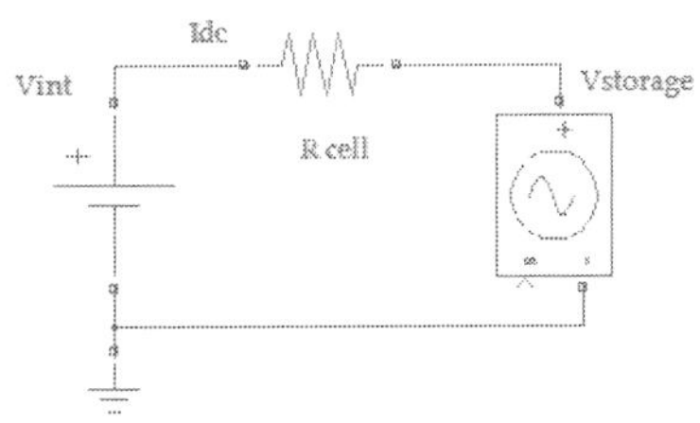

Fig. 15. Electrochemical equivalent.

2.58 Volt per cell, with a slight dependence on the SOC. The resistor takes into account the internal Joule losses and is assumed to be function of SOC and temperature.
The controlled generator, V_{batt}, models the behaviour of the dc/ac converter. Its task is to close the circuit and set, by the means of the control system, the current value to have the desired value of power flowing in the circuit. The current is assumed positive if it flows from V_{oc} to V_{batt} implying, hence, a discharge action.
The value of V_{oc} depends on the SOC as depicted by the characteristic (reported in physical values and per unit on a 2.58 V base) shown in Fig. 16 (Bossi et al, 2005).

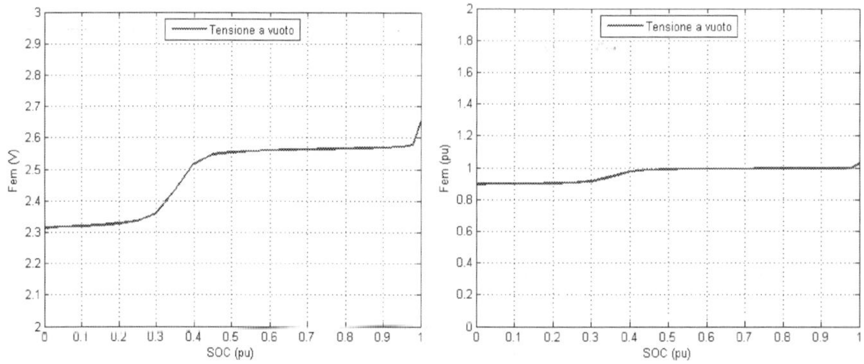

Fig. 16. Open cell voltage characteristic in function of the SOC in physical value (left picture) and per unit (right picture).

The behaviour of the internal cell resistance is displayed in Fig. 17.

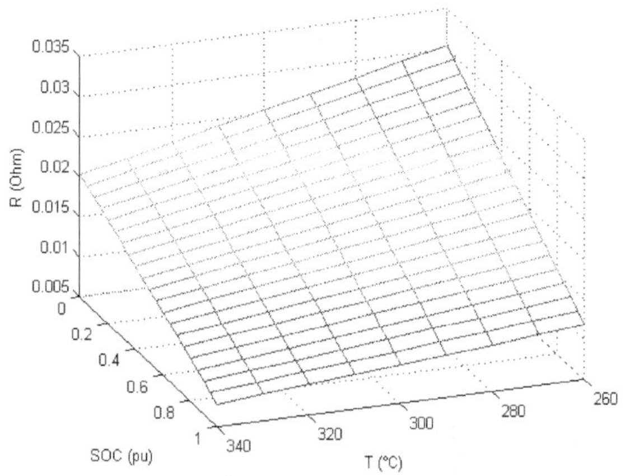

Fig. 17. Na-$NiCl_2$ cell internal resistance in function of SOC and temperature

The dependance of the resistance is not linear on the temperature and on the SOC. It can be noticed that the resistance increases with the decrease of the temperaure, this fact must kept in mind when setting the temperature thresholds that trigger the cooling fan.

3.4 Current control loop

The control of the power output is demanded to the current control loop. The power reference is converted in a current reference by the knowledge of the voltage at the connection point of the battery. The reference current is therefore compared with the current flowing in the battery system and the difference, the error, is the input for the PI (Proportional-Integral) controller. The output of the controller sets the values for the voltage source, that models the electronic converter; therefore the current flowing in the battery cells is evaluated by comparing this voltage with the open circuit voltage of the battery and by knowning the internal resistance. The integral action of the controller grants that the current error is zero in the steady state conditions. Fig. 18 summarizes graphically what above mentioned.

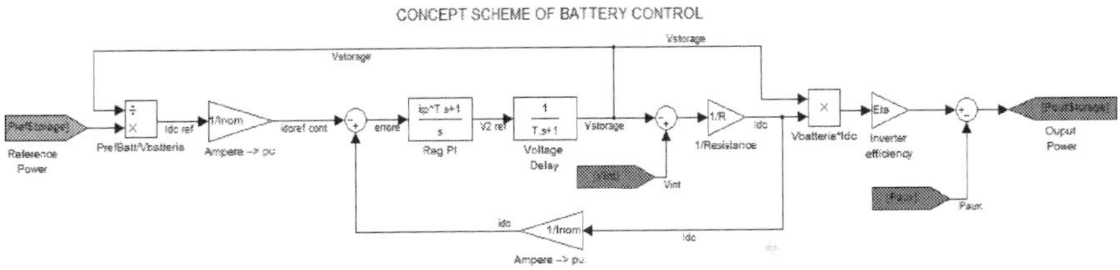

Fig. 18. Block diagram of the battery control.

3.5 Thermal dynamic

The thermal dynamic is very crucial when high power stress is required—this kind of management requires high currents leading to high thermal dissipation on internal resistance—and, moreover, for hot-temperature battery devices because they require working in a specific temperature range (e.g. 260 °C ÷ 360 °C for a Ni-$NaCl_2$ chemistry based battery). The battery model is equipped with auxiliary systems (fans and heaters) in order to enable the control system to keep temperature within the allowed range.

A possible representation is reported in Fig. 19. Using the electric components, the differential equations, reported subsequently, can be easily understood.

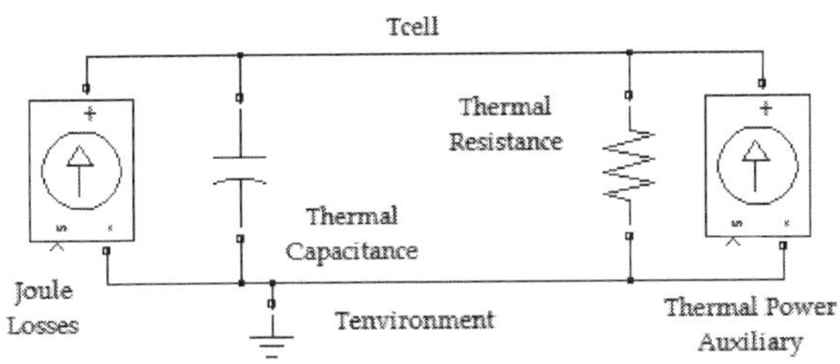

Fig. 19. Thermal dynamic described by the mean of electric equivalent.

If it is assumed that the temperature, T (°C), behaves like the voltage and the thermal power, Q (W), like the current, then the capacitor and the resistor represents the thermal capacitance, C_T (J/K), and the thermal resistance, R_T (K/W), of the battery. The two current generators are used to model the thermal power generated by Joule effect, Q_{Pr}, and the thermal contribution of both cooling fan and heater devices, Q_{Paux}.

$$
\begin{cases}
Q_{Pr} + Q_{Paux} + Q_R = Q_C \\
Q_R = \Delta T / R_T \\
Q_C = C_T * dT/dt \\
T(0) = T_0
\end{cases}
\quad (9)
$$

3.6 Protection and limitation

The battery control system has the task to stop the battery if the SOC level approaches 0.2 or 1. It is also sensible to the temperature and for example it has to turn on the cooling fan when temperature goes above a threshold, that is function of SOC, or turn on heater devices if the battery is cooling down. The protection actions are summarized in Table 2.

Action	Pilot Signal	Thresholds	Hysteresis band	Protection Output
Charge	SOC, I_{ref}	SOC > 1 & I_{ref} < 0 (charge)	0,01 pu SOC	[0, 1]
Discharge	SOC, I_{ref}	SOC < $SOC_{minimum}$ & I > 0 (discharge)	0,01 pu SOC	[0, 1]
Charge/Discharge	Temperature	360° C	10 °C	[0, 1]
Charge	Temperature, SOC	T_MAX= f(SOC)	10 °C	[0, 1]

Table 2. Protections Table

Moreover two limitation thresholds are present. The first one has to reduce the reference current whenever the, during a charge, the SOC is above 0.8 pu; while the second reduces the current if the temperature, during a discharge, reaches 340 °C.

Action	Pilot Signal	Thresholds	Hysteresis band	Protection Output
Charge	SOC	0.8 pu	0.1 pu SOC	Max current 2 → 0.5 pu
Discharge	Temperature	340 °C	10 °C	Max current 2 → 0.5 pu

Table 3. Limitation Table

4. Power plant layout and controllers

As foretold, the idea is to control the battery charging and discharging in order to control the whole plant output at the Point of Common Coupling (PCC). An overview of the layout of the park is shown Fig. 20, the four 2 MW wind turbines are identified by the blocks on the right side of the picture while the battery is located in the central block.

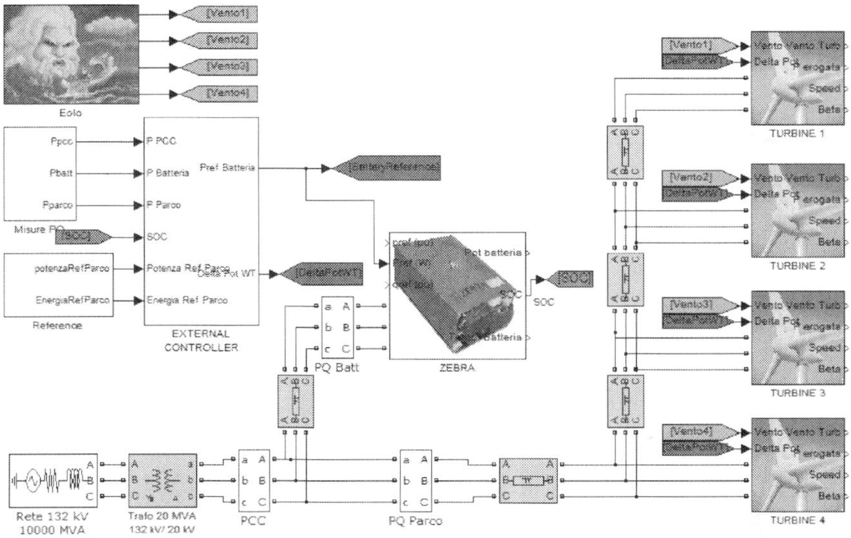

Fig. 20. Power plant layout

4.1 External battery controller

The external battery controller sets the reference power that the storage system has to accomplish. The controller, shown in Fig. 21, is equipped with a PI regulator and is sensible to the error between the power produced by all the wind turbines and the expected reference power. Moreover it is present in the control loop another contribution sensible to the SOC level of the battery (Yoshimoto et al., 2006). This control reduces or increases the battery reference power with the purpose to keep the SOC in an adequate range so that storage is always available.

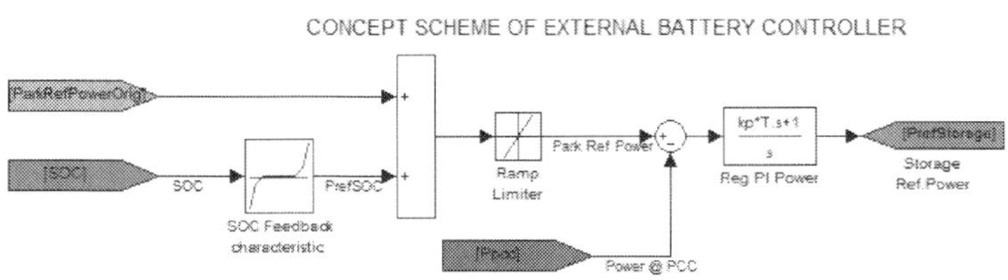

Fig. 21. Block diagram of the battery external controller

4.2 External wind turbine controller

The external wind turbine controller, shown in Fig. 22, forces the turbines to reduce their output, by means of a signal, generated by a PI controller sensible to the difference between reference power and the power flowing at the PCC, which overrides the one generated by maximum power tracking curve. In fact, whenever there is the mandatory order from the Distribution System Operator to grant zero power transit at the PCC and the battery is not able to store the whole energy produced by the turbines, in order to avoid the complete shut down of the turbines, they can be forced to reduce their output. This situation can happen obviously if the storage is full, but also if it is above the threshold that reduces the charge intensity (i.e. SOC 0.8) and thus the battery is able to store just a fraction of wind production.

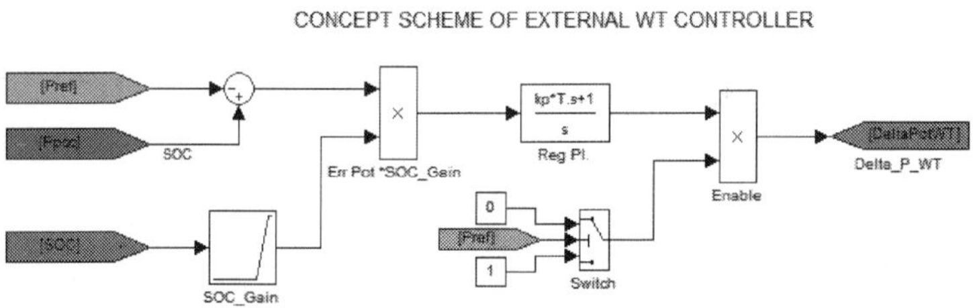

Fig. 22. Block diagram of the external wind turbine controller

5. Simulations & scenarios analyzed

The behaviours of the stand alone models are firstly tested. Thus the dynamic response of the wind turbine is tested by a sequence of wind steps starting from the cut-in speed (3 m/s) to the cut-out speed (25 m/s). The storage is tested by means of a sequence of both charge/discharge actions, with different magnitudes.

After that the models are coupled and two scenarios are studied: the first one regards the levelling of the overall wind park output in order to have a smoother power output at the PCC, the second one foresees the task to grant no power transit at the PCC for a certain time because of DSO request.

In order to perform these simulations it is assumed that the forecasted power output of the park is correct, that means that each ten minutes a reference power, which depends on the mean wind speed, is sent to the plants. It is clear that at each time step the wind turbines will not produce the forecasted power because of the turbulence. The main task is hence to smooth the fast fluctuation induced in the wind by the local terrain roughness.

5.1 Models testing: wind turbine

First of all the turbine model is tested by means of a series of wind steps, shown in the first diagram of Fig. 23 along with the pitch angle curve. The output power can be seen in the second diagram, while the rotational speed in the latter one. As long as the wind speed increases so does the rotational speed and therefore the output power. Once the wind reaches the nominal value (12 m/s) the power does not go above the nominal power (2 MW) and at each wind step the pitch controller keeps under control the rotational speed (1 pu) by increasing the pitch angle.

Fig. 24 offers an overview of the aero dynamical behaviour of the machine. The first diagram shows the increase of wind power with the increase of wind speed and the portion produced by the machine.

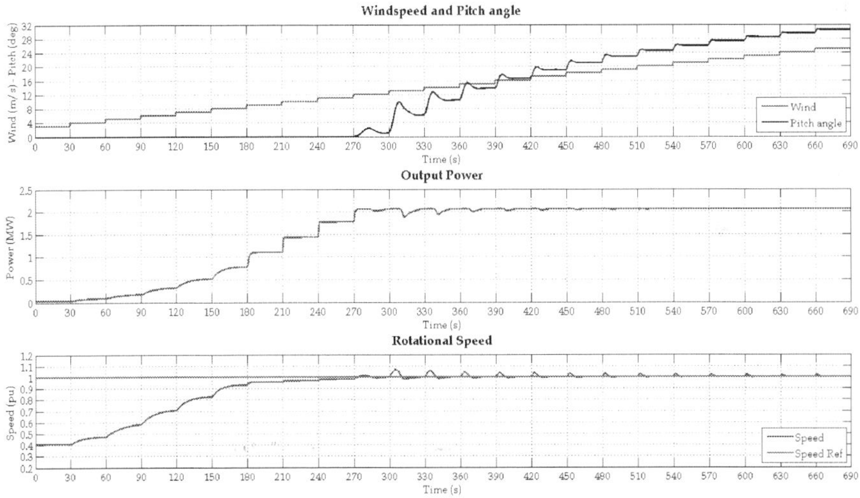

Fig. 23. Wind speed and pitch angle; Output power; Rotational speed

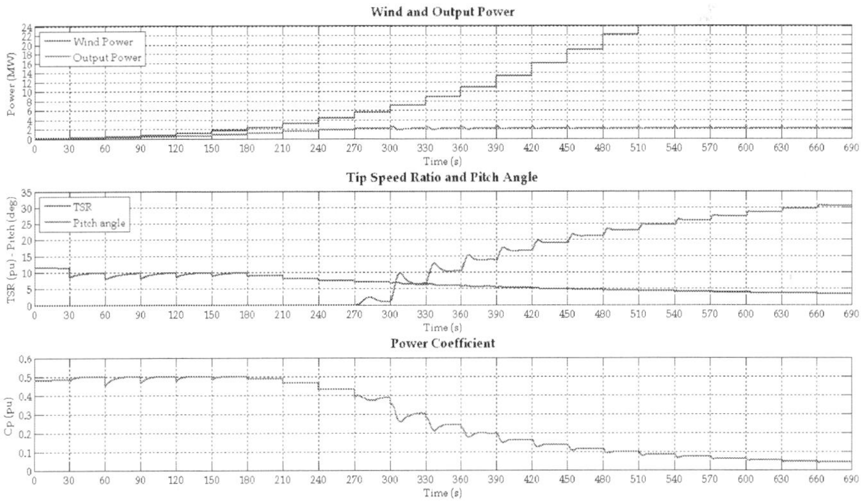

Fig. 24. Wind and output power; Tip speed ratio and pitch angle; Power coefficient

Because all the components of the machine are rated for 2 MW it is clear that all the amount of power present in strong winds cannot be harvested. The second diagram shows the Tip Speed Ratio (TSR or lambda), i.e. the ratio between the peripheral speed of the blades and the wind speed, and the pitch angle. As explained previously it is role of the MPT control to tune the rotational speed in order to pursue the optimal TSR (equal to 10 for the aerodynamic characteristic implemented in this turbine model) for light winds. As the wind

starts blowing stronger and the rotational speed reaches the limit (1 pu), the TSR is reduced contributing to the reduction of the power coefficient (cp, shown in the third diagram). The reduction of the TSR alone (see Fig. 8), however, is not enough to reduce the cp and, hence, it is necessary also to pitch the blades in order to reduce the lift of the blades.
The next step is to analyze the response of the four wind turbines, each fed by its own wind profile. Fig. 25 reports the simulation results for a 2-hours window simulation.

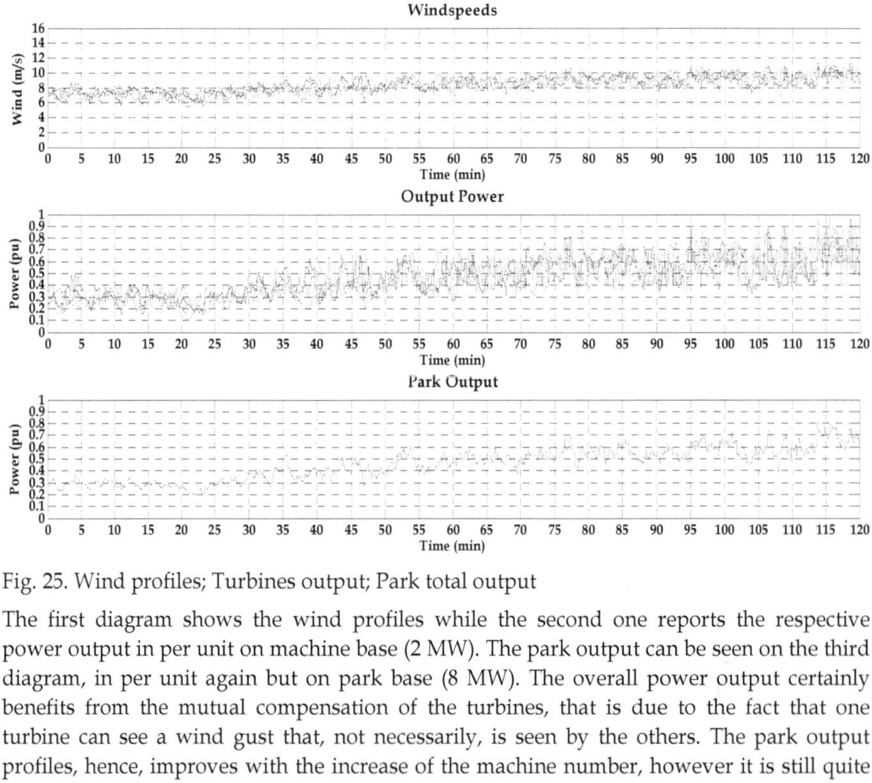

Fig. 25. Wind profiles; Turbines output; Park total output

The first diagram shows the wind profiles while the second one reports the respective power output in per unit on machine base (2 MW). The park output can be seen on the third diagram, in per unit again but on park base (8 MW). The overall power output certainly benefits from the mutual compensation of the turbines, that is due to the fact that one turbine can see a wind gust that, not necessarily, is seen by the others. The park output profiles, hence, improves with the increase of the machine number, however it is still quite turbulent. The measure of the power turbulence intensity, defined, similarly to the wind turbulence, as the ratio between the standard deviation and the average power in a 10-minutes basis, is about 19% (equal to 0.38 MW) for the single turbines while it is reduced to 11% (equal to 0.88 MW) at the PCC for the 2-hours timeframe considered.

5.2 Models testing: Storage

The storage size is 2.5 MW – 2 MWh and is chosen in order to compensate ± 3 times the PCC "power turbulence" intensity. The initial conditions are 60% SOC and 290 °C. Firstly a charge-discharge cycle at nominal power is performed followed by a more intense and shorter cycle at two time nominal power. The results are reported in Fig. 26, the first diagram shows the reference power that the battery is supposed to supply and the power that can be provided. At the 19th minute the SOC level reaches 0.8 and that triggers the limitation in the charge action. The power that can be stored is reduced by half. The battery voltage is higher than the open circuit because the current has to flow inside the battery. At the 30th minute the charge is stopped and a discharge is required, the battery inverter thus sets a lower voltage in order to make the current flowing outside the battery. The controller has to progressively reduce the voltage in order to provide the required power, this because the lower is the SOC the higher gets the battery resistance (see Fig. 17).

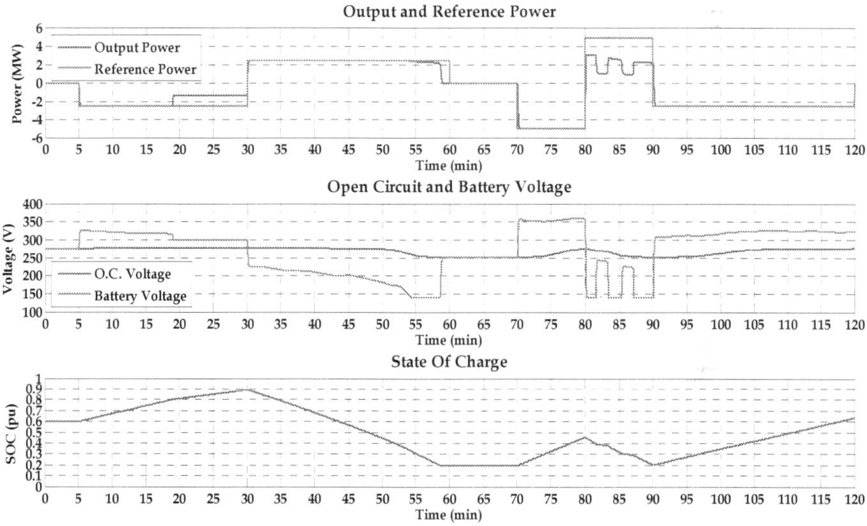

Fig. 26. Output and reference powers; Open circuit and battery voltage, SOC

When the SOC reaches 0.2 (58th minute) the minimum SOC protection stops the discharge. Afterward a more intense charge is required and successfully performed. At the 80th minute a deeper discharge is requested and the battery controller has to reduce the intensity several times due to the intervention of the thermal protection, as can be better appreciated in the Fig. 27. In the second diagram, in fact, at the 82nd minute the temperature reaches 340 °C that is the maximum allowed value. The power is hence reduced and as a result the joule losses drop (blue curve first diagram) and the thermal power removed by the fans (red curve) determines the reduction of the temperature. At the 90th minute the discharge action is terminated and a charge at nominal power is performed until the end of the simulation. In this last timeframe, the first diagram of Fig. 27 shows a sequence of turn on/off of the cooling fan (red curve) triggered by the thermal control system. The increase of the SOC, in fact, determines a reduction of the maximum allowed temperature.

5.3 Smoothing action

In this scenario is analyzed the coupling of the storage system with the wind farm. The mission of the battery is to smooth the wind farm output and the results are depicted in Fig. 28.

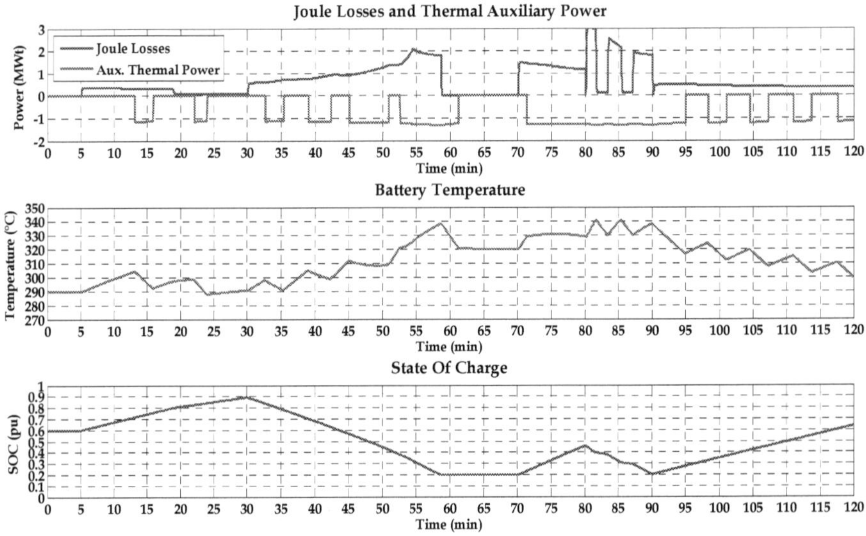

Fig. 27. Joule losses and auxiliary thermal power; Cell temperature; SOC

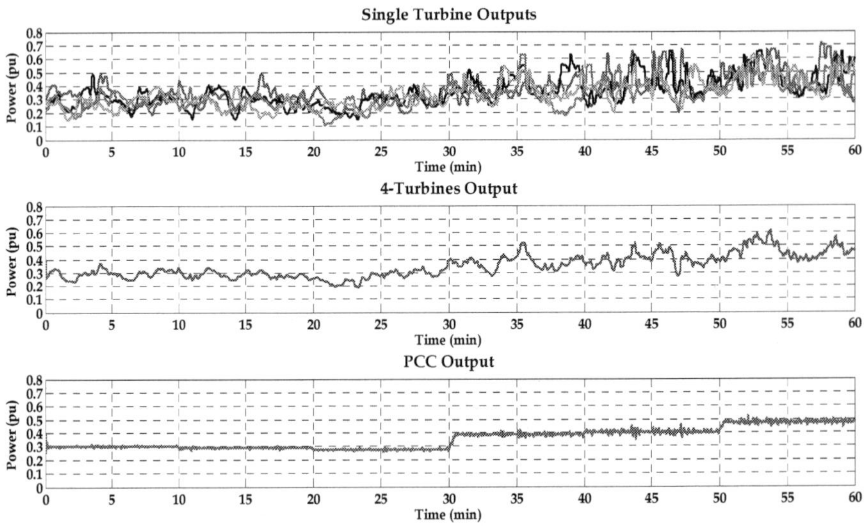

Fig. 28. Single turbine outputs; 4-Turbines overall output; PCC output

The different turbine outputs are reported in the first diagram, while the overall wind production is depicted in the second diagram. The output power at the PCC, shown in the third diagram, presents a much smoother profile due to the compensation action of the battery. The storage system in fact, controlled by the reference set point sent by the external controller, by the means of charge/discharge cycles is able to grant the expected reference power. Thus the wind turbulence is compensated by the storage. The "power turbulence" intensity, whose value is 11% for the power measured at the PCC, drops below the 3% for the 2-hours timeframe considered. The benefit of the storage is thus proved, however a warning must be raised concerning the stress that these charging/discharging cycles can have on the battery lifetime.

Fig. 29 offers an overview of the main storage variables: the power profile required and the one realized are shown in the first diagram, while the voltages are reported in the second one. The SOC is slightly decreasing because the battery has a non-null internal resistance and thus all these charge/discharge cycles deplete energy. The temperature instead is quite constant, meaning that the joule losses are matching the battery natural cooling.

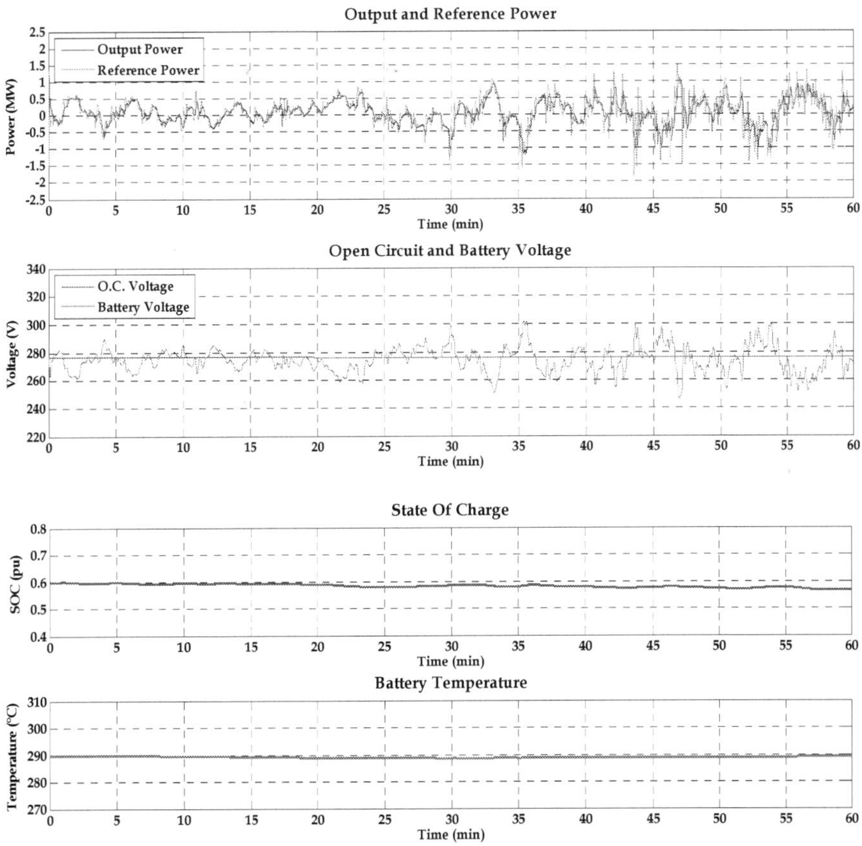

Fig. 29. Output and reference powers; O.C. and battery voltage; SOC; Cell temperature

5.4 Zero power transit at the PCC

In this last scenario the combined control of wind turbines and storage system is envisaged. It is supposed that the DSO requires no power transit at the PCC for half an hour, due for instance to congestions of the medium voltage network.

The third diagram of Fig. 30 reports the results for the power transit at the PCC: at the 2nd minute the battery begins to store the turbines' output in order to grant no power transit. Meanwhile the SOC, shown in Fig. 31, increases and when it reaches 0.8 (17th minute) the battery control system reduces the charging current and thus the power that the battery can store, because such an intensive charge cannot be born. The first diagram of Fig. 31 reports the actual level that the battery is allowed to manage and the request from the external battery controller that is trying to grant zero transit at the PCC. At this point the external controller of the turbines is enabled (see Fig. 22) and, by overriding the MPT controller, forces them to reduce their output.

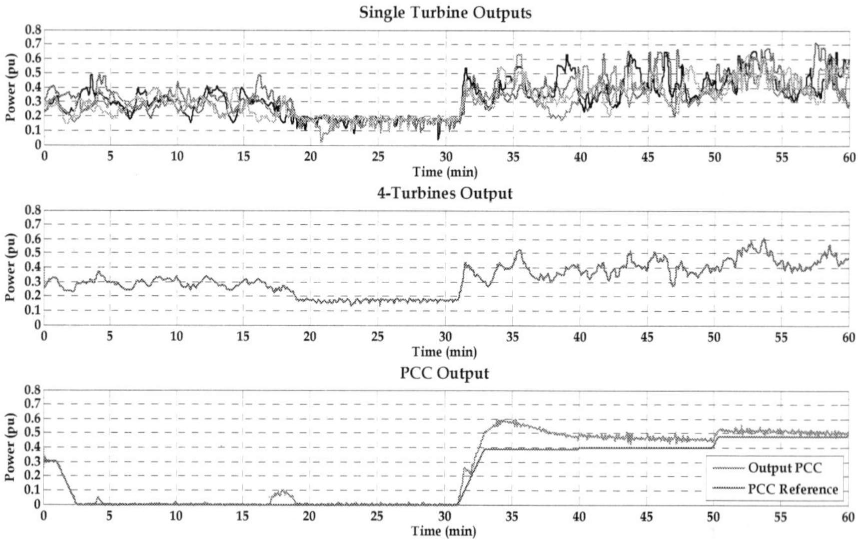

Fig. 30. Single turbine outputs; 4-Turbines overall output; PCC output and reference

This signal forces the reference power to a value that implies the acceleration of the turbines rotor, due to the unbalance between the mechanical torque (the accelerating one) and the electromagnetic torque of the generator (the braking one). The pitch control begins to increase the blade pitch angle to reduce the aero-dynamic efficiency and thus to contrast the rotor speed deviation. The wind turbine dynamics can be appreciated in Fig. 32 and Fig. 33.

When the DSO allows again power transit (31st minute) the MPT control of the turbines is fully restored and hence the turbines produce again the maximum power. The third diagram of Fig. 30 highlights an extra power production compared to the one that was programmed (blue line). This is due to the SOC feedback action of the battery external controller that forces the storage to release this amount of energy in order to take the SOC level far from the maximum level (see third diagram Fig. 31). As foretold this SOC feedback is needed to keep the battery within a reasonable range of values to avoid the block of the battery.

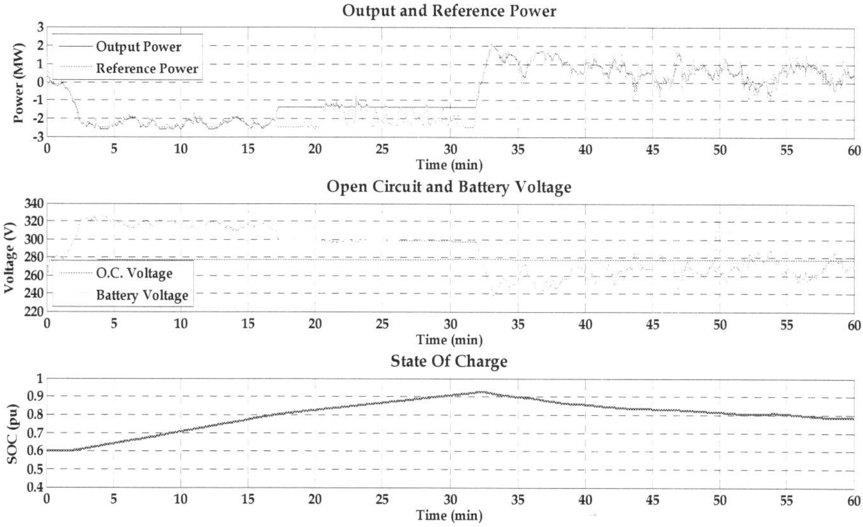

Fig. 31. Output and reference powers; Open circuit and battery voltage; SOC

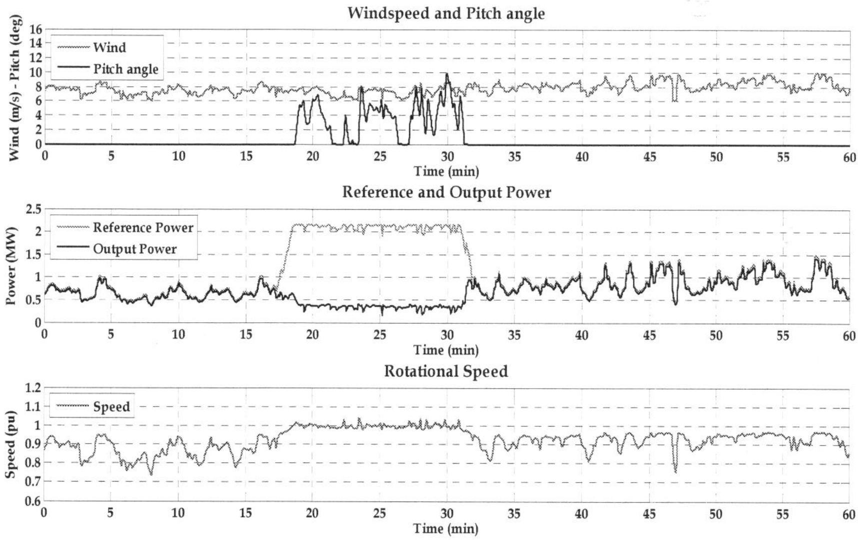

Fig. 32. Wind speed and pitch angle; Reference and output power; Rotational speed

It is interesting to analyze, in Fig. 32 and Fig. 33, the reduction of power in the turbines by the MPT override signal. This analysis can also help to better comprehend the wind turbine function. The power output reduction, in fact, is not realized by acting on the blades pitch angle but by the reduction of the reference power that the generator has to produce. The consequent torques unbalance causes the over speed which is counteracted by the pitch controller. The reduction in the rotor efficiency, and hence the reduction of the power coefficient, is not realized by the blade pitching alone but also by the fact that the turbine is going to work with a Tip Speed Ratio well above the optimal one (second diagram Fig. 33). These two factors both contribute to reduce the power coefficient, whose behaviour is depicted in the third diagram.

When the turbine is allowed to extract again the maximum power, the MPT control, by setting the appropriate electromagnetic torque, takes the turbine rotational speed in the range that determines the optimal TSR. Obviously this happens because the wind is blowing between 6 and 9 m/s, thus there is the interest to harvest all the energy available in the wind. If the wind had blown stronger, the MPT controller would not have taken the turbine in the working point that determines the optimal TSR (see Fig. 24).

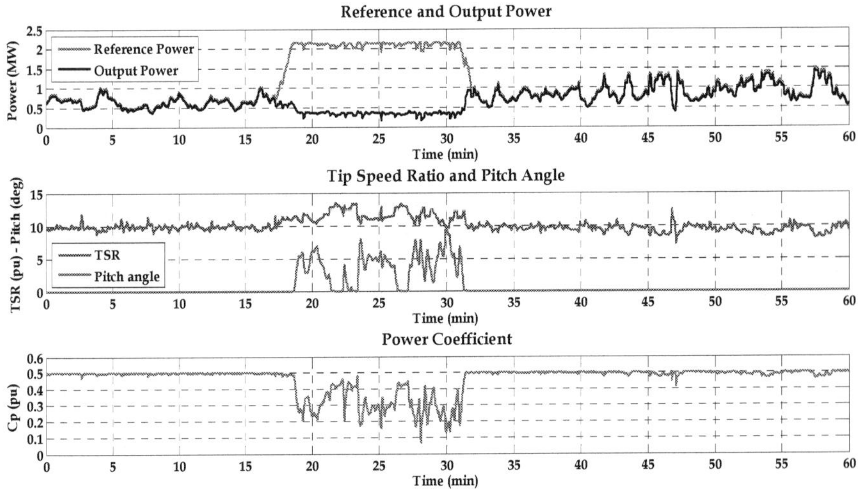

Fig. 33. Reference and output power; TSR and pitch angle; Power coefficient

6. Conclusions

The chapter focused on the development of models of wind turbines and storage systems, in Matlab-Simulink environment, for implementing integrated control strategies of the whole resulting system in order to describe the benefits that storage can provide.

A deep characterization of the wind profiles has been performed. The 2 MW turbine model was described from an electromechanical perspective, thus it provided: an analysis of the aerodynamic behaviour of the rotor including the pitch control system, the shaft dynamic and the maximum power tracking characteristic.

The storage model has been characterized from both the electrochemical and the thermal perspective. All the dynamics were built in the equivalent cell and the desired size of the battery (2.5 MW - 2 MWh) was obtained by multiplying the cell parameters for the number of series/parallel elements. The model presents a general validity and for this study has been tuned on the specifics of $Na\text{-}NiCl_2$ chemistry.

The control strategies analyzed had aimed at setting the battery charging and discharging phases in order to control the whole plant output.
The dynamics of the stand-alone models were studied by means of specific tasks: responses to wind steps and charge/discharge cycles at different intensity. Afterward the models have been coupled and two scenario were analyzed: the first one regarded the possibility to smooth the wind park turbulent output by charge/discharge control of the battery; the second one analyzed the possibility to control the storage to collect the power produced by the wind turbine in case of order by the DSO to reduce or to have zero power output at PCC and proved the possibility to reduce the wind farm output, without any turbine disconnection, in case of need.
It was thus highlighted how the storage system could grant benefits in term of controllability of the wind park output. However, the rigid power control, whose aim was to pursue a perfectly smoothed output, proved to be very stressful for the storage, threatening the battery lifetime. Nevertheless this control proved good performance in the zero transit scenario.
Further studies will regard the validation of the battery model, in order to insert proper values for the time constants and will envisage other control strategies as that sensible to an energy reference signal instead of the power one.

7. References

Thomas, R.J. (2009). Putting an action plan in place, *Power and Energy Magazine, IEEE*, vol. 7, no. 4, pp. 26-31, , July-Aug. 2009

Neural, A.; Kogan, V.I. & Schafer C.M. (2008). Load Leveling Reduces T&D Line Losses, *Power Delivery, IEEE Transactions on*, vol. 23, no. 4, pp. 2168–2173, Oct. 2008

Grillo, S.; Marinelli, M.; Pasca, E.; Petretto, G. & Silvestro, F. (2009). Characterization of Wind and Solar generation and their influence on distribution network performances, *Universities Power Engineering Conference (UPEC), 2009 Proceedings of the 44th International*, pp.1-6, 1-4 Sept. 2009, Glasgow

Fioravanti, R.; Vu, K.; Vu & Stadlin, W. (2008). Large scale solution, *Power and Energy Magazine, IEEE*, vol. 7, no. 4, pp. 48-57, July-Aug. 2009

Oudalov, A.; Chartouni, D.; Ohler, C. & Linhofer, G. (2006). Value Analysis of Battery Energy Storage Applications in Power Systems, *Power Systems Conference and Exposition, 2006. PSCE '06. 2006 IEEE PES*, pp.2206-2211, Oct. 29 2006-Nov. 1 2006

Di Rosa, D.; Fastelli, I.; Gigliucci, G.; Grillo, S.; Marinelli, M.; Massucco, S. & Silvestro, F. (2010). Generation and battery modelling and integrated control strategies for a better acceptance of intermittent renewable energy sources in the electric distribution system, *CIRED Workshop*. 7-8 June 2010, Lyon

Ackermann, T. et al. (2005). *Wind Power in Power Systems*, John Wiley and Sons Ltd, ISBN: 0470855088, New York

Marinelli, M.; Morini, A.; Pitto, A. & Silvestro, F. (2008). Modeling of doubly fed induction generator (DFIG) equipped wind turbine for dynamic studies, *Universities Power Engineering Conference, 2008. UPEC 2008. 43rd International*, pp. 1-6, 1-4 Sept. 2008, Padova

Achilles, S. & Pöller, M. (2004). Direct Drive Synchronous Machine Models for Stability Assessment of Wind Farm, *DIgSILENT Publications*, 2004

Sørensen, P.; Hansen, A.; Janosi, L.; Bech, J. & Bak-Jensen, B. (2001). Simulation of interaction between wind farm and power system. *Risø-R-1281, Risø National Laboratory, Roskilde*. December 2001

Hansen, A.; Iov, F.; Sørensen, P.; Cutululis, N.; Jauch, C. & Blaabjerg, F. (2007). Dynamic wind turbine models in power system simulation tool DIgSILENT. *Risø-R-1400 ed.2, Risø National Laboratory, Roskilde*. August 2007

Akhmatov V. (2003). Analysis of dynamic behaviour of dynamic power systems with large amount of wind power. *PhD Thesis, Orsted DTU,* 2003

Chen, M. & Rincon Mora, A. (2006). Accurate Electrical Battery Model Capable of Predicting Runtime and I-V Performance, *Energy Conversion, IEEE Transactions on,* vol. 21, no. 2

Bossi, C.; Buonarota, A. & Micolano, E. (2005). Risultati delle prove di laboratorio condotte su accumulatori avanzati. *Ricerca di Sistema, Technical Report,* Milano, 2005

Yoshimoto, K.; Nanahara, T.; Koshimizu, G. & Uchida, Y. (2006). New Control Method for Regulating State-of-Charge of a Battery in Hybrid Wind Power/Battery Energy Storage System, *Power Systems Conference and Exposition, 2006. PSCE '06. 2006 IEEE PES,* pp. 1244-1251, Oct. 29 2006-Nov. 1 2006

Part 3

Wind Turbine Generators and Drives

20

Wind Turbines with Permanent Magnet Synchronous Generator and Full-Power Converters: Modelling, Control and Simulation

Rui Melício[1], Victor M. F. Mendes[2] and João P. S. Catalão[3]
[1]*CIEEE – Center for Innovation in Electrical and Energy Engineering*
[2]*ISEL – Instituto Superior de Engenharia de Lisboa*
[3]*UBI – University of Beira Interior*
Portugal

1. Introduction

Research about dynamic models for grid-connected wind energy conversion systems is one of the challenges to achieve knowledge for the ongoing change due to the intensification of using wind energy in nowadays. This book chapter is an involvement on those models, but dealing with wind energy conversion systems consisting of wind turbines with permanent magnet synchronous generators (PMSG) and full-power converters. Particularly, the focus is on models integrating the dynamic of the system as much as potentially necessary in order to assert consequences on the operation of system.

In modelling the energy captured from the wind by the blades, disturbance imposed by the asymmetry in the turbine, the vortex tower interaction, and the mechanical eigenswings in the blades are introduced in order to assert a more accurate behaviour of wind energy conversion systems. The conversion system dynamic comes up from modelling the dynamic behaviour due to the main subsystems of this system: the variable speed wind turbine, the mechanical drive train, and the PMSG and power electronic converters. The mechanical drive train dynamic is considered by three different model approaches, respectively, one-mass, two-mass or three-mass model approaches in order to discuss which of the approaches are more appropriated in detaining the behaviour of the system. The power electronic converters are modelled for three different topologies, respectively, two-level, multilevel or matrix converters. The consideration of these topologies is in order to expose its particular behaviour and advantages in what regards the total harmonic distortion of the current injected in the electric network. The electric network is modelled by a circuit consisting in a series of a resistance and inductance with a voltage source, respectively, considering two hypotheses: without harmonic distortion or with distortion due to the third harmonic, in order to show the influence of this third harmonic in the converter output electric current. Two types of control strategies are considered in the dynamic models of this book chapter, respectively, through the use of classical control or fractional-order control.

Case studies were written down in order to emphasize the ability of the models to simulate new contributions for studies on grid-connected wind energy conversion systems.

Particularly, the contexts of possible malfunctions is an added contribution for the studies on: the pitch angle control of the turbine blades, malfunction characterized by momentarily imposing the position of wind gust on the blades; the power electronic converters control, malfunction characterized by an error choice assumed on the voltage vectors for the power electronic converter. Hence, simulation results of the dynamic models regarding the behaviours not only due to the fact that wind energy is not a controllable source of energy, but also due to possible malfunctions of devices that drive the control on wind energy conversion systems, are presented and discussed. Also, the simulation for the use of fractional-order control is a new contribution on the studies for controlling grid-connected wind energy conversion systems. A comparison between the classical control and the fractional-order control strategies in what regards the harmonic content, computed by the Discrete Fourier Transform, injected into the electric network is presented. The simulations of the mathematical models considered in this book chapter are implemented in Matlab/Simulink.

2. Modelling

2.1 Wind speed

The energy stored in the wind is in a low quality form of energy. Many factors influence the behaviour of the wind speed so it is modelised as a source intermittent and variable of energy and somehow characterized as random variable in magnitude and direction (Chen & Spooner, 2001). Although of the characterisation of wind energy as random variable, for the purpose of this book chapter the wind speed can be modelled as a deterministic sum of harmonics with frequency in range of 0.1–10 Hz (Xing et al., 2006), given by

$$u = u_0 \left[1 + \sum_n A_n \sin(\omega_n \, t) \right] \tag{1}$$

where u is the wind speed, u_0 is the average wind speed, A_n is the magnitude of the eigenswing n, ω_n is the eigenfrequency of the eigenswing n. The wind speed variation put into effect an action on the physical structure of a wind turbine (Akhmatov et al., 2000). This action causes a reaction of the physical structure, introducing mechanical effects perturbing the energy conversion. Hence, further consideration due to the wind speed variation was studied in order to better characterize the mechanical power associated with the energy conversion over the rotor due to the action on the physical structure.

2.2 Wind turbine

The mechanical power in the nonappearance of mechanical effects, influencing the energy conversion, on the physical structure of a wind turbine, is computed by

$$P_{tt} = \frac{1}{2} \rho \, A_t \, u^3 \, c_p \tag{2}$$

where P_{tt} is the mechanical power associated with the energy capture from the wind by the blades, ρ is the air density, A_t is the area covered by the rotor blades, and c_p is the power coefficient.

The power coefficient c_p is a function of the pitch angle θ of rotor blades and of the tip speed ratio λ, which is the ratio between blade tip speed and wind speed value upstream of the rotor, given by

$$\lambda = \frac{\omega_t \, R_t}{u} \tag{3}$$

where ω_t is the rotor angular speed at the wind turbine, R_t is the radius of the area covered by the blades. The computation of the power coefficient requires the use of blade element theory and the knowledge of blade geometry or the use of real wind turbine characteristic table. In this book chapter, the numerical approximation ascribed by (Slootweg et al., 2003) is followed. The power coefficient in this numerical approximation is given by

$$c_p = 0.73\left(\frac{151}{\lambda_i} - 0.58\theta - 0.002\theta^{2.14} - 13.2\right)e^{-\frac{18.4}{\lambda_i}} \tag{4}$$

$$\lambda_i = \frac{1}{\dfrac{1}{(\lambda - 0.02\theta)} - \dfrac{0.003}{(\theta^3 + 1)}} \tag{5}$$

The maximum power coefficient is at null pitch angle and for this numerical approximation it is equal to

$$c_{p\max}(\lambda_{opt}(0), 0) = 0.4412 \tag{6}$$

with the optimal tip speed ratio at null pitch angle equal to

$$\lambda_{opt}(0) = 7.057 \tag{7}$$

The maximum pitch angle for the numerical approximation is 55°, and the minimum power coefficient is equal to

$$c_{p\min}(\lambda\ (55), 55) = 0.0025 \tag{8}$$

with a tip speed ratio equal to

$$\lambda\ (55) = 3.475 \tag{9}$$

The conversion of wind energy into mechanical energy over the rotor of a wind turbine is influenced by various forces acting on the blades and on the tower of the wind turbine (e.g. centrifugal, gravity and varying aerodynamic forces acting on blades, gyroscopic forces acting on the tower), introducing mechanical effects influencing the energy conversion. Those mechanical effects have been modelled by eigenswings mainly due to the following phenomena: asymmetry in the turbine, vortex tower interaction, and mechanical eigenswing in the blades. The mechanical power over the rotor of a wind turbine has been modelled as sum of harmonic terms multiplied by the power given by (2), where the mechanical eigenswings define the harmonics. In this book chapter, the model developed in (Xing et al., 2006)-(Akhmatov et al., 2000) is followed, where the further consideration due to the action on the physical structure confers a mechanical power over the rotor given by

$$P_t = P_{tt}\left[1+\sum_{n=1}^{3} A_n\left(\sum_{m=1}^{2} a_{nm}\, g_{nm}(t)\right)h_n(t)\right] \tag{10}$$

$$g_{nm} = \sin\left(\int_0^t m\,\omega_n(t')\,dt' + \varphi_{nm}\right) \tag{11}$$

where P_t is the mechanical power of the wind turbine disturbed by the mechanical eigenswings, m is the order of the harmonic of an eigenswing, g_{nm} is the distribution of the m-order harmonic in the eigenswing n, a_{nm} is the normalized magnitude of g_{nm}, h_n is the modulation of eigenswing n, φ_{nm} is the phase of the m-order harmonic in the eigenswing n. The eigenfrequency range of the wind turbine model is considered to be from 0.1 to 10 Hz. The values used for computing P_t (Akhmatov et al., 2000) are given in Table 1.

n	Source	A_n	ω_n	h_n	m	a_{nm}	φ_{nm}
1	Asymmetry	0.01	ω_t	1	1	4/5	0
					2	1/5	π/2
2	Vortex tower interaction	0.08	3 ω_t	1	1	½	0
					2	½	π /2
3	Blades	0.15	9 π	1/2 ($g_{11}+g_{21}$)	1	1	0

Table 1. Mechanical eigenswings excited in the wind turbine

2.3 One mass drive train

In a one-mass drive train model all inertia components are lumped together, i.e., modelled as a single rotating mass. The equation for the one-mass model is based on the second law of Newton, deriving the state equation for the rotor angular speed at the wind turbine, given by

$$\frac{d\omega_t}{dt} = \frac{1}{J}\left(T_t - T_g\right) \tag{12}$$

where J is the moment of inertia for blades, hub and generator, T_t is the mechanical torque, T_g is the electric torque.

2.4 Two mass drive train

The equations for the two-mass model are based on the torsional version of the second law of Newton, deriving the state equation for the rotor angular speed at the wind turbine and for the rotor angular speed at the generator, given by

$$\frac{d\omega_t}{dt} = \frac{1}{J_t}\left(T_t - T_{dt} - T_{at} - T_{ts}\right) \tag{13}$$

$$\frac{d\omega_g}{dt} = \frac{1}{J_g}\left(T_{ts} - T_{dg} - T_{ag} - T_g\right) \tag{14}$$

where J_t is the moment of inertia for blades and hub, T_{dt} is the resistant torque in the wind turbine bearing, T_{at} is the resistant torque in the hub and blades due to the viscosity of the airflow, T_{ts} is the torque of torsional stiffness, ω_g is the rotor angular speed at the generator, J_g is the generator moment of inertia, T_{dg} is the resistant torque in the generator bearing, T_{ag} is the resistant torque due to the viscosity of the airflow in the generator. A comparative study of wind turbine generator system using different drive train models (Muyeen et al., 2006) has shown that the two-mass model may be more suitable for transient stability analysis than one-mass model. The simulations in this book chapter are in agreement with this comparative study in what regards the discarding of one-mass model.

2.5 Three mass drive train

With the increase in size of the wind turbines, one question arises whether long flexible blades have an important impact on the transient stability analysis of wind energy systems during a fault (Li & Chen, 2007). One way to determine the dynamic properties of the blades is through the use of finite element methods, but this approach cannot be straightforwardly accommodated in the context of studies of power system analysis. Hence, to avoid the use of the finite element methods it is necessary to approach the rotor dynamics in a compromising way of accessing its dynamic and preserving desirable proprieties for power system analysis programs. One straightforward way to achieve this compromise, where the blade bending dynamics is explained by a torsional system is illustrated in Fig. 1.

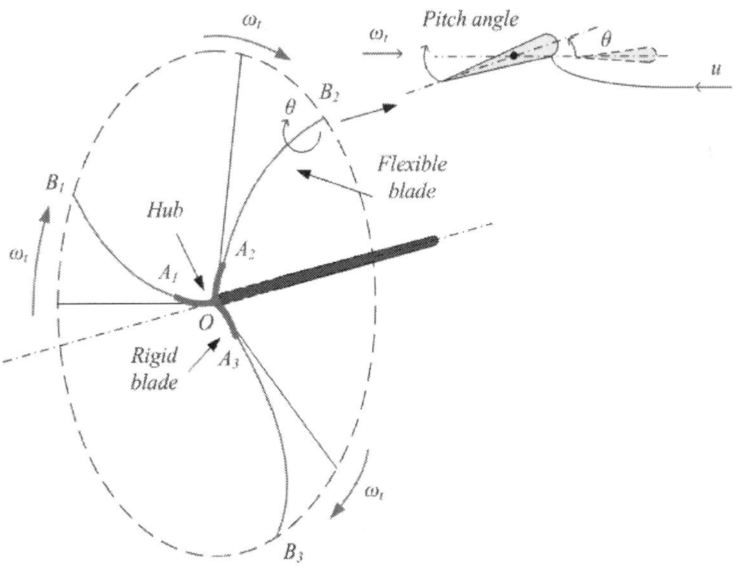

Fig. 1. Blade bending

Since the blade bending occurs at a significant distance from the joint between the blades and the hub, it is admissible to model the blades by splitting the blades in two parts: type OA parts, blade sections OA_1, OA_2 and OA_3; and type AB parts, blade sections A_1B_1, A_2B_2 and A_3B_3. Type OA parts have an equivalent moment of inertia associated with the inertia of the hub and the rigid blade sections. Type OB parts have an equivalent moment of inertia associated with the inertia of the rest of the blade sections. Type OB parts are the effective flexible blade sections and are considered by the moment of inertia of the flexible blade sections. Type OA and OB parts are joined by the interaction of a torsional element, but in addition a second torsional element connecting the rest of the inertia presented in the angular movement of the rotor is needed, i.e., it is necessary to consider the moment of inertia associated with the rest of the mechanical parts, mainly due to the inertia of the generator. Hence, the configuration of this model is of the type shown in Fig. 2.

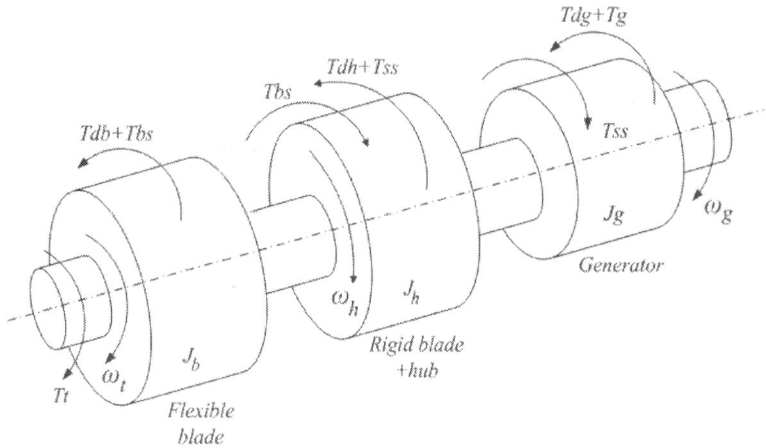

Fig. 2. Three-mass drive train model

The equations for the three-mass model are also based on the torsional version of the second law of Newton, given by

$$\frac{d\omega_t}{dt} = \frac{1}{J_b}(T_t - T_{db} - T_{bs}) \tag{15}$$

$$\frac{d\omega_h}{dt} = \frac{1}{J_h}(T_{bs} - T_{dh} - T_{ss}) \tag{16}$$

$$\frac{d\omega_g}{dt} = \frac{1}{J_g}(T_{ss} - T_{dg} - T_g) \tag{17}$$

where J_b is the moment of inertia of the flexible blades section , T_{db} is the resistant torque of the flexible blades, T_{bs} is the torsional flexible blades stifness torque, ω_h is the rotor angular speed at the rigid blades and the hub of the wind turbine, J_h is the moment of inertia of the hub and the rigid blades section, T_{dh} is the resistant torque of the rigid blades and the hub, T_{ss} is the torsional shaft stifness torque, T_{dg} is the resistant torque of the generator. The moments of inertia for the model are given as input data, but in their absence an estimation of the moments of inertia is possible (Ramtharan & Jenkins, 2007).

2.6 Generator

The generator considered in this book chapter is a PMSG. The equations for modelling a PMSG, using the motor machine convention (Ong, 1998), are given by

$$\frac{di_d}{dt} = \frac{1}{L_d}[u_d + p\omega_g L_q i_q - R_d i_d] \tag{18}$$

$$\frac{di_q}{dt} = \frac{1}{L_q}[u_q - p\omega_g (L_d i_d + M i_f) - R_q i_q] \tag{19}$$

where i_d, i_q are the stator currents, u_d, u_q are the stator voltages, p is the number of pairs of poles, L_d, L_q are the stator inductances, R_d, R_q are the stator resistances, M is the mutual inductance, i_f is the equivalent rotor current. In order to avoid demagnetization of permanent magnet in the PMSG, a null stator current associated with the direct axis is imposed (Senjyu et al., 2003).

2.7 Two-level converter

The two-level converter is an AC-DC-AC converter, with six unidirectional commanded insulated gate bipolar transistors (IGBTs) used as a rectifier, and with the same number of unidirectional commanded IGBTs used as an inverter. There are two IGBTs identified by *i*, respectively with *i* equal to *1* and *2*, linked to the same phase. Each group of two IGBTs linked to the same phase constitute a leg k of the converter. Therefore, each IGBT can be uniquely identified by the order pair *(i, k)*. The logic conduction state of an IGBT identified by *(i, k)* is indicated by S_{ik}. The rectifier is connected between the PMSG and a capacitor bank. The inverter is connected between this capacitor bank and a second order filter, which in turn is connected to the electric network. The configuration of the simulated wind energy conversion system with two-level converter is shown in Fig. 3.

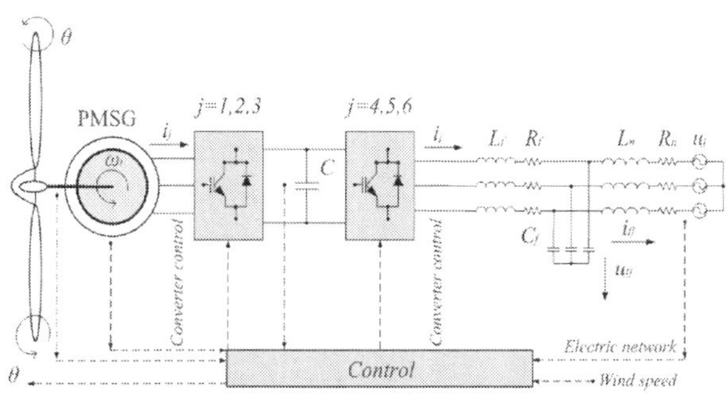

Fig. 3. Wind energy conversion system using a two-level converter

For the switching function of each IGBT, the switching variable γ_k is used to identify the state of the IGBT i in the leg k of the converter. Respectively, the index k with $k \in \{1,2,3\}$ identifies a leg for the rectifier and $k \in \{4,5,6\}$ identifies leg for the inverter. The switching variable a leg in function of the logical conduction states (Rojas et al., 1995) is given by

$$\gamma_k = \begin{cases} 1, & (S_{1k} = 1 \text{ and } S_{2k} = 0) \\ 0, & (S_{1k} = 0 \text{ and } S_{2k} = 1) \end{cases} \quad k \in \{1, \dots ,6\} \tag{20}$$

but logical conduction states are constrained by the topological restrictions given by

$$\sum_{i=1}^{2} S_{ik} = 1 \quad k \in \{1, \dots ,6\} \tag{21}$$

Each switching variable depends on the conducting and blocking states of the IGBTs. The voltage v_{dc} is modelled by the state equation given by

$$\frac{dv_{dc}}{dt} = \frac{1}{C}\left(\sum_{k=1}^{3} \gamma_k i_k - \sum_{k=4}^{6} \gamma_k i_k\right) \tag{22}$$

Hence, the two-level converter is modelled by (20) to (22).

2.8 Multilevel converter

The multilevel converter is an AC-DC-AC converter, with twelve unidirectional commanded IGBTs used as a rectifier, and with the same number of unidirectional commanded IGBTs used as an inverter.

The rectifier is connected between the PMSG and a capacitor bank. The inverter is connected between this capacitor bank and a second order filter, which in turn is connected to an electric network. The groups of four IGBTs linked to the same phase constitute a leg k of the converter. The index i with $i \in \{1,2,3,4\}$ identifies a IGBT in leg k. As in the two-level converter modelling the logic conduction state of an IGBT identified by the pair (i, k) is indicated by S_{ik}. The configuration of the simulated wind energy conversion system with multilevel converter is shown in Fig. 4.

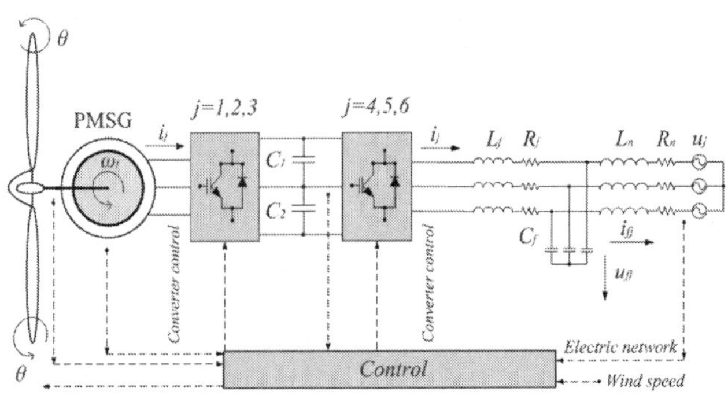

Fig. 4. Wind energy conversion system using a multilevel converter

For the switching function of each IGBT, the switching variable γ_k is used to identify the state of the IGBT i in the leg k of the converter. The index k with $k \in \{1,2,3\}$ identifies the leg for the rectifier and $k \in \{4,5,6\}$ identifies the inverter one. The switching variable of each leg k (Rojas et al., 1995) are given by

$$\gamma_k = \begin{cases} 1, (S_{1k} \text{ and } S_{2k}) = 1 \text{ and } (S_{3k} \text{ or } S_{4k}) = 0 \\ 0, (S_{2k} \text{ and } S_{3k}) = 1 \text{ and } (S_{1k} \text{ or } S_{4k}) = 0 \\ -1, (S_{3k} \text{ and } S_{4k}) = 1 \text{ and } (S_{1k} \text{ or } S_{2k}) = 0 \end{cases}$$

$$k \in \{1, \dots, 6\} \tag{23}$$

constrained by the topological restrictions given by

$$(S_{1k} . S_{2k}) + (S_{2k} . S_{3k}) + (S_{3k} . S_{4k}) = 1 \quad k \in \{1, \dots, 6\} \tag{24}$$

With the two upper IGBTs in each leg k (S_{1k} and S_{2k}) of the converters it is associated a switching variable Φ_{1k} and also with the two lower IGBTs (S_{3k} and S_{4k}) it is associated a switching variable Φ_{2k}, respectively given by

$$\Phi_{1k} = \frac{\gamma_k(1+\gamma_k)}{2} \; ; \; \Phi_{2k} = \frac{\gamma_k(1-\gamma_k)}{2} \quad k \in \{1, \dots, 6\} \tag{25}$$

The voltage v_{dc} is the sum of the voltages v_{C1} and v_{C2} in the capacitor banks C_1 and C_2, modelled by the state equation

$$\begin{aligned} \frac{dv_{dc}}{dt} &= \frac{1}{C_1}\left(\sum_{k=1}^{3} \Phi_{1k} i_k - \sum_{k=4}^{6} \Phi_{1k} i_k\right) + \\ &+ \frac{1}{C_2}\left(\sum_{k=1}^{3} \Phi_{2k} i_k - \sum_{k=4}^{6} \Phi_{2k} i_k\right) \end{aligned} \tag{26}$$

Hence, the multilevel converter is modelled by (23) to (26).

2.9 Matrix converter

The matrix converter is an AC-AC converter, with nine bidirectional commanded insulated gate bipolar transistors (IGBTs). The logic conduction state of an IGBT is indicated by S_{ij}. The matrix converter is connected between a first order filter and a second order filter. The first order filter is connected to a PMSG, while the second order filter is connected to an electric network. The configuration of the simulated wind energy conversion system with matrix converter is shown in Fig. 5.

The IGBTs commands S_{ij} are function of the on and off states, given by

$$S_{ij} = \begin{cases} 1, \text{ (on)} \\ 0, \text{(off)} \end{cases} \quad i, j \in \{1, 2, 3\} \tag{27}$$

For the matrix converter modelling, the following restrictions are considered

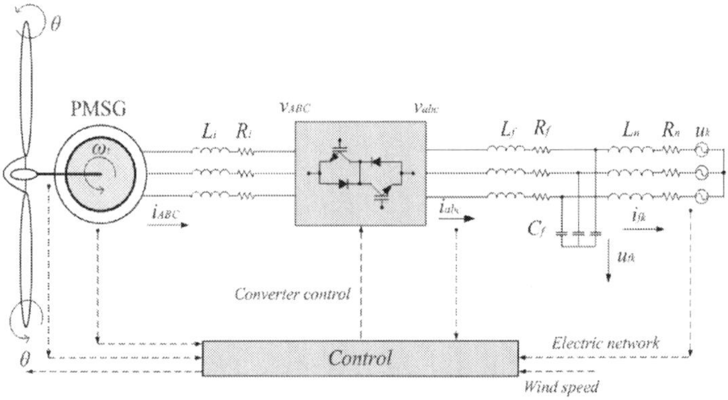

Fig. 5. Wind energy conversion system using a matrix converter

$$\sum_{j=1}^{3} S_{ij} = 1 \quad i \in \{1, 2, 3\} \tag{28}$$

$$\sum_{i=1}^{3} S_{ij} = 1 \quad j \in \{1, 2, 3\} \tag{29}$$

The vector of output phase voltages is related to the vector of input phase voltages through the command matrix. The vector of output phase voltages (Alesina & Venturini, 1981) is given by

$$\begin{bmatrix} v_A \\ v_B \\ v_C \end{bmatrix} = \begin{bmatrix} S_{11} & S_{12} & S_{13} \\ S_{21} & S_{22} & S_{23} \\ S_{31} & S_{32} & S_{33} \end{bmatrix} \begin{bmatrix} v_a \\ v_b \\ v_c \end{bmatrix} = [S] \begin{bmatrix} v_a \\ v_b \\ v_c \end{bmatrix} \tag{30}$$

The vector of input phase currents is related to the vector of output phase currents through the command matrix. The vector of input phase currents is given by

$$[i_a \quad i_b \quad i_c]^T = [S]^T \, [i_A \quad i_B \quad i_C]^T \tag{31}$$

where

$$[i_a \quad i_b \quad i_c] = [i_4 \quad i_5 \quad i_6] \tag{32}$$

$$[v_a \quad v_b \quad v_c] = [v_4 \quad v_5 \quad v_6] \tag{33}$$

Hence, the matrix converter is modelled by (27) to (33). A switching strategy can be chosen so that the output voltages have the most achievable sinusoidal waveform at the desired frequency, magnitude and phase angle, and the input currents are nearly sinusoidal as possible at the desired displacement power factor (Alesina & Ventirini, 1981). But, in general terms it can be said that due to the absence of an energy storage element, the matrix converter is particular sensitive to the appearance of malfunctions (Cruz & Ferreira, 2009).

2.10 Electric network

A three-phase active symmetrical circuit given by a series of a resistance and an inductance with a voltage source models the electric network. The phase currents injected in the electric network are modelled by the state equation given by

$$\frac{di_{fk}}{dt} = \frac{1}{L_n}(u_{fk} - R_n i_{fk} - u_k) \quad k = \{4,5,6\} \tag{34}$$

where R_n and L_n are the resistance and the inductance of the electric network, respectively, u_{fk} is the voltage at the filter and u_k is the voltage source for the simulation of the electric network.

3. Control strategy

3.1 Fractional order controllers

A control strategy based on fractional-order PI^{μ} controllers is considered for the variable-speed operation of wind turbines with PMSG/full-power converter topology. Fractional-order controllers are based on fractional calculus theory, which is a generalization of ordinary differentiation and integration to arbitrary non-integer order (Podlubny, 1999). Applications of fractional calculus theory in practical control field have increased significantly (Li & Hori, 2007), regarding mainly on linear systems (Çelik & Demir, 2010).

The design of a control strategy based on fractional-order PI^{μ} controllers is more complex than that of classical PI controllers, but the use of fractional-order PI^{μ} controllers can improve properties and controlling abilities (Jun-Yi et al., 2006)-(Arijit et al., 2009). Different design methods have been reported including pole distribution, frequency domain approach, state-space design, and two-stage or hybrid approach which uses conventional integer order design method of the controller and then improves performance of the designed control system by adding proper fractional order controller. An alternative design method used is based on a particle swarm optimization (PSO) algorithm and employment of a novel cost function, which offers flexible control over time domain and frequency domain specifications (Zamani et al., 2009).

Although applications and design methods regard mainly on linear systems, it is possible to use some of the knowledge already attained to envisage it on nonlinear systems, since the performance of fractional-order controllers in the presence of nonlinearity is of great practical interest (Barbosa et al., 2007). In order to examine the ability of fractional-order controllers for the variable-speed operation of wind turbines, this book chapter follows the tuning rules in (Maione & Lino, 2007). But, a more systematic procedure for controllers design needs further research in order to well develop tuning implementation techniques (Chen et al., 2009) for a ubiquitous use of fractional-order controllers.

The fractional-order differentiator denoted by the operator ${}_aD_t^{\mu}$ (Calderón et al., 2006) is given by

$$ {}_aD_t^{\mu} = \begin{cases} \dfrac{d^{\mu}}{dt^{\mu}}, & \Re(\mu) > 0 \\ 1, & \Re(\mu) = 0 \\ \int_a^t (d\tau)^{-\mu}, & \Re(\mu) < 0 \end{cases} \tag{35}$$

where μ is the order of derivative or integral, which can be a complex number, and $\Re(\mu)$ is the real part of the μ. The mathematical definition of fractional derivative and integral has been the subject of several approaches. The most frequently encountered one is the Riemann–Liouville definition, in which the fractional-order integral is given by

$$ {}_aD_t^{-\mu} f(t) = \frac{1}{\Gamma(\mu)} \int_a^t (t-\tau)^{\mu-1} f(\tau)\, d\tau \tag{36} $$

while the definition of fractional-order derivative is given by

$$ {}_aD_t^{\mu} f(t) = \frac{1}{\Gamma(n-\mu)} \frac{d^n}{dt^n} \left[\int_a^t \frac{f(\tau)}{(t-\tau)^{\mu-n+1}}\, d\tau \right] \tag{37} $$

where

$$ \Gamma(x) \equiv \int_0^\infty y^{x-1} e^{-y}\, dy \tag{38} $$

is the Euler's Gamma function, a and t are the limits of the operation, and μ identifies the fractional order. In this book chapter, μ is assumed as a real number that for the fractional order controller satisfies the restrictions $0 < \mu < 1$. Normally, it is assumed that $a = 0$. In what follows, the following convention is used ${}_0D_t^{-\mu} \equiv D_t^{-\mu}$. The other approach is Grünwald-Letnikov definition of fractional-order integral given by

$$ D_t^{-\mu} f(t) = \lim_{h \to 0} h^{\mu} \sum_{r=0}^{\frac{t-a}{h}} \frac{\Gamma(\mu+r)}{r!\,\Gamma(\mu)} f(t-rh) \tag{39} $$

while the definition of fractional-order derivative is given by

$$ D_t^{\mu} f(t) = \lim_{\substack{h \to 0 \\ nh \approx t-a}} h^{-\mu} \sum_{r=0}^{n} (-1)^{\mu} f(t-rh) \tag{40} $$

An important property revealed by the Riemann-Liouville and Grünwald-Letnikov definitions is that while integer-order operators imply finite series, the fractional-order counterparts are defined by infinite series (Calderón et al., 2006), (Arijit et al., 2009). This means that integer operators are local operators in opposition with the fractional operators that have, implicitly, a memory of the past events. The differential equation for the fractional-order PI^{μ} controller $0 < \mu < 1$ is given by

$$ u(t) = K_p\, e(t) + K_i\, D_t^{-\mu} e(t) \tag{41} $$

where K_p is the proportional constant and K_i is the integration constant. Taking $\mu = 1$ in (41) a classical PI controller is obtained. The fractional-order PI^{μ} controller is more flexible than the classical PI controller, because it has one more adjustable parameter, which reflects the intensity of integration. The transfer function of the fractional-order PI^{μ} controller, using the Laplace transform on (41), is given by

$$G(s) = K_p + K_i \, s^{-\mu} \tag{42}$$

A good trade-off between robustness and dynamic performance, presented in (Maione & Lino, 2007), is in favour of a value for μ in the range [0.4, 0.6].

3.2 Converters control

Power electronic converters are variable structure systems, because of the on/off switching of their IGBTs. Pulse width modulation (PWM) by space vector modulation (SVM) associated with sliding mode (SM) is used for controlling the converters. The sliding mode control strategy presents attractive features such as robustness to parametric uncertainties of the wind turbine and the generator, as well as to electric grid disturbances (Beltran et al., 2008). Sliding mode controllers are particularly interesting in systems with variable structure, such as switching power electronic converters, guaranteeing the choice of the most appropriate space vectors. Their aim is to let the system slide along a predefined sliding surface by changing the system structure.

The power semiconductors present physical limitations that have to be considered during design phase and during simulation. Particularly, they cannot switch at infinite frequency. Also, for a finite value of the switching frequency, an error exists between the reference value and the control value. In order to guarantee that the system slides along the sliding surface, it has been proven that it is necessary to ensure that the state trajectory near the surfaces verifies the stability conditions (Rojas et al., 1995) given by

$$S(e_{\alpha\beta},t)\,\frac{dS(e_{\alpha\beta},t)}{dt} < 0 \tag{43}$$

in practice a small error $\varepsilon > 0$ for $S(e_{\alpha\beta},t)$ is allowed, due to power semiconductors switching only at finite frequency. Consequently, a switching strategy has to be considered given by

$$-\varepsilon < S(e_{\alpha\beta},t) < +\varepsilon \tag{44}$$

A practical implementation of this switching strategy at the simulation level could be accomplished by using hysteresis comparators. The output voltages of matrix converter are switched discontinuous variables. If high enough switching frequencies are considered, it is possible to assume that in each switching period T_s the average value of the output voltages is nearly equal to their reference average value. Hence, it is assumed that

$$\frac{1}{T_s}\int_{nT_s}^{(n+1)T_s} v_{\alpha\beta}\, dt = v_{\alpha\beta}^{*} \tag{45}$$

Similar to the average value of the output voltages, the average value of the input current is nearly equal to their reference average value. Hence, it is assumed that

$$\frac{1}{T_s}\int_{nT_s}^{(n+1)T_s} i_q\, dt = i_q^{*} \tag{46}$$

The output voltage vectors in the $\alpha\beta$ plane for the two-level converter are shown in Fig. 6.

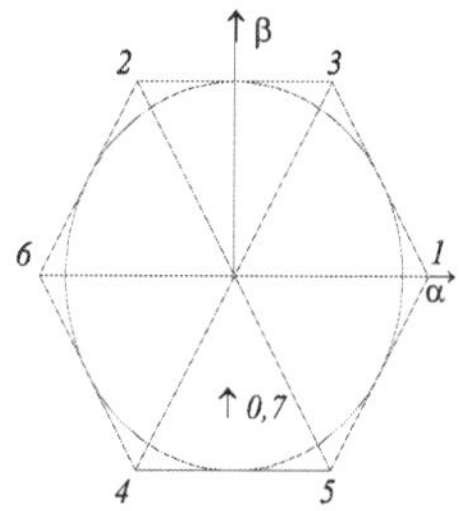

Fig. 6. Output voltage vectors for the two-level converter

Also, the integer variables $\sigma_{\alpha\beta}$ for the two-level converter take the values

$$\sigma_{\alpha\beta} \text{ with } \sigma_{\alpha}, \sigma_{\beta} \in \{-1, 0, 1\} \tag{47}$$

The output voltage vectors in the $\alpha\beta$ plane for the multilevel converter are shown in Fig. 7.

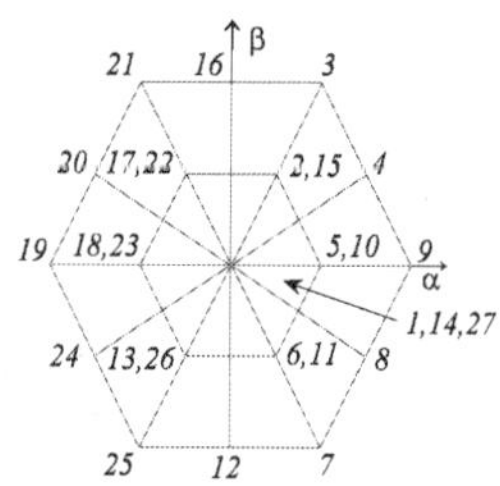

Fig. 7. Output voltage vectors for the two-level converter

The integer variables $\sigma_{\alpha\beta}$ for the multilevel converter take the values

$$\sigma_{\alpha\beta} \text{ with } \sigma_{\alpha}, \sigma_{\beta} \in \{-2, -1, 0, 1, 2\} \tag{48}$$

If $v_{C1} \neq v_{C2}$, then a new vector is selected. The output voltage vectors and the input current vectors in the $\alpha\beta$ plane for the matrix converter are shown respectively in Fig. 8 and Fig. 9.

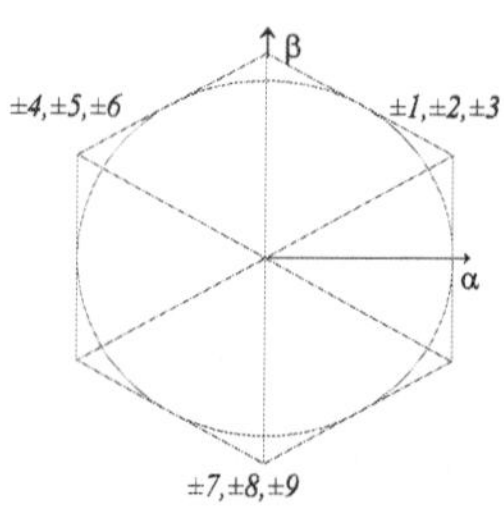

Fig. 8. Output voltage vectors for the matrix converter

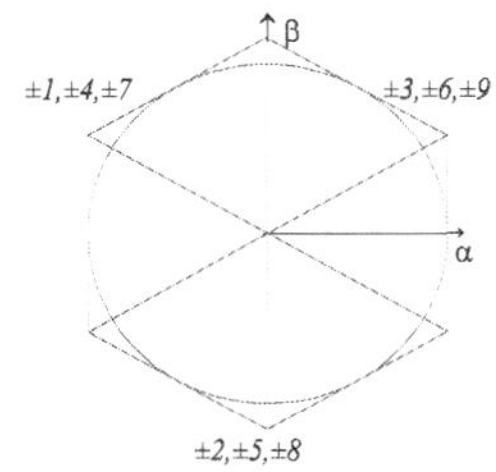

Fig. 9. Input current vectors for the matrix converter

The outputs of the hysteresis comparators are integer variables (Melício et al., 2010a). The voltage integer variables $\sigma_{\alpha\beta}$ for the matrix converter take the values

$$\sigma_{\alpha\beta} \text{ with } \sigma_{\alpha}, \sigma_{\beta} \in \{-1,0,1\} \tag{49}$$

The current integer variables σ_q for the matrix converter in dq coordinates take the values

$$\sigma_q \in \{-1,1\} \tag{50}$$

Hence, the proposed control strategy for the power electronic converters is given by the consideration of (43) to (50). Design of PI^{μ} controllers on the wind energy conversion systems is made over the respective configurations. The design of PI^{μ} controllers follows the tuning rules in (Maione & Lino, 2007). Power electronic converters are modelled as a pure delay (Chinchilla et al., 2006) and the left-over dynamics are modelled with a second order equivalent transfer function, following the identification of a step response.

4. Power quality evaluation

The harmonic behaviour computed by the Discrete Fourier Transform is given by

$$X(k) = \sum_{n=0}^{N-1} e^{-j2\pi kn/N}\, x(n) \text{ for } k = 0,\ldots,N-1 \tag{51}$$

where $x(n)$ is the input signal and $X(k)$ is a complex giving the amplitude and phase of the different sinusoidal components of $x(n)$. The total harmonic distortion THD is defined by the expression given by

$$\text{THD}\,(\%) = 100\,\frac{\sqrt{\sum_{H=2}^{50} X_H^2}}{X_F} \tag{52}$$

where X_H is the root mean square (RMS) value of the non-fundamental H harmonic, component of the signal, and X_F is the RMS value of the fundamental component harmonic. Standards such as IEEE-519 (Standard 519, 1992) impose limits for different order harmonics and the THD. The limit is 5% for THD is considered in the IEEE-519 standard and is used in this book chapter as a guideline for comparison purposes.

5. Simulation results

Consider a wind power system with the moments of inertia of the drive train, the stiffness, the turbine rotor diameter, the tip speed, the rotor speed, and the generator rated power, given in Table 2.

Blades moment of inertia	400×10^3 kgm^2
Hub moment of inertia	19.2×10^3 kgm^2
Generator moment of inertia	16×10^3 kgm^2
Stiffness	1.8×10^6 Nm
Turbine rotor diameter	49 m
Tip speed	17.64-81.04 m/s
Rotor speed	6.9-30.6 rpm
Generator rated power	900 kW

Table 2. Wind energy system data

The time horizon considered in the simulations is 5 s.

5.1 Pitch angle control malfunction

Consider a pitch angle control malfunction starting at 2 s and lengthen until 2.5 s due to a total cut-off on the capture of the energy from the wind by the blades (Melício et al., 2010b), and consider the model for the wind speed given by

$$u(t) = 15\left[1 + \sum_k A_k \sin(\omega_k\, t)\right] \quad 0 \le t \le 5 \tag{53}$$

This wind speed in function of the time is shown in Fig. 10.

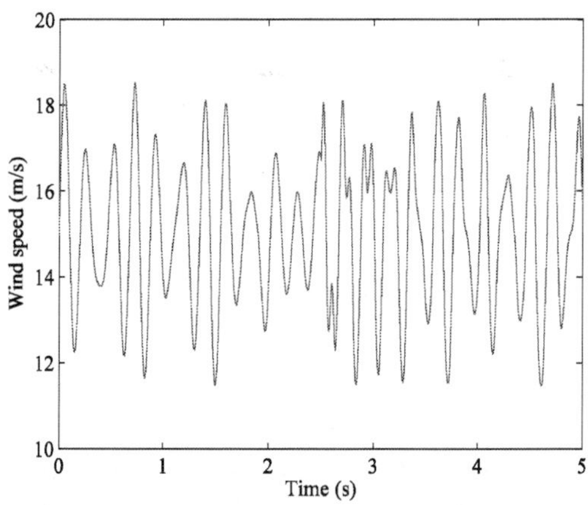

Fig. 10. Wind speed

In this simulation after some tuning it is assumed that $\mu = 0.5$. The mechanical power over the rotor of the wind turbine disturbed by the mechanical eigenswings, and the electric power of the generator is shown in Fig. 11.
Fig. 11 shows an admissible drop in the electrical power of the generator, while the mechanical power over the rotor is null due to the total cut-off on the capture of the energy from the wind by the blades. The power coefficient is shown in Fig. 12.

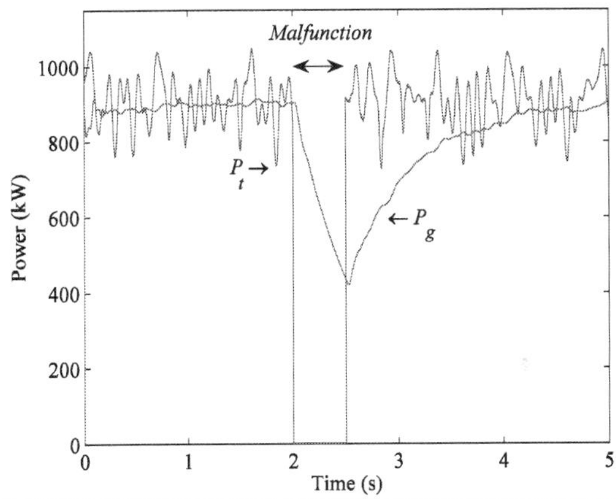

Fig. 11. Mechanical power over the rotor and electric power

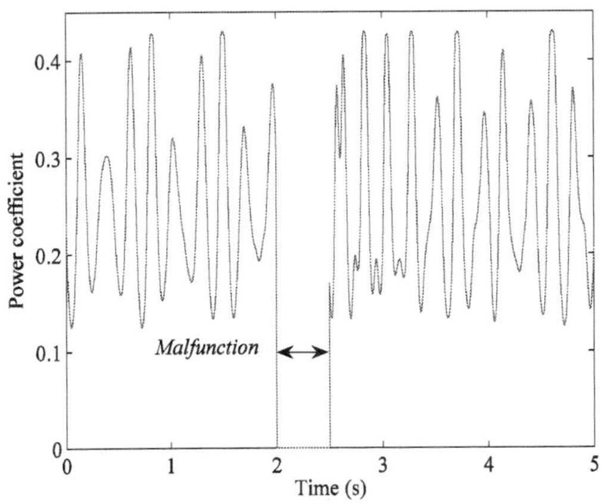

Fig. 12. Power coefficient

The power coefficient is at the minimum value during the pitch control malfunction, induced by the mistaken wind gust position. The voltage v_{dc} for the two-level converter with the fractional order controller, respectively for the three different mass drive train models are shown in Fig. 13.

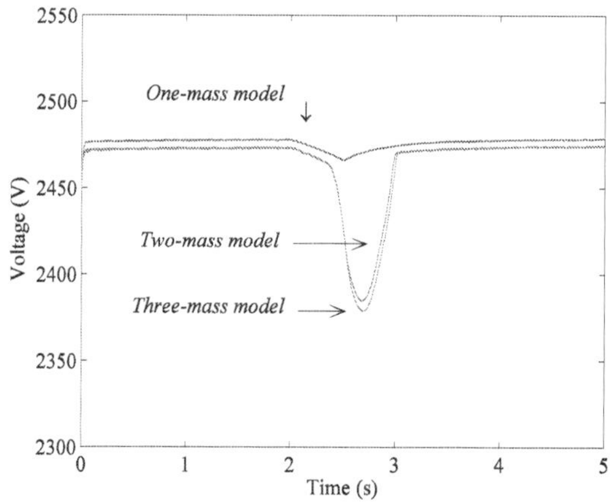

Fig. 13. Voltage at the capacitor for two-level converter using a fractional-order controller

The voltage v_{dc} for the two-level converter presents almost the same behaviour with the two-mass or the three-mass models for the drive train. But one-mass model omits significant dynamic response as seen in Fig. 13. Hence, as expected there is in this case an admissible use for a two-mass model; one-mass model is not recommended in confining this behaviour. Nevertheless, notice that: the three-mass model is capturing more information about the behaviour of the mechanical drive train on the system. The increase on the electric power of wind turbines, imposing the increase on the size of the rotor of wind turbines, with longer flexible blades, is in favour of the three-mass modelling.

5.2 Converter control malfunction

Consider a wind speed given by

$$u(t) = 20\left[1 + \sum_k A_k \sin(\omega_k\, t)\right] \quad 0 \le t \le 5 \tag{54}$$

The converter control malfunction is assumed to occur between 2.00 s and 2.02 s, imposing a momentary malfunction on the vector selection for the inverter of the two-level and the multilevel converters and on the vector selection for the matrix converter. This malfunction is simulated by a random selection of vectors satisfying the constraint of no short circuits on the converters. In this simulation after some tuning it was assumed $\mu = 0.7$. Simulations with the model for the matrix converter were carried out, considering one-mass, two-mass and three-mass drive train models in order to establish a comparative behaviour (Melício et al., 2010c).

The mechanical torque over the rotor of the wind turbine disturbed by the mechanical eigenswings and the electric torque of the generator, with the one-mass, two-mass and three-mass drive train models, are shown in Figs. 14, 15 and 16, respectively. As shown in these figures, the electric torque of the generator follows the rotor speed at the PMSG, except when it is decreased due to the malfunction. For the same fault conditions, the transient response of the three-mass drive train model is significantly different than that of the two-mass model, since it captures more information about the behaviour of the system. The results have shown that the consideration of the bending flexibility of blades influences the wind turbine response during internal faults (Melício et al., 2010d).

The voltage v_{dc} for the two-level and the multilevel converters with a three-mass drive train model is shown in Fig. 17. As expected during the malfunction this voltage suffers a small increase, the capacitor is charging, but almost after the end of the malfunction voltage recovers to its normal value (Melício et al., 2010e).

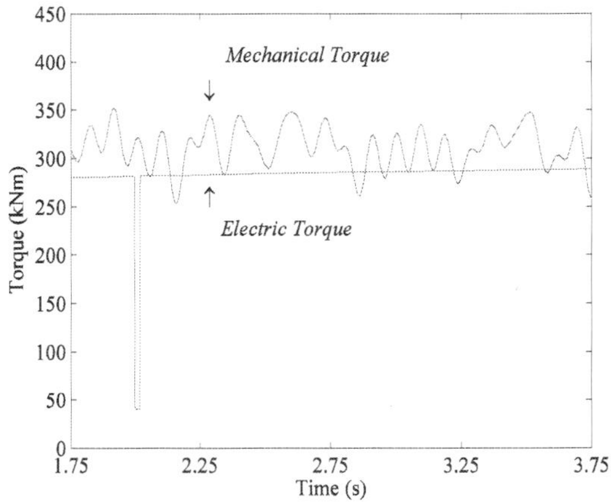

Fig. 14. Mechanical and electric torque with the one-mass drive train model and using a matrix converter

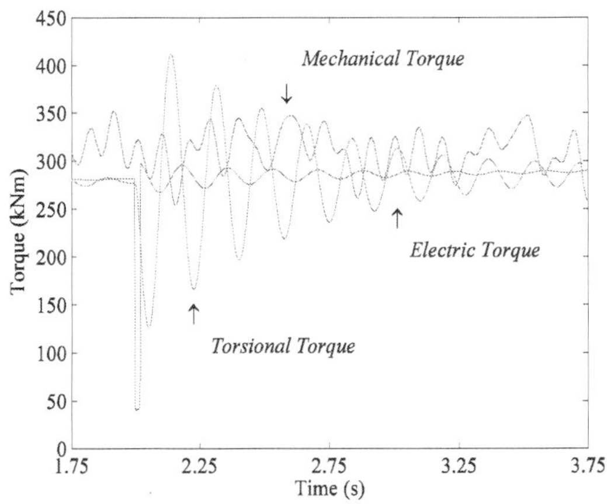

Fig. 15. Mechanical and electric torque with the two-mass drive train model and using a matrix converter

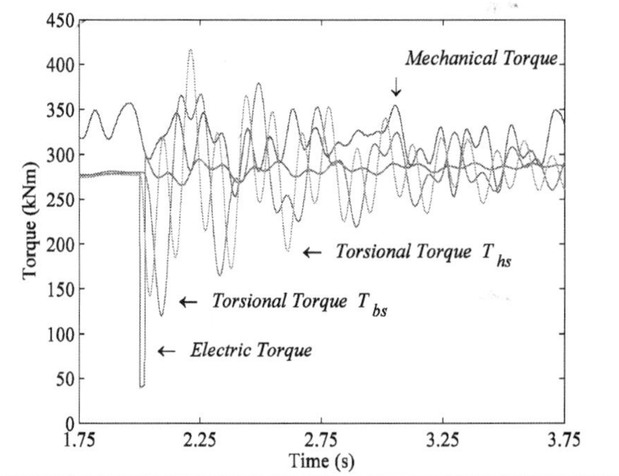

Fig. 16. Mechanical and electric torque with the three-mass drive train model and using a matrix converter

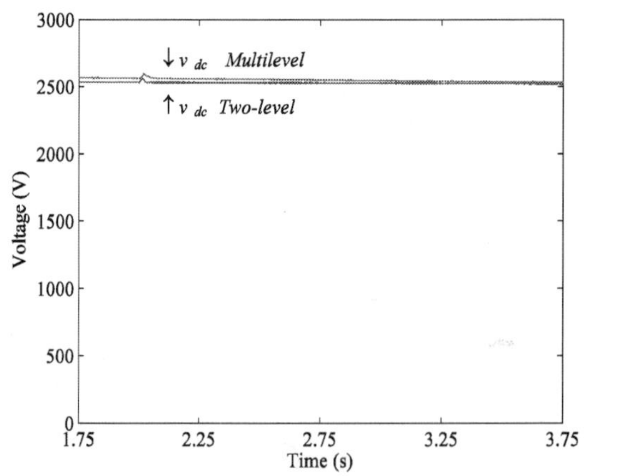

Fig. 17. Voltage v_{dc} for the two-level and the multilevel converter using a three-mass drive train model

The currents injected into the electric grid by the wind energy system with the two-level converter and a three-mass drive train model are shown in Fig. 18.
The currents injected into the electric grid for the wind energy system with a multilevel converter and a three-mass drive train model are shown in Fig. 19.
The currents injected into the electric grid for the wind energy system with matrix converter and a three-mass drive train model are shown in Fig. 20.

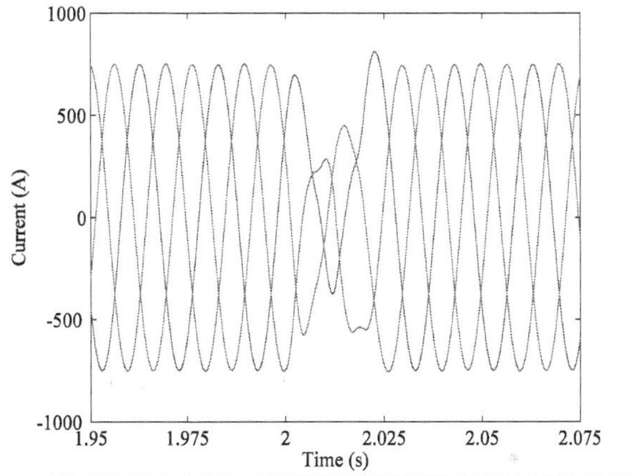

Fig. 18. Currents injected into the electric grid (two-level converter and a three-mass drive train model)

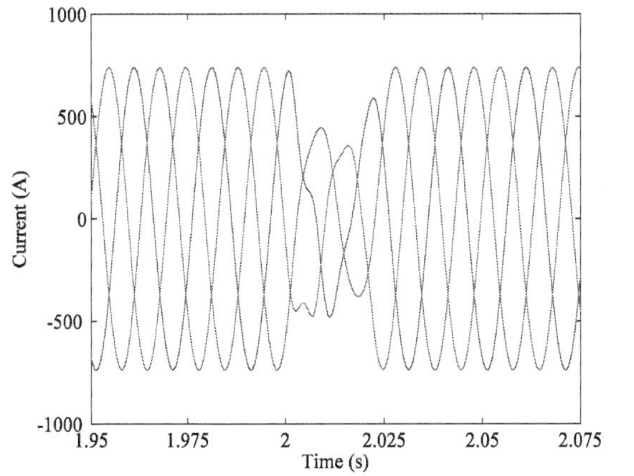

Fig. 19. Currents injected into the electric grid (multilevel converter and three-mass drive train model)

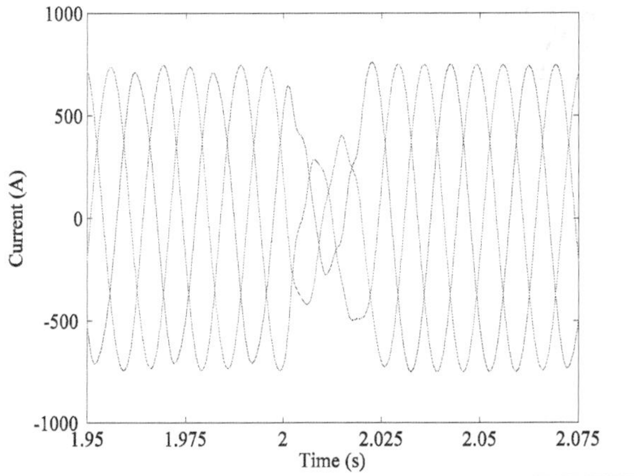

Fig. 20. Currents injected into the electric grid (matrix converter and three-mass drive train model)

As shown in Figs. 18, 19 or 20, during the malfunction the current decreases, but almost after the end of the malfunction the current recovers its normal behaviour.

5.3 Harmonic assessment considering ideal sinusoidal voltage on the network

Consider the network modelled as a three-phase active symmetrical circuit in series, with 850 V at 50 Hz. In the simulation after some tuning it is assumed that $\mu = 0.5$. Consider the wind speed in steady-state ranging from 5-25 m/s. The goal of the simulation is to assess the third harmonic and the THD of the output current.

The wind energy conversion system using a two-level converter in steady-state has the third harmonic of the output current shown in Fig. 21, and the THD of the output current shown in Fig. 22. The wind energy conversion system with multilevel converter in steady-state has

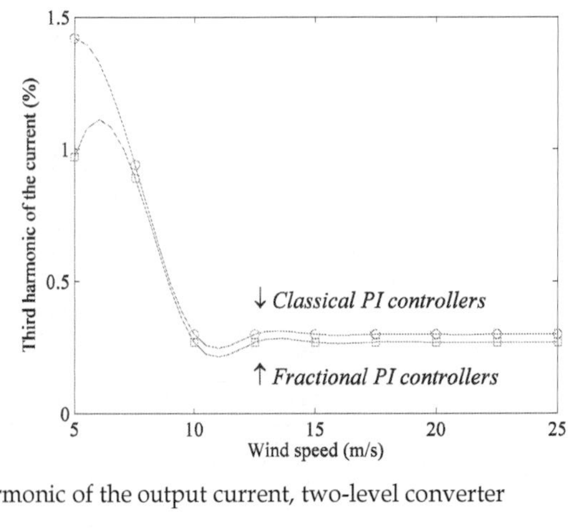

Fig. 21. Third harmonic of the output current, two-level converter

the third harmonic of the output current shown in Fig. 23, and the THD of the output current shown in Fig. 24. The wind energy conversion system with matrix converter in steady-state has the third harmonic of the output current shown in Fig. 25, and the THD of the output current shown in Fig. 26.
The fractional-order control strategy provides better results comparatively to a classical integer-order control strategy, in what regards the third harmonic of the output current and the THD.

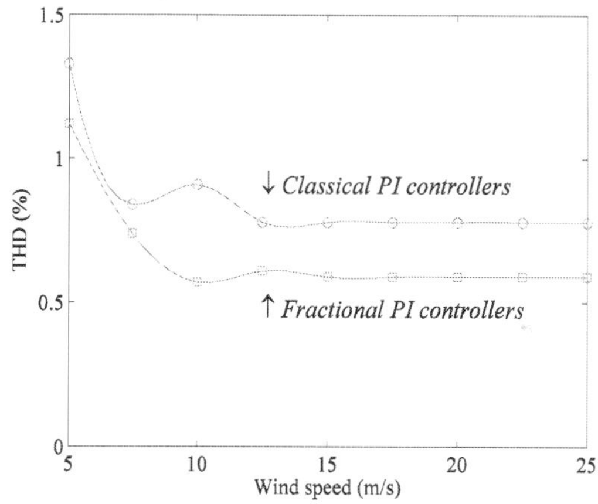

Fig. 22. THD of the output current, two-level converter

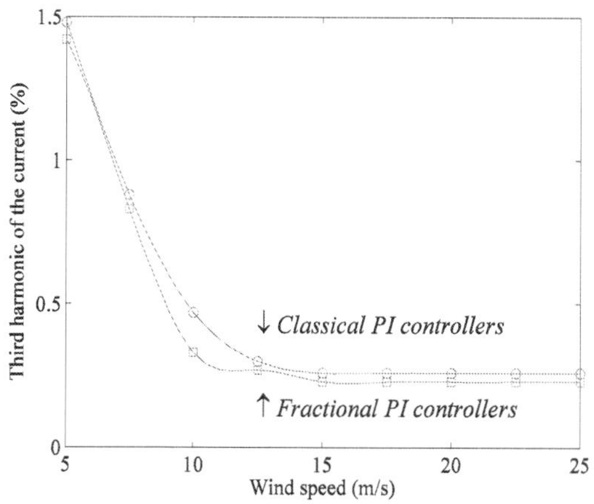

Fig. 23. Third harmonic of the output current, multilevel converter

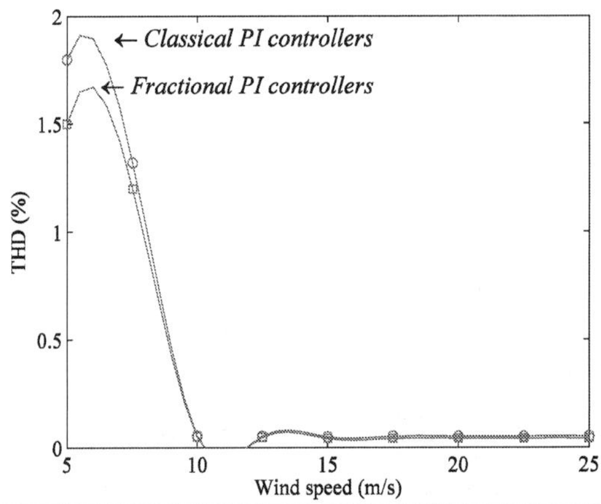

Fig. 24. THD of the output current, multilevel converter

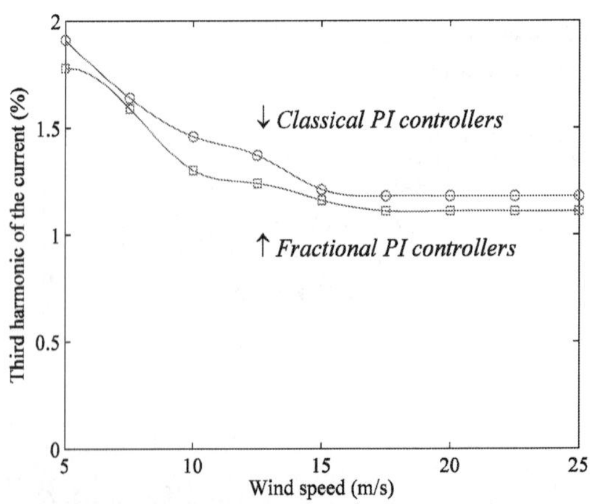

Fig. 25. Third harmonic of the output current, matrix converter

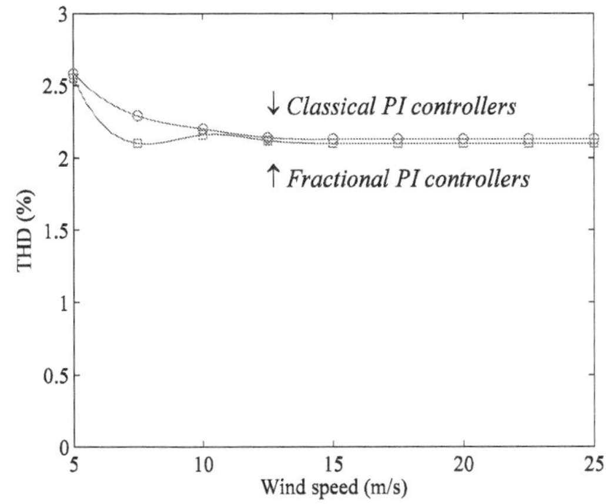

Fig. 26. THD of the output current, matrix converter

5.4 Harmonic assessment considering non-ideal sinusoidal voltage on the network

Consider the network modelled as a three-phase active symmetrical circuit in series, with 850 V at 50 Hz and 5 % of third harmonic component. In this simulation after some tuning it is assumed that $\mu = 0.5$. Consider the wind speed in steady-state ranging from 5-25 m/s.
The wind energy conversion system using a two-level converter in steady-state has the third harmonic of the output current shown in Fig. 27, and the THD of the output current shown in Fig. 28. The wind energy conversion system with multilevel converter in steady-state has the third harmonic of the output current shown in Fig. 29, and the THD of the output current shown in Fig. 30. The wind energy conversion system with matrix converter in steady-state has the third harmonic of the output current shown in Fig. 31, and the THD of the output current shown in Fig. 32.

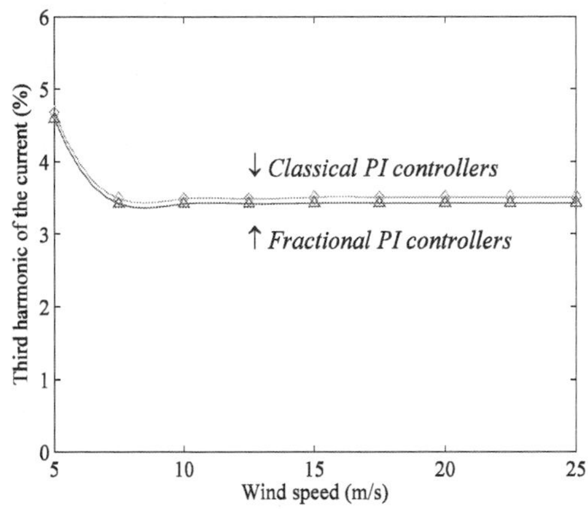

Fig. 27. Third harmonic of the output current, two-level converter

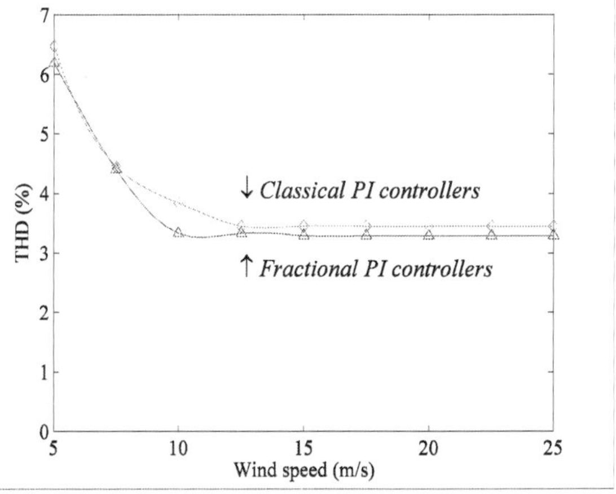

Fig. 28. THD of the output current, two-level converter

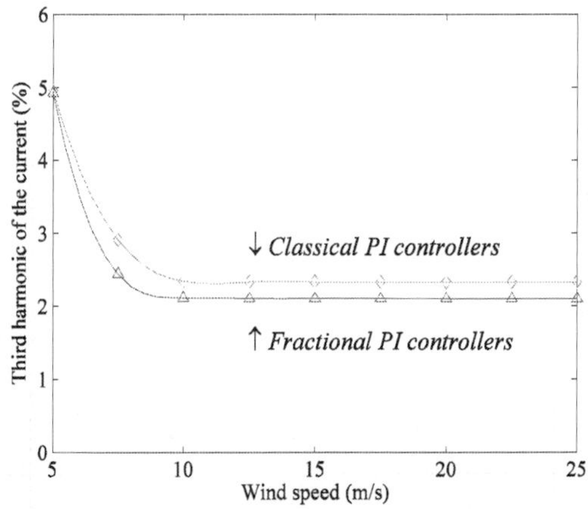

Fig. 29. Third harmonic of the output current, multilevel converter

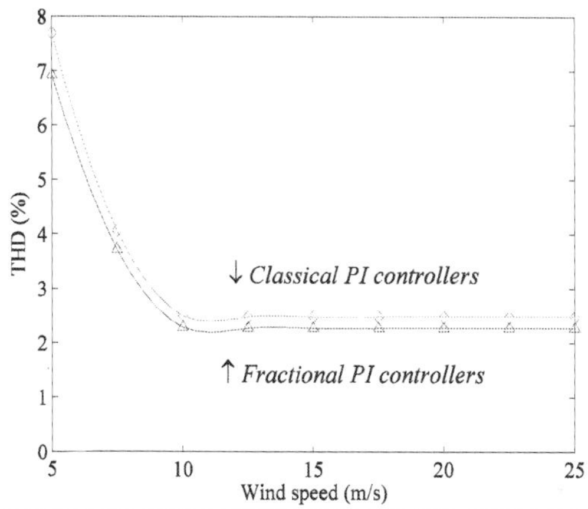

Fig. 30. THD of the output current, multilevel converter

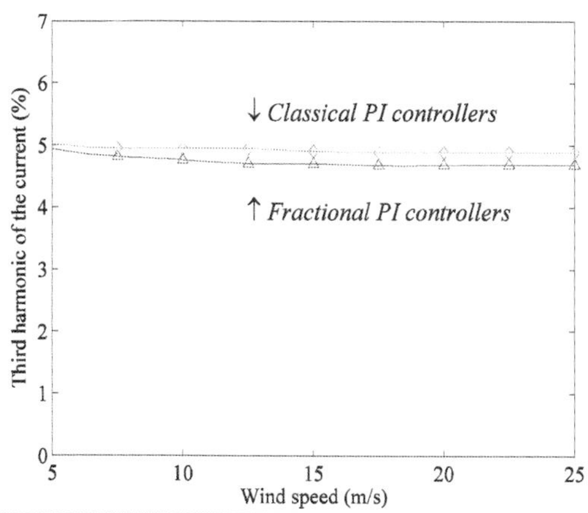

Fig. 31. Third harmonic of the output current, matrix converter

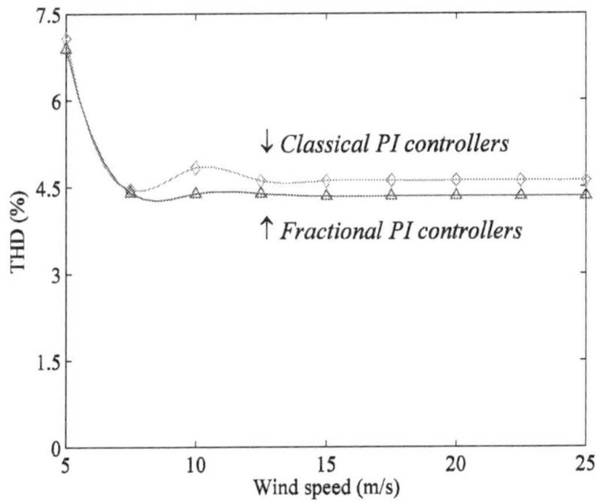

Fig. 32. THD of the output current, matrix converter

The THD of the current injected into the network is lower than the 5% limit imposed by IEEE-519 standard for the two-level and multilevel converters, and almost near to this limit for the matrix converter.

The fact that a non-ideal sinusoidal voltage on the network has an effect on the current THD of the converters can be observed by comparison of Fig. 21 until Fig. 26 with Fig. 27 until Fig. 32.

Also, from those figures it is possible to conclude that the fractional-order controller simulated for the variable-speed operation of wind turbines equipped with a PMSG has shown an improvement on the power quality in comparison with a classical integer-order control strategy, in what regards the third harmonic of the output current and the THD.

6. Conclusion

The assessment of the models ability to predict the behaviour for the drive train is in favour of the two-mass model or the three-mass model for wind energy conversion systems with electric power in the range of the values used in the simulations.

But, with the increase on the electric power of wind turbines, imposing the increase on the size of the rotor of wind turbines, with longer flexible blades, the study on the transient stability analysis of wind energy conversion systems will show a favour for a three-mass modelling.

The fractional-order controller simulated for the variable-speed operation of wind turbines equipped with a PMSG has shown an improvement on the power quality in comparison with a classical integer-order control strategy, in what regards the level for the third harmonic on the output current and the THD.

Accordingly, it is shown that the THD of the current for the wind energy conversion system with either a two-level or a multilevel converter is lower than with a matrix converter.

Finally, the results have shown that the distortion on the voltage of the electric network due to the third harmonic influences significantly the converter output electric current.

Particularly, in what concerns to the use of the two-level, multilevel and matrix converters the last one can be the first to be out of the 5% limit imposed by IEEE-519 standard, if a major harmonic content on the voltage of the electric network is to be considered.

7. References

Akhmatov, V.; Knudsen, H. & Nielsen, A. H. (2000). Advanced simulation of windmills in the electric power supply. *Int. J. Electr. Power Energy Syst.*, Vol. 22, No. 6, August 2000, pp. 421-434, ISSN: 0142-0615.

Alesina, A. & Venturini, M. (2000). Solid-state power conversion: a Fourier analysis approach to generalized transformer synthesis. *IEEE Trans. Circuits Syst.*, Vol. 28, No. 4, April 1981, pp. 319-330, ISSN: 0098-4094.

Arijit, B.; Swagatam, D.; Ajith, A. & Sambarta, D. (2009). Design of fractional-order PI-lambda-D-mu-controllers with an improved differential evolution. *Eng. Appl. Artif. Intell.*, Vol. 22, No. 2, March 2009, pp. 343-350, ISSN: 0952-1976.

Barbosa, R. S.; Machado, J. A. T. & Galhano, M. (2007). Performance of fractional PID algorithms controlling nonlinear systems with saturation and backlash phenomena. *J. Vibration and Control*, Vol. 13, No. 9-10, September 2007, pp. 1407-1418, ISSN: 1077-5463.

Beltran, B.; Ahmed-Ali, T. & Benbouzid, M. E. H. (2008). Sliding mode power control of variable-speed wind energy conversion systems. *IEEE Trans. Energy Convers.*, Vol. 23, No. 2, June 2008, pp. 551-558, ISSN: 0885-8969.

Calderón, A. J.; Vinagre, B. M. & Feliu, V. (2006). Fractional order control strategies for power electronic buck converters. *Signal Processing*, Vol. 86, No. 10, October 2006, pp. 2803-2819, ISSN: 0165-1684.

Chen, Y. Q.; Petráš, I. & Xue, D. (2009). Fractional order control–a tutorial, *Proceedings of 2009 American Control Conference*, pp. 1397-1411, ISBN: 978-1-4244-4523-3, USA, June 2009, St. Louis.

Chen, Z. & Spooner, E. (2001). Grid power quality with variable speed wind turbines. *IEEE Trans. Energy Convers.*, Vol. 16, No. 2, (June 2001) pp. 148-154, ISSN: 0885-8969.

Çelik, V. & Demir, Y. (2010). Effects on the chaotic system of fractional order PI-alfa controller. *Nonlinear Dynamics*, Vol. 59, No. 1-2, January 2010, pp. 143-159, ISSN: 0924-090X.

Chinchilla, M.; Arnaltes, S. & Burgos, J. C. (2006). Control of permanent-magnet generators applied to variable-speed wind energy systems connected to the grid. *IEEE Trans. Energy Convers.*, Vol. 21, No. 1, March 2006, pp. 130-135, ISSN: 0885-8969.

Cruz, S. M. A. & Ferreira, M. (2009). Comparison between back-to-back and matrix converter drives under faulty conditions, *Proceedings of EPE 2009*, pp. 1-10, ISBN: 978-1-4244-4432-8, Barcelona, September 2009, Spain.

Jun-Yi, C. & Bing-Gang, C. (2006). Design of fractional order controllers based on particle swarm, *Proceedings of IEEE ICIEA 2006*, pp. 1-6, ISBN: 0-7803-9514-X, Malaysia, May 2006, Singapore.

Li, H. & Chen, Z. (2007). Transient stability analysis of wind turbines with induction generators considering blades and shaft flexibility, *Proceedings of 33rd IEEE IECON 2007*, pp. 1604-1609, ISBN: 1-4244-0783-4, Taiwan, November 2007, Taipei.

Li, W. & Hori, Y. (2007). Vibration suppression using single neuron-based PI fuzzy controller and fractional-order disturbance observer. *IEEE Trans. Ind. Electron.*, Vol. 54, No. 1, February 2007, pp. 117-126, ISSN: 0278-0046.

Maione, G. & Lino, P. (2007). New tuning rules for fractional PI-alfa controllers. *Nonlinear Dynamics*, Vol. 49, No. 1-2, July 2007, pp. 251-257, ISSN: 0924-090X.

Melício, R.; Mendes, V. M. F. & Catalão, J. P. S. (2010a). Modeling, control and simulation of full-power converter wind turbines equipped with permanent magnet synchronous generator. *International Review of Electrical Engineering*, Vol. 5, No. 2, March-April 2010, pp. 397-408, ISSN: 1827-6660.

Melício, R.; Mendes, V. M. F. & Catalão, J. P. S. (2010b). A pitch control malfunction analysis for wind turbines with permanent magnet synchronous generator and full-power converters: proportional integral versus fractional-order controllers. *Electric Power Components and Systems*, Vol. 38, No. 4, 2010, pp. 387-406, ISSN: 1532-5008.

Melício, R.; Mendes, V. M. F. & Catalão, J. P. S. (2010c). Wind turbines equipped with fractional-order controllers: stress on the mechanical drive train due to a converter control malfunction. *Wind Energy*, 2010, in press, DOI: 10.1002/we.399, ISSN: 1095-4244.

Melício, R.; Mendes, V. M. F. & Catalão, J. P. S. (2010d). Harmonic assessment of variable-speed wind turbines considering a converter control malfunction. *IET Renewable Power Generation*, Vol. 4, No. 2, 2010, pp. 139-152, ISSN: 1752-1416.

Melício, R.; Mendes, V. M. F. & Catalão, J. P. S. (2010e). Fractional-order control and simulation of wind energy systems with PMSG/full-power converter topology. *Energy Conversion and Management*, Vol. 51, No. 6, June 2010, pp. 1250-1258, ISSN: 0196-8904.

Muyeen, S. M.; Hassan Ali, M.; Takahashi, R.; Murata, T.; Tamura, J.; Tomaki, Y.; Sakahara, A. & Sasano, E. (2006). Transient stability analysis of grid connected wind turbine generator system considering multi-mass shaft modeling. *Electr. Power Compon. Syst.*, Vol. 34, No. 10, October 2006, pp. 1121-1138, ISSN: 1532-5016.

Ong, P.-M. (1998). *Dynamic Simulation of Electric Machinery: Using Matlab/Simulink*, Prentice-Hall, ISBN: 0137237855, New Jersey.

Podlubny, I. (1999). Fractional-order systems and PI-lambda-D-mu-controllers. *IEEE Trans. Autom. Control*, Vol. 44, No. 1, January 1999, pp. 208-214, ISSN: 0018-9286.

Ramtharan, G. & Jenkins, N. (2007). Influence of rotor structural dynamics representations on the electrical transient performance of DFIG wind turbines. *Wind Energy*, Vol. 10, No. 4, (March 2007) pp. 293-301, ISSN: 1095-4244.

Rojas, R.; Ohnishi, T. & Suzuki, T. (1995). Neutral-point-clamped inverter with improved voltage waveform and control range. *IEEE Trans. Industrial Elect.*, Vol. 42, No. 6, (December 1995) pp. 587-594, ISSN: 0278-0046.

Salman, S. K. & Teo, A. L. J. (2003). Windmill modeling consideration and factors influencing the stability of a grid-connected wind power-based embedded generator. *IEEE Trans. Power Syst.*, Vol. 16, No. 2, (May 2003) pp. 793-802, ISSN: 0885-8950.

Senjyu, T.; Tamaki, S.; Urasaki, N. & Uezato, K. (2003). Wind velocity and position sensorless operation for PMSG wind generator, *Proceedings of 5th Int. Conference on Power Electronics and Drive Systems 2003*, pp. 787-792, ISBN: 0-7803-7885-7, Malaysia, January 2003, Singapore.

Slootweg, J. G.; de Haan, S. W. H.; Polinder, H. & Kling, W. L. (2003). General model for representing variable speed wind turbines in power system dynamics simulations. *IEEE Trans. Power Syst.*, Vol. 18, No. 1, February 2003, pp. 144-151, ISSN: 0885-8950.

Standard 519 (1992). *[52] IEEE Guide for Harmonic Control and Reactive Compensation of Static Power Converters*, IEEE Standard 519-1992.

Timbus, A.; Lisserre, M.; Teodorescu, R.; Rodriguez, P. & Blaabjerg, F. (2009). Evaluation of current controllers for distributed power generation systems. *IEEE Trans. Power Electron.*, Vol. 24, No. 3, March 2009, pp. 654-664, ISSN: 0885-8993.

Xing, Z. X.; Zheng, Q. L.; Yao, X. J. & Jing, Y. J. (2005). Integration of large doubly-fed wind power generator system into grid, *Proceedings of 8th Int. Conf. Electrical Machines and Systems*, pp. 1000-1004, ISBN: 7-5062-7407-8, China, September 2005, Nanjing.

Zamani, M.; Karimi-Ghartemani, M.; Sadati, N. & Parniani, M. (2009). Design of fractional order PID controller for an AVR using particle swarm optimization. *Control Eng. Practice*, Vol. 17, No. 12, December 2009, pp. 1380-1387, ISSN: 0967-0661.

21

Reactive Power Control of Direct Drive Synchronous Generators to Enhance the Low Voltage Ride-Through Capability

Andrey C. Lopes, André C. Nascimento, João P. A. Vieira, Marcus V. A. Nunes and Ubiratan H. Bezerra
Federal University of Pará
Brazil

1. Introduction

The development of wind power generation has grown considerably during the last years. The use of wind generators forming groups denominated wind farms, operating together with conventional sources of energy in weak grids has also increased [1]. The increased penetration of wind energy into the power system over the last few years is directly reflected in the requirements for grid connection of wind turbines. These codes are becoming more and more demanding, requiring wind farms to behave more and more as conventional power plants in the power system. Therefore it is essential to analyze the characteristics of wind generators during the network disturbances [2].

Currently, most of the grid requirements address low voltage ride-through (LVRT) and grid support capabilities of the wind generators. The LVRT requirement specifies that wind generators need to remain connected to the grid when an abnormal grid voltage is detected (e.g. during short-circuit faults). The grid support capability specifies wind generators to assist the power system by supplying ancillary services, such as voltage control, to assure a safe and reliable grid operation. Power electronics based solutions for grid interfacing of wind turbines seem to be a very promising technology that can cope with these grid requirements.

The configurations of variable speed wind generation that employ direct drive synchronous generators with permanent magnet and rotor field excitation, present noticeable advantages such as the decoupled control of the generators active and reactive power, the improvement of system efficiency and the fact that the machine stator frequency is decoupled from the grid frequency. The stator converter uses a high frequency switching PWM converter to achieve high control performance with low harmonic distortion, [3].

Due to those aspects in a worldwide basis and specifically in Brazil, there is a trend to install a large amount of wind power based on that technology. The interaction with the grid becomes increasingly important then. This can be understood as follows. When all wind generators would be disconnected in case of a grid failure, these renewable generators will—unlike conventional power plants—not be able to support the voltage and the grid frequency during and immediately following the grid failure. This would cause major problems for the system stability [3].

With the perspective of integration of more wind parks in Brazil the Grid National Operator (ONS) already has set requirements for the behavior of the wind generators protection. Instead of disconnecting them from the grid, the wind generators should be able to follow the characteristic shown in Fig. 1.

Only when the grid voltage goes below the curve (in duration or voltage level), the wind generators is allowed to disconnect. When the voltage is in the gray area, the wind generators should supply reactive power. In this paper a method is proposed that makes it possible for wind generators using direct drive synchronous generators to stay connected to the grid during faults.

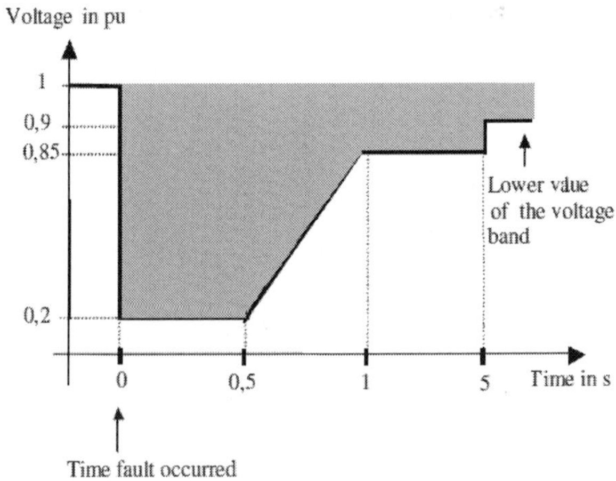

Fig. 1. ONS Requirements for Wind Parks Behaviour during Faults.

The main attention of the chapter is focused on the control strategies of the grid-side converter to provide reactive power support in case of grid disturbances. This control strategy can improve the voltage level during the fault and also contribute to the LVRT requirement.

For strong grids there is a decoupling between the active and reactive powers. Thus the total reactive power injection during a fault is a good solution for this control strategy without to compromissing the power system transient stability. However if the wind farm is connected to a weak electrical grid (i.e. voltage and frequency fluctuations conditions), there may be some power surge problems due to technical constraints related to the weak grid. In this case the proposed control strategy is based in the reactive injection curve defined by the grid code requirement improving the power system transient stability.

In this chapter, a variable speed wind turbine with a power electronic interface (a full power converter system) is considered. It is assumed that the wind turbines are equipped with a voltage dip ride-through facility and have a rapid current controller. Based on these assumptions, the wind park is modeled as a current injection source with current limitation determined by the converter capacity constraint. A similar approach of modeling a wind park was also adopted in [4] and [5]. The reference current is generated in accordance with the E.ON grid code which is then injected into the grid.

2. Grid requirements and fault ride-through capability

For this study, a set of minimum technical requirements concerning the installation of new wind farm was defined. The interconnection requirements, active and reactive power control and fault ride-through capability are based on the E.ON grid codes and ONS requirements.
The Wind Farm covered by the ONS requirements should present Fault Ride Through Capability in order to stay connected to the grid if during and in the moments subsequent to a fault or system disturbance the wind farm terminal voltage is above the value defined in the Time-Voltage characteristic curve presented in Fig. 1. Below the line, the wind park is allowed to trip.
For OEN grid code, it is required that during voltage dips a demand of reactive power injection must follow a specified curve. This requirement, besides improving the voltage levels in the electric grid in a defect condition, allows the wind park not to be removed of the system by the trip of the under voltage relay, increasing the ride-through capacity.
The generating plants must support the grid voltage with additional reactive current during a voltage dip. To do this, the voltage control must be activated as shown in Fig. 2 in the event of a voltage dip of more than 10% of the rms generator voltage. The voltage control must take place within 20 ms after fault recognition by providing a reactive current on the voltage side of the generator transformer amounting to at least 2% of the rated current for each percent of the voltage dip. A reactive power output of at least 100% of the rated current must be possible if necessary [EON]. The reactive injection curve is generated in accordance with the E.ON grid code as shown in Fig 2.

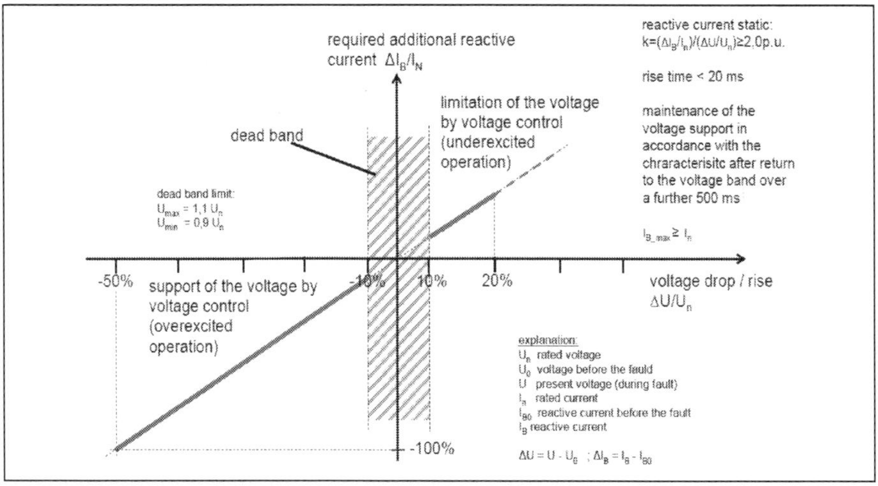

Fig. 2. The Principle of Voltage Support in Event of Grid Faults

When the voltage dip goes larger than 10% of the rated generator voltage the control action should act 20 ms after the identification of the fault, supplying reactive current on the low voltage side of the generator transformer at least 2% of the nominal current for each 1% of voltage dip.

3. Direct drive PMSG wind generators and controls

3.1 Modeling assumptions

The first stage in the simulation process is to model individual system components with an appropriate degree of complexity. The structure of the direct drive PMSG wind power system is shown in Fig. 3, in which the generator-side converter is a diode rectifier, the grid-side converter is a Pulse Width Modulation (PWM) inverter used to sustain the dc-bus voltage and regulate the grid-side power factor controlled independently through the decoupled d-q vector control approach for modern PMSG wind generators designs. The DC-link created by the capacitor in the middle decouples the operation of the two converters, as shown in Fig. 3.

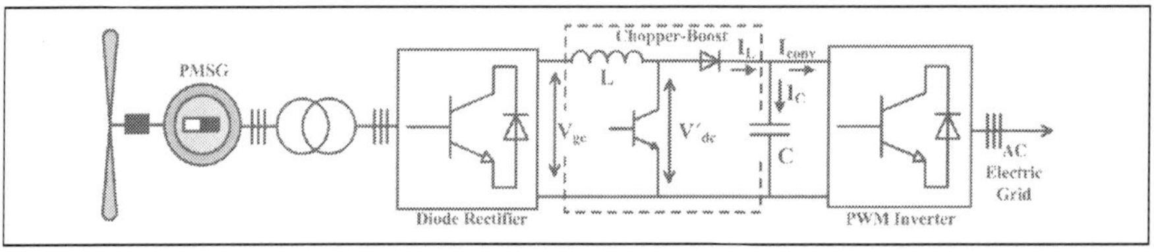

Fig. 3. Configuration of a PMSG Wind Generators

It has been demonstrated that the PMSG speed changes during voltage dips are negligible [6]. This is because the power supply to the DC circuit from the machine-side converter is approximately constant. So in terms of the grid connecting point (PCC) results, the PMSG mechanical behavior is not particularly relevant. So the model of the PMSG and the machine-side converter, can be omitted. A constant DC power current source can be used to represent these instead (thus neglecting the machine side power fluctuations), and the simulation results at the PCC will not be greatly compromised [6].

3.2 Simplified model

The model introduced in this section is a simplification of the model used in the previous section. In this model, the DC link voltage is controlled with a boost-chopper. The grid-side inverter is controlling the reactive power as usual. Fig. 4 shows the structure of this simplified model with the modified grid-side inverter controller.

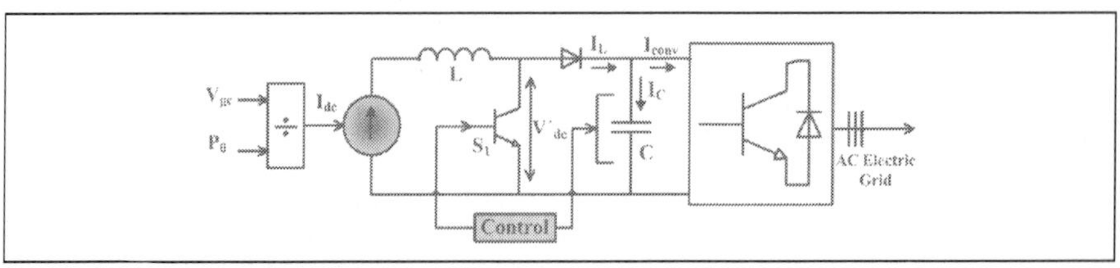

Fig. 4. Structure of the Simplified Model

In Fig. 4, P_0 represents the initial active power at the beginning of the simulation, based on the assumption of constant wind speed (wind power) in the wind farm during the short simulation time frame.

4. Models of the converters

Wind generators based on synchronous machines are connected to the electrical grid through static converters as illustrated previously in Fig. 3. In these cases it is required to model both converters, that is, the converter connected to the electrical grid and the one connected to the generator stator, which are both self-commutated PWM converters.

4.1 The grid-side converters models

The grid side converter as showed in Fig. 5 is controlled using a synchronous reference frame. The active part of the complex current becomes Iq and the reactive part is Id. The ouput power of the converter may be written as:

$$S = P_s + jQ_s = V_q\left(I_q - jI_d\right) \tag{1}$$

The converter terminal voltage, V´, is defined as:

$$V' = \frac{\sqrt{3}}{2\sqrt{2}} m V_{DC} \tag{2}$$

The pulse-width modulation index m is the control variable of the PWM converter. Equation (2) is valid for $0 \le m < 1$.
Reactive power is controlled directly by the reactive current Iq. If the converter is to operate with unity power factor, the reference for the reactive current must be set equal to zero. The output signals in the converter model are the pulse modulation.

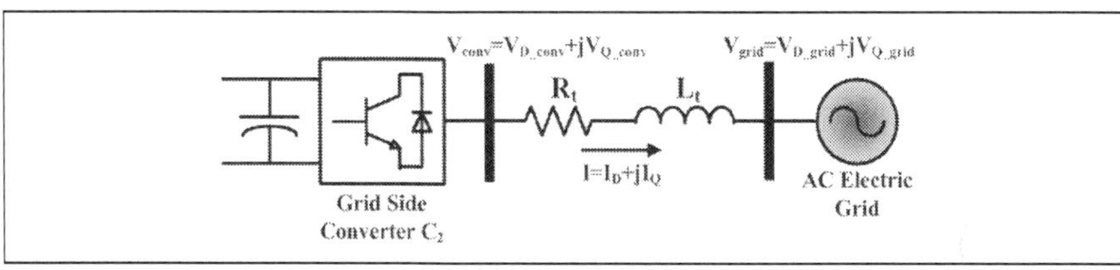

Fig. 5. Grid-Side Converter Schematic

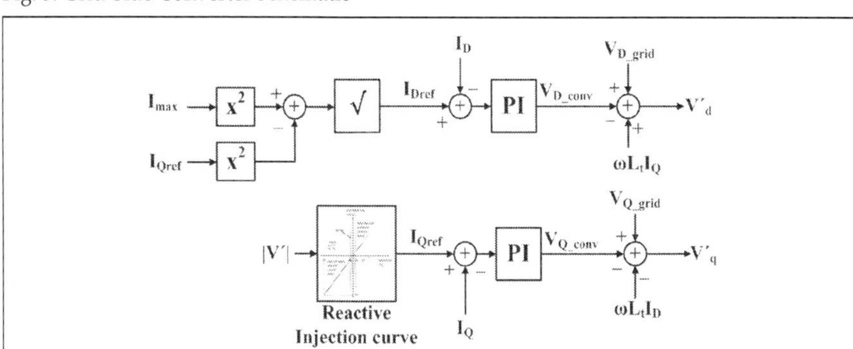

Fig. 6. Control Loop for the Grid-Side Converter

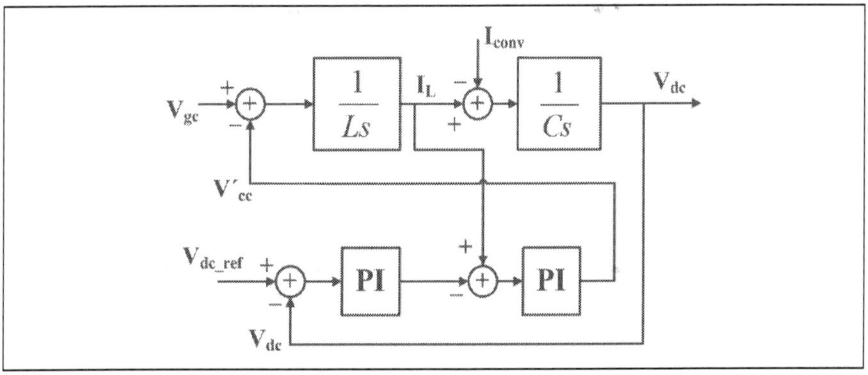

Fig. 7. Control Loop for the DC-Link Voltage

5. The sample power system

The electrical network used as a basis for this investigation is similar to that in [7]. For this system, a wind park is planned to be installed at bus 2 as illustrated in Fig. 8. The wind park to be connected is considered in this study as a dynamic equivalent, this way, an equivalent wind generator of 16MW and 400 V is considered. Each machine has a rated capacity of 2 MW and is designed to operate with rated terminal voltage of 400 V. The wind park is to be connected to the distribution electrical grid by 0.4/13.8 kV transformers. Besides the wind farm generation, two conventional diesel electric plants of 75 MVA and 36 MVA respectively connected to buses 1 and 3 are rated to supply the electric load of this system.

The model parameters of the speed and voltage regulators and synchronous generators of the diesel units were obtained from [9] and [10]. The equivalent automatic voltage regulator used is an IEEE Type 1 model. The equivalent primary machine of the synchronous generator and its speed governor are first-order models with proportional/integral frequency control [11], [12].

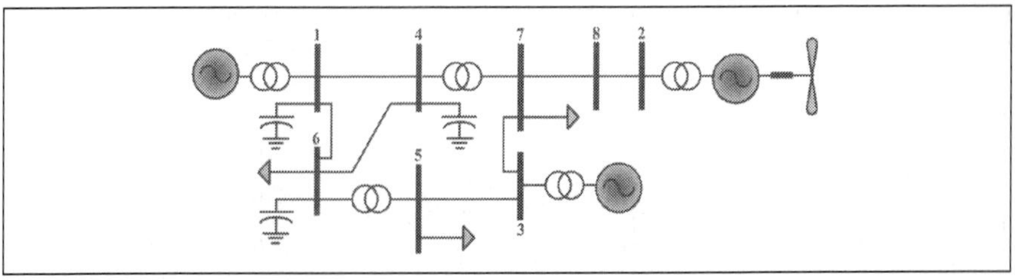

Fig. 8. Electrical Grid with Wind and Diesel Generation

6. Results

The simulation studies were performed considering a new operational practice that recommends to keep the wind generators in operation even during the fault period. The evaluation of reactive power injection during the fault will be performed on two conditions: the first is the adopted by Brazilian grid code which does not require the reactive power injection. The second injects reactive power following the EON curve.

6.1 Case I

The wind park generation at bus 2 was simulated using the full capacity of the park at 2 MW rated power on each synchronous generator. The short circuit was located near bus 1, starting at t = 1s, and lasting for 500 ms.

Fig. 9 shows the voltage profile on PCC during the fault period. It is observed that voltage dip is below the minimum of the ONS curve adopted in Brazil, consequently in this case the wind park is disconnect of the grid due the under voltage relay trip. For this case the voltage at the PCC does not return to nominal value due to instability problem on the system. The system transient stability was also affected due to wind park to be disconnected as shown in Fig. 10.

With the adoption of the reactive injection criteria it is observed that the voltage profile is over the limit of the ONS curve. Thus the wind park is maintained connected to the grid increasing the ride-through capacity.

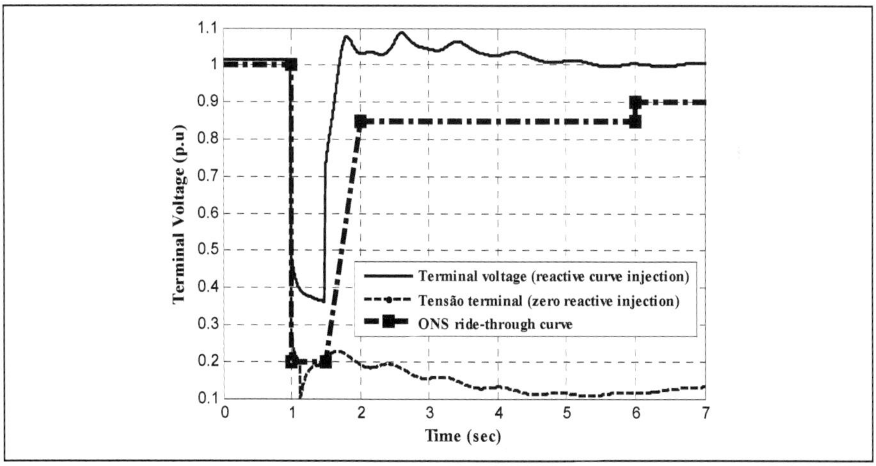

Fig. 9. Voltage Behaviour at PCC

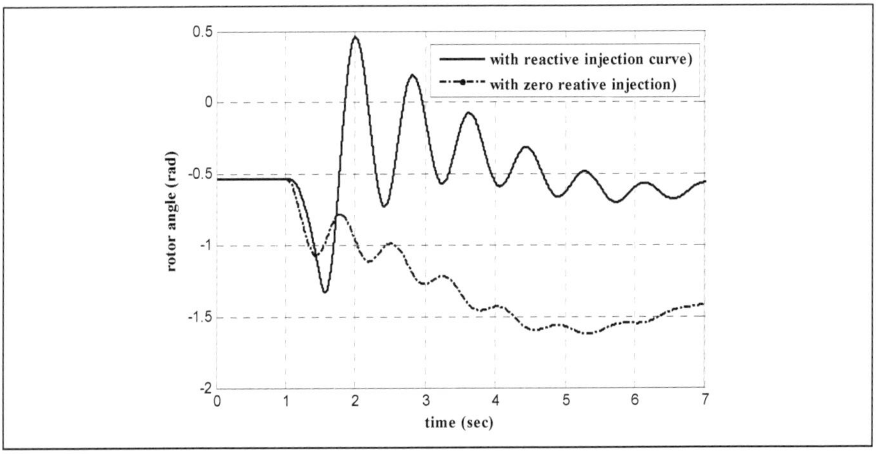

Fig. 10. Rotor Angle of the Synchronous Generator at bus 1.

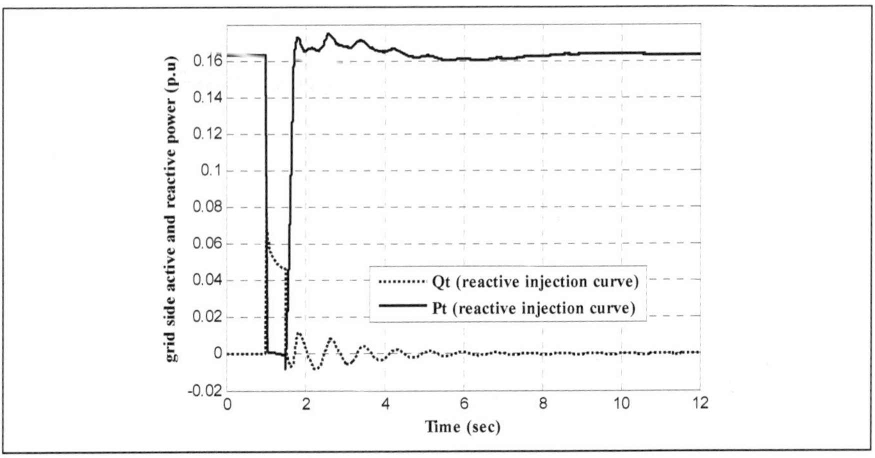

Fig. 11. Reactive and Active Power at PCC

6.2 Case II

The wind park generation at bus 6 was simulated using the full capacity of the wind park at 2 MW rated power on each synchronous generator. The short circuit was located near bus 6, starting at t = 1s, and lasting for 200 ms. Fig. 12 shows the voltage profile on PCC during the fault period. It is observed that the voltage dip is 0.65 p.u for the case of zero reactive injection, whereas for the case of following the reactive power injection curve the voltage dip is 0.8 p.u. and the reactive current support is provided within 20 ms after the fault detection in accordance with the E.ON regulation requirement. Thus the reactive power injection curve improves voltage profile and the ride-through capability.

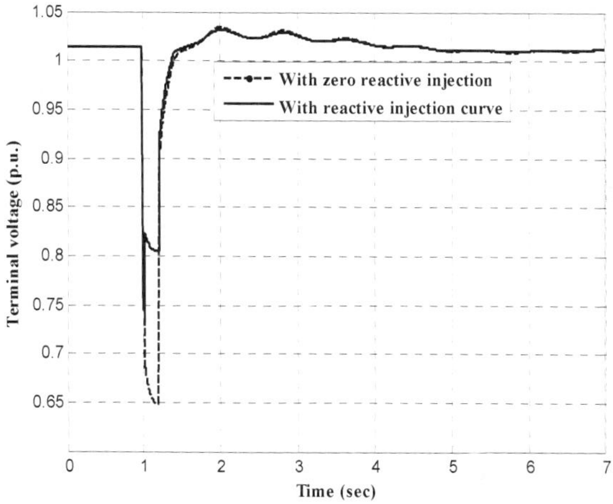

Fig. 12. Voltage Behaviour at PCC

Fig. 13 shows the converter reactive power which follows the reactive injection curve during the fault, improving the voltage profile as shown in Fig. 12.

The E.ON grid code requirements allow the active power injection by the converter when the voltage dip is above 0.5 p.u. as shown in Fig. 14. It is also observed after fault cleaning time a power injection ramp until the reference power. This method increases the transient stability margin as shown in Fig. 15, in which it is observed a small rotor angle oscillation with the reactive power injection curve.

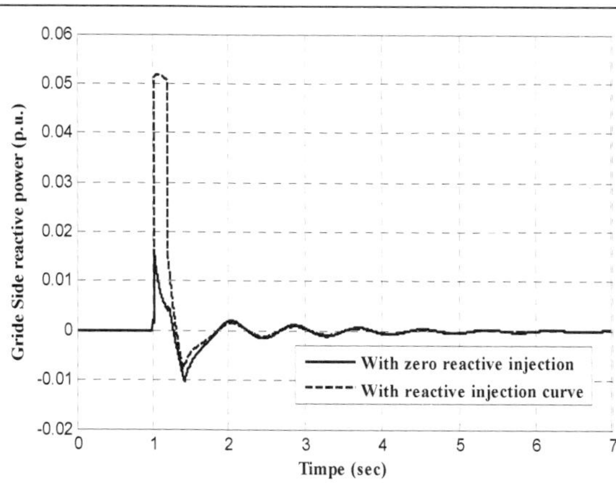

Fig. 13. Reactive Power at PCC

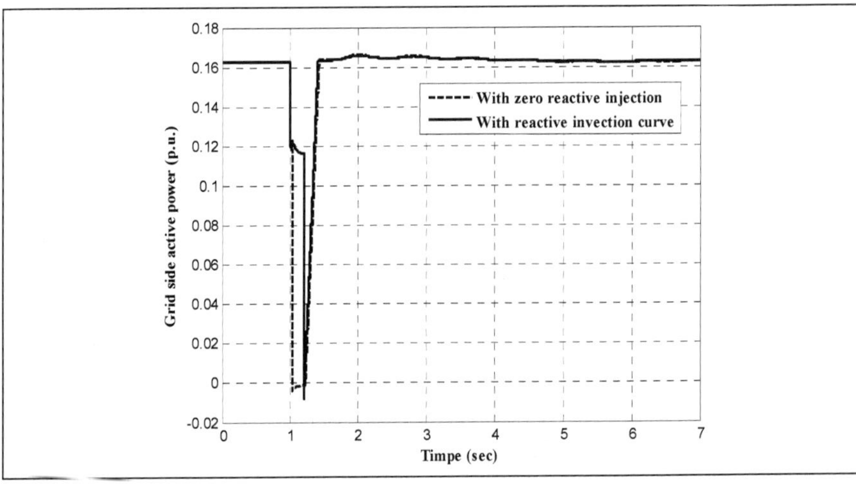

Fig. 14. Active Power at PCC

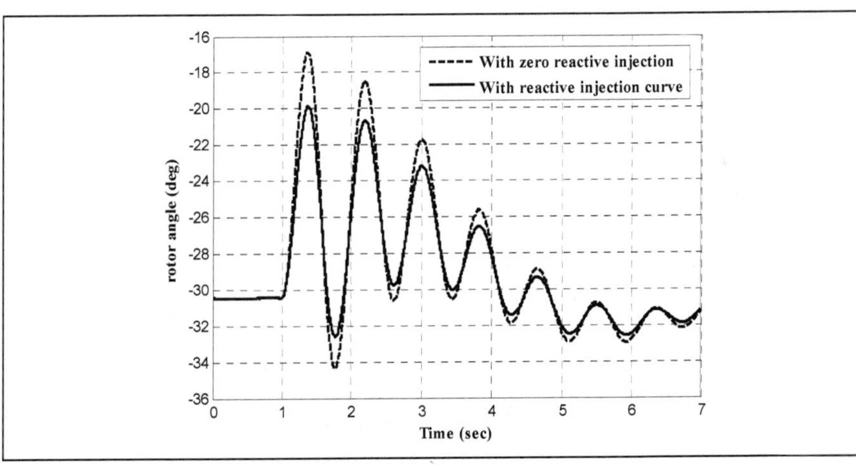

Fig. 15. Rotor Angle of the Synchronous Generator at Bus 1

7. Conclusion

In this chapter, it was explored the performance of alternative voltage control strategies applied to direct drive synchronous wind generators, more specifically with permanent magnetic (PMSG). The reactive power control of the grid-side converter is investigated for voltages purposes. The E.ON fault response code for wind farms is taken as the base case for the study. The simulated results presented in this chapter have considered that the proposed operational procedure has kept running during the fault period (ride-through capability) the wind generators, and also offers the possibility to supply reactive power during the voltage dip in order to facilitate voltage restoration. This is possible with the control of the grid side converter. The results have demonstrated that the consequence of this new approach is positive in the sense of maintaining transient voltage and rotor angle stability, once a variable speed wind generator technology, as direct drive synchronous generator is used.

8. References

[1] A.D. Hansen, G. Michalke, "Multi-pole permanent magnet synchronous generator wind turbines," grid support capability in uninterrupted operation during grid faults. IET Renew. Power Gener., 2009, Vol. 3, Iss. 3, pp. 333–348.

[2] A.D. Hansen, L.H. Hansen, "Wind turbine concepts market penetration over ten years (1995 to 2004)', Wind Energy, 2007, 10, (1), pp. 81–97

[3] J. P. A. Vieira, M. V. A. Nunes, A. C. Nascimento, S. R. Silva, U. H. Bezerra, M. F. Medeiros Júnior, "Analysis of Ride-Through With the Integration of Direct Drive Synchronous Wind Generators in Power Systems" In: VII INDUSCON, 2006, Recife. 2006. v1. p1-6.

[4] N. R. Ullah, T. Thiringer, "Variable speed wind turbines for power system stability enhancement" IEEE Trans. Energy Conv., vol. 22, no. 1, pp. 52-60, mar. 2007.

[5] A. Larson, A. Petersson, N. R. Ullah, O. Carlsson, "Krieger's flak wind farm" in Proc. Nordic Wind Power Conf. NWPC 2006, Helsinki, Finland, May 2006.

[6] J. Conroy, R. Watson, "Aggregate modelling of wind farms containing full-converter wind turbine generators with permanent magnet synchronous machines: transient stability studies" IET Renewable Power Generation. 2009, vol 3, N. 1, pp. 39-52.

[7] M. V. A. Nunes, J. A. Peças Lopes, H. H. Zurn, U. H. Bezerra, R. G. Almeida, "Influence of the Variable-Speed Wind Generators in Transient Stability Margin of The Conventional Generators Integrated in Electrical Grids" IEEE Transactions on Energy Conversion, Vol. 19, NO. 4, December 2004.

[8] Grid Code: High and Extra High Voltage, E. ON Netz GmbH Tech. Rep., 2006, Status: 1.

[9] P. Kundur, Power system stability and contol, Electric Power Research Institue McGrall-Hill, 1994.

[10] P. M. Anderson, A. A. Fouad, "Analysis of Faulted Power Systems" Ames, Iowa: Iowa State Univ. Press, 1995.

[11] G. N. Kariniotakis, G. S. Stavrakakis, "A general simulation for the accurate assessment of isolated diesel-wind turbines interaction, part I: a general multimachine power system model," IEEE Trans. Energy Conversion, vol. 10, pp. 577–583, Sept. 1995.

[12] N. Jenkins, J.B. Ekanayake, L. Holsworth, X. Wu, "Dynamic modeling of doubly fed induction generator wind turbines," IEEE Trans. Power Syst., vol. 18, pp. 803-809, May 2003.

22

Electromagnetic Calculation of a Wind Turbine Earthing System

Yasuda Yoh[1] and Fujii Toshiaki[2]
[1]*Kansai University*
[2]*Otowa Electric Co., Ltd.*
Japan

1. Introduction

Lightning protection systems (LPS) for wind power generation is becoming an important public issue since installations of wind turbines (WT) have greatly increased worldwide and their generating capacities are approaching 150 GW (GWEC 2010). WTs are often struck by lightning because of their open-air locations, special shapes and very high construction heights. As well as seriously damaging blades, accidents where low-voltage and control circuit breakdowns frequently occur in many wind farms. An earthing (grounding) system is one of the most important components required for appropriate LPSs in WTs and wind farms.

Japan, in particular, suffers from frequent and heavy lightning strikes, an example being the notorious „winter lightning" found in coastal areas of the Sea of Japan (Yokoyama 2002). Indeed, many turbines in Japan have been hit by lightning, and winter lightning poses a specific threat due to its intense power and electric current that are much higher than the world average (Shindo 2009, Natsuno 2010). Furthermore, due to its narrow landmass, wind turbines in Japan tend to be constructed in mountainous areas with high resistivity soil. Thus, earthing is one of the most important issues in Japan for protecting wind turbines from lightning.

Although IEC 61400-24:2010 („Wind turbine Generator System - Part 24: Lightning Protection") indicates a methodology for protection against such accidents, a standardised solution remains to be established. The earthing issue for a WT described in the IEC 61400-24 is completely subject to IEC 62305-3:2006 („Protection against lightning - Part 3: Physical damage to structures and life hazard"), which handles lightning protection for a general structure including a house and a building. The foundation and earthing system of a WT are generally much smaller than those of an ordinary building that is equivalent in height to a wind turbine. Furthermore, the lightning level for a WT is much higher than that of a normal house that has an equivalent foundation to a turbine. The smaller earthing system of a WT may not have enough capacity for lightning protection compared to conventional equipment.

In this chapter, the authors will discuss, using electromagnetic calculations, possible unexpected cases for the requirement regarding „the minimum length" of an earth electrode defined in IEC 61400-24 and IEC 62305-3.

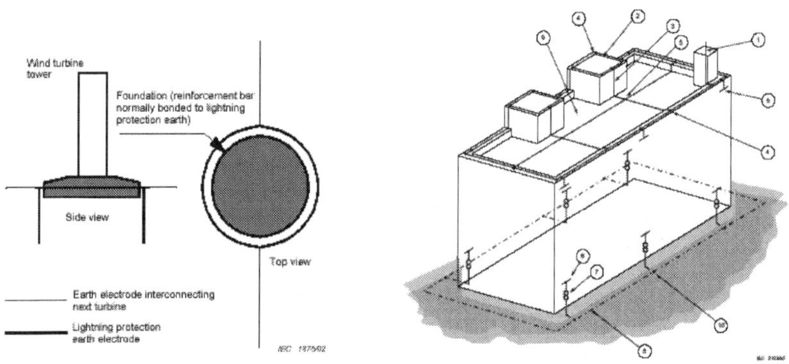

Fig. 1. Conceptual figure of a ring earth electrode. Left: for wind turbine (IEC 2010); Right: for ordinary building (IEC 2006)

Symbols and abbreviations

l_r [m]:	horizontal length of buried electrode
l_v [m]:	vertical length of buried electrode
l_1 [m]:	minimum length defined in IEC 62305-3
r [m]:	radius of hemispherical electrode
r_e [m]:	equivalent radius of assumed hemispherical electrode
t [s]:	time
x, y, z [m]	lengths in 3-dimensional space
n:	number of iterations in the time domain
i, j, k:	number of iterations in the space domain
$\mathbf{E}$ [V/m]:	electric field (3-dimensional vector)
$\mathbf{H}$ [A/m]:	magnetic field (3-dimensional vector)
$\mathbf{D}$ [C/m^2]:	magnetic flux density (3-dimensional vector)
$\boldsymbol{B}$ [T]:	magnetic flux density (3-dimentional vector)
ε [F/m]	permittivity
σ [S/m]	conductivity (inverse of resistivity)
μ [H/m]:	permeability
ρ [Ω m]:	resistivity of soil (inverse of conductivity)
R [Ω]:	resistance to earth (in this Chapter, simply „earth resistance")
d [m]:	depth of buried ring earth electrode, or length of vertical rods

CIGRE	International Council on Large Electric Systems
EM:	Electro-Magnetic (*adj.*)
EPR:	Earth Potential Rise
FDTD:	Finite Difference Time Domain method
IEC:	International Electrotechnical Commission
LPS:	Lightning Protection System
WT:	Wind Turbine

2. Basic concept of an earthing system for lightning protection

2.1 What is "earthing"?

Before starting a discussion on wind turbine earthing, we need to review briefly the meaning of „earthing" to ensure a common understanding.

When people use the word „earthing" (or „grounding") in every day conversation, they probably mean that an electric device should be connected to the earth (ground) via a cable or a rod to achieve „zero voltage" at an earth terminal (or a chassis). This understanding may be quite acceptable for „normal cases" like electric leakage protection or an electric noise reduction. However, strictly speaking, „zero voltage" cannot be realised by earthing. This fact would be very evident in an „abnormal" case such as surge invasion due to lightning.

Theoretically, the earth resistance of an electrode can be determined by the integration of a function including the resistivity of the surrounding soil and the size (i.e. radius) of the electrode. As the resistivity of the actual soil can never be zero, it is a natural consequence that the earth resistance can never be zero. Incidentally, it is obvious that the earth voltage V [V] of the terminal of the earthing electrode is solved using the simple Ohm's law of $V = IR$, where I [A] is an inrush current such as a lightning surge. Suppose the earth resistance is 10 Ω, which is normally the safe value as defined in many standards and regulations, then if a lightning current of 30 kA is the inrush current to the earthing electrode, the earthing voltage of the electrode, namely an earthing potential rise (EPR), would rise to 300 kV! This is a basic concept in earthing for a LPS.

It follows from the above description that the best way to implement a LPS is to reduce the earth resistance to as low as possible. However, in an actual WT site, in addition to the conventional electrical equipment the cost of reducing the earth resistance may be excessive because of the need for additional construction work involving large or long conductive materials, a significant amount of labour and special heavy machinery. Thus, many international or domestic standards and regulations normally set the minimum safe value for the steady resistance to be 10 Ω. It is therefore important to estimate the resistance of the earthing system as accurately as possible before WT construction.

2.2 Theory of earth resistance

Theoretically, earth resistance is a function of the resistivity of the surrounding soil and the size of the electrode. As a normal resistance is required, earth resistance R is defined as follows (note that the range of integration is from the origin infinity):

$$R = \int_0^{\infty} \frac{\rho}{S(r)} dr. \tag{1}$$

Assuming the simple case of a hemispherical electrode with radius r [m] and the surrounding soil is homogeneous, with a resistivity ρ [Ωm] as shown Fig. 2, the following can solve the earth resistance:

$$\begin{aligned} R &= \int_0^{\infty} \frac{\rho}{2\pi r^2} dr \\ &= \frac{\rho}{2\pi r}. \end{aligned} \tag{2}$$

The earth resistance therefore is a function of the resistivity of the surrounding soil and the size (i.e. radius) of the electrode. As shown in the above equation, there must be an earth resistance with a non-zero value.

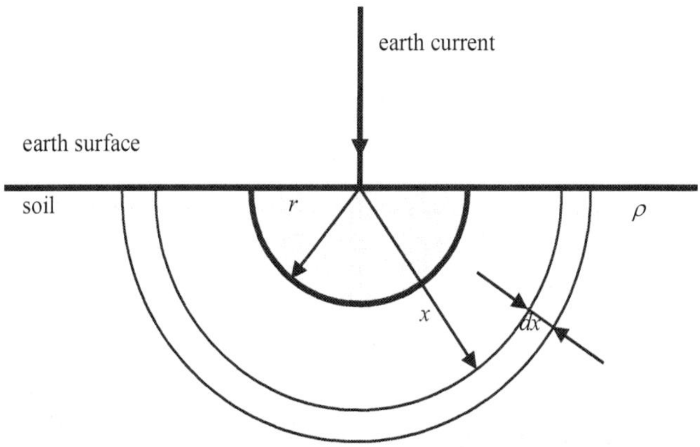

Fig. 2. Conceptual illustration of earth resistance in case of a hemispherical electrode

Similarly, the earth resistance of a vertical conductor of radius r [m] and length (depth) d [m] buried in the soil is approximately defined as

$$R = \frac{\rho}{2\pi d} \ln\left(\frac{2d}{r}\right), \tag{3}$$

and that of a ring earth electrode whose outer radius, inner radius and buried depth is r_o, r_i, and d [m] is proposed by Sunde (Sunde 1949) as follows:

$$R = \frac{\rho}{2\pi r_o} \ln\left(\frac{8r_o}{\sqrt{2r_i d}}\right). \tag{4}$$

Note that the above equations are just theoretical or approximate calculating equations for the typical shapes of various types of electrode. As it is normal that varying types of electrodes are combined in practical use, it is difficult to estimate correctly an earthing value of an arbitrary electrode with a complex shape. Although there are several theories concerning a combination effect or an adjacent effect, no equations have been proposed to universally express a complex shape. Furthermore, it is not always satisfactory to consider only the steady value of the earth resistance. For EPR, it is important to consider transient voltage under a lightning impulse lasting up to 10 μs. This is why numerical calculations including the FDTD method are important to calculate the earthing electrode and design an accurate earthing system for LPS.

2.3 Wind turbine earthing system described in IEC standards

According to IEC 61400-24, a „Type A arrangement" (with vertical and/or horizontal electrode) and „Type B arrangement" (with ring earth electrode) are recommended for wind turbine earthing. The type B arrangement is described as „*this type of arrangement comprises a ring earth electrode external to the structure in contact with the soil for at least 80 % of its total length or a foundation earth electrode. Additional vertical and horizontal earth electrodes may be used in combination with the ring electrode. The electrode should be buried to a depth of at least 0.5 m.*" This arrangement was originally defined as an earthing method for ordinary houses or buildings in IEC 62305-3:2006 (originally IEC 61024-1-2, which was abolished and the revised version was re-numbered as the current standard). The concept of the earth electrode is to create equipotential bonding surrounding a house or a building to avoid values of step and touch voltages that conventionally are considered dangerous.

On the other hand, IEC 62305-3 states that, "*For the ring earth electrode (or foundation earth electrode), the mean radius r_e of the area enclosed by the ring earth electrode (or foundation earth electrode) shall be not less than the value l_1:*

$$r_e \geq l_1 \tag{5}$$

where l_1 is represented in Figure 2 (note by authors: in this text, the figure is denoted as Fig. 3) *according to LPS class I, II, III and IV.*"

IEC 62305-3 also continues with, "*When the required value of l_1 is larger than the convenient value of r_e, additional horizontal or vertical (or inclined) electrodes shall be added with individual lengths l_r (horizontal) and l_v (vertical) given by the following equations:*

$$l_r = l_1 - r_e \tag{6}$$

and

$$l_v = (l_1 - r_e) / 2. \tag{7}$$

It is recommended that the number of electrodes shall be not less than the number of the down-conductors, with a minimum of two."

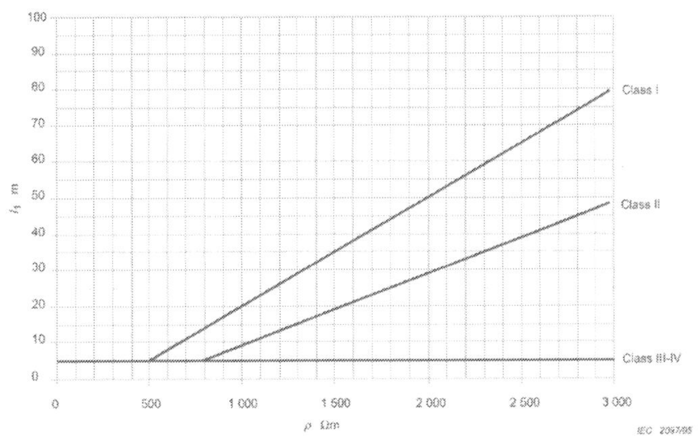

Fig. 3. Minimum length l_1 defined in IEC 62305-3 (from IEC 2006)

Here, note that there is no information about the installing location of the additional electrodes attaching to the ring earth electrode evident in IEC 62305-3. It is considered that

the reason for this is that it does not matter whether the additional electrodes are attached to a conventional structure such as a building because the structure has relatively wide foundations. Moreover, it is normal for the ring earth electrode to be installed relatively close to the foundations because of land area limitations. In contrast, the foundation of a wind turbine is comparatively small and therefore the ring earth electrode must be installed as far away from the original foundation as possible. In this situation, it is possible that the installed location of the additional electrodes could be very sensitive.
In this chapter, the minimum length of the electrodes will be researched in detail. A numerical calculation using the FDTD (Finite Difference Time Domain) method is employed to clarify how the size and the location of the attachment of the additional electrodes will affect the earth resistance. This study not only shows the unexpected inappropriate cases but also proposes an improved recommendation, particularly for a wind turbine earthing system.

3. FDTD electromagnetic calculation

A Finite Difference Time Domain (FDTD) method is a computing calculation algorithm in which Maxwell's electromagnetic equations are computationally treated as difference equations in both the time and space domains. While the FDTD method was initially applied to electromagnetic field analysis around an antenna (Yee 1966, Kunz 1993), with the increased CPU power in PC machines, various investigations into high voltage engineering including lightning surge and earth system analysis have also employed the algorithm.

3.1 Theory of Fine Difference Time Domain (FDTD) method

In the FDTD method, an analysis domain surrounding a wave source and the measured objects is assumed. The domain is divided into a small rectangular solid, which is called a „cell". The following Maxwell differential equations, Eqs. (8) and (9), are directly applied to all the cells.

$$\mathrm{rot}\mathbf{E}(\mathbf{r},t) = -\frac{\partial \mathbf{B}(\mathbf{r},t)}{\partial t}, \tag{8}$$

$$\mathrm{rot}\mathbf{H}(\mathbf{r},t) = -\frac{\partial \mathbf{D}(\mathbf{r},t)}{\partial t} + \mathbf{J}(\mathbf{r},t). \tag{9}$$

In the actual calculation, Maxwell's equations are arranged as a first-order central difference approximation called Yee's algorithm (Yee 1966) and the magnetic and electric fields are calculated step by step as shown in Fig. 4. For example, an electric field E^n is calculated from E^{n-1} at $t = (n-1)\Delta t$ and a magnetic field $\boldsymbol{H}^{n-1/2}$ at $t = (n-1/2)\Delta t$. In addition, $\mathbf{H}^{n+1/2}$ is generated from $\mathbf{H}^{n-1/2}$ and E^n. Using Maxwellian constitutive equations $\mathbf{B} = \mu\mathrm{H}$, $\mathrm{D} = \varepsilon\mathrm{E}$ and $\mathrm{J} = \sigma\mathrm{E}$ under the assumption of isotropic and non-dispersion media, Eqs. (8) and (9) can be transformed to the following equations:

$$\frac{\partial \mathbf{H}}{\partial t} = -\frac{1}{\mu}\nabla\times\mathbf{E}, \tag{10}$$

$$\frac{\partial \mathbf{E}}{\partial t} = -\frac{\sigma}{\varepsilon}\mathbf{E} + \frac{1}{\varepsilon}\nabla\times\mathbf{H}. \tag{11}$$

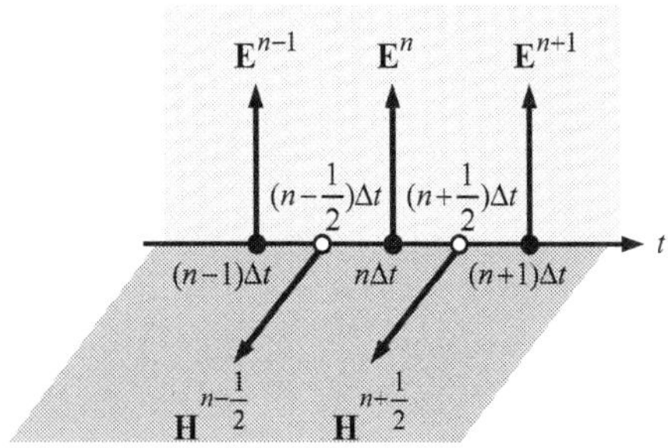

Fig. 4. Arrangement of the electric field **E** and magnetic field **H** in the time difference domain

Eqs. (10) and (11) can be converted to Eqs. (12) and (13) by difference approximation.

$$\frac{\mathbf{H}^{n+\frac{1}{2}}-\mathbf{H}^{n-\frac{1}{2}}}{\Delta t}=-\frac{1}{\mu}\nabla\times\mathbf{E}^{n}, \tag{12}$$

$$\frac{\mathbf{E}^{n}-\mathbf{E}^{n-1}}{\Delta t}=-\frac{\sigma}{\varepsilon}\mathbf{E}^{n-\frac{1}{2}}+\frac{1}{\varepsilon}\nabla\times\mathbf{H}^{n-\frac{1}{2}}. \tag{13}$$

So, a recurrence formula for the magnetic field **H** can be simply expressed as follows:

$$\mathbf{H}^{n+\frac{1}{2}}-\mathbf{H}^{n-\frac{1}{2}}=-\frac{\Delta t}{\mu}\nabla\times\mathbf{E}^{n}. \tag{14}$$

On the other hand, in the electric field, $\mathbf{E}^{n-1/2}$ cannot exist because each electric field is defined only at integer time. Consequently, approximating it as the average of $\mathbf{E}^{n}$ and $\mathbf{E}^{n-1}$, a recurrence formula for the electric field *E* is given as follows:

$$\mathbf{E}^{n}-\frac{1-\frac{\sigma\Delta t}{2\varepsilon}}{1+\frac{\sigma\Delta t}{2\varepsilon}}\mathbf{E}^{n-1}=\frac{\frac{\Delta t}{\varepsilon}}{1+\frac{\sigma\Delta t}{2\varepsilon}}\nabla\times\mathbf{H}^{n-\frac{1}{2}}. \tag{15}$$

As can be seen from Eqs. (14) and (15), in the FDTD methods, the electric field E^{n} is generated from the previous half step of the electric field $\mathbf{E}^{n-1}$ at $t = (n-1)\Delta t$ and the magnetic field $\mathbf{H}^{n-1/2}$ at $t = (n-1/2)\Delta t$. Likewise, the magnetic field $\mathbf{H}^{n+1/2}$ is calculated from the previous half step of the electric field E^{n} and the magnetic field $\mathbf{H}^{n-1/2}$.

The next step of Yee's Algorithm is a difference formulation and an arrangement of an electric field and a magnetic field in the space domain as shown in Fig. 5. The alternate arrangement of the electric field and the magnetic field replicates exactly the concept of the original Maxwell's equation whose physical meaning is that „a rotation of an electric field forms a magnetic field and a rotation of a magnetic field forms the electric field". Consequently, a magnetic field at an arbitrary point at an arbitrary time can be expressed by electric fields at a neighbour point as follows:

$$H_x^{n+\frac{1}{2}}\left(i, j+\frac{1}{2}, k+\frac{1}{2}\right) - H_x^{n-\frac{1}{2}}\left(i, j+\frac{1}{2}, k+\frac{1}{2}\right)$$
$$= -\frac{\Delta t}{\mu}\left[\frac{E_z^n\left(i, j+1, k+\frac{1}{2}\right) - E_z^n\left(i, j, k+\frac{1}{2}\right)}{\Delta y} - \frac{E_y^n\left(i, j+\frac{1}{2}, k+1\right) - E_y^n\left(i, j+\frac{1}{2}, k\right)}{\Delta z}\right]. \tag{16}$$

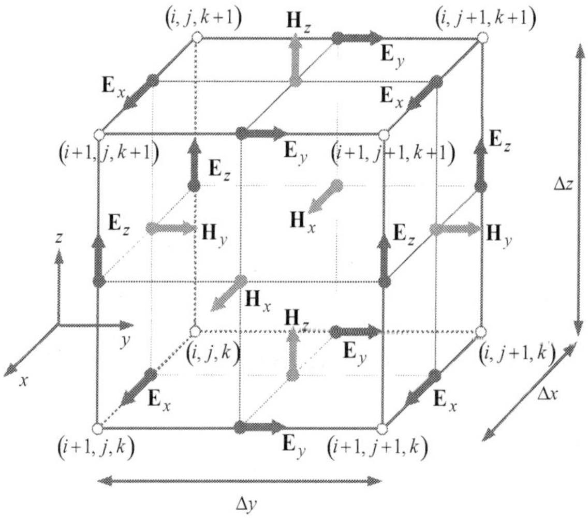

Fig. 5. Arrangement in the difference space domain of an electric field **E** and a magnetic field **H**

In the same manner, it is possible to calculate a magnetic field at an arbitrary point at an arbitrary time as:

$$E_z^n\left(i, j, k+\frac{1}{2}\right) - \frac{1-\frac{\sigma\Delta t}{2\varepsilon}}{1+\frac{\sigma\Delta t}{2\varepsilon}} E_z^{n-1}\left(i, j, k+\frac{1}{2}\right)$$
$$= -\frac{1-\frac{\sigma\Delta t}{2\varepsilon}}{1+\frac{\sigma\Delta t}{2\varepsilon}}\left[\frac{H_y^{n-\frac{1}{2}}\left(i+\frac{1}{2}, j, k+\frac{1}{2}\right) - H_y^{n-\frac{1}{2}}\left(i-\frac{1}{2}, j, k+\frac{1}{2}\right)}{\Delta x}\right. \tag{17}$$
$$\left. - \frac{H_x^{n-\frac{1}{2}}\left(i, j+\frac{1}{2}, k+\frac{1}{2}\right) - H_x^{n-\frac{1}{2}}\left(i, j-\frac{1}{2}, k+\frac{1}{2}\right)}{\Delta y}\right].$$

3.2 Applications of the FDTD method for lightning protection of a wind turbine

As mentioned above, early applications of the FDTD method focused on the electromagnetic analysis of antenna. Some of the earliest reports on applying it to electric power apparatus were published by Tanabe, which discussed the electromagnetic field propagation in soil from a buried vertical rod when a lightning surge was imposed (Tanabe 2000, Tanabe 2001). As the calculation power of PCs has dramatically increased since early 2000, many reports and papers on surge analysis using the FDTD method have been published, especially by Japanese researchers. Noda proposed a novel method, which described a thin wire, such as an overhead wire, and an underground cable in distribution lines (Noda 2002). Moreover, a thin-wire representation in a non-quadric grid (Taniguchi 2008a), and an algorithm to calculate a circular object in cylindrical coordinates (Taniguchi 2008b) were initially proposed and discussed. A research group in the Central Research Institute of Electric Power Industry (CRIEPI), Japan developed a commercial software named „VSTL" (Virtual Surge Test Lab.), which specialised in the surge analysis of electric power apparatus based on the FDTD method (Noda 2005).

Only a few papers on applications to WT-LPS have been reported since the start of 2000. Yamamoto investigated the state of surge propagation in a wind turbine including blades, down-conductors, the nacelle, the tower, and an earthing system, comparing actual measurements using a downsized WT model and FDTD calculations (Yamamoto 2009, Yamamoto 2010). An investigation group of Doshisha University, Japan, calculated the surge propagation from a turbine when struck by lightning, to another turbine via a buried interconnecting earthing wire (Nagao 2009). The authors' investigation group also stored knowledge and results on surge analysis and an earthing design of a WT (Yasuda 2007a, Yasuda 2007b, Fujii 2009). In the following sections, the latest results by the authors will be presented.

Numerical calculations on WT-LPS are not limited to the FDTD method. Several early researchers have raised important questions about WT earthing and its LPS. A report from a joint research group at UMIST (University of Manchester Institute of Technology) and the National Technical University of Athens is one of the earliest on the subject (Hatziargyriou 1997, Cotton 1997, Cotton 1999, Lorenzou 2000), where they compared the results by EMTP and a software package named CEDGS based on a numerical integration method in the frequency domain. Good examples of later investigations included the use of the MoM (moment method) (Lewke 2006), the FEM (finite element method) (Muto 2010), CEDGS (Kontargyri 2005, Elmghairbi 2009), EMTP (electromagnetic transient program) and its related software (Yasuda 2008), and an algebraic analysis based on a travelling-wave theory (Hermoso 2006, Sekioka 2010).

4. Analysis of Wind Turbine Earthing using FDTD Calculation I: (Evaluation of the effect of a ring earth electrode)

As mentioned in the Introduction, a ring earth electrode was originally installed for use with conventional buildings and households to reduce a touch and a step voltage mainly for human safety. Though the original purpose was effective for a WT earthing system, the ring earth electrode was expected to have the effect of reducing not only a touch and a step voltage but also a steady resistance and an earth potential rise (EPR). This is because a ring electrode for WT is normally installed in a much wider area than for conventional buildings and households (see Fig. 1 and relative discussion in Sec. 1). In this section, the evaluation of the effect of the ring earth electrode for WT is discussed especially from the aspect of a steady resistance and an EPR. The results introduced in this section are mainly the outcomes from the authors' paper (Fujii 2009).

4.1 Models of the WT foundation for FDTD calculation

For this modeling, initially, a simplified ideal foundation as shown in Fig. 6(a) is adopted with the following assumptions:

i. a WT tower is not considered and a lightning current is simulated as a direct inrush at the top surface of the foundation,
ii. the foundation is made of several blocks of rectangular solids and the area of the base is 12 m × 12 m (which simulates the normal foundation of a 2 MW class WT), and
iii. a reinforced bar in the foundation is simulated by a copper frame surrounding the foundation.

Other details of conditions in the present FDTD calculations are shown in Table 1. In addition, Fig. 6(b) and (c) show the case with four vertical rods, and the case with an outer ring earth electrode, x m on a side.

domain of space	50 m × 50 m × 110 m	
space step size	0.50 m (x and y = 100 splits, z = 220 splits)	
time step size	4.5×10^{-10} s (satisfying Courant's stable condition)	
relative permittivity	air	1
	soil	10
	concrete	6
conductivity	air	0
	soil	3.33×10^{-3} S/m
	concrete	58×10^{-4} S/m
	conductor (copper)	58×10^{6} S/m
permeability	air	$4\pi\times10^{-7}$ H/m
velocity of electromagnetic wave	3×10^{8} m/s	
lightning: lump wave	crest width	1 μs
	wave tail	70 μs
	crest peak	30 kA
soil resistivity (varying parameter)	100, 500, 1000, 1500, 2000 Ωm	
equivalent radius of thin wire for thin-wire approximation	0.1149 m = 0.2298 Δx (Baba 2005)	
boundary condition	the second-order Liao's absorbing condition (Liao 1984)	
measurement method of EPR	integration of electric field	

Table 1. Parameters for the present FDTD calculation

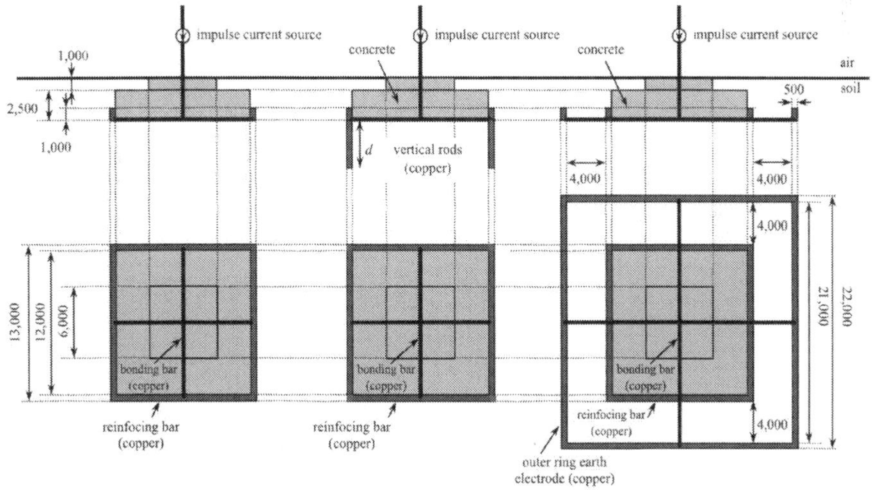

Fig. 6. Analysis models of various WT earthing systems

4.2 Qualitative observation of the FDTD calculation

Since the FDTD calculation has a significant advantage for calculating Maxwell's equation directly in the time domain, it is very easy to check how an electric field is distributed at an arbitrary time. Figure 7 illustrates contour plots of an electric field around each foundation model, using the FDTD calculation. In Fig. 7(a), the immediate area around the original foundation is painted red, which means the electric field is up to 1 × 10^5 V/m. This result indicates that it is not enough to use only the foundation and reinforcing bars as an earth system for lightning protection. In contrast, Figs. 7(b) and (c) show the relatively lower

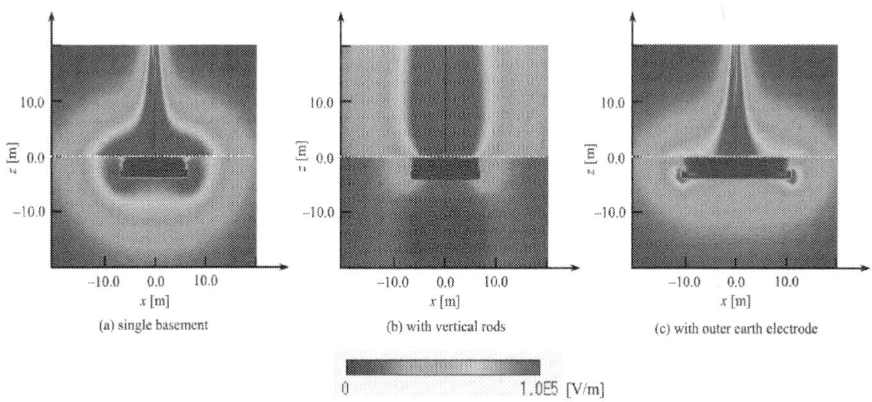

Fig. 7. Contour plots of electric field with each earthing model (ρ = 2000 Ωm, t = 2.00 μs)

electric field in the soil due to the auxiliary vertical rods or the outer ring earth electrode (Note that the buried vertical rods are invisible in Fig. 7(b) because the graph is a cross-section at $y = 0$). This suggests effective suppression of EPR during a lightning surge can be expected when an auxiliary electrode is employed. Note that, from the original view point of the touch and step voltage, the result of the case with the ring earth electrode (Fig. 7(c)) shows that the potential difference on the earth surface clearly improved because of the ring electrode.

4.3 Transient analysis of the various earthing systems

4.3.1 Transient analysis for the original foundation case

To observe the FDTD results in more detail, the calculated waveforms of EPR (curves of simultaneous potentials on the top surface of the foundation) and the simultaneous impedance (curves of simultaneous quotients of simultaneous potentials and simultaneous input current) of the earthing system are as illustrated in Figs. 8, 9 and 10. In this analysis, the parameter of soil resistivity is assumed to be 100, 500, 1000, 1500 and 2000 Ωm.
Although the waveforms in Fig. 8(b) have steep peaks up to 60 Ω at about $t = 0.1$ μs, they are not considered essential in the present discussion. This is because they are determined by division calculations whose denominators are almost zero at that time and may have almost no influence to the potential rise at the time. In fact, it is evident that the transient elevation of the EPR waveform at $t = 0.1$ μs cannot be seen in Fig. 8(a). The problem should be considered as the special case of a very steep current rise in sub-micro seconds at a subsequent lightning stroke. Disregarding the steep peak at about $t = 0.1$ μs, it is shown that curves in cases of less than 2000 Ωm have creeping inductive characteristics, and that the case of 2000 Ωm has a slightly capacitive characteristics before $t = 1.0$ μs. After $t = 1.0$ μs, all the curves clearly have resistive characteristics with a flat and stable behaviour.

4.3.2 Transient analysis for the case with vertical rods

In contrast, Fig. 9 shows the results from the case with four vertical rods as shown in Fig. 6(b). In the present analysis, the varying parameter was set to be d [m] which was the buried depth of each vertical rod and the calculations were performed under the condition of a soil resistivity of 2000 Ωm. The calculation was performed to clarify the effect of vertical rods. As can be seen in Fig. 9(a), with every parameter EPR curves indicate the effect of vertical rods compared with the case without rods. It is also found that the simultaneous impedance curves have moderate peaks and show slight inductivity.
However, in the case of rods of more than 30 m, the effect of holding the EPR down is not seen any more and steady values of impedance *i.e.* earth resistance converge to a standard value. This is very similar to a result from a conventional transmission tower foundation with vertical rods. Thus, it becomes clear that vertical rods are effective for a WT-LPS for moderating both steady resistance and a rise in inductive potential. However, rods that are too long may not be cost-effective or realistic for reducing EPR and the steady earth resistance.

4.3.3 Transient analysis in the case with a ring earth electrode

The result in the case with a ring earth electrode as shown in Fig. 6(c) is presented in Fig. 10. In this analysis, the varying parameter is a side of the square electrode. Comparing the curves in the two graphs, it is evident that the larger the ring electrode, the lower the EPR,

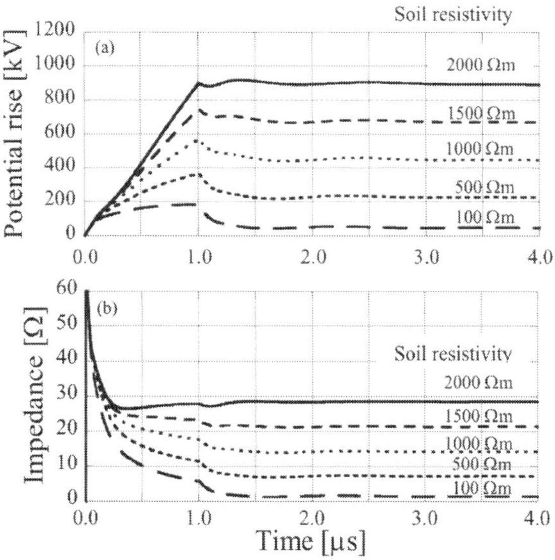

Fig. 8. Transient waveforms of Model (a): standard foundation. (upper graph: earth potential rise, lower graph: simultaneous impedance)

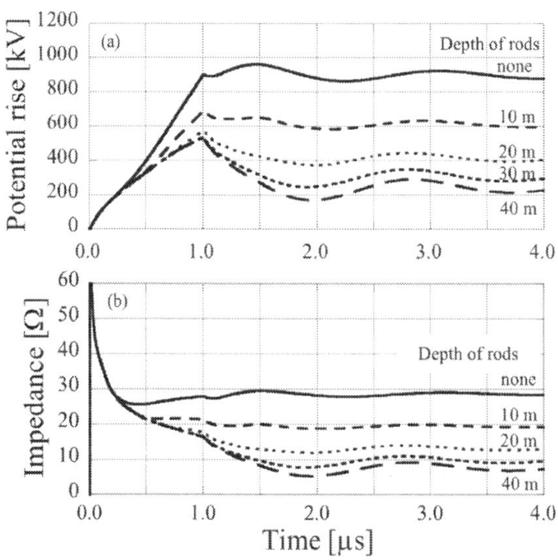

Fig. 9. Transient waveforms of Model (b): foundation with vertical rods. (upper graph: earth potential rise, lower graph: simultaneous impedance)

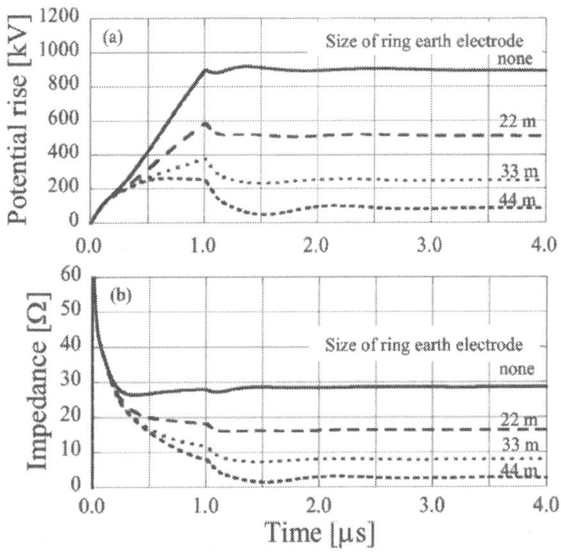

Fig. 10. Transient waveforms of Model (c): foundation with ring earth electrode. (upper graph: earth potential rise, lower graph: simultaneous impedance)

effectively suppressing the steady impedance. This indicates that a ring earth electrode has sufficient influence to prevent an EPR due to a lightning strike, and its effect can be considered as almost equal to that of vertical rods.

From Fig. 10(b), it is evident that curves in cases smaller than 22 m have moderate inductive characteristics, while a curve in the case of 22 m has almost resistive or slightly capacitive characteristics. The potential rise shown in Fig. 10(a) is much lower than originally speculated and can be considered of no concern when protecting electrical and electronic devices in a WT.

Furthermore, to examine in detail the EPR suppression effect of the ring earth electrode, a summarised result of all the parameters of soil resistivity is set out in Fig. 11. The graph shows that the higher the soil resistivity, the greater the effect that can be expected from the ring electrode. This is considered to be because the effect of the inductivity tends to directly affect the EPR in low resistivity soil. Figure 12 shows another summarised graph of the reduction ratio of EPR and earth impedance compared with that of the original foundation. As can clearly be seen in the graph, the reduction ratio of the steady value of earth resistance always stays around 40 %; therefore, it is evident that an almost-half suppressing effect can be expected in every resistivity case using a 22 m square ring earth electrode. By contrast, the higher the resistivity, the lower the effective suppression of the reduction ratio of the EPR. This can also be explained as due to the relative strength of the inductive element in cases of lower resistivity. However, in all cases the EPR due to the inductive characteristic of a ring earth electrode tends to be moderate, and can be adequately suppressed. Thus, it is confirmed that a ring earth electrode as well as the vertical rods can be expected to prevent any potential rise due to a lightning surge.

Note that other shaped earth electrodes, *i.e.* hexagonal and octagonal, were evaluated by the authors, and their results show almost the same tendency to suppress EPR and provide steady resistance.

4.4 Comparison of vertical rods and the ring earth electrode

As mentioned above, vertical rods are an effective auxiliary electrode, but might be a much more expensive method when an installation goes to several tens of meters in depth. By contrast, a ring earth electrode is a relatively inexpensive method that can be buried horizontally in shallow soil, spreading the electrode bonding to the reinforcing bars as part of the construction system. Figure 13 shows two graphs comparing a case with vertical rods and one with a ring earth electrode at 2000 Ωm soil resistivity. From these graphs, the ring earth electrode of about 20 m square is expected to have an equivalent or superior effect to vertical rods going to about 20 m in depth, in reducing the temporal EPR. It should be noted is that there is a tendency for EPR to reduce according to the depth of the rods. As shown in the left line graph in Fig. 13(a), few reduction effects on EPR can be expected if the rods are longer than 30 m. On the other hand, the straight bar graph in Fig. 13(b) shows the reduction effect by the ring earth is more likely over a long period. Therefore, it can be concluded that a ring earth electrode installed in a wide area is very effective for suppressing EPR.

Likewise, a similar conclusion on the steady earth resistance can be deduced from Fig. 13(b). The reduction effect of the steady resistance using a 22 m square electrode is equal to that with 15 m rods. Furthermore, the reduction effect with a 33 m square is just same as that with 45 m rods! Although it is not easy to compare construction costs per meter, it is evident that a ring earth electrode is one of the best solutions for suppressing steady resistance safely and cost-effectively.

Although both auxiliary electrodes are individually calculated in the present analysis, parallel usage is expected to be more effective in reducing the steady resistance of the earthing system of a WT. Thus, a practical combination of a wide ring earth electrode and relatively short vertical electrodes holds promise as a cost-effective earthing system for wind turbine lightning protection. Detailed discussion follows.

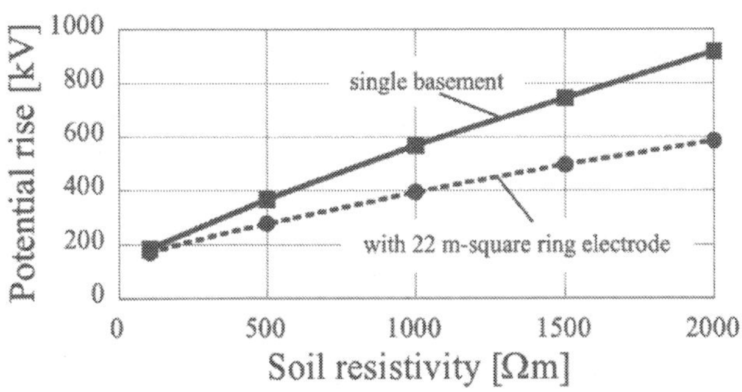

Fig. 11. Comparison of the peak voltage with and without a ring earth electrode.

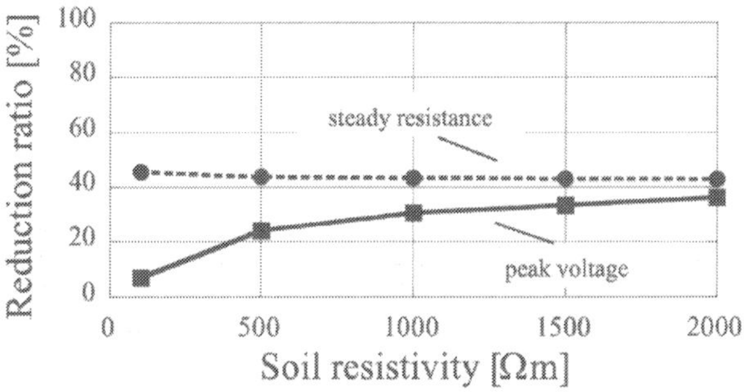

Fig. 12. Reduction ratio of the peak voltage and steady resistance.

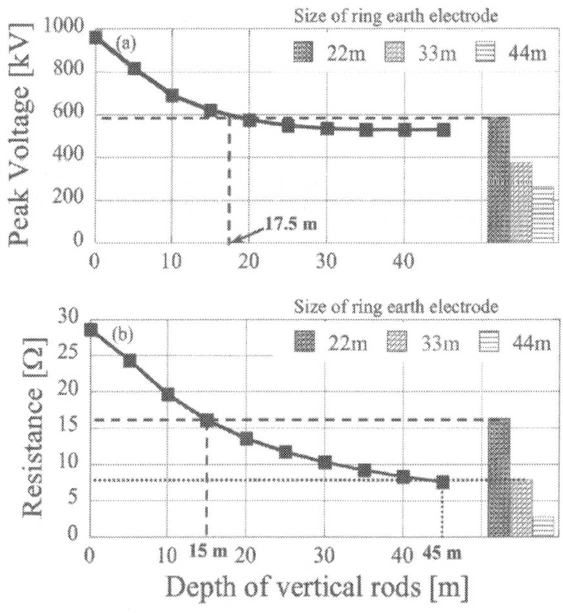

Fig. 13. Comparison between vertical rods and ring earth electrode. ((a) Peak voltage, (b) Steady resistance).

5. Analysis of wind turbine earthing using FDTD Calculation II (Combining the effect of a ring earth electrode with vertical rods)

As discussed in Section 2, the earthing system of a WT is defined in IEC 61400-24:2010, almost all of which was followed by IEC 62305-3:2006 which covered all electrical apparatus. Both the IEC standards state „the minimum length" of the Type A arrangement (*i.e.* vertical rods) when it is combined with the Type B arrangement (*i.e.* ring earth electrode) as shown in Fig. 3, Section 2. Interestingly, although the conductors are limited in number as in „*with a minimum of two"* to determine the minimum length of the Type A arrangement, the location of the conductors is not stated in the standards at all. This would not be a serious problem for conventional buildings and households or power apparatus such as power stations and power plants, but, is it really suitable for WTs even with new apparatus with a special shape and the special characteristic of a LPS?

Though a few reports have noted the problem, the essential questions do not appear to have been resolved. As mentioned previously, a WT sometimes has a wide ring earth electrode compared to its small foundation. This is because the area occupied by a WT is normally much larger than that of a conventional building and it is relatively easy to bury electrodes horizontally. Moreover, compared with the conventional power apparatus, which normally has a mesh electrode, it is not realistic for the total site of the WT or a wind farm to be covered with a mesh electrode due to the high construction costs. The question arises as to whether the location of the installed conductors really affects the determination of the minimum length and what combination of the ring electrode and vertical rods has the most effect on LPS for a WT. These questions need to be resolved to establish safe countermeasures against lightning accidents. The discussion in this section is mainly based on the authors' paper (Yasuda 2007b).

5.1 Models for a WT foundation

Figure 14 shows models of a simplified WT earthing system with a square foundation. Conditions for the FDTD calculations in this section are almost the same as those in the previous section as shown in Table 1. In addition, Fig. 14 shows the various combinations of vertical rods and a ring earth electrode. It has already been confirmed that few differences were apparent between the results for the square ring electrode and those of other shaped rings such as hexagonal, octagonal or circular. In the present investigation, two models varying the attaching location are prepared as follows;

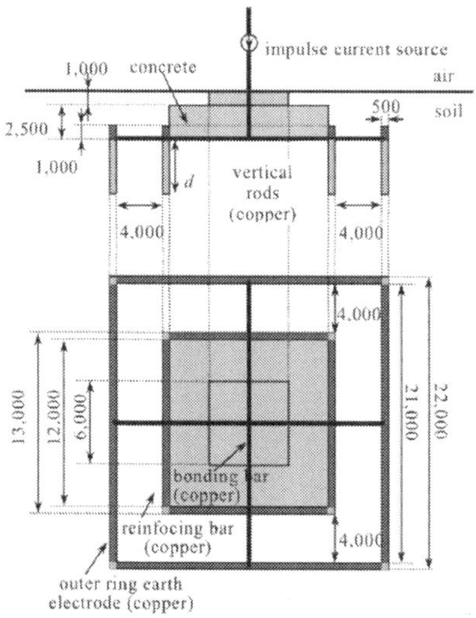

Fig. 14. FDTD model of wind turbine foundation and earthing system

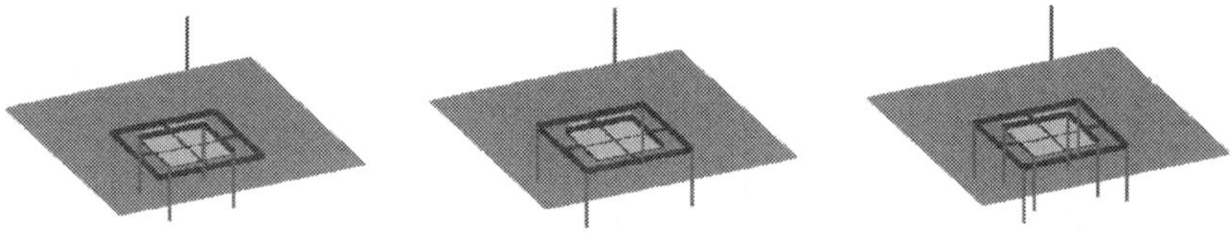

Fig. 15. Three case studies for the present FDTD calculation.

- Case I: four (4) vertical rods are installed on the bottom of the foundation's four corners,
- Case II: four (4) rods are installed on the bottom of the four corners of the outer ring earth electrode.
- Case III: eight (8) rods are installed on the bottom of every corners of both the inner and outer ring earth electrodes.

5.2 Results for the FDTD calculations

The results in each case are shown in Fig. 16. Comparing the curves for Case I and Case II, it is clear that the inner vertical rods (Case I) have less effect than the outer rods (Case II) in reducing earth resistance. For example, to reduce the earth resistance of a WT foundation with a 25 m square ring earth electrode to less than 10 Ω, only 20 m outer rods or 35 m inner rods are needed. Thus, the auxiliary vertical rods are better installed on the bottom of four corners of the *outer* ring electrode.

Considering Case III which is a combined configuration of Case I and Case II, it is evident that the effect of the auxiliary vertical electrodes has further increased. In this case, the vertical rods are required to be no more than 15 m to achieve the 10 Ω earth resistance. The cost of material for Case III with 8 rods would be twice that for Case II with 4 rods. However, because of heavy construction equipment and the construction period, the deeper a vertical rod is installed, the higher the installation cost. Thus, the results for Case III suggest that it is possible to install vertical rods cost effectively.

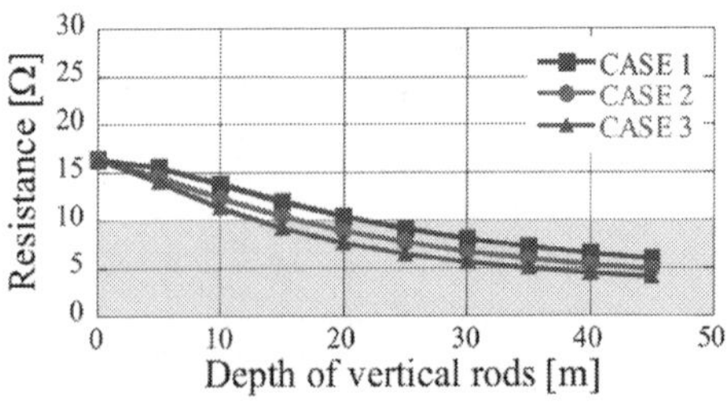

Fig. 16. Three case studies for the present FDTD calculation.

Detailed calculations using the above three models obtained the results shown in Fig. 17. To obtain the curves, a significant number of FDTD calculations were performed. Initially, a certain length l_v for the vertical electrodes was assumed for the case of a certain resistivity ρ. If the earth resistance results from the FDTD calculation as a function of the assumed length l_v is higher than 10 Ω, the next calculation with the longer length is performed. If a required result where the resistance is equal to, or lower than, 10 Ω is found, the minimum length for the case of the given resistivity can be obtained according to Eq. (7). In the next step, by varying the resistivity ρ, the above calculation was succeeded. Meanwhile, r_e is an equivalent radius of the earth electrode, which is calculated as 12.4 m against the 22 m square ring earth electrode in the present analysis.

According to Eq. (7), the plot for each case can be put down in Fig. 17. If Case I is chosen (with four rods on the bottom of the foundation), the individual length of the vertical rods is required to be more than 60 m to achieve 10 Ω earth resistance in the present condition. This goes far beyond the required minimum length defined in IEC TR61400-24 and IEC 62305-3. Case II (with four rods on the bottom of the outer ring) seems to barely satisfy the IEC requirement and Case III (with eight rods) clearly stays within the recommended area.

In the graph, a solid line denotes the minimum length defined in IEC 62305-3 (similar to the top line „Class I" as shown in Fig. 3). Comparing the solid line and the dotted results of Case I, it is clear that an unexpected inappropriate result was obtained in Case I, which is the worst case where the additional vertical electrodes are installed below the inner foundation rather than the outer ring electrode.

In a conventional building, the above problem occurs merely because it is not normal to install a ring electrode away from the foundation and, inevitably, few differences in the results would occur between Case I and Case II. In contrast, the construction situation may be completely different for a WT, where the ring earth electrode should be installed as far away from the foundation as possible. Furthermore, turbine developers tend not to dig new deep shafts below the outer ring earth electrode but use the foundation piles of the turbine as vertical electrodes to reduce their additional construction costs for LPS. Consequently, as can be seen in Fig. 17, there is a real possibility that an unexpected inappropriate case (i.e. Case I) would occur, even if the required condition in IEC 62305-3 is completely satisfied.

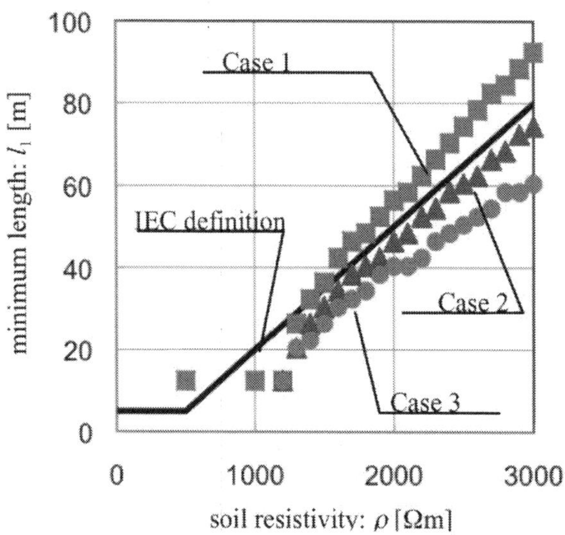

Fig. 17. Comparison between IEC definition and the FDTD results.

5.3 Final discussion

In this section, the authors show the possibility of unexpected inappropriate cases for wind turbine earthing systems with regard to the minimum length of additional electrodes defined in IEC 62305-3. After the electromagnetic calculation using the FDTD method, it became clear that an inappropriate case might possibly occur in the wind turbine earthing system.
Though the case shown in Fig. 17 is a relatively extreme case, a high-resistivity soil of over 1500 Ωm is often seen in mountainous areas of Japan, in the flat lands of Brazil and other countries with complex geophysical characteristics. Note that this does not mean there are serious defects in the present IEC 61400-24 and IEC 62305 series. But, in some special cases such as a WT, which is considered to be a special shape compared to conventional electrical apparatus, careful attention must be paid to the design of an earthing system according to site observations, when constructed on very high resistivity soil.

6. Conclusions

In this chapter, the authors showed the importance of an earthing system for lightening protection of a wind turbine. As an example of a numerical calculation for the earthing design of a WT, the Finite Difference Time Domain (FDTD) method was introduced. Furthermore, the latest examples of the FDTD method applied to WT earthing analysis were shown and discussed. The effect of a ring earth electrode was also evaluated from the perspective of an EPR and a steady resistance.
Earthing systems and lightning protection are both old and new problems. While a recognised theory that was proposed in the first half of the 20th century is still available, new computing methods are now being applied to this field. While much has been solved by developments in technology, a lot remains unsolved. As a wind turbine is a relative newcomer in the field of conventional buildings and power apparatus, the proliferation in its use seems to add another layer of confusion to the issue. One of the authors is a member of the „Investigation Committee on Earthing Systems of Wind Turbines" (convener: Prof. Sekioka, Shonan Institute of Technology, Japan) in the Institute for Electrical Engineers, Japan, where numerous papers and reports that have been published worldwide are researched and summarised (Sekioka 2007). The committee plans to release its final report written in Japanese in the autumn of 2011 and an English translation (or an abridged translation) is expected to be published. Moreover, one of the authors also attends an international committee in CIGRE named „WG C4.409 (Lightning Protection of Wind Turbine Blades)" (convener: Dr. Yokoyama, CRIEPI, Japan), whose main research area is not WT earthing but the WT blade (Hermoso 2010). The final report is planned for completion in the Spring of 2011 and its summarised paper will be published in the Electra Journal.

7. References

Baba, Y.; Nagaoka, N. & Ametani A. (2005). Modeling of Thin Wires in a Lossy Medium for FDTD Simulations, *IEEE Trans. on Electromagnetic Compatibility*, Vol.47, No.1, pp. 54-60

Cotton, I. & Jenkins, N. (1997). The Effects of Lightning on Structures and Establishing the Level of Risk, *Proceedings of IEE Half-Day Colloquium on Lightning Protection of Wind Turbines*, No.3

Cotton, I. & Jenkins, N. (1999). Windfarm Earthing, *Proceedings of European Wind Energy Conference (EWEC1999)*, pp. 725-728

Elmghairbi, A.; Haddad, A. & Griffiths, H. (2009). Potential Rise and Safety Voltages of Wind Turbine Earthing Systems under Transient Conditions, *Proceedings of 20th International Conference on Electricity Distribution (CIRED2009)*, pp. 8-11, June 2009.

Fujii, T.; Yasuda, Y. & Ueda, T. (2009). Electromagnetic Analysis of Ring Earth Electrode of Wind Turbine, *IEEJ Transaction on Power and Energy*, Vol.129, No.8, pp.1047-1055 (in Japanese)

GWEC (2010). *Global wind power boom continues despite economic woes*, http://www.gwec.net/fileadmin/documents/PressReleases/PR_2010/Annex%20stats%20PR%202009.pdf , Global Wind Energy Council

Hatziargvriou, N.; Lorentzou, M.; Cotton, I. & Jenkins, N. (1997). Wind Farm Earthing, *Proceedings of IEE Half-Day Colloquium on Lightning Protection of Wind Turbines*, No.6

Hermoso, B. (2006). Wind Farm Earthing Installations: Rated and Lightning Frequencies Behaviour, *Proceedings of International Conference on Grounding and Earthing (GROUND'2006)*, Maceió, November 2006, pp.411-414

Hermoso, B. & Yokoyama, S. (2010). A Review of Research Methods for Lightning Protection in Wind Turbine Blades and Activity of CIGRE WG C4.409, *Proceedings of 30th International Conference on Lightning Protection (ICLP2010)*, No.9A-1157, Cagliari, September 2010

IEC (2006). *Protection against lightning – Part 3: Physical damage to structures and life hazard*, IEC 62305-3, Ed. 1.0(b), International Electro-technical Commission, Geneva

IEC (2010). *Wind Turbine Generation System - 24: Lightning Protection*, IEC61400-24, International Electro-technical Commission, Geneva

Kontargyri V.T., Gonos I.F., Stathopulos I.A. (2005). Frequency Response of Grounding Systems for Wind Turbine Generators, *Proceedings of the 14th International Symposium on High-Voltage Engineering (ISH 2005)*, Beijing, August 2005, No.B-13

Kunz, K. S. & Luebbers, R. J. (1993). *The Finite Difference Time Domain Method for Electromagnetics*, CRC Press, ISBN 978-0849386572

Lewke, B.; Krug, F. & Kindersberger, J. (2006). Risk of Lightning Strike to Wind Turbines for Maintenance Personnel Inside the Hub, *Proceedings of 28th International Conference on Lightning Protection (ICLP2006)*, No.XI-9, Kanazawa, September 2006.

Liao, Z. P.; Wong, H. L.; Yang, B.-P. & Yuan, Y.-F. (1984). A transmitting boundary for transient wave analysis, *Science Sinica*, Vol.27, No.10, pp. 1063-1076

Lorentzou, M.; Hatziargyriou, N. & Papadias, B. (2000). Analysis of Wind Turbine Grounding Systems, *Proceeding of 10th Mediterranean Electrotechnical Conference (MELECON2000)*, pp. 936-939, Cyprus, May 2000, ISBN 0-7803-6290-X, IEEE, New York.

Muto, A.; Suzuki, J. & Ueda, T. (2010). Performance Comparison of Wind Turbine Blade Receptor for Lightning Protection, *Proceedings of 30th International Conference on Lightning Protection (ICLP2010)*, No.9A-1263, Cagliari, September 2010

Nagao, M.; Baba, Y.; Nagaoka, N. & Ametani, A. (2009). FDTD Electromagnetic Analysis of a Wind Turbine Generator Tower struck by Lighting, *IEEJ Transaction of Electric and Power*, Vol.129, No.10, pp. 1181-1186

Natsuno, D.; Yokoyama, S.; Shindo, T.; Ishii, M. & Shiraishi, H. (2010). Guideline for Lightning Protection of Wind Turbines in Japan, *Proceedings of 30th International Conference on Lightning Protection (ICLP2010)*, No.SSA-1259, Cagliari, September 2010

Noda, T. & Yokoyama, S. (2002). Thin Wire Representation in Finite Difference Time Domain Surge Simulation, *IEEE Transactions on Power Delivery*, Vol.17, No.3, pp. 840-847

Noda, T.; Tatematsu, A. & Yokoyama, S. (2002). Code and Its Application to the Lightning Overvoltage Calculation of a Transmission Tower, *Proceedings of the International Conference on Power Systems Transients (IPST'05)*, Montreal, June 2005, Paper No. IPST05-138

Sekioka S.; Yamamoto K.; Minowa M. & Yokoyama S. (2007). Damages in Japanese Wind Turbine Generator Systems due to Winter Lightning, *Proceedings of IX International Symposium on Lightning Protection (IX SIPDA)*, Foz do Iguaçu, November 2007

Sekioka, S. & Funabashi, T. (2010). A Study on Effective Length for Practical Design of Grounding System in a Wind Turbine, *Proceedings of 30th International Conference on Lightning Protection (ICLP2010)*, No.5B-1085, Cagliari, September, 2010,

Shindo, T.; Suda, T. (2009). Lightning Risk of Wind Turbine Generator System, *IEEJ Transaction on Power and Energy*, Vol.129, No.10, pp.1219-1224

Sunde, E. D. (1949). *Earth Conduction Effects in Transmission Systems*, D. Van Nostrand Co., New York.

Tanabe, K. (2000). Calculation Results for Dynamic Behavior of Grounding Systems Obtained Using the FD-TD Method, *Proceeding of 25th International Conference on Lightning Protection (ICLP2000)*, Athens, September 2000, pp. 452-457

Tanabe, K. (2001). Novel method for analyzing the transient behavior of grounding systems based on the finite-difference time-domain method, *Proceedings of IEEE Power Engineering Society Winter Meeting*, Vol. 3, pp. 1128–1132

Tanabe, K. & Asakawa A. (2003). Computer Analysis of Transient Performance of Grounding Grid Element Based on the Finite-Difference Time-Domain Method, *IEEE Transaction on Power Energy*, Vol.120-B, No.8/9, pp. 209-212

Taniguchi, Y.; Baba, Y.; Nagaoka, N. & Ametani, A. (2008a). Modification on a Thin-Wire Representation for FDTD Calculations in Nonsquare Grids, *IEEE Transactions on Electromagnetic Compatibility*, Vol.50, No.2, pp. 427 - 431, ISSN 0018-9375

Taniguchi, Y.; Baba, Y.; Nagaoka, N. & Ametani, A. (2008b). Representation of an Arbitrary-Radius Wire for FDTD Calculations in the 2-D Cylindrical Coordinate System, *IEEE Transactions on Electromagnetic Compatibility*, Vol.50, No.4, pp. 1014 - 1018, ISSN 0018-9375

Yamamoto, K.; Yokoyama, S. & Ametani, A. (2009). Experimental and analytical studies of lightning overvoltages in wind turbine generator systems, *Electric Power Systems Research*, Vol.79, No.3, pp. 436-442

Yamamoto, K.; Yanagawa, S.; Yamabuki, K.; Sekioka, S. & Yokoyama, S. (2010). Analytical Surveys of Transient and Frequency-Dependent Grounding Characteristics of a Wind Turbine Generator System on the Basis of Field Tests, *IEEE Transactions on Power Delivery*, Vol.25, No.4, pp. 3035-3043, ISSN 0885-8977

Yasuda, Y.; Fuji, T. & Ueda, T. (2007a). Transient Analysis of Ring Earth Electrode for Wind Turbine, *Proceedings of European Wind Energy Conference (EWEC2007)*, No.BL3.212, Milan, May 2007

Yasuda, Y.; Fuji, T. & Ueda, T. (2007b). How does Ring Earth Electrode effect to Wing Turbine?, *Proceeding of 42st International Universities Power Engineering Conference (UPEC2007)*, Brighton, September 2007, pp. 796-800.

Yasuda, Y.; Uno, N.; Kobayashi, H. & Funabashi, T. (2008). Surge Analysis on Wind Farm When Winter Lightning Strikes, *IEEE Transactions on Energy Conversion*, Vol.23, No.1, pp. 257-262, ISSN 0885-8969

Yee, K. S. (1966). Numerical solution of initial boundary value problems involving Maxwell's equations in isotropic media, *IEEE Transaction on Antennas and Propagation*, Vol. 14, No.3, pp. 302-307.

Yokoyama, S. (2002). Lightning detection and lightning protection of power systems in Japan, *Proceedings of IEEE/PES Asia Pacific Transmission and Distribution Conference and Exhibition 2002*, Vol.1, pp. 546-551

23

Rotor Speed Stability Analysis of a Constant Speed Wind Turbine Generator

Mitalkumar Kanabar and Prof. Srikrishna Khaparde
Indian Institute of Technology Bombay
India

1. Introduction

As wind turbine generator (WTG) technology is one of the fastest growing renewable energy technologies, the focus is given towards the cost-benefit analysis (Agalgaonkar et al., 2006); as well as, study of its specific grid integration issues (Zavadil et al., 2005). Many countries have their own grid codes (rules and regulations) to integrate WTG into the utility grid. Most common grid codes for WTG include low voltage ride through (LVRT) capability, voltage control, power quality, and protection requirements. AWEA recommended adoption of an LVRT requirement developed by E.ON Netz as shown in Fig. 1. Whereas, WECC (Western Electricity Coordinating Council) has put effort to lenient this stringent requirement in May 2005.

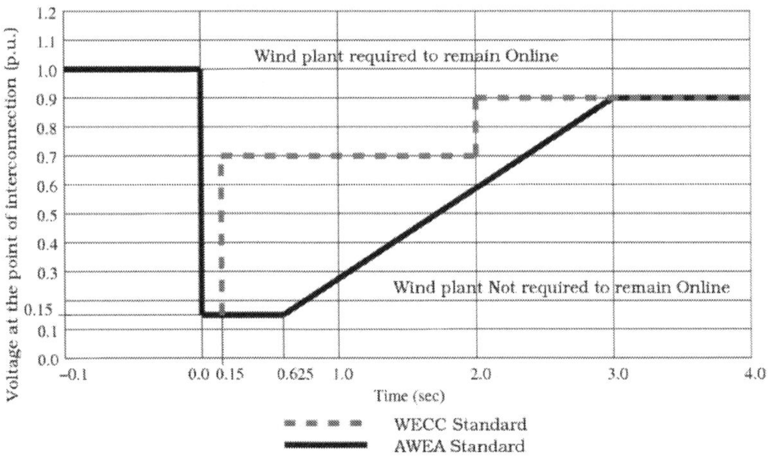

Fig. 1. LVRT requirements for wind generation facilities [4].

There are two major types of WTG technology: constant speed WTG and variable speed WTG (Ackermann, 2005). The major earlier installations in developing countries are based on the constant speed WTG technology because it is robust, economical, and simple in design (Kanabar & Khaparde, 2008). India ranks fifth in the world with a total installed capacity of more than 11,500 MW by the end of March, 2010 (Wind Power, 2010). Most of the WTGs currently installed in India are fixed or constant speed induction generators. However, unlike variable speed WTG, the constant speed WTG does not satisfy LVRT requirements that are because during a nearby fault, rotor accelerates to very high value and hence the WTG becomes unstable (Kanabar et al., 2006). This phenomenon is referred to as rotor speed instability (Samuelsson & Lindahl, 2005).

Several literatures are available on stability analysis of a constant speed WTG. The study of transient stability of a constant speed WTG using dynamic simulation has been presented in (Rodriguez et al., 2002) with case study on Spanish system. The comparison of transient stability margin between constant speed and variable speed WTGs has been discussed in (Nunes et al., 2003). It has been shown that a constant speed WTG has a much lesser transient stability margin as compared with a variable speed WTG. Reference (Chompoo-inwai et al., 2005) examines the response of a constant speed WTG during faults and the possible impacts on the system stability when the percentage of wind generation increases. Moreover, literature also suggests controlling wind turbine blade angles to stabilize the generator during fault conditions. A comparative study between active stall control and pitch control of a constant speed WTGs are discussed in (Ackermann, 2005). It can be concluded from the literature that to meet the LVRT requirements (according to grid codes), rotor speed stability margin of a constant speed WTG should be improved. There are two different methods to control a constant speed WTG: 1) providing additional reactive power support for improving the terminal voltage, this in turn, will increase the electromagnetic torque and hence, the rotor acceleration can be reduced; 2) mechanical torque control by pitching the blades during a system disturbance can be used to reduce the rotor acceleration. There are several already installed constant speed WTGs without this blade pitch control functionality. Therefore, first method of providing additional reactive power would be suitable for them. On the other hand, many modern wind turbine blades have pitch control mechanism. And hence, second method would be suitable for them.

Section-2 of the chapter presents the detailed analytical formulae derived for rotor speed stability margin to determine the exact amount of additional reactive power support required to achieve the LVRT capabilities. Using simulation of the dynamic model in MATLAB, it has been shown that the calculated value of reactive power is sufficient for the WTG to comply with the LVRT requirements by enhancing rotor speed stability. Section-3 demonstrates the implementation of the active stall controller for the constant speed WTG enhances, and explains how this method can enhance rotor speed stability to meet the LVRT requirements with the help of simulation using DIgSILENT software.

2. Enhancement of rotor speed stability margin of a constant speed WTG using additional reactive power compensation

A constant speed WTG consumes reactive power and hence, shunt capacitor banks are connected to it to supply reactive power locally. Conventionally, the nominal rating of a capacitor bank selected is to compensate for no-load reactive power demand (Jenkinds et al., 2000). However, as real power is exported, additional reactive power is drawn from the network. This reactive power consumption ramps up drastically during faults due to acceleration of the rotor. Consequently, it has to be disconnected from the grid due to rotor speed instability. Therefore, a constant speed WTG without pitch control does not possess LVRT capability. During a fault, if additional shunt capacitor banks are connected, then the recovery of voltage and electromagnetic torque can be improved. Thereby, rotor acceleration can be reduced. This will increase the critical clearing time of a constant speed WTG. This section will quantify the amount of additional reactive power support required to improve the critical clearing slip and time such that it can meet the LVRT requirements. Further, the effect of wind velocity and rotor inertia constant on the critical clearing slip and time have also been discussed.

2.1 Steady state analysis of a constant speed WTG

Fig. 2 shows a single line diagram of a sample system with a constant speed WTG and capacitor banks connected to an infinite bus through a step-up power transformer.

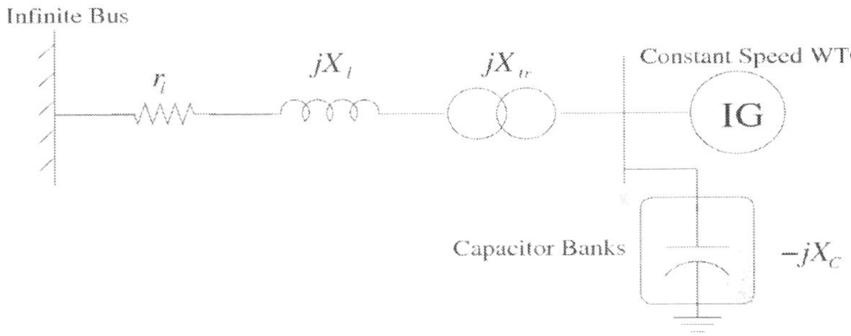

Fig. 2. Single line diagram of a sample system.

To calculate the exact amount of reactive power support required to satisfy the LVRT requirement, it is necessary to obtain a relation between the critical slip (scr) and the reactive power support in steady state.

Let us consider the steady-state equivalent model of a constant speed induction generator as shown in Fig. 3. In this figure, r_l and X_l are the line resistance and reactance respectively; Xtr is the transformer reactance; X_C is the capacitive reactance; r_s and X_s are the stator resistance and reactance respectively; X_m is the magnetizing reactance; r'_r and X'_r are the rotor resistance and reactance referred to the stator side respectively and s is the rotor slip. All these above quantities are in per unit.

To obtain the torque-slip characteristics of a constant speed WTG, a Thevenin equivalent has been derived across points A and B, as shown in Figs. 3 and 4.

The formula to calculate the Thevenin's voltage is indicated below:

$$V_{th} = \frac{V X_C X_m}{(r_l + jX_{lt})[r_s + j(X_s + X_m - X_C)] + X_C(X_s + X_m) - jX_C r_s} \quad (1)$$

Where,

$$X_{lt} = X_l + X_{tr}$$

The values of the Thevenin resistance (rth) and Thevenin reactance (xth) are obtained as follows:

$$r_{th} = \frac{a \cdot c + b \cdot d}{c^2 + d^2} \quad (2)$$

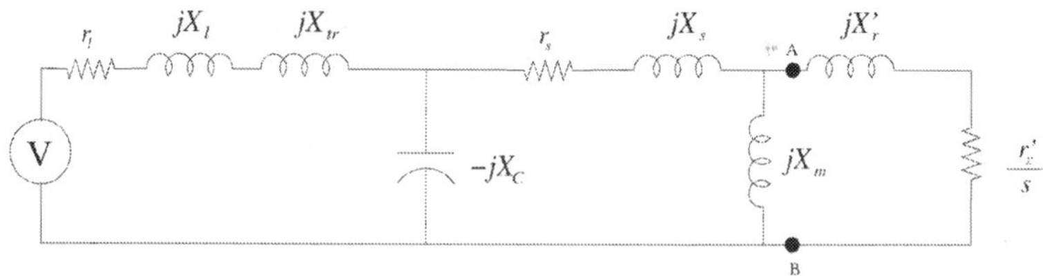

Fig. 3. Equivalent circuit diagram of a sample system.

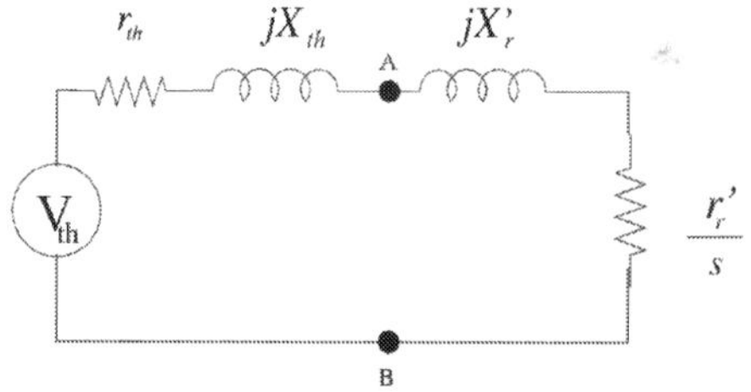

Fig. 4. Thevenin's equivalent circuit.

$$X_{th} = \frac{b \cdot c - a \cdot d}{c^2 + d^2} \tag{3}$$

Where,

$$B_C = \frac{1}{X_C} = \text{Total susceptance of the capacitor banks}$$

$$a = B_C[X_s X_{lt} X_m - r_s r_l X_m] - X_m(X_s + X_{lt})$$

$$b = -B_C[r_l X_s X_m + r_s X_{lt} X_m] + X_m(r_s + r_l)$$

$$c = -B_C[r_l(X_s + X_m) + r_s X_{lt}] + r_s + r_l$$

$$d = B_C[r_s r_l - X_{lt}(X_s + X_m)] + X_s + X_m + X_{lt}$$

From Fig. 4, the electromagnetic torque (in p.u.) of a constant speed WTG can be determined as follows:

$$T_e = \frac{|V_{th}|^2 \cdot \frac{r'_r}{s}}{[\frac{r'_r}{s} + r_{th}]^2 + [X'_r + X_{th}]^2} \tag{4}$$

Using (6.4), the torque versus slip characteristic has been obtained, and is shown in Fig. 4.4. In the normal operating condition, the electrical and mechanical torques will be equal; hence, the WTG will operate at slip s_0 (point Q). When a severe fault occurs close to the WTG, the terminal voltage of the WTG falls drastically. This will reduce the electrical torque to almost zero. Consequently, the rotor will oscillate, and the slip of the WTG will increase gradually. Once the fault is cleared, the terminal voltage and electrical torque will again increase to its nominal value and thereby, the rotor will decelerate. If the fault is cleared after the critical clearing time (t_{cr}), the rotor may accelerate to a higher than critical slip (s_{cr}) value. In this case, although the fault is cleared and the terminal voltage is recovered back, the rotor will continue to accelerate (beyond s_{cr}), and therefore, the WTG will enter into the unstable region. This phenomenon implies that if the rotor slips crosses the point P (as shown in Fig 5), the WTG will be disconnected from the grid due to over-speed protection. In practice, over-speed protection circuit disconnects the WTG from the grid when the speed of the WTG exceeds 1.2 p.u.

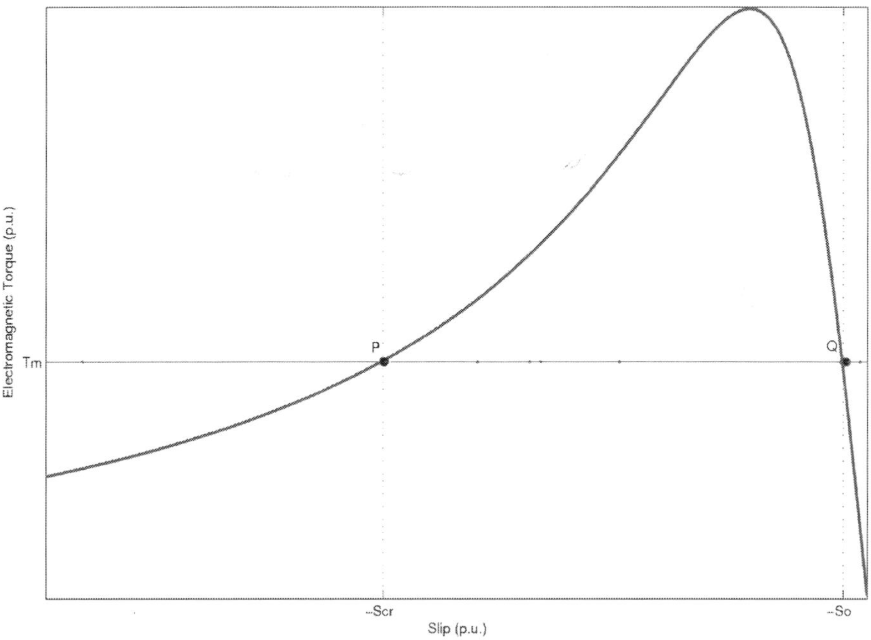

Fig. 5. Torque-slip characteristic of a constant speed WTG.

2.1.1 Evaluation of critical slip

As shown in Fig. 5, the critical slip can be obtained by equating the electrical torque of the WTG with its mechanical torque.
From (4),

$$T_e = \frac{|V_{th}|^2 \cdot \frac{r_r'}{s}}{[\frac{r_r'}{s} + r_{th}]^2 + [X_r' + X_{th}]^2} = T_m \tag{5}$$

This leads to,

$$\frac{|V_{th}|^2}{T_m}\cdot\frac{r_r'}{s} = (\frac{r_r'}{s})^2 + 2r_{th}\frac{r_r'}{s} + r_{th}^2 + (X_r' + X_{th})^2 \tag{6}$$

$$\Rightarrow [r_{th}^2 + (X_r' + X_{th})^2]\cdot s^2 + [2r_{th}r_r' - r_r'\frac{|V_{th}|^2}{T_m}]\cdot s + (r_r')^2 = 0$$

This is a quadratic equation in 's', and following are its roots (scr,s0) (as depicted in Fig. 5)

$$s_{cr} = \frac{-[2r_{th}r_r' - r_r'\frac{|V_{th}|^2}{T_m}] + \sqrt{[2r_{th}r_r' - r_r'\frac{|V_{th}|^2}{T_m}]^2 - 4(r_r')^2[r_{th}^2 + (X_r' + X_{th})^2]}}{2[r_{th}^2 + (X_r' + X_{th})^2]} \tag{7}$$

$$s_0 = \frac{-[2r_{th}r_r' - r_r'\frac{|V_{th}|^2}{T_m}] - \sqrt{[2r_{th}r_r' - r_r'\frac{|V_{th}|^2}{T_m}]^2 - 4(r_r')^2[r_{th}^2 + (X_r' + X_{th})^2]}}{2[r_{th}^2 + (X_r' + X_{th})^2]} \tag{8}$$

Equation (7) shows that s_{cr} is mainly a function of the parameters such as r_{th}, X_{th} and V_{th}. The value of these parameters also depends on B_C (the amount of reactive power compensation in per unit).

2.2 Calculation of critical clearing time

Let us consider rotor dynamics to obtain the critical clearing time,

$$\frac{ds}{dt} = \frac{T_m - T_e}{2H} \tag{9}$$

where, s is the slip in p.u., Tm is the mechanical torque in p.u., Te is the electromagnetic torque in p.u., H is the combined inertia constant of the WTG system in sec. Integration of (9) leads to,

$$\int_0^{t_{cr}} dt = \frac{2H}{T_m - T_e}\int_{s_0}^{s_{cr}} ds \tag{10}$$

It has been assumed that during fault $T_m - T_e$ remains constant.
Finally,

$$t_{cr} = \frac{2H}{T_m - T_e}(s_{cr} - s_0) \tag{11}$$

From the above equation, it can be observed that the critical clearing time is directly proportional to the inertia constant and the difference of the critical and initial slip, and inversely proportional to the difference between the mechanical and electromagnetic torque.

2.3 Simulation results and discussions

Modelling of the constant speed WTG, capacitor banks and grid has been carried out in MATLAB/SIMULINK software tool. Currents from the WTG have been added to the currents from the capacitor banks. The total current has been injected into the grid. From

this injected current, the terminal voltage has been calculated which is given as an input to the WTG and capacitor banks.
A 600 kW constant speed WTG has been considered for this analysis. The machine parameters are listed in Table 1. This WTG is stall controlled and hence, it does not possess blade-pitch control.

Parameters	Value
Rated Power	600 kW
Rated Phase Voltage	690 V
Rated Frequency	50 Hz
Number of Poles	4
Stator Resistance (rs)	0.016 p.u.
Stator Leakage Reactance (Xls)	0.15 p.u.
Rotor Resistance (rr)	0.01 p.u.
Rotor Leakage Reactance (Xlr)	0.11 p.u.
Magnetizing Reactance (Xm)	7.28 p.u.
Inertia constant (J)	18.029 $kg \cdot m^2$

Table 1. Parameters of a Constant Speed WTG

2.4.1 Effect of additional reactive power support

With the help of simulation, it has been shown that this 600 kW constant speed WTG does not comply with the LVRT requirements (as shown in Fig. 8). Hence, the WTG has to be disconnected from the grid due to rotor speed instability whenever a fault occurs in its vicinity. However, by providing additional reactive power support, rotor speed stability of the WTG can be enhanced such that it can satisfy the LVRT requirements. This phenomenon has been discussed in this section with the help of simulation results. Further, the exact quantification of reactive support required achieving particular values of critical slip and critical clearing time has been obtained theoretically. Finally, using the dynamic model of the system, it has been shown that dynamic simulation results (critical slip and time) match with the analytically calculated results using equations (7) and (11). Fig. 6 shows the torque-slip characteristics of a 600 kW WTG with two different value of reactive power injection.
Equation (3) shows that the electromagnetic torque is a function of V_{th}, r_{th} and X_{th}. For a given set of machine parameters, V_{th}, r_{th} and X_{th} are functions of B_C (the value of reactive power compensation) as shown in equations (1), (2) and (3) respectively. For nominal reactive power support, the value of B_C is 0.23 p.u., and with an additional capacitor bank of 0.22 p.u., the value of B_C will be 0.45 p.u. As indicated in Fig. 6, an additional value of B_C will shift the torque-slip characteristic upwards. Consequently, the value of critical slip will increase from -0.118 p.u. to -0.15 p.u. Improvement of s_{cr} because of additional B_C can also be obtained numerically using (7). Similarly, using (11), t_{cr} has been calculated as 0.12 s with the nominal capacitor bank, and 0.155 s with an additional capacitor bank. The equations for s_{cr} and t_{cr} (as a function of B_C) have been verified using a dynamic-simulation model of the sample system. A severe three phase-to-ground fault has been created on the system such that the terminal voltage at the constant speed WTG should remain as per LVRT requirements. To consider the worst condition, the wind velocity has been kept constant at its rated value in the simulation. Therefore, the mechanical torque of the turbine will remain

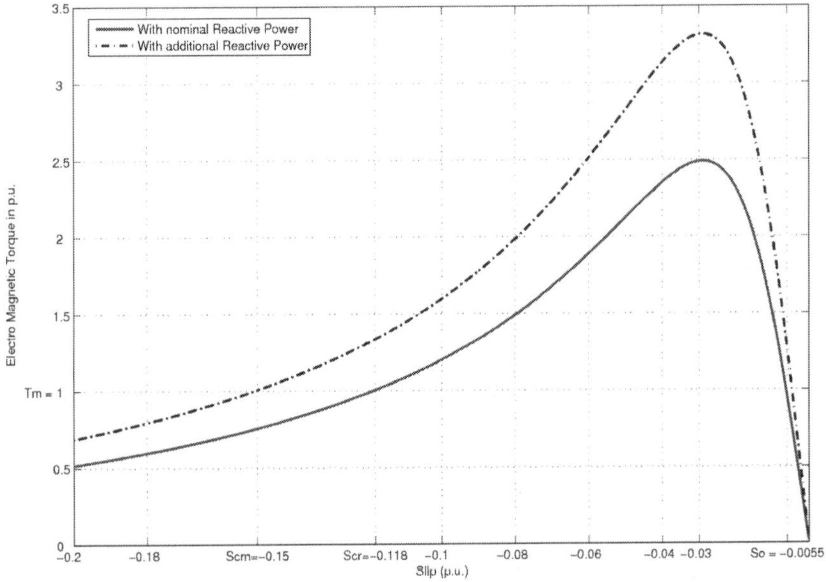

Fig. 6. Torque-slip characteristics of the WTG with nominal and additional reactive power.

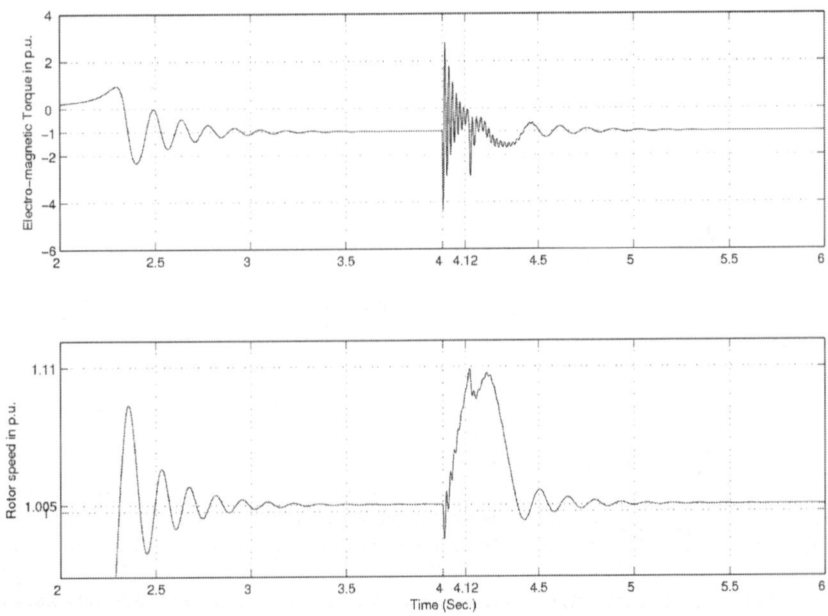

Fig. 7. Electromagnetic torque and rotor speed without additional capacitor bank.

at 1 p.u. during the fault. As per LVRT requirements (refer Fig. 1), a WTG should remain stable for 0.15 s with a terminal voltage of 0.15 p.u. It means that to satisfy the LVRT requirement, a WTG should have a critical clearing time of at least 0.15 s.
Fig. 7 shows the electromagnetic torque and rotor speed in per unit. During the fault, the value of Te will reduce to almost zero. Hence, the rotor will accelerate to -0.11 p.u. slip within 0.12 s. The numerical values of the critical slip and time are calculated to be -0.118 p.u. and 0.12 s from equations (7) and (11) respectively. If the fault persist for more than the critical clearing time (0.12 s), the rotor will continue to accelerate and theWTG becomes unstable. Fig. 8 shows the electromagnetic torque and rotor speed for a fault of duration 0.15 s.

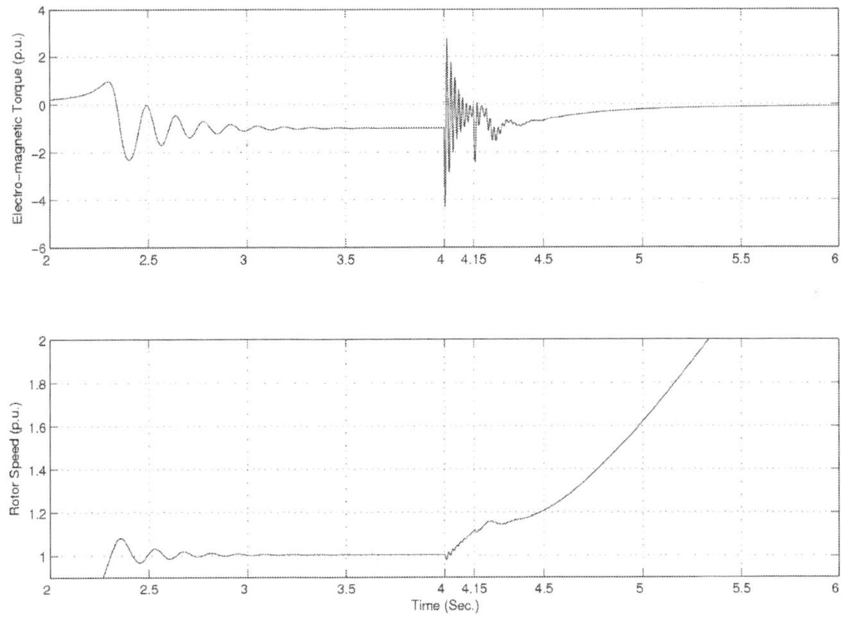

Fig. 8. Electromagnetic torque and rotor speed without additional capacitor bank for a fault of duration 0.15 s.

The speed of the WTG ramps up gradually and it has to be disconnected from the grid before the mechanical constraint on the rotor speed (1.2 p.u.) is reached. However, as per LVRT requirements, a constant speed WTG should remain connected for at least 0.15 s, which is not satisfied for this 600 kW WTG with a nominal value of reactive power compensation. To meet the LVRT requirements, an additional capacitor bank has been employed in parallel with the WTG. With this, executing the dynamic model again, it has been observed that the WTG remains stable even for 0.15 s. The rotor accelerates to -0.15 p.u. slip, and after the fault is cleared, the rotor comes back to its nominal value of -0.005 p.u. slip as shown in Fig. 9.

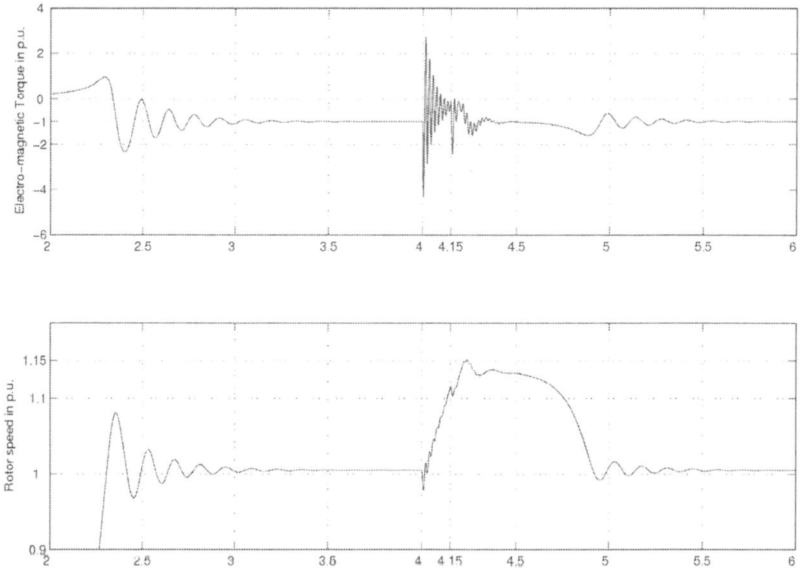

Fig. 9. Electromagnetic torque and rotor speed with additional capacitor bank.

3. Enhancement of rotor speed stability margin using active stall control

3.1 About active stall control

The rotor speed stability of a constant speedWTG can be improved by active stall control. Under transient conditions, the active stall controller controls the blade pitch angle (β) in the negative direction to reduce the turbine torque. This action helps to reduce the acceleration of the rotor, and improves the rotor speed stability. Using this phenomenon, the enhancement of rotor speed stability of a constant speed WTG with active stall is explained as follows. The active stall control based constant speed WTG has control of pitch in the negative direction (i.e. between −90° to 0°) with respect to pitch control based WTG (Type-A1). The rate of negative pitch control is normally less than 8° per second.

Fig. 10 shows the difference in the direction of blade rotation between pitch controlled and active stall controlled constant speed WTG. In the figure, the chord line is the straight line connecting the leading and trailing edges of an airfoil. The plane of rotation is the plane in which the blade tips lie as they rotate. The pitch angle (β) is the angle between the chord line of the blade and the plane of rotation. And, the angle of attack (α), is the angle between the chord line of the blade and the relative wind or the effective direction of air flow.

In active stall control, at low wind speeds, the machine is usually controlled by pitching its blades similar to a pitch controlled machine, to track maximum power gen-eration. When the machine reaches its rated power value, the blades are pitched in the direction opposite from what a pitch controlled machine does, in order to reduce output power. This needs pitch angle β to be decreased, typically, by a small amount only. Hence, the rating of pitch drives is less for active stall control as compare with pitch control (Ackermann, 2005). Therefore, the cost and complexity are comparatively less for active stall controlled WTGs.

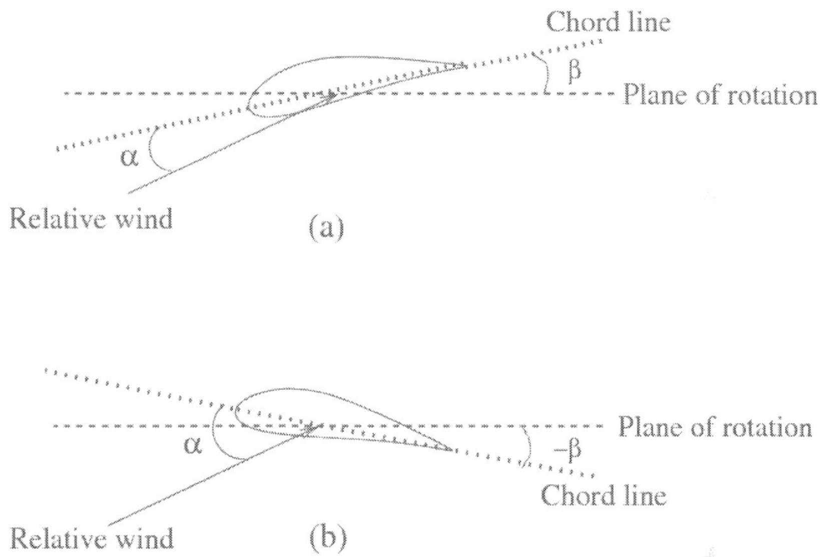

Fig. 10. Power generation control methods: (a) pitch control, (b) active stall control.

3.2 Design of active stall controller for constant speed WTG

The block diagram of active stall based WTG is shown in Fig. 11. To enhance rotor speed stability, pitch angle controller controls the generator speed by means of pitch angle of the blades. Accordingly, the mechanical torque (input to the induction generator) changes and it controls the rotor speed to lie within the allowable range.

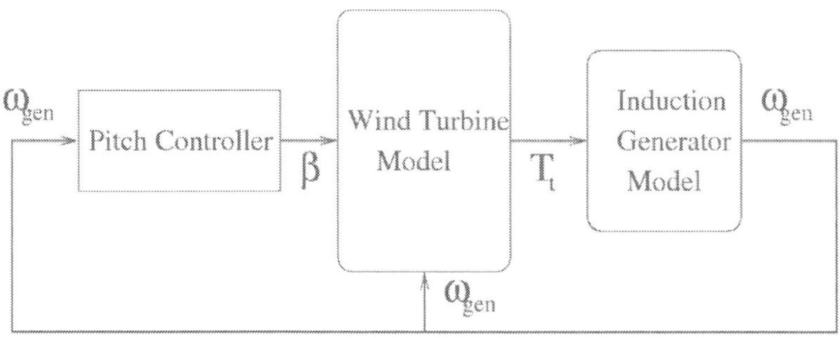

Fig. 11. Block diagram of active stall control based WTG.

As shown in Fig. 12, the pitch angle controller controls the pitch angle of the wind turbine blades. This would change the power coe cient Cp, which changes the mechanical torque input to the induction generator. Finally, the pitch angle controller controls the speed of the generator and enhances rotor speed stability. In the figure, T represents the time constant of the servo mechanism. The values of the controller parameters are given in Table 3.

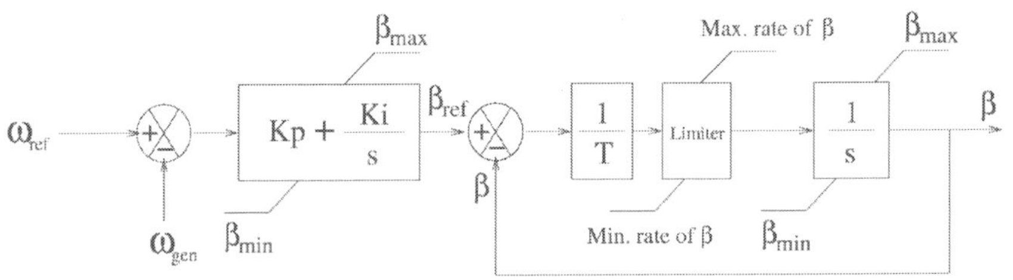

Fig. 12. Active stall pitch controller.

3.2.1 Aerodynamic model of the wind turbine

The aerodynamic torque developed on the main shaft of a wind turbine is give by:

$$T_a = \frac{1}{2\omega_m}\rho\pi R^2 V_w^3 C_p(\beta,\lambda) \tag{12}$$

where, Ta is the aerodynamic torque developed in N · m; ωm is the speed of the wind turbine in rad/s; ρ is air density in kg/m2; R is the radius of wind turbine blade rotation area in m; Vw is the average wind speed in m/s; Cp is the power coeffcient of the wind turbine.

3.3 Simulation results and discussions

As shown in Fig. 13, a 1.5 MW constant speed WTG has been connected to a medium voltage (MV) distribution network. Modeling of WTG with capacitor bank connected to a sample system has been simulated using DIgSILENT PowerFactory.

Parameters	Value
Rated Power	1.5 MW
Rated Phase Voltage	690 V
Rated Frequency	50 Hz
Number of Poles	4
Stator Resistance (rs)	0.01 p.u.
Stator Leakage Reactance (Xls)	0.1 p.u.
Rotor Resistance (rr)	0.01 p.u.
Rotor Leakage Reactance (Xlr)	0.1 p.u.
Magnetizing Reactance (Xm)	3.0 p.u.
Inertia constant (J)	20.0 kg m^2

Table 2. Parameters of a 1.5 MW constant speed WTG

For rotor speed stability analysis, a 3-phase severe fault has been created on line-1 at time t=5 s. The fault has been cleared by removing that line from the system at time t=5.15 s. To analyze the behavior of the WTG during this grid disturbance, the quantities of induction generator (at bus-6) such as active power generation (in MW), rotor speed (in p.u.), reactive power generation (in Mvar), and generator terminal voltage (in p.u.) have been plotted. The effect of the operating point and active stall control on rotor speed stability is illustrated in the next sub-sections. If a fault occurs close to a constant speed WTG, the voltage at the generator terminals of the wind turbine drops, which results in the reduction of active power. If a wind turbine controller does not attempt to reduce the mechanical power input, the turbine accelerates during the fault. If a wind turbine has no means of controlling its power, then critical clearing time will be very short (Ledesma et al., 2003).

Figs. 14 and 15 show the induction generator quantities when a 3-phase line fault is created on Line-1 at 5 s and cleared at 5.15 s by disconnecting the line. During the fault, the voltage dips and active power generator reduce to almost zero. As the torque of wind turbine is not controlled, the generator rotor continue to accelerate. This result into the rotor speed instability. Finally, the over-speed relay removes the generator from the grid. Because of this phenomenon, it can be said that a constant speed (squirrel cage) induction generator does not have LVRT capability.

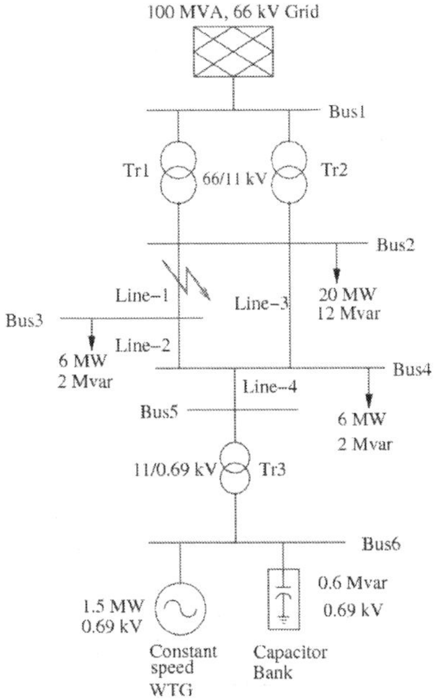

Fig. 13. A sample system with a 1.5 MW constant speed WTG.

Parameters	Value
Proportional gain (Kp)	500
Integral gain Ki	20
Time constant (T)	0.5s
Min. limit of β (βmin)	−30°
Max. limit of β (βmax)	10°
Rate of β	8°/s

Table 3. Parameters of active stall controller

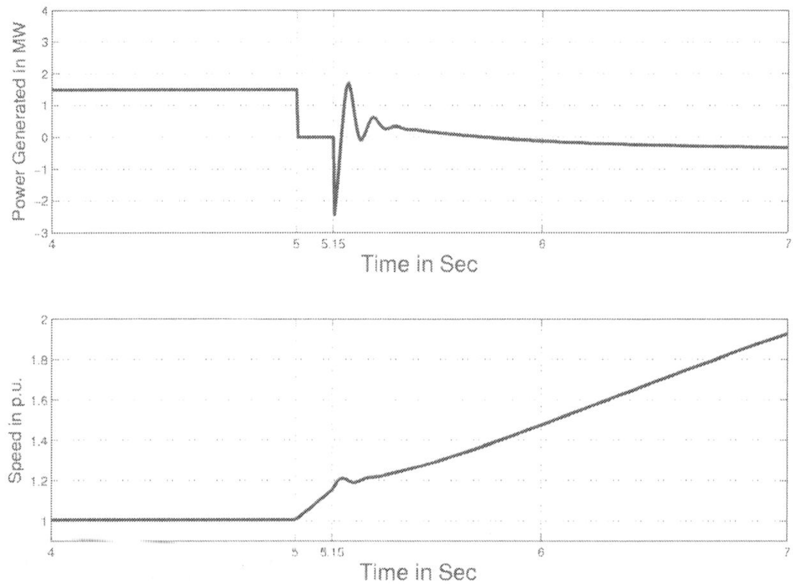

Fig. 14. Generator active power and speed without active stall control.

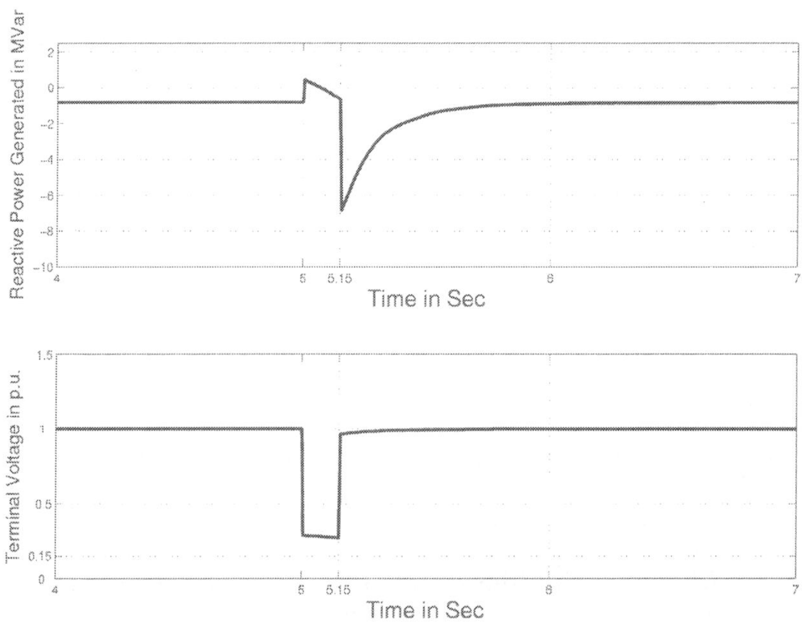

Fig. 15. Generator reactive power and terminal voltage without active stall control.

Parameters	Value
WTG mechanical inertial (J)	4x10^6 kgm^2
Stiff coefficient(k)	100x10^6 Nm/rad
Damping coe cient(D)	0 Nms/rad
Radius of blade rotation area (R)	35 m
Air density (ρ)	1.225kg/m^3

Table 4. Parameters of Wind Turbine

3.4 Effect of active stall control on the rotor speed stability

With the help of active stall control, the input mechanical torque of a constant speed WTG can be reduced. During a grid disturbance, the system voltage sags, and hence, the active power supplied by the WTG decreases. As a result, the rotor speed increases and the WTG draws a very high reactive power, as shown in Figs. 16 and 17. In this case, the WTG power output is 1.5 MW (at constant wind speed). Using active stall pitch controller, the WTG will remain stable even though the fault is cleared at 5.4 s. That is because the mechanical input power, and hence, the active power generated is reduced by controlling the pitch angle in negative direction. This results into reduction in mechanical input torque. Consequently, the speed and hence, the reactive power drawn reduces and this improves the rotor speed stability, as shown in Figs. 16 and 17. A constant speed WTG continues to supply power to the grid and satisfy the LVRT requirements of the WTG.

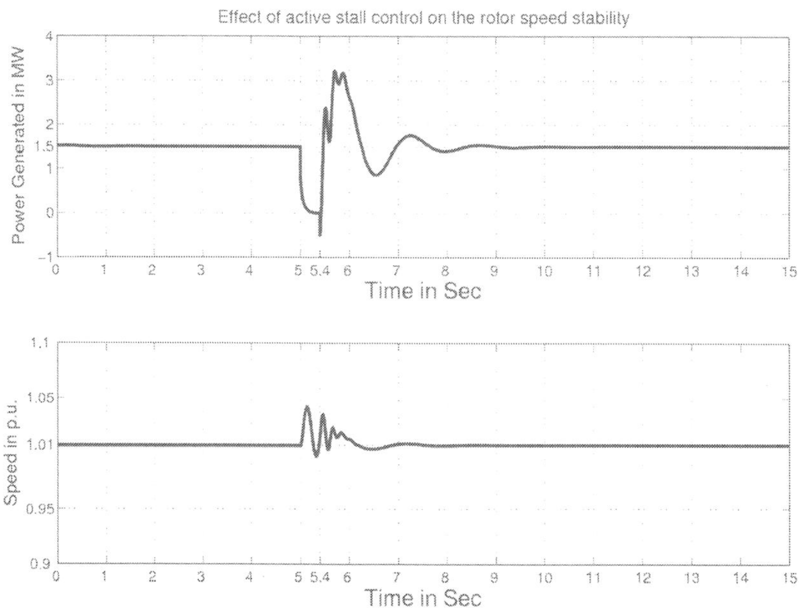

Fig. 16. Effect of active stall control on rotor speed stability: active power and speed.

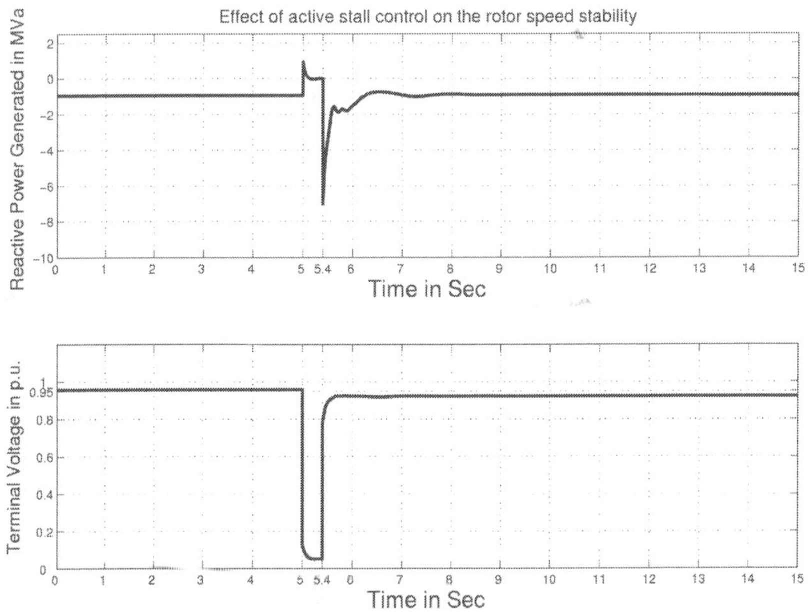

Fig. 17. Effect of active stall control on rotor speed stability: reactive power and voltage.

4. Conclusion

It has been shown that a constant speed WTG cannot meet LVRT grid code due to rotor speed instability. Consequently, the analysis has been aimed towards developing techniques to enhance the rotor speed stability such a way that a constant speed WTG can satisfy LVRT grid code.

The first technique involves providing additional reactive power support during disturbances. This technique is useful for a constant speed WTG without active stall control. This extra reactive power would recover the voltage and hence, the electromagnetic torque. Therefore, the rotor acceleration reduces and the value of t_{cr} increases. For this analysis, a 600 kW constant speed WTG connected to a sample network with nominal reactive power support (B_C =0.22 p.u.) has been simulated using MATLAB. Using the simulation results, it has been shown that with this nominal value of reactive power support, t_{cr} is 0.12 s. However, according to the LVRT requirements t_{cr} should be at least 0.15 s. Therefore, additional reactive power support of the same value (B_C =0.22 p.u.) has been connected. This extra reactive power increases the value of t_{cr} to 0.15 s, which is in compliance with the LVRT requirements.

Second technique is based on pitch angle control of turbine blades to control the mechanical torque. For this analysis, an active stall controller for use with turbine dynamic model has been designed using DIgSILENT simulation tool. From this simulation model, it has been shown that during a network disturbance, the active stall controller reduces the turbine torque and the rotor acceleration. Consequently, this would enhance the rotor speed stability margin. For this analysis, a 1.5 MW constant speed WTG connected to a sample system has been simulated using DIgSILENT. From the simulation results, it has been shown that with the help of active stall control, the value of t_{cr} increases from 0.1 s to 0.4 s in order to meet LVRT requirements.

During grid integration, it is mandatory for the owners of WTGs to ensure compliance with the grid codes. This thesis work, therefore, can be useful for a utility or a regulatory commission to determine whether a particular constant speed WTG can meet the grid codes. Further, this work also presents techniques to enhance the rotor speed stability margin. These techniques can be helpful to manufacturers of constant speed WTGs to achieve compliance with grid codes such as LVRT requirements.

In this chapter, the study of rotor speed stability has been carried out for a constant speed WTG. This work can be extended by considering analysis for a aggregate model of wind farm with a number of such constant speed WTGs. In practice, wind farms can also be equipped with FACTS devices, such as, SVC and STATCOM in addition to capacitor banks on each WTG. The sizing of such FACTS devices can be determined to satisfy LVRT requirements for the aggregated wind farm by modifying the analysis for the enhancement of rotor speed stability.

5. References

Agalgaonkar, A.; Dobariya, C.; Kanabar, M. & Khaparde, S. A. (2006). Optimal sizing of Distributed Generation in MicroGrid, *Proceedings of IEEE international Power India Conference*, ISBN: 10.1109/POWERI.2006.1632627, April 2006, IEEE, New Delhi.

Zavadil, R.; Miller, N.; Ellis, A. & Muljadi, E. (2005). Making connections, *IEEE Power Energy Magazine*, pp. 26-37, Nov. 2005.

Ackermann, T. (2005). *Wind Power in Power Systems*. John Wiley & Sons Ltd., ISBN 0470855088, England.

Kanabar, M. & Khaparde, S. A. (2008). Evaluation of Rotor Speed Stability Margin of a Constant Speed Wind Turbine Generator, *Proceedings of IEEE international Power India Conference*, ISBN: 10.1109/ICPST.2008.4745277, Oct. 2006, IEEE, New Delhi.

Wind Power India Website (2010). [Online]. Available: http://www.windpowerindia.com.

Kanabar, M. ; Dobariya, C. & Khaparde, S. A. (2006). Rotor Speed Stability Analysis of Constant Speed Wind Turbine Generators, *Proceedings of IEEE conference on Power Electronics Drives and Energy Systems*, ISBN: 10.1109/PEDES.2006.344270, Dec. 2006, IEEE, Delhi.

Samuelsson, O. & Lindahl, S. (2005). On speed stability, *IEEE Transaction on Power System*, Vol. 20, No. 2, May 2005, pp. 1179-1180, ISSN : 10.1109/TPWRS.2005.846194.

Rodriguez, J. M.; Fernandez, J. L.; Beato, D.; Iturbe, R.; Usaola, J.; Ledesma, P. & Wilhelmi, J. R. (2002). Incidence on power system dynamics of high penetration of fixed speed and doubly fed wind energy systems: study of the Spanish case, IEEE Transaction on Power System, Vol. 17, No. 4, Nov. 2002, pp. 1089-1095, ISSN : 10.1109/TPWRS.2002.804971.

Nunes, M. V. A.; . Bezerra, U. H. & Zurn H. H. (2003). Transient stability margin of variable versus fixed speed wind systems in electrical grids, *Proceedings of IEEE Power Tech.*, ISBN: 10.1109/PTC.2003.1304460, June 2003, IEEE, Bologna.

Chompoo-inwai, C.; Yingvivatanapong, C.; Methaprayoon, K. & Lee, W. J. (2005). Reactive compensation techniques to improve the ride-through capability of wind turbine during disturbance, *IEEE Transaction on Industry Applications*, Vol. 41, No. 3, May 2005, pp. 666-672 , ISSN : 10.1109/TIA.2005.847283.

Jenkinds, N.; Allan, R.; Crossley, P.; Kirschen, D. & Strbac, G. (2000). *Embedded Generation.* The Institution of Electrical Engineers, ISBN: 0852967748, England.

Ledesma, P.; Usaola, J. & Rodriguez, J. L. (2003). Transient stability of a fixed speed wind farm, *Renewable Energy*, Vol. 28, pp. 1341-1355, May 2003.

24

Power Quality in Grid-Connected Wind Turbines

J.J. Gutierrez, J. Ruiz, P. Saiz, I. Azcarate, L.A. Leturiondo and A. Lazkano
University of the Basque Country
Spain

1. Introduction

Electric power is a very unusual product. It must be constantly available for the customers but its massive storage is not possible. Therefore, electric power needs to be generated as it is used. Furthermore, it is usually generated far from the customers. All these factors make it difficult to control and assure the quality of electric power. There is no procedure that allows the removal of the poor electric power. Consequently, it is necessary to define preventive and monitoring procedures that guarantee a minimum level of power quality and hence the correct behavior of the equipment connected to the power distribution system. Perfect power quality means that the voltage is continuous and sinusoidal with a constant amplitude and frequency. Low-frequency conducted disturbances are the main defects that could compromise power quality. These are classified in the following groups: harmonics and interharmonics, voltage dips and short supply interruptions, voltage unbalance, power frequency variations and voltage fluctuations or flicker.

In the last years the use of wind farms and other distributed power generation systems has drastically increased. The question that needs to be raised is how those new generation systems will affect to the whole grid. A portion of the answer must be obtained from the impact that they have on the power quality. In principle, wind energy can be considered a risky source in terms of power quality. Moreover, when wind turbines are part of the grid the power, quality seems to be a complex issue which highly depends on the interaction between the grid and the wind turbines. The main impact on the grid by the wind turbines, concerning power quality, is related to voltage changes and fluctuations, harmonic content, power peaks and flicker. The presence of these disturbances is determined by meteorological conditions and by the technical features of the wind turbine: continuously variable output power due to wind shear, tower shadow or turbulences; performance of electrical components such as generators and transformers; aerodynamic and mechanical behavior of the rotor...

The power quality of the wind turbines must be certified on the basis of measurements performed according to international standards and guidelines. On one hand, the IEC 61400-21 standard is the reference normative for the certification of the power quality of wind turbines (IEC-61400-21 Ed. 2.0, 2008). The first edition was published in 2001 and it specifies the main relevant features of power quality that should be measured in a wind turbine. This standard establishes the procedures for the measurement and assessment of power quality characteristics of grid-connected wind turbines. According to it, measurements should be performed for harmonic content, flicker, voltage drops as well as active and reactive power, during normal and switching operations. To obtain those characteristics, the measurements should be made on the basis of long time-series of current and voltage, always depending on wind conditions. On the other hand, the international Measuring Network of Wind Energy Institute (MEASNET) has defined some guidelines based on the above-mentioned standard with the aim to adapt the procedures and hence the measurement results obtained by its members.

This chapter is organized in two main related sections. The first section provides a descriptive approach to the main factors that have an influence on the power quality of the grid-connected wind turbines. First we summarize the main rationale and objectives of the IEC 61400-21 standard. Then we detail the procedures specified by the standard for the measurement of the main parameters of the wind turbine power quality characteristics: harmonic content, flicker, voltage drops and power parameters. We also focus on the most relevant features that must be considered by a measurement system when trying to assess one of the most complex power quality parameters: flicker. In the second section we describe our own measurement system, a useful tool specifically developed for the assessment of the power quality of a grid-connected wind turbine, according to the IEC 61400-21 standard. To conclude the chapter, we provide some illustrative examples of power quality parameters measured on different wind turbines installed in a wind farm of Northern Spain.

2. Power quality characteristics of wind turbines

Power injection from grid-connected wind turbines affects substantially the power quality. The procedures for the measurement and assessment of the main parameters involved in the power quality characteristics of a wind turbine are described in the IEC 61400-21 standard. The tests are designed to be as non-site-specific as possible, so that power quality characteristics measured with the wind turbine connected at a test site can also be considered valid at other sites.

The validity of the measurement procedure is dependent upon the proper establishment of the test conditions. The wind turbine has to be directly connected to the MV-network and the measurements of the electrical characteristics have to be made at the wind turbine terminals. It is necessary to specify the rated data of the wind turbine including rated active power of wind turbine P_n, rated apparent power S_n, nominal phase-to-phase voltage U_n and the rated current I_n. Moreover, the location of the wind turbine terminals and the specific configuration of the assessed wind turbine including the relevant control parameter settings have to be clearly stated in the test report.

According to the standard there are seven parameters compromising the required power quality characteristics of a wind turbine: voltage fluctuations or flicker; harmonics and interharmonics; voltage drops; active power; reactive power; grid protection and reconnection time. In the following sections we will describe those parameters and the procedures specified for their measurement, stressing the most relevant issues affecting the assessment of harmonic and interharmonic content and flicker.

2.1 Current harmonics, interharmonics and higher frequency components

Voltage and current harmonics are usually present on the utility network. Non-linear and electronic loads, rectifiers and inverters, are some sources which produce harmonic content. The effects of the harmonics include overheating, faulty operation of protections, equipment failures or interferences with communication systems.

The standard specifically defines different procedures to assess the harmonics, interharmonics and higher frequency components for a wind turbine working under continuous conditions and operating with reactive power as close as possible to zero. This means that, if applicable, the reactive set-point control shall be set to zero. These parameters will not be considered under switching operations since the harmonic content is not harmful enough when the duration of the perturbation is limited to a short period of time.

The values of the individual current harmonics, interharmonics and higher frequency components and the Total Harmonic Current distortion (THC) must be provided in percentage of I_n and with the wind turbine operating within the active power bins 0, 10, 20,..., 100% of P_n, where 0, 10, 20,..., 100% are the bin midpoints. The harmonic current components must be specified as subgrouped RMS values for frequencies up to 50 times the fundamental grid frequency. The THC coefficient must be calculated from those values according to:

$$THC = \sqrt{\sum_{h=2}^{50} i_{sg,h}^2} \cdot 100 \qquad (1)$$

where $i_{sg,h} = \frac{I_{sg,h}}{I_n}$ and $I_{sg,h}$ is the subgrouped RMS current harmonic of harmonic order h.

The interharmonic current components must be specified as subgrouped RMS values $I_{isg,h}$ for frequencies up to 2 kHz in accordance to Annex A of the IEC 61000-4-7 standard (IEC-61000-4-7, 2002). The higher frequency current components must be specified as subgrouped values for frequencies between 2 kHz and 9 kHz in accordance to Annex B of the IEC 61000-4-7 standard.

At least nine 10 min time-series of instantaneous current measurements (three tests and three phases) must be collected for each 10% power bin. The 10 min averages of each frequency band must be calculated for each 10 min time-series, and subsequently the maximum 10 min averages of each frequency band in each 10% power bin must be reported.

2.1.1 Measurement of the subgrouped harmonic, interharmonic and higher frequency current components according to IEC 61000-4-7

The measurement of the harmonic current content is specified for a discrete signal obtained at a sampling rate of f_s. The basic tool for the measurement is the Discrete Fourier Transform (DFT) applied over a signal window of T_w seconds ($T_w \cdot f_s$ samples). This transformation provides the spectral components for the analyzed window with a spectral resolution of $f_w = \frac{1}{T_w}$ Hz.

The standard suggests the use of a rectangular window whose duration is 10 cycles of the fundamental frequency in 50 Hz systems and 12 cycles in 60 Hz systems (i.e. approximately 0.2 s). With these exact window lengths the spectral leakage has no influence on those spectral components that are a multiple number of the spectral resolution f_w = 5 Hz. To achieve this goal it is necessary to use a sampling rate locked to the fundamental frequency by means of a Phase Locked Loop system (PLL).

Finally, to measure the spectral components up to 9 kHz it is needed the use of a sampling rate over 18 kHz.

2.1.1.1 Calculation of the subgrouped harmonic, interharmonic and higher frequency current components.

The DFT applied to each window provides the spectral components, c_k, with a resolution of 5 Hz from the DC component up to $\frac{f_s}{2}$. Fig. 1 shows how the subgrouped harmonic and interharmonic components are grouped. The values of the components can be obtained by grouping the different spectral components from the DFT, according to the next equations:

$$I_{sg,h}^2 = \sum_{i=-1}^{1} c_{k+i}^2$$
$$I_{isg,h}^2 = \sum_{i=2}^{p} c_{k+i}^2 \qquad (2)$$

where the k index refers to the spectral line order provided by the DFT, corresponding to the h-th harmonic component ($k = 10 \cdot h$ for 50 Hz systems and $k = 12 \cdot h$ for 60 Hz systems). The value of p must be 8 for 50 Hz systems and 10 for 60 Hz systems.
The values of the higher frequency components are obtained by grouping the spectral lines from the DFT in 200 Hz bands from 2 to 9 kHz. By using k index for the spectral line corresponding to the band b=2100, 2300, ..., 8900 Hz ($k = \frac{b}{5}$), the higher frequency component I_b can be obtained as follows:

$$I_b^2 = \sum_{i=1-p}^{p} c_{k+i}^2 \qquad (3)$$

The value of p must be 20 for 50 Hz systems and 25 for 60 Hz systems.

2.1.1.2 Smoothing process.

For each 0.2 s window the equations (2) and (3) provide $I_{sg,h}$, $I_{isg,h}$ and I_b (this makes a total of 3000 values of each for every single value of h or b, in case of 10 min time-series). To avoid abrupt transitions between different windows, the subgrouped current components obtained for each value of h and b are smoothed by processing those 3000 values through a 1st low-pass filter with a time constant of 1.5 s. This filter must be designed for a sampling frequency of 5 $\frac{S}{s}$ because the $I_{sg,h}$, $I_{isg,h}$ and I_b values are available every 0.2 s. Moreover, it is necessary to eliminate the first 50 values corresponding to the filter transient in order to obtain an accurate average of the 3000 values at the filter output.

2.2 Response to voltage drops

One of the main objectives of the IEC 61400-21 standard is to provide a methodology to be used in wind energy generation systems so that they contribute to control and assess the quality of service of the electric power system, as conventional plants do. Moreover, one of the main concerns related to the massive insertion of renewable energy generation systems, such as wind turbines, is to maintain the reliability of the system despite the contingencies that may happen in the network.

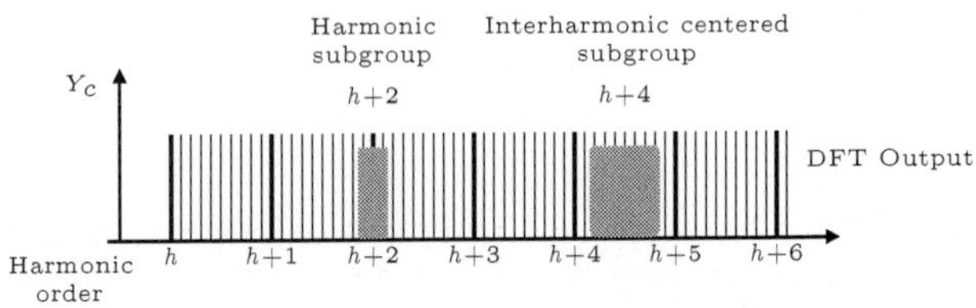

Fig. 1. Illustration of the harmonic subgroup and interharmonic centered subgroup.

A specific problem is related to the behavior of the wind farms in the presence of voltage drops in the electrical network. Voltage drops are sudden voltage dips mainly caused by faults in the network. These events are random in nature and can be characterized by their amplitudes and duration. Previous experiences generate doubts about the capacity wind power generation to remain connected, both during the fault and during the subsequent recovery. The standard tries to check that wind farms are able to actively contribute to grid stability in case of voltage drops, and to that end a specific test is included in the standard.
This test is defined for off-line conditions, i.e. when the turbine under test is disconnected from the grid and therefore does not contribute to modify the voltage shape. The test verifies the response of a wind turbine to voltage drops, with the wind turbine operating at two different situations concerning the rated active power P_n: between 10% and 30% of P_n and above 90%. A number of six voltage drops are defined, specifying the magnitude and duration of the rectangular voltage drop (see Table 1).

Case	Magnitude of voltage phase to phase	Magnitude of positive sequence voltage	Duration(s)
VD1[1]	0.90 ± 0.05	0.90 ± 0.05	0.5 ± 0.02
VD2[1]	0.50 ± 0.05	0.50 ± 0.05	0.5 ± 0.02
VD3[1]	0.20 ± 0.05	0.20 ± 0.05	0.2 ± 0.02
VD4[2]	0.90 ± 0.05	0.95 ± 0.05	0.5 ± 0.02
VD5[2]	0.50 ± 0.05	0.75 ± 0.05	0.5 ± 0.02
VD6[2]	0.20 ± 0.05	0.60 ± 0.05	0.2 ± 0.02

[1] Symmetrical three-phase voltage drop
[2] Two-phase voltage drop

Table 1. Specification of the test of voltage drops.

These test signals are used in the measurement procedure to obtain time-series of active power, reactive power, current and voltage at the wind turbine terminals for the time shortly prior to the voltage drop and until the effect of the voltage drop has extinguished. The test can be carried out using a set-up, such as in Fig. 2, in which the voltage drops are created by a short-circuit emulator that connects the three phases or two phases to ground via an impedance, or connecting the three or two phases together through an impedance.

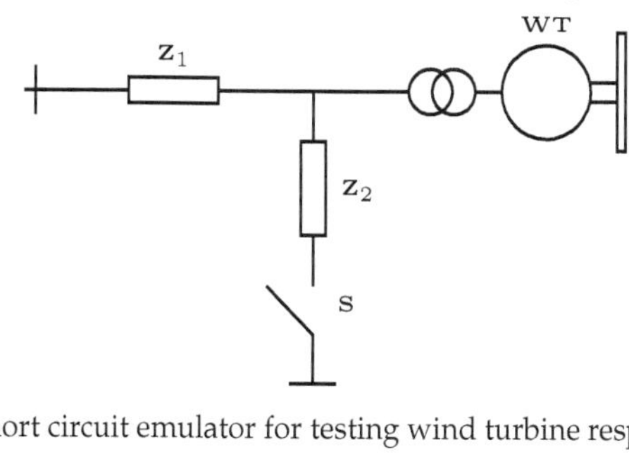

Fig. 2. System with short circuit emulator for testing wind turbine response to temporary voltage drops.

The voltage drop is created by connecting the impedance Z_2 by the switch S, which shall be able to accurately control the time between connection and disconnection of Z_2. The impedance value of Z_2 must be set to obtain the voltage magnitudes specified in the standard when the wind turbine is not connected. The function of impedance Z_1 is to limit the effect of the short-circuit on the up-stream grid. The magnitude of this impedance should be selected so that the voltage drop tests do not cause an unacceptable situation at the upstream grid, and at the same the impedance does not affect the transient response of the wind turbine in a significant manner.

2.3 Active and reactive power

The standard tries to assess the capability of the wind turbine concerning the active and reactive powers. The assessment must be done by means of different types of tests, some of them based on the wind speed and others considering both the wind speed and the wind turbine regulation system.

2.3.1 Active power

For the assessment of the active power three different tests are considered. First, the maximum power must be measured from at least 5 time-series of 10 min, collected for each 1 $\frac{m}{s}$ wind speed bin between the cut-in wind speed and 15 $\frac{m}{s}$. The measured power must be transferred to 0.2 s average data and 60 s average data by a block averaging:

- $P_{0.2}$ will be determined as the highest value obtained from 0.2 s windows, recorded during the 10 min period.
- P_{60} will be determined as the highest valid 60 s value calculated by averaging the 0.2 s values, recorded during the 10 min period.
- P_{600} will be determined as the highest 600 s value calculated by averaging the 0.2 s values, recorded during the 10 min period.

On the other hand, the ability of the wind turbine to operate in active power set-point control mode and to operate in ramp rate limitation control mode must be tested. For both tests, the results will be the active power calculated from 0.2 s average data, the wind speed and the available active power. The available active power must be obtained from the control system of the wind turbine. If the wind turbine control system does not provide it, an approximate value can be used based on measured wind speed combined with the power curve of the wind turbine.

In the case of ramp rate limitation, the wind turbine must be started from stand still and the ramp rate must be set to 10% of rated power per minute. Moreover, the available active power output must be at least 50% of rated power. In the case of set-point control, the test must be carried out during a test period of 10 min. The ramp rate limitation must be deactivated during this test and the set-point signal must be reduced from 100% to 20% in steps of 20% during 2 min at each set-point value. Moreover, the available active power output must be at least 90% of rated power.

2.3.2 Reactive power

For the assessment of the reactive power two different tests are considered. Both tests must be done considering the regulation system of the wind turbine.

The first test tries to assess the capability of the wind turbine concerning the maximum inductive reactive power and the maximum capacitive reactive power. For each of the two settings, the measurements must be taken so that at least 30 time-series of 1 min of active and

reactive power are collected at each 10% power bin from 0% to 100%. The sampled data will be calculated as 1 min average data by applying 0.2 s block averaging for each 1 min period.
On the other hand, the reactive power control by set-point value must also be measured, considering two cases: the measurement at a set-point of reactive power at zero and the measurement during the step change of reactive power. For the first case, the procedure is the same as that one used to assess the capability of the wind turbine concerning the maximum reactive power. For the second case, the test must be of 6 min period and the set-point of reactive power must be regulated for 2 min intervals corresponding to reactive power of zero, maximum capacitive reactive power and maximum inductive reactive power. The active power output, measured as 1 min average values, must be approximately 50% of rated power. The reactive power must be 0.2 s average data.
The results of the test must be the reactive power from 0.2 s windows, together with the set-point value of reactive power.

2.4 Voltage fluctuations (Flicker)

The impression of unsteadiness of visual sensation induced by variation in the intensity of a light source due to fluctuations of the supply voltage is known as flicker.
As a result of the subjective nature of the perception of annoyance related to the sensitivity of each person to light fluctuations, the precise measurement of flicker is not an easy task. IEC 61000-4-15 provides a detailed description of the structure and functional specifications of flicker measuring device called flickermeter. This measurement tool represents the relationship between voltage fluctuations and the human discomfort providing a short-term, P_{st}, and a long-term, P_{lt}, indicator. The P_{st} is the flicker severity evaluated over a short period (10 minutes is used in practice) and the conventional threshold of irritability is set in $P_{st} = 1$. The P_{lt} term is the flicker severity evaluated over a long period of two hours and it is obtained by using successive P_{st} values.
Fluctuating loads in the electrical power system, e.g. welding machines, arc furnaces or electric boilers, are the main sources of these perturbations in the electrical power system. Moreover, from the point of view of power generation, the connection of wind turbines to the grid can affect the ideal form of the voltage signal. Among the perturbations generated by the wind turbines, the fluctuations in voltage are the most notable (Ackerman, 2005). Rapid variations in wind speed produce fluctuating power, which can lead to voltage fluctuations at the Point of Common Coupling (PCC), which in turn generate flicker. The standard specifies a test for the voltage fluctuations with the aim of obtaining the measurements independently of the characteristics and conditions of the network to which the wind turbine is connected. Furthermore, the standard requires the characterization of the voltage fluctuations for two situations, namely continuous operation and switching operations.
The following paragraphs will describe the test procedures for both types of functional conditions.

2.4.1 Continuous operation

It is described as the normal operation of the wind turbine excluding start-up and shut-down operations. The standard establishes a processing and statistical evaluation scheme to obtain the flicker coefficients (see Fig. 3). These coefficients must be estimated from the current and voltage time-series measured during the continuous operation.
The specification establishes a specific test procedure with the aim of obtaining a normalized measure of the flicker emission. The phase-to-neutral voltage and the line current, $u_m(t)$ and

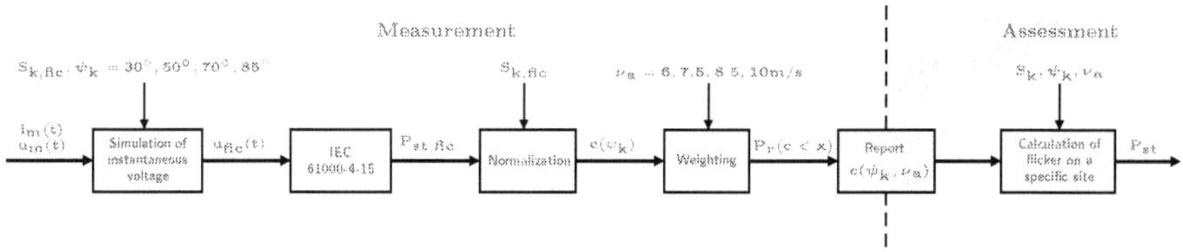

Fig. 3. Scheme of the measurement and assessment procedures for flicker during continuous operation of the wind turbines in accordance with IEC 61400-21.

$i_m(t)$, need to be processed for at least 15 registers of 10 min duration for each 1 $\frac{m}{s}$ wind speed bin between the cut-in wind speed and 15 $\frac{m}{s}$. For each time-series and for each network impedance, specified in the standard with the values of 30°, 50°, 70° and 85°, the fictitious voltage $u_{fic}(t)$ is calculated from the circuit of Fig. 4. This model represents a fictitious grid that enables the assessment of flicker caused exclusively by the wind turbine. In compliance with IEC 61000-4-15 and using the fictitious voltage $u_{fic}(t)$ as the input to the flickermeter, a flicker emission value, $P_{st,fic}$, can be obtained.

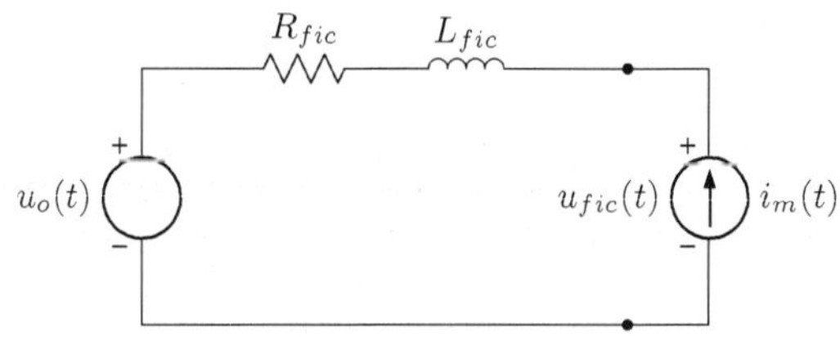

Fig. 4. Fictitious grid used for flicker assessment in wind turbines.

The flicker coefficient has to be determined for each of the calculated flicker emission values by applying the equation (4):

$$c(\psi_k) = P_{st,fic} \cdot \frac{S_{k,fic}}{S_n} \tag{4}$$

where S_n is the rated apparent power of the wind turbine and $S_{k,fic}$ is the short-circuit apparent power of the fictitious grid.
For each network impedance phase angle ψ_k, a weighting procedure calculates the weighted accumulated distribution functions of the flicker coefficients, $P_r(c < x)$, assuming four different Rayleigh distributed wind speeds of mean v_a = 6, 7.5, 8.5 and 10 $\frac{m}{s}$. For each accumulated distribution, the 99% percentile, $c(\psi_k, v_a)$, of the flicker coefficient is then reported.
The assessment procedure specifies how the reported flicker coefficients can be used to estimate the flicker emission from a single wind turbine or a group of wind turbines. The short and long-term flicker emission from the wind turbine installation must be compared with the short and long-term flicker emission limits for the relevant PCC, and with that purpose, these flicker emission terms must be obtained as follows:

$$P_{st} = c(\psi_k, v_a) \cdot \frac{S_n}{S_k} \tag{5}$$

where $c(\psi_k, v_a)$ is the flicker coefficient of the wind turbine, S_n is the rated apparent power of the wind turbine and S_k is the short-circuit apparent power at the PCC.
In case more wind turbines are connected to the PCC, the flicker emission due to the sum of them can be estimated as:

$$P_{st\Sigma} = \frac{1}{S_k} \cdot \sqrt{\sum_{i=1}^{N_{wt}} [c_1(\psi_k, v_a) \cdot S_{n,i}]^2} \quad (6)$$

where $c_i(\psi_k, v_a)$ is the flicker coefficient of the individual wind turbine, $S_{n,i}$ is the rated apparent power of the individual wind turbine and N_{wt} is the number of wind turbines connected to the PCC.

2.4.2 Switching operations

The standard establishes an alternative processing and statistical evaluation scheme during start-up or switching between generators (see Fig. 5). Four different parameters must be obtained to assess the consequences of the start-up and shut-down maneuvers of the wind turbine: the maximum number of switching operations within a 10 min and 2 hour period, N_{10_m} and N_{120_m} respectively, the flicker step factor $k_f(\psi_k)$ and the voltage change factor $k_u(\psi_k)$.
The specification establishes a procedure of measurements and subsequent calculations to determine $k_u(\psi_k)$ and $k_f(\psi_k)$ for each type of switching operation. The phase-to-neutral voltage and the line current, $u_m(t)$ and $i_m(t)$, need to be processed for at least 15 registers of a period T_p long enough to pass the transient of the switching operation. As in the case of the continuous operation, the fictitious voltage, $u_{fic}(t)$, and the flicker emission values, $P_{st,fic}$, are calculated. Flicker step factor and voltage change factor can be obtained by applying the expressions (7) and (8) respectively, and finally they are determined as the average result of the 15 calculated values.

$$k_f(\psi_k) = \frac{1}{130} \frac{S_{k,fic}}{S_n} \cdot P_{st,fic} \cdot T_p^{0,31} \quad (7)$$

$$k_u(\psi_k) = \sqrt{3} \cdot \frac{U_{fic,m\acute{a}x} - U_{fic,min}}{U_n} \cdot \frac{S_{k,fic}}{S_n} \quad (8)$$

where $U_{fic,m\acute{a}x}$ and $U_{fic,min}$ are the maximum and minimum one period root mean square (RMS) value of the voltage on the fictitious grid during the switching operation.
The assessment procedure for switching operations specifies how to estimate the flicker emission and voltage changes during switching operations on any specified site, using the reported flicker step factors and voltage change factors.

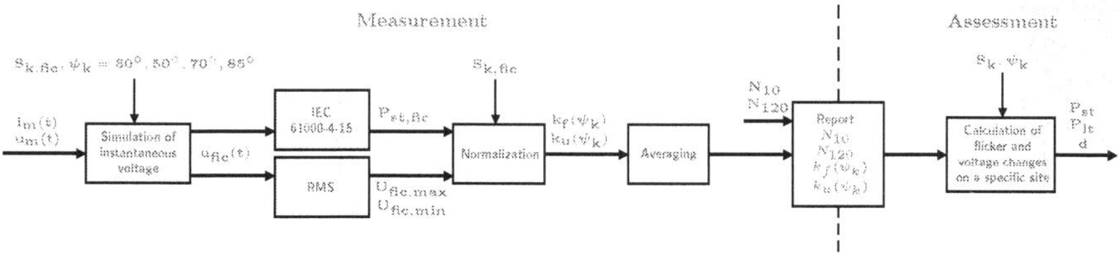

Fig. 5. Measurement and assessment procedures for flicker during switching operations of the wind turbines in accordance with IEC 61400-21.

During the assessment procedure, the established flicker emission limits must be compared with the short and long-term flicker parameters that can be obtained from the next expressions:

$$P_{st} = 18 \cdot N_{10_m}^{0.31} \cdot K_f(\psi_k) \cdot \frac{S_n}{S_k} \tag{9}$$

$$P_{lt} = 8 \cdot N_{120_m}^{0.31} \cdot K_f(\psi_k) \cdot \frac{S_n}{S_k} \tag{10}$$

In the case that more wind turbines are connected to the PCC, the flicker emission from the sum of them can be estimated from equation (11) and equation (12):

$$P_{st\Sigma} = \frac{18}{S_k} \cdot \left(\sum_{i=1}^{N_{wt}} N_{10_{m,i}} \cdot [k_{f,i}(\psi_k) \cdot S_{n,i}]^{3.2} \right)^{0.31} \tag{11}$$

$$P_{lt\Sigma} = \frac{18}{S_k} \cdot \left(\sum_{i=1}^{N_{wt}} N_{120_{m,i}} \cdot [k_{f,i}(\psi_k) \cdot S_{n,i}]^{3.2} \right)^{0.31} \tag{12}$$

where $N_{10_{m,i}}$ and $N_{120_{m,i}}$ are the number of switching operations of the individual wind turbine within a 10 min and 2 hour period respectively, and $k_{f,i}(k)$ is the flicker step factor of the individual wind turbine.

2.4.3 Relevant issues for the flicker test implementation

There are two relevant aspects that should be considered when implementing the test procedure under both functional conditions, normal operation and switching operations. First, the estimation of the fictitious voltage obtained from the resolution of the fictitious grid specified by the IEC 61400-21 standard. Second, the implementation of the IEC flickermeter, according to the functional specifications defined by the IEC 61000-4-15 standard.

2.4.3.1 Estimation of the fictitious voltage.

IEC 61400-21 standard specifies a method that uses current and voltage time-series measured at the wind turbine terminals to simulate the voltage fluctuations on a fictitious grid with no source of voltage fluctuations other than the wind turbine. The fictitious grid is shown in Fig 4. The turbine is represented by a current generator with a value of $i_m(t)$, and the network to which it is connected is represented by its Thevenin equivalent circuit and an ideal phase-to-neutral voltage source with the instantaneous value $u_0(t)$. The network impedance is composed of a resistance R_{fic} in series with an inductance L_{fic}.
The ideal voltage source $u_0(t)$ models a network free of fluctuations and is defined as:

$$u_0(t) = \sqrt{\tfrac{2}{3}} \cdot U_n \sin \alpha_m(t) \tag{13}$$

The electrical angle $\alpha_m(t)$ of the fundamental can be described as:

$$\alpha_m(t) = 2\pi \int_0^t f(t)dt + \alpha_0 \tag{14}$$

where $f(t)$ is the fundamental frequency, which may vary over time, and α_0 is the electrical angle of the fundamental frequency at $t = 0$.

With this model a fictitious voltage, $u_{fic}(t)$, at the wind turbine terminals can be obtained using the expression (15):

$$u_{fic}(t) = u_0(t) + R_{fic} \cdot i_m(t) + L_{fic} \cdot \frac{di_m(t)}{dt} \tag{15}$$

The main error source in the calculation of $u_{fic}(t)$ appears from the estimation of $u_0(t)$, which must fulfill the following two conditions:

1. Flicker on the voltage $u_0(t)$ should be zero.
2. The ideal voltage source $u_0(t)$ should have the same electrical angle $\alpha_m(t)$ as the fundamental frequency of the measured voltage $u_m(t)$.

A small error in the estimation of the phase of the fundamental frequency of $u_m(t)$ can generate important changes in $u_{fic}(t)$ that significantly affect the $P_{st,fic}$ value calculated (Gutierrez et al., 2008). To obtain an accurate estimation of $u_0(t)$ fulfilling the previous conditions, the selection of a proper signal processing technique turns out to be a key factor. First, it is important to understand that $u_m(t)$ is a band-limited signal and most of its power is concentrated around its fundamental frequency, which is equal or very close to 50 Hz. As it has been demonstrated in previous works (Gutierrez et al., 2008), the obtention of a precise estimation of $u_0(t)$ entails necessarily the combination of two processes, applied to $u_m(t)$:

- Filtering the fundamental frequency of $u_m(t)$. We propose the implementation of a narrow band adaptive filter, whose results can be improved by an anticausal zero-phase filter implementation.
- Calculation of the instantaneous phase of the fundamental frequency of $u_m(t)$ by implementing a classical zero-crossing method.

Next, we will describe the main technical considerations for a proper implementation of those processes.

1. *Narrow band filter and anticausal zero-phase filter implementation.*

   The typical method of eliminating a narrow band interference consists of filtering the signal using a notch filter. Our case is the inverse, given that the objective is the fundamental component of the signal $u_m(t)$. Working in the discrete domain, a very narrow band-pass filter needs to be designed around the discrete pulsation corresponding to the fundamental frequency $\Omega_0 = 2\pi\frac{f_0}{f_s}$ with f_0 = 50 Hz. A proper solution could be a narrow band-pass filter implemented through an adaptive scheme based on the Least Mean Square algorithm (LMS). This design makes it possible to obtain the fundamental component at 50 Hz without distortion and without any delay at the output with respect to the input (Widrow & Stearns, 1985).

   From the transfer function of the noise-canceller, the transfer function of the adaptive filter output can be calculated as:

$$H(z) = 1 - H_1(z) = 2\mu C^2 \cdot \frac{z\cos(\Omega_0) - 1}{z^2 - 2(1-\mu C^2)z\cos(\Omega_0) + 1 - \mu C^2} \tag{16}$$

   corresponding to a 3 dB bandwidth $BW = 2\mu C^2$ rad = $\frac{f_s \mu C^2}{\pi}$ Hz.

   The frequency response $H(\Omega)$ corresponds to a narrow band-pass filter that enables obtaining the fundamental component of $u_m[n]$. When working with $C = 1$, $f_s = 3200\frac{S}{s}$

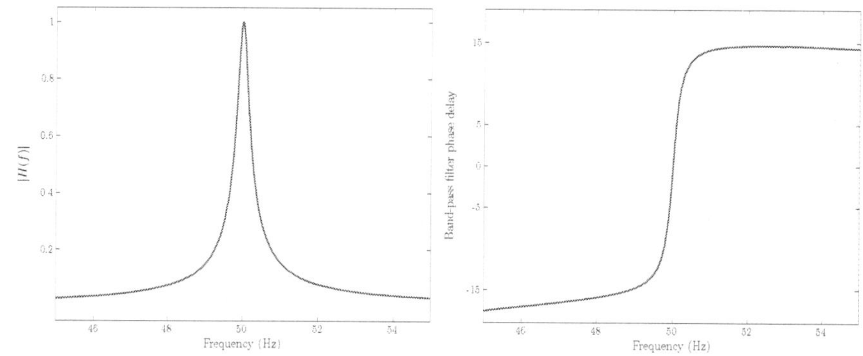

(a) Module of the band-pass filter frequency response. (b) Phase delay of the band-pass filter.

Fig. 6. Frequency responses of the band-pass filter.

and $\mu = 0.0003$, a bandwidth of approximately 0.3 Hz is found around the 50 Hz component. Fig. 6 shows the module and the phase delay $\tau_f(\Omega) = \frac{-\phi(\Omega)}{\Omega}$ of $H(\Omega)$ scaling the axis of frequency in Hz.

The main problem that $H(z)$ presents to obtain the fundamental component of $u_m[n]$ is the abrupt behavior of the phase delay around 50 Hz, that produces displacements of several samples in the output due to eventual small variations of the fundamental frequency around 50 Hz. This causes an appreciable error in the P_{st} of $u_{fic}(t)$. To solve this problem, the phase distortion can be eliminated using the Anticausal Zero-Phase Filter Implementation.

Considering the processing scheme in Fig. 7, after filtering in the forward direction, the filtered sequence is reversed and run back through the filter. The result has exactly zero-phase distortion. In fact, in the frequency domain $Y(\Omega) = U_m(\Omega) \cdot |H(\Omega)|^2$. The magnitude is the square of the filter's magnitude response, and the filter order is double the order of $H(z)$.

This implementation can only be used in cases in which $u_m[n]$ is a finite duration signal known before being filtered. From the signal obtained, $y[n]$, it is necessary to eliminate the transitory at both ends.

2. *Zero-Crossing Method.*

The estimation of the frequency of the power system using the zero-crossing technique has been well known for a long time (Lee & Devaney, 1994). Constructing the instantaneous phase of the signal $u_m(t)$ from the frequency or period of each cycle of $u_m(t)$, is straightforward.

Working in the discrete domain, the algorithm searches for the positions of the contiguous samples of $u_m(t)$ that mark a transition of values from positive to negative. To achieve a

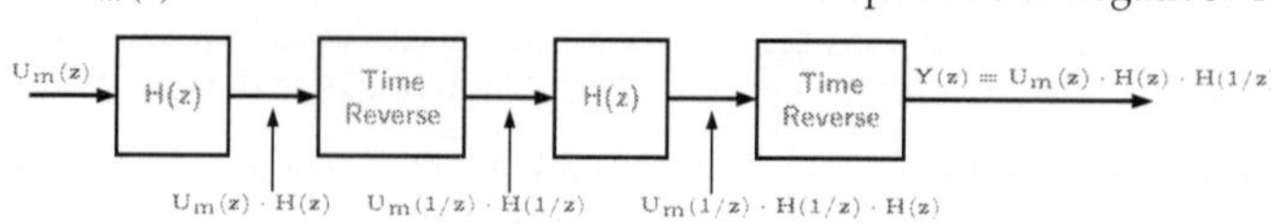

Fig. 7. The anticausal zero-phase filter scheme.

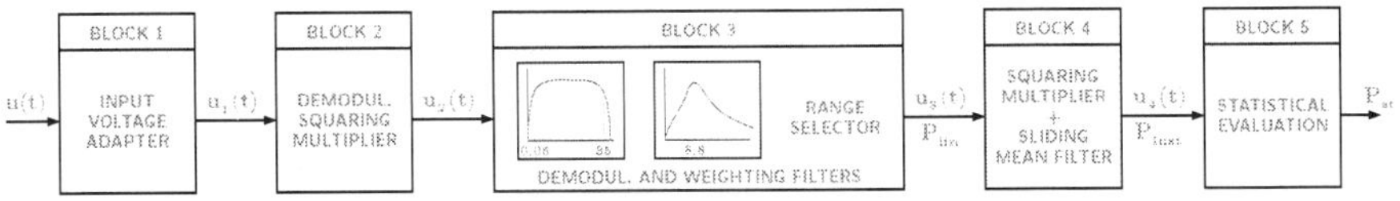

Fig. 8. Block diagram of the IEC flickermeter according to IEC 61000-4-15 (IEC-61000-4-15 Ed. 2.0, 2010).

more precise approximation to the zero-crossing point, a linear interpolation between the points of the transition can be used.

Knowing the number and the fraction of the samples that make up a period, reconstruction of the instantaneous phase of the fundamental component is done, sharing the 2π radians uniformly for each sampling instant.

2.4.3.2 Estimation of $\mathbf{P_{st,fic}}$

Once the fictitious voltage $u_{fic}(t)$ has been estimated, the standard specifies the calculation of the short-term flicker severity P_{st} produced by that voltage. The measurement of the flicker severity is a complex procedure whose functional specifications are detailed in the IEC 61000-4-15 standard, namely IEC flickermeter, and it is worth providing some details and relevant considerations about its implementation. Fig. 8 shows the block diagram of the IEC flickermeter.

Block 1 of the flickermeter scales the input voltage to an internal reference value. The objective of this block is to make flicker measurements independent of the input voltage level.

Block 2 recovers the voltage fluctuations by squaring the scaled input voltage, thereby simulating the behavior of an incandescent lamp.

Block 3 of the flickermeter consists of three cascaded filters, followed by a range selector that determines the sensitivity of the device. The first two filters are part of the demodulation process and consist of a first-order high-pass filter (cutoff frequency = 0.05 Hz) and a sixth-order low-pass Butterworth filter (cutoff frequency = 35 Hz). The third filter models the behavior of the *lamp–eye* system. The analogue response of this band-pass filter is defined in the standard for 230 V and 120 V reference lamps.

Block 4 implements an eye–brain model. It includes a squaring multiplier followed by a low-pass filter that is specified to be a sliding-mean filter, having the transfer function of a 1st-order low-pass resistance–capacitance filter with a time constant of 300 ms. The output of Block 4 represents the instantaneous flicker sensation P_{inst}. A unit output from Block 4 corresponds to the reference human flicker perceptibility threshold. In Block 5, P_{st} is calculated by performing a statistical classification of P_{inst} over a short period of time (usually 10 min). The method for obtaining the P_{st} value is a multipoint algorithm that uses the percentiles obtained from the cumulative probability distribution of P_{inst}, namely

$$P_{st} = \sqrt{0.0314 \cdot P_{0.1} + 0.0525 \cdot P_{1s} + 0.0657 \cdot P_{3s} + 0.28 \cdot P_{10s} + 0.08 \cdot P_{50s}} \quad (17)$$

For the measurements performed in this work, we have implemented a highly accurate IEC flickermeter. This reference flickermeter is the complete digital MatLab implementation previously used in other studies (Ruiz et al., 2007; 2010). Its main features are:

1. The input can be either analytically generated signals, as rectangular voltage fluctuations, or actual registered signals that have been digitally formatted.

2. It uses any input sampling selected from these two sets: $f_s = 1600, 3200, 6400, 12800$ and $25600 \frac{S}{s}$ and $f_s = 1280, 2560, 5120, 10240$ and $20480 \frac{S}{s}$.
3. It performs a decimation process at the output of the sixth-order low-pass Butterworth filter to allow a constant sampling rate of $1600 \frac{S}{s}$ for the first set of sampling frequencies and $1280 \frac{S}{s}$ for the second, in the following blocks, independently of the input sampling rate.
4. To avoid errors coming from the classification process in terms of number of classes, type of classification or type of interpolation between the classes, our reference flickermeter does not classify the P_{inst} signal. It stores all the samples of P_{inst} during the short-term period of 10 min and calculates the percentiles of (17) in an absolutely accurate way.
5. Finally, it is important to remark that the complete precision of the reference flickermeter has been contrasted with other well-known implementations (Key et al., 1999; Mombauer, 1998), also used in previous works (Gallo et al., 2006; WG2CIGRÉ, 2004).

3. A system for the measurement of the power quality characteristics of grid-connected wind turbines

The assessment of the power quality characteristics of a wind turbine requires the obtention of several voltage and current time-series for different wind speeds. These time-series must be obtained for two types of functional status of the wind turbine: continuous and switching operations.

Moreover, all that information must be processed to measure several parameters, which requires, in principle, the connection of different power quality analyzers. There are not so many commercial analyzers particularly designed to fulfill the requirements of the IEC 61400-21 standard. Furthermore, there are very few research works about integrated systems to assess the power quality of grid-connected wind turbines (Gherasim et al., 2006).

We developed a measurement system for the acquisition, storage and processing of the voltage, current and wind speed time-series required by the standard. In order to provide more flexibility to the measurement system, instead of using a commercial equipment, we chose the implementation of our own system. The main rationale of our system is to divide the whole measurement into independent processes:

1. The recording and storage of the wind, voltage and current time-series.
2. The off-line measurement and assessment of the power quality characteristics by post-processing the stored time-series.

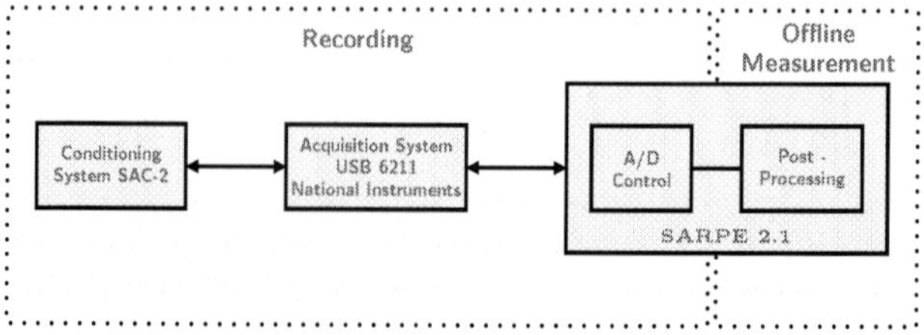

Fig. 9. Scheme of the measurement system.

To perform those processes we developed two interconnected sub-systems: a signal conditioning system (SAC-2) and a control system (SARPE 2.1). The scheme of the overall measurement system is shown in Fig. 9. The conditioning system is a hardware module transforming the three-phase voltage and current, as well as the wind speed, to the appropriate levels for the post-processing. This operation is managed by the control system. This is a software tool that controls the acquisition and stores the voltage, current and wind speed time-series. The control system also includes a post-processing module that works off-line by reading the recorded time-series and calculates the parameters of the power quality characteristics of the wind turbine.

3.1 Conditioning system SAC-2

This system converts the voltage, current and wind speed input levels to the appropriate levels for the final measurement. Fig. 10 shows a photograph of the developed hardware platform, SAC-2.

For a precise conditioning of the input levels, this hardware platform provides four voltage channels and four different scales per channel (see Table 2 (a)).

There are also four current channels, and four different types of current sensors can be used (see Table 2 (b)). There are two additional analog channels to register wind characteristics.

The system provides several clocks to use as external sampling frequencies in the acquisition by the control module. The first set of sampling frequencies allows the use of a number of samples per cycle of 50 Hz always corresponding to a power of 2 (f_S = 1600, 3200, 6400, 12800 and 25600 $\frac{S}{s}$). A second set provides a number of samples per 10 cycles of 50 Hz or 12 cycles of 60 Hz corresponding to a power of 2 (f_S = 1280, 2560, 5120, 10240 and 20480 $\frac{S}{s}$). This second group of sampling frequencies makes possible the implementation of the harmonics and interharmonics measurement method specified by the standard. Each channel includes a fifth order Butterworth anti-aliasing filter with adjustable cutoff frequency. A phase-locked loop (PLL) synchronizes the sampling rate to the first channel grid frequency, either 50 or 60 Hz.

The system also includes four digital inputs, activated by dry contacts, to trigger the start of the acquisition by the control system.

3.2 Control system SARPE 2.1

The control system is a MatLab tool running on a PC laptop and consists of two modules: the acquisition module and the post-processing module.

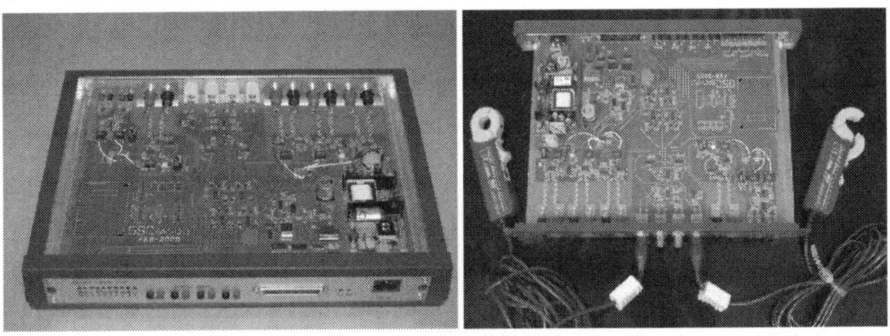

Fig. 10. Layout of the conditioning system, SAC-2.

(a) Working ranges for the voltage channels.

Scale (V_{rms})	Range (V_{rms})
60	16-78
150	40-195
300	78-390
600	156-780

(b) Working ranges for the current sensors.

Sensitivity (mV/A)	Scale (A_{rms})	Range (A_{rms})
0.1	5000	1300-6500
	500	130-650
1	500	130-650
	50	13-65
10	50	13-65
	5	1.3-6.5
100	50	1.3-6.5
	0.5	0.13-0.65

Table 2. Conditioning system working range.

The acquisition module manages both the acquisition and the recording operations. Its main functions are:

1. Acquisition of the conditioned signals using the DAQ 6062E card from National Instruments (12-bit resolution).
2. Selection of different parameters that configure the register:
   a. Internal clock from the acquisition card or external clock from the conditioning system.
   b. Sampling frequency.
   c. Use of the anti-aliasing filters.
   d. Activation of the PLL.
   e. Channels to be recorded.
   f. Scale of each selected voltage and current channels.
   h. Duration of the register.
3. Checking the functional status of the wind turbine to validate the storage of the corresponding time-series.
4. Selection of the trigger type:
   a. Delayed start.
   b. Digitally controlled start.
5. Communication by GSM/GPRS system to remotely control the status of the registering process.

On the other hand, the post-processing module recovers the recorded sampled data and processes them according to the procedures specified by the IEC 61400-21 standard.
The off-line processing provides several advantages for the wind turbine certification process. Since the power quality standards may change, the off-line processing makes it easier to incorporate these changes by software modifications. In this sense, this method of post-processing allows the analysis of the waveforms that have produced a specific power quality characteristic. It is not necessary to retest a specific wind turbine to calculate the characteristics. Another advantage is that all parties involved in the certification process have access to the stored information.

4. Case study of power quality characteristic of grid-connected wind turbines

There are not too many works assessing the power quality in wind farms according to the IEC 61400-21 standard (Foussekis et al., 2003; Srensen, 2001; Srensen et al., 2007). This section shows the results of measurements performed on two wind turbines with different constructive characteristics and located in an experimental wind farm in the northwest of Spain[1]. The first tested wind turbine (WT1) corresponds to a machine with a double speed asynchronous generation system (4 and 6 poles), fixed sail passage and fixed generator speed, with a nominal power of 660 Kw and nominal voltage of 690 V. In this wind turbine, a total of 135 records were registered under different wind speed and power conditions, each record containing a current and voltage 10 min time-series. The second wind turbine (WT2) has a 4-pole synchronous generating system and electronic power control, variable sail passage and variable generator speed, and it provides a nominal power of 800 KW and a nominal voltage of 1000 V. In this wind turbine, a total of 75 voltage and current 10 min time-series were recorded under different wind speed and power conditions.
Voltage fluctuations and harmonic and interharmonic content of the registered records are analyzed hereunder.

4.1 Voltage fluctuations

Each set of measured voltage-current time-series pair $u_m(t), i_m(t)$ is used as input to calculate the fictitious voltage $u_{fic}(t)$ on the fictitious grid. This has been done for the four different network impedance phase angles (ψ_k) specified in the IEC 61400-21 standard, using a short-circuit apparent power of $S_{k,fic} = 20 \cdot S_n$.

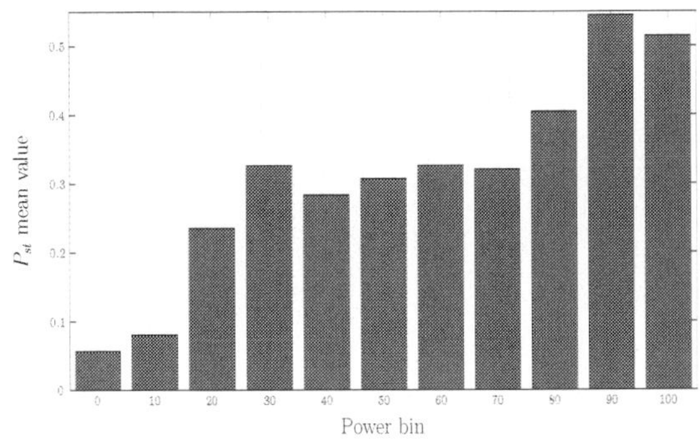

Fig. 11. Mean values of $P_{st,fic}$ in terms of the active power bins for WT1.

In order to calculate the voltage $u_0(t)$ required to obtain $u_{fic}(t)$, $u_m(t)$ has been filtered with an IIR band-pass notch filter; the filter implementation is based on anticausal zero-phase technique. Each of the obtained $u_{fic}(t)$ (one for each ψ_k for each $u_m - i_m$ pair) has been input to an IEC 61000-4-15 compliant flickermeter to obtain the flicker severity value $P_{st,fic}$.
Fig. 11 shows the results obtained for the first wind turbine working with a network impedance phase angle $\psi_k = 85°$. The 135 $P_{st,fic}$ values obtained for this wind turbine have been grouped according to the power of the machine, for the operation of the wind turbine within the active power bins 0, 10, 20, ..., 100% of P_n. The first bin corresponds to $P < 5\%$

[1] The authors would like to thank SOTAVENTO GALICIA S.A. (Spain) for making the signals available free of charge for the purpose of this work.

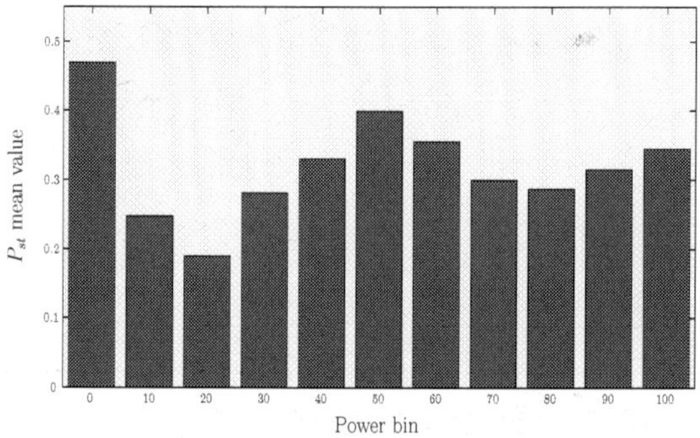

Fig. 12. Mean values of $P_{st,fic}$ in terms of the active power bins for WT2.

of P_n; the last one to $P > 95\%$ of P_n; the intermediate bins correspond to 10%-wide ranges centered in the bin midpoints. It can be observed that $P_{st,fic}$ values increase as the working power increases.

In the same way, Fig. 12 shows the $P_{st,fic}$ values obtained for the 75 u_{fic} associated to the second wind turbine; the values are also given for $\psi_k = 85°$, as in the previous case. In this case the $P_{st,fic}$ values are high when the machine operates at low power.

4.2 Current harmonics and interharmonics

The harmonic and interharmonic content of each current signal has been obtained according to the IEC 61000-4-7 standard. A 10-cycle (in a 50 Hz system, approximately 200 ms) rectangular window has been used. For each window within a 10 min period, the Discrete Fourier Transform (DFT) has been calculated and the resulting spectral lines have been grouped to obtain the harmonic subgroups (HS) and the interharmonic centered subgroups (ICS).

The resulting HS and ICS time-series have been smoothed using a first order low-pass filter with a time constant of 1.5 s. The smoothing filter introduces a transient; in order to eliminate this transient the first 10 seconds of each HS and ICS have been suppressed. Afterwards the HS and ICS RMS value has been calculated. Therefore given a 10 min current time-series, a single value is obtained for each HS and ICS. In order to illustrate the effect of the smoothing process, Fig. 13 shows the time evolution of the 7th harmonic subgroup $I_{sg,7}$ (a) and the 7th interharmonic centered subgroup $I_{isg,7}$ (b) of phase1-current in a 10 min register recorded in the first wind turbine while working at 53% of nominal power. The evolution of the RMS value of $I_{sg,7}$ and $I_{isg,7}$ along the 10-cycle intervals, i.e. the input to the smoothing low-pass filter, is represented in blue; the output from the smoothing low-pass filter is shown in red. It can be observed that smoothing reduces the abrupt changes produced between two adjacent windows.

Fig. 14 shows in more detail the initial 20 s of Fig. 13. The transient produced by the smoothing process is clearly observed. This transient has to be suppressed so that it is not taken into consideration in the calculation of the HS and ICS of the analyzed record.

Fig. 15 shows the HS and ICS content (up to 12th order harmonic) of phase1-current in the same 10 min record of the first wind turbine in which Fig. 13 and 14 are based. A high contribution of 5th, 7th and 11th harmonics can be observed, whereas the rest of harmonics and all the interharmonics take very low values.

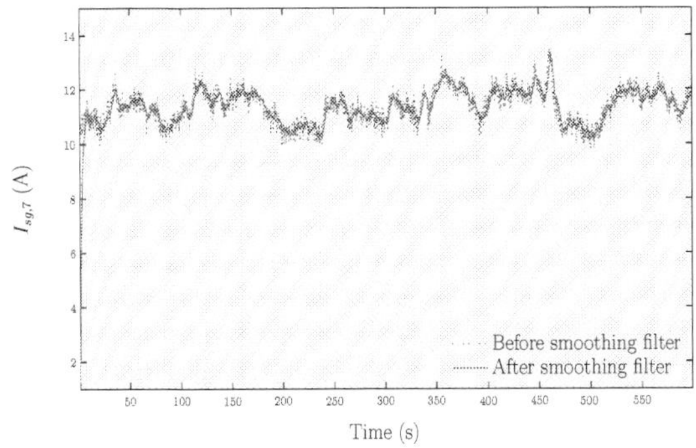

(a) RMS value of 7th harmonic subgroup.

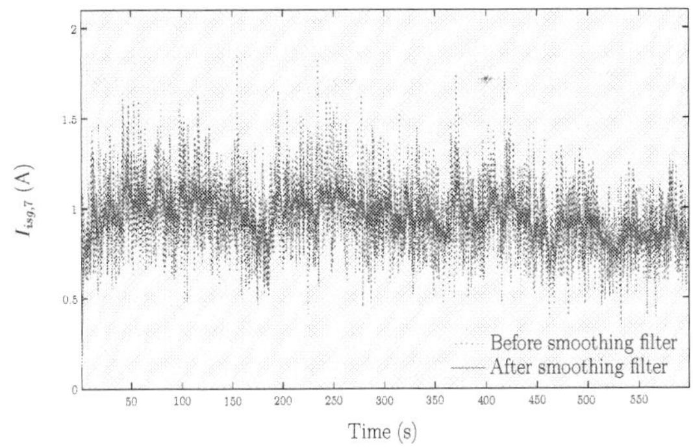

(b) RMS value of 7th interharmonic centered subgroup.

Fig. 13. Effect of smoothing.

Fig. 16 shows the same information as Figure 15, but for a 10 min record of the second wind turbine. The mean power of the selected record is also 53% of S_n, as in the first turbine. The comparison between Fig. 15 and 16 shows that harmonic and interharmonic content distribution is completely different in both turbines. In the second one, harmonic content is more uniformly distributed among all the HS and ICS.

The 10 min averages of each subgrouped harmonic and interharmonic have been calculated for each of the 135 10-min time-series recorded in the first wind turbine. Fig. 17 represents the maximum 10-min averages of 7th harmonic (a) and 7th interharmonic (b) in each 10% power bin. The first bin corresponds to $P < 5\%$ of P_n; the last one to $P > 95\%$ of P_n; the intermediate bins correspond to 10%-wide ranges centered in the bin midpoints (10, 20, ..., 90). The results are shown in percentage of I_n.

The same process is performed with the phase1-current of the 75 records of the second turbine. Fig. 18 shows the maximum 10-min averages of 7th harmonic (a) and 7th interharmonic (b) in percentage of I_n in each 10% power bin.

The comparison between Fig. 17 and 18 shows that the maximum value of 7th harmonic content is higher in the first wind turbine than in the second one for nearly all the power

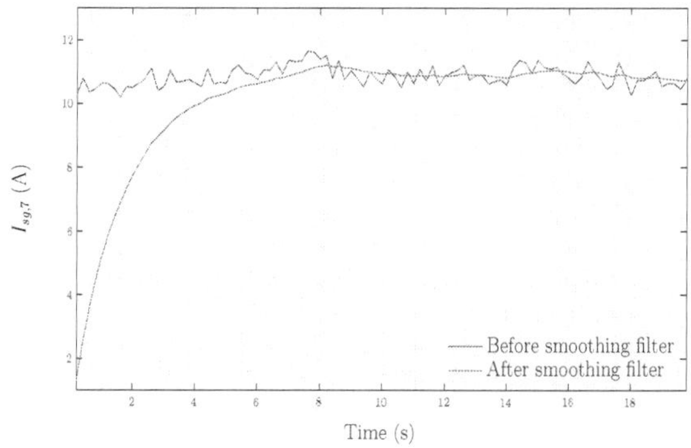

(a) RMS value of 7th harmonic subgroup.

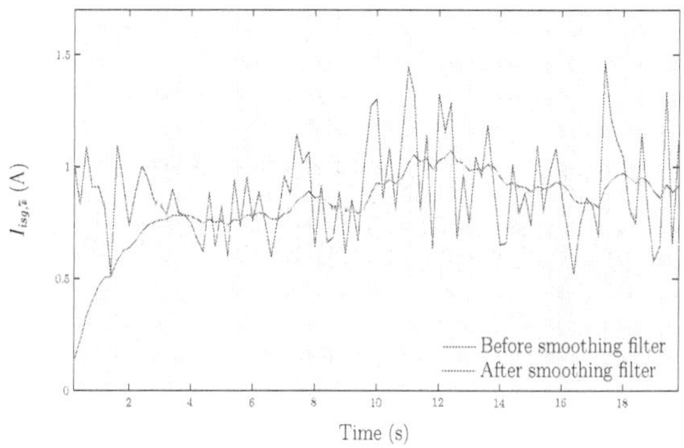

(b) RMS value of 7th interharmonic centered subgroup.

Fig. 14. Transient time due to smoothing.

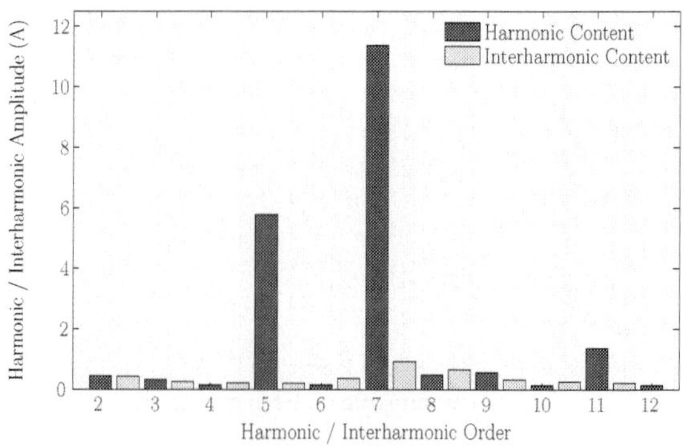

Fig. 15. Harmonic and Interharmonic content: WT1, $\frac{P}{S_n} = 53\%$.

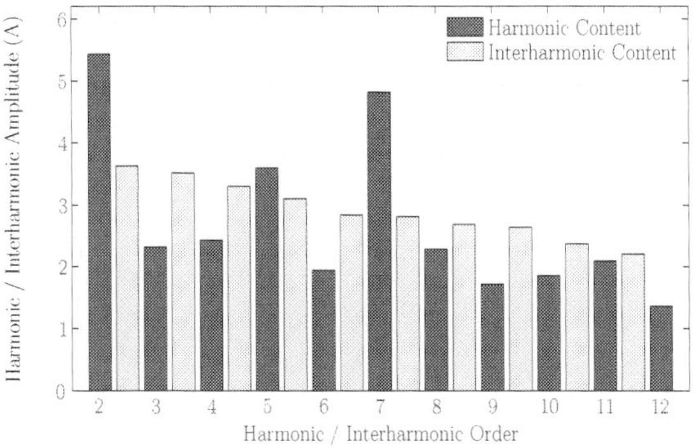

Fig. 16. Harmonic and Interharmonic content: WT2, $\frac{P}{S_n} = 53\%$.

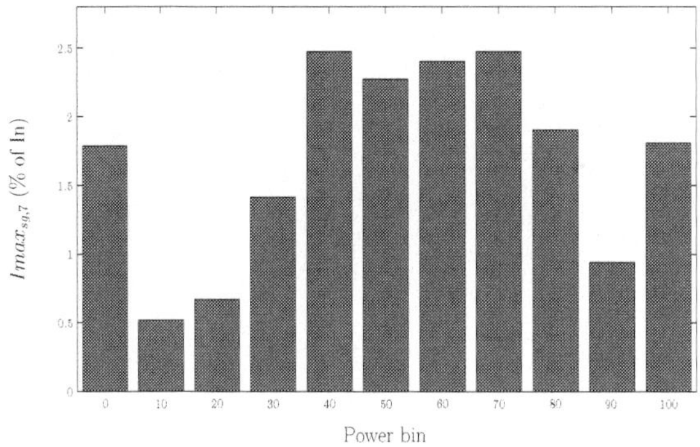

(a) 7th harmonic subgroup.

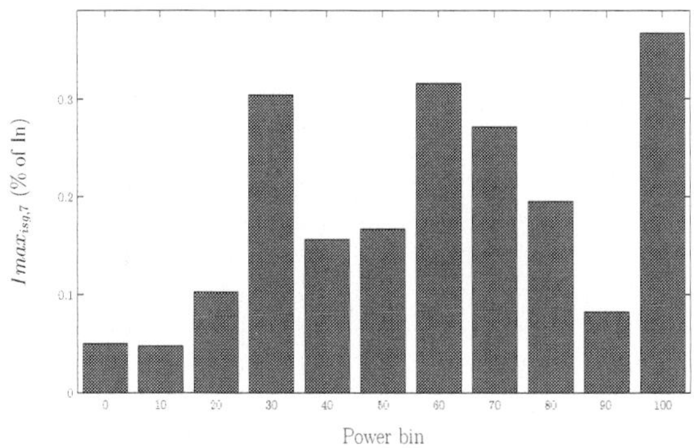

(b) 7th interharmonic centered subgroup.

Fig. 17. WT1: Maximum 10-min average of a given harmonic/interharmonic (in % of I_n) in each power bin.

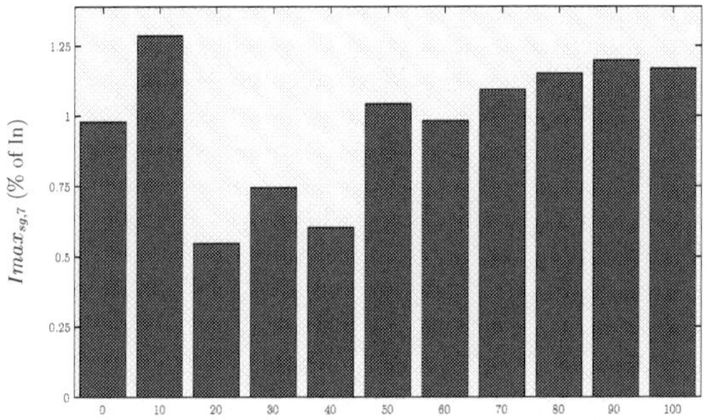

(a) 7th harmonic subgroup.

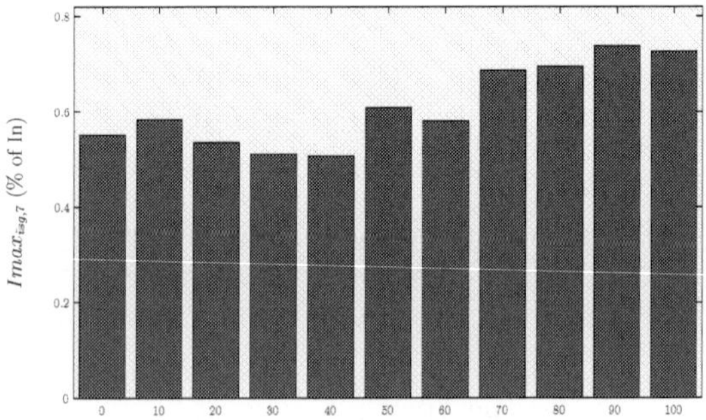

(b) 7th interharmonic centered subgroup.

Fig. 18. WT2: Maximum 10-min average of a given harmonic/interharmonic (in % of I_n) in each power bin.

bins. The reverse consideration is true for 7th interharmonic: its content is lower in the first wind turbine than in the second one for all the power bins.

5. Conclusions

Power injection from wind turbines affects substantially the power quality. This chapter described the main parameters involved in the assessment of the power quality of grid-connected wind turbines. The definition of those parameters and the procedures and methods for their assessment are compiled in the IEC 61400-21 standard. This text has become the reference normative for the certification of the grid-connected wind turbines in terms of power quality. According to it there are seven parameters compromising the required power quality characteristic of a wind turbine: voltage fluctuations or flicker; harmonics and interharmonics; voltage drops; active power; reactive power; grid protection and reconnection time.

The implementation of the measurement and assessment procedures specified by the standard requires a deep knowledge and experience on power quality issues. Moreover, all the procedures require the storage and processing of a great amount of data from voltage, current and wind time-series. For that purpose we developed our own measurement system, a useful tool specifically designed for the assessment of the power quality of a grid-connected wind turbine. This system acquires, stores and processes the voltage, current and wind speed time-series required by the standard, providing the output parameters specified by the standard for the assessment of the power quality characteristics.

The methods to assess the voltage fluctuations and harmonic content are particularly demanding. Their measurements are based on complex functional systems and advanced signal processing techniques. We have extensively detailed some of these techniques and we have applied them to a case study based on the analysis of both parameters on two wind turbines with different constructive characteristics and located in an experimental wind farm in the northwest of Spain. Since there are not too many experimental works assessing the power quality in wind farms according to the IEC 61400-21 standard, this work can be useful to enlarge the knowledge about the influence of the wind turbines on the power quality.

6. References

Ackerman, T. (2005). Wind Power in Power Systems., John Wiley & Sons, Ltd.

Foussekis, D., Kokkalidis, F., Tentzerakis, S. & Agoris, D. (2003). Power quality measurements on different types of wind turbines operating in the same wind farm, *Proceedings on European Wind Energy Conference and Exhibition* .

Gallo, D., Landi, C. & Pasquino, N. (2006). Design and Calibration of an Objective Flickermeter, *IEEE Transactions on Instrumentation and Measurement* 55(6): 2118–2125.

Gherasim, C., Croes, T., den Keybus, J. V., Driesen, J. & Belmans, R. (2006). Development of a flickermeter for grid-connected wind turbines using a DSP-based prototyping system, *Instrumentation and Measurement, IEEE Transactions on* 55(2): 550–556.

Gutierrez, J., Ruiz, J., Leturiondo, L. & Lazkano, A. (2008). Filcker Measurement System for Wind Turbine Certification, IEEE Transactions on Instrumentation and Measurement 57(12): 375–382.

IEC-61000-4-15 Ed. 2.0 (2010). Electromagnetic compatibility (emc) Ű part 4: Testing and measurement techniques - section 15: Flickermeter functional and design specifications.

IEC-61000-4-7 (2002). Electromagnetic compatibility (emc) Ű part 4: Testing and measurement techniques - section 7: General guide on harmonics and interharmonics measurements and instrumentation, for power supply systems and equipment connected thereto.

IEC-61400-21 Ed. 2.0 (2008). Wind turbine generator systems. Part 21: Power quality requirements for grid connected wind turbines.

Key, T., Nastasi, D., Sakulin, H., Harding, J. & Cooke, T. (1999). System Compatibility Research Project Final Report, Task 21: Power Line Monitors, Part II: Flickermeters, EPRI PEAC Corporation .

Lee, J. & Devaney, M. (1994). Accurate measurement of line frequency in the presence of noiseusing time domain data, Proceedings of the 10th IEEE Instrumentation and Measurement Technology Conference., pp. 1016–1019.

Mombauer, W. (1998). Calculating a new reference point for the IEC-flickermeter, *European Transactions on Electrical Power* 8(6): 429–436.

Ruiz, J., Gutierrez, J. & Irusta, U. (2007). Singular Frequencies in Rectangular Fluctuations in the IEC Flickermeter, *IEEE Transactions on Power Delivery* 22(2): 1255–1256.

Ruiz, J., Gutierrez, J., Lazkano, A. & Ruiz de Gauna, S. (2010). A Review of Flicker Severity Assessment by the IEC Flickermeter, *IEEE Transactions on Instrumentation and Measurement* 59(8): 2037–2047.

Srensen, P. (2001). European Wind Turbine Testing Procedure Developments. Task 2: Power Quality, Ris National Laboratory .

Srensen, P., Cutululis, N., Lund, T., Hansen, A., Srensen, T., Hjerrild, J., Donovan, M., Christensen, L. & Nielsen, H. (2007). Power quality issues on wind power installations in denmark, *Power Engineering Society General Meeting, 2007. IEEE*, pp. 1–6.

WG2CIGRÉ (2004). Test Protocol for IEC Flickermeter used in Power System Voltage Monitoring.

Widrow, B. & Stearns, S. (1985). Adaptive Signal Processing, Prentice Hall.

25

Optimal Selection of Drive Components for Doubly-Fed Induction Generator Based Wind Turbines

Dr. Davide Aguglia[1,2], Prof. Philippe Viarouge[1], Prof. René Wamkeue[3] and Prof Jérôme Cros[1]
[1]*LEEPCI Laboratory, Laval University, G1K 7P4 Quebec (QC)*
[2]*CERN – European Organization for Nuclear Research, Electric power converter Group, CH- 1211 Geneva 23*
[3]*UQAT, Abitibi-Témiscamingue Quebec University, J9X 5E4 Quebec (QC)*
[1,3]*Canada*
[2]*Switzerland*

1. Introduction

This chapter presents recent advancements and global design methods, as well as new results, regarding the electromechanical drive of wind turbines equipped with Doubly-Fed Induction Generators (DFIG). Currently DFIG is the most popular topology used by the large wind turbine manufacturers, because a wide variable speed range with a reduced static converter size can be obtained with a suitable control of the real and reactive power flows (Pena et al., 1996). The variable speed operation of wind generation systems presents several advantages that have been well established in the literature (Zigner et al., 1997) and improves the annualized wind energy yield. However the design process of such an electromechanical system needs suitable compromises between the performances of the wind turbine on the whole speed range and the respective characteristics of the induction machine, the gearbox and the static converters. The optimal solution in terms of performance and cost must be derived from a global design approach. For a given turbine, a given DFIG and a given annual wind distribution, a suitable compromise must be found between the gearbox ratio and the sizing of the power converters to maximize the annual energy yield. The choice of the gearbox ratio for variable speed DFIG is not detailed in the literature. This chapter presents a methodology that can be used for the optimal determination of at least these two important variables of the design process in terms of annual energy yield and initial cost. To demonstrate the advantages of the use of a global approach to define the best compromises in terms of relative size between the drive components and control laws is one of this chapter's goals.

Figure 1 presents the structure of the topology under study, which is using a four quadrants variable frequency static converter.

The purpose of this global approach is to determine the optimal dimensioning, or sizing, compromises to be adopted in order to optimize the overall performances of the

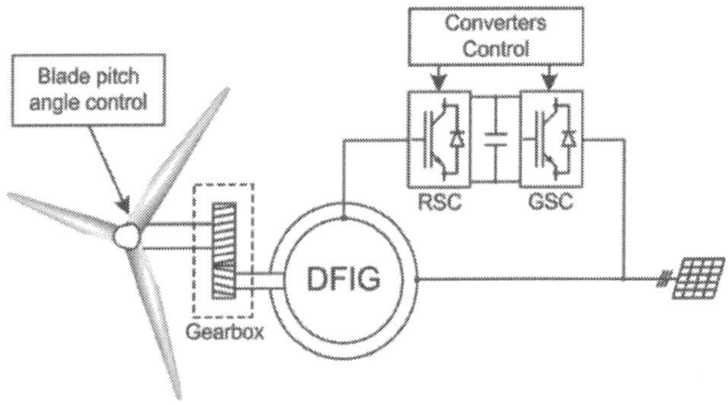

Fig. 1. Structure of the DFIG based wind turbine under study.

electromechanical drive. For instance one has to define the gearbox ratio, the power converter rating and the steady-state control laws which are maximizing the wind turbine annual production at minimal cost and minimal mass installed into the nacelle (Aguglia et al., 2007a). A global analysis is needed to achieve this objective. In the industry, each drive component is offered by a different supplier, and the final assembly doesn't necessarily represent a globally optimized solution in terms of sizing or cost. Except recent publications from the authors of this chapter, such an approach is not presented in the literature. The annual production of a wind turbine is one of its most important global performances. The higher the energy production, the higher the investment returns. To determine the annual production of a wind turbine one has to adopt a global analysis, by obtaining all the necessary models which allow the prediction of the losses of all sub-components of the drive. Furthermore, because the annual production depends on the wind statistics of a given site, the models used for its computation should be flexible enough to allow the computation of losses for any operation point of the plant, characterized by the generator torque, the rotating speed and the electrical power factor. The chapter will present the method allowing taking into account a statistical and cyclical operation of such a plant (Aguglia et al., 2007b). For each operating point of the plant, the rotor phasor voltage and the current imposed, or controlled, by the Rotor Side Converter (RSC) (refer to Fig. 1) must be imposed by a suitable control law to optimize the performance. To obtain a fast computation of the overall performances of the system on the whole variable speed range, one needs to obtain the RSC control laws for each operating point. A novel analytical method to derive such laws is also presented in the chapter (Aguglia et al., 2008). Once all the performances can be easily determined on the whole variable speed range, it is possible to perform a sizing analysis. For instance, it is possible to determine the annual production versus the dimensional constraints or the total mass installed into the nacelle. This chapter presents guidelines to answer to the following questions:

- How to select the gearbox ratio in order to optimize the annual production?
- What is the influence of the maximal voltage delivered by the RSC on the annual production? (DC-Bus voltage limit imposed by the switches technology).
- What is the influence of the maximal power converter apparent power selection on the variable speed range and the annual production?
- What are the compromises between the annual production and the total mass installed into the nacelle?

2. Models of the wind energy conversion system

2.1 Wind and wind turbine models

The wind site model is based on a *Weibull* probability distribution function of the wind speed (Heier, 2006). Figure 2(a) presents an example of such wind speed distribution for a site with an annual average wind speed of 6 [m/s]. Expression (1) is used to compute the mechanical power delivered by the wind turbine. A typical characteristic of the maximal power against wind speed is presented in Fig. 2(b).

$$P_t = \frac{1}{2} C_p(\lambda,\beta) \rho_{air} \pi r_{turb}^2 v^3 \quad (1)$$

ρ_{air}, r_{turb} and v are the air density, the wind turbine radius and the wind speed respectively. The factor C_p is the power coefficient that depends on the tip speed ratio (2) and the blade pitch angle β.

$$\lambda = \frac{r_{turb}\Omega_{turb}}{v} \quad (2)$$

Fig. 2(d) presents the variations of C_p against the tip speed ratio λ for different values of pitch angle β. An analytical expression for C_p can be found in (Heier, 2006). The C_p factor must be maintained to its maximal value to optimize the output mechanical power of the turbine on the whole wind speed range. For this purpose, the tip speed ratio must be kept constant at its optimal value. According to (2), the rotational speed of the turbine must remain proportional to the wind speed as presented in Fig. 2(c).

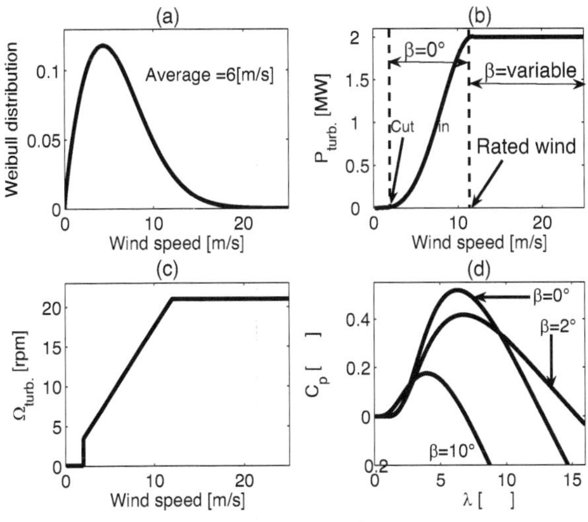

Fig. 2. Wind statistics and wind turbine steady-state characteristics. (*Weibull* distribution (a), Maximal power (b), optimal rotating speed profile (c), and power coefficient C_p (d) against wind speed).

2.2 Gearbox losses and mass models

The gearbox losses can be approximately modeled by constant viscous losses. Since gearboxes for wind turbines equipped with DFIGs are usually composed of three stages, the oil friction losses are higher in the second and third stages because their rotating speed is higher. A viscous loss proportion of 1% of rated power per stage is a reasonable assumption (Cotrell, 2002). The gearbox efficiency and its mechanical losses can be determined by (3).

$$\eta_{gearbox} = \frac{P_{Into\,gearbox} - q \cdot 0.01 \cdot P_{rated\,gearbox}}{P_{Into\,gearbox}}$$
$$P_{gearbox\,losses} = P_{\max\,gearbox\,losses} \frac{n}{n_{rated}} \tag{3}$$

q is the number of gearbox stages, n is the actual rotor speed (rpm) and n_{rated} is the rated rotor speed (rpm). The gearbox mass is also an important factor of the design of a DFIG drive system. For DFIG wind turbines in the MW range, the gearbox mass can be twice the DFIG mass. According to the data available in (Flender, 2006), the gearbox mass depends on the torque on the low speed shaft and on the number of stages. For example, the mass of a three-stage gearbox of 2 [*MW*] with a low rotating speed of 19 [*rpm*] and a torque of 1.06 [*MNm*], is equal to 8.5 Tons. Its speed ratio G_r can be selected in a range between $45<G_r<125$. For lower speed ratios, the gearbox has two stages only. In this case, the mass is 7.8 Tons and the ratio range is $25<G_r<40$. For single stage gearboxes in the *MW* power range, the speed ratio is between $3<G_r<8$. The mass model of single stage gearboxes is described by the following equation (Li et al., 2009):

$$M_{Gear} = \frac{3.2 T_n F_s F_m}{1000} \tag{4}$$

Where T_n is the nominal torque on the high speed output, F_s the service factor (F_s=1.25 is used here) and F_m is the mass factor defined as in (5).

$$F_m = \frac{1}{Z} + \frac{1}{Z r_w} + r_w + r_w^2 + 0.4 \frac{1+r_w}{Z} (G_r - 1)^2 \tag{5}$$

Were Z is the number of satellites (Z=6 is used here) and r_w is a factor defined as $r_w=(G_r/2)-1$.

2.3 Doubly-Fed induction generator model

The DFIG steady-state model is based on the classical wounded rotor induction machines equivalent circuit depicted in Fig. 3. From this circuit the overall performances of the DFIG can be derived, if both stator and rotor power supplies are defined. Compared to the classical induction machine, the DFIGs electromagnetic torque derivation is slightly different and is presented in section 3.1.

2.4 Static power converter losses model

The total losses of the back-to-back PWM VSI converter can be derived from (6). This loss model is based on the parameters of conventional 1700*V* IGBT modules, with a switching frequency of 5 [*kHz*], and a third harmonic injection process (Peterson, 2005).

$$P_{losses,conv} = P_{losses,RSC} + P_{losses,GSC}$$
$$= 3\left(3.88\frac{\sqrt{2}\cdot 2}{\pi}I_{RSC} + \frac{1.76}{2I_{RSC,\max}}I_{RSC}^{2}\right)$$
$$+3\left(3.88\frac{\sqrt{2}\cdot 2}{\pi}I_{GSC} + \frac{1.76}{2I_{GSC,\max}}I_{GSC}^{2}\right) \quad (6)$$

I_{RSC} and I_{GSC} are the respective *rms* values of the line currents of the rotor and grid side converters.

3. Wind power plant control

3.1 DFIG torque control

The analytical determination of the control laws presented in this paper doesn't take into account the mechanical and magnetic losses of the drive. The steady-state equivalent circuit of the induction generator is presented in Fig. 3.

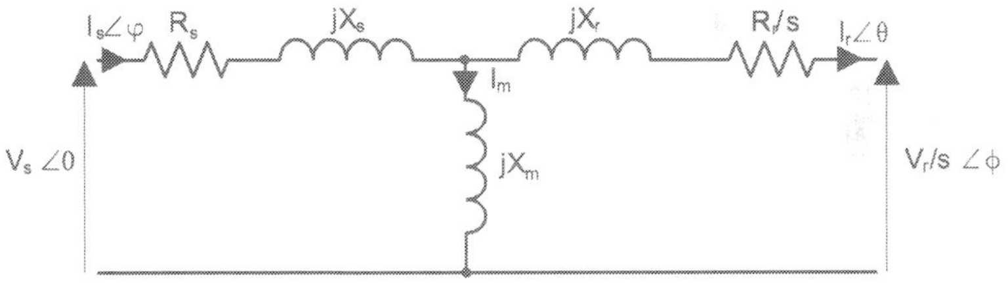

Fig. 3. DFIG Equivalent circuit.

The generator air gap power P_{ag} is:

$$P_{ag} = 3\frac{R_r}{s}I_r^2 + 3\frac{V_r}{s}I_r\cos(\phi-\theta) \quad (7)$$

All the phase angles are referred to the stator voltage V_s, except for the rotor phase angle $\psi = \phi - \theta$ that is the angle between the rotor current and the rotor voltage. Therefore, the electromagnetic torque T_e can be written as:

$$T_e = \frac{3p}{\omega_s}\left[\frac{R_r}{s}I_r^2 + \frac{V_r}{s}I_r\cos(\psi)\right]$$
$$= \frac{p}{\omega_s s}\left[3R_rI_r^2 + P_r\right] \quad (8)$$

Equation (8) shows the influence of rotor voltage V_r and phase angle ψ on the electromagnetic torque. From the same equation, one can notice that the active power flow P_r in the RSC controls the electromagnetic torque of the DFIG. Because there are two control variables, the same torque T_e value (or power P_r) can be obtained with different combinations of V_r-ψ pairs but with different performances in terms of efficiency and power factor.

3.2 Novel analytical determination of converter control laws

The rotor side power converter control strategy has a direct influence on the DFIG performance. Several approaches have been proposed in the literature to determine specific rotor converter control laws that minimize some objective function like the total drive losses for example. Non-linear constrained optimization techniques have been proposed for this purpose (Çadirici et al., 1992) (Smith et al., 2005). But these approaches can be problematic if the determination of the control law must be performed for each step of a DFIG global design process that is also based on the use of a non-linear constrained optimization procedure. Two embedded optimization methods in the same CAD environment can lead to convergence problems and drastically increase the total number of iterations (see Fig. 4).

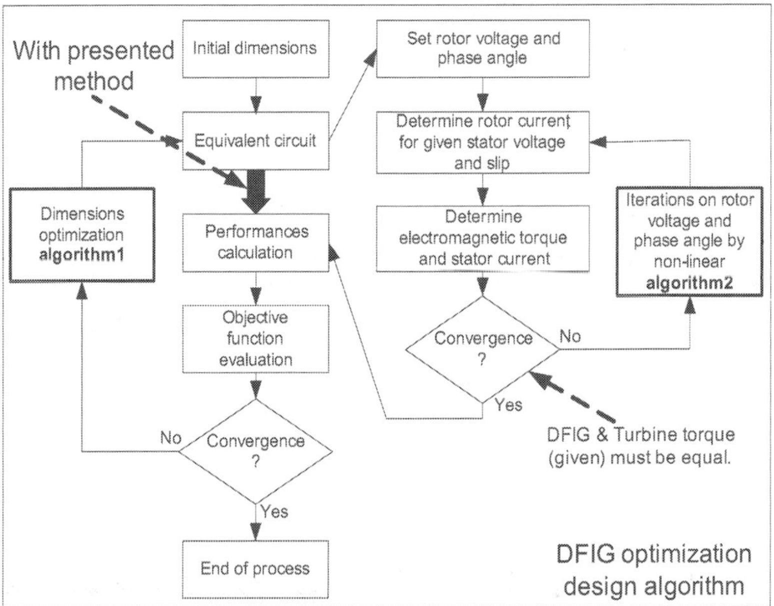

Fig. 4. DFIG global design process with two embedded non linear optimization methods

It is possible to find an analytical solution for the determination of the rotor control laws to avoid the use of the non-linear algorithm 2 in Fig. 4. A single solution for the rotor control variables V_r and ψ exists for each steady-state operation point because the electromagnetic torque and the generator stator power factor are imposed by the wind turbine torque-speed characteristic and the grid specifications.

Fig. 5 illustrates the method that is proposed to derive the analytical formulation of the rotor control laws. For each steady-state operation point, the generator electromagnetic torque T_e is considered to be equal to the output mechanical torque T_t of the turbine since the mechanical losses are neglected. The stator voltage V_s and the stator power factor SPF=$cos\varphi$ (where φ is the phase angle between the fundamental components of the stator current and voltage) are imposed because the grid power factor GPF is fixed by the grid specifications and the grid side converter power factor is equal to unity.

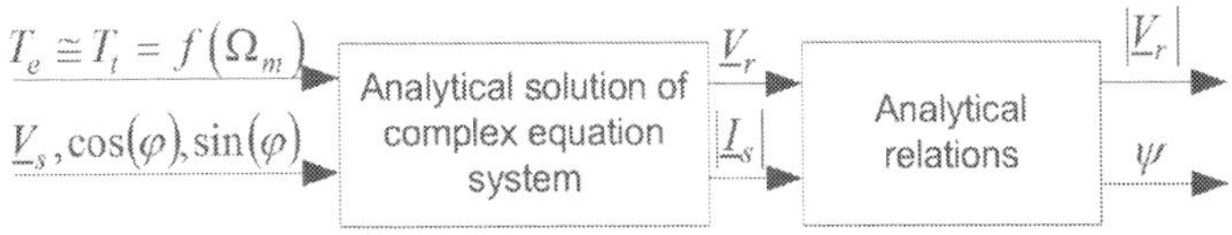

Fig. 5. Diagram of the procedure for the analytical determination of converter control laws.

It is common to operate the rotor side converter to deliver the reactive power absorbed by the DFIG and to keep the grid side converter at unity power factor. This strategy is preferable because the voltage delivered by the rotor side converter is multiplied by a factor 1/s before being applied to the magnetizing inductance of Fig. 3, which is absorbing the main part of the total DFIG reactive power. If the voltage drop across the rotor resistance and leakage inductance are neglected, the total DFIG reactive power can be expressed as in (9).

$$Q_m \cong 3\left(\frac{V_r}{s}\right)^2 \frac{1}{X_m} \cong 3\frac{1}{s}V_r I_r \sin(\psi) \cong \frac{1}{s}Q_{RSC} \tag{9}$$

Since the value of the slip s is always less than unity, it is preferable to compensate the DFIG reactive power consumption with the RSC. A system of three analytical equations can be established to derive the three intermediate variables listed on Fig. 5 (*rms* value of the stator current I_s, real and imaginary components of $\underline{V}_r$) from the input parameters T_t, V_s, $\cos\varphi$ and $\sin\varphi$. The Thévénin equivalent circuit of the DFIG depicted in Fig. 6 is used for this purpose.

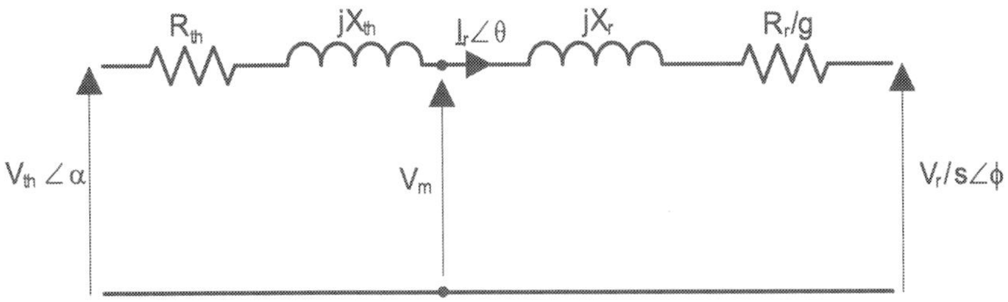

Fig. 6. DFIG Thévénin equivalent circuit.

The parameters R_{th} and X_{th}, and the voltage V_{th} can be derived from the stator voltage V_s, the resistance R_s, the reactances X_s and X_m, by applying the Thévénin transformation on the equivalent circuit of Fig. 3. Several intermediate variables are listed in (10):

$$\begin{aligned}
&\underline{V}_s = V_s + j0 \\
&\underline{V}_{th} = V_{thr} + jV_{thi} = V_{th}^{\angle\alpha} \\
&\underline{V}_r = a + jb = V_r^{\angle\phi} \\
&\underline{I}_s = c\cdot\cos(\varphi) + jc\cdot\sin(\varphi) = c\cdot x + jc\cdot y = I_s^{\angle\varphi}
\end{aligned} \tag{10}$$

The three unknown variables a, b and c in (10) can be determined with a set of three equations. The first two equations can be derived by computing the rotor complex current $\underline{I}_r$ in the equivalent circuits of Fig. 3 and Fig. 6. Equation (11) leads to a system of two equations, one for the real part and another for the imaginary part of the rotor current.

$$\underline{I}_r = \underline{I}_s - \underline{I}_m = \frac{V_{thr} + jV_{thi} - \dfrac{a+jb}{s}}{R_{th} + \dfrac{R_r}{s} + j\left(X_{th} + X_r\right)} = c\cdot x + jc\cdot y - \frac{V_s - \left(c\cdot x + jc\cdot y\right)\left(R_s + jX_s\right)}{jX_m} \tag{11}$$

The third equation is derived from the DFIG torque equation that must be equal to the turbine torque under steady-state operation. In (12) the electromagnetic torque is a function of the stator current, the rotor voltage and the DFIG parameters only. The system of equations (11) and (12) can be solved to determine the variables a, b and c and derive the rotor voltage and stator current.

$$T_e = T_t = \frac{3p}{\omega_s}\left(\begin{array}{c}\dfrac{R_r}{s}\left(\left(c\cdot x + \dfrac{c\cdot xX_s + c\cdot yR_s}{X_m}\right)^2 + \left(c\cdot y + \dfrac{V_s - c\cdot xR_s + c\cdot yX_s}{X_m}\right)^2\right) + \\ \dfrac{c\cdot xa + c\cdot yb}{s} + \dfrac{bV_s - c\cdot xR_sb + c\cdot xX_sa + c\cdot yR_sa + c\cdot yX_sb}{X_m s}\end{array}\right) \tag{12}$$

The symbolic analytical solution of such a complex system of equations was performed by use of *Mathematica*®. The complex rotor current $\underline{I}_r$ and the phase angle ψ can be finally derived from equations (13) and (14) obtained with the equivalent circuit of Fig. 3.

$$\underline{I}_r = \frac{-\left(V_s - \underline{I}_s\left(R_s + jX_s\right)\right)}{jX_m} + \underline{I}_s \tag{13}$$

$$\psi = \arctan\left(\frac{\mathrm{Im}\left(\underline{I}_r\right)}{\mathrm{Re}\left(\underline{I}_r\right)}\right) - \arctan\left(\frac{\mathrm{Im}\left(\underline{V}_r\right)}{\mathrm{Re}\left(\underline{V}_r\right)}\right) \tag{14}$$

The efficiency of the DFIG is:

$$\eta = \frac{3\left[V_r I_r \cos(\psi) - V_s I_s \cos(\varphi)\right]}{3\left[V_r I_r \cos(\psi) - V_s I_s \cos(\varphi) + R_s I_s^2 + R_r I_r^2\right] + P_{mags} + P_{magr}} \tag{15}$$

One can notice that the magnetic losses are included in the efficiency formulation These losses are computed from the DFIG dimensions and magnetic induction levels. The stator magnetic losses P_{mags} are nearly constant because a constant voltage/frequency ratio is imposed by the grid at the stator terminals. The rotor magnetic losses P_{magr} that are not negligible because of large slip operation are also computed.

3.3 Variable speed range limitations

To maximize the energy production of the power plant, the turbine rotating speed must vary with the wind speed in order to maintain the tip speed ratio λ at its optimal value (see eq. (1) and (2)). As visualized in Fig. 2 the turbine rotating speed should vary linearly with the wind starting from very low wind speeds values up to the nominal wind and rotating speeds. One has to verify that the generator is able to guarantee such a variable speed range. A limiting factor is provided by the voltage limitation introduced by the rotor power converter. To understand this phenomena suppose that stator and rotor windings resistances and leakages inductances are negligible (which is a legitimate hypothesis, especially for large generators). Then, the DFIG equivalent circuit is reduced to impedance represented by the magnetizing inductance only as shown in Fig. 7.

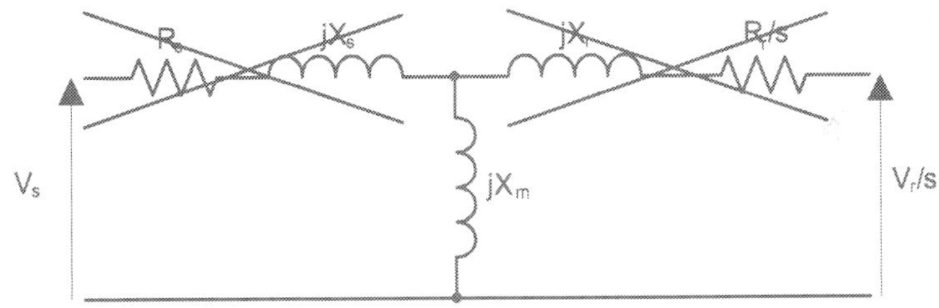

Fig. 7. DFIG equivalent circuit assuming no voltage drop across winding resistances and leakage inductances.

Based on this assumption one is allowed to write the following identity:

$$V_s = \frac{V_r}{s} \tag{16}$$

The voltage V_r is referred to the stator side. If we define *sr* as the stator to rotor windings turns ratio, or the stator to rotor voltages ratio when the DFIG is at standstill (neglecting the windings pitch and distribution factors), we can rewrite (16) as:

$$V_s = \frac{V_r^{'} sr}{s} \tag{17}$$

$V_r^{'}$ represents the voltage at the rotor side converter terminals, which has an amplitude limitation imposed by the switches technology. For instance, the use of 1700V IGBT technology limits the power converter DC-Bus voltage to approximately 1000V to 1200V. Therefore, the maximal *rms* phase voltage which can be handled by the rotor side converter is in the order of $1000V/(\sqrt{2}\sqrt{3}) \cong 400V$. Considering eq. (17) one notices that since the stator voltage V_s is fixed and imposed by the grid, and since the stator to rotor turns ratio *sr* cannot vary, when the power converter voltage V_r' has reached its maximal value, the slip *s* is not allowed to vary anymore (respect of eq. (17)). This means that as soon as the maximal rotor voltage is reached, the wind turbine must operate at constant speed even if the wind speed is varying. This situation can appear during low wind speeds operations where the slip *s* and the rotor voltage V_r' are very large (eq. (17)). This equation reveals another important issue; the selection of stator to rotor turns ratio *sr*. Usually the maximal rotor voltage at the converter terminals is equal to the stator voltage V_s. In steady-state operation of the wind turbine the rotor voltage is always kept below its maximal value in order to guarantee a voltage margin for torque-speed correction (e.g. wind gusts). It is usual to keep a safety rotor voltage margin of 25% (Petersson, 2005), which means that the maximal rotor side converter voltage during steady state operation is about 300V. For instance, supposing that the maximal slip is *s=0.3*, one can approximately derive the stator to rotor turns ratio as shown in (17).

$$sr = \frac{sV_s}{V_r'} = \frac{0.3 \cdot 400V}{300V} = 0.4\ [-] \qquad (18)$$

Here it is easy to understand that due to the voltage limitation, and due to the objective of maximizing the variable speed range, the selection of the power converter maximal voltage (rating), the placement of the synchronous speed with respect to the wind speed, i.e. the maximal slip s which can be adjusted with the selection of the gearbox ratio, and the selection of the stator to rotor turns ratio sr, in not a trivial process and needs a global optimization approach.

3.4 2.5 MW wind turbine performances illustration by numerical simulation

The parameters of a typical 2.5 MW wind generator using a DFIG system have been used to illustrate the proposed analytical method to derive the control law. All needed parameters are presented in Table 1.

Rated power	2.5	MW
Rated voltage	690	V
Base frequency	50	Hz
Stator resistance	0.045	p.u.
Rotor resistance	0.028	p.u.
Stator leakage inductance	0.077	p.u.
Rotor leakage inductance	0.15	p.u.
Mutual inductance	3.3	p.u.
Number of poles-pairs	3	-
Rotor diameter	80	m
Rated wind speed	15	m/s
Turbine speed range	10.9-19.1	rpm
Gearbox ratio	1:68.1	-

Table 1. 2.5 MW DFIG and wind turbine parameters used for simulations.

The mechanical power collected by a wind turbine can be derived from the kinetic energy of the air flow facing the swept area of the blades as expressed in (1). Figure 8 shows an example of the C_p characteristic as a function of the tip speed ratio, for three different blade pitch angles.

The maximum power characteristic of a real 2.5 MW wind turbine (Nordex, 2007) is presented in Fig. 9(a). The variation of the power coefficient of this turbine versus wind speed is reported in Fig. 9(b) (Nordex, 2007).

One can identify four different wind speed regions of operation on Fig. 9(a). Region 2 coincides to a nearly constant and maximum power coefficient C_p where the blade pitch angle β and the tip speed ratio λ are constant and set to their optimal value. In this region, the output power of the turbine is proportional to the third power of the wind speed (1). At a wind speed of approximately 11 m/s, the maximum rotating speed of the turbine is reached and must be limited to respect the mechanical and aerodynamic constraints. In region 3, the limitation of the rotating speed is provided by a suitable control of the DFIG torque, and the blade pitch angle is controlled in order to track a maximum power coefficient C_p. However, the tip speed ratio λ and the power coefficient C_p are decreasing as

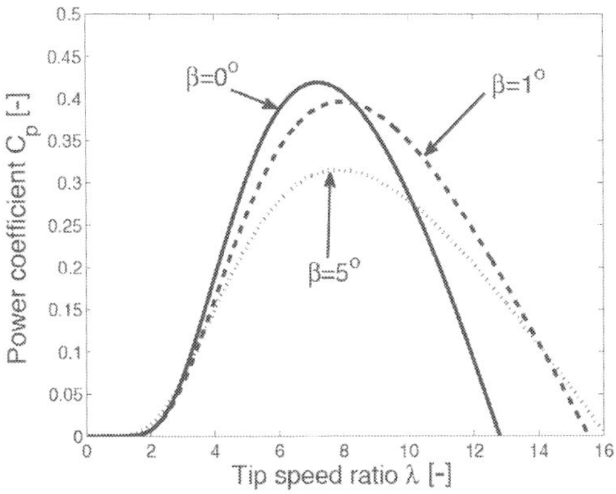

Fig. 8. Typical C_p factor characteristic for MW range wind turbines.

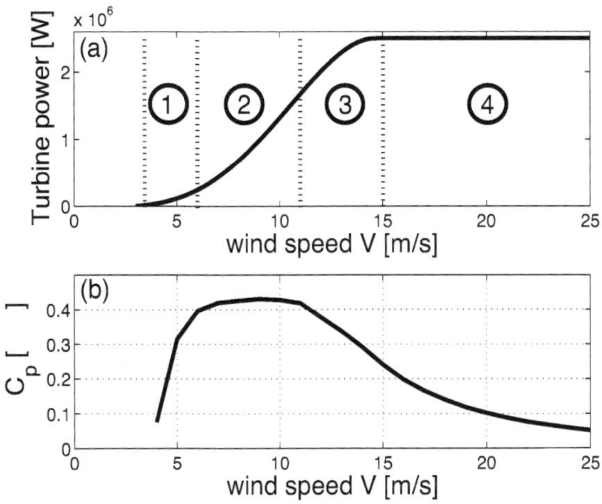

Fig. 9. 2.5 MW Wind turbine power characteristics (Nordex, 2007).

the wind speed increases. The turbine output power is still increasing with the wind speed but with a smaller rate of change *dPt/dV*. When the rated power is reached at v=*15 m/s*, one enters the region 4 where the blade pitch angle β only is controlled in order to limit the turbine output to its rated maximal value. In region 1 the rotating speed is limited to a minimal value that depends on the maximal rotor voltage that can be produced (Analysis of section 3.3). The wind speed is varying while the rotating speed is fixed, leading to a decrease of the C_p factor (see Fig. 9(b) in region 1). By using the data of Table 1, it is possible to determine the rotating speed versus wind speed control characteristic of the DFIG (Fig. 10(a)) and the turbine output power versus rotating speed characteristic (Fig. 10(b)) which corresponds to the maximum power characteristic of the 2.5 MW wind turbine of Fig. 9(a).

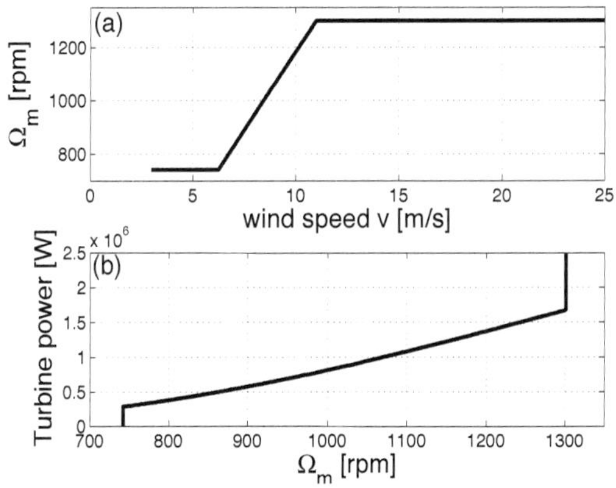

Fig. 10. DFIG rotation speed versus wind speed and turbine power versus rotation speed characteristics.

The DFIG torque characteristic on the whole speed range can be easily derived from the characteristics of Fig. 10 ($T=P/\Omega$).
In this case study, the stator power factor of the generator is set to unity on the whole operation range and the grid side converter (GSC) is operated at unity power factor as well. This leads to a unity power factor operation of the whole system, seen by the grid. With such a strategy, the rotor side converter must deliver the reactive power consumption of the generator only. The steady-state control laws of the rotor side converter, presented in Fig. 11, have been determined by the proposed analytical method. These laws are represented by the characteristics of the two control variables $V_r(\Omega_m)$ and $\Psi(\Omega_m)$. Figure 12 to 15 illustrate the steady-state performances of the whole system on the whole operation range: generator and converter apparent powers, DFIG active power flow, DFIG currents (referred to stator) and DFIG efficiency.

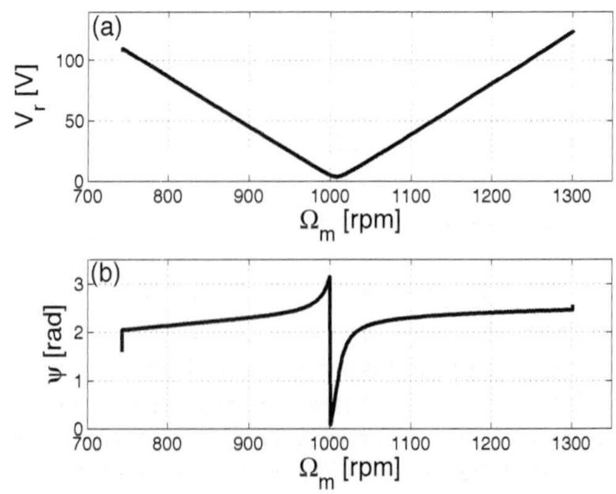

Fig. 11. Rotor voltage and phase angle control law for GPF=1.

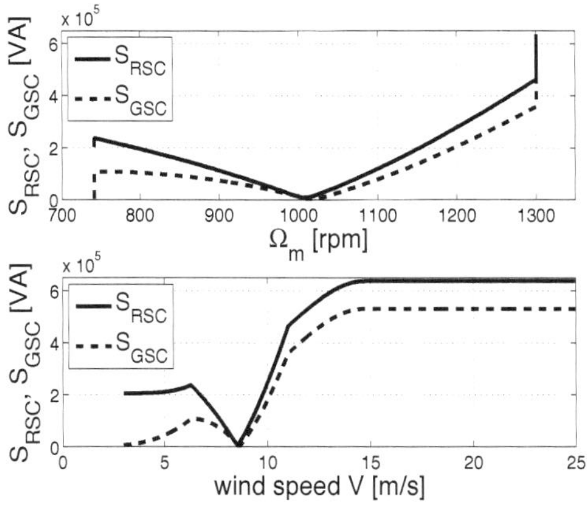

Fig. 12. RSC and GSC apparent power for GPF=1.

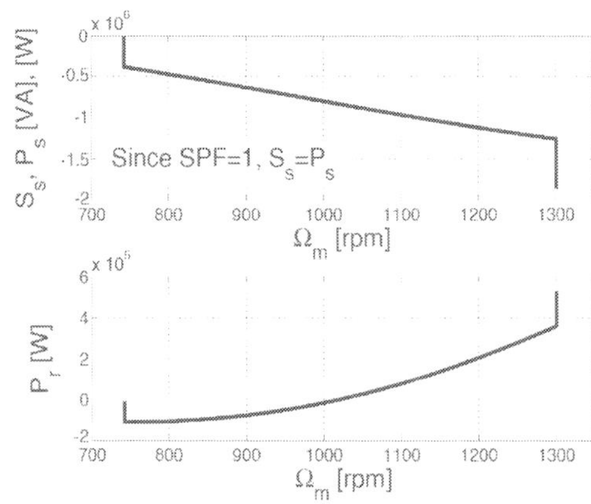

Fig. 13. DFIG powers flows for GPF=1.

The maximal apparent power of the RSC is higher than the maximal apparent power of the GSC since the reactive current is supplied by the RSC. The apparent and active powers of the DFIG stator are identical since the stator power factor is set to unity. One can notice on Fig. 13 that the rotor real power flow P_r is reversed at the synchronous operation speed (1000 [rpm]). The rotor current in Fig. 14 is relatively high in low wind speed regions since the reactive current must always be supplied. In this region, the power extracted from the wind is very low and the rotor copper losses are high; the DFIG efficiency of Fig. 15 is then drastically decreased.

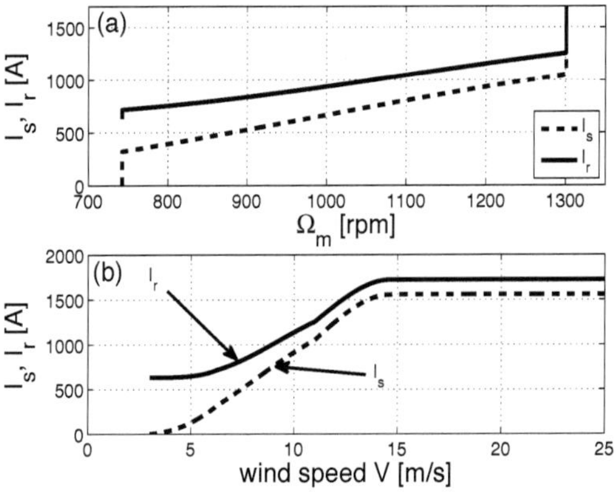

Fig. 14. DFIG stator & rotor RMS currents for GPF=1.

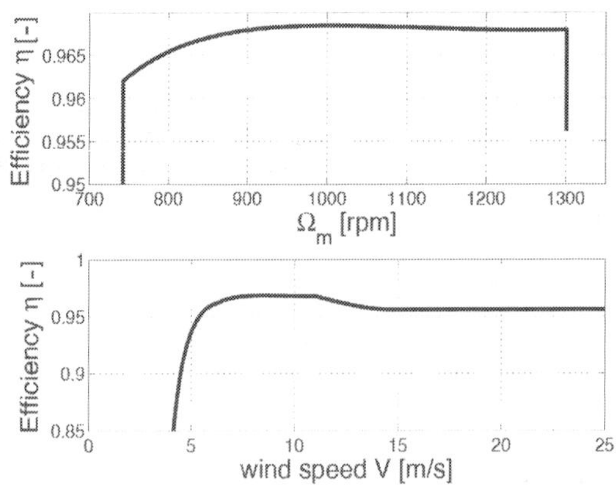

Fig. 15. DFIG Efficiency for GPF=1.

4. Annual energy production

An optimal design must take into account the annual energy production of the plant. It depends on the annual wind distribution of the plant site, on the wind turbine characteristics, on the DFIG, gearbox and power converter losses, and on the control strategy. One computes, for each wind speed, the number of hours of operation and the amount of power extracted by the turbine from the wind flow. The annual mechanical energy production of the turbine can be derived from eq. (19)(Grauers, 1996). $P_t(v)$ is the output power of the turbine and $p(v)$ is the wind probability distribution while the constant 8760 is the number of hours per year.

$$E_{an,turbine} = 8760 \int_{cut-in}^{cut-out} P_t(v)p(v)dv \tag{19}$$

For a given wind turbine, the annual electrical energy production of the plant is determined by introducing the whole system efficiency. The total losses of the drive system (gearbox, DFIG and power converter) can be computed from the models presented in section 2, for each operation point that corresponds to each wind speed value. One can derive the total efficiency of the drive $\eta_{tot}(v)$ for each wind speed value. The expression for the annual energy yield is expressed in (21).

$$E_{an,grid} = 8760 \int_{cut-in}^{cut-out} P_t(v)p(v)\eta_{total}(v)dv \tag{21}$$

From equation (21) one can notice that the DFIG dimensions, the selected gearbox ratio, the power converter technology or the selection of the stator to rotor turns ratio highly influence the annual energy production. This means that the annual production performance is a suitable performance to be maximized, since it is dependent on all the choices of the drive components.

5. Optimal selection of drive components

5.1 Gearbox ratio selection

The annual energy yield varies with the gearbox ratio since the efficiency of the total wind generation system depends on the exact location of the DFIG synchronous speed operation on the effective wind speed range of the site. The super-synchronous and sub-synchronous speed ranges are changing with the gearbox ratio. This concept is illustrated in Fig. 16 for a given wind turbine and a given DFIG with two different values of the gearbox ratio *Gr1* and *Gr2* (*Gr1*<*Gr2*). The two DFIG rotating speed characteristics are presented on Fig. 16. Since the synchronous speed Ω_s of the DFIG is imposed by the grid frequency and the DFIG magnetic poles number, different sub-synchronous and super-synchronous speed ranges are obtained with different gearbox ratios. The wind speed which coincides to the synchronous speed of the DFIG is changed. If the *Weibull* distribution, describing the wind speeds probability, is taken into account (on top of Fig. 16), one notices that for different gearbox ratios the probability of sub-synchronous or super-synchronous operation is different. The choice of the gearbox ratio is directly related to the annual energy production because the total efficiency of the generation system is different for every rotating speed (e.g. sub-synchronous region efficiency different from super-synchronous region efficiency). Furthermore, the power converter voltage rating is a function of the maximum slip operation (eq. (17)). The choice of the gearbox ratio has direct influence on both the annual energy production and the power converter rating.
A very important aspect to be considered is the gearbox mass, which depends on its speed ratio. As presented in section 2.2, the gearbox mass tends to increase with the gearbox ratio. This has to be considered during the gearbox ratio selection, since it is technically and economically interesting to minimize the mass installed into the nacelle of the wind turbine.

5.2 Power converter apparent power selection

One has to bear in mind that the most important advantage of a DFIG based wind turbine is the low initial cost given by a power converter rated at a fraction of the nominal power only. Therefore, during the design of the DFIG drive, a special attention must be paid to the cost minimization of the power converter in order to maintain the advantage of such topology.

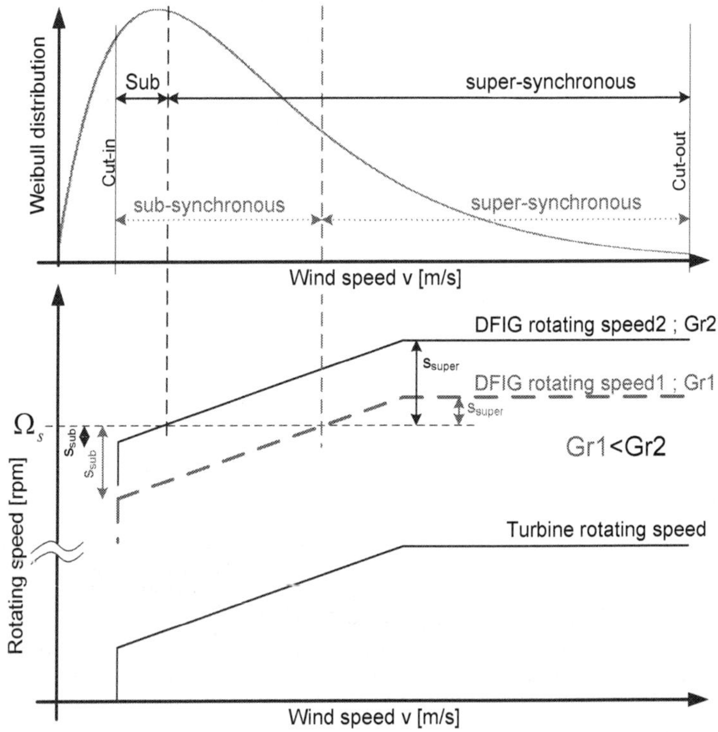

Fig. 16. Effects of the gearbox ratio selection on sub and super-synchronous probabilities of operation.

There are several ways of selecting the power converter rating. To derive the power converter rating one has to define its maximal voltage and its maximal current. The maximal voltage is directly related to the switches technology choice. The maximal current flowing through the power converters is directly related to the maximal active and reactive power flows, which are strongly dependent on the maximal DFIG slip excursion (which in turn is related to the selection of the gearbox ratio). The problematic is that the selection of the gearbox ratio influence the selection of the power converter rating, furthermore all these choices affect the annual production. Because of such complexity it is useful to proceed with an illustration of the influences of every choice on the performances of the wind turbine, this is what is presented in the next section 5.3.

5.3 Illustration of optimal DFIG drive components selection

The selection of the optimal DFIG drive components is an interactive search for optimal compromises between initial cost, mass and energy production. There is no unique and global optimal solution for every specification since all choices have consequences in both technical and economic aspects. As in all engineering domains, technical-economical compromises must be found. The models presented in this chapter can easily be integrated in a non-linear optimization environment which can be used to determine the optimal selection of the gearbox ratio, the power converter rating and global mass, to optimize an objective function composed by economical and technical performances. However, for the sake of illustration and understanding, this section presents the sensitivity of some global wind turbine performances versus the selection variables. This illustration is based on the wind turbine characterized by the parameters presented in Table 1. Considering a wind site characterized by an average wind speed of μ_v=*8m/s* and a *Weibull* shape factor k_w=*1.8* (Heier, 2006), the annual energy production versus the gearbox ratio selection is presented in Figure 17.

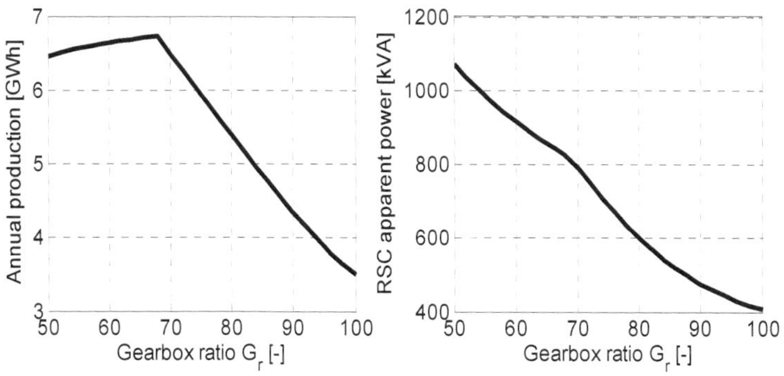

Fig. 17. Annual production and11 rotor side converter maximal apparent (dimensioning) power versus gearbox ratio - 2.5MW wind turbine & average wind speed site of 8 m/s.

The optimal gearbox ratio that is maximizing the annual production is G_r=*68.1 [-]*. The performances of this configuration are illustrated on figures 8 to 15. However, looking at the power converter (RSC) maximal apparent power (dimensioning power) on the right side of Figure 17, one notices that the higher G_r, the lower the power converter rating. For gearbox ratios higher than G_r=*68.1 [-]* both the annual production and the power converter rating are drastically decreasing. The considered voltage for the calculation of the RSC apparent power was kept fixed to 690V (line), since the IGBT technology was fixed (1700V IGBTs). To understand the behavior depicted in Figure 17, the visualization of some wind turbine performances versus the wind speed for different gearbox ratios is necessary. This analysis is presented in Figure 18, where turbine power, rotor voltage (rotor side), turbine power coefficient and turbine rotating speed are plotted versus the wind speed for three different gearbox ratios G_r=*50*, G_r=*68.1* and G_r=*90*.

Solution with G_r=*68.1*: One can notice that the annual energy is maximized (Fig. 17). From Fig. 18 (b) one notice that the lowest voltage is reached when the DFIG is operating at synchronous speed. An increase of the DFIG slip *s* is associated to a linear increase of the rotor voltage, as showed by the simplified equation (16). The rotor voltage V_r' can increase until it reaches the maximal steady state value (a margin of 25% is kept). Once this limit is reached, either in super-synchronous or in sub-synchronous regions, the DFIG slip *s* is not allowed to vary anymore (eq. (16)) even though the wind speed is changing. Due to mechanical coupling the DFIG speed limitations can be seen on the turbine speed Ω_{turb} of Fig. 18 (d) (10.9 *rpm* and 19.1 *rpm*). This limitation produces regions where the turbine rotating speed is not changing with the wind speed, therefore, recalling eq. (1) and (2) and Fig. 2 (d), such regions present a degradation of the turbine power coefficient C_p (Fig. 18 (c))

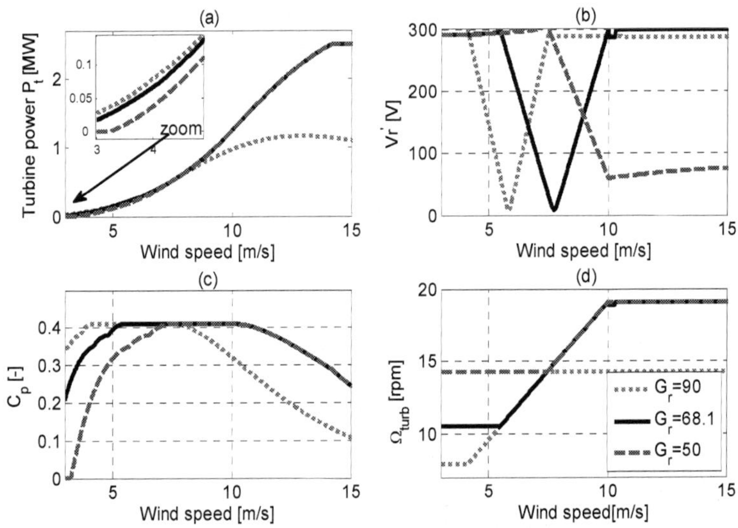

Fig. 18. Turbine power P_t, rotor voltage V_r'(rotor side), turbine power coefficient C_p and turbine rotating speed Ω_{turb} versus wind speed for three different gearbox ratios.

and of the turbine power P_t (zoom of Fig. 18 (a)). The turbine power degradation in the super-sychronous region, starting at a wind speed of 10 *m/s* is noticeable by the fact that Pt is not increasing with the third power of the wind speed anymore, as described by eq. (1). This is the symptom of a decrease of the power coefficient C_p.

Solution with G_r=90: In this case the gearbox ratio is higher, and the DFIG synchronous speed coincides with a lower wind speed (6 *m/s*), as depicted in Fig. 18 (b). As the wind speed increases beyond 6 m/s, the rotor voltage increases with the same rate of change as in the solution with G_r=*68.1*, since the stator to rotor turns ratio *sr* has not changed (17). Therefore the rotor voltage limit is reached prematurely (7.5 *m/s*) drastically reducing the maximal DFIG and turbine rotating speed (14 *rpm*) as depicted in Fig 18 (d). Consequently the power coefficient C_p experiences a serious degradation (Fig. 18 (c)), causing a dramatic reduction of the wind turbine power P_t in Fig. 18 (a). This is why in Fig. 17 the RSC apparent power is sharply decreasing for high gearbox ratios, since the currents flowing into it decrease with the turbine power decrease.

Solution with G_r=50: Here the gearbox ratio is very low and the DFIG is operating in sub-synchronous only. Therefore the rotor voltage limitation is occurring in the sub-synchronous region in a wide wind speed region (Fig. 18 (b)). This is drastically limiting the minimum turbine rotating speed (fig. 18 (d)), which in turn is degrading the Cp coefficient at low wind speeds. Therefore the turbine power degradation takes place in the lower wind speed region as depicted in the zoom of Fig. 18 (a). This explains why in Fig. 17 the annual energy production is slightly decreasing for low gearbox ratios. It might be surprising that such a slight decrease in the turbine power characteristic, in the low wind speed region, is producing such a loss in the annual production. However one has to consider that at such low wind speeds the *Weibull* distribution can present very high wind probabilities; it is the case in this example.

In this chapter the design of the generator is not discussed, but the simple models previously presented can be used to analyze the importance of the DFIG stator to rotor turns ratio *sr* selection. Figure 19 presents the annual production and the RSC maximal apparent power versus the stator to rotor turns ratio *sr*. In this analysis the gearbox ratio is kept at its optimal value G_r=*68.1*.

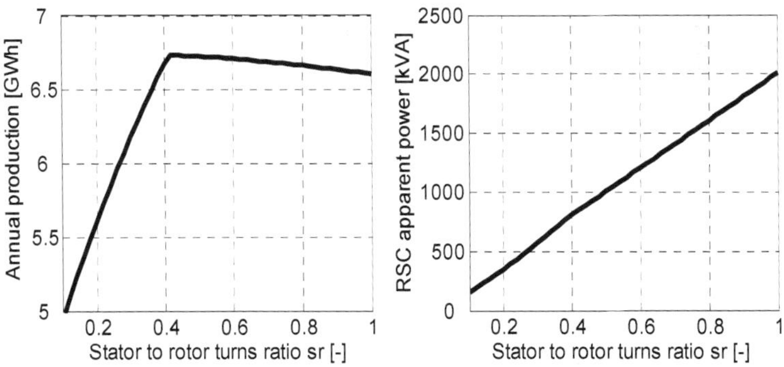

Fig. 19. Annual production and rotor side converter maximal apparent (dimensioning) power versus stator to rotor turns ratio (G_r=*68.1*) - 2.5MW wind turbine & average wind speed site of 8 m/s.

Even in this case one notices an optimal value of *sr* which maximizes the annual production, whereas the RSC maximal apparent power is pseudo-linearly increasing with *sr*. As in the previous analysis of the gearbox ratio selection, the explanation of the curves of Fig. 19 can be given thank to an illustration of some wind turbine performances, versus the wind speed, for several choices of the stator to rotor turns ratio, as proposed in Fig. 20.

Solution with *sr=0.41*: This choice corresponds to the optimal turns ratio that is maximizing the annual energy production as showed in Fig. 19. This case has already been described in the "Solution with G_r=*68.1*" paragraph.

Solution with *sr=0.15*: The value of *sr* is too low in this case. Referring to (17) it is clear that a small variation in the DFIG slip *s*, in the region close to *s=0*, produces a remarkable increase in the rotor voltage. Therefore the rotor voltage limit is reached for very small DFIG slips *s*., as shown in Fig. 20 (b). Thus, the variable speed range is highly reduced (Fig. 20 (d)), producing a degradation of the turbine power coefficient Cp over a wide wind speed region as shown in Fig. 20 (c). Since the rotating speed limitation is occurring in the sub-synchronous and super-synchronous regions, the wind turbine power degradation occurs in the low and high wind speeds regions, explaining the severe decrease of the annual production illustrated in Fig. 19.

Solution with *sr=0.8*: The *sr* value is so low that the rotor voltage increases slowly with the DFIG slip, as shown in Fig. 20 (b). For low wind speeds, where the DFIG is operating in sub-synchronous mode, no rotor voltage limitation is reached. This extends the variable speed range to the low wind speeds region, increasing the turbine power above the values reached in the case with *sr=0.41*, as shown in the zoom of Fig. 20 (a). In the super-synchronous region the rotor voltage never reaches its limitation since the turbine mechanical speed limitation is reached first. This means that this solution maximizes the turbines power to its

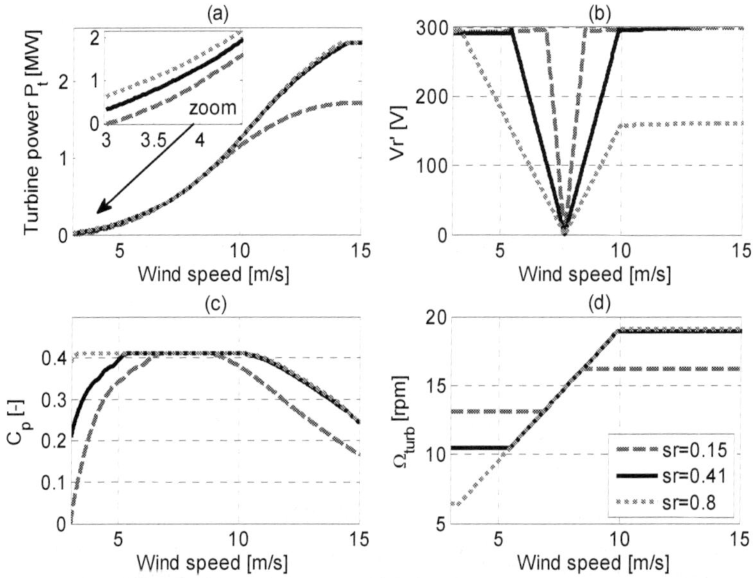

Fig. 20. Turbine power P_t, rotor voltage V_r'(rotor side), turbine power coefficient C_p and turbine rotating speed Ω_{turb} versus wind speed for three different stator to rotor turns ratios (G_r=68.1).

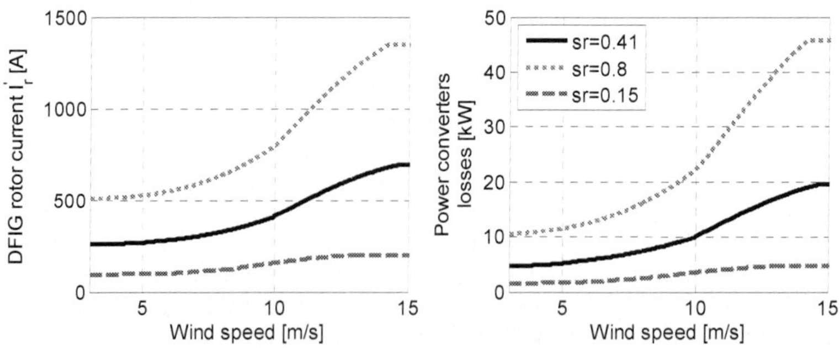

Fig. 21. DFIG rotor current and total power converters losses (RSC+GSC) versus wind speed, for three different stator to rotor turns ratios.

From simple analytical relations, neglecting stator and rotor losses and mechanical losses, it is possible to derive the well known rotor active power balance:

$$P_r = P_t \frac{s}{1-s} \tag{22}$$

With P_r the DFIG rotor active power. Since for the stator to rotor turns ratio analysis the gearbox ratio is maintained fixed, the super-synchronous maximal slip is the same between

solution with *sr=0.41* and *sr=0.8*, as it can be understood from Fig. 20 (d) (same maximal rotating speed). Therefore, referring to eq. (22), the rotor active power flow P_r is the same for both solutions on a wide operating wind speed (a difference can be noticed in the wind speeds region below 6 m/s as shown in Fig. 20 (d)). Neglecting the reactive power, the active power is the product of the rotor voltage and rotor current. However in one case (Gr=0.41) the rotor voltage is higher than in the other case (Gr=0.8), leading to a remarkable difference between the rotor currents behaviors, as presented in Fig. 21. This difference in currents produces a difference in the power converter losses which are strongly dependent on the currents. Therefore the global efficiency of the system decreased when sr is increased, explaining the annual production degradation illustrated in Fig. 19 for *sr* values above *sr=0.41*. Furthermore the increase in the RSC currents with the increase of *sr*, explains the increase of the power converter ratings presented in Fig. 19.
The gearbox mass issue was not considered here because its value is constant in the range of ratios used in these examples (according to the model presented in (4) and (5)). However the model can be very useful when the designers are exploring new topologies considering a wide range of gearbox ratios and stages (Aguglia et al., 2009). The objective of the analysis presented in this section was mainly to give a flavor of the complex selection process of some key variables of wind turbines DFIG drive systems. This complex process can be handled by use of a non-linear constrained optimization program, which can be used to select the optimal compromises between DFIG, power converter and gearbox performance/cost to maximize the annual production. The dimensional design of the DFIG itself, which consists in finding the optimal mechanical dimensions of the active materials (iron & copper), can be easily integrated in this global environment as presented by the authors in (Aguglia et al., 2009).

6. Conclusion

The DFIG drive components selection process, or design process, needs a global approach of the system in order to optimize its global performances. In the case of a wind turbine plant, such global performances are represented by the annual production, the overall mass and the initial cost. For this purpose the designer needs a model of each sub-component.
In this chapter only a few key variable for the DFIG drive design were considered for a sensitivity analysis with respect to the annual production and size of the converter. It is demonstrated that every choice of drive components (gearbox, DFIG and power converter) has an influence on the annual energy production and power converter cost. This powerful design methodology can be used to design the DFIG mechanical geometry as well. With this approach it is possible to integrate into the design process the wind probability distribution. Therefore, the plant is not optimized for a given operating point only, but for the global operation spectrum. This methodology is very useful for every electrical drive design in the variable speed application area, where all operating points must be considered and weighted with a certain probability of operation (e.g. typical torque vs. time behavior of an electric vehicle or typical cyclic operation of a fan).
The selection, or design, process can be coupled with a non-linear optimization algorithm, which can help in the complex task of selecting the optimal variables. In such an iterative process it is essential to have efficient models which allow to quickly obtaining all global performances. Analytical formulations of these models are well adapted for this purpose.
The presented results for DFIG wind turbines drives have been obtained thank to the proposed analytical determination of the rotor power converter control laws. The most

important variable influencing the annual production are the gearbox ratio and the DFIG stator to rotor turns ratio. It is important to mention that this sensitivity to these two variables is strongly dependent on the rotor side converter voltage limit. Therefore it is of extreme importance to take into account this limitation during the design process.

7. References

Aguglia D., Viarouge P., Wamkeue R., Cros J. (2007a). Selection of Gearbox Ratio and Power Converters Ratings for Wind Turbines Equipped With Doubly-Fed Induction Generators, IEEE conference "Electrical Machines and Drives - IEMDC", Vol. 1, 3-5 May 2007, pp. 447-452.

Aguglia D., Wamkeue R., Viarouge P., Cros J. (2007b). Optimizing the Annual Energy Production of Doubly-Fed Induction Generator Based Wind Turbines, IEEE Conference "Electrical Power Conference–EPC", Montreal, 25-26 October 2007, pp. 248-255.

Aguglia D., Viarouge P., Wamkeue R., Cros J. (2008). Analytical determination of steady-state converter control laws for wind turbines equipped with doubly fed induction generators, IET. Journal on Renewable Power Generation, Vol. 2, no 1, March 2008, pp. 16 -25.

Aguglia D., Viarouge P., Wamkeue R., Cros J. (2009). Doubly-Fed Induction Generator Drive Optimal Design for Wind Turbines with Reduced Gearbox Stages Number, "European Wind Energy Conference (EWEC)", 16-19 March 2009, pp. 1-10.

Çadirici I., and Ermis M.: 'Double-output induction generator operating at sub synchronous and super synchronous speeds: steady-state performances optimization and wind-energy recovery', *IEE Proceedings-B*, Vol. 139, No. 5, 1992

Cotrell J.R., "A preliminary evaluation of a multiple-generator drivetrain configuration for wind turbines," in *Proc. 21st ASME Wind Energy Symp.*, 2002, pp. 345-352.

Flender "Planurex® 2, Planetary gear units", Brochure, [Online]. Available: http://www.flender.com/_upload/k256en.pdf

Generation Using Doubly Fed Wound Rotor Induction Machine - A comparison With Alternative Schemes," *IEEE Trans. Energy Convers.*, Vol. 17, No. 3, 2002.

Grauers A., "Efficiency of three wind energy generator systems," *IEEE Trans. Energy Convers.*, Vol. 11, No. 3, September 1996.

Heier S.: "Grid integration of Wind Energy Conversion Systems, *Second edition,*" John Whiley & Sons, Ltd, 2006, pp. 31-44.

Li H., Chen Z., Polinder H. : 'Optimization of multibrid permanent-magnet wind generator systems', *IEEE Trans. on energy conv.*, Vol. 24, No. 1, March 2009, pp 82-92.

Pena R., Clare J. C., Asher G. M. (1996). Doubly fed induction generator using back-to-back PWM converters and its application to variable-speed wind–energy generation, *IET Journal on Electric Power Applications*, pp. 231-241, Vol. 43, no. 3

Petersson A., 'Analysis, modeling and control of doubly-fed induction generators for wind turbines', *Ph.D. Thesis, Chalmers University of Technology*, Sweden 2005.

Smith S., Todd R., Barnes M., and Tavner P. J.: 'Improved Energy Conversion for Doubly-Fed Wind Generators', *IEEE IAS Conference*, Vol. 4, 2005 Nordex N80/2500kW wind turbine Brochure, Online, Available: http://www.nordex-online.com/en/nordex/downloads.html, accessed November 2007

Zinger D. S. and Mulijadi E.: "Annualized Wind Energy Improvement Using Variable Speeds," *IEEE Trans. Ind. Appl.*, Vol. 33, No. 6, 1997.

26

Wind Turbine Model and Maximum Power Tracking Strategy

Hengameh Kojooyan Jafari and Ahmed Radan
Islamic Azad University-Islamshahr Branch,
K.N. University of Technology
Iran

1. Introduction

Today doubly fed induction generators (DFIG) are used for modern wind turbines to deliver electrical power to the grid. A speed variation of ±30% around synchronous speed can be obtained by the use of power converter of ±30% of nominal power. Furthermore, it is possible to control active and reactive power, which gives a better performance, and the power electronics enables the wind turbine to act as a more dynamic power source to the grid. The DFIG does not need either a soft starter or a reactive power compensator. This system is naturally a little bit more expensive compared to the classical systems; however, it is possible to save money on the safety margin of gear and reactive power compensation units, and it is also possible to capture more energy from the wind (Blaabjerg & Chen, 2006). A wind turbine with maximum power tracking is a very suitable power source to the grid.
This new model, as a dynamic power source to the grid, comprises a maximum power tracking wind turbine, a doubly fed induction machine with winding configuration, external rotor resistance and external rotor source which has a variable phase and amplitude. In this chapter its simulation, effects of important parameters, design of a special kind of voltage controller and a new combined controller for it and comparison of these controllers are presented.

2. Key words

Doubly fed machine, Wind turbine, Voltage controller, Combined controller

3. Maximum power tracking wind turbine

Maximum power tracking wind turbine can deliver maximum power to the grid in low and high wind speeds.
Turbine torque via wind is inferred from following equations (1) to (3):

$$\lambda = \frac{\omega_M \times R}{V_{wind}} \tag{1}$$

$$P_M = \frac{1}{2}\rho\pi R^5 C_P \frac{\omega_M^{\ 3}}{\lambda^3} \tag{2}$$

$$T_M = \frac{P_M}{\omega_M} = \frac{1}{2}\rho\pi R^5 C_P \frac{{\omega_M}^2}{\lambda^3} \tag{3}$$

Where, V_{wind}, the wind speed, is measured in m/s, R, the blade radius is measured per m, ρ (1.24kg/m³[4]), air density is measured in kg/m³, ω_M, turbine mechanical speed, is measured in rad/sec, λ is tip-speed ratio (TSR) and C_p is power coefficient, i.e. ratio of turbine power (power extracted) to wind power (power available) and it depends on aerodynamics specifications of blades (Hoseinpur, 2001), (Burter et al., 2001).
C_p is function of λ (Burter et al., 2001):

$$C_P = 0.22\left[116\left(\frac{1}{\lambda+0.08\beta} - \frac{0.035}{\beta^3+1}\right) - 0.4\beta - 5\right] e^{-12.5\left(\frac{1}{\lambda+0.08\beta} - \frac{0.035}{\beta^3+1}\right)} \tag{4}$$

Where β is blade pitch angle.
Simulation of turbine for two typical wind speed, 4 and 5m/s that are in valid range of speed between low-shutdown speed and high stopped speed, has been performed for improved turbine parameters according to table1 (Hoseinpur, 2001):

Nominal power	15kw
Blade radius	5.5m
Blade pitch angle	0°

Table 1. Turbine parameters

In a fixed wind speed, maximum power of turbine can be achieved from C_{Pmax} function considering improved λ. Improved parameters from equation (8) are presented in table2 (Hoseinpur, 2001).

λ_i or Improved TSR	C_{pmax} or Maximum power coefficient
8.636	0.48

Table 2. Improved parameters of turbine

Then, by using equations (1), (2), maximum turbine power is calculated.

4. Doubly fed induction machine

Most of wind turbine generators are induction generators that are very reliable and costs of them are low (Ehernberg et al., 2001).
Induction generators are not complicated. These generators can give active power to grid however they take reactive power from it.
In these generators at 50HZ frequency, the angular frequency is usually among 1200rpm to 1800rpm (relative to number of poles) and gear ratio is among 30 to 50 (Burter et al., 2001).
Recently use of doubly-fed induction generators in wind turbines has become more common; however, they are more complicated than ordinary induction machines.
Voltage equations of an induction generator in ABC system are given by equation (5) (Krause, 1986):

$$V_{S,r} = R_{S,r} \bullet i_{S,r} + \frac{d}{dt}\left(L \bullet i_{S,r}\right) \tag{5}$$

And n, the ratio of equivalent stator turns to equivalent rotor turns is unit (Krause, 1986):

$$\begin{aligned} n &= 1 \\ L_{ms} &= \frac{2}{3} L_m \\ L_{ms} &= L_{mr} \end{aligned} \tag{6}$$

And electromagnetic torque is according to equation (7) (Krause, 1986):

$$T_e = \left(i_{abcS}\right)^T \frac{d}{d\theta_m}\left(L'_{sr}\right) \bullet i'_{abcr} \tag{7}$$

And rotor mechanical speed can be obtained from equation (8) (Krause, 1986):

$$T_e - T_m = J\frac{d\omega_m}{dt} + D\omega_m \tag{8}$$

Where T_m is mechanical torque, T_e is generator torque, D is system drag (friction) coefficient and J is total inertia.
In induction machine with rotor configuration that is referred to as a winding rotor, rotor external resistance is used to increase slip and its amount is usually low and is nearly one over ten percent of rotor resistance per phase.
In doubly-fed induction generator, an external source with adjustable amplitude and phase is used to control induction generator speed and power (Ehernberg et al., 2001).
According to table 3 and by using induction machine model of MATLAB-SIMULINK the simulation has been performed.

Nominal power	15 kW
Line to line nominal voltage	460 V
Nominal frequency	60 HZ
Number of pair poles	4
Stator resistance, R_S	0.2761 Ω
Stator inductance, L_{ls}	2.2 mH
Rotor resistance, R_r	0.1645 Ω
Rotor inductance, L_{lr}	2.2 mH
Magnetizing inductance, L_m	76.14 mH
Inertia, J	0.1 kg.m^2
Friction coefficient, F	0.018 N.m.s

Table 3. Induction machine parameters in side of stator (Hoseinpur et al., 2001)

5. Machine simulation results

Results of simulation of fig.1 are presented in table 4 (Kojooyan Jafari & Radan, 2008). Simulation has been performed for 2 seconds, using MATLAB-SIMULINK.
In table4, the polarity of input power to machine is considered negative and that of output from machine is considered positive.

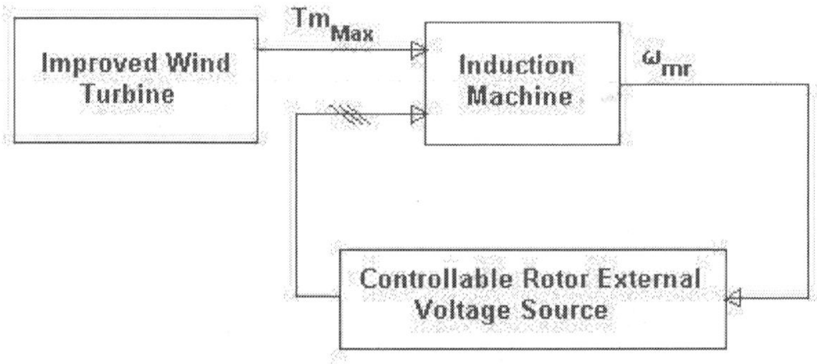

Fig. 1. Model of doubly-fed machine with improved wind turbine

In simulation, the gearbox effect is considered and output torque of gearbox is multiplied by inverse of gear ratio where gear ratio is the ratio of generator shaft speed to low-speed shaft speed in relation to equation (9):

$$N_{GB} = \frac{\omega_{mr}}{\omega_T} \tag{9}$$

V_{wind}	k	θ	r_{ex}	ω_T	ω_{mr}	N_g	$P_{T3\Phi}$	$P_{S3\Phi}$
4	5	-0.02	0.016	6.28	94.5	15	-1.8k	1.4k
4	5	-0.78	0.016	6.28	94.5	15	-1.8k	1.4k
4	10	-0.02	0.016	6.28	94.5	15	-1.8k	1.4k
4	10	-0.78	0.016	6.28	94.5	15	-1.8k	1.4k
5	15	-0.02	0.016	7.85	95	12	-3.53k	3k
5	15	-0.78	0.016	7.85	95	12	-3.53k	3k
5	20	-0.02	0.016	7.85	95	12	-3.53k	3k
5	20	-0.78	0.016	7.85	95	12	-3.53k	3.53k

Table 4. Simulation results of wind turbine and doubly-fed generator

V_{wind}	k	r_{ex}	$P_{r3\Phi}$	P_{loss}	$Q_{S3\Phi}$	$Q_{r3\Phi}$
4	5	0.016	6	394	-7.5k	11.5
4	5	0.016	6	394	-7.5k	11.5
4	10	0.016	10	390	-8.5k	47
4	10	0.016	10	390	-8.5k	47
5	15	0.016	7	523	-8k	21
5	15	0.016	7	523	-8k	21
5	20	0.016	10	480	-8.5k	40
5	20	0.016	10	480	-8.5k	40

Table 4. (continue)

Where k and θ are amplitude and phase of external rotor source, r_{ex} is external rotor resistance, $Q_{S3\Phi}$ and $Q_{r3\Phi}$ are 3-phased reactive power of rotor and stator in VAR, $P_{T3\Phi}$ is maximum turbine power, $P_{r3\Phi}$ and $P_{S3\Phi}$ are 3-phased active power of rotor and stator and P_{loss} is power losses of machine that all are in watt, ω_T is turbine speed in rad/sec, V_{wind} is wind speed in m/s, ω_{mr} is mechanical speed of rotor in rad/sec and Ng is ratio of gear

Table5 shows the results of simulation for two amounts of external rotor resistance.

V_{wind}	k	θ	r_{ex}	ω_T	ω_{mr}	Ng	$Q_{r3\Phi}$
4	10	-0.78	0.016	6.28	94.5	15	47
4	10	-0.78	3	6.28	96.7	15.4	-1
5	15	-0.78	0.016	7.85	95	12	21
5	15	-0.78	3	7.85	99	12.6	-0.5

Table 5. Simulation results of wind turbine with doubly-fed generator for two different r_{ex}

V_{wind}	k	$Q_{S3\Phi}$	$P_{T3\Phi}$	$P_{S3\Phi}$	$P_{r3\Phi}$	P_{loss}
4	10	-8.5k	-1.8k	1.4k	10	390
4	10	-7.2k	-1.8k	1.5k	-0.5	300.5
5	15	-8k	-3.53k	3k	7	523
5	15	-7.3k	-3.53k	3.1k	2	428

Table 5. (continue)

The curves of simulation are presented in the following figs.2 to 17 (Kojooyan Jafari & Radan, 2009).

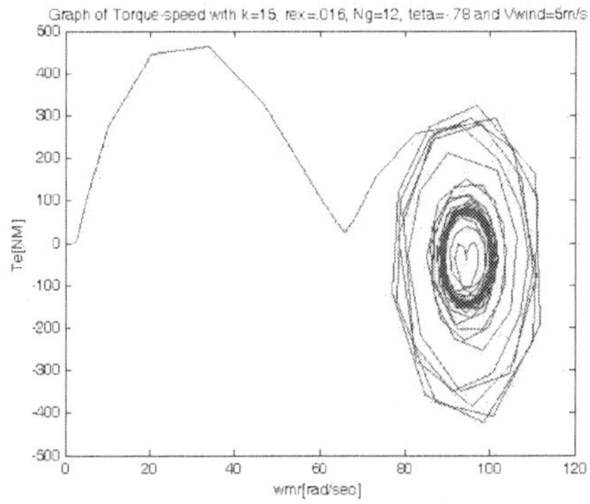

Fig. 2. Curve of torque-speed for k=15, r_{ex}=0.016, V_{wind}=5, θ=-0.78 and Ng=12

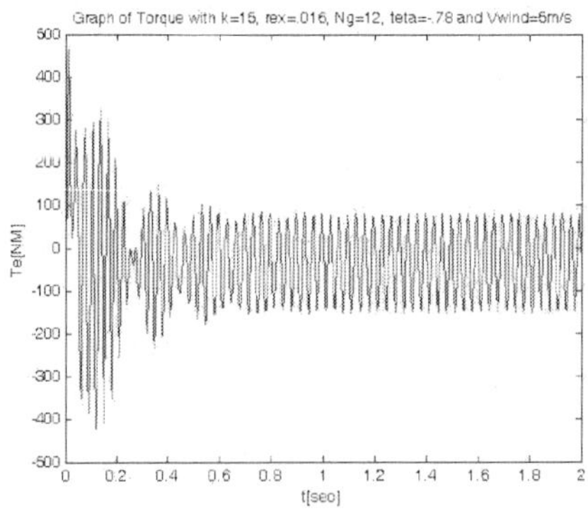

Fig. 3. Curve of electromagnetic torque-time for k=15, r_{ex}=0.016, V_{wind}=5, θ=-0.78 and Ng=12

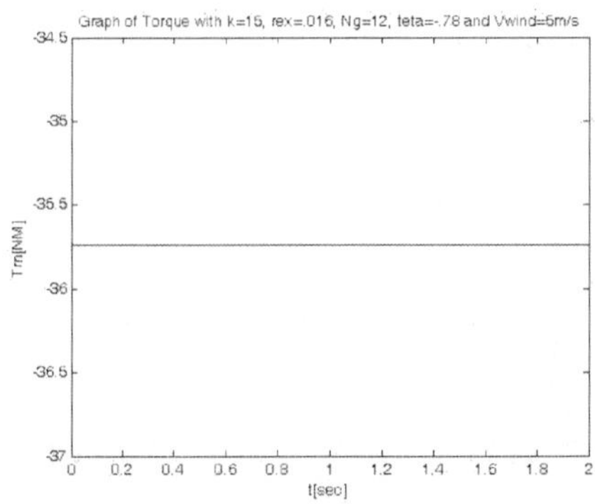

Fig. 4. Curve of mechanical torque-time for k=15, r_{ex}=0.016, V_{wind}=5, θ=-0.78 and Ng=12

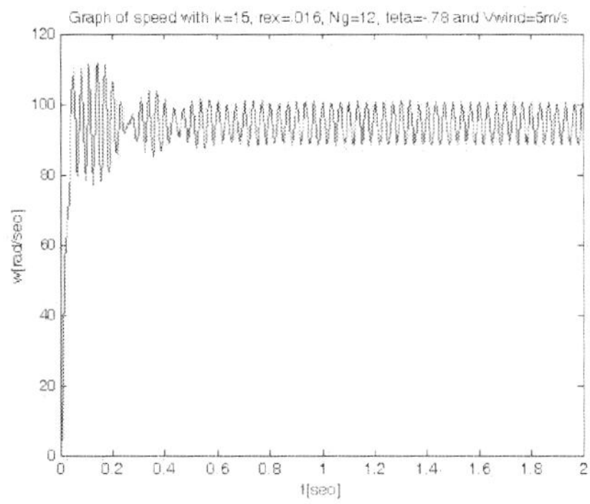

Fig. 5. Curve of mechanical rotor speed-time for k=15, r_{ex}=0.016, V_{wind}=5, θ=-0.78 and Ng=12

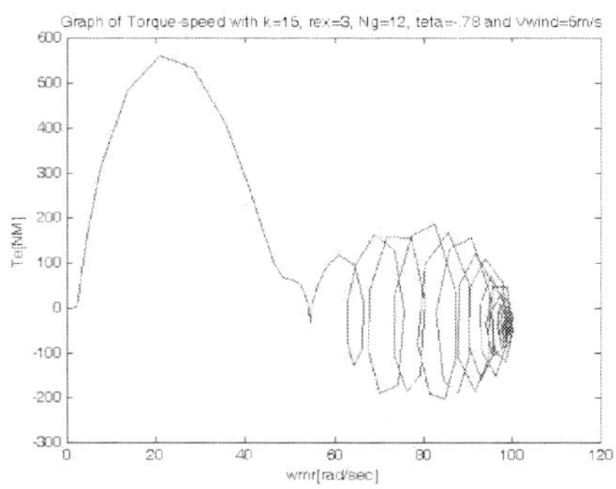

Fig. 6. Curve of torque-speed for k=15, r_{ex}=3, V_{wind}=5, θ=-0.78 and Ng=12

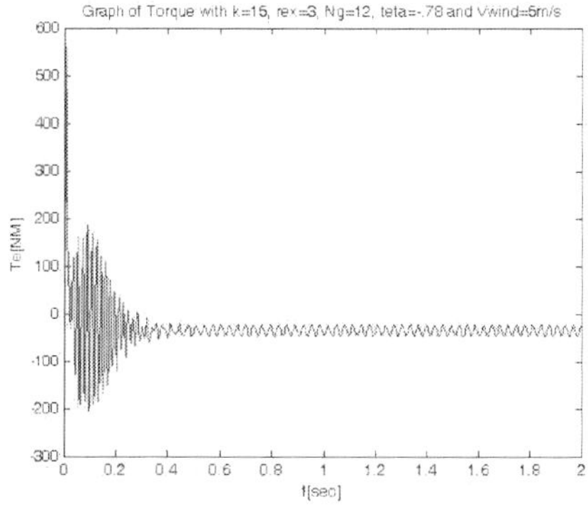

Fig. 7. Curve of electromagnetic torque-time for k=15, r_{ex}=3, V_{wind}=5, θ=-0.78 and Ng=12

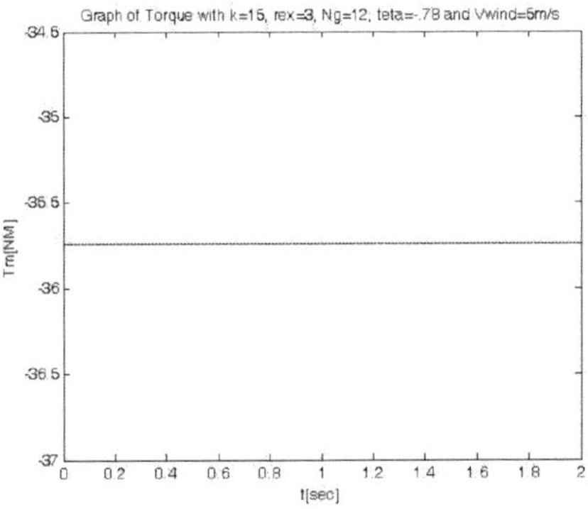

Fig. 8. Curve of mechanical torque-time for k=15, r_{ex}=3, V_{wind}=5, θ=-0.78 and Ng=12

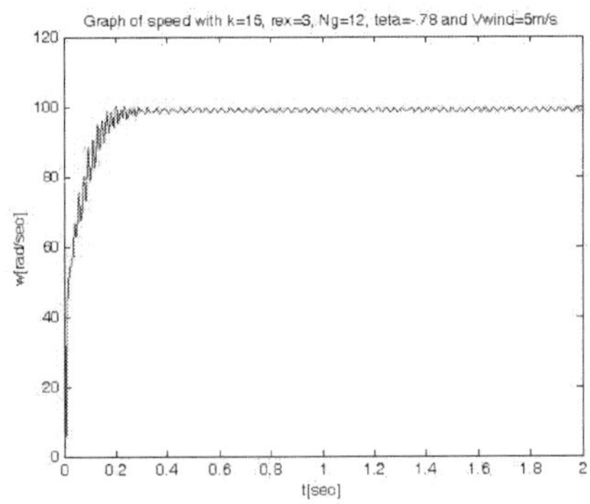

Fig. 9. Curve of mechanical rotor speed-time for k=15, r_{ex}=3, V_{wind}=5, θ=-0.78 and Ng=12

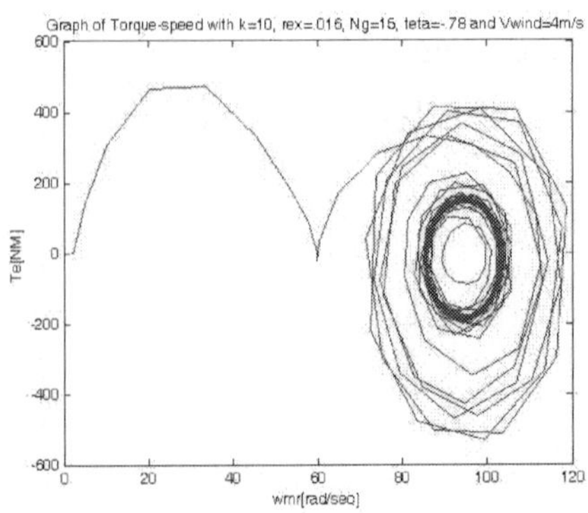

Fig. 10. Curve of torque-speed for k=10, r_{ex}=0.016, V_{wind}=4, θ=-0.78 and Ng=15

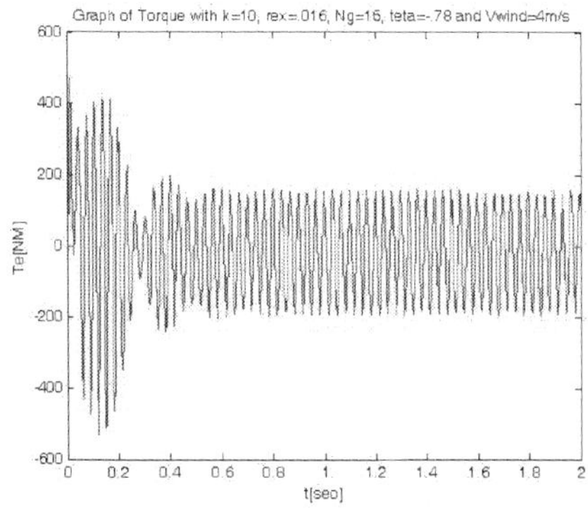

Fig. 11. Curve of electromagnetic torque-time for k=10, r_{ex}=0.016, V_{wind}=4, θ=-0.78 and Ng=15

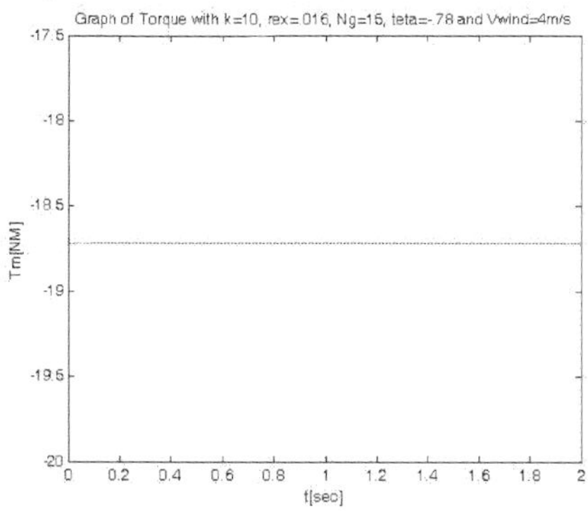

Fig. 12. Curve of mechanical torque-time for k=10, r_{ex}=0.016, V_{wind}=4, θ=-0.78 and Ng=15

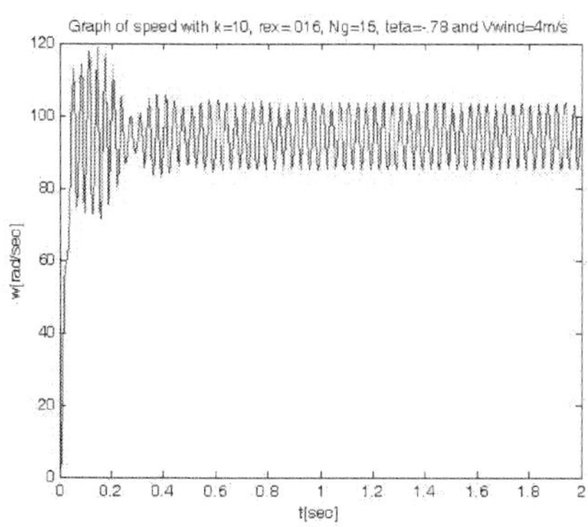

Fig. 13. Curve of mechanical rotor speed-time for k=10, r_{ex}=0.016, V_{wind}=4, θ=-0.78 and Ng=15

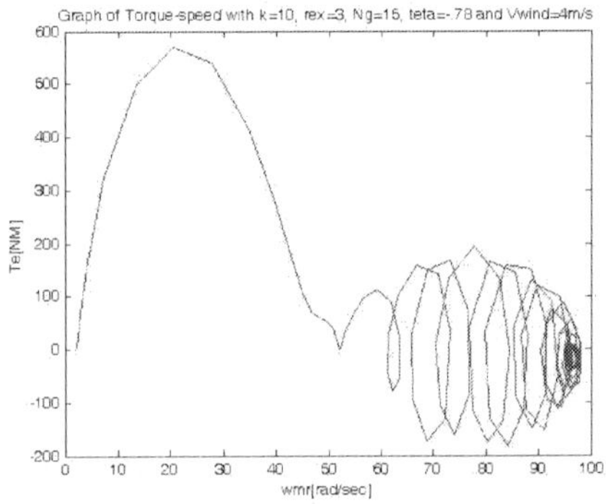

Fig. 14. Curve of torque-speed for k=10, r_{ex}=3, V_{wind}=4, θ=-0.78 and Ng=15

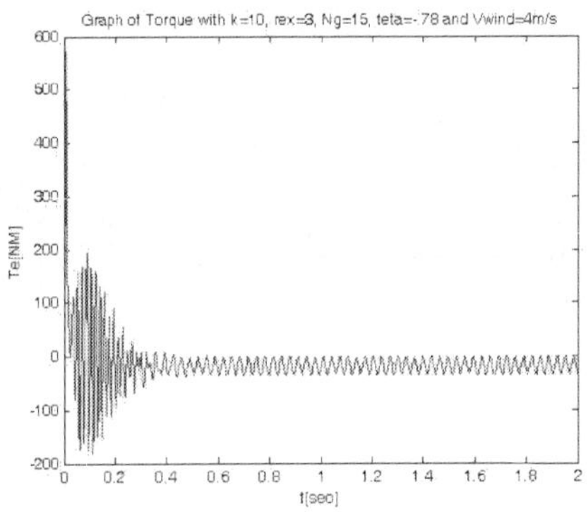

Fig. 15. Curve of electromagnetic torque-time for k=10, r_{ex}=3, V_{wind}=4, θ=-0.78 and Ng=15

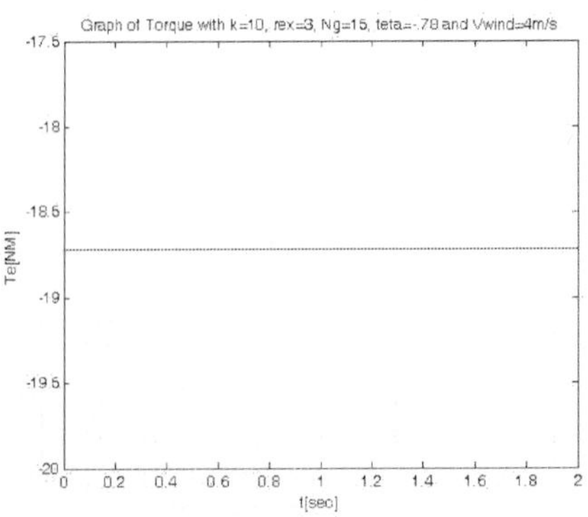

Fig. 16. Curve of mechanical torque-time for k=10, r_{ex}=3, V_{wind}=4, θ=-0.78 and Ng=15

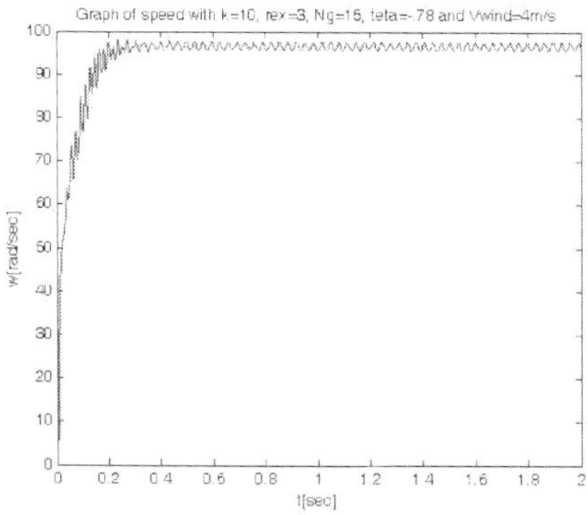

Fig. 17. Curve of mechanical rotor speed-time for k=10, r_{ex}=3, V_{wind}=4, θ=-0.78 and Ng=15

6. PI self tuning voltage controller

One of the most important subjects is control of output power when rotor external voltage source domain drops down.A PI self tuning voltage controller, shown in Fig.18, controls stator output power through adjusting the voltage at rotor terminals. V_{wind} in diagram is related to ω_M of turbine according to the equation (1). For self tuning control, P parameter of controller is adopted by k; domain of external rotor source according to fig. 19 and equation (9) when I parameter is constant and 0.0001, in every low and high wind speed (Kojooyan Jafari & Radan, 2010).

$$p = 8 / k \quad (9)$$

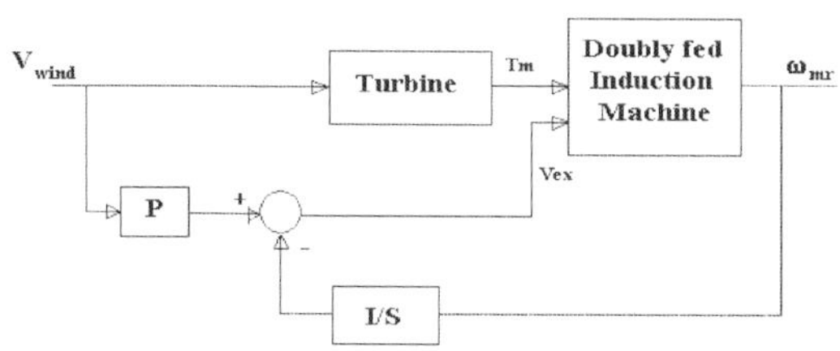

Fig. 18. Block diagram of PI self tuning voltage controller

7. PI self tuning combined voltage and pitch controller

A P self tuning voltage and pitch controller controls the system proportionally according to fig.20 then torque compensation is exerted whereas maximum turbine torque according to equation (3) can be achieved and stator output power is controlled consequently.
In this system self tuning combined control is designed by constant parameters of P_1 and P_2 according to table 10.

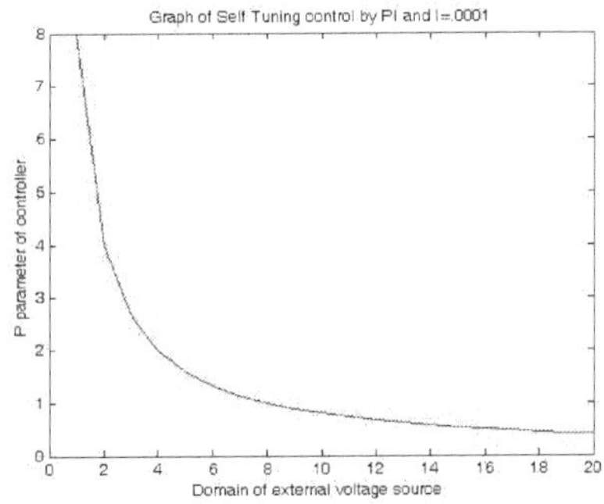

Fig. 19. Curve of P parameter when domain of external rotor source changes while I parameter is 0.0001

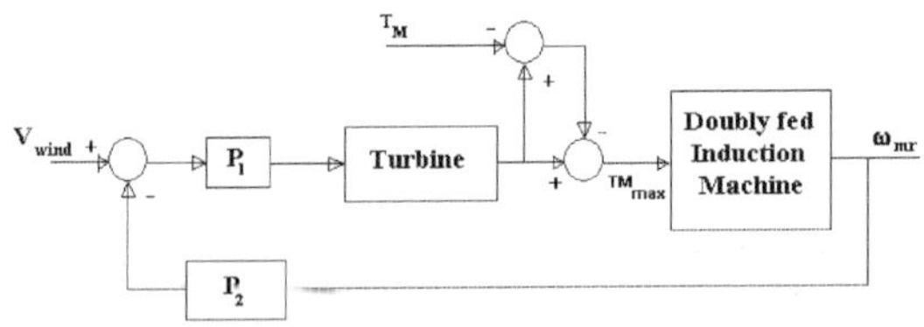

Fig. 20. Block diagram of P combined controller

8. Controllers simulation results

The results of simulation for reference points of tables 6 to 9 with parameters of PI and P controllers according to table 10 are presented in figs. 21 to 68 for a typical wind speed; 6m/s, for self tuning PI and P controllers.

In table 6, the polarity of input power to machine is considered negative and that of output from machine is considered positive for set points of tables 6 to 9. Table 10 shows parameters of controllers.

In simulation, the gearbox effect is considered in such a way that output torque of gearbox is multiplied by inverse of gear ratio.

#Points	$\omega_{mr[rad/s]}$	$P_{s[W]}$	$Q_{s[VAR]}$
1	96.6	1.5k	-7.2k
2	98.9	3k	-7.3k
3	117	15.2k	-9.4k

Table 6. Three reference point characteristic

$V_{wind[m/s]}$	N_g
4	15.4
5	24
6	34
12	140

Table 7. N_G differences for the first reference point

$V_{wind[m/s]}$	N_g
4	8
5	15.4
6	17
12	74

Table 8. N_G differences for the second reference point

$V_{wind[m/s]}$	N_g
4	1.7
5	2.6
6	3.8
12	15.4

Table 9. N_G differences for third reference point

When k; rotor external voltage domain of system with PI controller, drops down from 10 to 2 in relation to figs. 21 to 68, P parameter of self tuning PI controller can control the system according to equation (9) and table 10; however, constant parameters of self tuning P controller control the system without change according to table 10, when domain of rotor external voltage source drops down.

P of PI	I of PI	P_1 of P	P_2 of P
8/k	.0001	.8	.1

Table 10. Parameters of Controllers

Simulation has been done for both PI, as shown in figs. 21 to 44 and P according to Figs. 45 to 68. Stator active and reactive powers delivered to the grid are controlled by both P and PI controllers; however it is seen that output responses of the system with P controller has less swing in relation to figs. 45 to 68 furthermore when k drops down from 10 to 2 according to the figs. 21 to 68 it is inferred that P controller can control the stator active and reactive powers with decreasing swing of them; however, PI controller controls the stator active and reactive powers without any change in them. Also torque speed curves show stability of machine and rotor currents are in sinusoidal form.

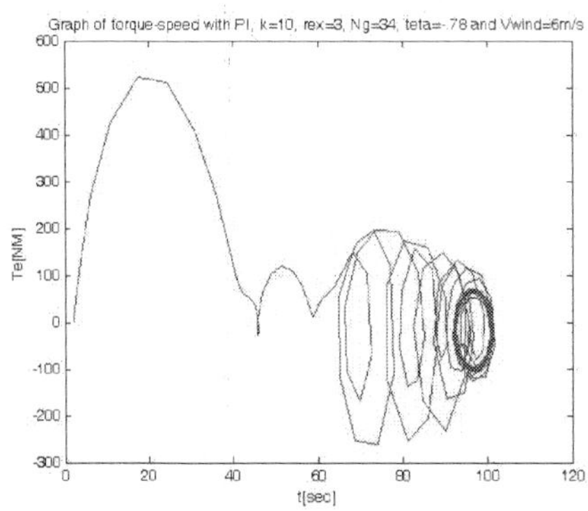

Fig. 21. Curve of torque-speed for typical v_{wind}=6m/s, k=10, Ng=34 and PI controller.

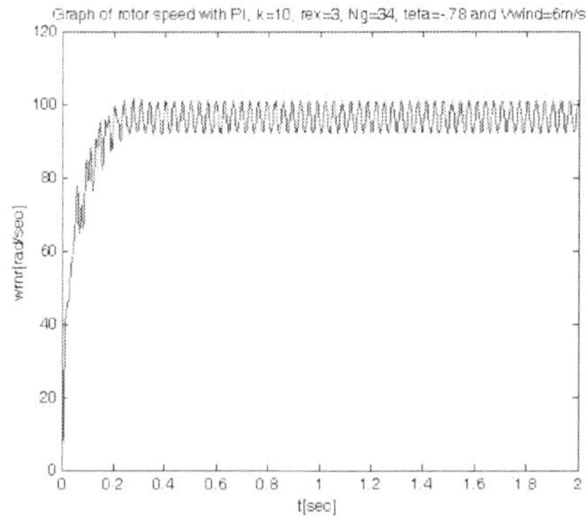

Fig. 22. Curve of rotor speed for typical v_{wind}=6m/s, k=10, Ng=34 and PI controller.

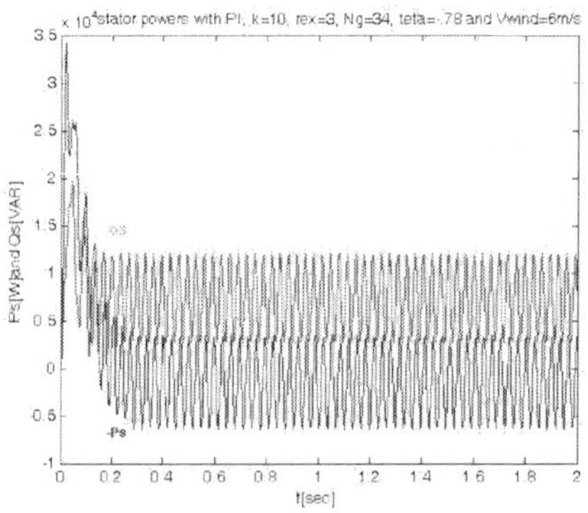

Fig. 23. Curve of stator powers for typical v_{wind}=6m/s, k=10, Ng=34 and PI controller.

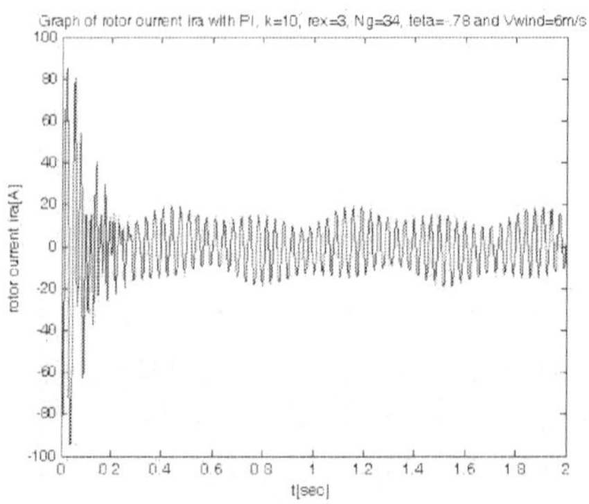

Fig. 24. Curve of rotor current i_{ra} for typical v_{wind}=6m/s, k=10, Ng=34 and PI controller.

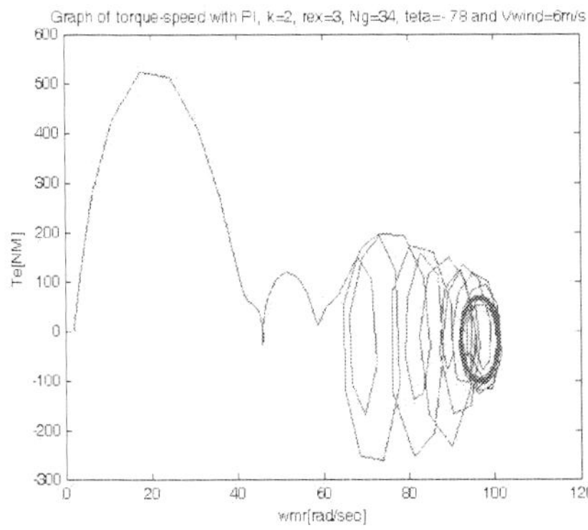

Fig. 25. Curve of torque-speed for typical v_{wind}=6m/s, k=2, Ng=34 and PI controller.

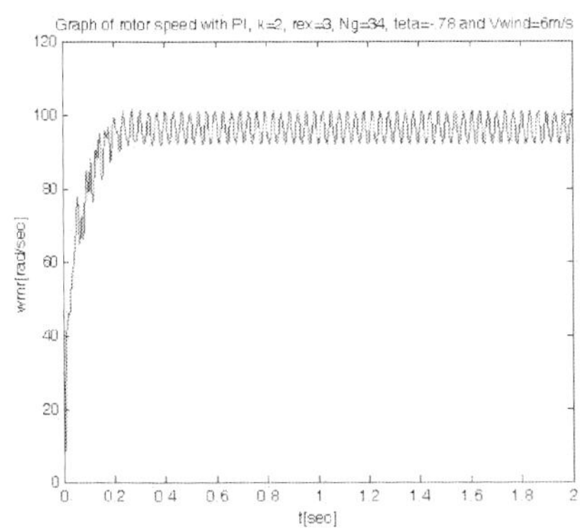

Fig. 26. Curve of rotor speed for typical v_{wind}=6m/s, k=2, Ng=34 and PI controller.

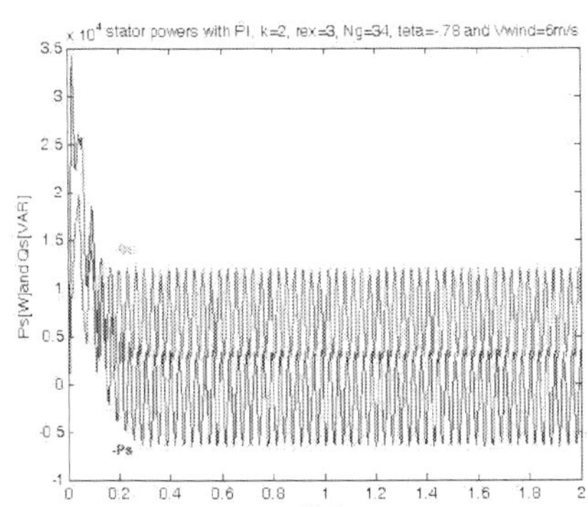

Fig. 27. Curve of stator powers for typical v_{wind}=6m/s, k=2, Ng=34 and PI controller.

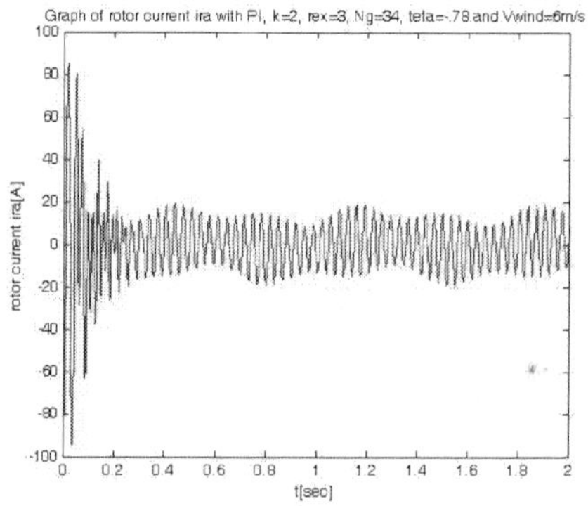

Fig. 28. Curve of rotor current i_{ra} for typical v_{wind}=6m/s, k=2, Ng=34 and PI controller.

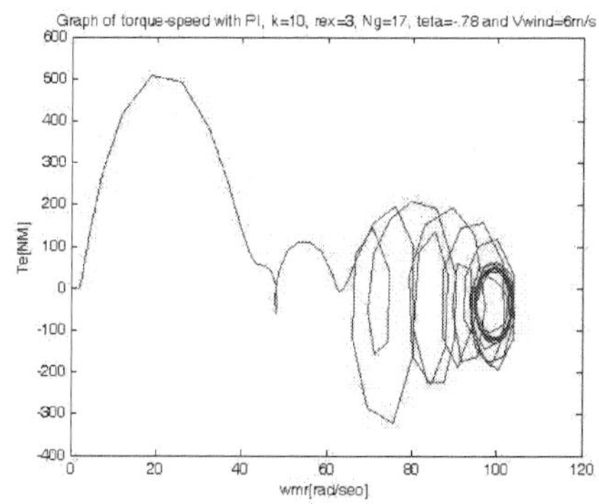

Fig. 29. Curve of torque-speed for typical v_{wind}=6m/s, k=10, Ng=17 and PI controller.

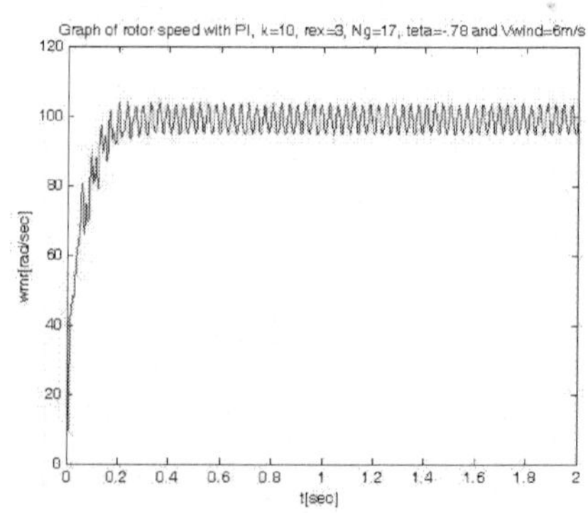

Fig. 30. Curve of rotor speed for typical v_{wind}=6m/s, k=10, Ng=17 and PI controller.

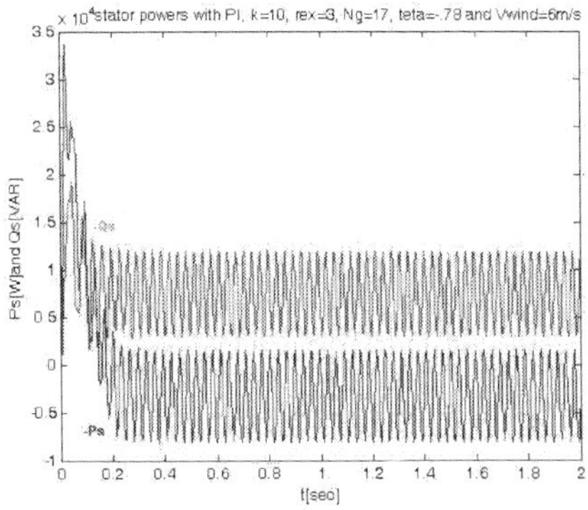

Fig. 31. Curve of stator powers for typical v_{wind}=6m/s, k=10, Ng=17 and PI controller.

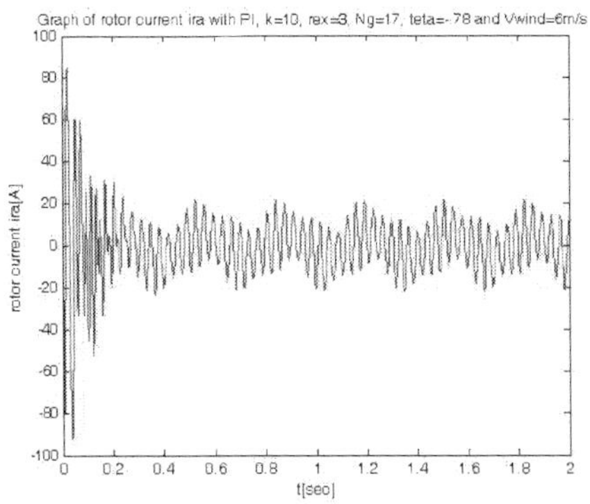

Fig. 32. Curve of rotor current i_{ra} for typical v_{wind}=6m/s, k=10, Ng=17 and PI controller.

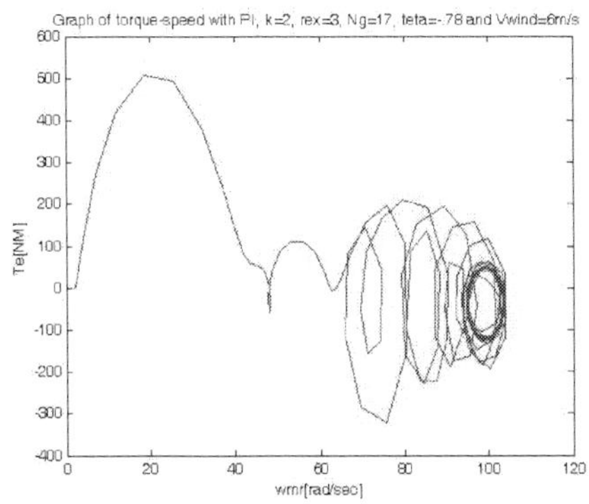

Fig. 33. Curve of torque-speed for typical v_{wind}=6m/s, k=2, Ng=17 and PI controller.

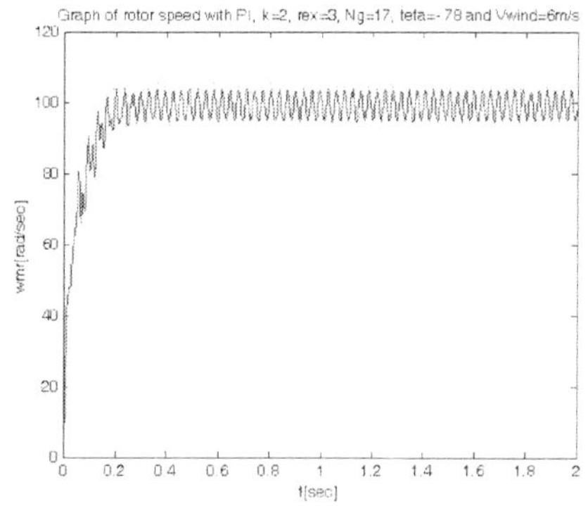

Fig. 34. Curve of rotor speed for typical v_{wind}=6m/s, k=2, Ng=17 and PI controller.

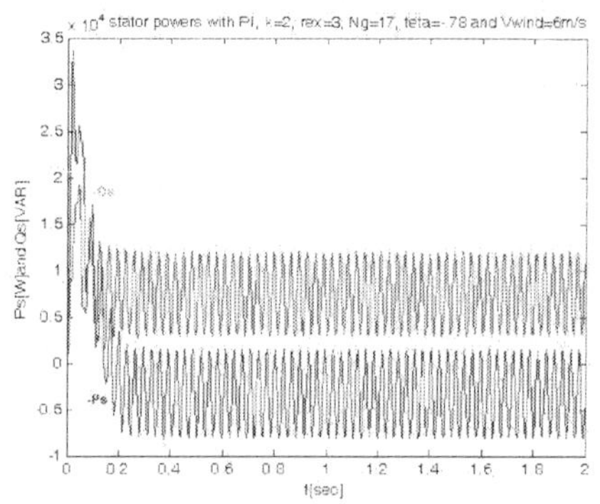

Fig. 35. Curve of stator powers for typical v_{wind}=6m/s, k=2, Ng=17 and PI controller.

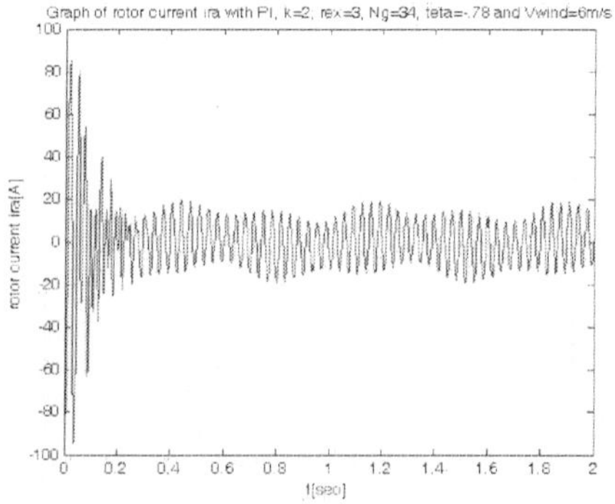

Fig. 36. Curve of rotor current i_{ra} for typical v_{wind}=6m/s, k=2, Ng=17 and PI controller.

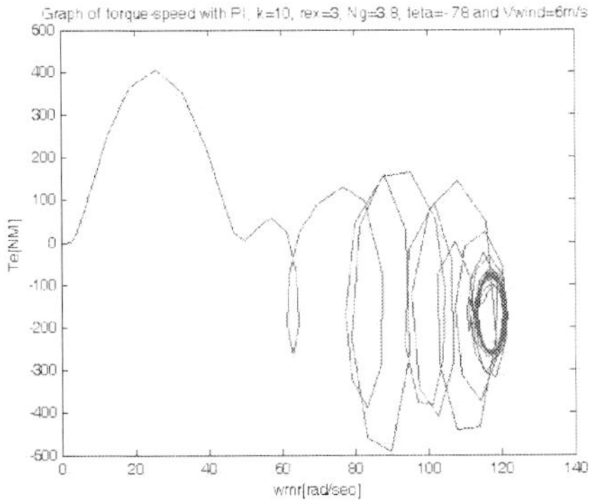

Fig. 37. Curve of torque-speed for typical v_{wind}=6m/s, k=10, Ng=3.8 and PI controller.

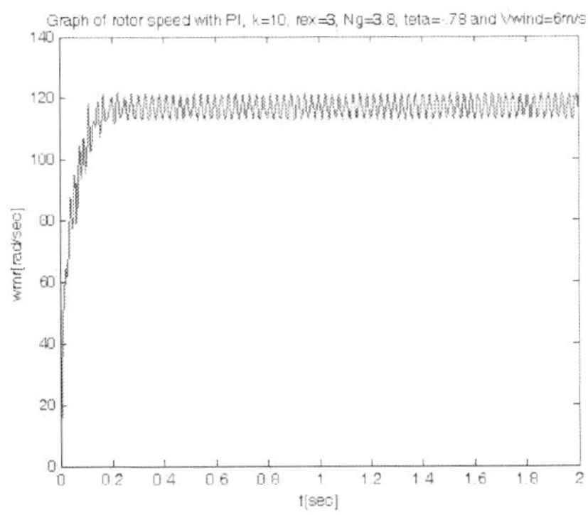

Fig. 38. Curve of rotor speed for typical v_{wind}=6m/s, k=10, Ng=3.8 and PI controller.

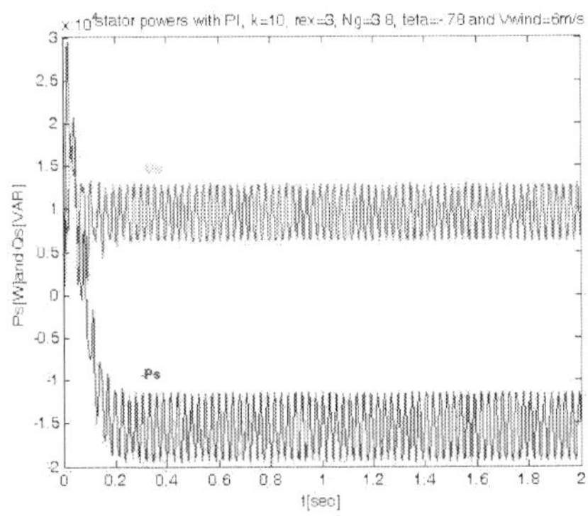

Fig. 39. Curve of stator powers for typical v_{wind}=6m/s, k=10, Ng=3.8 and PI controller.

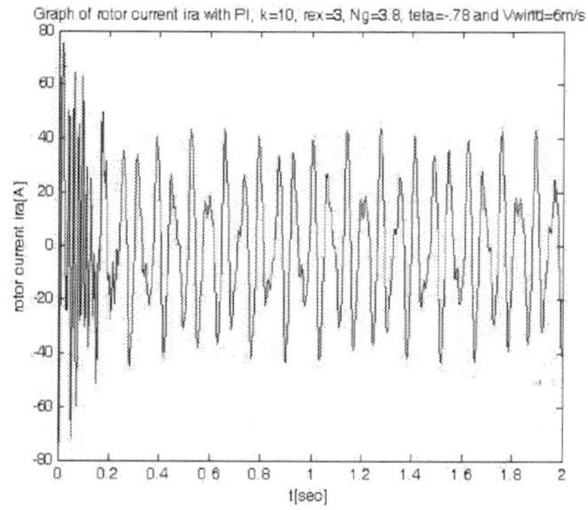

Fig. 40. Curve of rotor current i_{ra} for typical v_{wind}=6m/s, k=10, Ng=3.8 and PI controller.

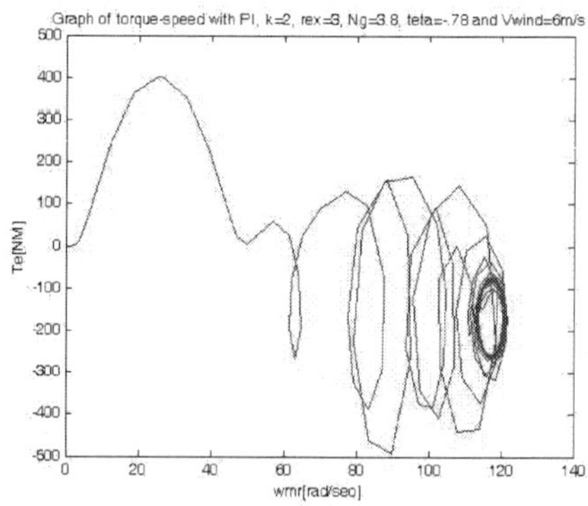

Fig. 41. Curve of torque-speed for typical v_{wind}=6m/s, k=2, Ng=3.8 and PI controller.

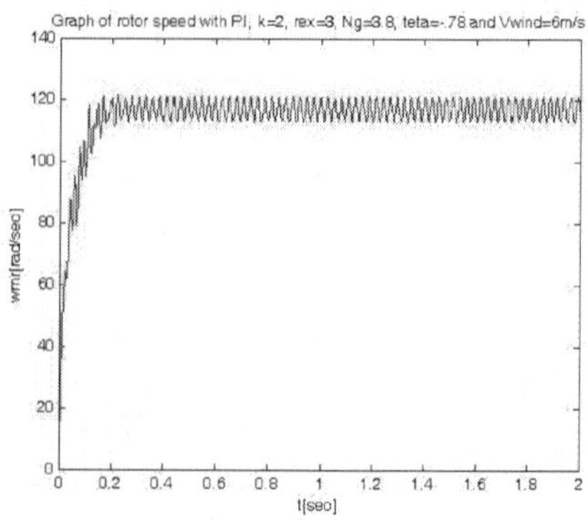

Fig. 42. Curve of rotor speed for typical v_{wind}=6m/s, k=2, Ng=3.8 and PI controller.

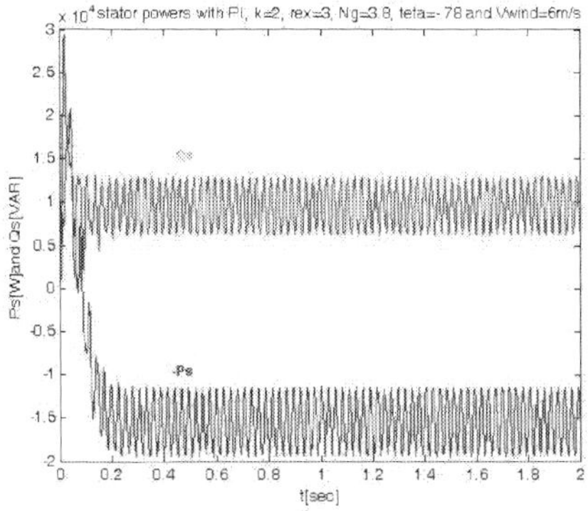

Fig. 43. Curve of stator powers for typical v_{wind}=6m/s, k=2, Ng=3.8 and PI controller.

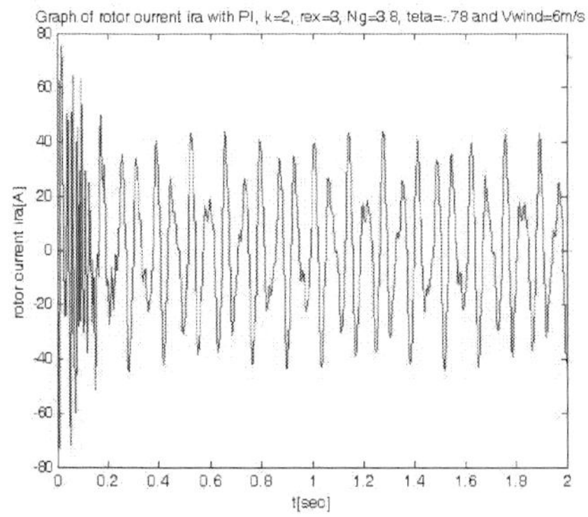

Fig. 44. Curve of rotor current i_{ra} for typical v_{wind}=6m/s, k=2, Ng=3.8 and PI controller.

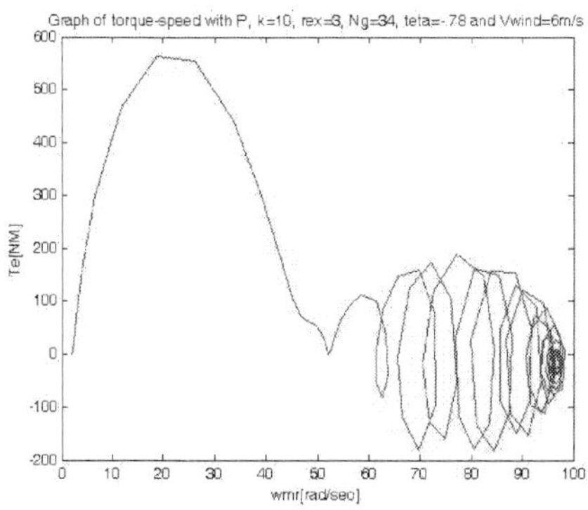

Fig. 45. Curve of torque-speed for typical v_{wind}=6m/s, k=10, Ng=34 and P controller.

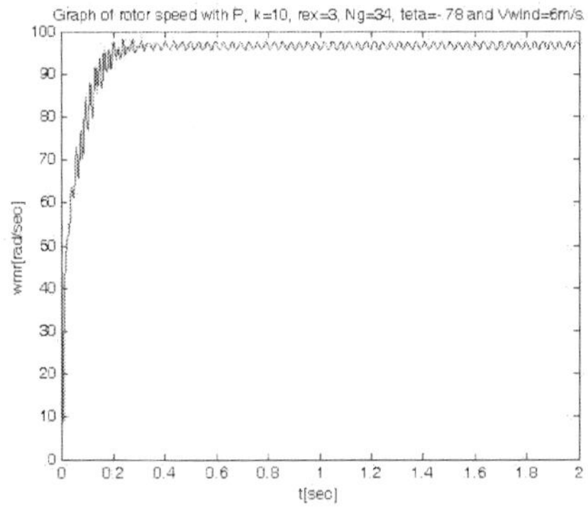

Fig. 46. Curve of rotor speed for typical v_{wind}=6m/s, k=10, Ng=34 and P controller.

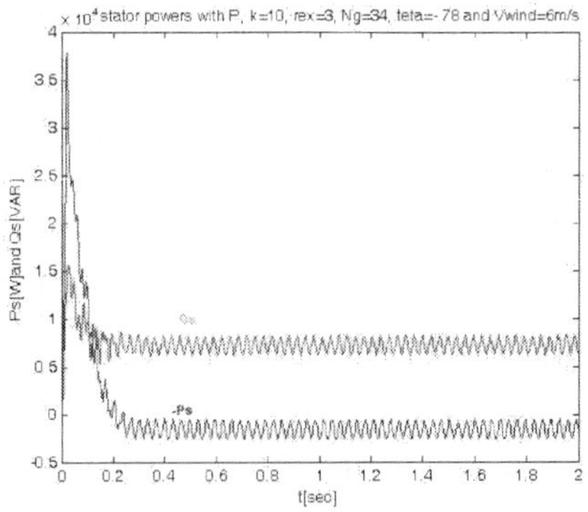

Fig. 47. Curve of stator powers for typical v_{wind}=6m/s, k=10, Ng=34 and P controller.

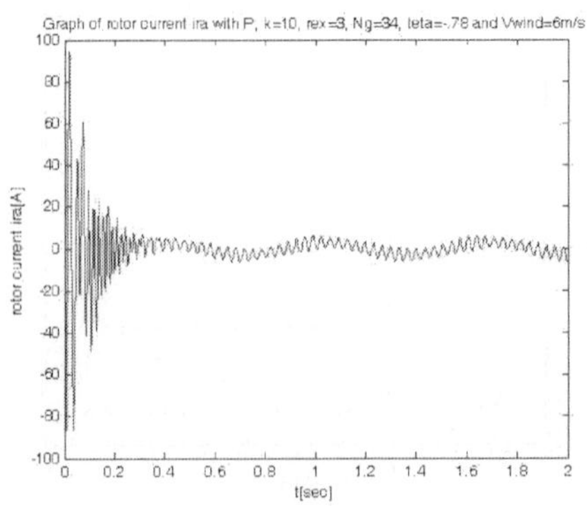

Fig. 48. Curve of rotor current i_{ra} for typical v_{wind}=6m/s, k=10, Ng=34 and P controller.

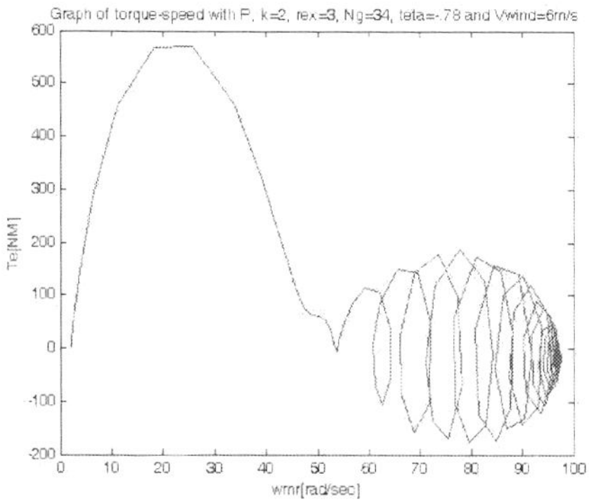

Fig. 49. Curve of torque-speed for typical v_{wind}=6m/s, k=2, Ng=34 and P controller.

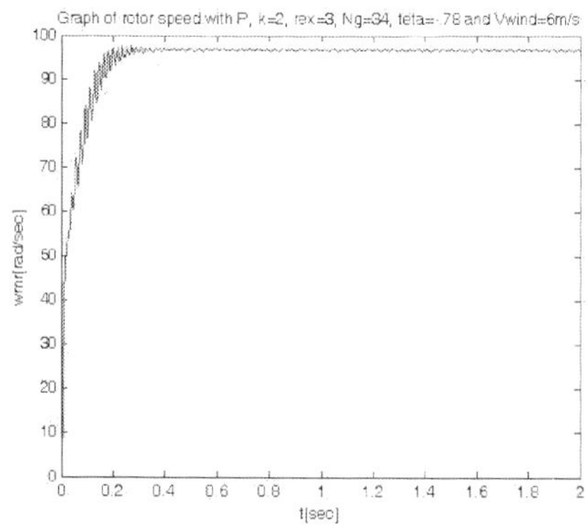

Fig. 50. Curve of rotor speed for typical v_{wind}=6m/s, k=2, Ng=34 and P controller.

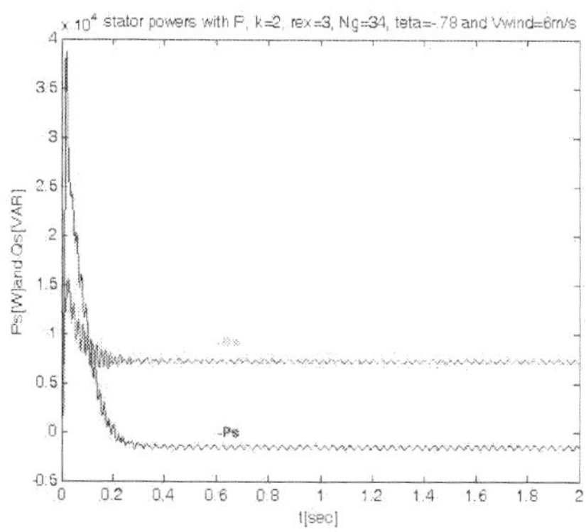

Fig. 51. Curve of stator powers for typical v_{wind}=6m/s, k=2, Ng=34 and P controller.

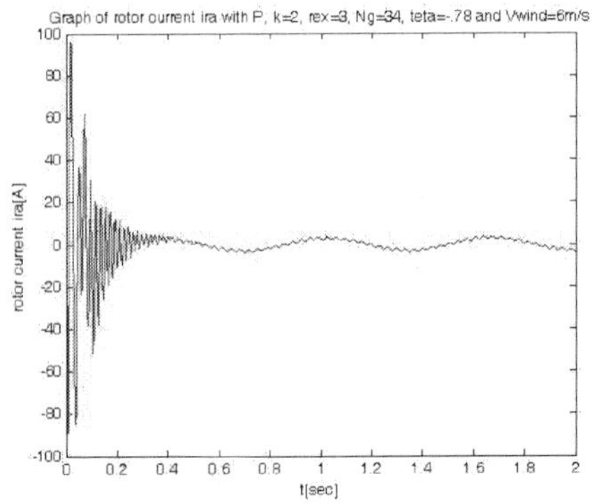

Fig. 52. Curve of rotor current i_{ra} for typical v_{wind}=6m/s, k=2, Ng=34 and P controller.

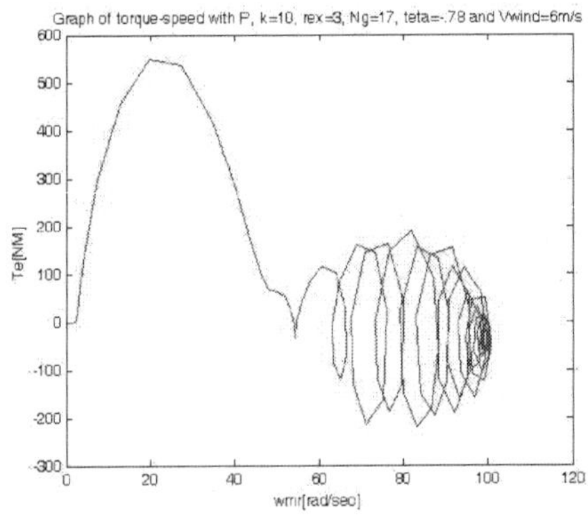

Fig. 53. Curve of torque-speed for typical v_{wind}=6m/s, k=10, Ng=17 and P controller.

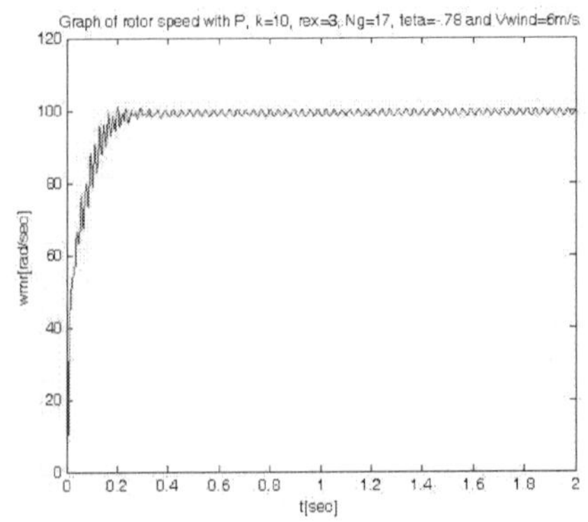

Fig. 54. Curve of rotor speed for typical v_{wind}=6m/s, k=10, Ng=17 and P controller.

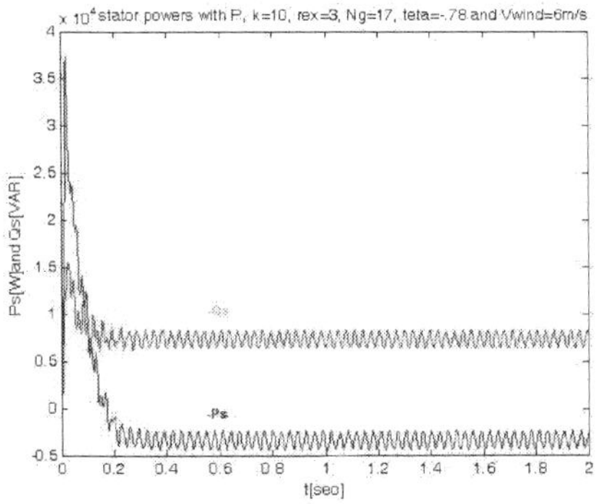

Fig. 55. Curve of stator powers for typical v_{wind}=6m/s, k=10, Ng=17 and P controller.

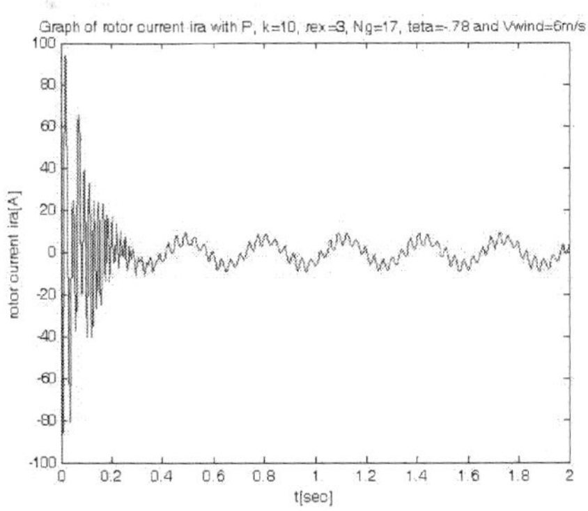

Fig. 56. Curve of rotor current i_{ra} for typical v_{wind}=6m/s, k=10, Ng=17 and P controller.

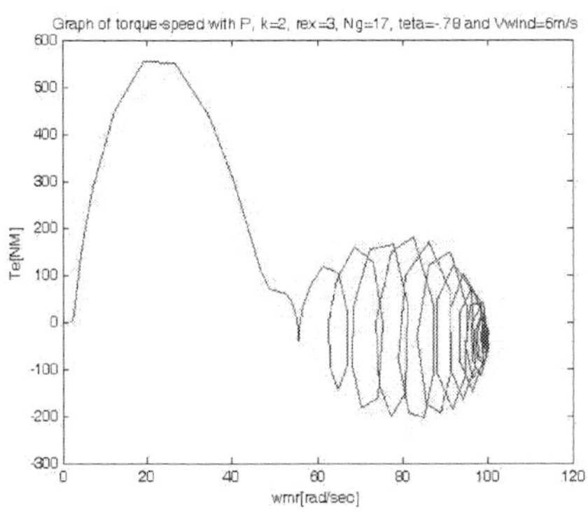

Fig. 57. Curve of torque-speed for typical v_{wind}=6m/s, k=2, Ng=17 and P controller.

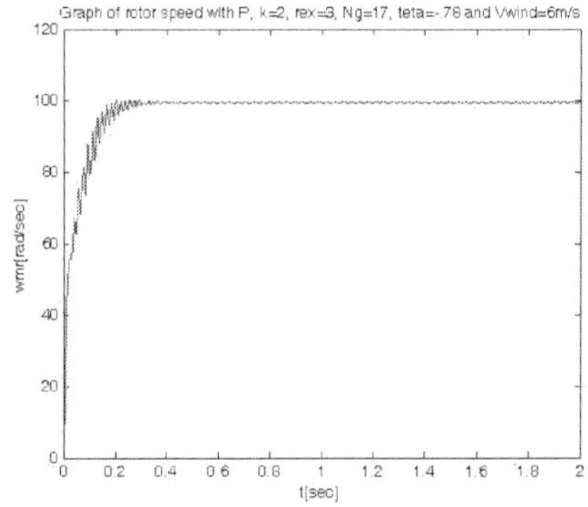

Fig. 58. Curve of rotor speed for typical v_{wind}=6m/s, k=2, Ng=17 and P controller.

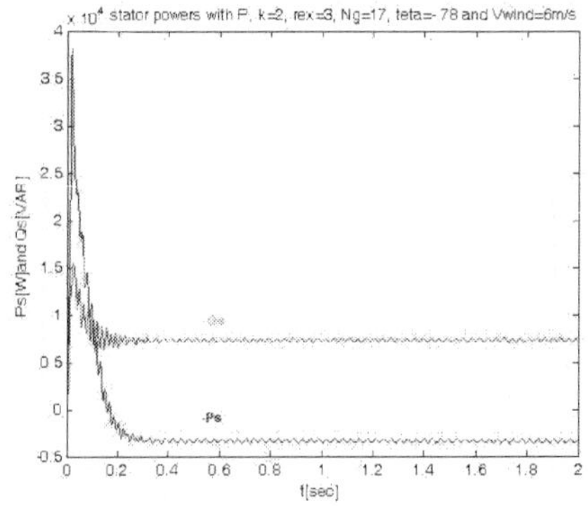

Fig. 59. Curve of stator powers for typical v_{wind}=6m/s, k=2, Ng=17 and P controller.

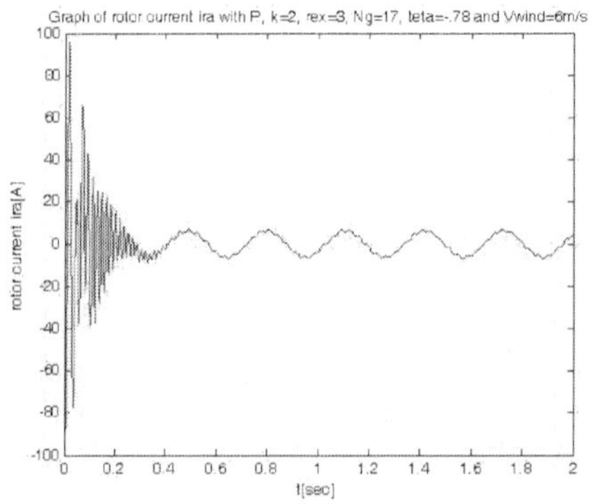

Fig. 60. Curve of rotor current i_{ra} for typical v_{wind}=6m/s, k=2, Ng=17 and P controller.

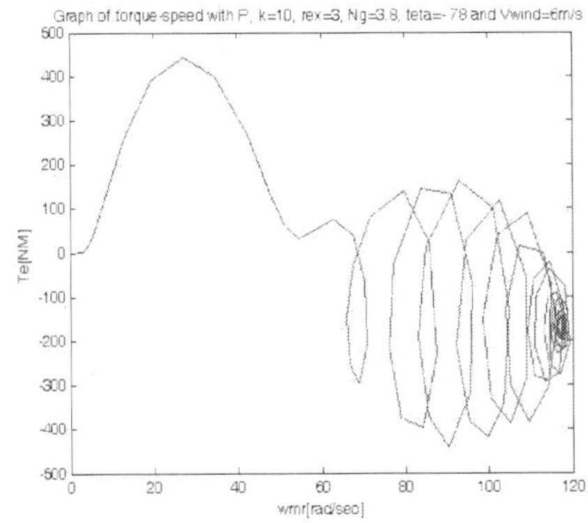

Fig. 61. Curve of torque-speed for typical v_{wind}=6m/s, k=10, Ng=3.8 and P controller.

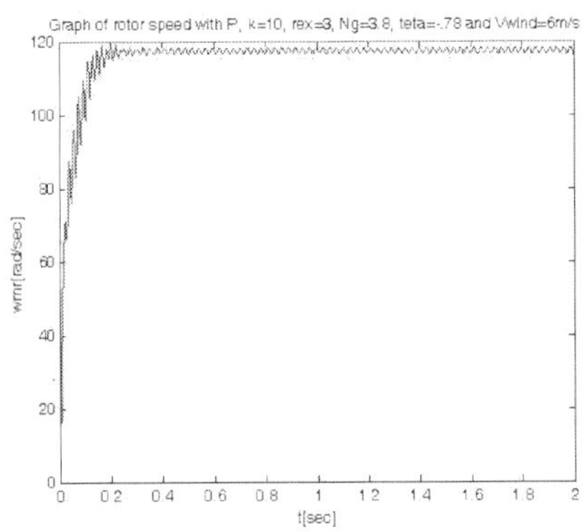

Fig. 62. Curve of rotor speed for typical v_{wind}=6m/s, k=10, Ng=3.8 and P controller.

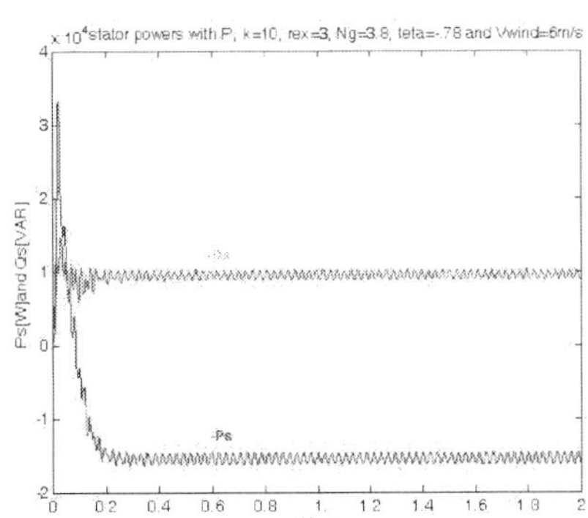

Fig. 63. Curve of stator powers for typical v_{wind}=6m/s, k=10, Ng=3.8 and P controller.

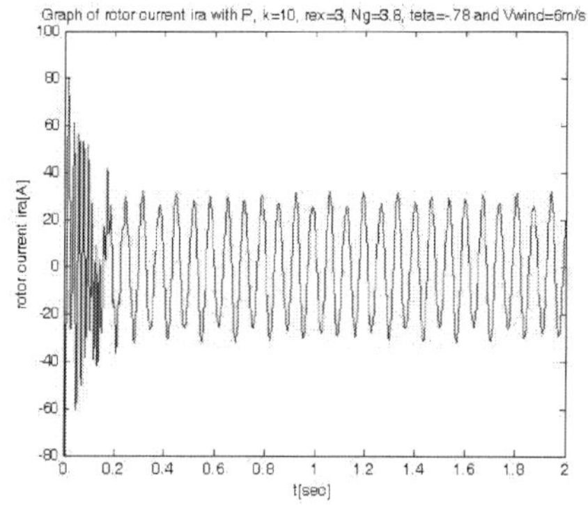

Fig. 64. Curve of rotor current i_{ra} for typical v_{wind}=6m/s, k=10, Ng=3.8 and P controller.

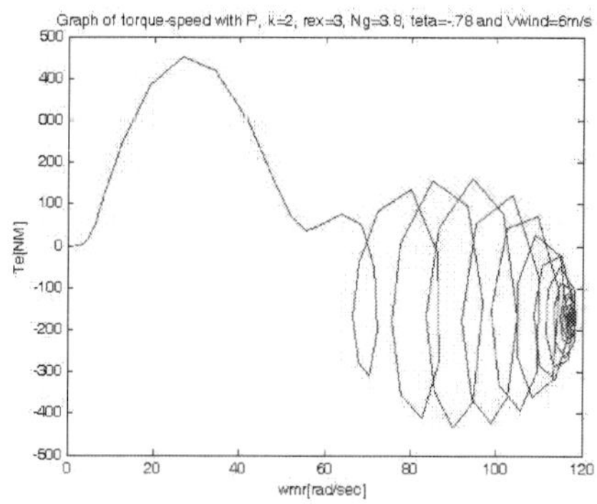

Fig. 65. Curve of torque-speed for typical v_{wind}=6m/s, k=2, Ng=3.8 and P controller.

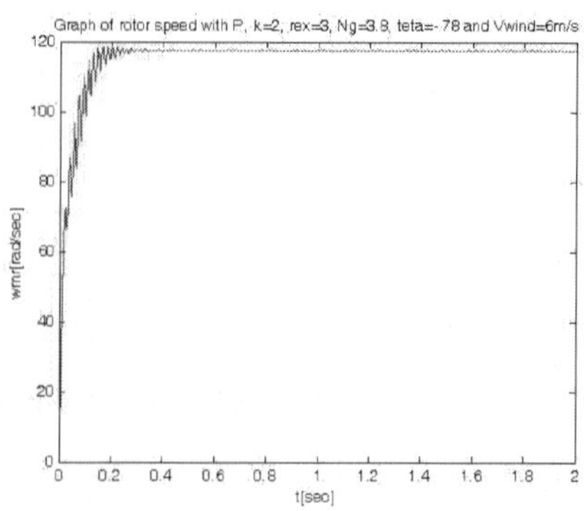

Fig. 66. Curve of rotor speed for typical v_{wind}=6m/s, k=2, Ng=3.8 and P controller.

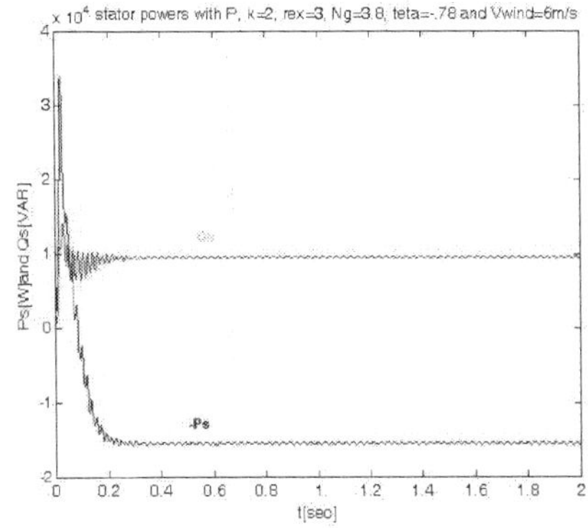

Fig. 67. Curve of stator powers for typical v_{wind}=6m/s, k=2, Ng=3.8 and P controller.

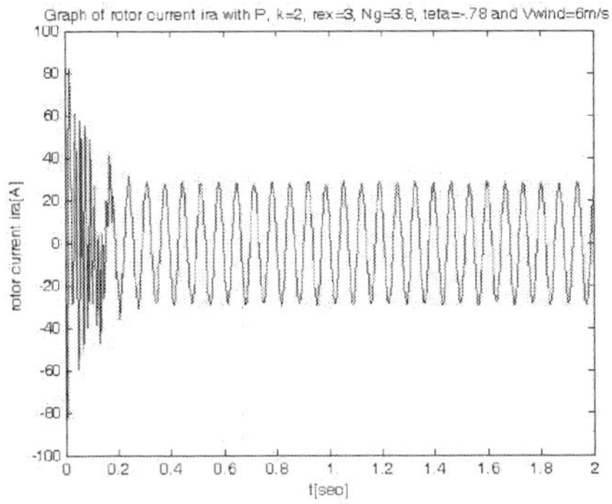

Fig. 68. Curve of rotor current i_{ra} for typical v_{wind}=6m/s, k1=2, Ng=3.8 and P controller.

9. Conclusion

As in this chapter a maximum power wind turbine is modeled with doubly fed induction machine and two P and PI controllers are designed to control stator active and reactive powers of simulated machine, according to the results, it can be concluded that:

1. Sensitivity of Q to K is more than sensitivity of P (especially in that of rotor).
2. The simulated doubly-fed induction machine with improved wind turbine, in generator mode, gives active power to the grid and takes reactive power from it.
3. In this system, gear ratio differs for different speeds and this variation is obvious especially in high speeds as adapting of turbine speed to induction generator speed in high speeds is more difficult.
4. By increasing r_{ex}, slope of torque-speed curve rises and speed of induction machine increases.
5. Sensitivity of P,Q and ω_{mr} to θ is very low.
6. Sensitivity of ω_{mr} to k is low and increasing of k, raises swing of ω_{mr}.
7. In a set point, with increasing k, machine losses declines.

8. PI voltage and P combinational self tuning controllers can both automatically control the system when rotor external voltage source domain of system drops down.
9. In a PI self tuning voltage controller, Stator active and reactive powers are controlled with the phase and amplitude of rotor external voltage but in P self tuning combinational controller those are also controlled by pitch angle and torque compensation whereas maximum turbine torque and power can be achieved anyway.
10. In PI self tuning voltage controller p parameter is tuned proportionally with the amplitude of modeled external rotor voltage but in P self tuning combinational controller proportional parameters are constant.
11. According to output curves, output responses of P self tuning controller are better with lower swing in comparison to PI self tuning output results.
12. A PI voltage controller is more economic for a new wind power plant.
13. A P combinational controller is suitable for wind power plants with pitch angle controller especially with power delivery more than 3000[w] which are promoted.
14. P controller can control the stator active and reactive powers with decreasing swings of them; however, PI controller controls the stator active and reactive powers without any change in them.
15. Simulated doubly fed induction generator delivers active power to the network grid and takes reactive power from it.

10. References

Blaabjerg, F. & Chen, Z., (2006). Power electronics for modern wind turbine, *Proceedings of* 1598290320, ISBN1598290339, Morgan & Claypool, Aalborg

Burter, T. & Bossanyi, E. & Jenkins N. & Sharpe, D. (2001). Wind energy handbook, John Wiley and sons

Ehernberg, J. & Andersen B. & Rebsdorf A., (2001). Windkraf tan lagen fur den megawattbereich, Digitale steuerungeines dopplet gespei sten asynchron generators ohne lagegeber, *Magazine Electronik,* Issue 19, page numbers(78-87)

Hosseinpur, M. & Yazdiyan, A. & Mohamadiyan, M. (2001). Maximum power control of wind system, variable speed connected to grid using matrix converter. *ICEE Transaction*

Kojooyan Jafari H. & Radan, A. (2008). Open loop control of doubly fed machine with improved wind turbine, *Proceedings of 6th WSEAS conference on environment, ecosystems and development,* page numbers (22-27), WSEAS, ISBN978-960-474-045-1, Cairo, Egypt

Kojooyan Jafari H. & Radan, A. (2009). Open loop control of doubly fed machine with improved wind turbine, *WSEAS Transaction on circuit and systems,* Vol.8, No.2 (Feb.2009) page numbers (217-226), WSEAS, ISSN1109-2734

Kojooyan Jafari H. & Radan, A. (2010). Comparison between Self Tuning PI Voltage Control of DFIG and a Combinational Control for Improved Wind Turbines, *IREACO International Control,* (Sep.2010), IREACO Journal of Praise Worthy Prize, ISSN1974-6059 , Cd-Rom ISSN1974-6067

Krause, P.C., (1986). Analysis of electrical machinery, McGraw-Hill Inc

27

High-Temperature Superconducting Wind Turbine Generators

Wenping Cao
Newcastle University Upon Tyne
United Kingdom

1. Introduction

It is well acknowledged that wind represents a clean, renewable and reliable source of energy for electricity generation. The past two decades have seen a rapid growth in wind energy utilization throughout the world and a particular trend towards ever increasing turbine ratings. Indeed, the exploitation of wind energy has been high on many governments' and organizations' agenda. However, for onshore wind farms the interference with neighboring residents and low wind speeds may limit the number of acceptable sites available for development. On the contrary, high wind speeds and low interference make offshore wind farms attractive which are expected to be a significant growth area to bridge the energy gap, with many wind farms rated as highly as 1000 MW being currently under development. Yet, the drawbacks of the offshore installation are generally associated with the foundation of wind turbines, grid connection and maintenance. Thus, it is economically preferable for offshore wind farms to have fewer but larger wind turbines instead of many small ones. Currently, the wind turbine generators of 5-7.5 MW are commercially available in the marketplace (UK Wind Power, 2008) and these of 10 MW are under development (Windpower Engineering, 2010). When the mast-top weight of mega-watts generators exceeds 100 tonnes, installation would become increasingly difficult and costly. In fact, the existing offshore installation vessels have a lifting capability of around 300 tonnes (Scots, 2008). The nacelle mass of some large wind turbines currently available can easily exceed this (Lewis & Muller, 2007) so that it is not economically viable to develop heavier wind turbines for offshore applications. Not surprisingly, therefore, the possibility of utilizing superconductors in wind turbine generators to reduce the mass and volume has come under the spotlight, primarily aided by the continuous improvement of superconducting materials.

2. Development of superconductors

The phenomenon of superconductivity was first discovered in 1911 by Onnes (later won a Nobel Prize in 1913) when he observed that the resistance of the mercury dropped to zero at a temperature of 4.2 K (-269°C). Since then, earlier developed metallic low-temperature superconductors (LTS) always operated at a temperature of 4.2 K or below, which requires the employment of costly liquid helium refrigeration for cooling purposes. The first commercial LTS were developed at Westinghouse in 1962. Typical LTS materials include NbTi (at 4 K), Nb_3Sn (at 4 K) and magnesium diboride MgB_2 tapes (at 20 K) (Gieras, 2009).

Due to the complexity and cost of liquid helium refrigeration equipment, the superconductors were rarely used in applications. A breakthrough was made in 1986 when a cuprate-perovskite ceratic material lanthanum barium copper oxide (LBaCuO, or LBCO) was discovered with critical temperatures in excess of 77 K. This marked the beginning of high-temperature superconductors (HTS), paving the way for a gradual proliferation of superconductivity. Having been further explored over two decades, the family of HTS continues to expand and today includes many rare-earth ceramic materials such as bismuth strontium calcium copper oxide (BiSrCaCuO, or BSCCO) (first generation) and yttrium barium copper oxide (YBaCuO, or YBCO) (second generation). The chronological discovery of superconductors is shown in detail in Fig. 1. The 1G HTS are produced as multiple filaments in a silver matrix, involving multiple rolling processes followed by controlled heat treatment in the manufacturing. In contrast, the 2G HTS are coated on a buffer layer and substrate. Main specifications for typical BSCCO and YBCO wires are listed in Table 1.

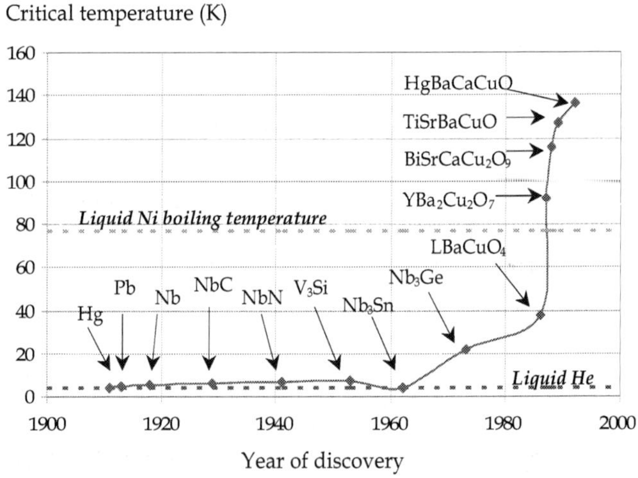

Fig. 1. Chronological discovery of superconductors

Specifications	BSCCO (Bi-2223)	YBCO (348C)
Thickness (avg)	0.19-0.23 mm	0.18-0.22 mm
Width (avg)	3.9-4.3 mm	4.70-4.95 mm
Min critical current	115 A *	≥100 A *
Min current density	135 A/mm² *	93 A/mm² *
Max stress	75 MPa **	150 MPa **
Max strain	0.15% **	0.3% **
Min bend diameter	100 mm **	30 mm **
Note	* at 77 K, self-field, 10 mV/mm ** With 95% Ic retention	

Table 1. Properties for typical BSCCO and YBCO wires (AMSC, 2008)

In principle, the HTS have neither resistance nor magnetic properties under their critical conditions. More importantly, they have non-linear *J-E* characteristics, which are governed by the following equation.

$$\vec{E} = E_c\left(\frac{J}{J_c}\right)^{N-1}\frac{\vec{J}}{J_c} \tag{1}$$

where E and E_c are the electrical field and electrical critical field, J_c is the critical current density, and N is an empirical factor chosen to match the experimental results.

In order to maintain superconductors in the superconducting state, it is required to keep their temperature, current density and magnetic field all below their critical values. Otherwise, the superconductors would return to the resistive mode, termed as "quenching". As a whole, HTS materials are characterized by their high transition (or critical) temperatures. For example, transition temperatures for the second generation superconductors are typically between 77 K and 125 K (i.e. -196°C and -148°C). This feature makes it possible to use cheap liquid nitrogen for cooling purposes. In fact, it is this refrigeration advantage that has spurred an immediate resurrection of interest in superconducting applications. At the present time, the development of HTS is nearing viable commercialization. A cost comparison between HTS and copper conductors is depicted in Fig. 2(a).

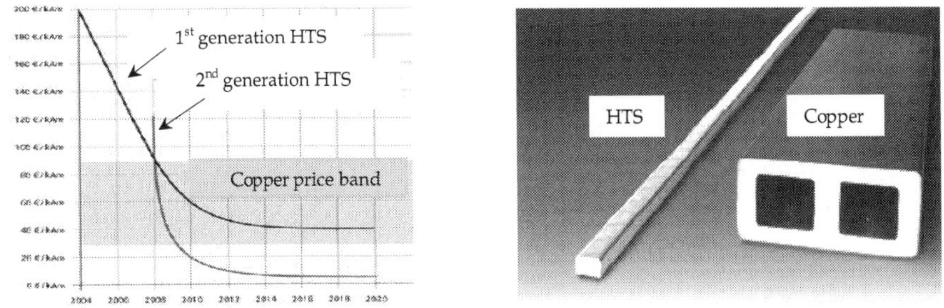

(a) (b)

Fig. 2. (a) HTS price in €/kA over time along with copper price band; (b) Comparison of 1kA insulated HTS stack with copper bar. (Lewis & Muller, 2007)

However, the quest for higher-temperature superconducting materials or even room-temperature superconductors (RTS) (precisely, temperatures above 0°C) has never ceased. A recent breakthrough was achieved in 2008 by a group of UK-US scientists who successfully identified a key component to unravelling the mystery of room-temperature superconductivity (Suchitra et al., 2008). This takes us a big step closer to the dream of RTS.

The fact that HTS could transport electricity with virtually no resistance at relatively high temperatures holds vast potentials and opens many doors for applications. These include electrical machines (Bumby, 1983; Superczynski & Waltman, 1997; Ship et al., 2002; Sykulski et al., 2002; Fee et al., 2003; Fogarty, 2004; Hull & Murakami, 2004; Kalsi et al., 2004; Masson & Luongo, 2005; Snitchler et al., 2005; Nakamura et al., 2006; Al-Mosawi et al., 2007; Klaus et al., 2007; Wilke et al., 2008; Abrahamsen et al., 2009; Wen et al., 2009), transformers (Kamijo et al., 2007; Choi et al., 2008; Liu et al., 2010), power transmission lines (Wang & Lancaster, 2006; Sasaki et al., 2007; Johnson et al., 2009), energy storages (de Andrade et al., 2005; Hirabayashi et al., 2008; Dai et al., 2010) and so forth. It becomes clear that the superconducting apparatus have to be sufficiently large to justify the costs of superconductors and the associated cooling system.

In the application of superconducting rotating machines, superconductors can be used as either HTS wires (Kalsi et al., 2004), bulk HTS materials (Hull & Murakami, 2004), or a combination of the two (Masson & Luongo, 2005). This chapter considers HT superconductors to be used as coils rather than bulks, which are commonplace in practice. In the literature, most large HTS machines are of synchronous type which uses HTS wires in the dc field winding at cryogenic temperatures and copper wires in the stator winding at ambient temperatures. The largest manufactured and tested to date is a 36.5 MW 120 rpm HTS propulsion motor for the US Navy, designed and supplied by American Superconductor Corporation (AMSC) and Northrop Grumman (AMSC, 2007).

Overall, it is estimated that the multi-MW machines would be highly efficient (over 98%) and capable of reducing the weight and volume by 50% and the capital cost by 25% in the general power application. obviously, large wind turbine generators (with MW rating) will fit the bill perfectly.

3. High temperature superconducting wind turbine generators (HTSWTGs)

The ability of superconductors to increase current density allows for high magnetic fields, leading to a significant reduction in mass and size for superconducting machines. This feature is shown in Fig. 2(b) for comparison with copper bars. In the economic perspective, it is estimated that the superconducting technology could achieve an efficiency improvement of 1% in large electrical machines, which offers substantial savings to utilities and end-users. A cost comparison for 300 MW generators is given in Table 2. Another distinct feature of superconducting machines is their part-load efficiency advantages. Namely, the HTS machine efficiency at partial loads is significantly higher when compared to conventional machines, as is illustrated in Fig. 3. This is particularly relevant to wind power generation since the wind turbines operate mostly at part-load conditions. As a result, HTS wind turbine generators can extract more wind energy than other types of machines even though they have the same nominal efficiency on the nameplates.

Machine types	Efficiency (%)	Capital cost (MD)	Annual losses (MD)	Total annual costs (MD)
Conventional	98.6	880	1400	1565
HTS generator	99.7	740	280	420

(MD: millions of dollars)

Table 2. Cost comparison for 300 MW generators (Giese et al., 1992)

In wind turbine generators, there are several competing topologies. Currently the mature technology for large wind turbines is the doubly-fed induction generator (DFIG) (Pena et al., 1996; Muller et al., 2002; Petersson et al., 2005; Ramtharan et al., 2007). Their power rating is typically between 0.5-5 MW. The prime advantage of the DFIGs is the reduced cost of the partially rated power electronic converter. However, the generator still uses brushes and

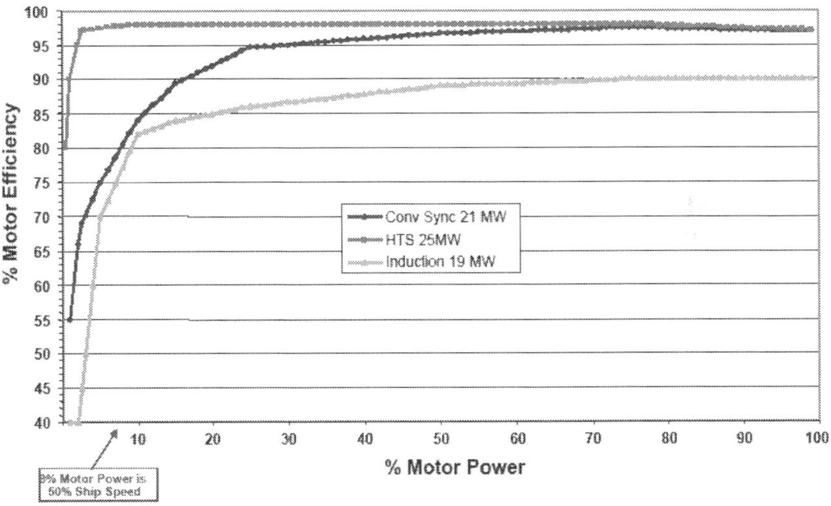

Fig. 3. Higher efficiency at full and part loads for superconducting machines (AMSC, 2010)

slip-rings which require regular maintenance. Additionally, the reactive power drawn by the machine must be compensated using power electronic converters or other measures. Nowadays, an emerging and competing technology is the permanent magnet (PM) generators. They are gradually penetrating into the wind energy market and their power rating is typically between 2-6 MW. Clearly, the use of PMs provides much promise in reduced losses and machine weight. But the flux density and power density are still limited by the existing materials (PM, copper and iron). If further improvements are demanded, superconducting generators will be an advanced technology because they eliminate conductor joule losses and provide superior performance and compactness. The likely power rating for HTSWTGs is 5-10 MW and above. Up to date a number of HTSWTG projects have been initiated worldwide. AMSC commenced a project on developing a 10 MW HTS offshore wind turbine generator in 2007 (Marsh, 2009), which is still ongoing to date. It is estimated that this machine weights around 120 tonnes and costs a US $ 6.8 million. It can be seen that the weight is nearly a third of a similarly-rated conventional direct-drive turbine generator.

4. Topologies of HTSWTGs

Since superconducting machine design is still evolving and is highly dependent on the superconducting materials used, it should not rule out any viable topology at the stage of feasibility study. In the literature, a variety of HTS machine types have been reported. These include homopolar machines (Superczynski & Waltman, 1997), induction machines (Nakamura et al., 2006) and in the majority, synchronous machines (Ship et al., 2002; Sykulski et al., 2002; Al-Mosawi et al., 2007; Klaus et al., 2007; Wilke et al., 2008).

In essence, the existing HTS are prone to ac losses when experiencing field variations. In this regard, the synchronous machine may become a natural choice of machine type since, in synchronous machines, the HTS in the rotation would see a standstill and constant field generated from the stator winding. For the same reason, superconductors are rarely used in the stator winding. Also, it is difficult to build the stator winding by flat HTS coils for an effective use of space. Clearly a radical design of the whole superconducting arrangement (both stator and rotor) is not realistic before a new type of superconductors is developed to overcome these ac losses. As a result, a feasible design of the HTSWTG is based on the synchronous generator with the copper stator and the superconductor rotor (Bumby, 1983, Al-Mosawi et al., 2007). HTS coils are generally wound in the form of very thin race-track tapes due to their ceramic features. When installed on the rotor, they provide a constant magnetic field similar to the PMs.

Regarding the HTS synchronous machines, they can be further broken down into four different types (Lewis & Muller, 2007):

Type 1. Conventional stator and HTS rotor with magnetic pole bodies
This type offers gains in efficiency due to minimized rotor loss but does not offer substantial reduction in weight and dimension.

Type 2. Conventional stator and HTS rotor with non-magnetic pole bodies
This type offers gains in efficiency due to minimized rotor loss, and reduces costly cold magnetic materials and complex thermal isolation. However, it needs more HTS wires to establish the necessary flux density.

Type 3. Airgap stator winding and HTS rotor with magnetic pole bodies
This type produces significantly higher flux density at the airgap than a conventional stator and thus decreases the mass and size of the machine. However, the rotor iron can operate highly saturated so that the efficiency is reduced.

Type 4. Airgap stator winding and HTS rotor with non-magnetic pole bodies
This type allows significant reduction in weight and dimension, and also minimizes potential high cost cold magnetic materials. But it requires more HTS wires in use.

Since wind energy is a cost sensitive market, the capital cost is paramount in producing large machines. For offshore wind turbines, the mass saving is more important than size reduction. Therefore, the actual compromise is made between the low cost, low mass and high efficiency. For the direct-drive wind generator, types 3 and 4 both can offer the advantage of lowest mass. The cost balance between the two largely relies on the relative cost of HTS wires against other materials (cold and warm irons, copper etc). Based on the existing production pricing for iron and 2G HTS wires, type 4 may be a more cost-effective solution. Converteam has undertaken the design of an 8 MW 12 rpm HTSWTG following type 4 topology (Lewis & Muller, 2007).

From the above analysis, a multi-MW, low speed, direct-drive HTSWTG system can be sketched, as shown in Fig. 4. The machine consists of the stator back iron, stator copper winding, HTS field coils, rotor core, rotor support structure, rotor cooling system, cryostat and external refrigerator, electromagnetic (EM) shield and damper, bearing, shaft and housing. Among all the components, key to the machine design are the arrangements for the stator, rotor, cooling and gearbox.

Stator arrangements

Although in theory the whole superconducting arrangement is possible, the excessive ac losses in the stator and physical difficulty in bending and twisting the fragile HTS to fit into the stator slots pose particular challenges. Thus it is not surprising for some designs to choose the existing stator in availability (Ship et al., 2002).

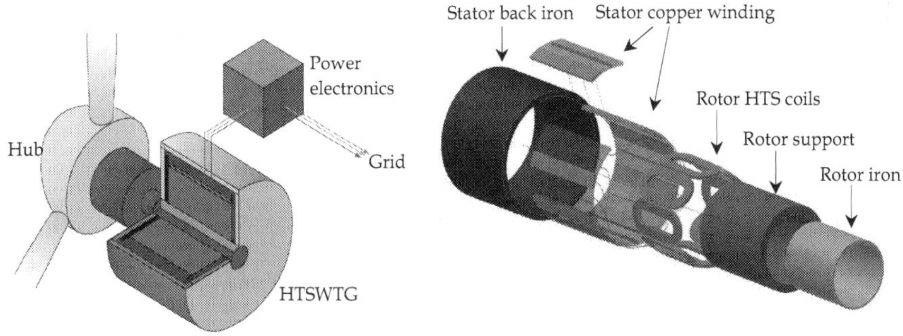

Fig. 4. Schematic of a multi-MW, low speed, direct drive HTSWTG system (Courtesy of Converteam)

In the literature, there are three different stator winding topologies reported for HTS machines. As shown in Fig. 5(a), the copper windings are placed in the slots between magnetic teeth and an iron yoke is used to guide the magnetic flux through the stator.

Obviously, this configuration can take advantage of existing stator and stator winding design but has a limit on the maximum flux density in relation to the saturation of the iron teeth. The second winding topology is designed to overcome this problem by replacing the iron teeth with no-magnetic teeth (e.g. fiber reinforced plastics). Consequently, a higher level of flux density can be achieved, as well as a higher power generated from the machine. Nonetheless, replacing the magnetic teeth would change the flux distribution in the stator, giving rise to transverse flux components. In order to mitigate the resultant eddy currents, the windings should be modified using smaller-diameter insulated and transposed copper strands (Litz wires). The difference between (b) and (c) lies in the width of the non-magnetic stator teeth. The thinner teeth in (c) implies that more copper can be inserted in the slots and higher ampere-turn loading can be obtained without causing saturation in the stator teeth. But this topology is subjected to mechanical stability and cooling capability.

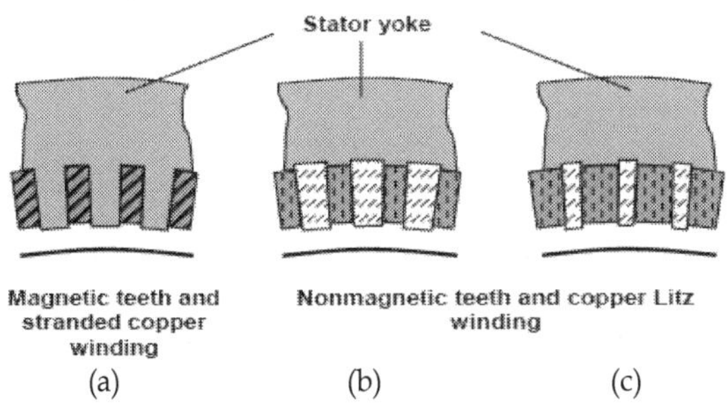

Fig. 5. Three different stator winding arrangements. (a) conventional stator, (b) high field stator, (c) high field, high magnetomotive force (mmf) loading stator (Klaus et al., 2007)

Rotor arrangements

In conventional copper wires, the maximum current density achieved is typically 5 A/mm^2 while, in 2G HTS, this can be as much as 135 A/mm^2. The significant increase in current density leads to a much compact HTS rotor for the HTSWTGs.

In terms of magnetic characteristics, the rotor core can be either magnetic or non-magnetic. The use of magnetic irons can reduce the mmf required to establish the same field since core material forms part of magnetic circuits (much better than air). Clearly, the rotor mass would be increased accordingly and so is the rotor inertia. However, the latter does not cause problems since in direct-drive wind turbines the actual rotation speed is quite low. In practice, it is very difficult to twist the HTS coils to align with the field for the purposes of minimizing ac losses so that iron (and the flux diverters) should be used to guide the flux in the desired direction and away from the HTS. But the fact that iron saturates at approximately 2 T puts a limit on the maximum flux density.

In theory, the high current density in superconductors makes it possible to produce sufficient air-gap flux density without a rotor core. Therefore, the rotor can be of air-cored type (coreless rotor) (Ship & Sykulski, 2004; Lukasik et al., 2008). This configuration provides a significant reduction in the weight of the rotor and the associated eddy current losses. Nevertheless, it may increase the amount of superconductors used and the current level in the superconductor so as to produce the required flux density. Similarly, because there is no iron core, the support structure should be strong to transmit the high torque, which is the case of direct-drive wind turbines.

With regard to the rotor cooling arrangement, the HTSWTG can use either warm or cold rotors, as demonstrated in Fig. 6. In Fig. 6(a), only HTS coils are cooled at cryogenic temperature so that the so-called "cold mass" is low. This results in short cool-down periods and reduced eddy current losses. But the supporting structure would be complicated to hold the HTS and also to prevent heat leakage. In contrast, in Fig. 6(b), the cold rotor structure is relatively simple and the whole rotor is cooled at cryogenic temperature, requiring additional cooling capacity to remove the heat inside the rotor. Moreover, an auxiliary torque transmission element is needed to connect the rotor and the shaft. Since the two are operated at different temperatures, heat leakage arises via the intermediate element. Besides, cooling the rotor core to a very low temperature gives rise to eddy current losses when exposed to mmf harmonics. This effect can be significant and requires a careful design of the rotor EM shield to prevent the harmonics from entering the cold part. In large wind turbines, warm rotor topology may be preferred due to the minimized cooling requirement and eddy current losses.

Cooling arrangements

Cooling arrangements play a crucial role in the success of the HTS machines. When designing the cryogenic system, one should consider its ease of operation and maintenance, minimum complexity and cost, and integration with the superconducting machines. Early LTS designs used liquid helium to achieve a temperature of 4.2 K whereas the latest HTS use liquid nitrogen or even inexpensive liquid hydrogen to cool the superconductors down to 77-125 K. The cost of cryogenic cooling systems depends more on operating temperatures than anything else. Therefore, the overall cost constantly drops as the critical temperatures of HTS increase.

When the operating temperature decreases, the critical temperature and critical current in HTS wires increase. For instance, when the operating temperatures reduce from 77 K to 50 K, the critical current in the HTS is doubled but the cooling power required only increases by 15% (Jha, 1998).

The cryogenic cooling systems generally use counter-current streams for optimum economy. In this respect, the conductors with a high surface-to-volume ratio can lead to a high cooling

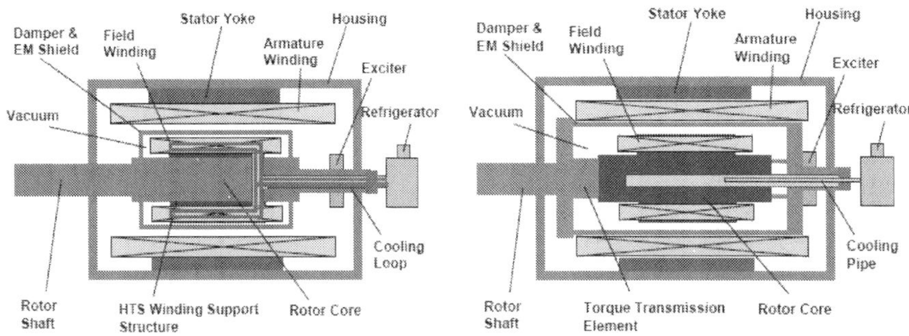

(a) Warm rotor (b) Cold rotor

Fig. 6. Two different rotor arrangements (Klaus et al., 2007)

efficiency. It is easily understood that cooling efficiency is also dependent on the thermal insulation of HTS. In reality, to remove 1 W of heat generated at 77 K requires 10 W of electricity (Giese et al., 1992). Thus a key aspect of the cooling design is to minimize the power losses in the support structure and EM shields.

Selection of gearbox

Historically, gearbox failures are proven to be major challenges to the operation of wind farms (Robb, 2005; Ribrant & Bertling, 2007). This is especially true for offshore wind turbines which are situated in harsh environments and which may be realistically accessed once per year.

Obviously, direct drive configuration removes the necessity for gears, slip-rings and the associated reliability problems. A comparison of different drive train configurations is presented in Table 3. As a result, some wind turbine manufacturers are now moving toward direct-drive generators to improve reliability. However, a drawback of the direct drive is associated with the low operating speed of the turbine generator. Low speed operation implies a high torque required for a given power output, i.e., a physically large machine. As the nominal speed of the machine reduces, the volume and weight would increase approximately in inverse proportion. This may offset some of the weight savings from using the HTS. Nevertheless, the system as a whole can still benefit from reduced mass and size, taking account of savings made from removing gearboxes. For example, a direct-drive 6 MW HTSWTG is estimated to be approximately 20% of the mass of an equivalent conventional synchronous generator, half of the mass of an optimized PM direct-drive generator, and a similar mass of a conventional geared high-speed generator (Lewis & Muller, 2007).

Drive trains	Turbine speed	Gearing	Generator speed	Problems
Conventional	15 rpm	1:100 gear	1500 rpm	Heavy & problematic gearbox
Hybrid	15 rpm	1:6 gear	90 rpm	In between
Direct drive	15 rpm	No	15 rpm	Large & heavy generator

Table 3. Three types of drive train configurations

5. Design considerations and challenges

A good design of electrical machines should allow for better use of materials and space while meeting electrical, mechanical, thermal, economic and reliability requirements. In the design of HTSWTGs, typical optimization parameters in the consideration are: low mass and size, minimum use of superconductors, low capital cost, high efficiency, high levels of reliability and stability. However, it is highly likely that they are conflicting in practice and a compromise has to be made based on personal experiences. For instance, the working point of the machine is dictated by the critical current of the HTS coils and the maximum flux density at the conductor, which are both dependent on the operating temperature. When machine compactness is achieved by increasing the flux density, iron losses in magnetic iron parts will be increased, thus reducing the efficiency. When the operating temperature of HTS is reduced, electrical performance improves but cooling power required increases.

Without a doubt, firstly, the mechanical properties of the HTS place some constraints on the machine design. Physically, they are limited in the shape and coil arrangement. The difficulty in the cryogenic design arises from the difference in thermal contraction between the superconductors and the core, which must be taken into consideration. In the rotor design, the supporting structure must be mechanically strong to carry the loads imposed by the centrifugal forces and thermally arranged by appropriate thermal insulation to prevent the heat leak from the warm part of the rotor entering into the cryostat.

At first glance, it may be tempting to view HTS as conventional conductors with zero resistance. But this is not the case in the machine design for the *J-E* characteristics are highly non-linear, depending on the magnetic field intensity and orientation, the temperature and current allowances for safety margin. If any one of these parameters reaches its thresholds, the superconductivity can be lost.

It is widely accepted that existing superconductors work best with dc currents and constant fields. When experiencing ac field variations, hysteresis and eddy current losses are induced in the conductors. Magnetically, the superconductors are anisotropic and particularly vulnerable to magnetic fields in perpendicular direction. When used as superconducting tapes, care should be exercised in the design to accommodate the constraints resulting from their anisotropic properties. The magnetic fields (especially perpendicular to the HTS tape's broad face) should be kept below certain limits to avoid significant power losses. Another source of power losses in the cold part of the rotor is associated with eddy currents (Sykulski et al., 2002). They can result in a significant load on the cryogenic system and therefore put a constraint on the machine design. As a consequence, electromagnetic shields should be used to protect the rotor from ac flux components. Electrically, divert rings and metal screen can also act as separate damping windings to improve the machine's transient responses.

In the stator design, a challenge is the centrifugal forces which act on the stator conductors and which are highly cycle fatigue loads. Therefore, stator copper coils need to be made from stranded Litz wire to eliminate eddy current loss and to provide physical flexibility. When the non-magnetic teeth are used, electromagnetic forces need to be transmitted to the back iron and frame via non-magnetic elements. In addition, some problems are associated with harmonic contents in the stator voltage. The output voltage harmonics are determined by the configuration of the stator winding and the air-gap flux density waveform produced by the field winding (Lukasik et al., 2008). Since HTS machines' synchronous reactance is low, the voltage harmonics have an exaggerated impact on the external circuits. It is found that the fifth harmonic is the dominant harmonic component and should be mitigated in the design of the pole face (Ship et al., 2002).

The design of a 10 kW direct-drive HTSWTG is described in (Abrahamsen et al., 2009) and the main specifications are tabulated in Table 4 for reference.

Items	Value	Items	Value
Rating	10 kW	Critical current density	110 A/mm^{-2}
Pole No.	8	Stator max flux density	0.96 T
Type of HTS	BSCCO-2223	Rotor max flux density	1.79 T
Working temperature	50 K	Stator line voltage	400 V
Stator diameter	0.32 m	Stator phase current	14.4 A
Rotor diameter	0.25 m	HTS wire length	7539 m
Rotor length	0.4 m	HTS wire weight	91 kg

Table 4. Main specifications of a 10 kW direct-drive HTSWTG. (Abrahamsen et al., 2009)

6. Integrating HTSWTGs into the power network

Power system stability relies on large wind turbines that remain connected when undergoing voltage surges and short-circuits at local or remote distances. Fig. 7 shows a simplified representation of the HTSWTG in a power system. Equivalent circuits for the *d*- and *q*-axis representations of superconducting generators are developed in (Liese et al., 1984), which comprise a large number of series connected *T*-networks (Kulig et al., 1984). An important feature in the modeling of the superconducting machine is the rotor EM shield, which in effect distorts the radial and tangential flux densities and affects the machine dynamic performance and output power.

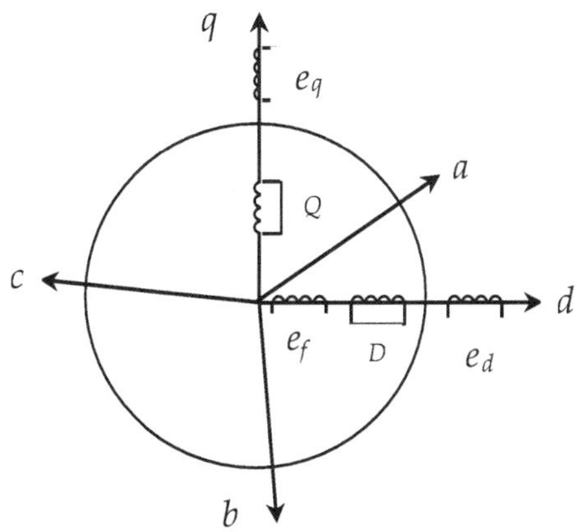

Fig. 7. Representation of the HTSWTG

When integrating large HTSWTGs into the power network, considerations of their impacts are twofold. Firstly, there is an impact of the HTSWTG on the power network and, secondly, there is an impact of the power grid faults on the HTSWTG system.

If the power network is strong, it may be able to accept more wind generation within normal power quality criteria. Nonetheless, most large wind power sites are remote where the adjacent distribution networks or substations are low in their capacity. For analysis purposes, a weak network can be represented by a short-circuit ratio (SCR) of less than 6 (Abbey et al., 2005). Calculating a local network's SCR can help optimize the wind farm design in handling the weakest point of the system. The intermittent power output of a wind farm can result in voltage fluctuations on these networks, known as "flickers". These would be significant for small numbers of large wind turbines connected at low voltages, as is the case for offshore wind turbines. Moreover, variable-speed wind turbines can also induce harmonic voltages to appear on the network, causing equipment to malfunction or overheat. Compared to the conventional wind turbine generators, HTSWTGs may have lower synchronous and sub-transient reactances. Therefore, their dynamic responses tend to be faster despite a greater L/R time constant they have. Although HTSWTGs may provide a larger dynamic stability limit, their dynamic behaviors are largely dictated by the transformer-transmission line reactance. Clearly, with the increased proliferation of wind power generation in the network, the power system may become weaker and power system stability may be of great concern.

On the other hand, it is equally important to examine the fault-ride-through (FRT) capability of the HTSWTG system responding to grid faults. Nowadays, many power network codes require wind turbines to ride through voltage sages (E.ON, 2003; Denmark, 2004; FERC, 2005; Ireland, 2007; UK, 2008). In addition to voltage fluctuations caused by varying loads connected on the network, power faults at local or remote buses of the power network are also the sources of problem. HTSWTGs may be able to provide better damping resulting from rotor electromagnetic shield and/or damping screen than conventional generators. Consequently, real power fluctuations following a grid fault should be smaller and HTSWTGs are considered to be more resistant to the transient system faults. In particular, when equipped with power electronics and low voltage ride-through-capability, large HTSWTGs may be incorporated into remote networks without compromising power system stability.

7. Conclusions

The implementation of superconducting technology in electrical machines offers significant reductions in mass and size, as well as superior performance and reliability, and potentially competitive costs. In the offshore wind power generation, the dominant DFIG configuration suffers from regular maintenance associated with slip-rings and gearboxes. Development of HTS materials has made superconductivity technically and economically viable to fill the gap.

This chapter has overviewed the historical development of superconductivity and considered the potential merits of applying HTS coils to large wind turbine generators. A number of machine topologies and design issues have been discussed. It is found that: 1) HTS provide potential benefits for wind turbine development in lowering the overall cost of wind energy while improving energy efficiency; 2) synchronous generators with the HTS field coils promise to be a favorable configuration for next generation wind turbine generators. This is so far a proven technology in large electrical machines and may still need some time to develop its economic competitiveness; and 3) used in combination, direct-drive arrangement can reduce the reliability problems associated with the gearbox but it comes at a price in terms of machine size. An increase in system efficiency would have significant economic implications since the machines considered are multi-MW and above. Improved fault ride-through capacity of the HTSWTG would help minimize the need for maintenance and the likelihood of machine breakdowns. Further work is currently underway to model a 10 MW direct-drive HTSWTG using 3D finite-element tools.

Looking to the future, it would be highly desired that the next generation room-temperature superconductors be developed in commercial availability. If such a day comes, superconductivity would offer unprecedentedly significant benefits in cost saving and performance improvement, and would undoubtedly revolutionize every aspect of electrical machine design.

8. Acknowledgment

The author gratefully acknowledges the helpful discussions with Prof. G. Asher of Nottingham University and Prof. B. Mecrow of Newcastle University.

9. References

Abbey, C., Khodabakhchian, B., Zhou, F. (2005). "Transient modeling and comparison of wind generator topologies", the International Conference on Power System Transients (IPST'05), 10-23, June 2005

Abrahamsen, A.B., Mijatovic, N., Seiler, E., Sorensen, M.P., Koch, M., Norgard, P.B., Pedersen, N.F., Traeholt, C., Andersen, N.H., Ostergard, J. (2009). "Design study of 10 kW superconducting generator for wind turbine applications", *IEEE Trans. Applied Superconductivity*, Vol. 19, Issue: 3, Part: 2, pp. 1678-1682

Al-Mosawi, M.K., Goddard, K., Beduz, C., Yang, Y. (2007). "Coreless HTS synchronous generator operating at liquid nitrogen temperatures", *IEEE Trans. Applied Superconductivity*, Vol. 17, Issue 2, Part 2, June 2007, pp. 1599-1602

American Superconductor Corporate (AMSC), (2007), "American Superconductor complete successful test of 36.5MW HTS ship propulsion motor", *Materials, Online*: http://www.azom.com/news.asp?newsID=8114

American Superconductor Corporate (AMSC), (2008). "Bi-2223 high current density wire", *Materials, Online*: http://www.amsuper.com/products/library/002-multifactfs01-02.pdf

American Superconductor Corporate (AMSC), (2010). *Materials,* Online: http://www.amsc.com/products/library/HTS_efficiency_advantage.pdf

Bumby, J.R. (1983). "Superconducting rotating electrical machines", *Clarendon Press*, Oxford

Choi, J., Lee, S., Choi, S., Park, M., Kim, W., Lee, K., Choi, K., Lee, H., Hahn, S. (2008). "Conceptual design of a 5 MVA single phase high temperature superconducting transformer", *IEEE Trans. Applied Superconductivity*, Vol. 18, Issue: 2, pp. 636-639

Dai, T., Tang, Y., Shi, J., Jiao, F., Wang;L. (2010). "Design of a 10 MJ HTS superconducting magnetic energy storage magnet", *IEEE Trans. Applied Superconductivity*, Vol. 20, Issue: 3, pp. 1356-1359

de Andrade, R. Jr., Ferreira, A.C., Sotelo, G.G., Neto, J.L.S., Rolim, L.G.B., Suemitsu, W.I., Bessa, M.F., Stephan, R.M., Nicolsky, R. (2005). "Voltage sags compensation using a superconducting flywheel energy storage system", *IEEE Trans. Applied Superconductivity*, Vol. 15, Issue: 2, Part: 2, pp. 2265-2268

Denmark, (2004). "Wind turbines connected to grids with voltages below 100 kV-Technical Regulations TF 3.2.62", *Transmission Lines Department*, May 2004, Online, http://www.energinet.dk 2010

E.ON, (2003). "Grid code high and extra high voltage," *Tech. Rep., Netz GmgH Bayreuth*, Online: http://eon-netz.com

Gieras, J.F. (2009). "Advancements in electric machines", Springer, Milton Keynes

Giese, R.F., Sheahen, T.P., Wolsky, A.M., Sharma, D.K. (1992). "High-temperature superconductors: their potential for utility applications", *IEEE Trans. Energy Conversion*, Vol. 7, No. 3, September 1992, pp. 589-597

Fee, M., Staines, M.P., Buckley, R.G., Watterson, P.A., Zhu, J.G. (2003). "Calculation of AC loss in an HTS wind turbine generator", *IEEE Trans. Applied Superconductivity*, Vol. 13, Issue: 2, Part: 2, pp. 2193-2196

FERC, (2005). "Docket no. RM05-4-001; Order no. 661-A, Interconnection for Wind Energy", *Federal Energy Regulatory Commission (FERC)*, United States of America, Dec. 12, 2005

Fogarty, J.M. (2004). "Development of a 100 MVA high temperature superconducting generator", *the 2004 IEEE Power Engineering Society General Meeting*, Vol. 2, pp. 2065-2067

Jha, A.R. (1998). "Superconductor technology", *John Wiley & Sons*, USA

Johnson, B.K., Appikonda, J.M., Venkataramanan, G. (2009). "Reactive compensation needs for superconducting transmission systems", *IEEE Trans. Applied Superconductivity*, Vol. 19, Issue: 3, Part: 2, pp. 2066-2069

Hirabayashi, H., Makida, Y., Nomura, S., Shintomi, T. (2008). "Liquid hydrogen cooled superconducting magnet and energy storage", *IEEE Trans. Applied Superconductivity*, Vol. 18, Issue: 2, pp. 766-769

Hull J.R., Murakami, M. (2004). "Applications of bulk high temperature superconductors", *IEEE Proceedings*, Vol. 92 No. 10, October 2004, pp. 1705-1718

Ireland, (2007). "Wind Farm Power Station Grid Code Provisions, WFPS1", *Ireland National Grid, Grid Code Version 2*, pp. 213–216, Jan. 2007

Kalsi, S., Weeber, K., Takesue, H., Lewis, C., Neumueller, H.W., Blaugher, R.D. (2004). "Development status of rotating machines employing superconducting field windings", *IEEE Proceedings*, Vol. 92 No. 10, October 2004, pp 1688-1704

Klaus, G., Wilke, M., Frauenhofer, J., Nick, W., Neumuller, H.W. (2007). "Design challenges and benefits of HTS synchronous machines", *the 2007 IEEE Power Engineering Society General Meeting*, pp. 1- 8

Kamijo, H., Hata, H., Fujimoto, H., Inoue, A., Nagashima, K., Ikeda, K., Iwakuma, M., Funaki, K., Sanuki, Y., Tomioka, A., Yamada, H., Uwamori, K., Yoshida, S. (2007). "Tests of superconducting traction transformer for railway rolling stock", *IEEE Trans. Applied Superconductivity*, Vol. 17, Issue: 2, Part: 2 pp. 1927-1930

Kulig, T.S.; Lambrecht, D.; Liese, M. (1984). "Investigation of the transient performance of superconducting generators with an advanced Network-Torsion-Machine program", *IEEE Trans. Power Apparatus and Systems*, Vol. PAS-103, Issue 7, July 1984, pp. 1764-1772

Lewis, C., & Muller, J. (2007). "A direct drive wind turbine HTS generator", IEEE Society Power Engineering General Meeting, 24-28 June 2007, Florida, USA, pp. 1-8

Liese, M.; Kulig, T.S.; Lambrecht, D. (1984). "Superconducting generator modeling by iterative T-network equivalent circuits for investigation of transient performance", *IEEE Trans. Power Apparatus and Systems*, Vol. PAS-103, Issue 7, July 1984, pp. 1756-1763

Liu, H.J., Wu, Y., Ren, Z.B., Wu, S.T., Shi, Y., Peng, J.Q., Chen, J.L., Long, F., Yu, M., Qian, L. (2010). "Manufacturing of 50 kA superconducting transformer for ITER correction coil conductor test", *Review of Scientific Instruments*, Vol. 81, Issue: 4, pp. 044701-044701-3

Lukasik, B., Goddard, K.F., Sykulski, J.K. (2008). "Finite-element assisted method to reduce harmonic content in the air-gap flux density of a high-temperature superconducting coreless rotor generator", *IET Science, Measurement & Technology*, Vol. 2 , Issue: 6 pp. 485-492

Marsh, G. (2009). "Rise of the superconductor", *Renewable Energy Focus*, Jul./Aug. 2009, pp. 38-42

Masson, P.J., Luongo, C.A. (2005). "High power density superconducting motor for all-electric aircraft propulsion", *IEEE Trans. Applied Superconductivity*, Vol. 15, no. 2, June 2005, pp 2226-2229

Muller, S., Deicke, M., and De Doncker, R.W. (2002) "Doubly fed induction generator systems for wind turbines," *IEEE Ind. Appl. Mag.*, Vol. 8, no.3, May/Jun. 2002, pp. 26–33

Nakamura, T., Miyake, H., Ogama, Y. , Morita, G., Muta, I., Hoshino, T. (2006). "Fabrication and characteristics of HTS induction motor by the use of B1-2223/Ag squirrel-cage rotor", *IEEE Trans. Applied Superconductivity*, Vol. 16, no. 2, June 2006, pp. 1469-1472

Pena, R., Clare, J.C. and Asher, G.M. (1996). "A doubly fed induction generator using back-to-back PWM converters and its application to variable-speed wind-energy generation," *Proc. Inst. Elect. Eng.*, Vol. 143, no. 5, May 1996, pp. 231–241

Petersson, A., Harnefors, L., and Thiringer, T. (2005). "Evaluation of current control methods for wind turbines using doubly-fed induction machines," *IEEE Trans. Power Electron.*, vol. 20, no. 1, Jan. 2005, pp. 227-235

Ramtharan, G., Ekanayake, J.B., and Jenkins, N. (2007). "Frequency support from doubly fed induction generator wind turbines," *IET Renewable Power Generation*, Vol.1, no. 1, pp. 3–9

Ribrant, J., Bertling,L. (2007). "Survey of failures in wind power systems with focus on Swedish wind power plants during 1997–2005", *IEEE Trans. Energy Conversion*, Vol. 22, No. 1, March 2007, pp. 167-173

Robb, D. (2005). "The gearbox challenge-the role of bearings in gearbox failure", *Windpower Monthly Magazine*, November 2005, pp. 53-60

Sasaki, A., Hamabe, M., Famakinwa, T., Yamaguchi, S., Radovinsky, A., Okumura, H., Emoto, M., Toyota, T. (2007). "Cryogenic fluid dynamics for DC superconducting power transmission line", *IEEE Trans. Applied Superconductivity*, Vol. 17 , Issue: 2, Part: 2, pp. 1748-1751

Scots, M. (2008). "Offshore wind installation ships: gaming the market", *New Energy Finance*, 11 April 2008, pp. 5

Ship, K.S., Goddard, K.F., Sykulski, J.K. (2002). "Field optimisation in a synchronous generator with high temperature superconducting field winding and magnetic core", IEE Proc. Sci. Meas. Technol. Vol. 149, No. 5, September 2002, pp. 194–198

Ship, K.S., Sykulski, J.K. (2004). "Field simulation studies for a high temperature superconducting synchronous generator with a coreless rotor", IEE Proc. Sci. Meas. Technol. Vol. 151, No. 6, November 2004, pp. 414–418

Snitchler, G., Gamble, B., Kalsi, S. (2005). "The performance of a 5 MW high temperature superconducting superconductor ship propulsion motor", *IEEE Trans. Applied Superconductivity,* Vol. 15, no. 2, June 2005, pp 2206-2209

Suchitra, E., Sebastian, N., Harrison, E., Palm, T., Murphy, C.H., Liang, R., Bonn, D.A., Hardy, W.N., and Lonzarich, G.G. (2008). "A multi-component Fermi surface in the vortex state of an underdoped high-Tc superconductor", *Nature,* Vol. 454, pp. 200-203

Superczynski, M.J., Waltman, D.J. (1997). "Homopolar motor with high temperature superconductor field windings", *IEEE Trans. Applied Superconductivity,* Vol. 7, no. 2, June 1997, pp 513-518

Sykulski, J.K., Goddard, K.F., and Ship. K.S. (2002). "Modelling and evaluations of eddy-current loss in high temperature superconducting synchronous generator", in *Studies in Applied Electromagnetics and Mechanics,* pp. 142-147

UK, (2008). "The grid code: Revision 31," *Grid Electricity Transmission,* No. 3, Oct 2008, Online: http://www.nationalgrid.com/uk

UK Wind Power, (2008). "World's biggest 7.5MW wind turbine for UK", August 11, 2008, *Materials,* Online: http://www.energy-business-review.com/article_news.asp?guid=9F250E84-4F17-41AC-845F-3484E3040767

Wang, Y., Lancaster, M.J. (2006). "High-temperature superconducting coplanar left-handed transmission lines and resonators", *IEEE Trans. Applied Superconductivity,* Vol. 16, pp. 1893-1897

Wen, H., Bailey, W., Goddard, K., Al-Mosawi, M., Beduz, C., Yang, Y. (2009). "Performance test of a 100 kW HTS generator operating at 67 K–77 K", *IEEE Trans. Applied Superconductivity,* Vol. 19, Issue: 3, Part: 2, pp. 1652-1655

Wilke, M., Schleicher, K., Klaus, G., Nick, W., Neumuller, H.W., Frauenhofer, J., Kahlen, K., Hartig, R. (2008). "Numerical calculations for high-temperature superconducting electrical machines", *18th International Conference on Electrical Machines (ICEM 2008),* pp. 1-6

Windpower Engineering, (2010). "Breaking the 9-MW barrier", May 2010, *Materials, Online:* http://www.clipperwind.com/pdf/wpe_Britannia.pdf

Small Scale Wind Energy Conversion Systems

Mostafa Abarzadeh, Hossein Madadi Kojabadi[1] and Liuchen Chang[2]
[1]*Sahand University of Technology*
[2]*University of New Brunswick*
[1]*Iran*
[2]*Canada*

1. Introduction

Electricity generation using wind energy has been well recognized as environmentally friendly, socially beneficial, and economically competitive for many applications. Because of crucial fossil energy resources shortage and environmental issues the wind energy is very important resource for electricity production. Small wind turbines, photovoltaic systems, full cells and pump as turbines (PAT) in small scale are main resources for distributed generation systems. Meanwhile, for remote areas wind energy beside photovoltaic system can combine as a hybrid system to provide necessary electric power of users. This system should be designed in such a way that the load demand of remote areas be provided with maximum reliability. Usually Direct coupled axial flux permanent magnet synchronous generator (AFPMSG), self-excited induction generator with gear box and permanent magnet synchronous generator(PMSG) with gear box can be used to connect to small wind turbine.
In the past few years, there have been many studies on small scale wind energy conversion systems. Authors of (Jia Yaoqin et al., 2002), (Nobutoshi Mutoh et al., 2006), (T.Tafticht et al., 2006), (Ch.Patsios et al., 2008) and (M.G.Molina et al., 2008) presented maximum power point tracking(MPPT) methods for small scale wind turbines. (Etienne Audierne et al., 2009), (M.G.Molina et al., 2008), (Boubekeur Boukhezzar et al., 2005), (Md.Arifujjaman et al., 2005) and (Jan T.Bialasiewicz, 2003) described small scale wind turbine furling system and modeled small scale wind turbines.
In this chapter we reviewed the working principles, over speed, output power control and MPPT control methods of small scale wind energy conversion system.

2. Wind turbine characteristics

The kinetic energy of the air stream available for the wind turbine given by

$$E = \frac{1}{2}\rho_a v V^2 \tag{1}$$

where ρ_a is air density, v is the volume of air available to the wind turbine rotor and V is the velocity of wind stream in m/s. The air parcel interacting with the rotor per second has a cross-sectional area equal to that of the rotor ($A_T(m^2)$) and thickness equal to the wind velocity ($V(m/s)$). Hence power of air stream available for wind turbine given by

$$P = \frac{1}{2}\rho_a A_T V^3 \tag{2}$$

However, wind turbine can not convert power of air stream completely. When the power stream passes the turbine, a part of its kinetic energy is transferred to the rotor and the air leaving the turbine carries the rest way. The actual power produced by wind turbine, usually, describe by power coefficient (C_P). C_P is the ratio of available power from wind stream and the power transferred to wind turbine. Hence

$$C_P = \frac{2P_T}{\rho_a A_T V^3} \tag{3}$$

where P_T is the power available from wind stream. According to Betz's law, no turbine can capture more than 59.3 percent of the kinetic energy in wind. The ideal or maximum theoretical efficiency (also called power coefficient, C_P) of a wind turbine is the ratio of maximum power obtained from the wind to the total power available in the wind. The factor 0.593 is known as Betz's coefficient. It is the maximum fraction of the power in a wind stream that can be extracted.

The C_P of a wind turbine depends on the profile of rotor blades, blade arrangement and setting etc. A designer would try to fix these parameters at its optimum level so as to attain maximum C_P at a wide range of wind velocities.

The thrust force experienced by the rotor(F) and rotor torque(T) are given by

$$F = \frac{1}{2}\rho_a A_T V^2 \tag{4}$$

$$T = \frac{1}{2}\rho_a A_T V^2 R \tag{5}$$

where R is the radius of the rotor. The ratio between the actual torque developed by the rotor and theoretical torque is termed as the torque coefficient (C_T). Thus,

$$C_T = \frac{2T_T}{\rho_a A_T V^2 R} \tag{6}$$

where T_T is the actual torque developed by the rotor.

The ratio between the velocity of the rotor tip and the wind velocity is termed as the tip speed ratio (λ). The power developed by the rotor at a certain wind speed greatly depends on tip speed ratio (λ). Thus,

$$\lambda = \frac{R\Omega}{V} = \frac{2\pi NR}{V} \tag{7}$$

where Ω is the angular velocity and N is the rotational speed of the rotor. The power coefficient and torque coefficient of a rotor vary with the tip speed ratio. The tip speed ratio is given by the ratio between the power coefficient and torque coefficient of the rotor.

$$\frac{C_P}{C_T} = \frac{R\Omega}{V} = \lambda \quad (8)$$

The efficiency with which a rotor can extract power from the wind depends on the dynamic matching between the rotor and wind stream. The $C_P - \lambda$ curve will represent the rotor performance irrespective of the rotor size and site parameters. The $C_P - \lambda$ curve represent the performance of the turbine irrespective rotor size and site parameters.

Typical $C_P - \lambda$ curves for different rotors are presented in Fig.1. in general, initially the power coefficient of the turbine increases with the tip speed ratio, reaches a maximum at a typical λ, and then decreases with further increase in the tip speed ratio.

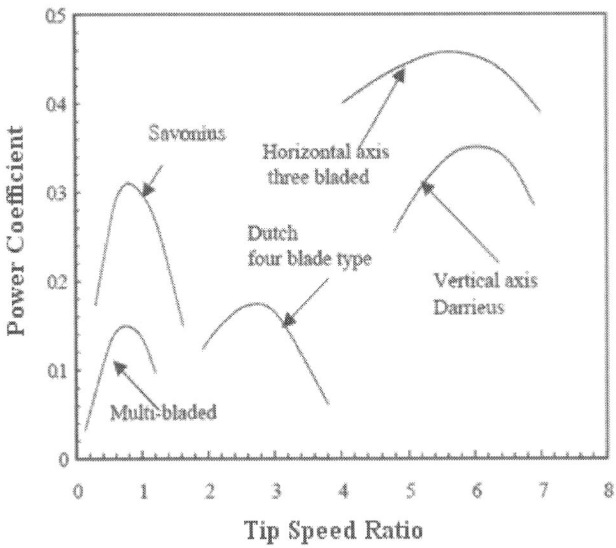

Fig. 1. Performance characteristics of wind rotors

The variations in C_P with λ depend on several design features of the rotor. American multi-bladed rotors show the lowest power coefficient and work at low speed ratio with the wind. However they have high solidity and hence high starting torque which make them attractive for water pumping. Two and three blade propeller turbines and the darrieus design turbine work at higher tip speed ratios. Hence they are suitable for wind electric generators.(Sathyajith Mathew)

3. Wind energy conversion systems

The main components of a wind turbine system are illustrated in Fig.2, including a turbine rotor, gearbox, generator, power electronic system and transformer.

Wind turbines convert the power from wind to mechanical power. It is important to be able to control and limit the converted mechanical power during higher wind speeds. The power

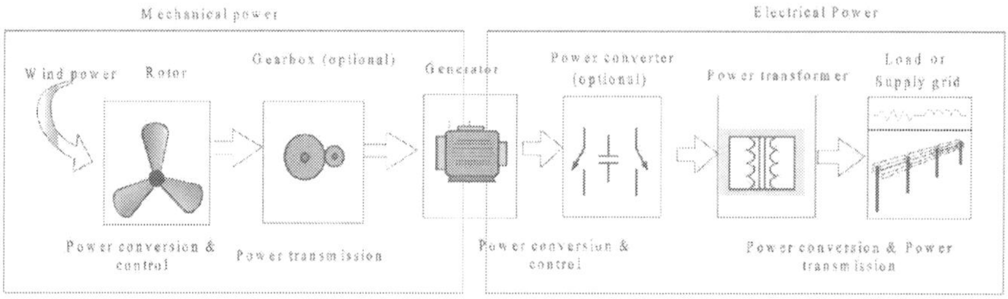

Fig. 2. Component of a wind turbine system.

limitation may be done either by stall control, active stall, or pitch control whose power curves are shown in Fig.3.

It can be seen that the power may be smoothly limited by rotating the blades either by pitch control or active stall control while the power from a stall-controlled turbines show a small overshoot and lower power output for higher wind speed.(Zhe Chen et al, 2009)

All of three methods for wind turbine power limitation usually used in large scale wind turbines, hence the power limitation during higher wind speeds in small scale wind turbines may be done by furling control or soft-stall control.

Many small wind turbines use an upwind rotor configuration with a tail vane for passive yaw control. Typically, the tail vane is hinged, allowing the rotor to furl (turn) in high winds, providing both power regulation and over-speed protection. Most the today's small wind turbines are operated using a variable speed generator. At higher wind speeds, the generated power of the wind turbine can go above the limit of the generator or the wind turbine design. When this occurs, small wind turbines use mechanical control or furling to turn the rotor out of the wind resulting in shedding the aerodynamic power or a steep drop in the power curve. Often, small turbine rotors furl abruptly at a wind speed only slightly above their rated wind speed, resulting in a very "peaky" power curve and poor energy capture at higher wind speeds. This energy loss is compounded by the furling hysteresis, in which the wind speed must drop considerably below the rated wind speed before the rotor will unfurl and resume efficient operation.

One way to improve the performance of furling wind turbines is to design the rotor to furl progressively, causing the power output to remain at or near rated power as the wind speed increases beyond it's rated value. This approach has two drawbacks: wind turbine rotors operating at high furl angles tend to be very noisy and experience high flap loads.

Fig.4 is the free body diagram of the system. It illustrates the simplified description of the furling mechanism. In a normal condition, the effective wind speed $V_n = V$ is the useful wind directed to the plane of rotation. The thrust is the force perpendicular to the plane of rotation. It is proportional to the square of the effective wind speed. The in-plane force, P_{force}, which is parallel to the plane of rotation, does not exist in the normal condition. When the wind speed increases, both the thrust and the P_{force} on the blade create moments due to the offset d_1 and d_2. **As** a result, the angle θ increases thus reducing the normal component of the wind speed V_n. **As** V_n decreases, the thrust and the wind energy converted to aerodynamic power also decreases.

The forces contributing to the moment around the pivot point are the thrust and the P_{force}. The thrust can be computed by considering the normal component of the wind speed.

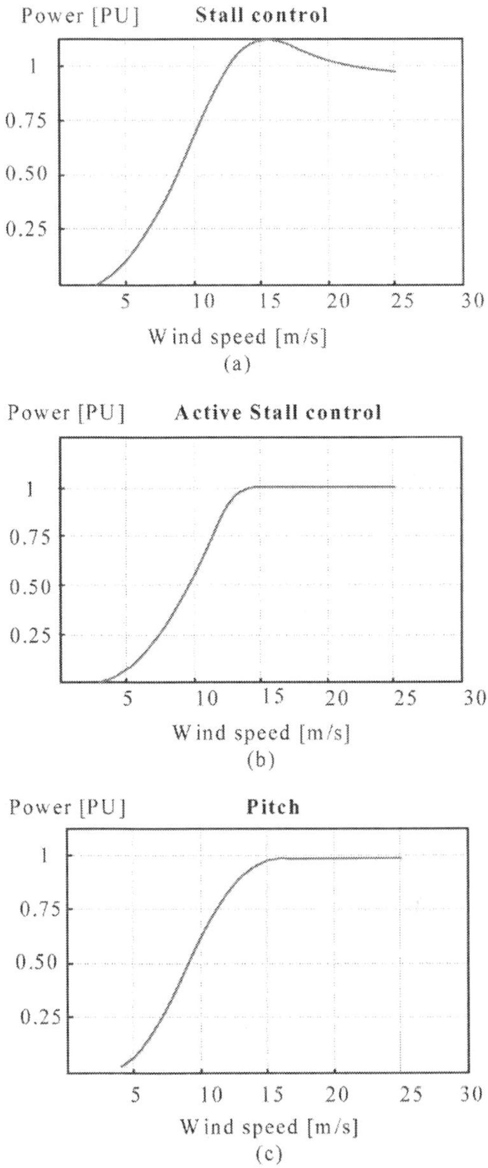

Fig. 3. Power characteristics of wind turbines. (a)stall control, (b)active stall control, (c)pitch control

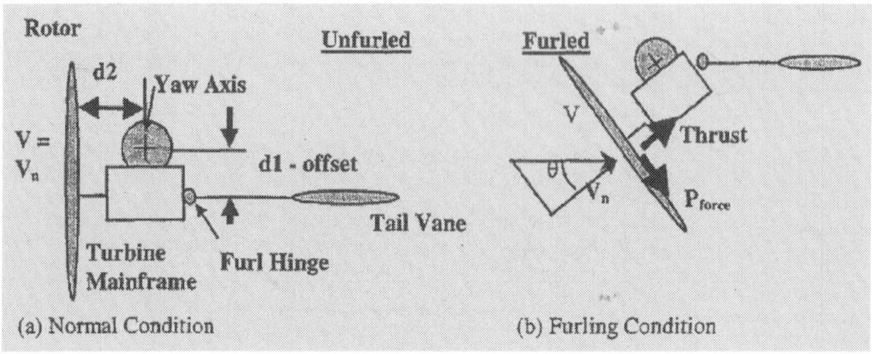

Fig. 4. Free body diagram of the furling system

$$Thrust = 0.5 \rho C_T A V_n^2 \tag{9}$$

The furling moment created by the blade can be written as

$$M_B = Thrust \times d1 + P_{force} \times d2 \tag{10}$$

And the restraining moment from the tail vane can be approximately represented by

$$M_T = K1 + K2 \times \theta \tag{11}$$

where K1 and K2 are the parameters of the wind turbine for the furling mechanism.
The equilibrium is governed by the following equation:

$$M_B - M_T = J\theta'' \tag{12}$$

where J is the moment inertia of the turbine with respect to the yaw axis, θ is the furling angle, and θ'' is the acceleration of furling rotation.

Note that manufactured wind turbines use a damper to reduce the furling loop hysteresis. Damping is necessary to keep the wind turbine from cycling or chattering in and out of furling. The damping plus the gyroscopic effect of turning wind turbine blades add to the unproductive time of entering and leaving the furling condition creating a hysteresis during transition. All of these delays reduce the wind turbine energy production.

The soft-stall concept is to control the generator rotations per minute (rpm) and achieve optimum operation over a wide range of rotor rpm. In order to control the generator rpm, the soft-stall concept regulates the stall mode of the wind turbine, thus furling can be delayed in normal operation. Furling is still used in the soft-stall concept during very high winds and emergency conditions. Potential advantages of soft-stall control are listed as follows:

- Delays furling as long as possible, which increases energy production
- Controls the wind turbine rotational speed to achieve the maximum power coefficient
- Operates the wind turbine at a low tip-speed ratio during high wind speeds to reduce noise and thrust loads. (E. Muljadi. et al, 1998 and Bialasiewicz, J.T., 2003)

The only difference between furling and soft-stall control is the addition of the DC-DC converter that allows the power to be controlled. With the DC-DC Converter between the rectifier and load, the transmitted power to the load can be controlled according to prescribed power/rpm schedule. Generated power curve for furling and soft stall control methods are shown in Fig.5.

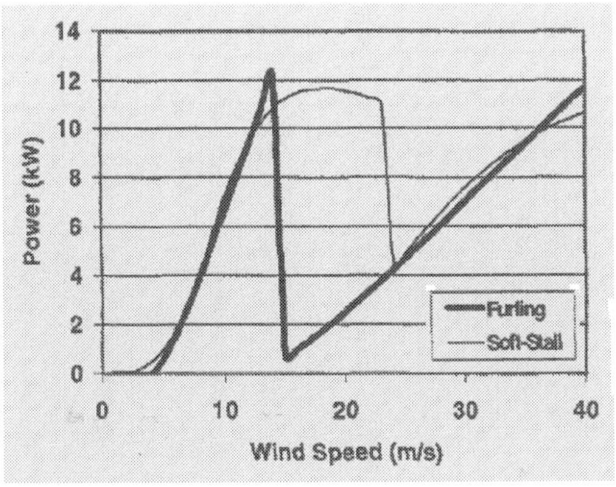

Fig. 5. Generated power for furling and soft-stall control

The two common types of electrical machines used in small scale wind turbines are self-excited induction generators (SEIG) and permanent magnet synchronous generators (PMSG). In these cases, the common way to convert the low-speed mechanical power to electrical power is a utilizing a gearbox and a SEIG or PMSG with standard speed. The gearbox adapts the low speed of the turbine rotor to the high speed of generators, though the gearbox may not be necessary for multiple-pole generator systems. The generator converts the mechanical power into electrical power.

In the self-excited induction generators, the reactive power necessary to energize the magnetic circuits must be supplied from parallel capacitors bank at the machine terminal. In this case, the terminal voltage or reactive power may not be directly controlled, and the induction generators may suffer from voltage instability problem.

There is considerable interest in the application of the multiple-pole Axial Flux Permanent Magnet Synchronous Generators (AFPMSG) driven by a wind-turbine shaft without gearbox.

Small scale wind conversion system may be integrated into loads or power systems with full rated power electronic converters. The wind turbines with a full scale power converter between the generator and load give the added technical performance. Usually, a back-to-back voltage source converter (VSC) is used in order to achieve full control of the active and reactive power. But in this case, the control of whole system would be a difficult task. Since the generator has been decoupled from electric load, it can be operated at wide range frequency (speed) condition and maximum power extract.

Fig.6 shows two most used solutions with full-scale power converters. All two solutions have almost the same controllable characteristics since the generator is decoupled from the load by a dc link.

The configuration shown in Fig.6(a) is characterized by having a gearbox. The wind turbine system with a self-excited induction generator and full rated power electronic converters is shown in Fig.6(a). Multipole systems with the axial flux permanent magnet synchronous generator without a gearbox is shown in Fig.6(b).

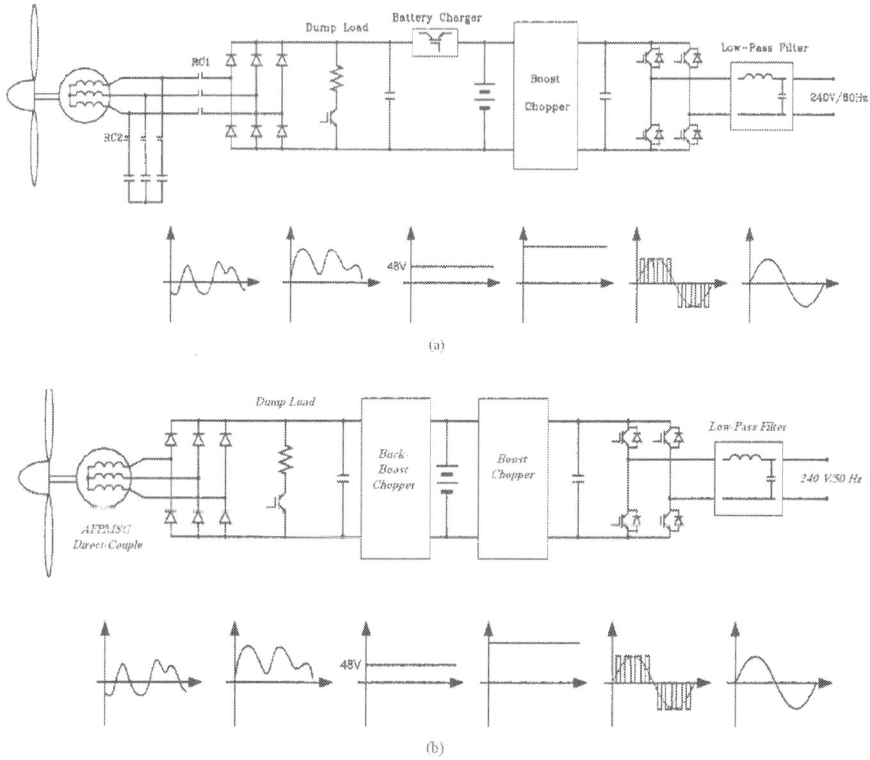

Fig. 6. Small scale wind generation system. (a)self-excited induction generator with gearbox (b)direct coupled axial flux permanent magnet synchronous generator

4. Maximum power point tracking systems for wind generators

Variable-speed wind turbines are able to operate at an optimal rotation speed as a function of the wind speed. The power electronic converter may control the turbine rotation speed to get the maximum possible power by means of a maximum power point tracking (MPPT) algorithm. In this way, it is also possible to avoid exceeding the nominal power if the wind speed increases. At the same time, the dc-link capacitor voltage is kept as constant as possible, achieving a decoupling between the turbine-side converter and the grid-side converter. The grid-connected inverters will inject the active power to the grid with minimum total harmonic distortion (THD) of output current and voltage. The grid voltage and inverter output voltage will be synchronized by zero-crossing circuit.

The rotor efficiency curve $C_p(\lambda)$ is a nonlinear function of the TSR, λ, which is determined by the blade design, and the pitch angle. From Fig.7(a), it is clear that there is a value of λ for which C_P is maximized, thus maximizing the power for a given wind speed. Because of the relationship between C_P and λ, for each wind velocity, there is a turbine speed that gives a maximum output power. The peak power points for various wind speeds are shown in Fig.7(b). Normally, a variable-speed wind turbine follows the $C_{P\max}$ to capture the maximum power up to the rated speed by varying the rotor speed to keep the system at the optimum TSR, λ_{opt}. (Zhe Chen et al, 2009)

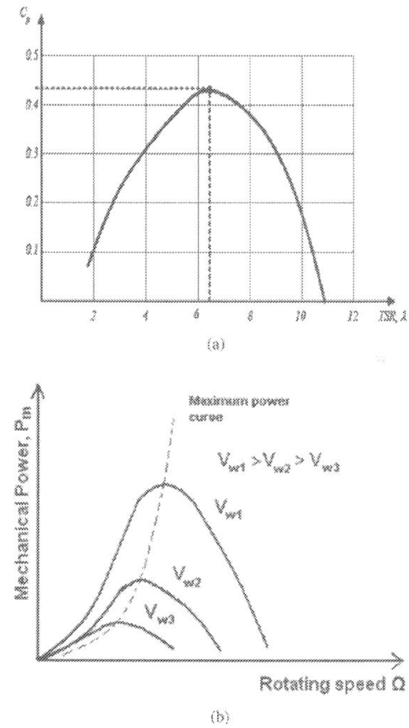

Fig. 7. (a)relationship between the TSR and the power coefficient. (b)wind turbine $P-\Omega$ characteristics and maximum power curve different wind speeds.

A typical example of the relationship between the wind speed and the power generated by the wind turbine is shown in Fig.8. The blades start to move around 4 m/s, and optimal aerodynamic efficiency is achieved up to the rated wind speed, about 15 m/s. Between the rated wind speed and 25 m/s, the power delivered is limited in order to avoid overloading on the wind turbine system. Above the cutout wind speed, the turbine has to be stopped in order to avoid damages.(Zhe Chen et al, 2009)

Two of the most commonly applied trends in the MPPT processes namely: The tracking method based on the optimum power versus speed characteristic and the perturbation and observation (P&O) of the output power method.(Ch.Patsios et al, 2008)

During the optimal condition of wind speed, the wind generator may be adjusted to follow the various methods to perform MPPT algorithm that will be summarized as follows.

1. *TSR Control:* Fig.9 shows this kind of MPPT controller, which needs the wind speed measured by an anemometer. The controller regulates the wind turbine speed to maintain an optimal TSR. However, the accurate wind speed may be difficult to obtain.

In addition, the use of an external anemometer increases the complexity and cost of the system.
In addition, the use of wind speed sensor to measure the wind speed adds to a system a cost and presents some difficulties in practical implementation. These MPPT methods described in the current literature are too expensive compared with generator whose rated capacity is small.(T.Tafticht et al., 2006)

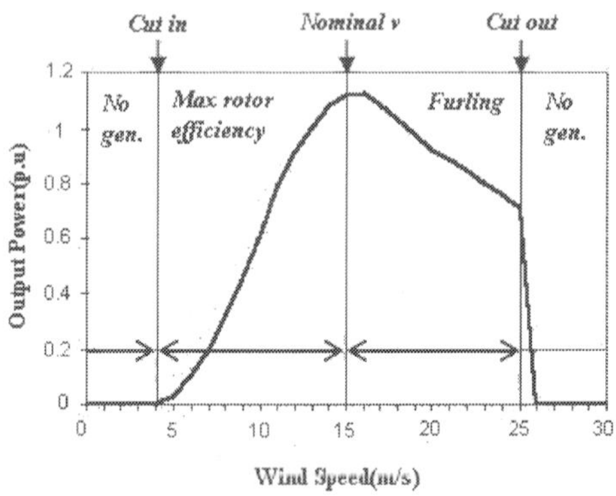

Fig. 8. output power of wind turbine as a function of wind speed.

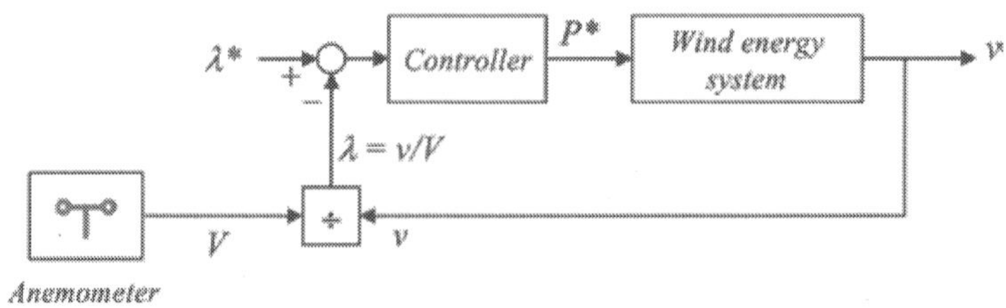

Fig. 9. Block diagram of TSR Control

2. *Power Signal Feedback (PSF) Control:* This control, depicted in Fig.10, requires the knowledge of the maximum power curves of the turbine, which may be obtained through simulations and practical tests. The speed of the wind turbine is used to select the stored power curve, which gives the target power to be tracked by the system. In many cases, this power curve may be substituted by a predictor or an observer of the wind speed as a function of the power and the wind-turbine speed.(Zhe Chen et al, 2009)
3. *Perturbation and observation(P & O) Control:* This MPPT process is based on monitoring the wind-generator (WG) output power using measurements of the WG output voltage and current and directly adjusting the dc/dc converter duty cycle according to the result of comparison between successive WG-output-power values. Although the wind speed varies highly with time, the power absorbed by the WG varies relatively slowly, because of the slow dynamic response of the interconnected wind-turbine/generator system. Thus, the problem of maximizing the WG output power using the converter

duty cycle as a control variable can be effectively solved using the steepest ascent method according to the following control law:

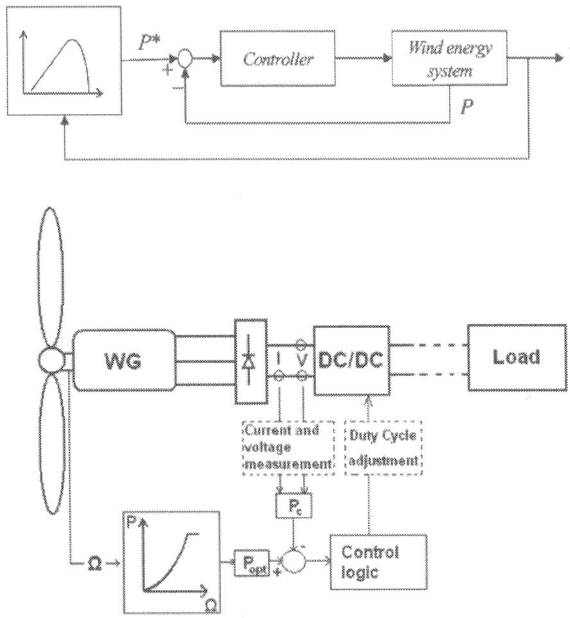

Fig. 10. Block diagram of PSF Control

$$D_k = D_{k-1} + C_1 \frac{\Delta P_{k-1}}{\Delta D_{k-1}} \tag{13}$$

where D_k and D_{k-1} are the duty-cycle values at iterations k and k − 1, respectively ($0 < D_k < 1$); $\frac{\Delta P_{k-1}}{\Delta D_{k-1}}$ is the WG power gradient at step k − 1; and C_1 is the step change.(Shirazi, M. et al, 2009) A version of the P&O algorithm is shown in Fig.11

4. *Hill Climbing Searching (HCS) Control:* In the HCS method, a controller compares the output power of the turbine with the previous power and based on the comparison it controls the load. Using a hill-climbing algorithm the controller tries to extract the maximum power from the wind, while the generator output is observed as the furl angle increase or decreases. In this strategy, the controller will try to adjust the load by measuring the consecutive power and thus the extraction of maximum power from the wind.(Md.Arifujjaman et al., 2005)
When the wind-turbine speed increases, the output power should normally increase as well, otherwise the speed should be decreased (see Fig.12). However, this method could be ineffective for large wind turbines, since the large turbines are difficult to adjust the speed fast. (Zhe Chen et al, 2009)

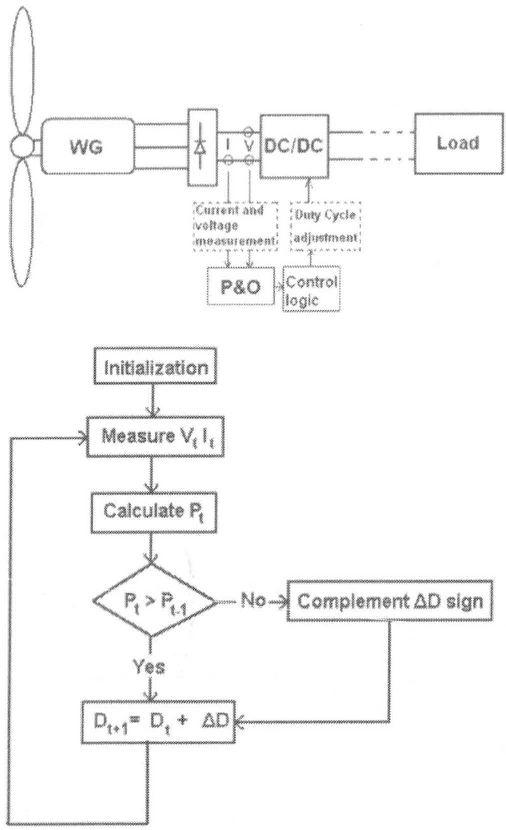

Fig. 11. Block diagram of P&O control

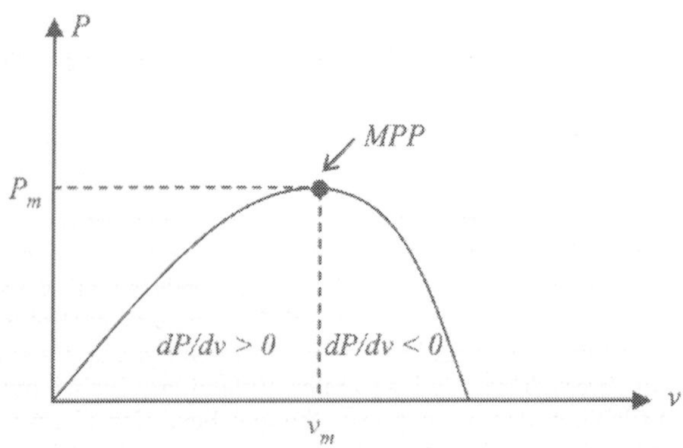

Fig. 12. Block diagram of HCS Control.

5. *MPPT by a maximum-efficiency control and a maximum-torque Control:* Based on the turbine characteristics of a selected turbine, the relationship between the optimum generator torque and the generator speed is established. This relationship determines the behaviour of the maximum torque control. For any particular wind speed the generator torque balances the mechanical torque so that they will be equivalent at the optimum operating point. Since the generator torque is controlled in such a way that it tracks the optimum torque curve.

Fig.13 shows the responces of MPPT control. Let it be assumed that the wind speed is V_{w3} and the generator torque T_g balances the mechanical torque T_m at the optimum point A as shown in Fig.13(a). When wind speed changes to V_{w2}, T_m jumps to point B, but T_g remains at point A. The generator speed increases according to the difference of torque (T_m - T_g) (see Fig.13(b)). Then T_g increases on the optimum torque curve and T_m decreases. After all, they settle down to the optimum torque T_{opt} for V_{w2} (point C in Fig. 13(a)). This means that, the generator torque is controlled to track the optimum torque curve for various wind speed by MPPT controller without a wind speed detector.

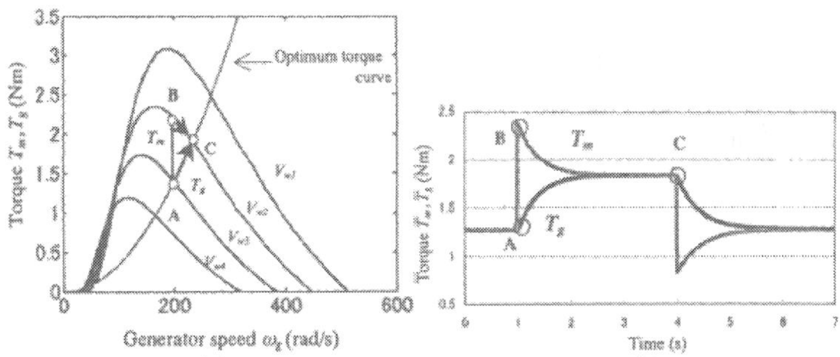

(a) (b)

Fig. 13. (a) Torque-speed curves. (b) Dynamic response of MPPT control.

An advantage of this method is that it does not require a wind speed detector. A drawback of this method is that to select the proportional constant that describes the relationship between the generator torque and speed is based on the turbine characteristics. This dependency hinders its ability to be used for various wind turbines, since different turbines have different characteristics. (Shirazi, M. et al, 2009)

5. Conclusion

This chapter has reviewed the small scale wind energy conversion systems. Various arrangements of small scale wind generators with different generators and control systems are described. The power limitation during higher wind speeds in small scale wind turbines may be done by furling control or soft-stall control. The soft-stall control method is better approach than furling control method. In compare with wind generators with gearbox, the main advantages of direct-drive wind generator systems are higher overall efficiency, reliability, and availability due to omitting the gearbox. Considering the improved performance and reduced cost of PM materials over recent years, direct drive AFPMSG have come more attractive for small scale wind generation systems.

The various MPPT algorithms have described in this chapter. Some methods use the changes in power (ΔP) and the changes in generator speed (Δω) to adjust the generator speed towards the optimum operating point. These methods are independent of turbine characteristics, so they are flexible and can be applied to various turbines. However, they would be slower than maximum-torque control method. The maximum-torque control method is fast and efficient, but having prior knowledge of the turbine characteristics is required.

6. References

Sathyajith Mathew, Wind Energy Fundamentals, Resource Analysis and Economics, Springer Verlag Berlin Heidelberg, 2006.

Zhe Chen.; Josep M. Guerrero.; Frede Blaabjerg.(2009). A Review of the State of the Art of Power Electronics for Wind Turbines, IEEE TRANSACTIONS ON POWER ELECTRONICS, VOL. 24, NO. 8, pp 1859-1874, AUGUST 2009.

E. Muljadi.; T. Forsyth.; C.P. Butterfield.(1998). SOFT-STALL CONTROL VERSUS FURLING CONTROL FOR SMALL WIND TURBINE POWER REGULATION, Windpower '98 Bakersfield, CA April 27-May 1 ,1998

E. Muljadi; K. Pierce; P. Migliore; *Soft-stall control for variable-speed stall-regulated wind turbines,* Journal of Wind Engineering and Industrial Aerodynamics, Volume 85, Issue 3, 24 April 2000, Pages 277-291.

Bialasiewicz, J.T. ; Furling control for small wind turbine power regulation , Industrial Electronics, 2003. ISIE '03. 2003 IEEE International Symposium on, pp. 804-809 vol.2, 9-11 June 2003

Patsios, C.; Chaniotis, A.; Kladas, A.; A Hybrid Maximum Power Point Tracking System for Grid-Connected Variable Speed Wind-Generators, Power Electronics Specialists Conference, 2008. PESC 2008. IEEE, pp- 1749 - 1754, 15-19 June 2008

Shirazi, M.; Viki, A.H.; Babayi, O.; A Comparative Study of Maximum Power Extraction Strategies in PMSG Wind Turbine System, : Electrical Power & Energy Conference (EPEC), 2009 IEEE, pp.1-6, 22-23 Oct. 2009

Arifujjaman, M. Iqbal, M.T. Ouaicoe, J.E. Khan, M.J.; Modeling and control of a small wind turbine , Electrical and Computer Engineering, 2005. Canadian Conference on, pp. 778 - 781, 1-4 May 2005

Tafticht, T.; Agbossou, K.; Cheriti, A.; DC bus control of variable speed wind turbine using a buck-boost converter, Power Engineering Society General Meeting, 2006. IEEE, 5 pp., 16 October 2006

Jia Yaoqin; Yang Zhongqing; Cao Binggang; A new maximum power point tracking control scheme for wind generation, Power System Technology, 2002. Proceedings. PowerCon 2002. International Conference on, pp. 144 - 148 vol.1, 13-17 Oct 2002

Etienne Audierne.; Jorge Elizondo.; Leonardo Bergami.; Humberto Ibarra.; Oliver Probst.; Analysis of the furling behavior of small wind turbines, Applied Energy Volume 87, Issue 7, July 2010, Pages 2278-2292

Molina, M.G.; Mercado, P.E.; A new control strategy of variable speed wind turbine generator for three-phase grid-connected applications, Transmission and Distribution Conference and Exposition: Latin America, 2008 IEEE/PES, pp. 1 - 8, 13-15 Aug. 2008